VOLUME ONE HUNDRED AND SEVEN

VITAMINS AND HORMONES

Ovarian Cycle

VOLUME ONE HUNDRED AND SEVEN

VITAMINS AND HORMONES

Ovarian Cycle

Series Editor

GERALD LITWACK, PhD
Toluca Lake, North Hollywood, California

ELSEVIER

ACADEMIC PRESS
An imprint of Elsevier

Cover photo credit:
JoAnne S. Richards
The Ovarian Cycle
Vitamins and Hormones (2018) **107**, pp. 1–26.

Academic Press is an imprint of Elsevier
50 Hampshire Street, 5th Floor, Cambridge, MA 02139, United States
525 B Street, Suite 1800, San Diego, CA 92101-4495, United States
The Boulevard, Langford Lane, Kidlington, Oxford OX5 1GB, United Kingdom
125 London Wall, London, EC2Y 5AS, United Kingdom

First edition 2018

ISBN: 978-0-12-814359-9
ISSN: 0083-6729

Publisher: Zoe Kruze
Acquisition Editor: Ashlie Jackman
Editorial Project Manager: Joanna Collett
Production Project Manager: Vignesh Tamil
Cover Designer: Matthew Limbert

Typeset by SPi Global, India

CONTENTS

CONTRIBUTORS

George Anifandis
Faculty of Medicine, School of Health Sciences, University of Thessaly, Larissa, Greece

Frank Arfuso
Stem Cell and Cancer Biology Laboratory, School of Pharmacy and Biomedical Sciences, Curtin Health Innovation Research Institute, Curtin University, Perth, WA, Australia

Martina Belli
University of California San Diego, School of Medicine, La Jolla, CA, United States

Jill Bennett-Toomey
Ohio Northern University, Ada, OH, United States

Zeev Blumenfeld
Reproductive Endocrinology, Faculty of Medicine, Technion—Israel Institute of Technology, Haifa, Israel

Mei-Jou Chen
National Taiwan University Livia Shangyu Wan Scholar, Department of Obstetrics and Gynecology, National Taiwan University Hospital, Taipei, Taiwan

Chia-Hong Chou
Department of Obstetrics and Gynecology, National Taiwan University Hospital, Taipei, Taiwan

Lique M. Coolen
University of Mississippi Medical Center, Jackson, MS, United States

Maria Paola Costi
University of Modena and Reggio Emilia, Modena, Italy

Alexandros Daponte
Faculty of Medicine, School of Health Sciences, University of Thessaly, Larissa, Greece

Arun Dharmarajan
Stem Cell and Cancer Biology Laboratory, School of Pharmacy and Biomedical Sciences, Curtin Health Innovation Research Institute, Curtin University, Perth, WA, Australia

Stefania Ferrari
University of Modena and Reggio Emilia, Modena, Italy

Graça Ferreira-Dias
C.I.I.S.A., Faculty of Veterinary Medicine, University of Lisbon, Lisbon, Portugal

Fabrizio Fontana
Università degli Studi di Milano, Milano, Italy

António M. Galvão
C.I.I.S.A., Faculty of Veterinary Medicine, University of Lisbon, Lisbon, Portugal; Institute of Animal Reproduction and Food Research of PAS, Olsztyn, Poland

Antonios Garas
Faculty of Medicine, School of Health Sciences, University of Thessaly, Larissa, Greece

Robert L. Goodman
West Virginia University, Morgantown, WV, United States

Thippeswamy Gulappa
The University of Michigan Medical School, Ann Arbor, MI, United States

Phil G. Knight
School of Biological Sciences, Hopkins Building, University of Reading, Reading, United Kingdom

Michael N. Lehman
University of Mississippi Medical Center, Jackson, MS, United States

Yee Leung
Western Australian Gynaecologic Cancer Service, King Edward Memorial Hospital for Women, Perth, WA, Australia

Patrizia Limonta
Università degli Studi di Milano, Milano, Italy

Tao Liu
Pacific Northwest National Laboratory, Richland, WA, United States

Lorena Losi
University of Modena and Reggio Emilia, Modena, Italy

Roberto Maggi
Università degli Studi di Milano, Milano, Italy

Gaetano Marverti
University of Modena and Reggio Emilia, Modena, Italy

Monica Marzagalli
Università degli Studi di Milano, Milano, Italy

Bindu Menon
The University of Michigan Medical School, Ann Arbor, MI, United States

K.M.J. Menon
The University of Michigan Medical School, Ann Arbor, MI, United States

Christina I. Messini
Faculty of Medicine, School of Health Sciences, University of Thessaly, Larissa, Greece

Ioannis E. Messinis
Faculty of Medicine, School of Health Sciences, University of Thessaly, Larissa, Greece

Marina Montagnani Marelli
Università degli Studi di Milano, Milano, Italy

Roberta Moretti
Università degli Studi di Milano, Milano, Italy

Fumio Otsuka
Okayama University Graduate School of Medicine, Dentistry and Pharmaceutical Sciences, Okayama, Japan

Vasantha Padmanabhan
University of Michigan, Ann Arbor, MI, United States

Glauco Ponterini
University of Modena and Reggio Emilia, Modena, Italy

Cecilia Pozzi
University of Siena, Siena, Italy

Muraly Puttabyatappa
University of Michigan, Ann Arbor, MI, United States

Antonio Quotadamo
University of Modena and Reggio Emilia, Modena, Italy

Doodipala Samba Reddy
College of Medicine, Texas A&M University Health Science Center, Bryan, TX, United States

Sheena L.P. Regan
Stem Cell and Cancer Biology Laboratory, School of Pharmacy and Biomedical Sciences, Curtin Health Innovation Research Institute, Curtin University, Perth, WA, Australia

JoAnne S. Richards
Dan L. Duncan Cancer Center; Center for Reproductive Medicine, Baylor College of Medicine, Houston, TX, United States

Karin D. Rodland
Pacific Northwest National Laboratory, Richland, WA, United States

Leda Severi
University of Modena and Reggio Emilia, Modena, Italy

Shunichi Shimasaki
University of California San Diego, School of Medicine, La Jolla, CA, United States

Dariusz Skarzynski
Institute of Animal Reproduction and Food Research of PAS, Olsztyn, Poland

Richard D. Smith
Pacific Northwest National Laboratory, Richland, WA, United States

Carlos Stocco
College of Medicine, University of Illinois at Chicago, Chicago, IL, United States

Inger Sundström-Poromaa
Department of Women's and Children's Health, Uppsala University, Uppsala, Sweden

Peyton W. Weems
Graduate Program in Neuroscience, University of Mississippi Medical Center, Jackson, MS, United States

John L. Yovich
Stem Cell and Cancer Biology Laboratory, School of Pharmacy and Biomedical Sciences, Curtin Health Innovation Research Institute, Curtin University; PIVET Medical Centre, Perth, WA, Australia

PREFACE

The system governing the generation of the ovum for reproduction is exquisitely regulated by a host of different hormones. In the ovarian cycle, there develops the ovarian graafian follicle. The follicle becomes ruptured so that the ovum can be discharged. Following this, a corpus luteum is formed and regresses, unless fertilization occurs. Also part of this cycle is the phasing of the uterus. Menstruation occurs in the absence of fertilization and this is followed by a proliferative phase and a secretive phase. An overview of the system is eloquently described in the opening chapter by Dr. JoAnne S. Richards of Baylor University.

The chapters in this volume are arranged as follows. First, as already mentioned, "The ovarian cycle" is reviewed by J.A.S. Richards. This is followed by "GnRH in the human female reproductive axis" by P. Limonta, M. Montagnani Marelli, R. Moretti, M. Marzagalli, F. Fontana, and R. Maggi. Next, K.M.J. Menon, B. Menon, and T. Gulappa report on "Regulation of luteinizing hormone receptor mRNA expression in the ovary: The role of miR-122." "The roles of neurokinins and endogenous opioid peptides in control of pulsatile LH secretion" is authored by P.W. Weems, M.N. Lehman, L.M. Coolen, and R.L. Goodman. F. Otsuka discusses "Interaction of melatonin and BMP-6 in ovarian steroidogenesis." "The effect of steroid hormones on ovarian follicle development" is the topic developed by C.-H. Chou and M.-J. Chen. D.S. Reddy follows with "GABA-A receptors mediate tonic inhibition and neurosteroid sensitivity in the brain." J. Bennett-Toomey and C. Stocco report on "GATA regulation and function during the ovarian life cycle." "Involvement of bone morphogenetic proteins (BMP) in the regulation of ovarian function" is the work of S.L.P. Regan, P.G. Knight, J. L. Yovich, Y. Leung, F. Arfuso, and A. Dharmarajan. Following this, I. E. Messinis, C.I. Messini, G. Anifandis, A. Garas, and A. Daponte report on "Gonadotropin surge-attenuating factor: A nonsteroidal ovarian hormone controlling GnRH-induced LH secretion in the normal menstrual cycle." "Luteolysis and the auto-, paracrine role of cytokines from tumor necrosis factor α and transforming growth factor β superfamilies" is described by A.M. Galvão, D. Skarzynski, and G. Ferreira-Dias.

Following these chapters, based mainly in the basic research aspects of the ovarian cycle, there are chapters relating to clinical aspects. The first of these is "Molecular aspects and clinical relevance of GDF9 and BMP15 in ovarian function" by M. Belli and S. Shimasaki. I. Sundström-Poromaa writes on "The menstrual cycle influences emotion but has limited effect on cognitive function." "Developmental programming of ovarian functions and dysfunctions" is the topic of M. Puttabyatappa and V. Padmanabhan. Z. Blumenfeld discusses "The ovarian hyperstimulation syndrome (OHSS)." J.A.S. Richards follows with "From follicular development and ovulation to ovarian cancers: An unexpected journey." "Human thymidylate synthase inhibitors halting ovarian cancer growth" is reviewed by S. Ferrari, L. Severi, C. Pozzi, A. Quotadamo, G. Ponterini, L. Losi, G. Marverti, M.P. Costi. Finally, concluding this section, T. Liu, K.D. Rodland, and R.D. Smith report on the "Characterization of the ovarian tumor peptidome."

I gratefully acknowledge the work of Joanna Collett and the cooperation of Ashlie Jackman of Elsevier, Oxford, UK in the production of this volume.

The illustration on the cover is derived from Figure 6 of the opening chapter entitled "The ovarian cycle" by Dr. JoAnne S. Richards.

Gerald Litwack
Toluca Lake, North Hollywood, CA
27 November, 2017

CHAPTER ONE

The Ovarian Cycle

JoAnne S. Richards[1]

Baylor College of Medicine, Houston, TX, United States
Dan L. Duncan Cancer Center, Baylor College of Medicine, Houston, TX, United States
Center for Reproductive Medicine, Baylor College of Medicine, Houston, TX, United States
[1]Corresponding author: e-mail address: joanner@bcm.edu

Contents

Abstract

The "ovarian cycle" is an exquisite and dynamic endocrine system that includes ovarian events, hypothalamic–pituitary interactions, uterine endometrial and myometrial changes during implantation and pregnancy, cervical alterations in structure, and breast development. The ovarian cycle and the steroid hormones produced by the ovary also impact epithelial cancer development in the ovary, uterus, cervix, and breast. This chapter provides a personal view of recent developments that occur in this complex endocrine environment.

1. INTRODUCTION

The ovarian cycle encompasses many ovarian events: the dynamic formation, growth, and ovulation of ovarian follicles and their transition to terminally differentiated structures called corpora lutea (Edson, Nagaraja, &

Vitamins and Hormones, Volume 107
ISSN 0083-6729
https://doi.org/10.1016/bs.vh.2018.01.009

Matzuk, 2009; Richards, 1979; Richards & Pangas, 2010). It also includes events that determine the selection and demise of nearly 99% of all oocytes and follicles that are present in the ovaries of mammals at birth but do not survive into adulthood. The ovarian cycle is an exquisite, highly orchestrated, and complex developmental process that depends on functional interactions and critical signaling pathways that communicate among the oocyte, granulosa cells, theca cells, and surrounding stromal components, including the vascular and immune systems. Within the ovary, each follicle is like an individual community of cells that moves through the ovarian stroma and eventually either dies by apoptosis or ovulates and delivers an oocyte for fertilization.

The ovarian cycle (Fig. 1), in turn, (1) impacts and is regulated by hypothalamic and pituitary functions, neurotransmitters, and hormones (Fig. 2), (2) directs specific functions of the oviduct (oocyte and embryo transport), uterus (implantation, pregnancy, and parturition), and cervix (pregnancy and parturition), (3) influences activities of many other organs, including adipocytes, vascular cells, and immune cells (Physiology of Reproduction, 2015), and (4) responds to factors in the environment including the length of day (Karsch, 1980).

Premature disruption of the normal ovarian cycle leads to infertility; ovarian aging and the natural cessation of the ovarian cycle lead to

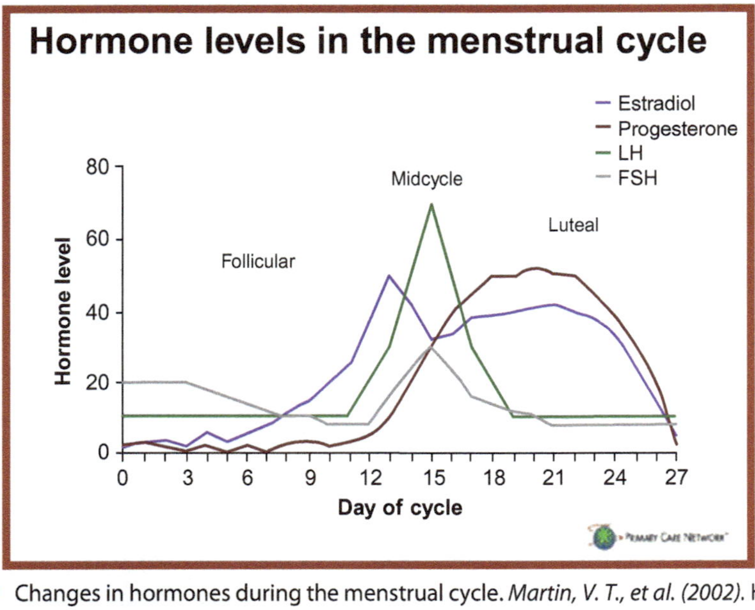

Fig. 1 Changes in hormones during the menstrual cycle. *Martin, V. T., et al. (2002).* Headache, 42, *438. Abstract OR19.*

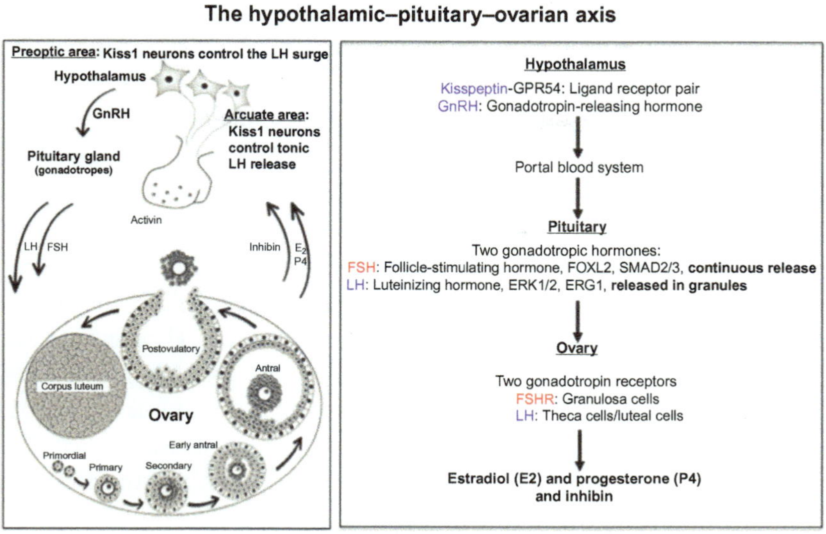

Fig. 2 Schematic and text describing the hypothalamic, pituitary, and ovarian events that occur during the ovarian cycle. *Adapted from Monsivais, D., Matzuk, M.M., & Pangas, S.A. (2017). The TGF-β family in the reproductive tract.* Cold Spring Harbor Perspectives in Biology, 9, *a022251, PMID: 28193725.*

menopause. Altering the hypothalamic–pituitary–ovarian axis can predispose women to polycystic ovarian syndrome (PCOS) and other infertile conditions. Contraceptives, once controversial, provide a voluntary and specific mechanism to disrupt the ovarian cycle. Hormone replacement therapies, also controversial, can alleviate symptoms of menopause but may also enhance the risks of cancer. Ovarian cancer (epithelial or granulosa cell) caused by genomic instability and oncogenic factor expression is the leading cause of gynecologic death in women. Other types of mutations that disrupt follicular cell functions and growth can also lead to premature ovarian failure (Physiology of Reproduction, 2015).

Because the scope of the "ovarian cycle" is vast and the complexities of factors that control each reproductive tissue and cell are overwhelming and still to be resolved, this chapter is a selected personal view, but hopefully, will provide a reasonable overview of current concepts.

2. OVARIAN EVENTS

2.1 Follicle Formation

During embryo development, primordial germ cells (PGCs) migrate to the genital ridge, proliferate and survive in response to specific bone morphogenic proteins (BMPs) (Pangas & Rajkovic, 2015) and the cKIT–KIT ligand signaling pathway (Driancourt, Reynaud, Cortvrindt, & Smitz, 2000). In the female embryonic gonad, these PGCs eventually cease dividing and enter prophase I of meiosis, a process dependent on the expression of *Stra8* (Falender, Shimada, Lo, & Richards, 2005; Menke & Page, 2002). Once arrested in this phase of meiosis, oocytes remain quiescent in meiotic prophase until ovulation (Fig. 3).

The formation of the female gonad depends on specific signaling pathways that prevent formation of the male gonad. These include the WNT4/FDZ/β-catennin (CTNNB1) and RSPO1 pathway as well as specific transcription factors, including WT1, FOXL2, and SF1 (NR5A1) (Morohashi, Yoshioka, & Kawajiri, 2001; Pangas & Rajkovic, 2015; Shima et al., 2011). Importantly, mice null for SF1 lack not only gonads but also adrenal glands and pituitary gonadotropes (Luo, Ikeda, & Parker, 1994) making SF1 a master gene of steroidogenic and hormone-secreting cells.

In the embryonic gonad, PGCs form syncytia (or nests) that breakdown by the migration of progenitor somatic cells, "pregranulosa cells" that surround the oocytes, leading to the formation of primordial follicles, known as the ovarian reserve (Pangas & Rajkovic, 2015). Nest breakdown is regulated, in part, by estradiol and Notch signaling pathways

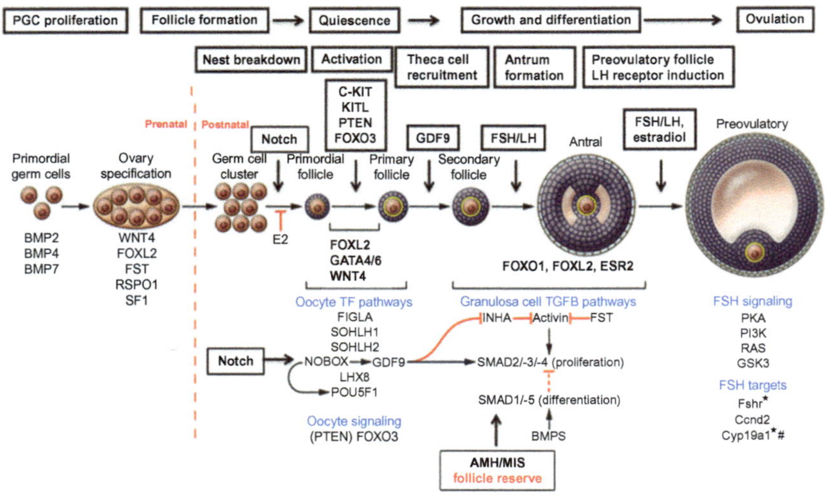

Fig. 3 Schematic of events and factors that regulate ovarian follicular formation, development, and ovulation. *Adapted from Richards, J. S., & Pangas, S. A. (2010). The ovary: Basic biology and clinical implications.* The Journal of Clinical Investigation, 120, *963–972. PMID: 20364094; PMCID:PMC22846061.*

(Vanorny & Mayo, 2017). Based on recent studies, there are two waves of primordial follicle development. In the first wave, granulosa cells are derived from somatic cells migrating from the mesonephros into the embryonic gonad. In the second wave, granulosa cells appear to be derived from cells in the cortical region of the postnatal ovary (Zheng, Zhang, Gorre, et al., 2014; Zheng, Zhang, & Liu, 2014). Whether or not these pregranulosa cells represent two distinct populations of progenitor granulosa cells remains to be firmly established (Hummitzsch et al., 2013).

Recent studies have shown that theca cells of ovarian follicles are derived from two progenitor cell types in the embryonic gonad: mesenchymal cells that migrate into the ovary from the mesonephros region and WT1 + stromal cells that are indigenous to the embryonic ovarian medullary region (Liu, Peng, Matzuk, & Yao, 2015). These fetal progenitor cells differentiate into theca cells after birth when the mesonephros-derived progenitors become the steroidogenic androgen-producing cells within the theca, whereas the ovarian WT1 + cells appear to give rise to theca fibroblasts, perivascular smooth muscle cells, and interstitial ovarian tissue (Nicol & Yao, 2014).

Early formation of primordial follicles is also regulated by the oocyte and oocyte factors that regulate the integrity of the oocyte. These factors include the transcription factors FIGLA, LHX8, SOHLH1, and SOHLH2 (Pangas & Rajkovic, 2015; Richards & Pangas, 2010). Whereas FIGLA controls the expression of the zona pellucida proteins, SOHLH1/2 regulate the expression of NOBOX. NOBOX in turn controls the expression of the TGFβ family protein, growth differentiation 9 (GDF9). GDF9 is secreted and regulates somatic cell functions (Liu, Peng, et al., 2015). NOTCH signaling also impacts oocyte functions as well as early stages of follicle growth (Vanorny & Mayo, 2017).

2.2 Early Stages of Follicular Growth

The release of primordial follicles from a quiescent state to a growing phase is initiated by intraovarian factors, such as Kit ligand, coming from somatic cells, that activates the c-KIT receptor (Driancourt et al., 2000) and PI3 kinase pathway in oocytes (Driancourt et al., 2000). Activated PI3K phosphorylates and thereby inhibits the functions of the transcription factor FOXO3 leading to primordial follicle activation (Zheng, Nagaraju, Liu, & Liu, 2012). These conclusions are based primarily on specific mutant mouse models in which disruption of *Foxo3* or *Pten* (that inhibits PI3K) in oocytes leads to global primordial follicle activation, depletion of all oocytes and follicles, and premature ovarian failure (Castrillon, Miao, Kollipara, Horner, & DePinho, 2003; Liu et al., 2006; Reddy et al., 2008; Sullivan & Castrillon, 2011; Fig. 3).

One key event in the early stages of follicle growth involves the recruitment and differentiation of theca cells. Based on the phenotype of *Gdf9* knockout mice (Dong et al., 1996), oocyte expression and secretion of GDF9 play a key role in this process. Recent studies show further that the oocyte-derived factor GDF9 induces the expression of Desert hedgehog (DHH) and Indian Hedgehog (IHH) in granulosa cells (Liu, Peng, et al., 2015). These Hedgehog pathway ligands then activate the PATCHED/SMOOTHENED receptor complex present in theca progenitor cells leading to activation of the GLI transcription factors and theca cell differentiation into the steroidogenic and fibroblast stromal cell lineages, respectively. These recent studies provide an explanation for why the disruption of GDF9 prevents theca cell recruitment, proliferation, steroidogenesis, and hence follicle growth, leading to premature ovarian failure (Liu, Peng, et al., 2015). C-KIT and KITL also impact theca cell differentiation and androgen production throughout follicular growth and may prevent follicular atresia (Driancourt et al., 2000).

Following theca cell recruitment and differentiation in response to HH signaling events, primary and secondary follicles grow in response to other TGFβ family members including antimullerian hormone a marker of the follicle reserve that reduces follicle growth and activin that enhances follicle growth (Pangas & Rajkovic, 2015). Follicular growth and maintenance of the female gonad specification is also dependent on specific transcription factors (FOXL2, GATA4/6, AR) as well as the WNT4/CTNNB1 signaling pathway (Ottolenghi et al., 2007; Pangas & Rajkovic, 2015). FOXL2, in particular, is critical not only for ovary specification by suppressing expression of the male-determining gene SOX9 but also for enhancing expression of aromatase (*Cyp19a1*). FOXO1 is also a critical regulator of follicle growth, apoptosis, and cell-fate decisions (Liu, Ren, et al., 2015). Depletion of *Foxo1* and *Foxo3* in granulosa cells leads to ovarian granulosa cell tumors (GCTs) and persistent expression of FOXL2, GATA4, WNT4, and activin and activin-mediated phosphorylation of SMAD2/3 (Liu, Ren, et al., 2015). It is possible that FOXO1 and FOXL2 balance the expression of specific genes in granulosa cells that impact growth, differentiation, and apoptosis.

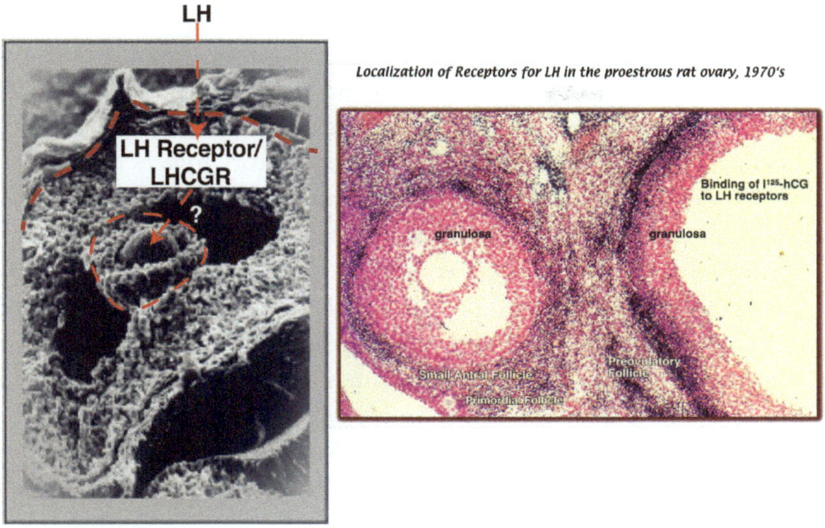

Fig. 4 LH induction of ovulation and COC expansion depends on LH receptor induction in granulosa cells of preovulatory follicles.

2.2.1 Later Stages of Follicular Growth

At the secondary and tertiary stages of follicular growth, the pituitary hormones FSH and LH begin to play more major roles. FSH regulates granulosa cell functions by binding and activating the FSH receptor, enhancing proliferation, inducing *Cyp19a1* expression and estradiol production. LH regulates theca cells by binding to the LH receptor leading to the production of androgens that are converted to estrogens in granulosa cells. Estradiol, in turn, via its receptor plays critical roles in granulosa cell proliferation, the regulation of FSH receptor activation, and the induction of the LH receptor in granulosa cells in response to FSH (Richards, 1994; Fig. 4).

2.2.2 Ovulation

Preovulatory follicles are characterized by increased responsiveness to FSH, the induction of LH receptors in granulosa cells and hence their acquisition of responsiveness to LH leading to ovulation. The LH surge activates not only the LH receptor stimulation of cAMP production but also the induction of EGF-like factors (*Areg*, *Ereg*, and *Btc*) that activate the EGF receptor and the ERK1/2 pathway leading to the induction of many inflammatory-related genes (Fan et al., 2009). The LH surge induces ovulation, a process that is like a "hit-and-run" reaction because the hormonal and molecular events mediating the ovulation process are induced rapidly and persist only transiently (Richards, Liu, & Shimada, 2015; Fig. 5). Ovulation is a complex process by which the preovulatory follicle ruptures and releases an expanded cumulus cell–oocyte complex (COC) from the surface of the follicle. Central to successful LH-induced ovulation are major changes in the expression of specific genes mediated, in part, by CITED4/CBP-mediated histone acetylation (Zhang et al., 2014). Specific ovulation-induced genes include proteases and targeted proteolysis, progesterone receptor (*Pgr*) and its activation by progesterone and the prostaglandin synthase 2 (*Ptgs2*) leading to the synthesis and action of prostaglandins, PGE2 (Kim & Duffy, 2016; Richards & Pangas, 2010; Xu et al., 2011). The inflammatory-related events of ovulation include the recruitment

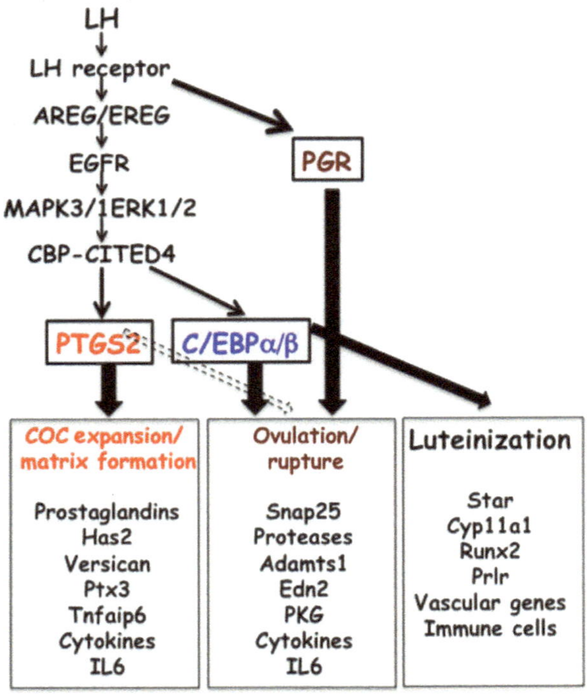

Fig. 5 Schematic of signaling pathways and genes that are activated and induced downstream of the LH surge in preovulatory follicles that control COC expansion, ovulation, and luteinization. *Adapted from Richards, J. S., & Pangas, S. A. (2010). The ovary: Basic biology and clinical implications.* The Journal of Clinical Investigation, 120, *963–972. PMID: 20364094; PMCID:PMC22846061.*

and functions of peripheral immune cells (Adams et al., 2016; Al-Alem et al., 2015; Richards & Pangas, 2010; Schmidt et al., 2014).

Based on in vivo perfusion studies and the conditional knockout of the progesterone receptor (*Pgr*), progesterone signaling is essential for successful ovulation (Brännström & Janson, 1989; Lydon et al., 1995; Robker et al., 2000; Tanaka, Espey, Kawano, & Okamura, 1991). Both forms of the progesterone receptor (PGRA and PGRB) are rapidly but transiently induced in granulosa cells of preovulatory follicles by the LH surge. One target of PGR (PGRA) action in granulosa cells of ovulating follicles is the protease, ADAMTS1, that targets and degrades the proteoglycan versican, a major component of the expanded cumulus COC (Robker et al., 2000). ADAMTS1 likely also controls other events leading to follicle rupture (Curry, 2010). Induction of PGRA/B by the LH surge is dependent on the EGF-like factors and EGFR activation. Importantly, PGRA and PGRB regulate the expression of distinct genes in granulosa cells, including, *Ctsl, Snap25, Rgs2, Egfr, Edn1, Cbp/p300 transactivator*, and *Adam8* (Shimada et al., 2007; Sriraman et al., 2008; Sriraman & Richards, 2010).

Prostaglandins are also essential for ovulation (Edelman, Jensen, Doom, & Hennebold, 2013; Richards et al., 2015; Richards & Pangas, 2010). Their synthesis is promoted as a consequence of LH induction of prostaglandin synthase 2 (*Ptgs2*; also known as cyclooxygenase 2) (Kim & Duffy, 2016; Sirois, Levy, Simmons, & Richards, 1993; Sirois, Simmons, & Richards, 1992; Wong & Richards, 1991, 1992). Prostaglandins, PGE2 by binding to the EP2/4 receptors, impact the process of COC expansion and genes that control COC matrix formation and stabilization.

The transcription factors C/EBPα and C/EBPβ are critical not only for ovulation but also for the formation of corpora lutea and progesterone biosynthesis (Fan, Liu, Johnson, & Richards, 2011). Based on molecular and immunohistochemical analyses, these transcription factors appear to impact the inflammatory response that is essential for the recruitment of immune cells and vascular components obligatory for the formation of corpora lutea.

Peripheral immune cells are essential for ovulation (Care et al., 2013; Cohen-Fredarow et al., 2014). This has been documented in mice by selective removal immune cells or dendritic cells prior to hormonal stimulation of ovulation using the compound chlodronate or diphtheria toxin-mediated disruption of immune cells, respectively. Ovulation is a highly conserved inflammatory process in mammals (Boerboom & Sirois, 1998; Peluffo, Murphy, Talcott Baughman, Stouffer, & Hennebold, 2011), and therefore, immune cells likely impact ovulation in humans as well.

2.2.3 Luteinization

The terminal differentiation of granulosa cells to luteal cells and the formation of the corpus luteum complete the "ovarian cycle." One of the major processes that occurs during corpus luteum formation is the invasion of blood vessels into the avascular ovulated follicles (Kim, Trau, & Duffy, 2017). Vascular development and luteal formation in response to gonadotropins are regulated, in part, by HIF1a-dependent VEGF expression as indicated recently in mice expressing a mutant *Vegf* gene lacking HIF response elements (Rico et al., 2012). These mice lack corpora lutea and exhibit a diminished vascular network as illustrated by CLARITY technology (Feng et al., 2017). Vascularization of the ovulated follicle is also dependent on PGEs (Kim & Duffy, 2016) and immune cells (Care et al., 2013). As mentioned earlier, the transcription factors C/EBPα and C/EBPβ are also essential for vascular development during luteinization (Fan et al., 2011). Progesterone produced by the corpus luteum is important for implantation and maintenance of pregnancy. Unlike follicular development that is a highly conserved process in all mammals, the functions and maintenance of corpora lutea are regulated by many luteotrophic hormones that are species specific and include LH, hCG, prolactin, and estrogen (Stocco, Telleria, & Gibori, 2007). Luteolysis, or the demise of the corpus luteum, can be induced by prostaglandins, involves immune cell activity (Walusimbi & Pate, 2013), and may be mediated by specific TGFβ family members (Goodman & Inskeep, 2015; Talbott et al., 2017).

3. PERIPHERAL IMPACT OF THE OVARIAN CYCLE

Completion of the "ovarian cycle" in the ovary from the primordial to ovulatory stages of follicle development and terminal differentiation of follicles to corpora lutea is associated with cyclical changes in the release of ovarian-derived steroids (estradiol and progesterone) and proteins (inhibin) that impact the female reproductive system (Herbison, 2016). In each organ, the ovarian steroid and peptides hormones impact changes in cell functions by binding to specific receptors and activating (or inhibiting) specific cellular and developmental processes.

3.1 Hypothalamus–Pituitary

Steroids are potent regulators of hypothalamic and pituitary cell functions (Goodman & Inskeep, 2015; Levine, 2015; Zeleznik & Plant, 2015). Tonic levels of estradiol and progesterone negatively regulate basal GnRH pulsatile release from neurons in the arcuate nucleus and thereby maintain basal levels of serum LH. The effects of steroids on basal LH release are mediated, in part, by their actions on kisspeptin neurons in the arcuate nucleus (Herbison, 2016; Fig. 2). By contrast, elevated levels of estradiol coming from preovulatory follicles induce increased pulsatile release of GnRH from neurons in the preoptic area leading to increased pulsatile release and the ovulatory surge of LH from the pituitary. The surge mode of LH release is controlled, in part by estradiol-mediated action on distinct kisspeptin neurons present in the preoptic area (Herbison, 2016). The effects of GnRH on LH secretion from gonadotropes are mediated, in part, by activation of the ERK1/2 pathway and ERG1 binding to the LH beta (*Lhb*) promoter. LH is packaged and released from the gonadotropes in secretion granules (Wang et al., 2014). Thus, not only is LH biosynthesis dependent on ERK1/2 signaling, but also, at the ovarian level, the LH surge induces EGF-like factors that activate ERK1/2 that are essential for ovulation. Progesterone is primarily a negative regulator of the basal and surge modes of GnRH and LH release, acting in part on kisspeptin1 neurons in the arcuate and preoptic areas (Herbison, 2016).

The biosynthesis and release of FSH are also modulated by estradiol and progesterone regulation of GnRH release, but, in addition, FSH is uniquely suppressed by the ovarian and pituitary-derived factor inhibin. Inhibin mediates its effects by blocking activin-induced phosphorylation and activation of SMAD2/3, transcription factors that bind and control transcriptional activation of the FSH beta (*Fshb*) promoter. FSH, unlike LH, is continuously released from the gonadotropes and is not secreted in granules (Wang et al., 2014). FSH beta expression is also regulated by the transcription factors FOXL2 and SF1. Thus, not only is FSH biosynthesis dependent on SMADs and FOXL2, but also FSH action in granulosa cells is associated with activin/SMAD2/3, FOXL2, and SF1 signaling events.

Steroids can also impact neuronal functions. Mating behavior in mice (lordosis) is dependent on estradiol and progesterone and their receptors as well as the phosphorylation and activation of the progesterone receptor by dopamine (Mani & Blaustein, 2012; Mani & Oyola, 2012). Importantly, theca-derived androgens impact ovarian follicular development by regulating FSH receptor expression and with FSH impact expression of aromatase and the LH receptor in granulosa cells (Fitzpatrick & Richards, 1991; Sen & Hammes, 2010; Sen et al., 2014). Elevated androgens in women with PCOS also appear to mediate both a systemic proinflammatory condition and a distinct ovarian inflammatory response to gonadotropins by mechanisms that appear to involve immune cell recruitment and functions and possible vascular changes (Adams et al., 2016; Adams, Polson, & Franks, 1986; Al-Alem et al., 2015; Diamanti-Kandarakis & Dunaif, 2012; Dumesic & Richards, 2013; Legro, Driscoll, Strauss, Fox, & Dunaif, 1998).

Resident immune/stem cells, called folliculostellate cells, also appear to be critical for pituitary functions (Castinetti, Davis, Brue, & Camper, 2011; Devnath & Inoue, 2008).

3.2 Oviduct and Uterus

Ovarian steroids orchestrate oviductal, uterine, and cervical functions that impact oocyte and embryo transport, implantation, pregnancy, and parturition (Kinder, Winuthayanon, Hewitt, Couse, & Korach, 2015; Mazur, Large, & DeMayo, 2015; Fig. 6). Hormonal control of these physiological processes is complex, dependent on precisely timed expression of specific regulatory factors and involves multiple signaling cascades, many of which are similar to those regulating follicular development. In addition, many events controlling pregnancy are species specific rather than being highly conserved as is observed in follicle development.

In the oviduct, estradiol from preovulatory follicles induces cilia formation in ciliated cells and mucin secretion from the glandular cells. The cilia are critical for promoting the movement of the ovulated COC and embryo (if fertilization occurs) down the oviduct (Lyons, Saridogan, & Djahanbakhch, 2006). Specific mucins, mainly MUC1, are secreted from the glandular cells and presumably provide nutrient, lubricant, and immune protective functions to the embryo. Loss of *Muc1* is associated with ectopic pregnancies (Al-Azemi, Refaat, Aplin, & Ledger, 2009).

In the uterine endometrium, estradiol and progesterone act on luminal epithelial cells and/or stromal cells to control the window of receptivity, embryo attachment, implantation, decidualization, and parturition (Mazur et al., 2015; Fig. 7). The effects of steroids depend on their interactions and cross talk with other signaling cascades in the endometrial uterine compartments for attachment, implantation, and decidualization. Estradiol acting via ESR in the luminal cells promotes the expression of the mucin *Muc1*. Estradiol via ESR and FGF signaling in stromal cells promotes the proliferation of the luminal cells by enhancing expression of *Ccne1* and *Ccne2* and *Igf1* (Clementi et al., 2013).

Attachment of the embryo to the uterine luminal epithelial cells and subsequent placental development is dependent on adhesion molecules present

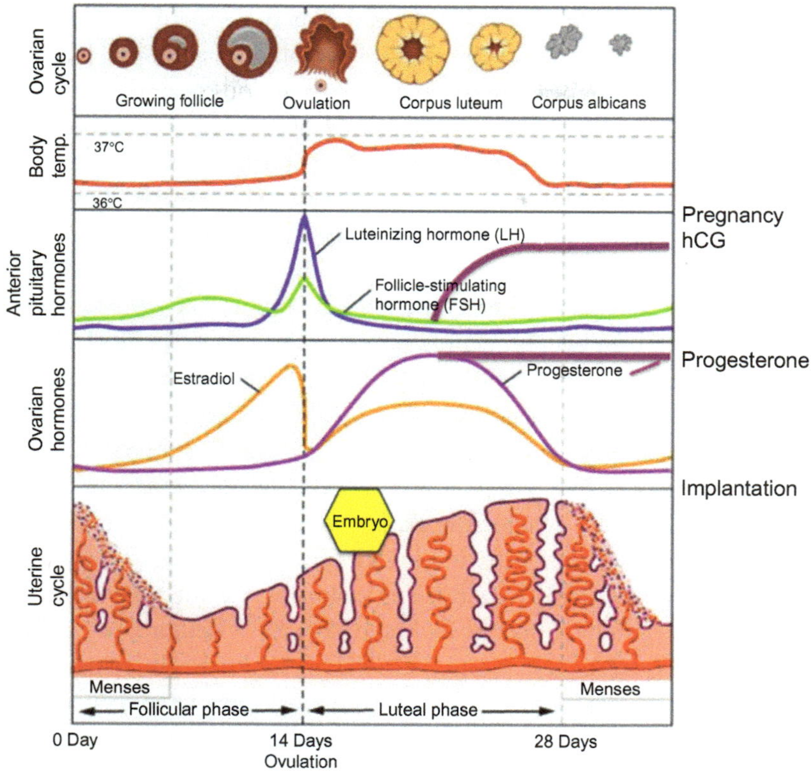

Fig. 6 Ovarian, hormonal and uterine events, including embryo attachment and implantation, during the reproductive cycle and pregnancy.

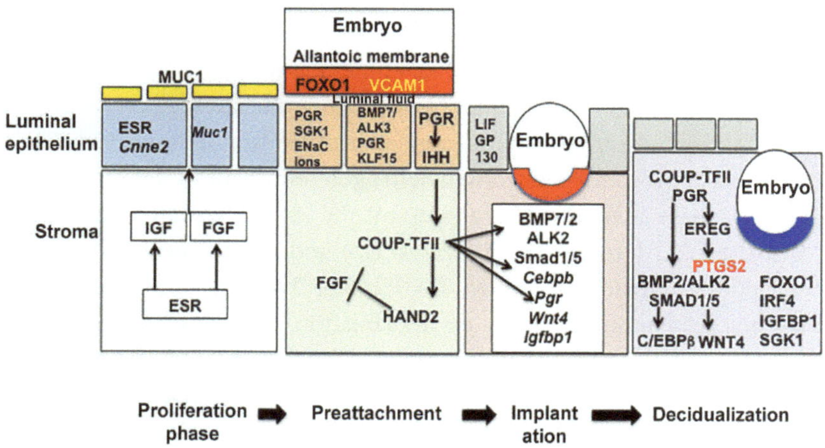

Fig. 7 Molecular events associated with embryo attachment, implantation, and the decidual response.

on the allantoic membrane of the embryo (Bai et al., 2014; Ferdous et al., 2011). Specifically, allantoic expression of vascular cell adhesion molecule 1 (VCAM1) is essential for embryo attachment (Ferdous et al., 2011; Gurtner et al., 1995). *Vcam1*-null mice die early in embryogenesis. Although heart defects were thought to mediate embryonic mortality in the *Vcam1*-null mice (Kwee et al., 1995), recent analyses of the *Foxo1*-null mice that also die early in embryogenesis indicate that embryo attachment via VCAM1 is the primary critical event (Ferdous et al., 2011). Moreover, FOXO1 is an upstream regulator of the *Vcam1* gene expression (Ferdous et al., 2011). Likely, similar events occur in other species (Bai et al., 2014). In addition, attachment is dependent on proper expression of ion channels and luminal fluid levels of potassium and sodium (Wetendorf & DeMayo, 2012) and a kinase SGK1 that regulates ion channel (ENaC) expression in the luminal cells (Fisher & Giudice, 2011). Overexpression of SGK1 in luminal cells prevents attachment (Fisher & Giudice, 2011; Salker et al., 2011). Overexpression of PGR in luminal cells of the $mPgrA^{LsL/+}$ mice also prevents attachment and alters the expression of specific genes, including *Sgk1* and *Vcam1* (Wetendorf & DeMayo, 2012). Whether or not there is a direct link between SGK1 and VCAM1 in luminal cells remains to be determined. Elevated SGK1 in human luminal cells or reduced SGK1 in the stromal compartment is associated with infertility and repeated pregnancy loss, indicating that SGK1 also plays critical roles in uterine function and pregnancy in women (Fisher & Giudice, 2011). Of note, the expression of the TGFβ family member *LEFTY2* is regulated by SGK1 and may play a role in regulating Nodal signaling events in the uterus (Salker et al., 2011).

Implantation is complex and characterized not only by changes in luminal epithelial cells to permit embryo attachment but also changes to support implantation and decidualization. These include the reduced responsiveness of luminal cells to estradiol and decreased proliferation (Mazur et al., 2015). In the mouse, the transition from the estrogen dominant to the progesterone dominant phase is mediated, in part, by the expression and activation of BMPs, the BMP1 receptors (ALK2/ALK3) and downstream SMAD1/5/4 transcription factors. Specifically, BMP2/ALK2 control the expression of genes involved in early stages of implantation, including *Cebpb, IGfbp1, Pgr*, and *Wnt4* in the stroma (Clementi et al., 2013). The BMP/ALK3 pathway and progesterone (PGR) induce *Klf15* that blocks epithelial cell proliferation (Mazur et al., 2015). Other mechanisms that suppress epithelial cell proliferation include PGR regulation of IHH in epithelial cells that impacts the Hedgehog (HH) (PTCH/SMO/GLI) signaling pathway in stromal cells. PGR also interacts with the transcription factor COUP-TFII (NR2F2) that impacts the expression of HAND2 that blocks FGF and its proliferative effects (Mazur et al., 2015).

Decidualization begins in response to embryo attachment and is an extension of implantation that is facilitated by LIF/GP130 activation and depends on the differentiation of stromal mesenchymal cells and fibroblasts that express specific genes (Aghajanova, Horcajadas, Esteban, & Giudice, 2010; Mazur et al., 2015; Spitzer et al., 2012). In mice, the mechanisms that promote decidualization include COUP-TFII and PGR regulation of BMP2 and WNT4/CTNNB1 signaling pathways and the induced expression of *Ptgs2* and PGE2 action. Induction and activation of C/EBPβ regulated by BMP/ALK2 (Clementi et al., 2013), and the EGF-like factor EREG also impact decidualization (Large & DeMayo, 2012). EGFR activation, in turn, can impact expression of WNT4 and BMP2. Uterine expression of FOXO1 also impacts the decidualization process by regulating endometrial expression of IRF4, a factor that impacts expression of WNT4 and IGFBP1 (Vasquez et al., 2015). SGK1 and LEFTY2 impact and regulate stress responses in decidual cells (Salker et al., 2011). Thus, many complex signaling pathways from the embryo to the uterus and within the uterine endometrium control attachment, implantation, and decidualization. Of note, many of these pathways are similar to those that impact ovulation and luteinization in the mouse. In humans there is evidence the bone marrow-derived mesenchymal stem cells, like endothelial mesenchymal stem cells, can differentiate into different cell lineages indicating great plasticity in these cells (Aghajanova et al., 2010; Spitzer et al., 2012).

In addition, the placenta acquires an immune-privileged environment to prevent immune recognition and rejection of the embryo, a topic to vast to cover here (Care et al., 2013).

In the uterine myometrium, estradiol and progesterone also play major roles to control uterine contractility (Renthal et al., 2015). During pregnancy progesterone promotes quiescence by suppressing proinflammatory response pathways (expression of *Ptgs2* and *Nfkb*) and "contractile" genes (oxytocin, *Oxtc*; connexin 43, *Cx43*). However, the transition to labor is associated with increased estradiol and uterine contractility and the suppression of progesterone-mediated events. In particular, this transition involves decreased expression of specific microRNAs that block *Ptgs2* expression and increase 20αHSD that metabolizes progesterone to an inactive metabolite. As a consequence of increased estradiol, *Ptgs2* expression and PGEs production and progesterone conversion to its inert metabolite increase and thereby lead to sensitizing the myometrium to estradiol and increased contractility.

3.3 Cervical Structure and Function

The cervix plays a critical role in maintaining pregnancy and eventually facilitating birth. As in other tissues, the ovarian steroid hormones estradiol and progesterone play distinct and complementary roles that impact structural components (collagen and elastin) of the cervix and their reorganization in the cervix during pregnancy and at parturition (Nallasamy et al., 2017).

4. BREAST DEVELOPMENT

The steroid hormones promote the development of the mammary gland and ultimately alveolar formation in preparation for providing milk for the offspring (Brisken & Ataca, 2015; Visvader & Stingl, 2014). As in other tissues, estradiol promotes proliferation of epithelial cells in breast ductal tissues, whereas progesterone in general promotes differentiation. Other regulators of breast development and function include prolactin (PRL), the WNT/CTNNB1, and FOXO1/IGF1 signaling cascades.

5. OVARIAN CANCER

Cancer in humans is largely related to the transformation of epithelial cells that undergo genomic instability, oncogene expression, proliferation, metastasis, and immune evasion. This is also true of ovarian cancer with the unique feature that ovarian epithelial cancer cells metastasize to the peritoneal cavity and home to the omentum, a metastatic site that does not depend on dissemination through the peripheral blood. There is quite strong epidemiological evidence that steroid hormones can impact ovarian cancer: the incidence of ovarian cancer is lower in women on contraceptives for as few as 5 years and higher in women on hormone replacement therapy (King & Wong, 2011; Mullany et al., 2014; Schildkraut, Calingaert, Marchbanks, Moorman, & Rodriguez, 2002). However, many other factors are also involved in this complex disease.

5.1 Epithelial Carcinomas

Epithelial ovarian cancer (EOC) continues to be the most lethal among all gynecologic malignancies and the fifth leading cause for cancer-associated deaths among women in the United States (Cancer Genome Atlas Research Network, 2011). EOC is most frequently associated with the transformation of epithelial cells derived either from the ovarian surface epithelium or from the fimbria of the oviduct. Based on morphological criteria, EOC has been divided into subtypes: serous, endometrial, mucinous, and clear cell, representing tissues of the reproductive tract (Cho & Shih, 2009). Of these subtypes, high-grade serous carcinomas (HGSC) are the most common and malignant. Based on molecular signatures HGSCs have been divided into additional subtypes: mesenchymal, immune reactive, epithelial, and vascular that exhibit distinct differences in morphology, responses to drugs and long-term survival outcome (Tothill et al., 2008).

Mutations in the tumor protein 53 (TP53, p53) coding region occur in over 96% of HGSC (Cancer Genome Atlas Research Network, 2011; Cole et al., 2016). While some of these mutations contribute to cancer progression as a result of loss of wild-type p53 activity, many result in the gain of an oncogenic function. These gain-of-function oncogenic p53 mutant proteins accumulate to high levels in cells, form stable protein aggregates, activate alternative gene expression programs, and contribute to carcinogenesis as well as development of drug resistance (Hoadley et al., 2014; Vousden & Privies, 2009).

5.2 Granulosa Cell Tumors

GCTs are quite rare and are divided into juvenile and adult types (Jamieson & Fuller, 2012). Mutant mouse models in which GCTs develop include activating mutations in β-catenin (*Ctnnb1*) combined with either the loss of *Pten* or the expression of active *Kras*^{G12D} (Boerboom et al., 2005; Richards et al., 2011). GCTs also develop in mice following granulosa cell-specific disruption of *Foxo1/Foxo3* (Liu, Ren, et al., 2015), depletion of *Smad1/5* (Middlebrook, Eldin, Li, Shivasankaran, & Pangas, 2009) and depletion of *Inha*. GCTs in the *Foxo1/3*, *Smad1/5*, and *Inha* (Matzuk, Finegold, Su, Hsueh, & Bradley, 1992) mutant mice exhibit persistent phosphorylation of SMAD2/3 that can be explained, in part, by the apparent unopposed proliferative actions of activin with sustained expression of FOXL2, GATA4, and WNT4 (Liu, Ren, et al., 2015).

6. UTERINE, BREAST, AND CERVICAL CANCERS

The steroid hormones impact the development, progression, and metastasis of cancer in other reproductive organs via their respective receptors (Brisken & Ataca, 2015). In the uterus, breast, and cervix epithelial cells and stromal cells contribute to tumorigenesis not only via activation of the cognate steroid receptors but also by aberrant activation or inactivation of many other signaling cascades, some of which, include mutations in *p53*, *Pten*, and *Brca1*, are common to EOC. Unique to cervical cancer are the interactions of steroids and their receptors with human papilloma virus 16 (HPV16) (Chung, 2015; Mehta, Baik, & Chung, 2016). Hopefully, novel therapies will soon emerge to help control or eliminate all of these malignancies.

REFERENCES

Adams, J., Liu, Z., Ren, Y. A., Wun, W.-S., Zhou, W., Kenigsberg, S., et al. (2016). Enhanced inflammatory transcriptome in granulosa cells of women with polycystic ovarian syndrome. *The Journal of Clinical Endocrinology and Metabolism, 101*, 3459–3468. 27228368.

Adams, J., Polson, D. W., & Franks, S. (1986). Prevalence of polycystic ovaries in women with anovulation and idiopathic hirsutism. *British Medical Journal (Clinical Research Ed.), 293*(6543), 355–359. 3089520.

Aghajanova, L., Horcajadas, J. A., Esteban, F. J., & Giudice, L. C. (2010). The bone marrow-derived human mesenchymal stem cell: Potential progenitor of the endometrial stromal fibroblast. *Biology of Reproduction, 82*(6), 1076–1087. PMCID: PMC2874495. https://doi.org/10.1095/biolreprod.109.082867.

Al-Alem, L., Puttabyatappa, M., Rosewell, K., Brännström, M., Akin, J., Boldt, J., et al. (2015). Chemokine ligand 20: A signal for leukocyte recruitment during human ovulation? *Endocrinology, 156*(9), 3358–3369. PMID: 26125463. https://doi.org/10.1210/en.2014-1874.

Al-Azemi, M., Refaat, B., Aplin, J., & Ledger, W. (2009). The expression of MUC1 in human fallopian tube during the menstrual cycle and in ectopic pregnancy. *Human Reproduction, 24*(10), 2582–2587. PMID: 19561343. https://doi.org/10.1093/humrep/dep233.

Bai, R., Bai, H., Kuse, M., Ideta, A., Aoyagi, Y., Fujiwara, H., et al. (2014). Involvement of VCAM1 in the bovine conceptus adhesion to the uterine endometrium. *Reproduction, 148*(2), 119–127. 24803492.

Boerboom, D., Paquet, M., Hsieh, M., Liu, J., Jamin, S. P., Behringer, R. R., et al. (2005). Misregulated Wnt/b-catenin signaling leads to ovarian granulosa cell tumor development. *Cancer Research, 65*, 9206–9215. 16230381.

Boerboom, D., & Sirois, J. (1998). Molecular characterization of equine prostaglandin G/H synthase-2 and regulation of its messenger ribonucleic acid in preovulatory follicles. *Endocrinology, 139*, 1662–1670.

Brännström, M., & Janson, P. O. (1989). Progesterone is a mediator in the ovulatory process of the in vitro-perfused rat ovary. *Biology of Reproduction, 40*(6), 1170–1178.

Brisken, C., & Ataca, D. (2015). Endocrine hormones and local signals during the development of the mouse mammary gland. *Wiley Interdisciplinary Reviews: Developmental Biology, 4*(3), 181–195. PMID: 25645332. https://doi.org/10.1002/wdev.172.

Cancer Genome Atlas Research Network. (2011). Integrated genomic analyses of ovarian carcinoma. *Nature, 474*, 609–615. 21720365.

Care, A. S., Diener, K. R., Jasper, M. J., Brown, H. M., Ingman, W. V., & Robertson, S. A. (2013). Macrophages regulate corpus luteum development during embryo implantation in mice. *The Journal of Clinical Investigation, 123*(8), 3472–3487. PMID: 23867505. https://doi.org/10.1172/jci60561.

Castinetti, F., Davis, S. W., Brue, T., & Camper, S. A. (2011). Pituitary stem cell update and potential implications for treating hypopituitarism. *Endocrine Reviews, 32*(4), 453–471. PMCID: PMC3369576. https://doi.org/10.1210/er.2010-0011.

Castrillon, D. H., Miao, L., Kollipara, R., Horner, J. W., & DePinho, R. A. (2003). Suppression of ovarian follicle activation in mice by the transcription factor Foxo3a. *Science, 301*, 215–218. 12855809.

Cho, K. R., & Shih, I.-M. (2009). Ovarian cancer. *Annual Review of Pathology, 4*, 287–313. 18842102. PMCID: PMC12679364.

Chung, S.-H. (2015). Targeting female hormone receptors as cervical cancer therapy. *Trends in Endocrinology & Metabolism, 26*(8), 399–401. PMID: 26163756. https://doi.org/10.1016/j.tem.2015.06.004.

Clementi, C., Tripurani, S. K., Large, M. J., Edson, M. A., Creighton, C. J., Hawkins, S. M., et al. (2013). Activin-like kinase 2 functions in peri-implantation uterine signaling in mice and humans. *PLoS Genetics, 9*(11), e1003863. https://doi.org/10.1371/journal.pgen.1003863.

Cohen-Fredarow, A., Tadmor, A., Raz, T., Meterani, N., Addadi, Y., Nevo, N., et al. (2014). Ovarian dendritic cells act as a double-edged pro-ovulatory and anti-inflammatory sword. *Molecular Endocrinology, 28*(7), 1039–1054. PMID: 24825398. https://doi.org/10.1210/me.2013-1400.

Cole, A. J., Dwight, T., Gill, A. J., Dickson, K.-A., Zhu, Y., Clarkson, A., et al. (2016). Assessing mutant p53 in primary high-grade serous ovarian cancer using immunohistochemistry and massively parallel sequencing. *Scientific Reports, 6*, 26191 PMID: 27189670. https://doi.org/10.1038/srep26191 http://www.nature.com/articles/srep26191. supplementary-information.

Curry, T. E., Jr. (2010). ADAMTS1 and versican: Partners in ovulation and fertilization. *Biology of Reproduction, 83*, 505–506.

Devnath, S., & Inoue, K. (2008). An insight to pituitary folliculo-stellate cells. *Journal of Neuroendocrinology, 20*(6), 687–691. PMID: 18601690 https://doi.org/10.1111/j.1365-2826.2008.01716.x.

Diamanti-Kandarakis, E., & Dunaif, A. (2012). Insulin resistance and the polycystic ovary syndrome revisited: An update on mechanisms and implications. *Endocrine Reviews, 33*(6), 981–1030. 23065822.

Dong, J., Albertini, D. F., Nishimori, K., Kumar, T. R., Lu, N., & Matzuk, M. M. (1996). Growth differentiation factor-9 is required during early ovarian folliculogenesis. *Nature, 383*, 531–535.

Driancourt, M. A., Reynaud, K., Cortvrindt, R., & Smitz, J. (2000). Roles of KIT and KIT LIGAND in ovarian function. *Reviews of Reproduction, 5*, 143–152. https://doi.org/10.1530/ror.0.0050143.

Dumesic, D. A., & Richards, J. S. (2013). Ontogeny of the ovary in polycystic ovary syndrome. *Fertility and Sterility, 100*(1), 23–38. PMID: 23472949 https://doi.org/10.1016/j.fertnstert.2013.02.011.

Edelman, A. B., Jensen, J. T., Doom, C., & Hennebold, J. D. (2013). Impact of the prostaglandin synthase-2 inhibitor celecoxib on ovulation and luteal events in women. *Contraception, 87*(3), 352–357. https://doi.org/10.1016/j.contraception.2012.07.004.

Edson, M. A., Nagaraja, A. K., & Matzuk, M. M. (2009). The mammalian ovary from genesis to revelation. *Endocrine Reviews, 30*, 624–712. 19776209. PMCID: PMC12761115.

Falender, A. E., Shimada, M., Lo, Y., & Richards, J. S. (2005). TAF4b, a TBP associated factor is required for oocyte development and function. *Developmental Biology, 288*, 405–419.

Fan, H. Y., Liu, Z., Johnson, P. F., & Richards, J. S. (2011). CCAAT/Enhancer-binding proteins (C/EBP)-{alpha} and -{beta} are essential for ovulation, luteinization and expression of key target genes. *Molecular Endocrinology, 25*, 253–268. 21177758. PMCID: PMC23386543.

Fan, H. Y., Liu, Z., Shimada, M., Sterneck, E., Johnson, P. F., Hedrick, S. M., et al. (2009). MAPK3/1 (ERK1/2) in ovarian granulosa cells are essential for female fertility. *Science, 324*, 938–941. 19443782. PMC12847890.

Feng, Y., Cui, P., Lu, X., Hsueh, B., Möller Billig, F., Zarnescu Yanez, L., et al. (2017). CLARITY reveals dynamics of ovarian follicular architecture and vasculature in three-dimensions. *Scientific Reports*, 7, 44810. PMID: 28333125. https://doi.org/10.1038/srep44810.

Ferdous, A., Morris, J., Abedin, M. J., Collins, S., Richardson, J. A., & Hill, J. A. (2011). Forkhead factor FoxO1 is essential for placental morphogenesis in the developing embryo. *Proceedings of the National Academy of Sciences of the United States of America, 108*(39), 16307–16312. 21930913.

Fisher, S. J., & Giudice, L. C. (2011). SGK1: A fine balancing act for human pregnancy. *Nature Medicine, 17*(11), 1348–1349. https://doi.org/10.1038/nm.2549.

Fitzpatrick, S. L., & Richards, J. S. (1991). Regulation of cytochrome P450 aromatase messenger ribonucleic acid and activity by steroids and gonadotropins in rat granulosa cells. *Endocrinology, 129*(3), 1452–1462. 1651851.

Goodman, R. L., & Inskeep, E. K. (2015). Control of the ovarian cycle of the sheep. In T. M. Plant & A. J. Zeleznik (Eds.), *Knobil and Neill's physiology of reproduction: Vol. 2* (pp. 1259–1305). Waltham, MA: Academic Press.

Gurtner, G. C., Davis, V., Li, H., McCoy, M. J., Sharpe, A., & Cybulsky, M. I. (1995). Targeted disruption of the murine VCAM1 gene: Essential role of VCAM-1 in chorio-allantoic fusion and placentation. *Genes and Development, 9*(1), 1–14. 7530222.

Herbison, A. E. (2016). Control of puberty onset and fertility by gonadotropin-releasing hormone neurons. *Nature Reviews. Endocrinology, 12*(8), 452–466. PMID: 27199290. https://doi.org/10.1038/nrendo.2016.70.

Hoadley, K. A., Yau, C., Wolf, D. M., Cherniack, A. D., Tamborero, D., Ng, S., et al. (2014). Multiplatform analysis of 12 cancer types reveals molecular classification within and across tissues of origin. *Cell, 158*(4), 929–944. PMID: 25109877. https://doi.org/10.1016/j.cell.2014.06.049.

Hummitzsch, K., Irving-Rodgers, H. F., Hatzirodos, N., Bonner, W., Sabatier, L., Reinhardt, D. P., et al. (2013). A new model of development of the mammalian ovary and follicles. *PLoS One, 8*(2), e55578. https://doi.org/10.1371/journal.pone.0055578.

Jamieson, S., & Fuller, J. P. (2012). Molecular pathogenesis of granulosa cell tumors of the ovary. *Endocrine Reviews, 33*(1), 109–144. PMID: 22240241. https://doi.org/10.1210/er.2011-0014.

Karsch, F. J. (1980). Seasonal reproduction: A saga of reversible fertility. *The Physiologist, 23*, 29–39. 7208642.

Kim, S. O., & Duffy, D. M. (2016). Mapping PTGERs to the ovulatory follicle: Regional responses to the ovulatory PGE2 signal1. *Biology of Reproduction, 95*(2), 33, 31–11, PMID: 27307073. https://doi.org/10.1095/biolreprod.116.140574.

Kim, S. O., Trau, H. A., & Duffy, D. M. (2017). Vascular endothelial growth factors C and D may promote angiogenesis in the primate ovulatory follicle. *Biology of Reproduction, 96*(2), 389–400. PMID: 28203718. https://doi.org/10.1095/biolreprod.116.144733.

Kinder, A. K., Winuthayanon, W., Hewitt, S. C., Couse, J. F., & Korach, K. S. (2015). Steroid receptors in the uterus and ovary. In T. M. Plant & A. J. Zeleznik (Eds.), *Knobil and Neill's physiology of reproduction: Vol. 1* (pp. 1099–1193). Waltham, MA: Academic Press.
King, E. R., & Wong, K. K. (2011). Steroid hormones and ovarian cancer. In *Steroids clinical aspect* (pp. 111–140). Intech. Open. https://doi.org/10.5772/27123.
Kwee, L., Baldwin, H. S., Shen, H. M., Stewart, C. L., Buck, C., Buck, C. A., et al. (1995). Defective development of the embryonic and extraembryonic circulatory systems in vascular cell adhesion molecule (VCAM-1) deficient mice. *Development, 121*(2), 489. 7539357.
Large, M. J., & DeMayo, F. J. (2012). The regulation of embryo implantation and endometrial decidualization by progesterone receptor signaling. *Molecular and Cellular Endocrinology, 358*(2), 155–165. https://doi.org/10.1016/j.mce.2011.07.027.
Legro, R. S., Driscoll, D., Strauss, J. F., Fox, J., & Dunaif, A. (1998). Evidence for a genetic basis for hyperandrogenemia in polycystic ovary syndrome. *Proceedings of the National Academy of Sciences of the United States of America, 95*, 1199–1257. https://doi.org/10.1073/pnas.95.25.14956.
Levine, J. E. (2015). Neuroendocrine control of the ovarian cycle of the rat. In T. M. Plant & A. J. Zeleznik (Eds.), *Knobil and Neill's physiology of reproduction: Vol. 2* (pp. 1199–1257). Waltham, MA: Academic Press.
Liu, C., Peng, J., Matzuk, M. M., & Yao, H. H. C. (2015). Lineage specification of ovarian theca cells requires multi-cellular interactions via oocyte and granulosa cells. *Nature Communications, 6*, 6934. PMCID: PMC4413950. https://doi.org/10.1038/ncomms7934.
Liu, L., Rajareddy, S., Reddy, P., Du, C., Jagarlamudi, K., Shen, Y., et al. (2006). Infertility caused by retardation of follicular development in mice with oocyte-specific expression of Foxo3a. *Development, 134*, 199–209.
Liu, Z., Ren, Y. A., Pangas, S. A., Adams, J., Zhou, W., Castrillon, D. H., et al. (2015). FOXO1/3 and PTEN depletion in granulosa cells promotes ovarian granulosa cell tumor development. *Molecular Endocrinology, 29*(7), 1006–1024. PMID: 26061565. https://doi.org/10.1210/me.2015-1103.
Luo, X., Ikeda, Y., & Parker, K. (1994). A cell-specific nuclear receptor is essential for adrenal and gonadal development and sexual differentiation. *Cell, 77*, 481–490.
Lydon, J. P., DeMayo, F., Funk, C. R., Mani, S. K., Hughes, A. R., Montgomery, C. A., et al. (1995). Mice lacking progesterone receptor exhibit reproductive abnormalities. *Genes & Development, 9*, 2266–2278.
Lyons, R. A., Saridogan, E., & Djahanbakhch, O. (2006). The reproductive significance of human fallopian tube cilia. *Human Reproduction Update, 12*(4), 363–372. PMID: 16565155. https://doi.org/10.1093/humupd/dml012.
Mani, S. K., & Blaustein, J. D. (2012). Neural progestin receptors and female sexual behavior. *Neuroendocrinology, 96*(2), 152–161. 22538437.
Mani, S. K., & Oyola, M. G. (2012). Progesterone signaling mechanisms in brain and behavior. *Frontiers in Endocrinology, 3*, 7. PMCID: PMC3355960.
Matzuk, M. M., Finegold, M., Su, J.-G. J., Hsueh, A. J. W., & Bradley, A. (1992). α-Inhibin is a tumour-suppressor gene with gonadal specificity in mice. *Nature, 360*, 313–319. 1448148.
Mazur, E. C., Large, M. J., & DeMayo, F. (2015). Human oviduct and endometrium: Changes over the menstrual cycle. In T. M. Plant & A. J. Zeleznik (Eds.), *Knobil and Neill's physiology of reproduction: Vol. 1* (pp. 1077–1097). Waltham, MA: Academic Press.
Mehta, F. F., Baik, S., & Chung, S.-H. (2016). Recurrence of cervical cancer and its resistance to progestin therapy in a mouse model. *Oncotarget, 8*, 2372–2380. 27911853.

Menke, D. B., & Page, D. C. (2002). Sexually dimorphic gene expression in the developing mouse gonad. *Gene Expression Patterns, 2*, 359–367. 12617826.

Middlebrook, B. S., Eldin, K., Li, X., Shivasankaran, S., & Pangas, S. A. (2009). Smad1-Smad5 ovarian conditional knockout mice develop a disease profile similar to the juvenile form of human granulosa cells tumors. *Endocrinology, 150*, 5208–5217. 19819941.

Morohashi, K., Yoshioka, H., & Kawajiri, K. (2001). In *Organogenesis of gonads and transcription factors. Program of the 83rd annual meeting of the endocrine society*. June 20–23, Abstract #TP1–1, pp. 22. Denver, Colorado.

Mullany, L. K., Liu, Z., Wong, K. K., Deneke, V., Ren, Y. A., Herron, A., et al. (2014). Tumor repressor protein 53 and steroid hormones provide a new paradigm for ovarian cancer metastases. *Molecular Endocrinology, 28*(1), 127–137. PMID: 24264574; PMCID: PMC23874458 https://doi.org/10.1210/me.2013-1308.

Nallasamy, S., Yoshida, K., Akins, M., Myers, K., Iozzo, R., & Mahendroo, M. (2017). Steroid hormones are key modulators of tissue mechanical function via regulation of collagen and elastic fibers. *Endocrinology, 158*(4), 950–962. PMID: 28204185 https://doi.org/10.1210/en.2016-1930.

Nicol, B., & Yao, H. H. (2014). Building an ovary: Insights into establishment of somatic cell lineages in the mouse. *Sexual Development, 8*(5), 243–251. 24480990.

Ottolenghi, C., Pelosi, E., Tran, J., Colombino, M., Douglass, E., Nedorezov, T., et al. (2007). Loss of Wnt4 and Foxl2 leads to female-to-male sex reversal extending to germ cells. *Human Molecular Genetics, 16*(23), 2795–2804. 17728319.

Pangas, S. A., & Rajkovic, A. (2015). Follicular development: Mouse, sheep and human models. In T. M. Plant & A. J. Zeleznik (Eds.), *Knobil and Neill's physiology of reproduction: Vol. 1* (pp. 947–995). Waltham, MA: Academic Press.

Peluffo, M. C., Murphy, M. J., Talcott Baughman, S., Stouffer, R. L., & Hennebold, J. D. (2011). Systematic analysis of protease gene expression in the rhesus macaque ovulatory follicle: Metalloproteinase involvement in follicle rupture. *Endocrinology, 152*(10), 3963–3974.

Physiology of Reproduction. (2015). (Vol. 1). Waltham, MA: Academic Press.

Reddy, P., Liu, L., Adhikari, D., Jagarlamudi, K., Rajareddy, S., Shen, Y., et al. (2008). Oocyte-specific deletion of Pten causes premature activation of the primordial follicle pool. *Science, 319*, 611–613.

Renthal, N. E., Williams, K. R. C., Montalbano, A. P., Chen, C.-C., Gao, L., & Mendelson, C. R. (2015). Molecular regulation of parturition: A myometrial perspective. *Cold Spring Harbor Perspectives in Medicine, 5*(11), 1–16. 26337112.

Richards, J. S. (1979). Hormonal control of ovarian follicular development: A 1978 perspective. *Recent Progress in Hormone Research, 35*, 343–373. 229527.

Richards, J. S. (1994). Hormonal control of gene expression in the ovary. *Endocrine Reviews, 15*, 725–751. 7705279.

Richards, J. S., Fan, H. Y., Liu, Z., Tsoi, M., Lague, M. N., Boyer, A., et al. (2011). Either Kras activation or Pten loss similarly enhances the dominant-stable CTNNB1-induced genetic program to promote granulosa cell tumor development in the ovary and testis. *Oncogene, 31*, 1504–1520. https://doi.org/10.1038/onc.2011.341. PMID: 21860425.

Richards, J. S., Liu, Z., & Shimada, M. (2015). Ovulation. In T. M. Plant & A. J. Zeleznik (Eds.), *Knobil and Neill's physiology of reproduction: Vol. 1* (4th ed., pp. 997–1021). Waltham, MA: Academic Press.

Richards, J. S., & Pangas, S. A. (2010). The ovary: Basic biology and clinical implications. *The Journal of Clinical Investigation, 120*, 963–972. 20364094. PMCID: PMC22846061.

Rico, C., Laguë, M.-N., Lefèvre, P., Tsoi, M., Dodelet-Devillers, A., Kumar, V., et al. (2012). Pharmacological targeting of mammalian target of rapamycin inhibits ovarian granulosa cell tumor growth. *Carcinogenesis, 33*(11), 2283–2292.

Robker, R. L., Russell, D. L., Espey, L. L., Lydon, J. P., O'Malley, B. W., & Richards, J. S. (2000). Progesterone-regulated genes in the ovulation process: ADAMTS-1 and cathepsin L proteases. *Proceedings of the National Academy of Sciences of the United States of America, 97*, 4689–4694. 10781075.

Salker, M. S., Christian, M., Steel, J. H., Nautiyal, J., Lavery, S., Trew, G., et al. (2011). Deregulation of the serum- and glucocorticoid-inducible kinase SGK1 in the endometrium causes reproductive failure. *Nature Medicine, 17*(11), 1509–1513. https://doi.org/10.1038/nm.2498. PMID: 22001908. http://www.nature.com/nm/journal/v17/n11/abs/nm.2498.html. supplementary-information.

Schildkraut, J. M., Calingaert, B., Marchbanks, P. A., Moorman, P. G., & Rodriguez, G. C. (2002). Impact of progestin and estrogen potency in oral contraceptives on ovarian cancer risk. *Journal of the National Cancer Institute, 94*, 32–38. 11773280.

Schmidt, J., Weijdegård, B., Mikkelsen, A. L., Lindenberg, S., Nilsson, L., & Brännström, M. (2014). Differential expression of inflammation-related genes in the ovarian stroma and granulosa cells of PCOS women. *Molecular Human Reproduction, 20*(1), 49–58. 23900753.

Sen, A., & Hammes, S. R. (2010). Granulosa cell-specific androgen receptors are critical regulators of ovarian development and function. *Molecular Endocrinology, 24*(7), 1393–1403. 20501640. PMCID: PMC22903904.

Sen, A., Prizant, H., Light, A., Biswas, A., Hayes, E., Lee, H.-J., et al. (2014). Androgens regulate ovarian follicular development by increasing follicle stimulating hormone receptor and microRNA-125b expression. *Proceedings of the National Academy of Sciences of the United States of America, 111*(8), 3008–3013. 24516121.

Shima, Y., Miyabayashi, K., Baba, T., Otake, H., Oka, S., Zubair, M., et al. (2011). Identification of an enhancer in the Ad4BP/SF-1 gene specific for fetal Leydig cells. *Endocrinology, 153*(1), 417–425. PMID: 22128023. https://doi.org/10.1210/en.2011-1407.

Shimada, M., Yanai, Y., Okazaki, T., Yamashita, Y., Sriraman, V., Wilson, M. C., et al. (2007). Synaptosomal associated protein 25 gene expression is hormonally regulated during ovulation and is involved in cytokine/chemokine exocytosis from granulosa cells. *Molecular Endocrinology, 21*, 2487–2502.

Sirois, J., Levy, L. O., Simmons, D. L., & Richards, J. S. (1993). Characterization and hormonal regulation of the promoter of the rat prostaglandin endoperoxide synthase 2 gene in granulosa cells. *The Journal of Biological Chemistry, 268*, 12199–12206.

Sirois, J., Simmons, D. L., & Richards, J. S. (1992). Hormonal regulation of messenger ribonucleic acid encoding a novel isoform of prostaglandin endoperoxide H synthase in rat preovulatory follicles. *The Journal of Biological Chemistry, 267*, 11586–11592.

Spitzer, T. L. B., Rojas, A., Zelenko, Z., Aghajanova, L., Erikson, D. W., Barragan, F., et al. (2012). Perivascular human endometrial mesenchymal stem cells express pathways relevant to self-renewal, lineage specification, and functional phenotype1. *Biology of Reproduction, 86*(2), 58, 51-16-58, 51-16. https://doi.org/10.1095/biolreprod.111.095885.

Sriraman, V., Eichenlaub-Ritter, U., Bartsch, J. W., Rittger, A., Mulders, S. M., & Richards, J. S. (2008). Regulated expression of ADAM8 (a disintegrin and metalloprotease domain 8) in the mouse ovary: Evidence for a regulatory role of luteinizing hormone, progesterone receptor, and epidermal growth factor-like growth factors. *Biology of Reproduction, 78*(6), 1038–1048.

Sriraman, V., & Richards, J. S. (2010). Progesterone receptor-induced gene expression in primary mouse granulosa cell cultures. *Biology of Reproduction, 82*, 402–412.

Stocco, C., Telleria, C., & Gibori, G. (2007). The molecular control of corpus luteum formation, function and regression. *Endocrine Reviews, 28*, 117–149.

Sullivan, S. D., & Castrillon, D. H. (2011). Insights into primary ovarian insufficiency through genetically engineered mouse models. *Seminars in Reproductive Medicine, 29*, 283–298.

Talbott, H., Hou, X., Qiu, F., Zhang, P., Guda, C., Yu, F., et al. (2017). Early transcriptome responses of the bovine midcycle corpus luteum to prostaglandin F2α includes cytokine signaling. *Molecular and Cellular Endocrinology, 452*, 93–109. PMID: 28549990. https://doi.org/10.1016/j.mce.2017.05.018.

Tanaka, N., Espey, L. L., Kawano, T., & Okamura, H. (1991). Comparison of inhibitory actions of indomethacin and epostane on ovulation in rats. *The American Journal of Physiology, 260*, E170–E174.

Tothill, R. W., Tinker, A. V., George, J., Brown, R., Fox, S. B., Lade, S., et al. (2008). Novel molecular subtypes of serous and endometrioid ovarian cancer linked to clinical outcome. *Clinical Cancer Research, 14*(16), 5198–5208. 18698038.

Vanorny, D. A., & Mayo, K. E. (2017). The role of Notch signaling in the mammalian ovary. *Reproduction, 153*(6), R187–R204. 28283672.

Vasquez, Y. M., Mazur, E. C., Li, X., Kommagani, R., Jiang, L., Chen, R., et al. (2015). FOXO1 is required for binding of PR on IRF4, novel transcriptional regulator of endometrial stromal decidualization. *Molecular Endocrinology, 29*(3), 421–433. PMID: 25584414 https://doi.org/10.1210/me.2014-1292.

Visvader, J. E., & Stingl, J. (2014). Mammary stem cells and the differentiation hierarchy: Current status and perspectives. *Genes and Development, 28*(11), 1143–1158. 24888586.

Vousden, K. H., & Privies, C. (2009). Blinded by the light: The growing complexity of p53. *Cell, 137*, 413–431. 19410540.

Walusimbi, S. S., & Pate, J. L. (2013). Physiology and endocrinology symposium: Role of immune cells in the corpus luteum. *Journal of Animal Science, 91*(4), 1650–1659. 23422006.

Wang, H., Larson, M., Jablonka-Shariff, A., Pearl, C. A., Miller, W. L., Conn, P. M., et al. (2014). Redirecting intracellular trafficking and the secretion pattern of FSH dramatically enhances ovarian function in mice. *Proceedings of the National Academy of Sciences of the United States of America, 111*(15), 5735–5740. PMID: 24706813. https://doi.org/10.1073/pnas.1321404111.

Wetendorf, M., & DeMayo, F. J. (2012). The progesterone receptor regulates implantation, decidualization, and glandular development via a complex paracrine signaling network. *Molecular and Cellular Endocrinology, 357*(1–2), 108–118. PMID: 22115959; PMCID: PMC23443857. https://doi.org/10.1016/j.mce.2011.10.028.

Wong, W. Y. W., & Richards, J. S. (1991). Evidence for two antigenically distinct molecular weight variants of prostaglandin H synthase in the rat ovary. *Molecular Endocrinology, 5*, 1269–1279.

Wong, W. Y., & Richards, J. S. (1992). Induction of prostaglandin H synthase in rat preovulatory follicles by gonadotropin-releasing hormone. *Endocrinology, 130*, 3512–3521.

Xu, F., Stouffer, R. L., Muller, J., Hennebold, J. D., Wright, J. W., Bahar, A., et al. (2011). Dynamics of the transcriptome in the primate ovulatory follicle. *Molecular Human Reproduction, 17*, 152–165. 21036944.

Zeleznik, A. J., & Plant, T. M. (2015). Control of the menstrual cycle. In T. M. Plant & A. J. Zeleznik (Eds.), *Knobil and Neill's physiology of reproduction: Vol. 2* (pp. 1307–1361). Waltham, MA: Academic Press.

Zhang, Y.-L., Xia, Y., Yu, C., Richards, J. S., Liu, J., & Fan, H.-Y. (2014). CBP-CITED4 is required for luteinizing hormone-triggered target gene expression during ovulation. *MHR: Basic Science of Reproductive Medicine, 20*(9), 850–860. PMID: 24878634. https://doi.org/10.1093/molehr/gau040.

Zheng, W., Nagaraju, G., Liu, Z., & Liu, K. (2012). Functional roles of the phosphatidylinositol 3-kinases (PI3Ks) signaling in the mammalian ovary. *Molecular and Cellular Endocrinology, 356*(1), 24–30. PMID: 21684319 https://doi.org/10.1016/j.mce.2011.05.027.

Zheng, W., Zhang, H., Gorre, N., Risal, S., Shen, Y., & Liu, K. (2014). Two classes of ovarian primordial follicles exhibit distinct developmental dynamics and physiological functions. *Human Molecular Genetics*, *23*(4), 920–928.

Zheng, W., Zhang, H., & Liu, K. (2014). The two classes of primordial follicles in the mouse ovary: Their development, physiological functions and implications for future research. *Molecular Human Reproduction*, *20*(4), 286–292.

CHAPTER TWO

GnRH in the Human Female Reproductive Axis☆

Patrizia Limonta, Marina Montagnani Marelli, Roberta Moretti, Monica Marzagalli, Fabrizio Fontana, Roberto Maggi[1]
Università degli Studi di Milano, Milano, Italy
[1]Corresponding author: e-mail address: roberto.maggi@unimi.it

Contents

Abstract

Gonadotropin-releasing hormone (GnRH) is recognized as the central regulator of the functions of the pituitary–gonadal axis. The increasing knowledge on the mechanisms controlling the development and the function of GnRH-producing neurons is leading to a better diagnostic and therapeutic approach for hypogonadotropic hypogonadisms and for alterations of the puberty onset. During female life span, the function of the

☆The paper does not require any permissions.

Vitamins and Hormones, Volume 107
ISSN 0083-6729
https://doi.org/10.1016/bs.vh.2018.01.003

GnRH pulse generator may be affected by a number of inputs from other neuronal systems, offering alternative strategies for diagnostic and therapeutic interventions. Moreover, the identification of a GnRH/GnRH receptor system in both human ovary and endometrium has widened the spectrum of action of the peptide outside its hypothalamic functions. The pharmacological use of GnRH itself or its synthetic analogs (agonists and antagonists) provides a valid tool to either stimulate or block gonadotropin secretion and to modulate the female fertility in several reproductive disorders and in assisted reproduction technology. The use of GnRH agonists in young female patients undergoing chemotherapy is also considered a promising therapeutic approach to counteract iatrogenic ovarian failure.

1. INTRODUCTION

Gonadotropin-releasing hormone (GnRH) is the main actor in the control of the reproductive axis. This decapeptide is released in a pulsatile way from hypothalamic neurons to reach the gonadotropes at the pituitary level and to stimulate its specific receptors (GnRHR), thus activating the pituitary–gonadal axis. GnRH release from hypothalamic neurons is locally regulated by different central signals, such as kisspeptin, neuropeptide Y, and norepinephrine, with stimulatory activity, or endogenous opioids and interleukin-1, with inhibitory activity. GnRH secretion is also controlled by gonadal steroids having both stimulatory and inhibitory effects according to the phase of the ovulatory cycle.

The pulsatility of GnRH release is crucial for the physiological maintenance of GnRHR functions and for the pattern of gonadotropin secretion. A pulsatile treatment with GnRH or GnRH agonists (GnRH-a) is used to mimic GnRH activity and maintain pituitary function (i.e., induction of puberty, fertility restoration in patients with GnRH deficiency), while continuous administration of GnRH analogs induces a downregulation/desensitization of GnRHRs and, after an initial flare effect, suppresses the release of pituitary gonadotropins; this therapeutic regimen is indicated for the treatment of central precocious puberty, endometriosis, hirsutism, polycystic ovarian disease, and steroid-dependent tumors.

It is well known that a GnRH/GnRHR system is expressed also in peripheral female reproductive tissues, such as the ovary and the endometrium, both in physiological and in pathological conditions. Interestingly, the GnRHR expressed in these tissues are coupled to a different intracellular signaling pathway when compared to pituitary GnRHR.

At the ovarian level, the GnRH/GnRHR system has been shown to be involved in the processes of follicular development/atresia and of luteinization/luteolysis, as well as in the control of ovarian epithelial cell proliferation in an autocrine/paracrine way. At the endometrial level, this system has been reported to play a crucial role in the cross talk between maternal endometrium and the embryo in the very first phases of gestation. Since in these peripheral tissues GnRHR are associated with an antiproliferative activity, they have been suggested to be an effective direct molecular target for GnRH-analog-based therapeutic approaches to treat steroid-dependent diseases, such as ovarian tumors.

More recently, GnRH analogs have been proposed as protective compounds against the side effects induced by cytotoxic drugs at the ovarian level in young female cancer patients undergoing chemotherapy. It is well known that cytotoxic drugs trigger the death of developing follicles, leading to ovarian failure and decreased fertility chances in these patients ("burnout" effect). By suppressing the pituitary–gonadal axis functions, GnRH-a keep the pool of primordial/primary follicles in a "dormant" condition, thus preventing the detrimental effects of chemotherapy. Most of the clinical data so far available suggest that GnRH-a, given in parallel with cytotoxic drugs, should be taken into consideration to preserve ovarian function and fertility in young female patients undergoing chemotherapy.

This review addresses the physiological functions of the GnRH/GnRHR system in the female reproductive tissues and its role as a molecular target for the treatment of different pathological conditions.

2. GnRH AND GnRH RECEPTORS

2.1 GnRH and GnRH Isoforms

GnRH belongs to the group of the hypothalamic releasing hormones. This neurohormone is a decapeptide discovered in 1971 by Schally's group (Schally et al., 1971). This discovery paved the way for the clarification of the key role of GnRH in the regulation of the reproductive functions (Conn & Crowley, 1994).

In humans, the GnRH gene, located on the short arm of chromosome 8 (8p11.2-p21), is organized into four exons and three introns (Seeburg, Mason, Stewart, & Nikolics, 1987). This gene is transcribed/translated into a 92 amino acid prohormone that is enzymatically cleaved and further modified into the secretory granules. This prohormone is composed of a signal peptide (23 amino acids) responsible for intracellular trafficking and secretion, the GnRH decapeptide, a three amino acid (Gly-Lys-Arg) proteolytic processing site, and GnRH-associated protein, a 56-amino acid peptide that is secreted with GnRH (whose function still has to be clarified).

GnRH is synthetized and secreted by hypothalamic preoptic parvocellular neurons that project their axons to the external zone of the median eminence where they deliver the decapeptide into the hypothalamo-hypophyseal portal vessels (Lehman, Robinson, Karsch, & Silverman, 1986); through this vascular system GnRH reaches the anterior pituitary. A unique neuronal network of hypothalamic neurons initiates and maintains reproductive functions in mammals (Wray & Hoffman, 1986).

These neurons originate in the embryonic nasal region, the olfactory placode, and undergo migration into the brain along the terminal and vomeronasal nerves during fetal development (around gestational weeks 9–12 in humans). The number of GnRH immunoreactive neurons in humans has been found significantly higher than previously thought; actually, about 10,000 GnRH neurons are present in the whole brain during fetal life with about 2000 neurons located in the hypothalamus and 8000 neurons distributed in widespread brain areas that are not involved in the control of the hypothalamic–pituitary–gonadal axis (Casoni et al., 2016) and that may account for the nonreproductive functions of GnRH (Maggi, 2016).

Genetic alterations of the development/migration of GnRH neurons are responsible for a large number of central idiopathic hypogonadotropic hypogonadisms and Kallmann's syndrome (Maggi et al., 2016).

At the pituitary level, GnRH binds to its specific receptors (gonadotropin-releasing hormone receptors, GnRHR) on gonadotropes and stimulates the biosynthesis/release of the two gonadotropins (luteinizing hormone, LH; follicle-stimulating hormone, FSH), thus ultimately regulating gonadal functions in both sexes (Conn & Crowley, 1994). The pulsatility of GnRH release is fundamental for the physiological maintenance of gonadotropin gene expression and for the pattern of gonadotropin secretion.

The amino acid sequence of native GnRH (pGlu-His-Trp-Ser-Tyr-Gly-Leu-Arg-Pro-Gly-NH_2) is reported in Fig. 1. It is now well known that the amino acid sequence of the C-terminal domain is necessary for the binding of the peptide to its receptors, while the amino acid sequence of the N-terminal domain is crucial for both the binding to and the activation of these receptors.

In addition to this classical form of GnRH (also called GnRH-I), other isoforms of the hormone have been identified in vertebrates. These decapeptides conserve in their structure both the N-terminal (Glp-His-Trp-Ser)

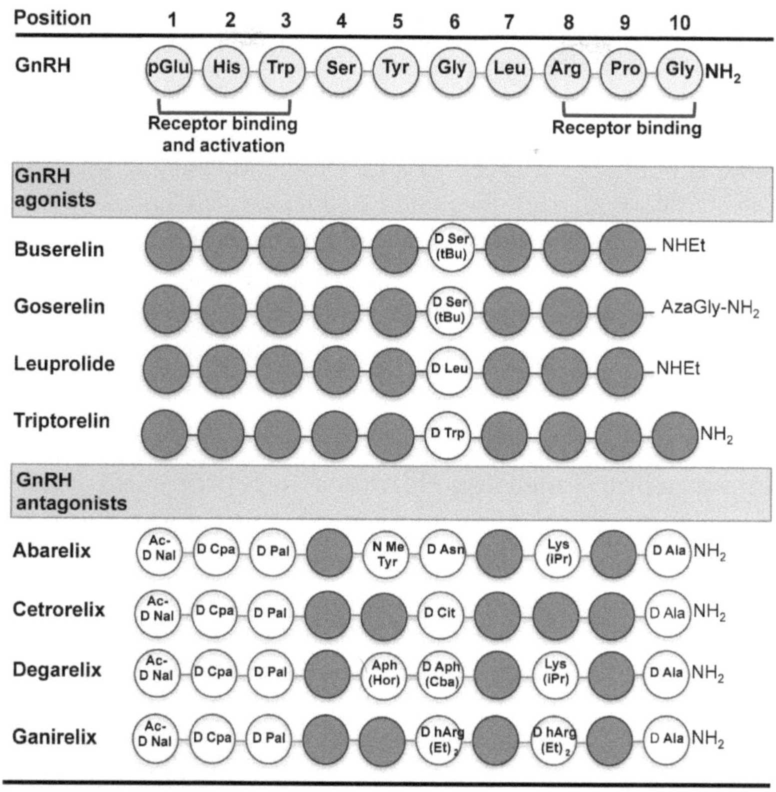

Ac, acetyl; *Aph*, 4-aminophenylalanine; *Cba*, carbamoyl; *Cit*, citrulline; *Cpa*, 4-chlorophenylalanine; *Hor*, L-hydroorotyl; *iPr*, isopropyl; *Nal*, 2-naphtylalanine; *N Me*, N Methyl; *Pal*, 3-pyridylalanine.

Fig. 1 Amino acid sequence of GnRH agonists and antagonists.

and the C-terminal (Pro-Gly-NH_2) amino acid sequences. In particular, a second form of GnRH has been identified in vertebrates, including humans (Millar, 2005; Neill, 2002). The structure of this form (GnRH-II) is highly conserved from fish to mammals. In humans, this peptide is encoded by a gene located on chromosome 20 and, at the amino acid level, it differs from the classical GnRH by three amino acids: His^5, Trp^7, Tyr^8 (Leung, Cheng, & Zhu, 2003; White, Eisen, Kasten, & Fernald, 1998). In the central nervous system, GnRH-II appears to be expressed in specific nuclei, such as the supraoptic, paraventricular, suprachiasmatic nuclei, and in the medial basal hypothalamus (Latimer et al., 2000); at this level, GnRH-II has been reported to act as a neuromodulator of the sexual behavior (Millar, 2005). GnRH-II is also expressed in tissues of the female reproductive system, such as the endometrium, ovary, and placenta (as well as in tumors derived from these tissues) (Cheon, Lee, Parhar, & Kang, 2001; Hong, Cheung, & Leung, 2008).

More recently, a third form of GnRH (GnRH-III) has been isolated from the brain of the sea lamprey (*Petromyzon marinus*); this decapeptide shares 60% of homology with the classical GnRH, differing in four amino acids: His^5, Asp^6, Trp^7, Lys^8 (Sower, Chiang, Lovas, & Conlon, 1993). In the lamprey, this peptide has been shown to play crucial roles in the control of the reproductive functions, such as gametogenesis and steroidogenesis (Deragon & Sower, 1994). In mammals, GnRH-III exerts a significant anti-tumor effect on different cancer cells, while possessing a very low activity on gonadotropin secretion (Kovacs et al., 2002).

2.2 Gonadotropin-Releasing Hormone Receptor

Hypothalamic GnRH exerts its hypophysiotropic activity through the binding to and activation of its specific receptors (GnRHR, also called "type I GnRHR") expressed on gonadotropes. The GnRHR is encoded by a gene located on chromosome 4 (4q13) and composed of three exons and two introns, that has been cloned in several species, including humans (Reinhart, Mertz, & Catt, 1992). This gene is transcribed/translated into a highly conserved 328-amino acid protein belonging to the GPCR (rhodopsin-like G protein-coupled receptors) receptor family. The GnRHR protein is characterized by a 35-amino acid extracellular N-terminus, with two putative glycosylation sites, followed by seven transmembrane domains (Millar, 2005; Stojilkovic, Reinhart, & Catt, 1994). The most peculiar feature of this receptor is the presence of a very short (1–2 amino acids) intracellular C-terminus. This intracellular part of the GPCR receptors is well known to be involved in their association with the specific G proteins as well as in the processes of receptor internalization and desensitization; this may be the reason why GnRHR, after ligand-induced activation, are slowly internalized and desensitized (Fan et al., 1994; Millar, 2005).

The intracellular signaling cascades associated with the GnRHR at the pituitary level are now well characterized (Fig. 2). Once activated by GnRH (or GnRH synthetic analogs), these receptors couple to the $G\alpha_{q/11}$ protein leading to the activation of the effector phospholipase Cβ (PLCβ) and the subsequent enzymatic hydrolysis of phosphatidylinositol 4,5 biphosphate (PIP_2) to form the two second messengers diacylglycerol (DAG) and inositol 1,4,5-triphosphate (IP_3). In turn, DAG activates protein kinase C (PKC) isoforms, while IP_3 triggers Ca^{2+} release from the endoplasmic reticulum

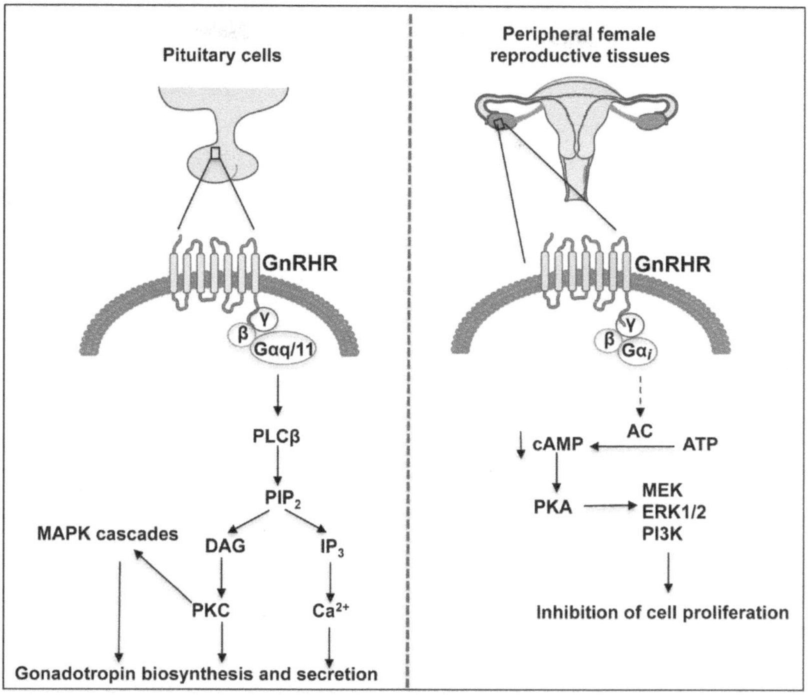

Fig. 2 Schematic representation of the intracellular signaling pathways coupled to the GnRHR in pituitary cells (*left panel*) and in peripheral reproductive tissues (*right panel*). In pituitary cells, the GnRHR is coupled to the $G\alpha_{q/11}$-PLCβ signaling pathway leading to PKC activation and increased cytoplasmic Ca^{2+} levels. PKC is responsible for the activation of downstream MAPK cascades. Ultimately, the activation of the MAPK cascades leads to phosphorylation/activation of transcription factors involved in the synthesis of gonadotropin subunits. In peripheral female reproductive tissues, GnRHR are coupled to a $G\alpha_i$ protein, leading to a decrease of cAMP levels, followed by the activation of different intracellular pathways, such as MEK, ERK1/2, and PI3K. Ultimately, these pathways lead to downregulation of gene transcription and antiproliferative/proapoptotic effects. *cAMP*, cyclic AMP; *DAG*, diacylglycerol; *ERK1/2*, extracellular signal-regulated kinase 1/2; *IP_3*, inositol 1,4,5-trisphosphate; *MAPK*, mitogen-activated protein kinase; *MEK*, MAPK/ERK kinase; *PI3K*, phosphoinositide 3-kinase; *PIP_2*, phosphatidylinositol 4,5-bisphosphate; *PKC*, protein kinase C; *PLCβ*, phospholipase Cβ.

(ER) stores by binding to its specific receptors located in the ER membrane that function as Ca^{2+} channels.

The accumulation of intracellular Ca^{2+} has been reported to be associated with GnRH-induced gonadotropin release (Tse, Tse, Hille, Horstmann, & Almers, 1997). On the other hand, PKC is responsible for the activation of downstream MAPK cascades, specifically ERK1/2 (through the Raf-1 kinase signal) and c-Jun N-terminal kinase (JNK; through the Src-CDC42/Rac pathway) and p38MAPK (Kraus, Naor, & Seger, 2001; Limonta, Moretti, Marelli, & Motta, 2003; Maggi et al., 2016; Naor, 2009). Phospholipase D and phospholipase A2, which in turn trigger the Raf-1/MAPK and PKC signaling pathways, respectively, are also activated by PLCβ (Kraus et al., 2001). Ultimately, the activation of the MAPK cascades will lead to the phosphorylation of transcription factors (such as c-Jun, c-Fos, Elk1) responsible for upregulating the gene expression of the gonadotropins α subunits and the FSHβ subunit (Fig. 2).

Thus, the GnRH-activated GnRHR/$G\alpha_{q/11}$/PLC/PKC axis in pituitary gonadotropes represents the major regulator of the synthesis and release of the two gonadotropin subunits (Kraus et al., 2001; Limonta et al., 2003; Maggi et al., 2016; Naor, 2009).

It is well known that different isoforms of PKC exist; they are produced by different genes and are differentially involved in signaling in health and disease conditions (Naor, Shearman, Kishimoto, & Nishizuka, 1988; Zeng, Webster, & Newton, 2012). Recently, it has been demonstrated that the PKCα, PKCβII, PKCδ, and PKCε isoforms, in particular, are specifically activated by GnRH in gonadotrope-derived cells and mediate MAPK activation in a ligand and cell context-dependent manner (Mugami, Dobkin-Bekman, Rahamim-Ben Navi, & Naor, 2017).

Although the $G\alpha_{q/11}$ remains the main G protein isoform to be associated with GnRHR, the $G\alpha_s$/adenylate cyclase/PKA signaling pathway has also been demonstrated to be involved in the physiological responses of pituitary gonadotropes to GnRHR stimulation. A specific and precise cross talk between all these intracellular pathways is fundamental for GnRH to exert its regulatory effects on gonadotropin synthesis and secretion (Kraus et al., 2001).

It is now well established that the features of the pulsatility of GnRH, both in terms of frequency and amplitude of pulses, can be responsible for a specific regulation of the synthesis and secretion of the two gonadotropin subunits (Haisenleder, Dalkin, Ortolano, Marshall, & Shupnik, 1991; Khadra & Li, 2006).

A pulsatile administration of GnRH (or GnRH agonists, GnRH-a) represents the most classical and effective therapeutic approach for those pathologies that are based on a defective pituitary–gonadal axis, such as hypogonadotropic hypogonadism and Kallmann's syndrome (see below). A prolonged administration of GnRH, after an initial stimulation of the pituitary GnRHR (flare), leads to receptor downregulation and desensitization, resulting in the suppression of gonadal steroid hormone secretion (medical castration) (Clayton, 1989; Kaiser, Conn, & Chin, 1997).

At the pituitary level, the expression of GnRHR is not constant but changes according to the levels of secretion of the two gonadotropins. For instance, in the pituitary of female rats, the number of GnRHR increases on the afternoon of proestrous, just before the gonadotropin surge responsible for ovulation. On the other hand, when low amounts of GnRH are released from hypothalamic neurons to reach the anterior pituitary, GnRHR levels on gonadotropes significantly decrease. Pituitary GnRHR expression is also regulated by circulating levels of gonadal steroids (i.e., estradiol and progesterone) (Yu, Ruman, & Christman, 2011).

As discussed earlier, a second isoform of GnRH (GnRH-II) has been identified in humans, suggesting the existence of a specific form of GnRHR for this peptide. However, it is now well accepted that a functional type II GnRHR is not expressed in human tissues (Millar, 2005; Neill, Musgrove, & Duck, 2004) and that GnRH-II exerts its biological functions through the activation of the classical form of GnRHR (Montagnani Marelli, Manea, Moretti, Marzagalli, & Limonta, 2015).

3. GnRH IN THE NEUROENDOCRINE CONTROL OF FEMALE REPRODUCTIVE FUNCTIONS

3.1 Control of GnRH Secretion

It has been found that only few (10%–30%) GnRH neurons are required to attain a normal fertility (Herbison, Porteous, Pape, Mora, & Hurst, 2008); however, they need complex control mechanisms to synchronize the release of the decapeptide in a pulsatile manner into the hypothalamo-hypophyseal portal vessels to induce an efficient secretion of pituitary gonadotropins (Marshall et al., 1991; Tsutsumi & Webster, 2009).

The relevance of the episodic release of GnRH is evident, considering that continuous infusion of GnRH exerts a suppressive action on both LH and FSH secretion readily reversed with a return to pulsatile stimulation. Actually, pulsatile administration of GnRH results in pituitary GnRH receptors (GnRHR) upregulation, whereas a loss of the GnRH response, due to a rapid uncoupling of the GnRH receptor from its intracellular signaling molecules followed by downregulation of receptor number, is induced by a continuous treatment (Cheng, Nathwani, & Leung, 2000). This feature of the GnRH system is exploited clinically to shut down the reproductive axis by the administration of long-acting GnRH-a (see below). The pulsatile secretion of GnRH has therefore the biological role to avoid the downregulation of the GnRH receptor in the gonadotrope cells.

In humans, the direct measurement of GnRH levels and pulsatility is not feasible and the low decapeptide peripheral blood levels do not accurately reflect hypothalamic GnRH secretion; therefore, the pulsatility of serum LH levels is accepted as a marker of the GnRH pulse generator activity; in particular, the measure of free α-subunit (FAS) levels, due to its shorter half-life than the LH β-subunit, is a reliable indicative measure of fast GnRH pulses (Hayes & Crowley, 1998; Landy, Schneyer, Whitcomb, & Crowley, 1990).

GnRH pulse generator may be cell autonomous, as determined by in vitro and in vivo observations showing that the activity of GnRH neurons is synchronized due to an extensive intercellular communication (Advis et al., 2003). Moreover, neurotransmitters and neuropeptides released by hypothalamic infundibular/arcuate nucleus (INF) and preoptic (POA) region (Piet, de Croft, Liu, & Herbison, 2015; Plant, 2015) may affect directly, or as intermediaries of circulating gonadal steroids, the GnRH pulsatility in humans. The different involvement of these two modalities in the physiological control of GnRH pulsatile secretion is still under discussion (Plant, 2015).

Since the first observation that mutations of the *KISS1* or *KISS1R* gene cause infertility (de Roux et al., 2003), it was proposed a role for the peptide kisspeptin, the product of *KISS1*, as a key regulator of the mammalian reproductive axis and, in particular, of GnRH release. Kisspeptin neuronal populations are mainly located in the INF region and less represented in POA; they project to GnRH neuron perikarya and fibers, which express kisspeptin receptors (Piet et al., 2015). Kisspeptin is a potent GnRH secretagogue and it is recognized to play a critical role in the onset of puberty, the regulation of sex steroid-mediated feedback, and the control of ovarian cycle and fertility in adulthood (Skorupskaite, George, & Anderson, 2014; Terasawa, Guerriero, & Plant, 2013). A complex system of neuroendocrine factors modulates the activity of kisspeptin neurons. In particular, INF kisspeptin neurons may coexpress neurokinin B and, even though rarely in humans, dynorphin, an endogenous opioid peptide (KNDy neurons) (Cheng, Coolen, Padmanabhan, Goodman, & Lehman, 2010; Lehman, Coolen, & Goodman, 2010; Ramaswamy, Seminara, & Plant, 2011; Skrapits et al., 2015). Such peptides are proposed to shape kisspeptin secretion through a paracrine/autocrine action. In nonhuman primates, neurokinin B was found to exert a kisspeptin-mediated stimulatory action on GnRH release, whereas dynorphin may inhibit the release of GnRH.

Likely to mutations of the kisspeptin receptor, the loss-of-function mutations in the gene for neurokinin B (TAC3) or for its receptor (TAC3R) are associated with hypogonadism and delayed puberty (Topaloglu et al., 2009). Other peptides, like substance P, galanin, NPY, agouti gene-related peptide, somatostatin, and proopiomelanocortin-derived peptides, have been identified in kisspeptin neurons in animals, but their presence and role in humans are still uncertain (Skrapits et al., 2015). GnRH neuron function may be modulated by a peptide recognized as gonadotropin-inhibitory hormone (Tsutsui et al., 2000) in the quail and identified in humans as RFamide-related peptide-3 (RFRP-3), but its modulatory effect on the regulation of human pubertal development, through the interaction with GPR147 receptors, has been provided only recently (Hinuma et al., 2000; Lima et al., 2014; Ubuka et al., 2009).

The release of the decapeptide from the hypothalamic neurons is therefore under the control of several central and peripheral signals which in turn may exert stimulatory (i.e., kisspeptin, norepinephrine, and neuropeptide Y) or inhibitory (i.e., endogenous opioids, interleukin-1, and progesterone) effects; GnRH itself modifies its own secretion through an ultra-short feedback loop (DePaolo, King, & Carrillo, 1987).

Finally, environmental factors, such as stress and changes in energy stores, were found to affect the release of GnRH and the activity of the pulse generator (Hrabovszky & Liposits, 2013).

3.2 GnRH in the Onset of Puberty and in the Adult Ovarian Cycle

3.2.1 Puberty

In primates, including humans, the GnRH neuronal system plays a key role in the onset of puberty.

It has been suggested that, during development, the activity of the neuroendocrine-reproductive axis may be distinguished in three phases (on–off–on) (Plant, Ramaswamy, Simorangkir, & Marshall, 2005) (Fig. 3). The first phase occurs during the fetal life (16th week of intrauterine life) and between the 4th and 10th weeks of postnatal life (infant) and it is characterized by a sustained GnRH pulsatility (on). This phase has been called "mini-puberty" since it is accompanied by a significant gonadotropin and steroid hormone secretion but, in contrast with the true puberty, either ovulation or spermatogenesis is not present. Mini-puberty is followed by a phase (juvenile) of relatively stable but low GnRH pulsatility and gonadal quiescence (off) that cover later infancy and juvenile development; GnRH

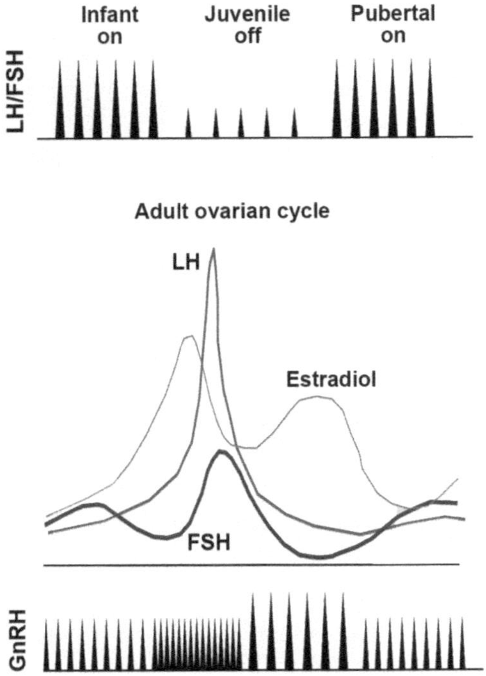

Fig. 3 Schematic representation of the frequency and amplitude of serum pulsatility of GnRH-induced gonadotropin release before and during puberty (*upper panel*) and of GnRH pulsatility in pituitary portal vessels during the human hormonal reproductive cycle (*lower panel*).

pulsatility is then reactivated (on) at the termination of the juvenile development (pubertal) to trigger the true puberty phase.

Different mechanisms control the changes of GnRH secretion along these three developmental phases. The activation of the GnRH pulse generator is determined by developmental events occurring in the brain and modulated by the gonadal steroids feedback control. In particular, INF KNDy neurons seem to form the pulse generation center during minipuberty and the pulsatile release of kisspeptin observed during this phase results in a corresponding pattern of GnRH as well as of gonadotropin secretion (Semple et al., 2005).

The decrease of GnRH secretion during late infancy and juvenile development is mainly due to neurobiological circuits impinging on the GnRH neurons (by either the decrease of stimulatory or the activation of inhibitory signals); it occurs also in the absence of gonadal signals and determines the characteristic delay in the puberty onset present in primates (Grumbach & Kaplan, 1990). Recent data show that the juvenile reproductive quiescence is characterized by a low pulsatile release of kisspeptin (Guerriero, Keen, & Terasawa, 2012). However, during this hypogonadotropic state, the hypothalamic content of GnRH is similar to that observed in adult animals and the reproductive axis keeps the responsivity to GnRH or its analogs (Plant, 2015).

On the basis of these observations, it appears that the limiting step of the puberty onset is the central *"brake"* on GnRH secretion. Moreover, in the primate female, the activity of the GnRH pulse generator during juvenile development seems to be only partially suppressed, and the negative feedback signals coming from the ovary may possibly contribute to the prepubertal restraint on gonadotropin secretion; this observation could therefore explain the earlier onset of puberty in females with respect to males and the higher frequency of GnRH-dependent precocious puberty in girls (Witchel & Plant, 2014).

The weakening of the central brake and the reactivation of high-amplitude and high-frequency release of kisspeptin and GnRH determine the puberty onset. GnRH pulse generation leads to an increase of gonadotropin release and to gonadal stimulation.

The consequent increased production of ovarian hormones can then exert a first negative feedback phase which characterizes the delay between the menarche and the first ovulation (Rapisarda, Bergman, Steiner, & Foster, 1983). Afterward, a positive feedback capability of estradiol develops which can trigger a larger release of GnRH, and LH, leading to ovulation.

The timing of the reactivation of GnRH pulses could be hypothesized to be due to both internal and external signals; actually, a hypothalamic pubertal clock, or brain sensors, able to monitor circulating signals of somatic development, may coordinate the reproductive axis reactivation according to the attainment of adult somatic size (Plant, 2015).

3.2.2 Adult Ovarian Cycle

In adult humans, an FAS/GnRH pulse is detected approximately every 60–100 min in the absence of sex steroids' negative feedback; in female, the pattern of pulsatile GnRH release is complicated by frequency variations during different reproductive stages and the ovulatory cycle.

A change in the frequency of pulsatile GnRH release is responsible for the rapid rise in estradiol secretion in the late follicular phase that triggers the preovulatory LH surge (Moenter, Caraty, Locatelli, & Karsch, 1991).

The GnRHRs (coupled to both $G_{\alpha q/11}$ and $G_{\alpha s}$ signaling pathways; see Section 2.2 and Fig. 2) present on the surface of the gonadotrope cells may decode the frequency and amplitude of GnRH pulses (Tsutsumi & Webster, 2009).

The activation of the different intracellular signal cascades (Fig. 2), as well as of transcription factors, may independently modulate the expression of LHβ and FSHβ genes (Lim et al., 2009; Tsutsumi & Webster, 2009). In particular, high-frequency GnRH pulses (1 pulse/60–90 min) induce LHβ expression and inhibit the expression of FSHβ. Low-frequency GnRH pulses (<1 pulse/2–3 h) lead to a repression of the LHβ promoter and to an increase of the expression of FSHβ. The two modalities of secretion of GnRH occur at different times during the ovarian cycle to characterize the hormonal cycle of the two gonadotropins (Fig. 3).

Sex steroids show a differential regulatory effect on GnRH secretion and pulsatility, with progesterone having an inhibitory action and estradiol (E2) both stimulatory and inhibitory effects, depending upon the stage of the menstrual cycle. A "tonic" GnRH pulsatile secretion is mainly present throughout the cycle and it is controlled by the negative feedback actions of ovarian steroids, with E2 inhibiting pulse amplitude and progesterone inhibiting pulse frequency and increasing pulse amplitude during the luteal phase; however, both steroids drive a diurnal variation in GnRH pulse frequency, with a lower frequency during the night (McCartney, 2010).

The maturation of the preovulatory follicle, at the end of the follicular phase, induces a rapid elevation of plasma E2 concentrations that increase, by a positive feedback action, GnRH pulse frequency and amplitude, and pituitary sensitivity to GnRH (i.e., by increasing GnRH receptor expression), leading to the LH preovulatory surge.

On the basis of the accepted model on the role of kisspeptin neurons, the production of ovarian E2 during the follicular phase of the ovulatory cycle induces the expression of progesterone receptors in the INF kisspeptin neurons which leads to override the normal negative feedback and to increase GnRH pulse frequency and amplitude to drive the preovulatory LH surge (Mittelman-Smith, Wong, Kathiresan, & Micevych, 2015; Stephens et al., 2015). The increase of progesterone production during the luteal phase, in the presence of lower levels of E2, downregulates kisspeptin expression in the INF neurons which slows GnRH pulse frequency with a decrease of LH and a progressive increase of FSH production (Rance, 2009). While the kisspeptin neuronal populations present in both INF and POA appear to differentially mediate the two types of steroid feedbacks in animal models, in humans only the INF population seems to play a main role in such process.

3.3 Dysregulation of Pulsatile GnRH Release

Many reproductive defects due to either physiological responses or pathological conditions may be associated to a dysregulation of the GnRH release.

Alteration of GnRH pulse generator may be present in functional hypothalamic amenorrhea (due to a prolonged fasting or exercise) (Liu & Patel, 2013) and in obesity (due to an altered insulin action on GnRH neurons) (DiVall et al., 2015). In women, the overall level of GnRH also increases with aging, possibly due to a pituitary resistance to GnRH or to the decrease of estrogen levels, with loss of the negative feedback. In younger postmenopausal women, gonadotropin, and therefore GnRH, pulses occur approximately every 50–55 min, likely to a normally cycling woman in the late follicular phase and midcycle surge; however, the pulse frequency decreases with age progression independently to ovarian hormonal milieu and by a possible impairment of the hypothalamic functions (Hall, 2007; Hall, Lavoie, Marsh, & Martin, 2000).

A number of reproductive disorders including hypogonadotropic hypogonadism, hypothalamic amenorrhea, hyperprolactinemia, and polycystic ovary syndrome (PCOS) are also associated with disruption of the normal pulsatile GnRH secretion. Regarding PCOS, the GnRH defect has been linked to an excess of fetal androgens which may permanently alter the steroid negative and positive feedback (Roland & Moenter, 2014) and to a possible involvement of neurokinin B, leptin, and kisspeptin (Yan, Yuan, Zhao, Cui, & Liu, 2014).

3.4 Therapeutic Use of GnRH

The induction of ovulation with a therapeutic regimen of pulsatile native GnRH (Reid, Leopold, & Yen, 1981) has been used since 1980 in women with disordered GnRH deficiency, such as PCOS, hyperandrogenic anovulation, and late-onset congenital adrenal hyperplasia (Scheiber & Liu, 2011).

The patients that show the highest response rates to a pulsatile regimen of GnRH administration are women affected either by the Kallmann's syndrome (i.e., primary GnRH deficiency) or by hypothalamic amenorrhea (decreased GnRH release in the presence of an intact hypothalamus)

(Crowley & McArthur, 1980; Leyendecker, Struve, & Plotz, 1980). Therefore, an accurate framework of the physiopathology of infertility is needed before the indication of a pulsatile GnRH therapy. Intravenous (IV) administration route was found to be more efficient to induce efficient episodic release of LH and FSH (Reid et al., 1981). Generally, IV pulses of about 1 μg of the peptide with a pulse frequency of 60–90 pulses/min induce a significant rise of GnRH (40–2000 pg/mL, a range of concentration found in human pituitary portal blood) within 4 min and higher rate of ovulation and pregnancy (Blacker, 1992; Filicori et al., 1991). At higher doses (10 μg/pulse), GnRH treatment may induce multiple (2–5) follicular development (Martin, 1993), supporting its indication in assisted reproductive technology (ART). On the other hand, subcutaneous (SC) administration may lead to a slower sustained rise in LH and FSH without a normal pulsatile pattern (Eckert et al., 1996).

The development of synthetic stable GnRH analogs led to a reduced use of native GnRH (see Section 5).

4. THE GnRH/GnRHR SYSTEM IN PERIPHERAL FEMALE REPRODUCTIVE TISSUES

4.1 Peripheral vs Central GnRHR

It is now well established that, in addition to the pituitary gonadotropes, GnRHR are expressed also in different nonpituitary tissues, specifically in tissues of the female reproductive tract, as well as in tumors derived from these tissues (Harrison, Wierman, Nett, & Glode, 2004; Maggi et al., 2016; Marelli, Moretti, Januszkiewicz-Caulier, Motta, & Limonta, 2006; Millar, 2005).

GnRHR were first shown to be expressed, both at the mRNA and at the protein level, in human placental cells; later, the expression of these receptors was also reported in the ovary and in the endometrium, both in physiological and in pathological conditions (Harrison et al., 2004; Limonta et al., 2012). The cDNA nucleotide sequence as well as the molecular size of these GnRHR corresponds to those of the receptors at the pituitary level (Kakar, Musgrove, Devor, Sellers, & Neill, 1992; Limonta et al., 1999; Peng, Fan, Ligier, Vaananen, & Leung, 1994); on the other hand, pituitary and peripheral receptors seem to possess a different pharmacological profile. For instance, the binding affinity of GnRH analogs is high for pituitary receptors, being lower for peripheral binding sites (Dondi et al., 1994; Iwashita, Evans, & Catt, 1986). Moreover, GnRH analogs known to be antagonists at the pituitary level have been consistently shown to behave as agonists in peripheral tissues and exert antiproliferative activity on tumor cells (Emons et al., 1993; Yano et al., 1994). The reason for these observations is still unclear. However, it has been postulated that some differences at the molecular level might exist between central vs peripheral receptors, such as different posttranslational modifications (i.e., glycosylation, phosphorylation) of the receptor protein (Rama & Rao, 2001). Moreover, Millar and coworkers (Millar, Pawson, Morgan, Rissman, & Lu, 2008) proposed that GnRHR may assume different conformations according to the cell type in which they are expressed; therefore, GnRH analogs can selectively bind to the receptors and specifically activate the intracellular signaling pathways according to the cell context. This hypothesis is known as "*Ligand-induced Selective Signaling*" (LiSS).

Peripheral GnRHR differ from pituitary receptors also in terms of their intracellular signaling cascade. In peripheral tissues (in particular in tumor tissues), GnRHR have been consistently shown to be coupled to a $G\alpha_i$ protein and to a subsequent decrease of cAMP levels. Decreased levels of cAMP are responsible for triggering the activation of different intracellular pathways, MAPK (p38 MAPK, ERK1/2, JNK), phosphatidylinositol-3-kinase (PI3K). $G\alpha_i$ is also responsible for the activation of the tyrosine phosphatase cascade, which in turn interferes with growth factor receptor signaling. Ultimately, decreased activity of MAPKs and growth factor receptor intracellular signaling pathways leads to downregulation of gene transcription at the nuclear level (Gunthert, Grundker, Hollmann, & Emons, 2002; Imai et al., 1996; Limonta et al., 2012; Marelli et al., 2006) (Fig. 2).

Similar to GnRHR, also the endogenous GnRH decapeptide has been found to be expressed in peripheral female reproductive tissues; data from the literature point out that this GnRH/GnRHR system exerts different autocrine/paracrine effects, both in physiological and in pathological conditions (Cheung & Wong, 2008; Grundker, Gunthert, Westphalen, & Emons, 2002; Limonta et al., 2012; Marelli et al., 2006; Ramakrishnappa, Rajamahendran, Lin, & Leung, 2005).

4.2 The GnRH/GnRHR System in the Ovary

4.2.1 Follicles and Corpora Lutea

GnRH and GnRHR have been widely demonstrated to be expressed at the ovarian level in different species, including humans (Kang, Choi, Yang, & Leung, 2003). As discussed earlier for pituitary GnRHR, the expression of these binding sites follows a dynamic pattern also at the ovarian level, according to the stage of follicular development. In rats, it has been shown that GnRHR levels are low in preantral/early antral follicles and increase over the course of follicular maturation, being highly expressed in the granulosa cells from graafian follicles and in atretic follicles (Clayton, Eccleston, Gossard, Thalbard, & Morel, 1992; Whitelaw, Eidne, Sellar, Smyth, & Hillier, 1995). In line with these observations, GnRHR were reported to be expressed in mouse ovaries, specifically in granulosa/cumulus cells (Torrealday, Lalioti, Guzeloglu-Kayisli, & Seli, 2013) as well as in porcine granulosa cells at varying follicular stages (Takekida, Matsuo, & Maruo, 2003).

In humans, both GnRH and GnRHR are not expressed in follicles in the early stages of development. On the other hand, they are present in the granulosa cell layer of graafian follicles as well as in granulosa-luteal cells (but not in theca-luteal cells) (Peng et al., 1994). Apoptosis of granulosa cells is known to play a crucial role in follicular atresia (Hsueh, Billig, & Tsafriri, 1994). The decapeptide and its receptors were found to localize also in the ovarian surface epithelium and in epithelial ovarian cancer cells and tissues (Grundker, Gunthert, et al., 2002; Imai et al., 1996). Ovarian GnRHR have been reported to correspond to the receptors at the pituitary level (Peng et al., 1994).

Finally, as pointed out earlier, the presence of the second form of GnRH (GnRH-II) has been demonstrated in human ovarian cells, supporting the notion that also this form may be involved in the control of the ovarian functions (Kang et al., 2003).

Taken together, these observations support that the GnRH/GnRHR system may be involved in the processes of follicle development/atresia and luteinization/luteolysis.

Apoptosis occurring both in granulosa and in theca interna cells is well known to play a role in follicular development. Apoptotic granulosa cells cannot be observed in primary follicles, while the number of granulosa apoptotic cells increases in antral follicles. Moreover, in the early stage of atresia of the antral follicles, apoptosis is present in granulosa cells, while it is apparent in the theca interna cells in the late stage of atresia of these follicles (Takekida et al., 2003).

Based on the effects of GnRH on the vitality of follicular cells, it has been suggested that the GnRH/GnRHR system might be involved in the proapoptotic process of follicular atresia (Harrison et al., 2004; Kang et al., 2003). In particular, GnRH was shown to decrease DNA synthesis and to induce apoptosis in rat and porcine granulosa cells from small and medium (but not from large) follicles (Takekida et al., 2003). Similar observations were reported by studies performed in human granulosa cells. The results obtained demonstrate that GnRH downregulates the proliferation of human granulosa cells and directly triggers the extrinsic apoptosis pathway by interfering with the IGF-I/Akt signaling pathway and through induction of caspase-8, -3, and -7; this effect was counteracted by cotreatment of the cells with FSH (Hong, Klausen, Cheung, & Leung, 2012). Taken together, these observations strongly support an important role of the GnRH/GnRHR system in regulating follicular development and atresia.

The proper growth and function of the ovary are related to a correctly orchestrated steroidogenesis and follicle maturation. In human granulosa cells, GnRH-a were shown to increase the expression of aromatase, thus stimulating estrogen production, and of anti-Müllerian hormone (AMH), known to be involved in the process of folliculogenesis. GnRH also induced the expression of genes involved in oocyte maturation and follicular rupture at the time of ovulation, including progesterone receptor gene and plasminogen activator (Kang et al., 2003; Ny, Liu, Ohlsson, Jones, & Hsueh, 1987). Finally, Miller and coworkers have recently reported that, in granulosa cells, GnRH-a increase the PEDF (pigment epithelium-derived factor, a potent antiangiogenic factor) to VEGF (vascular endothelial growth factor, the well-known key factor in angiogenesis) ratio, inversely to hCG (human chorionic gonadotropin). Based on this observation, the authors suggest that this GnRH-related antiangiogenic effect might be associated with the reduced risk of ovarian hyperstimulation syndrome (OHSS) obtained with GnRH-a compared to hCG in assisted reproduction protocols (Miller, Chuderland, Ron-El, Shalgi, & Ben-Ami, 2015). On the other hand, the antiangiogenic environment induced by GnRH-a might impair the functions of the corpus luteum that are dependent on a correct angiogenesis process; this may be related to the observed deficiency of corpus luteum function after GnRH-a trigger and the lower pregnancy rate obtained with this triggering protocol (Miller et al., 2015).

Experimental evidence has been reported indicating that the GnRH/GnRHR system may also be involved in the processes of luteinization and luteolysis. In particular, in human granulosa-luteal cells, GnRH-a decrease the expression of estrogen receptor alpha and beta (Chiang, Cheng, Igarashi, Nathwani, & Leung, 2000).

GnRH-a were reported to induce luteolysis in corpora lutea and this was associated with the stimulation of matrix metalloproteinases (i.e., MMP-2) and the membrane type 1 MMP, leading to the remodeling of the extracellular matrix in developed corpora lutea (Kang et al., 2003). Similar observations were also reported for the second isoform of GnRH; specifically, GnRH-II was shown to counteract progesterone secretion induced by hCG from human granulosa-luteal cells. In these cells, GnRH-II induced downregulation of both FSH and LH receptors (Kang et al., 2003).

Thus, locally expressed GnRH (and GnRH-II)/GnRHR system in the ovary is involved in the regulation of follicular development/atresia and of corpus luteum function.

4.2.2 Ovarian Surface Epithelium

As discussed earlier, GnRH, GnRH-II, and GnRHR localize also in the ovarian surface epithelium as well as in epithelial ovarian cancer cells and tissues (Emons et al., 1993; Imai et al., 1996). Ovarian surface epithelial (OSE) cells are involved in the physiological regulation of ovarian functions, contributing to ovulation. Specifically, they were suggested to contribute to ovulation through the secretion of matrix metalloproteinases, and to trigger ovulatory repair mechanisms by secreting some components of the epithelial basement membrane, such as collagen type IV and laminin (Kruk, Uitto, Firth, & Auersperg, 1994). GnRHR were shown to be expressed in human OSE cells and their activation, by means of GnRH-a, is associated with a significant (time- and dose-dependent) and specific suppression of cell proliferation (Kang, Tai, Cheng, & Leung, 2000). In line with these data, Choi and coworkers (Choi, Auersperg, & Leung, 2001) demonstrated that GnRH-II (as well as GnRH-I) significantly decreases the proliferation of immortalized OSE cells. Taken together, these observations strongly support the presence of a GnRH/GnRHR system in the OSE and that it is involved in an autocrine/paracrine way in the control of cell proliferation.

It is well known that most of all ovarian cancers are surface epithelial carcinomas. GnRH (both GnRH-I and -II) and GnRHR have been found to be expressed in several human ovarian cancer cell lines and tissues (Grundker, Gunthert, et al., 2002); at present, in primary ovarian tumors, the expression of the receptor is considered a favorable prognostic factor (Wilkinson et al., 2008). Activation of local GnRHR has been widely reported to be associated with a significant antitumor (antiproliferative, antimetastatic) activity (Grundker, Gunthert, Hellriegel, & Emons, 2004; Kang et al., 2003; Limonta et al., 2012; So, Cheng, Poon, & Leung, 2008). In line with these observations, clinical studies reported a partial remission and disease stabilization in ovarian cancer women who relapsed after chemotherapy and were treated with synthetic GnRH analogs

(Verschraegen et al., 2003). However, a recent review on randomized controlled trials comparing GnRH analogs with agonistic activity (see Section 5) with chemotherapeutic agents or placebo in women with epithelial ovarian cancer pointed out that there is no enough evidence to support the safety and effectiveness of GnRH analogs in treating ovarian cancer patients after platinum relapse.

More recently, targeted therapeutic strategies have been developed with the aim to improve the clinical outcome in cancer patients. For instance, GnRH analog-based cytotoxic hybrid compounds have been synthesized by linking a GnRH analog to a traditional chemotherapeutic drug. By binding to its receptors on cancer cells, the GnRH analog will specifically target the chemotherapeutic drug to tumor cells; after internalization, the hybrid compound is degraded at the lysosomal level to release the anticancer drug which can exert its cytotoxic activity (Engel et al., 2012; Schally & Nagy, 1999). AESZ-108, containing [D-Lys^6]-GnRH, which is linked to doxorubicin, was shown to inhibit the growth of GnRHR-positive ovarian cancer cells both in vitro and in vivo (preclinical studies) (Engel, Schally, Dietl, Rieger, & Honig, 2007; Grundker et al., 2002). Results from phase I and phase II clinical trials also supported a significant antitumor activity (median overall survival and time to progression) of this compound in women with ovarian cancer (Engel, Tinneberg, Rick, Berkes, & Schally, 2016). However, further studies are required to definitely confirm the complete effectiveness and safety of GnRH analog-based cytotoxic hybrids in late-stage ovarian cancers.

4.3 The GnRH/GnRHR System in the Endometrium

It is now well accepted that GnRH and GnRHR are expressed in the human endometrium, both in epithelial and in stromal cells, and that their levels of expression show a dynamic pattern, with an increase during the secretory phase of the menstrual cycle (Raga et al., 1998). The second isoform of GnRH (GnRH-II) has also been shown to be expressed in endometrial cell lines and to increase the secretion of enzymes involved in the remodeling of the extracellular matrix (matrix metalloproteinases MMP-2 and MMP-5, urokinase-type plasminogen activator uPA) from decidual endometrial stromal cells, thus promoting their motility behavior. Moreover, high levels of both GnRH and GnRHR are expressed in the endometrium during the luteal phase of the ovulatory cycle. Taken together, these observations indicate that the GnRH/GnRHR system may play a crucial role in promoting trophoblast invasion of the maternal endometrium during embryo implantation (Chou, MacCalman, & Leung, 2003; Wu et al., 2015).

Interestingly, endometrial GnRH has been reported to stimulate hCG secretion from the placenta, further supporting the crucial role of the cross talk between maternal endometrium and the embryo in the very first phases of gestation. Clinical observations further support this conclusion. It has actually been shown that, during in vitro fertilization (IVF), GnRH-a-induced ovarian stimulation is accompanied by a higher endometrial thickness (when compared with GnRH antagonist stimulation) and a consequent higher pregnancy rate (Al-Inany, Abou-Setta, & Aboulghar, 2006).

GnRH-I, GnRH-II, and GnRHR are known to be expressed in endometrial cancer cells and tumor tissues; however, contrasting results have been so far reported regarding their protective vs protumoral role in this gynecological cancer (Grundker et al., 2004; Maggi et al., 2016; Wu et al., 2015).

5. GnRH ANALOGS IN REPRODUCTION

Native GnRH has a half-life of 2–4 min mainly due to the degradation of the glycine–leucine bond between amino acids 6 and 7. To overcome this limitation, a number of synthetic GnRH analogs have been designed to be used to either restore fertility in patients with GnRH deficiency or suppress the pituitary–gonadal axis. Synthetic GnRH-a generally have a modification in Gly^6, usually replaced by a D-amino acid, to increase their plasma half-life, a deleted Gly^{10}-amide and the addition of a residue (ethylamide) to Pro^9 to increase the affinity for GnRHRs (Conn & Crowley, 1994; Engel & Schally, 2007). On the other hand, they preserve the amino acid sequence of the N-terminal and the C-terminal domains which are crucial for receptor binding (both N- and C-terminal domains) and activation (N-terminal domain). Fig. 1 reports the mostly used GnRH-a in clinical trials. Pulsatile administration of GnRH-a can be used to stimulate the pituitary–gonadal axis in substitution of native GnRH.

However, a prolonged administration of GnRH-a induces an initial stimulation of the pituitary GnRHR and of gonadotropin secretion (flare effect) followed by a receptor downregulation and reproductive axis desensitization resulting in the suppression of gonadotropins and gonadal steroid hormone secretion (medical castration) (Clayton, 1989; Kaiser et al., 1997).

GnRH-ant have been designed to obtain pharmacological compounds to block the pituitary–gonadal axis without the undesirable flare effect exerted by GnRH itself or by the GnRH-a (see Section 2.2). GnRH-ant competitively block the binding of GnRH to GnRHRs (Schally, 1999; Tan & Bukulmez, 2011); they present a very low amino acid similarity with the decapeptide and contain a Ac-D-Nal-D-Cpa-D-Pal motif in the N-terminal, D-Ala in position 10 and different amino acid substitutions in positions 5, 6, and 8 (Cook & Sheridan, 2000; Schally, 1999) (Fig. 1). Among different synthesized compounds, cetrorelix, ganirelix, and degarelix show the highest binding affinity and inhibitory activity. Abarelix was found to be highly active, but it was shown to induce systemic allergic reaction that restricted its marketing to European countries (Huhtaniemi, White, McArdle, & Persson, 2009).

The administration routes of GnRH analogs include SC, intramuscular (IM), nasal, or sublingual.

GnRH-a are used both to stimulate and to suppress reproductive functions according to their regimen of administration. Actually, a pulsatile treatment is used to mimic GnRH activity and maintain pituitary functions (i.e., induction of puberty, fertility restoration in patients with GnRH deficiency) (Abel et al., 2013; Zimmer, Ehrmann, & Rosenfield, 2010), while continuous (normally for 1–3 weeks) administration of GnRH-a induces a downregulation of GnRHRs and, after an initial flare effect, suppresses the release of pituitary gonadotropins (McArdle, 2012); its therapeutic regimen is indicated for central precocious puberty, endometriosis, hirsutism, and polycystic ovarian disease (Magon, 2011; Maheshwari, Gibreel, Siristatidis, & Bhattacharya, 2011).

GnRH-ant have several clinical applications in reproductive medicine and in gynecology and are particularly effective in the control of ovarian hyperstimulation protocols in assisted reproduction technology (Mancini et al., 2011; Tan & Bukulmez, 2011).

Possible side effects of a therapy with GnRH analogs are mainly related to sex hormone deficiency and similar to symptoms occurring in menopause, like vaginal dryness and hot flushes. Osteoporosis may also develop under prolonged treatment with GnRH analogs that should be therefore limited to 6 months, unless concomitant estrogen replacement therapy.

5.1 GnRH Analogs to Stimulate the Reproductive Axis

A pulsatile administration of GnRH (or GnRH-a) represents the most classical and effective therapeutic approach for those pathologies that are based on a defective pituitary–gonadal axis, such as hypogonadotropic hypogonadism and Kallmann's syndrome (Conn & Crowley, 1994; Raivio et al., 2007; Silveira & Latronico, 2013).

GnRH-a also provide a useful alternative to hCG treatment for the induction of LH release and ovarian stimulation for IVF, although applicable only in controlled ovarian stimulation protocol with GnRH-ant, which leaves the pituitary still responsive to the agonist (Orvieto et al., 2006). The advantages of the use of GnRH-a reside in a more physiological LH and FSH surge (with a better oocyte quality), an improved pregnancy, implantation, and live birth rates (van der Linden, Buckingham, Farquhar, Kremer, & Metwally, 2015; Youssef et al., 2014). It also reduces the risk of OHSS, consequent to the hCG treatment, because of its rapid induction of luteolysis (Casper, 2015; Mittelman-Smith et al., 2015).

5.2 GnRH Analogs to Block the Reproductive Axis

GnRH-a are used for treatment of girls (up to age 6) with central precocious puberty (Carel et al., 2009; Li, Li, & Yang, 2014). The treatment of early puberty with GnRH-a is safe and pubertal maturation is achieved usually within 1 year after treatment discontinuation, with no apparent abnormalities in reproductive functions (Thornton et al., 2014). An observed high prevalence of PCOs seems to be linked to the disease itself rather than to GnRH-a therapy and further studies are needed to address these possible limitations. GnRH-a treatments can be also used in preparation of female patients undergoing laparotomic, vaginal, or endoscopic surgery or in the woman approaching to menopause (Schmidt et al., 2009).

The ability of GnRH-ant to rapidly suppress the endogenous LH surge and gonadotropin levels without the initial stimulatory (flare) effect offers the main advantage of these compounds over GnRH-a in ART.

The GnRH-ant treatment has to be started soon after exogenous gonadotropin stimulation, but it requires less gonadotropin exposure, reducing OHSS incidence. Whether GnRH-ant protocols might lead to a slightly decreased pregnancy rates, compared with the GnRH-a protocols, is a still debated issue (Blumenfeld, 2001; Tan & Bukulmez, 2011). However, the shorter total treatment time and consequent reduced side effects improve patient's compliance for GnRH-ant medication.

5.3 GnRH Analogs for Fertility Preservation in Female Patients Undergoing Chemotherapy

The incidence of cancer in reproductive-aged women, together with the survival time of these patients, is increasing worldwide (Siegel, Miller, & Jemal, 2017). Both radiotherapy and chemotherapy treatments are associated with a failure of the reproductive functions. Radiotherapy is associated with ovarian failure and damage of the myometrium, which is responsible for endometrial dysfunction. On the other hand, cytotoxic drugs may induce apoptotic cell death in oocytes as well as in primordial/primary follicles and alterations in the ovarian vasculature leading to indirect loss of these follicles (Codacci-Pisanelli, Del Pup, Del Grande, & Peccatori, 2017; Lawrenz, Mahajan, & Fatemi, 2016; Meirow et al., 2007; Oktem & Oktay, 2007). These mechanisms induce a permanent amenorrhea, responsible for the suppression of the opportunity of pregnancy. Therapeutic and effective strategies aimed at preserving ovarian function and fertility in young women with cancer represent a major current clinical challenge (Blumenfeld, Zur, & Dann, 2015; Hickman, Valentine, & Falcone, 2016; Lambertini et al., 2017, 2015; Munster, 2013; Winkler-Crepaz, Bottcher, Toth, Wildt, & Hofer-Tollinger, 2017).

At present, different promising approaches are utilized with the aim to preserve the reproductive potential in chemotherapy-treated prepubertal girls and premenopausal women (Blumenfeld & Evron, 2015; Jensen et al., 2017; Maggi et al., 2016; Winkler-Crepaz et al., 2017).

One of these approaches is based on the *cryopreservation of oocytes*; however, the efficacy of this method largely depends on the number of oocytes that can be obtained (Harada & Osuga, 2016). *Embryo cryopreservation* has been proposed to avoid this limitation, particularly in adult female patients. According to this technology, embryos undergo cryopreservation after IVF. GnRH-a are utilized to trigger ovulation, after previous treatment with a GnRH-ant to reduce the risk of ovarian hyperstimulation (Benard et al., 2016). However, chemotherapy treatment must be postponed after oocyte retrieval and IVF. Moreover, it is suggested that women with estrogen-receptor cancers should be treated with an aromatase inhibitor during the process of ovarian stimulation, since high estrogen levels might be associated with detrimental effects at the tumor level (Linkeviciute, Boniolo, Chiavari, & Peccatori, 2014). It has been recently reported that, in women with breast cancer undergoing controlled ovarian hyperstimulation for fertility preservation, coadministration of letrozole (an aromatase inhibitor) suppresses estradiol levels without affecting oocyte recruitment or disease-free survival rates (Rodgers et al., 2017).

Another approach to safeguard fertility is based on the *cryopreservation of the ovarian cortical tissue*, which is rich in primordial follicles. At the end of the chemotherapeutic treatment, these tissues are autotransplanted either into the pelvis or outside the pelvis-abdominal wall (i.e., orthotopic or heterotopic transplantation) (Salama, Isachenko, Isachenko, Rahimi, & Mallmann, 2016). A recent review by Ladanyi and coworkers (Ladanyi, Mor, Christianson, Dhillon, & Segars, 2017) states that, so far, the autotransplantation of cryopreserved ovarian tissues has been able to restore fertility with a global pregnancy rate of 23%–37% and over 80 live births. On the other hand, it has been pointed out that this approach might not be appropriate in some types of cancer, where the risk of transplanting malignant cells present in the cryopreserved ovary may be high (Amorim & Shikanov, 2016). Moreover, ovarian tissue transplants are characterized by a relatively short life span (Salama et al., 2016). Thus, even if ovarian tissue cryopreservation may offer an advantage, in terms of fertility preservation, to young women, this approach is still considered experimental; further studies are required to definitely make it a standard care for these patients.

More recently, the "fertoprotective adjuvant" therapy has been developed with the aim to maintain ovarian reserve, through prevention of premature ovarian failure (POF), thus providing the basis for a long-term ovarian function. This therapy foresees the administration of GnRH-a prior to and during chemotherapy treatment (Blumenfeld, 2016; Blumenfeld et al., 2015; Mahajan, 2015). The rationale for this approach is based on different aspects of the activity of these compounds. First of all, through their classical mechanism of action, GnRH-a suppress the functions of the pituitary–ovarian axis, thus triggering a hypogonadotropic *milieu*, which is responsible for the inhibition of follicular recruitment and development. These compounds also act by counteracting the impairment of the ovarian stroma and vasculature associated with the low estrogen levels and possibly by protecting ovarian germinative stem cells (Blumenfeld, 2016; Mahajan, 2015).

However, to date, the most interesting mechanism of the protective effects of GnRH-a against cytotoxic drugs induced ovarian failure is based on their ability to counteract the so-called *burnout* effect (Fig. 4).

It is well known that primordial/primary follicles do not express FSH receptors and are gonadotropin unresponsive; on the other hand, antral follicles express FSH receptors and produce estrogens which, in turn, control FSH secretion. During the first phase of the ovarian cycle, a balance exists in which developing antral follicles inhibit primordial/primary follicle recruitment and are therefore responsible for their maintenance in a dormancy state. Actually, antral follicles are known to produce growth factors, such as transforming growth factor-β (TGF-β), AMH, bone morphogenic proteins (BMP) 2, 4, 5, 6, 7, and 15, and growth differentiation factor-9;

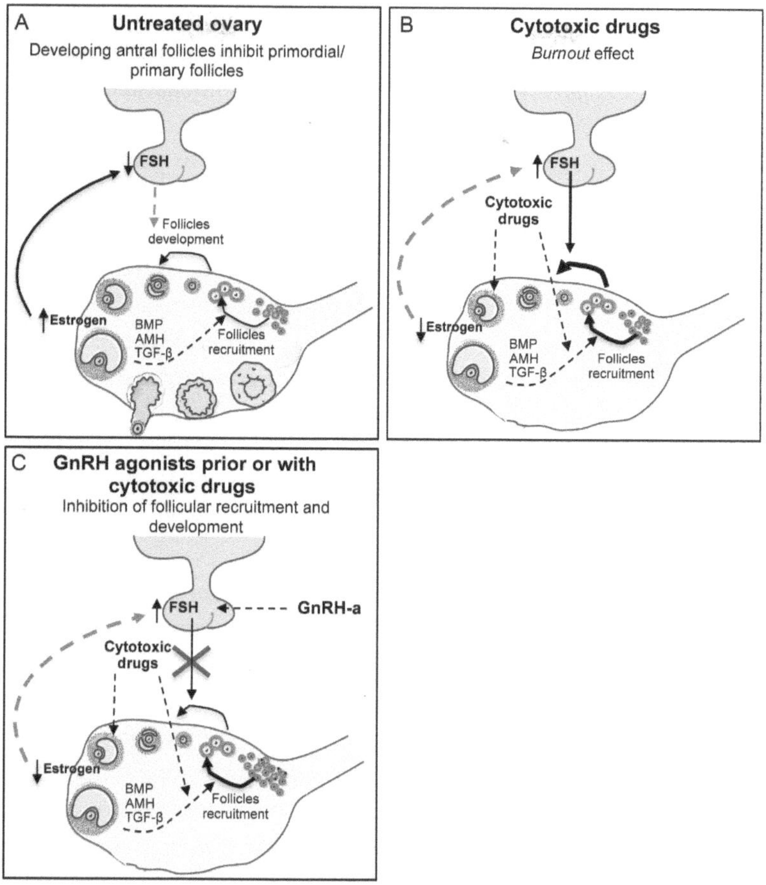

Fig. 4 Potential beneficial effects of GnRH-a in preserving ovarian function and fertility in young female cancer patients undergoing chemotherapy. (A) In the untreated ovary, antral (but not primordial/primary) follicles express FSH receptors and produce estrogens which, in turn, control FSH secretion. Antral follicles also produce growth factors (BMP, AMH, TGF-β) that act in a paracrine way to keep primordial/primary follicles in a dormant condition. Thus, during the first phase of the ovarian cycle, a balance exists in which developing follicles inhibit primordial/primary follicle recruitment. (B) Cytotoxic drugs trigger follicular damage leading to: decreased estrogen production and consequent increase in FSH levels; suppression of the negative paracrine control exerted by developing follicles on primordial/primary follicles. These two mechanisms are responsible for further recruitment of primordial/primary follicles that will also be destroyed by chemotherapy (*"burnout"* effect), thus depleting the ovarian reserve. (C) GnRH-a, given together with chemotherapy, suppress gonadotropin secretion, thus triggering a hypogonadotropic milieu which is responsible for the inhibition of follicular recruitment and development. In this way the primordial/primary follicles can remain in their dormant state (protection of the ovarian reserve) and will start developing again only after suspension of the treatment. *AMH*, anti-Müllerian-hormone; *BMP*, bone morphogenic proteins; *TGF-β*, transforming growth factor-β.

all these factors act in a paracrine way to keep primordial/primary follicles in a dormant condition (Blumenfeld & Evron, 2015; Knight & Glister, 2006). Cytotoxic drugs trigger the death of developing follicles and this leads to (a) decreased estrogen production and the consequent increase in FSH levels and (b) suppression of the negative paracrine control exerted by antral follicles on primordial/primary follicles. Together, these two mechanisms are responsible for the further recruitment of primordial follicles that will also be destroyed by the cytotoxic insults ("burnout" effect) (Roness, Gavish, Cohen, & Meirow, 2013) (Fig. 3). Through the desensitization of GnRHR at the pituitary level, GnRH-a suppress FSH secretion; this will lead to both inhibition of follicular growth and arrest of the recruitment of additional primordial follicles (Blumenfeld & Evron, 2015; Hasky et al., 2015).

Based on these observations, several clinical studies have been performed to confirm the beneficial effects of GnRH-a on fertility preservation in young female cancer patients undergoing chemotherapy.

In a retrospective cohort study in female patients affected by different types of tumors, Blumenfeld and coworkers (Blumenfeld et al., 2015) analyzed the effects of adding GnRH-a to chemotherapy on cyclic ovarian function and on fertility preservation, as evaluated by the number of pregnancies. The authors reported that the combination treatment resulted in a significantly higher percentage of women experiencing cyclic ovarian function and a lower percentage of patients developing ovarian failure. Adding GnRH-a to chemotherapy also significantly increased the occurrence of spontaneous conception. In another study performed in premenopausal breast cancer patients, the combination of a GnRH-a with chemotherapy resulted in a significantly higher pregnancy rate together with a significant improvement of disease-free and overall survival (Moore et al., 2015). In line with these observations, Lambertini and coworkers (Lambertini et al., 2015, 2017) reported that, in premenopausal women with breast cancer (either hormone receptor positive or hormone receptor negative) the concurrent administration of a GnRH-a and cytotoxic therapy is associated with a higher probability of ovarian function recovery and an increased chance of achieving pregnancy. In a systematic review and meta-analysis, a positive effect of GnRH-a in addition to chemotherapy has also been recently reported on the resumption of ovarian function in premenopausal women with breast cancer; however, the combination treatment did not appear to be associated with a protective effect in fertility (Shen et al., 2015).

On the other hand, these promising positive effects of GnRH-a in chemotherapy-treated patients could not be confirmed by other studies

(Vitek et al., 2014). A recent long-term analysis in lymphoma female patients reports that the GnRH-a does not prevent POF induced by chemotherapy and does not influence future pregnancy rate (Demeestere et al., 2016). Moreover, in in vitro studies, Bildik and coworkers (Bildik et al., 2015) demonstrated that, in human ovarian cortical specimens as well as in non-luteinized and in luteinized granulosa cells treated with cytotoxic drugs, GnRH-a do not prevent activation of apoptotic pathways, follicle loss, and DNA damage induced by chemotherapy drugs.

The protective effects of GnRH-a have also been suggested to be related to the specific mechanism of chemotherapy-induced ovarian damage. In an in vivo model (mice), Hasky and coworkers reported that GnRH-a can exert a protective effect on ovarian preservation and subsequent fertility in response to cyclophosphamide; on the other hand, when given in combination with doxorubicin, GnRH-a have an adverse effect on vascular recovery, leading to a sustained condition of ovarian failure (Hasky et al., 2015).

Thus, the efficacy of GnRH-a in preserving ovarian function and fertility in young cancer patients undergoing chemotherapy is still a matter of debate. However, the data so far reported suggest that, in addition to embryo/oocyte or ovarian tissue cryopreservation, GnRH-a should be taken into consideration to preserve ovarian function and fertility in these patients.

6. CONCLUSIONS AND FUTURE DIRECTIONS

The most recent data on the regulation of GnRH release as well as on the therapeutic use of its analogs offer interesting new perspectives in this field.

A GnRH/GnRHR system is expressed in the female reproductive axis. The pulsatile release of GnRH from the hypothalamic neurons is finely tuned by a series of neuropeptides that are under investigation as targets for future innovative therapeutic approaches for diseases of reproduction.

The functions of GnRHR at the ovarian level have not still completely clarified, but they could offer future new lines of intervention on ovarian dysfunctions.

Finally, the application of GnRH-a to preserve ovarian function in young female chemotherapy-treated cancer patients is an interesting approach that deserves further investigation

ACKNOWLEDGMENTS

Supported by Prin 2010/2011 2010NFEB9L_004; Interdisciplinary Research Project, Zukunftskolleg, University of Konstanz, Konstanz, Germany, 2012; PUR 2009, 2009-ATE-0228. We are also pleased to recognize the contribution of the Department of Pharmacological and Biomedical Sciences for the structural support.

REFERENCES

Abel, B. S., Shaw, N. D., Brown, J. M., Adams, J. M., Alati, T., Martin, K. A., et al. (2013). Responsiveness to a physiological regimen of GnRH therapy and relation to genotype in women with isolated hypogonadotropic hypogonadism. *The Journal of Clinical Endocrinology and Metabolism, 98*(2), E206–216.

Advis, J. P., Klein, J., Kuljis, R. O., Sarkar, D. K., McDonald, J. M., & Conover, C. A. (2003). Regulation of gonadotropin releasing hormone release by neuropeptide Y at the median eminence during the preovulatory period in ewes. *Neuroendocrinology, 77*(4), 246–257.

Al-Inany, H. G., Abou-Setta, A. M., & Aboulghar, M. (2006). Gonadotrophin-releasing hormone antagonists for assisted conception. *Cochrane Database of Systematic Reviews, 3*, CD001750.

Amorim, C. A., & Shikanov, A. (2016). The artificial ovary: Current status and future perspectives. *Future Oncology, 12*(20), 2323–2332.

Benard, J., Duros, S., El Hachem, H., Sonigo, C., Sifer, C., & Grynberg, M. (2016). Freezing oocytes or embryos after controlled ovarian hyperstimulation in cancer patients: The state of the art. *Future Oncology, 12*(14), 1731–1741.

Bildik, G., Akin, N., Senbabaoglu, F., Sahin, G. N., Karahuseyinoglu, S., Ince, U., et al. (2015). GnRH agonist leuprolide acetate does not confer any protection against ovarian damage induced by chemotherapy and radiation in vitro. *Human Reproduction, 30*(12), 2912–2925.

Blacker, C. M. (1992). Ovulation stimulation and induction. *Endocrinology and Metabolism Clinics of North America, 21*(1), 57–84.

Blumenfeld, Z. (2001). Gonadotropin-releasing hormone antagonists instead of agonists: A change for the better? *Fertility and Sterility, 76*(3), 443–444.

Blumenfeld, Z. (2016). Endocrine prevention of chemotherapy-induced ovarian failure. *Future Oncology, 12*(14), 1671–1674.

Blumenfeld, Z., & Evron, A. (2015). Preserving fertility when choosing chemotherapy regimens—The role of gonadotropin-releasing hormone agonists. *Expert Opinion on Pharmacotherapy, 16*(7), 1009–1020.

Blumenfeld, Z., Zur, H., & Dann, E. J. (2015). Gonadotropin-releasing hormone agonist cotreatment during chemotherapy may increase pregnancy rate in survivors. *The Oncologist, 20*(11), 1283–1289.

Carel, J. C., Eugster, E. A., Rogol, A., Ghizzoni, L., Palmert, M. R., Antoniazzi, F., et al. (2009). Consensus statement on the use of gonadotropin-releasing hormone analogs in children. *Pediatrics, 123*(4), e752–762.

Casoni, F., Malone, S. A., Belle, M., Luzzati, F., Collier, F., Allet, C., et al. (2016). Development of the neurons controlling fertility in humans: New insights from 3D imaging and transparent fetal brains. *Development, 143*(21), 3969–3981.

Casper, R. F. (2015). Introduction: Gonadotropin-releasing hormone agonist triggering of final follicular maturation for in vitro fertilization. *Fertility and Sterility, 103*(4), 865–866.

Cheng, G., Coolen, L. M., Padmanabhan, V., Goodman, R. L., & Lehman, M. N. (2010). The kisspeptin/neurokinin B/dynorphin (KNDy) cell population of the arcuate nucleus: Sex differences and effects of prenatal testosterone in sheep. *Endocrinology*, *151*(1), 301–311.
Cheng, K. W., Nathwani, P. S., & Leung, P. C. (2000). Regulation of human gonadotropin-releasing hormone receptor gene expression in placental cells. *Endocrinology*, *141*(7), 2340–2349.
Cheon, K. W., Lee, H. S., Parhar, I. S., & Kang, I. S. (2001). Expression of the second isoform of gonadotrophin-releasing hormone (GnRH-II) in human endometrium throughout the menstrual cycle. *Molecular Human Reproduction*, *7*(5), 447–452.
Cheung, L. W., & Wong, A. S. (2008). Gonadotropin-releasing hormone: GnRH receptor signaling in extrapituitary tissues. *The FEBS Journal*, *275*(22), 5479–5495.
Chiang, C. H., Cheng, K. W., Igarashi, S., Nathwani, P. S., & Leung, P. C. (2000). Hormonal regulation of estrogen receptor alpha and beta gene expression in human granulosa-luteal cells in vitro. *The Journal of Clinical Endocrinology and Metabolism*, *85*(10), 3828–3839.
Choi, K. C., Auersperg, N., & Leung, P. C. (2001). Expression and antiproliferative effect of a second form of gonadotropin-releasing hormone in normal and neoplastic ovarian surface epithelial cells. *The Journal of Clinical Endocrinology and Metabolism*, *86*(10), 5075–5078.
Chou, C. S., MacCalman, C. D., & Leung, P. C. (2003). Differential effects of gonadotropin-releasing hormone I and II on the urokinase-type plasminogen activator/plasminogen activator inhibitor system in human decidual stromal cells in vitro. *The Journal of Clinical Endocrinology and Metabolism*, *88*(8), 3806–3815.
Clayton, R. N. (1989). Gonadotrophin-releasing hormone: Its actions and receptors. *The Journal of Endocrinology*, *120*(1), 11–19.
Clayton, R. N., Eccleston, L., Gossard, F., Thalbard, J. C., & Morel, G. (1992). Rat granulosa cells express the gonadotrophin-releasing hormone gene: Evidence from in-situ hybridization histochemistry. *Journal of Molecular Endocrinology*, *9*(3), 189–195.
Codacci-Pisanelli, G., Del Pup, L., Del Grande, M., & Peccatori, F. A. (2017). Mechanisms of chemotherapy-induced ovarian damage in breast cancer patients. *Critical Reviews in Oncology/Hematology*, *113*, 90–96.
Conn, P. M., & Crowley, W. F., Jr. (1994). Gonadotropin-releasing hormone and its analogs. *Annual Review of Medicine*, *45*, 391–405.
Cook, T., & Sheridan, W. P. (2000). Development of GnRH antagonists for prostate cancer: New approaches to treatment. *The Oncologist*, *5*(2), 162–168.
Crowley, W. F., & McArthur, J. W. (1980). Simulation of the normal menstrual cycle in Kallman's syndrome by pulsatile administration of luteinizing hormone-releasing hormone (LHRH). *The Journal of Clinical Endocrinology and Metabolism*, *51*(1), 173–175.
Demeestere, I., Brice, P., Peccatori, F. A., Kentos, A., Dupuis, J., Zachee, P., et al. (2016). No evidence for the benefit of gonadotropin-releasing hormone agonist in preserving ovarian function and fertility in lymphoma survivors treated with chemotherapy: Final long-term report of a prospective randomized trial. *Journal of Clinical Oncology*, *34*(22), 2568–2574.
DePaolo, L. V., King, R. A., & Carrillo, A. J. (1987). In vivo and in vitro examination of an autoregulatory mechanism for luteinizing hormone-releasing hormone. *Endocrinology*, *120*(1), 272–279.
Deragon, K. L., & Sower, S. A. (1994). Effects of lamprey gonadotropin-releasing hormone-III on steroidogenesis and spermiation in male sea lampreys. *General and Comparative Endocrinology*, *95*(3), 363–367.
de Roux, N., Genin, E., Carel, J. C., Matsuda, F., Chaussain, J. L., & Milgrom, E. (2003). Hypogonadotropic hypogonadism due to loss of function of the KiSS1-derived peptide

receptor GPR54. *Proceedings of the National Academy of Sciences of the United States of America, 100*(19), 10972–10976.

DiVall, S. A., Herrera, D., Sklar, B., Wu, S., Wondisford, F., Radovick, S., et al. (2015). Insulin receptor signaling in the GnRH neuron plays a role in the abnormal GnRH pulsatility of obese female mice. *PLoS One, 10*(3), e0119995.

Dondi, D., Limonta, P., Moretti, R. M., Marelli, M. M., Garattini, E., & Motta, M. (1994). Antiproliferative effects of luteinizing hormone-releasing hormone (LHRH) agonists on human androgen-independent prostate cancer cell line DU 145: Evidence for an autocrine-inhibitory LHRH loop. *Cancer Research, 54*(15), 4091–4095.

Eckert, K. L., Wilson, D. M., Bachrach, L. K., Anhalt, H., Habiby, R. L., Olney, R. C., et al. (1996). A single-sample, subcutaneous gonadotropin-releasing hormone test for central precocious puberty. *Pediatrics, 97*(4), 517–519.

Emons, G., Ortmann, O., Becker, M., Irmer, G., Springer, B., Laun, R., et al. (1993). High affinity binding and direct antiproliferative effects of LHRH analogues in human ovarian cancer cell lines. *Cancer Research, 53*(22), 5439–5446.

Engel, J. B., & Schally, A. V. (2007). Drug insight: Clinical use of agonists and antagonists of luteinizing-hormone-releasing hormone. *Nature Clinical Practice. Endocrinology & Metabolism, 3*(2), 157–167.

Engel, J. B., Schally, A. V., Buchholz, S., Seitz, S., Emons, G., & Ortmann, O. (2012). Targeted chemotherapy of endometrial, ovarian and breast cancers with cytotoxic analogs of luteinizing hormone-releasing hormone (LHRH). *Archives of Gynecology and Obstetrics, 286*(2), 437–442.

Engel, J. B., Schally, A. V., Dietl, J., Rieger, L., & Honig, A. (2007). Targeted therapy of breast and gynecological cancers with cytotoxic analogues of peptide hormones. *Molecular Pharmaceutics, 4*(5), 652–658.

Engel, J. B., Tinneberg, H. R., Rick, F. G., Berkes, E., & Schally, A. V. (2016). Targeting of peptide cytotoxins to LHRH receptors for treatment of cancer. *Current Drug Targets, 17*(5), 488–494.

Fan, N. C., Jeung, E. B., Peng, C., Olofsson, J. I., Krisinger, J., & Leung, P. C. (1994). The human gonadotropin-releasing hormone (GnRH) receptor gene: Cloning, genomic organization and chromosomal assignment. *Molecular and Cellular Endocrinology, 103*(1–2), R1–6.

Filicori, M., Flamigni, C., Meriggiola, M. C., Cognigni, G., Valdiserri, A., Ferrari, P., et al. (1991). Ovulation induction with pulsatile gonadotropin-releasing hormone: Technical modalities and clinical perspectives. *Fertility and Sterility, 56*(1), 1–13.

Grumbach, M., & Kaplan, S. (1990). The neuroendocrinology of human puberty: An ontogenetic perspective. In M. M. Grumbach, P. C. Sizonenko, & M. Aubert (Eds.), *Control of the onset of puberty* (pp. 1–62). Baltimore, MD: Williams & Wilkins.

Grundker, C., Gunthert, A. R., Hellriegel, M., & Emons, G. (2004). Gonadotropin-releasing hormone (GnRH) agonist triptorelin inhibits estradiol-induced serum response element (SRE) activation and c-fos expression in human endometrial, ovarian and breast cancer cells. *European Journal of Endocrinology, 151*(5), 619–628.

Grundker, C., Gunthert, A. R., Westphalen, S., & Emons, G. (2002). Biology of the gonadotropin-releasing hormone system in gynecological cancers. *European Journal of Endocrinology, 146*(1), 1–14.

Grundker, C., Volker, P., Griesinger, F., Ramaswamy, A., Nagy, A., Schally, A. V., et al. (2002). Antitumor effects of the cytotoxic luteinizing hormone-releasing hormone analog AN-152 on human endometrial and ovarian cancers xenografted into nude mice. *American Journal of Obstetrics and Gynecology, 187*(3), 528–537.

Guerriero, K. A., Keen, K. L., & Terasawa, E. (2012). Developmental increase in kisspeptin-54 release in vivo is independent of the pubertal increase in estradiol in female rhesus monkeys (Macaca mulatta). *Endocrinology, 153*(4), 1887–1897.

Gunthert, A. R., Grundker, C., Hollmann, K., & Emons, G. (2002). Luteinizing hormone-releasing hormone induces JunD-DNA binding and extends cell cycle in human ovarian cancer cells. *Biochemical and Biophysical Research Communications*, *294*(1), 11–15.
Haisenleder, D. J., Dalkin, A. C., Ortolano, G. A., Marshall, J. C., & Shupnik, M. A. (1991). A pulsatile gonadotropin-releasing hormone stimulus is required to increase transcription of the gonadotropin subunit genes: Evidence for differential regulation of transcription by pulse frequency in vivo. *Endocrinology*, *128*(1), 509–517.
Hall, J. E. (2007). Neuroendocrine changes with reproductive aging in women. *Seminars in Reproductive Medicine*, *25*(5), 344–351.
Hall, J. E., Lavoie, H. B., Marsh, E. E., & Martin, K. A. (2000). Decrease in gonadotropin-releasing hormone (GnRH) pulse frequency with aging in postmenopausal women. *The Journal of Clinical Endocrinology and Metabolism*, *85*(5), 1794–1800.
Harada, M., & Osuga, Y. (2016). Where are oncofertility and fertility preservation treatments heading in 2016? *Future Oncology*, *12*(20), 2313–2321.
Harrison, G. S., Wierman, M. E., Nett, T. M., & Glode, L. M. (2004). Gonadotropin-releasing hormone and its receptor in normal and malignant cells. *Endocrine-Related Cancer*, *11*(4), 725–748.
Hasky, N., Uri-Belapolsky, S., Goldberg, K., Miller, I., Grossman, H., Stemmer, S. M., et al. (2015). Gonadotrophin-releasing hormone agonists for fertility preservation: Unraveling the enigma? *Human Reproduction*, *30*(5), 1089–1101.
Hayes, F. J., & Crowley, W. F. (1998). Gonadotropin pulsations across development. *Hormone Research*, *49*(3–4), 163–168.
Herbison, A. E., Porteous, R., Pape, J. R., Mora, J. M., & Hurst, P. R. (2008). Gonadotropin-releasing hormone neuron requirements for puberty, ovulation, and fertility. *Endocrinology*, *149*(2), 597–604.
Hickman, L. C., Valentine, L. N., & Falcone, T. (2016). Preservation of gonadal function in women undergoing chemotherapy: A review of the potential role for gonadotropin-releasing hormone agonists. *American Journal of Obstetrics and Gynecology*, *215*(4), 415–422.
Hinuma, S., Shintani, Y., Fukusumi, S., Iijima, N., Matsumoto, Y., Hosoya, M., et al. (2000). New neuropeptides containing carboxy-terminal RFamide and their receptor in mammals. *Nature Cell Biology*, *2*(10), 703–708.
Hong, I. S., Cheung, A. P., & Leung, P. C. (2008). Gonadotropin-releasing hormones I and II induce apoptosis in human granulosa cells. *The Journal of Clinical Endocrinology and Metabolism*, *93*(8), 3179–3185.
Hong, I. S., Klausen, C., Cheung, A. P., & Leung, P. C. (2012). Gonadotropin-releasing hormone-I or -II interacts with IGF-I/Akt but not connexin 43 in human granulosa cell apoptosis. *The Journal of Clinical Endocrinology and Metabolism*, *97*(2), 525–534.
Hrabovszky, E., & Liposits, Z. (2013). Afferent neuronal control of type-I gonadotropin releasing hormone neurons in the human. *Frontiers in Endocrinology*, *4*, 130.
Hsueh, A. J., Billig, H., & Tsafriri, A. (1994). Ovarian follicle atresia: A hormonally controlled apoptotic process. *Endocrine Reviews*, *15*(6), 707–724.
Huhtaniemi, I., White, R., McArdle, C. A., & Persson, B. E. (2009). Will GnRH antagonists improve prostate cancer treatment? *Trends in Endocrinology and Metabolism*, *20*(1), 43–50.
Imai, A., Takagi, H., Furui, T., Horibe, S., Fuseya, T., & Tamaya, T. (1996). Evidence for coupling of phosphotyrosine phosphatase to gonadotropin-releasing hormone receptor in ovarian carcinoma membrane. *Cancer*, *77*(1), 132–137.
Iwashita, M., Evans, M. I., & Catt, K. J. (1986). Characterization of a gonadotropin-releasing hormone receptor site in term placenta and chorionic villi. *The Journal of Clinical Endocrinology and Metabolism*, *62*(1), 127–133.
Jensen, A. K., Rechnitzer, C., Macklon, K. T., Ifversen, M. R., Birkebaek, N., Clausen, N., et al. (2017). Cryopreservation of ovarian tissue for fertility preservation in a large cohort of young girls: Focus on pubertal development. *Human Reproduction*, *32*(1), 154–164.

Kaiser, U. B., Conn, P. M., & Chin, W. W. (1997). Studies of gonadotropin-releasing hormone (GnRH) action using GnRH receptor-expressing pituitary cell lines. *Endocrine Reviews*, *18*(1), 46–70.

Kakar, S. S., Musgrove, L. C., Devor, D. C., Sellers, J. C., & Neill, J. D. (1992). Cloning, sequencing, and expression of human gonadotropin releasing hormone (GnRH) receptor. *Biochemical and Biophysical Research Communications*, *189*(1), 289–295.

Kang, S. K., Choi, K. C., Yang, H. S., & Leung, P. C. (2003). Potential role of gonadotrophin-releasing hormone (GnRH)-I and GnRH-II in the ovary and ovarian cancer. *Endocrine-Related Cancer*, *10*(2), 169–177.

Kang, S. K., Tai, C. J., Cheng, K. W., & Leung, P. C. (2000). Gonadotropin-releasing hormone activates mitogen-activated protein kinase in human ovarian and placental cells. *Molecular and Cellular Endocrinology*, *170*(1–2), 143–151.

Khadra, A., & Li, Y. X. (2006). A model for the pulsatile secretion of gonadotropin-releasing hormone from synchronized hypothalamic neurons. *Biophysical Journal*, *91*(1), 74–83.

Knight, P. G., & Glister, C. (2006). TGF-beta superfamily members and ovarian follicle development. *Reproduction*, *132*(2), 191–206.

Kovacs, M., Seprodi, J., Koppan, M., Horvath, J. E., Vincze, B., Teplan, I., et al. (2002). Lamprey gonadotropin hormone-releasing hormone-III has no selective follicle-stimulating hormone-releasing effect in rats. *Journal of Neuroendocrinology*, *14*(8), 647–655.

Kraus, S., Naor, Z., & Seger, R. (2001). Intracellular signaling pathways mediated by the gonadotropin-releasing hormone (GnRH) receptor. *Archives of Medical Research*, *32*(6), 499–509.

Kruk, P. A., Uitto, V. J., Firth, J. D., Dedhar, S., & Auersperg, N. (1994). Reciprocal interactions between human ovarian surface epithelial cells and adjacent extracellular matrix. *Experimental Cell Research*, *215*(1), 97–108.

Ladanyi, C., Mor, A., Christianson, M. S., Dhillon, N., & Segars, J. H. (2017). Recent advances in the field of ovarian tissue cryopreservation and opportunities for research. *Journal of Assisted Reproduction and Genetics*, *34*(6), 709–722.

Lambertini, M., Anserini, P., Fontana, V., Poggio, F., Iacono, G., Abate, A., et al. (2017). The PREgnancy and FERtility (PREFER) study: An Italian multicenter prospective cohort study on fertility preservation and pregnancy issues in young breast cancer patients. *BMC Cancer*, *17*(1), 346.

Lambertini, M., Ceppi, M., Poggio, F., Peccatori, F. A., Azim, H. A., Jr., Ugolini, D., et al. (2015). Ovarian suppression using luteinizing hormone-releasing hormone agonists during chemotherapy to preserve ovarian function and fertility of breast cancer patients: A meta-analysis of randomized studies. *Annals of Oncology*, *26*(12), 2408–2419.

Lambertini, M., Cinquini, M., Moschetti, I., Peccatori, F. A., Anserini, P., Valenzano Menada, M., et al. (2017). Temporary ovarian suppression during chemotherapy to preserve ovarian function and fertility in breast cancer patients: A GRADE approach for evidence evaluation and recommendations by the Italian Association of Medical Oncology. *European Journal of Cancer*, *71*, 25–33.

Landy, H., Schneyer, A. L., Whitcomb, R. W., & Crowley, W. F. (1990). Validation of highly specific and sensitive radioimmunoassays for lutropin, follitropin, and free alpha subunit in unextracted urine. *Clinical Chemistry*, *36*(2), 340–344.

Latimer, V. S., Rodrigues, S. M., Garyfallou, V. T., Kohama, S. G., White, R. B., Fernald, R. D., et al. (2000). Two molecular forms of gonadotropin-releasing hormone (GnRH-I and GnRH-II) are expressed by two separate populations of cells in the rhesus macaque hypothalamus. *Brain Research. Molecular Brain Research*, *75*(2), 287–292.

Lawrenz, B., Mahajan, N., & Fatemi, H. M. (2016). The effects of cancer therapy on women's fertility: What do we know now? *Future Oncology*, *12*(14), 1721–1729.

Lehman, M. N., Coolen, L. M., & Goodman, R. L. (2010). Minireview: Kisspeptin/neurokinin B/dynorphin (KNDy) cells of the arcuate nucleus: A central node in the control of gonadotropin-releasing hormone secretion. *Endocrinology, 151*(8), 3479–3489.
Lehman, M. N., Robinson, J. E., Karsch, F. J., & Silverman, A. J. (1986). Immunocytochemical localization of luteinizing hormone-releasing hormone (LHRH) pathways in the sheep brain during anestrus and the mid-luteal phase of the estrous cycle. *The Journal of Comparative Neurology, 244*(1), 19–35.
Leung, P. C., Cheng, C. K., & Zhu, X. M. (2003). Multi-factorial role of GnRH-I and GnRH-II in the human ovary. *Molecular and Cellular Endocrinology, 202*(1–2), 145–153.
Leyendecker, G., Struve, T., & Plotz, E. J. (1980). Induction of ovulation with chronic intermittent (pulsatile) administration of LH-RH in women with hypothalamic and hyperprolactinemic amenorrhea. *Archives of Gynecology, 229*(3), 177–190.
Li, P., Li, Y., & Yang, C. L. (2014). Gonadotropin releasing hormone agonist treatment to increase final stature in children with precocious puberty: A meta-analysis. *Medicine (Baltimore), 93*(27), e260.
Lim, S., Pnueli, L., Tan, J. H., Naor, Z., Rajagopal, G., & Melamed, P. (2009). Negative feedback governs gonadotrope frequency-decoding of gonadotropin releasing hormone pulse-frequency. *PLoS One, 4*(9), e7244.
Lima, C. J., Cardoso, S. C., Lemos, E. F., Zingler, E., Capanema, C., Menezes, L. D., et al. (2014). Mutational analysis of the genes encoding RFamide-related peptide-3, the human orthologue of gonadotrophin-inhibitory hormone, and its receptor (GPR147) in patients with gonadotrophin-releasing hormone-dependent pubertal disorders. *Journal of Neuroendocrinology, 26*(11), 817–824.
Limonta, P., Marelli, M. M., Mai, S., Motta, M., Martini, L., & Moretti, R. M. (2012). GnRH receptors in cancer: From cell biology to novel targeted therapeutic strategies. *Endocrine Reviews, 33*(5), 784–811.
Limonta, P., Moretti, R. M., Marelli, M. M., Dondi, D., Parenti, M., & Motta, M. (1999). The luteinizing hormone-releasing hormone receptor in human prostate cancer cells: Messenger ribonucleic acid expression, molecular size, and signal transduction pathway. *Endocrinology, 140*(11), 5250–5256.
Limonta, P., Moretti, R. M., Marelli, M. M., & Motta, M. (2003). The biology of gonadotropin hormone-releasing hormone: Role in the control of tumor growth and progression in humans. *Frontiers in Neuroendocrinology, 24*(4), 279–295.
Linkeviciute, A., Boniolo, G., Chiavari, L., & Peccatori, F. A. (2014). Fertility preservation in cancer patients: The global framework. *Cancer Treatment Reviews, 40*(8), 1019–1027.
Liu, J., & Patel, B. (2013). Central causes of amenorrhea. In L. De Groot, P. Beck-Peccoz, G. Chrousos, K. Dungan, A. Grossman, J. Hershman, et al. (Eds.), *Endotext [Internet]*. South Dartmouth, MA: Mdtext.com, Inc.
Maggi, R. (2016). Physiology of gonadotropin-releasing hormone (GnRH): Beyond the control of reproductive functions. *MOJ Anatomy and Physiology, 2*(5), 5.
Maggi, R., Cariboni, A. M., Marelli, M. M., Moretti, R. M., Andre, V., Marzagalli, M., et al. (2016). GnRH and GnRH receptors in the pathophysiology of the human female reproductive system. *Human Reproduction Update, 22*(3), 358–381.
Magon, N. (2011). Gonadotropin releasing hormone agonists: Expanding vistas. *Indian Journal of Endocrinology and Metabolism, 15*(4), 261–267.
Mahajan, N. (2015). Fertility preservation in female cancer patients: An overview. *Journal of Human Reproductive Sciences, 8*(1), 3–13.
Maheshwari, A., Gibreel, A., Siristatidis, C. S., & Bhattacharya, S. (2011). Gonadotrophin-releasing hormone agonist protocols for pituitary suppression in assisted reproduction. *Cochrane Database of Systematic Reviews, 8*, CD006919.

Mancini, F., Tur, R., Martinez, F., Coroleu, B., Rodriguez, I., & Barri, P. N. (2011). Gonadotrophin-releasing hormone-antagonists vs long agonist in in-vitro fertilization patients with polycystic ovary syndrome: A meta-analysis. *Gynecological Endocrinology, 27*(3), 150–155.

Marelli, M. M., Moretti, R. M., Januszkiewicz-Caulier, J., Motta, M., & Limonta, P. (2006). Gonadotropin-releasing hormone (GnRH) receptors in tumors: A new rationale for the therapeutical application of GnRH analogs in cancer patients? *Current Cancer Drug Targets, 6*(3), 257–269.

Marshall, J. C., Dalkin, A. C., Haisenleder, D. J., Paul, S. J., Ortolano, G. A., & Kelch, R. P. (1991). Gonadotropin-releasing hormone pulses: Regulators of gonadotropin synthesis and ovulatory cycles. *Recent Progress in Hormone Research, 47*, 155–187 [discussion 188–159].

Martin, J. B. (1993). Molecular genetics in neurology. *Annals of Neurology, 34*(6), 757–773.

McArdle, C. A. (2012). Gonadotropin-releasing hormone receptor signaling: Biased and unbiased. *Mini Reviews in Medicinal Chemistry, 12*(9), 841–850.

McCartney, C. R. (2010). Maturation of sleep-wake gonadotrophin-releasing hormone secretion across puberty in girls: Potential mechanisms and relevance to the pathogenesis of polycystic ovary syndrome. *Journal of Neuroendocrinology, 22*(7), 701–709.

Meirow, D., Dor, J., Kaufman, B., Shrim, A., Rabinovici, J., Schiff, E., et al. (2007). Cortical fibrosis and blood-vessels damage in human ovaries exposed to chemotherapy. Potential mechanisms of ovarian injury. *Human Reproduction, 22*(6), 1626–1633.

Millar, R. P. (2005). GnRHs and GnRH receptors. *Animal Reproduction Science, 88*(1–2), 5–28.

Millar, R. P., Pawson, A. J., Morgan, K., Rissman, E. F., & Lu, Z. L. (2008). Diversity of actions of GnRHs mediated by ligand-induced selective signaling. *Frontiers in Neuroendocrinology, 29*(1), 17–35.

Miller, I., Chuderland, D., Ron-El, R., Shalgi, R., & Ben-Ami, I. (2015). GnRH agonist triggering modulates PEDF to VEGF ratio inversely to hCG in granulosa cells. *The Journal of Clinical Endocrinology and Metabolism, 100*(11), E1428–1436.

Mittelman-Smith, M. A., Wong, A. M., Kathiresan, A. S., & Micevych, P. E. (2015). Classical and membrane-initiated estrogen signaling in an in vitro model of anterior hypothalamic kisspeptin neurons. *Endocrinology, 156*(6), 2162–2173.

Moenter, S. M., Caraty, A., Locatelli, A., & Karsch, F. J. (1991). Pattern of gonadotropin-releasing hormone (GnRH) secretion leading up to ovulation in the ewe: Existence of a preovulatory GnRH surge. *Endocrinology, 129*(3), 1175–1182.

Montagnani Marelli, M., Manea, M., Moretti, R. M., Marzagalli, M., & Limonta, P. (2015). Oxime bond-linked daunorubicin-GnRH-III bioconjugates exert antitumor activity in castration-resistant prostate cancer cells via the type I GnRH receptor. *International Journal of Oncology, 46*(1), 243–253.

Moore, H. C., Unger, J. M., Phillips, K. A., Boyle, F., Hitre, E., Porter, D., et al. (2015). Goserelin for ovarian protection during breast-cancer adjuvant chemotherapy. *The New England Journal of Medicine, 372*(10), 923–932.

Mugami, S., Dobkin-Bekman, M., Rahamim-Ben Navi, L., & Naor, Z. (2017). Differential roles of PKC isoforms (PKCs) in GnRH stimulation of MAPK phosphorylation in gonadotrope derived cells. *Molecular and Cellular Endocrinology*. https://doi.org/10.1016/j.mce.2017.04.004.

Munster, P. N. (2013). Fertility preservation and breast cancer: A complex problem. *Oncology (Williston Park), 27*(6), 533–539.

Naor, Z. (2009). Signaling by G-protein-coupled receptor (GPCR): Studies on the GnRH receptor. *Frontiers in Neuroendocrinology, 30*(1), 10–29.

Naor, Z., Shearman, M. S., Kishimoto, A., & Nishizuka, Y. (1988). Calcium-independent activation of hypothalamic type I protein kinase C by unsaturated fatty acids. *Molecular Endocrinology, 2*(11), 1043–1048.

Neill, J. D. (2002). GnRH and GnRH receptor genes in the human genome. *Endocrinology, 143*(3), 737–743.
Neill, J. D., Musgrove, L. C., & Duck, L. W. (2004). Newly recognized GnRH receptors: Function and relative role. *Trends in Endocrinology and Metabolism, 15*(8), 383–392.
Ny, T., Liu, Y. X., Ohlsson, M., Jones, P. B., & Hsueh, A. J. (1987). Regulation of tissue-type plasminogen activator activity and messenger RNA levels by gonadotropin-releasing hormone in cultured rat granulosa cells and cumulus-oocyte complexes. *The Journal of Biological Chemistry, 262*(24), 11790–11793.
Oktem, O., & Oktay, K. (2007). Quantitative assessment of the impact of chemotherapy on ovarian follicle reserve and stromal function. *Cancer, 110*(10), 2222–2229.
Orvieto, R., Rabinson, J., Meltzer, S., Homburg, R., Anteby, E., & Zohav, E. (2006). GnRH agonist versus GnRH antagonist in ovarian stimulation: Is the emperor naked? *Clinical and Experimental Obstetrics & Gynecology, 33*(4), 197–199.
Peng, C., Fan, N. C., Ligier, M., Vaananen, J., & Leung, P. C. (1994). Expression and regulation of gonadotropin-releasing hormone (GnRH) and GnRH receptor messenger ribonucleic acids in human granulosa-luteal cells. *Endocrinology, 135*(5), 1740–1746.
Piet, R., de Croft, S., Liu, X., & Herbison, A. E. (2015). Electrical properties of kisspeptin neurons and their regulation of GnRH neurons. *Frontiers in Neuroendocrinology, 36*, 15–27.
Plant, T. M. (2015). Neuroendocrine control of the onset of puberty. *Frontiers in Neuroendocrinology, 38*, 73–88.
Plant, T. M., Ramaswamy, S., Simorangkir, D., & Marshall, G. R. (2005). Postnatal and pubertal development of the rhesus monkey (Macaca mulatta) testis. *Annals of the New York Academy of Sciences, 1061*, 149–162.
Raga, F., Casan, E. M., Kruessel, J. S., Wen, Y., Huang, H. Y., Nezhat, C., et al. (1998). Quantitative gonadotropin-releasing hormone gene expression and immunohistochemical localization in human endometrium throughout the menstrual cycle. *Biology of Reproduction, 59*(3), 661–669.
Raivio, T., Falardeau, J., Dwyer, A., Quinton, R., Hayes, F. J., Hughes, V. A., et al. (2007). Reversal of idiopathic hypogonadotropic hypogonadism. *The New England Journal of Medicine, 357*(9), 863–873.
Rama, S., & Rao, A. J. (2001). Embryo implantation and GnRH antagonists: The search for the human placental GnRH receptor. *Human Reproduction, 16*(2), 201–205.
Ramakrishnappa, N., Rajamahendran, R., Lin, Y. M., & Leung, P. C. (2005). GnRH in non-hypothalamic reproductive tissues. *Animal Reproduction Science, 88*(1–2), 95–113.
Ramaswamy, S., Seminara, S. B., & Plant, T. M. (2011). Evidence from the agonadal juvenile male rhesus monkey (Macaca mulatta) for the view that the action of neurokinin B to trigger gonadotropin-releasing hormone release is upstream from the kisspeptin receptor. *Neuroendocrinology, 94*(3), 237–245.
Rance, N. E. (2009). Menopause and the human hypothalamus: Evidence for the role of kisspeptin/neurokinin B neurons in the regulation of estrogen negative feedback. *Peptides, 30*(1), 111–122.
Rapisarda, J. J., Bergman, K. S., Steiner, R. A., & Foster, D. L. (1983). Response to estradiol inhibition of tonic luteinizing hormone secretion decreases during the final stage of puberty in the rhesus monkey. *Endocrinology, 112*(4), 1172–1179.
Reid, R. L., Leopold, G. R., & Yen, S. S. (1981). Induction of ovulation and pregnancy with pulsatile luteinizing hormone releasing factor: Dosage and mode of delivery. *Fertility and Sterility, 36*(5), 553–559.
Reinhart, J., Mertz, L. M., & Catt, K. J. (1992). Molecular cloning and expression of cDNA encoding the murine gonadotropin-releasing hormone receptor. *The Journal of Biological Chemistry, 267*(30), 21281–21284.

Rodgers, R. J., Reid, G. D., Koch, J., Deans, R., Ledger, W. L., Friedlander, M., et al. (2017). The safety and efficacy of controlled ovarian hyperstimulation for fertility preservation in women with early breast cancer: A systematic review. *Human Reproduction*, *32*(5), 1033–1045.

Roland, A. V., & Moenter, S. M. (2014). Reproductive neuroendocrine dysfunction in polycystic ovary syndrome: Insight from animal models. *Frontiers in Neuroendocrinology*, *35*(4), 494–511.

Roness, H., Gavish, Z., Cohen, Y., & Meirow, D. (2013). Ovarian follicle burnout: A universal phenomenon? *Cell Cycle*, *12*(20), 3245–3246.

Salama, M., Isachenko, V., Isachenko, E., Rahimi, G., & Mallmann, P. (2016). Updates in preserving reproductive potential of prepubertal girls with cancer: Systematic review. *Critical Reviews in Oncology/Hematology*, *103*, 10–21.

Schally, A. V. (1999). LH-RH analogues: I. Their impact on reproductive medicine. *Gynecological Endocrinology*, *13*(6), 401–409.

Schally, A. V., Arimura, A., Baba, Y., Nair, R. M., Matsuo, H., Redding, T. W., et al. (1971). Isolation and properties of the FSH and LH-releasing hormone. *Biochemical and Biophysical Research Communications*, *43*(2), 393–399.

Schally, A. V., & Nagy, A. (1999). Cancer chemotherapy based on targeting of cytotoxic peptide conjugates to their receptors on tumors. *European Journal of Endocrinology*, *141*(1), 1–14.

Scheiber, M., & Liu, J. (2011). The use of gonadotropin-releasing hormone to induce ovulation. *The Global Library of Women's Medicine*. Retrieved from https://doi.org/10.3843/glowm.10340.

Schmidt, P. J., Steinberg, E. M., Negro, P. P., Haq, N., Gibson, C., & Rubinow, D. R. (2009). Pharmacologically induced hypogonadism and sexual function in healthy young women and men. *Neuropsychopharmacology*, *34*(3), 565–576.

Seeburg, P. H., Mason, A. J., Stewart, T. A., & Nikolics, K. (1987). The mammalian GnRH gene and its pivotal role in reproduction. *Recent Progress in Hormone Research*, *43*, 69–98.

Semple, R. K., Achermann, J. C., Ellery, J., Farooqi, I. S., Karet, F. E., Stanhope, R. G., et al. (2005). Two novel missense mutations in g protein-coupled receptor 54 in a patient with hypogonadotropic hypogonadism. *The Journal of Clinical Endocrinology and Metabolism*, *90*(3), 1849–1855.

Shen, Y. W., Zhang, X. M., Lv, M., Chen, L., Qin, T. J., Wang, F., et al. (2015). Utility of gonadotropin-releasing hormone agonists for prevention of chemotherapy-induced ovarian damage in premenopausal women with breast cancer: A systematic review and meta-analysis. *OncoTargets and Therapy*, *8*, 3349–3359.

Siegel, R. L., Miller, K. D., & Jemal, A. (2017). Cancer statistics, 2017. *CA: A Cancer Journal for Clinicians*, *67*(1), 7–30.

Silveira, L. F., & Latronico, A. C. (2013). Approach to the patient with hypogonadotropic hypogonadism. *The Journal of Clinical Endocrinology and Metabolism*, *98*(5), 1781–1788.

Skorupskaite, K., George, J. T., & Anderson, R. A. (2014). The kisspeptin-GnRH pathway in human reproductive health and disease. *Human Reproduction Update*, *20*(4), 485–500.

Skrapits, K., Borsay, B., Herczeg, L., Ciofi, P., Liposits, Z., & Hrabovszky, E. (2015). Neuropeptide co-expression in hypothalamic kisspeptin neurons of laboratory animals and the human. *Frontiers in Neuroscience*, *9*, 29.

So, W. K., Cheng, J. C., Poon, S. L., & Leung, P. C. (2008). Gonadotropin-releasing hormone and ovarian cancer: A functional and mechanistic overview. *The FEBS Journal*, *275*(22), 5496–5511.

Sower, S. A., Chiang, Y. C., Lovas, S., & Conlon, J. M. (1993). Primary structure and biological activity of a third gonadotropin-releasing hormone from lamprey brain. *Endocrinology*, *132*(3), 1125–1131.

Stephens, S. B., Tolson, K. P., Rouse, M. L., Jr., Poling, M. C., Hashimoto-Partyka, M. K., Mellon, P. L., et al. (2015). Absent progesterone signaling in kisspeptin neurons disrupts the LH surge and impairs fertility in female mice. *Endocrinology*, *156*(9), 3091–3097.

Stojilkovic, S. S., Reinhart, J., & Catt, K. J. (1994). Gonadotropin-releasing hormone receptors: Structure and signal transduction pathways. *Endocrine Reviews*, *15*(4), 462–499.

Takekida, S., Matsuo, H., & Maruo, T. (2003). GnRH agonist action on granulosa cells at varying follicular stages. *Molecular and Cellular Endocrinology*, *202*(1–2), 155–164.

Tan, O., & Bukulmez, O. (2011). Biochemistry, molecular biology and cell biology of gonadotropin-releasing hormone antagonists. *Current Opinion in Obstetrics & Gynecology*, *23*(4), 238–244.

Terasawa, E., Guerriero, K. A., & Plant, T. M. (2013). Kisspeptin and puberty in mammals. *Advances in Experimental Medicine and Biology*, *784*, 253–273.

Thornton, P., Silverman, L. A., Geffner, M. E., Neely, E. K., Gould, E., & Danoff, T. M. (2014). Review of outcomes after cessation of gonadotropin-releasing hormone agonist treatment of girls with precocious puberty. *Pediatric Endocrinology Reviews*, *11*(3), 306–317.

Topaloglu, A. K., Reimann, F., Guclu, M., Yalin, A. S., Kotan, L. D., Porter, K. M., et al. (2009). TAC3 and TACR3 mutations in familial hypogonadotropic hypogonadism reveal a key role for neurokinin B in the central control of reproduction. *Nature Genetics*, *41*(3), 354–358.

Torrealday, S., Lalioti, M. D., Guzeloglu-Kayisli, O., & Seli, E. (2013). Characterization of the gonadotropin releasing hormone receptor (GnRHR) expression and activity in the female mouse ovary. *Endocrinology*, *154*(10), 3877–3887.

Tse, F. W., Tse, A., Hille, B., Horstmann, H., & Almers, W. (1997). Local Ca2+ release from internal stores controls exocytosis in pituitary gonadotrophs. *Neuron*, *18*(1), 121–132.

Tsutsui, K., Saigoh, E., Ukena, K., Teranishi, H., Fujisawa, Y., Kikuchi, M., et al. (2000). A novel avian hypothalamic peptide inhibiting gonadotropin release. *Biochemical and Biophysical Research Communications*, *275*(2), 661–667.

Tsutsumi, R., & Webster, N. J. (2009). GnRH pulsatility, the pituitary response and reproductive dysfunction. *Endocrine Journal*, *56*(6), 729–737.

Ubuka, T., Morgan, K., Pawson, A. J., Osugi, T., Chowdhury, V. S., Minakata, H., et al. (2009). Identification of human GnIH homologs, RFRP-1 and RFRP-3, and the cognate receptor, GPR147 in the human hypothalamic pituitary axis. *PLoS One*, *4*(12), e8400.

van der Linden, M., Buckingham, K., Farquhar, C., Kremer, J. A., & Metwally, M. (2015). Luteal phase support for assisted reproduction cycles. *Cochrane Database of Systematic Reviews*, 7, CD009154.

Verschraegen, C. F., Westphalen, S., Hu, W., Loyer, E., Kudelka, A., Volker, P., et al. (2003). Phase II study of cetrorelix, a luteinizing hormone-releasing hormone antagonist in patients with platinum-resistant ovarian cancer. *Gynecologic Oncology*, *90*(3), 552–559.

Vitek, W. S., Shayne, M., Hoeger, K., Han, Y., Messing, S., & Fung, C. (2014). Gonadotropin-releasing hormone agonists for the preservation of ovarian function among women with breast cancer who did not use tamoxifen after chemotherapy: A systematic review and meta-analysis. *Fertility and Sterility*, *102*(3). 808–815.e1.

White, R. B., Eisen, J. A., Kasten, T. L., & Fernald, R. D. (1998). Second gene for gonadotropin-releasing hormone in humans. *Proceedings of the National Academy of Sciences of the United States of America*, *95*(1), 305–309.

Whitelaw, P. F., Eidne, K. A., Sellar, R., Smyth, C. D., & Hillier, S. G. (1995). Gonadotropin-releasing hormone receptor messenger ribonucleic acid expression in rat ovary. *Endocrinology*, *136*(1), 172–179.

Wilkinson, S. J., Kucukmetin, A., Cross, P., Darby, S., Gnanapragasam, V. J., Calvert, A. H., et al. (2008). Expression of gonadotrophin releasing hormone receptor I is a favorable prognostic factor in epithelial ovarian cancer. *Human Pathology*, *39*(8), 1197–1204.

Winkler-Crepaz, K., Bottcher, B., Toth, B., Wildt, L., & Hofer-Tollinger, S. (2017). What is new in 2017? Update on fertility preservation in cancer patients. *Minerva Endocrinologica*, *42*, 331–339.

Witchel, S., & Plant, T. (2014). Puberty: Gonardarche and adrenarche. In J. F. Strauss & R. L. Barbieri (Eds.), *Yen and Jaffe's reproductive endocrinology* (pp. 382–421). Philadelphia, PA: Saunders Elsevier.

Wray, S., & Hoffman, G. (1986). A developmental study of the quantitative distribution of LHRH neurons within the central nervous system of postnatal male and female rats. *The Journal of Comparative Neurology*, *252*(4), 522–531.

Wu, H. M., Huang, H. Y., Lee, C. L., Soong, Y. K., Leung, P. C., & Wang, H. S. (2015). Gonadotropin-releasing hormone type II (GnRH-II) agonist regulates the motility of human decidual endometrial stromal cells: Possible effect on embryo implantation and pregnancy. *Biology of Reproduction*, *92*(4), 1–11.

Yan, X., Yuan, C., Zhao, N., Cui, Y., & Liu, J. (2014). Prenatal androgen excess enhances stimulation of the GNRH pulse in pubertal female rats. *The Journal of Endocrinology*, *222*(1), 73–85.

Yano, T., Pinski, J., Halmos, G., Szepeshazi, K., Groot, K., & Schally, A. V. (1994). Inhibition of growth of OV-1063 human epithelial ovarian cancer xenografts in nude mice by treatment with luteinizing hormone-releasing hormone antagonist SB-75. *Proceedings of the National Academy of Sciences of the United States of America*, *91*(15), 7090–7094.

Youssef, M. A., Van der Veen, F., Al-Inany, H. G., Mochtar, M. H., Griesinger, G., Nagi Mohesen, M., et al. (2014). Gonadotropin-releasing hormone agonist versus HCG for oocyte triggering in antagonist-assisted reproductive technology. *Cochrane Database of Systematic Reviews*, *10*, CD008046.

Yu, B., Ruman, J., & Christman, G. (2011). The role of peripheral gonadotropin-releasing hormone receptors in female reproduction. *Fertility and Sterility*, *95*(2), 465–473.

Zeng, L., Webster, S. V., & Newton, P. M. (2012). The biology of protein kinase C. *Advances in Experimental Medicine and Biology*, *740*, 639–661.

Zimmer, C. A., Ehrmann, D. A., & Rosenfield, R. L. (2010). Potential diagnostic utility of intermittent administration of short-acting gonadotropin-releasing hormone agonist in gonadotropin deficiency. *Fertility and Sterility*, *94*(7), 2697–2702.

CHAPTER THREE

Regulation of Luteinizing Hormone Receptor mRNA Expression in the Ovary: The Role of miR-122

K.M.J. Menon[1], Bindu Menon, Thippeswamy Gulappa
The University of Michigan Medical School, Ann Arbor, MI, United States
[1]Corresponding author: e-mail address: kmjmenon@umich.edu

Contents

Abstract

The expression of luteinizing hormone receptor (LHR) in the mammalian ovary is regulated in response to changes in the secretion of follicle-stimulating hormone and luteinizing hormone by the anterior pituitary, at least in part, through posttranscriptional mechanisms. The steady-state levels of LHR mRNA are maintained by controlling its rate of degradation by an RNA-binding protein designated as LHR mRNA-binding protein (LRBP). LRBP forms a complex with LHR mRNA and targets it for degradation in the p bodies. miR-122, an 18 nucleotide noncoding RNA, regulates the expression of LRBP. Thus, the levels of miR-122 determine the cellular levels of LHR mRNA expression. This

Vitamins and Hormones, Volume 107
ISSN 0083-6729
https://doi.org/10.1016/bs.vh.2018.01.010

phenomenon has been examined during the induction of LHR mRNA expression that occurs during follicle maturation in response to rising levels of FSH. In this situation, miR-122 and LRBP levels decrease as LHR mRNA expression undergoes downregulation in response to preovulatory LH surge. miR-122 expression as well as LRBP levels show robust increases. The mechanism of induction of LRBP by miR-122 has also been discussed.

1. INTRODUCTION

The adult mammalian ovary undergoes cyclical changes throughout the reproductive period in response to constantly changing secretory patterns of the anterior pituitary hormones, LH and FSH, which are in turn regulated by ovarian steroids through feedback mechanisms. In the human, early events of follicular maturation consist principally of mitosis of granulosa cells and transformation of the surrounding stroma to layers of theca cells stimulated by gradually rising levels of FSH and low levels of LH. In rodents and humans, luteinizing hormone (LH) regulates androgen biosynthesis by the theca–interstitial cells during follicular development, and these androgens are converted to estrogens by the granulosa cells (Bjersing, 1968). The follicular growth during the second half of the follicular phase is accompanied by rising levels of estradiol and increased expression of the LH receptor (LHR). Although a large number of primordial follicles begin to develop in each cycle, all become atretic except the one that is destined to ovulate. The granulosa cells in the ovulatory follicles reach optimum size by mitotic activity supported by the increasing levels of estrogen produced in response to FSH and, indirectly, LH. The oocytes undergo meiotic division in the preovulatory follicles, and in most mammals, ovulation occurs when the oocyte has reached the metaphase of the second meiotic division. After ovulation, the cell types associated with corpus luteum secrete progesterone and estrogens to support the implantation of the blastocyst. In the absence of blastocyst implantation, corpus luteum regresses. The ovarian cycle repeats itself until menopause, which occurs after depletion of ovum results in the cessation of the cyclical changes in ovarian structure and function. The main driver of the ovarian cycle is the sequential and combined actions of FSH and LH. While the FSH action is confined to the granulosa cells within the ovarian follicles, LH acts on multiple targets including the theca–interstitial cells, FSH-stimulated granulosa cells, and the corpus luteum after ovulation. While FSH in concert with other paracrine factors is known to regulate the development of the primary follicles to preantral and antral stages, the LHR makes its appearance in significant amounts as a result of FSH stimulation.

The action of LH is mediated by LHR, which belongs to the glycoprotein subfamily of the large G protein-coupled receptor family with an unusually large extracellular domain, seven-membrane spanning helices, and an intracellular domain with two conserved palmitoylation sites (Kawate, Kletter, Wilson, Netzloff, & Menon, 1995; Kawate & Menon, 1994; Munshi, Clouser, Peegel, & Menon, 2005; Zhu, Wang, & Ascoli, 1995). Its expression increases during follicle growth and shows a dramatic decline in response to a bolus of LH that occurs prior to ovulation (Menon, Clouser, & Nair, 2005; Menon, Franzo-Romain, Damanpour, & Menon, 2011; Zeleznik, Midgley, & Reichert, 1974; Zeleznik, Schuler, & Reichert, 1981). During this period of transition, a refractory period of LH responsiveness occurs that is characterized by a transient loss of LHR (Hoffman, Peegel, Sprock, Zhang, & Menon, 1991; LaPolt, Oikawa, Jia, Dargan, & Hsueh, 1990; Lu, Peegel, Mosier, & Menon, 1993; Peegel, Randolph, Midgley, & Menon, 1994; Segaloff, Wang, & Richards, 1990) and consequent loss of the LH responsiveness (Azhar & Menon, 1979; Ghosh, Peegel, Dunham, Sands, & Menon, 1988; Hunzicker-Dunn, Jungmann, Derda, & Birnbaumer, 1979; Sen, Azhar, & Menon, 1979). The receptor levels then rise again, reaching maximum levels by mid-portion of the luteal phase to support progesterone production. The LHR expression then falls with the regression of the corpus luteum if pregnancy does not ensue (Azhar & Menon, 1979; Hoffman et al., 1991; LaPolt et al., 1990; Lu et al., 1993; Peegel et al., 1994). The changes in LHR expression are exquisitely regulated to ensure successful reproduction.

We have examined the mechanisms responsible for the rapid changes in the levels of LHR expression using both rodent and human models and found that a posttranscriptional mechanism, at least in part, accounts for these changes. Furthermore, we have discovered that a noncoding RNA, miR-122, plays a role in this process.

miR-122, a 22-nucleotide microRNA, was first discovered by cloning and sequencing of miRNAs from tissues in mice as a highly abundant liver-specific miRNA (Lagos-Quintana et al., 2002). It was subsequently found to be present in other tissues such as breast cancer tissues, ovaries, and granulosa cells (Fong et al., 2015; Li, Fang, Liu, & Yang, 2015; Salilew-Wondim et al., 2014; Sirotkin, Ovcharenko, Grossmann, Laukova, & Mlyncek, 2009). miR-122 is essential for the maintenance of liver homeostasis by regulating the expression of a large number of target mRNAs involved in diverse hepatic functions and also by suppressing nonhepatic genes (reviewed in Thakral & Ghoshal, 2015). Pharmacological inhibition and genetic knockout of miR-122 have been shown to induce dysregulation of systemic and hepatic lipid metabolism, iron homeostasis, and differentiation of hepatocytes. miR-122 regulates diverse functions such as cholesterol, glucose and iron homeostasis, lipid metabolism, and infection of hepatitis C virus. Furthermore, miR-122 inhibition leads to a reduction of serum cholesterol by downregulating genes involved in cholesterol biosynthesis, including the regulatory enzyme HMG-CoA reductase, thereby protecting animals from diet-induced hypercholesterolemia. The biogenesis of miR-122 follows the canonical pathway (Filipowicz & Grosshans, 2011; Jopling, 2012). *miR-122* gene is transcribed by RNA polymerase II (pol II) from a long noncoding RNA located on chromosome 18 in human and mouse, generating ~4.5 kb pri-miR-122, which is then processed to a 66 nt pre-miR-122 in the nucleus and subsequently to mature miR-122 in the cytoplasm. Our studies have shown that miR-122 regulates LHR mRNA expression post-transcriptionally by regulating the levels of LHR mRNA-binding protein designated as LRBP. Since miR-122 regulates LHR expression by altering LHR mRNA stability, a brief overview of our understanding of the post-transcriptional mechanism of LHR regulation is presented before discussing its regulation by miR-122.

2. POSTTRANSCRIPTIONAL REGULATION OF LHR mRNA EXPRESSION

Administration of hCG, a hormone with significant chemical identity to LH, produces a rapid decline in LHR mRNA and protein in rat and human ovaries. The loss of steady-state levels of receptor expression could also be due to a temporary pause in the synthesis of new receptor protein by decreased transcription, or an increase in degradation of mRNA. Treatment of superovulated rats with 50 IU hCG led to a rapid decline of LHR mRNA expression (Hoffman et al., 1991). The Northern blot analysis of the LHR mRNA transcripts showed that the decline started at 6 h following hCG treatment and reached undetectable levels by 24 h. The recovery from downregulation was seen after 24 h and almost reached control levels by 72 h. The in situ hybridization pattern of the receptor expression in the whole ovary was in full agreement with the qPCR results and Northern blot data (Peegel et al., 1994). The process of LHR mRNA loss was very specific since other mRNAs involved in ovarian function remained unchanged (Hoffman et al., 1991). This is further supported by the finding that LHR mRNA downregulation can be achieved by chronic elevation of cyclic AMP levels by repeated administration of inhibitors of type 2 phosphodiesterase (Peegel et al., 1994). The evidence for the decrease in the steady-state levels of LHR mRNA expression occurred through posttranscriptional mechanism came from the determination of transcription rates by performing nuclear run-on assays, which showed no decrease in transcription rates during the course of downregulation when compared to the control (Lu et al., 1993). In fact, the nuclei isolated from the hCG-treated ovaries showed a robust increase in the incorporation of [^{3}H] uridine into total RNA compared to the control. These findings supported the view that the loss of LHR mRNA during ligand-induced downregulation occurred posttranscriptionally by increasing LHR mRNA degradation.

2.1 LHR Expression During Ovarian Cycle Is Regulated Posttranscriptionally by an RNA-Binding Protein

It is now recognized that the expression of specific, highly regulated mRNAs is regulated, at least in part, at the level of mRNA degradation (Ross, 1995). In most such instances, the changes in the stability of the mRNA occur by the binding of proteins to specific sequences and/or structures of the mRNA. The specific regions that the mRNA-binding proteins interact with may be localized either on the 5′ untranslated region, the coding region, or the 3′ untranslated region of the mRNA (Ross, 1995). In general, the steady-state levels of the mRNA expression are regulated by either increasing or decreasing the degradation rate resulting from the interaction with specific RNA-binding proteins.

The possible existence of a binding protein that specifically recognizes LHR mRNA was examined by incubating ^{32}P-labeled LHR mRNA with a cytosolic fraction (100 × g supernatant) prepared from the ovary pretreated with hCG to downregulate LHR mRNA expression in the presence of ribonuclease inhibitors (Kash & Menon, 1998). After treating the reaction mixture with ribonuclease T1 to degrade the unreacted RNA probe, the ribonucleoprotein (RNP) complex was separated by electrophoresis on an 8% native acrylamide gel and subjected to autoradiography. Two RNP complexes were identified, one prominent band corresponding to 50 kDa, and a second less intense band of 45 kDa (Fig. 1). A threefold increase in the 50 kDa RNP complex was seen in samples derived from LHR downregulated ovaries, but no significant change in the intensity of the 45 kDa band was found. Because of this finding, further studies focused on the larger 50 kDa ribonucleoprotein complex, designated as LRBP

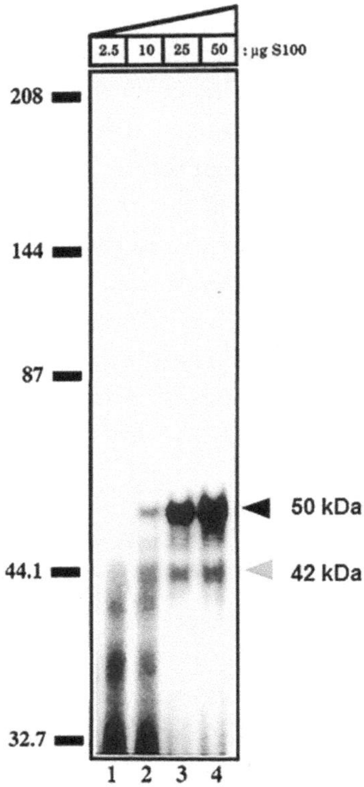

Fig. 1 LHR mRNA binding to cytosolic fraction of the ovary during LHR downregulation. Autoradiogram of RNA gel mobility shift analysis performed by incubating with different concentrations (2.5 µg (*lane 1*), 5 µg (*lane 2*), 10 µg (*lane 3*), 25 µg (*lane 4*), or 50 µg (*lane 5*)) of cytosolic proteins with ^{32}P-labeled (1×10^5 cpm) RNA. Two RNP complexes, corresponding to 50 and 45 kDa, are indicated by *arrows*. *Adapted from Kash, J. C., & Menon, K. M. J. (1998). Identification of a hormonally regulated luteinizing hormone/human chorionic gonadotropin receptor mRNA binding protein. Increased mrna binding during receptor down-regulation.* The Journal of Biological Chemistry, 273, *10658–10664. ©2002 by American Society for Biochemistry and Molecular Biology.*

(Kash & Menon, 1998). The binding of LHR mRNA to LRBP showed that the recognition site of the mRNA resided in the amino-terminal portion of the LHR corresponding to nucleotides 102–282. Further, truncation of this region showed that the contact site resided between nucleotides 188 and 228. RNA hydroxyl radical footprinting revealed a bipartite polypyrimidine-rich sequence (UCUCX7-UCUCCCU), corresponding to nucleotides 203–220 of LHR mRNA that specifically interacted with LRBP. Mutation of the C residues within the bipartite sequence revealed that all C residues participated in binding to LRBP (Kash & Menon, 1999). The interaction of LHR mRNA with LRBP exhibited high affinity with equilibrium dissociation constant (K_d) in the range of 4×10^{-9} M.

After establishing that the ovarian LRBP recognizes LHR mRNA with high affinity and specificity, the changes in LRBP activity in relation to tissue levels of LHR mRNA were examined. It has been established that LHR mRNA expression increases during maturation of the ovarian follicles in response to FSH. The levels then fall immediately after the preovulatory LH surge and rise again with the growth and development of the corpus luteum (Nair, Kash, Peegel, & Menon, 2002). The relationship between LHR mRNA expression and LRBP activity was examined in ovarian tissues after treating immature rats with pregnant mare serum gonadotropin to induce follicle growth and to increase LHR mRNA expression levels. At this time point, the LRBP in the cytosolic fractions prepared from the ovaries showed low levels of activity. However, when LHR expression was suppressed during ligand-induced downregulation by treatment with hCG, the LRBP activity was increased several fold in a time-dependent manner. Thus, when the LHR mRNA expression level was high, the RNA-binding protein activity was at the lowest level. Conversely, the LHR mRNA-binding activity was high when the mRNA levels were low. This inverse relationship (Fig. 2) suggested that LRBP is an endogenous regulator of LHR mRNA expression.

2.2 In Vitro mRNA Degradation by the LHR mRNA-Binding Protein

The direct role of the RNA-binding protein in LHR mRNA degradation was demonstrated by employing a cell-free mRNA decay system developed by Ross (1993) under in vitro conditions. The assay essentially determines the ability of ribosomes to degrade exogenous RNA in the presence and absence of the RNA-binding protein. Although all mRNAs are prone to degradation, the rate of degradation varies depending on the cellular environment. The rate of decay of LHR mRNA was very rapid in ribosomes isolated from the ovaries of rats treated with hCG to downregulate LHR expression compared to the degradation of LHR mRNA by ribosomes isolated from ovaries of the saline-treated control group. The rate of decay of exogenously added ovarian RNA by ribosomes isolated from saline-treated rats was accelerated by the addition of a partially purified ovarian LHR mRNA-binding protein (Nair et al., 2002), demonstrating that the LHR mRNA-binding protein plays a role in LHR mRNA degradation.

2.3 Characterization of LHR mRNA-Binding Protein

After exhausting several approaches to successfully purify LRBP by a variety of techniques, including affinity purification using covalently linked

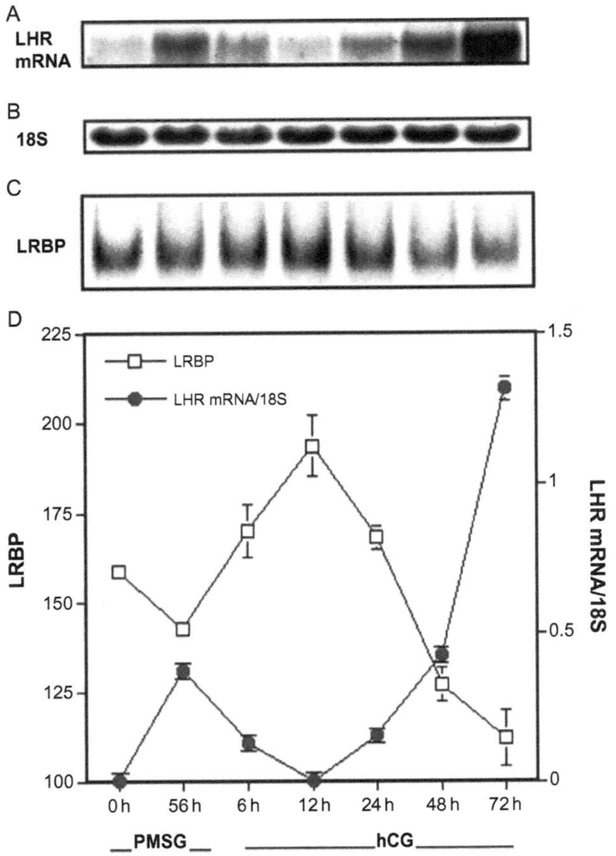

Fig. 2 LHR mRNA expression and RNA-binding activity of the cytosolic extracts from ovaries treated with PMSG and hCG. Twenty-three-day-old rats were treated with PMSG at 0 h. Ovaries were collected at 0 and 56 h later. hCG was administered at 56 h of PMSG administration, and ovaries were harvested at 6, 12, 24, 48, and 72 h. (A) Northern blots of LHR mRNA. (B) The Northern blots were normalized using 18S rRNA. (C) RNA electrophoretic mobility shift analysis performed at these time points after incubating 50 μg of cytosolic protein isolated from ovaries incubated with ^{32}P-labeled LHR mRNA probe. (D) 6.7-kb LHR mRNA transcript and RNA electrophoretic mobility shift analysis bands quantitated by densitometric scan and expressed in arbitrary units. *Adapted from Nair, A. K., Kash, J. C., Peegel, H., & Menon, K. M. J. (2002). Post-transcriptional regulation of luteinizing hormone receptor mRNA in the ovary by a novel mRNA-binding protein.* The Journal of Biological Chemistry, *277, 21468–21473. ©2002 by American Society for Biochemistry and Molecular Biology.*

RNA-binding sequence to sepharose, conventional techniques were used to purify the LHR mRNA-binding protein from the 100 × *g* supernatant fraction of the ovarian homogenates. The ovaries were initially downregulated by treatment with the ligand in order to increase the yield of the binding protein (Kash & Menon, 1999). The supernatants were subjected to chromatography on a strong cation exchange resin (Macro-Prep High S support) and eluted with 150 mM KCl. The eluates were concentrated and subjected to SDS-PAGE to separate the proteins. The ^{32}P-LHR mRNA-binding activity associated with the protein band on the gel was identified by an overlay assay (Northwestern blot) using ^{32}P-labeled LHR mRNA fragment (203 − 220) as the probe. After extensive standardization of the assay, the corresponding protein band showed the RNA-binding activity was cut, eluted, and renatured. The eluted protein was electrophoresed again to determine the purity of the preparation. The electrophoretically homogeneous protein band was then subjected to amino-terminal analysis, as well as MS-MALDI analysis, to establish its identity. Both analyses revealed the purified protein to be mevalonate kinase (Nair & Menon, 2004). The gene encoding the rat mevalonate kinase was then cloned and overexpressed in 293 T cells (Nair & Menon, 2004). The recombinant protein exhibited a concentration-dependent increase in the binding LHR mRNA probe (Nair & Menon, 2004). The binding exhibited all the characteristics of the expected LHR mRNA-binding protein with respect to specificity for binding to the previously identified contact site (nucleotides 203–220), competition by unlabeled LHR mRNA fragment (203–220), dependence on C residues in the ligand-binding site, and immunoreactivity of the recombinant protein similar to that seen for the electrophoretically purified rat mevalonate kinase. Mevalonate kinase is a metabolic enzyme involved in cholesterol biosynthesis which catalyzes the conversion of mevalonate to phosphomevalonate, the precursor of cholesterol and many natural products. In fact, we have shown that mevalonate kinase expression is regulated by LH in the corpus luteum both at the mRNA level and at the protein level (Wang & Menon, 2005). During LH/hCG-induced downregulation, mevalonate kinase expression was induced in the corpus luteum. The induction of expression of other sterol response element (SRE)-containing genes such as HMG-CoA reductase, farnesyl pyrophosphate synthase, and LDL receptor was also seen during this phase (Wang & Menon, 2005). These results suggest that mevalonate kinase expression plays a regulatory role in LHR mRNA expression in the corpus luteum.

Depletion of mevalonate kinase expression by 25 hydroxycholesterol resulted in the abrogation of LHR mRNA downregulation in cultured human granulosa cells (Wang, Nair, & Menon, 2007). The inhibitory effect of mevalonate kinase on LHR expression was documented independently by Ikeda et al. (2008), who showed that overexpression of mevalonate kinase can abrogate an FSH-induced increase in LHR mRNA expression in rat ovarian follicles.

The notion that mevalonate kinase, a metabolic enzyme, acts as an mRNA-binding protein is consistent with a similar role of several other metabolic enzymes that have been described (Menon et al., 2005). The iron-response element-binding protein, which plays a role in iron homeostasis, is an enzyme in the citric acid cycle (Hentze & Kuhn, 1996). Similarly, thymidylate synthase, dihydrofolate reductase, glyceraldehyde-3-phosphate dehydrogenase, glutamate dehydrogenase, and lactate dehydrogenase have all been shown to serve as RNA-binding proteins that regulate different aspects of RNA metabolism (Chu, Schmitz, Ju, & Copur, 1999; Nagy & Rigby, 1995; Pioli, Hamilton, Connolly, Brewer, & Rigby, 2002).

3. ROLE OF miR-122 IN LHR EXPRESSION

Since LRBP plays a crucial role in LHR expression during ligand-induced downregulation (Fig. 3), the regulatory factors controlling LRBP expression would be expected to have a major influence on LHR

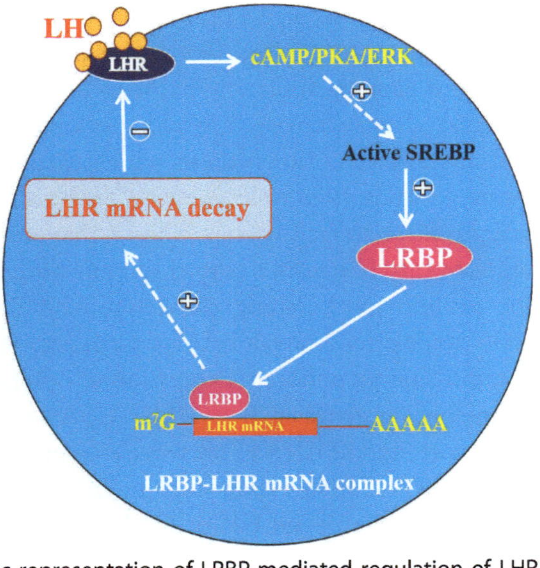

Fig. 3 Schematic representation of LRBP-mediated regulation of LHR expression. LH/hCG binding to the LHR activates the cAMP signal cascade and increases the transcription of LRBP through the activation of the transcription factor SREBP. LRBP binds to LHR mRNA, forming an untranslated mRNP complex, which is then targeted for degradation.

mRNA expression. While searching for possible regulatory mechanisms of LRBP expression, a study by Krutzfeldt et al. who tested the biological significance of silencing miR-122 using antagomirs in mice came to our attention (Krutzfeldt et al., 2005). They identified miR-122 as a positive regulator of many enzymes in the cholesterol biosynthetic pathway such as HMG-CoA reductase, phosphomevalonate kinase, and mevalonate kinase, among several other targets in the liver. Since miR-122 was shown to be a liver-specific regulator of mevalonate kinase in that study, we examined whether miR-122 is also expressed in the ovarian tissue and its potential role in regulating LHR mRNA expression. Since mevalonate kinase has been identified as the protein that binds LHR mRNA, we will be referring to mevalonate kinase as LRBP in the context of LHR mRNA regulation in this chapter.

3.1 LH/hCG Increases miR-122 Expression

In order to determine a possible role for miR-122 in LRBP and LHR regulation, we examined the changes in the levels of miR-122 during LHR downregulation using real-time PCR. As shown in Fig. 4A, there was a significant increase in the expression of miR-122 immediately after hCG treatment. The increase was initially evident at 30 min post-hCG and peaked around 1–2 h, but started to decline thereafter. The expression of miR-122 in the ovary and its upregulation in response to hCG were further confirmed by fluorescence in situ hybridization (FISH), followed by confocal microscopic analysis of frozen sections from normal and hCG-treated rat ovaries, using a 5′-FITC-labeled miR-122 probe (Menon et al., 2013). The results (Fig. 4) showed that while there was low or no expression of miR-122 in the control ovaries, its expression showed an increase in response to hCG treatment. It is noteworthy that control ovaries showed no detectable levels of miR-122 expression in real-time PCR assays or in FISH analyses. These findings suggest that miR-122 is hormonally regulated during the ovarian cycle. The failure to detect miR-122 expression in normal ovaries as reported in some studies (Landgraf et al., 2007) could be attributed to this fact, since the levels of miR-122 expression in the ovary is undetectable when LHR expression occurs maximally.

Analysis of the temporal relationship between miR-122 expression and ligand-induced downregulation of LHR mRNA showed a close relationship between the time frame of expressions of miR-122, LRBP, and

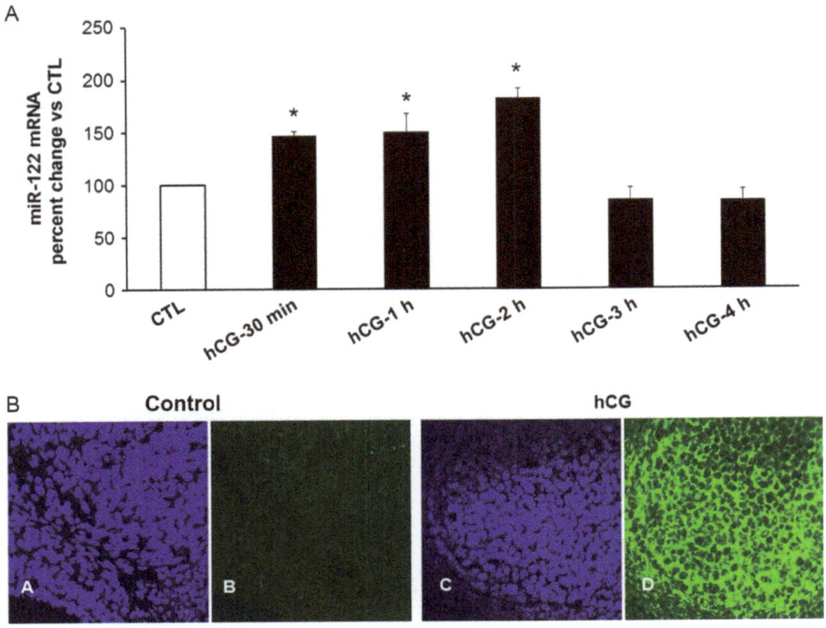

Fig. 4 Analysis of miR-122 expression in the ovaries using real-time PCR and fluorescent in situ hybridization analysis. (A) Superovulated rats were treated with hCG on day 5. Ovaries were collected and total RNA isolated at different time intervals. The levels of miR-122 and U6 snRNA were determined by real-time PCR. (B) Fluorescent in situ hybridization analysis (FISH) of control (CTL) and hCG-treated (1 h) frozen rat ovary sections were conducted using FITC-labeled miR-122 LNA probe. Each panel shows DAPI staining (a and c; *blue fluorescence*) on the *left* and FITC (miR-122) staining (b and d; *green fluorescence*) on the *right* (Menon, Sinden, Franzo-Romain, Botta, & Menon, 2013).

LHR mRNA (Fig. 5). miR-122 expression showed an increase at 30 min post-hCG and remained steady until 2 h. A rapid surge in LRBP mRNA levels was then observed. The levels of LRBP mRNA increased at 2 h after hCG treatment and remained high until 4 h. An increase in LRBP protein expression was then seen at 4 h and sustained until 6 h of hCG treatment. Most importantly, LHR mRNA levels started to decline at 4 h post-hCG, and a minimum decrease was observed by 12 h. Thus, an increase in the expression of miR-122 was followed by an increase in LRBP mRNA and protein expression, culminating in decreased levels of LHR mRNA regulation. This temporal relationship suggested that miR-122 might play a role in mediating an increase in LRBP levels during LHR mRNA downregulation.

The signaling pathway involved in hCG-induced expression of miR-122 showed the involvement of cAMP/PKA and ERK pathways

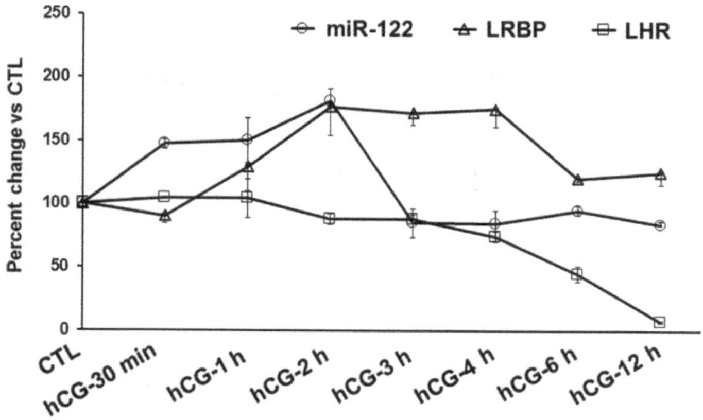

Fig. 5 Temporal relationship of miR-122, LRBP, and LHR mRNA expressions in the ovary following hCG treatment. Superovulated rats were treated with hCG on day 5, and ovaries were collected at 0, 30 min, 1, 2, 3, 4, 6, and 12 h later. Equal amounts of RNA from the control (CTL) or hCG-treated ovaries were subjected to real-time PCR using predesigned primers and probes for miR-122, LRBP, or LHR mRNA. The *graphs* represent the ratio of LRBP or LHR mRNA levels normalized to 18S RNA and that of miR-122 normalized to U6 snRNA, and are shown as percent change vs control (Menon et al., 2013).

(Menon, Gulappa, & Menon, 2015; Menon et al., 2013). For example, pretreatment with H-89, a PKA inhibitor, as well as inhibition of ERK1/2 using U0126, abrogated hCG-mediated increases in miR-122 in primary cultures of human granulosa cells. A significant increase in the expression of miR-122 was seen at 1 h following hCG treatment. This increase was abrogated when cells were pretreated with U0126 or H-89 1 h before hCG treatment. This is in agreement with our previous findings that the cAMP/PKA pathway plays a crucial role in the induction of LRBP expression and its binding to LHR mRNA during hCG-mediated LHR mRNA downregulation. These findings suggest that hCG treatment activates the cAMP/PKA/ERK1/2 signaling cascade and results in an increase in miR-122 levels, culminating in increased LRBP expression and leading to LHR mRNA downregulation.

3.2 Inhibition of miR-122 Abrogated hCG-Induced LRBP Induction and LHR mRNA Downregulation

The role of miR-122 as an upstream regulator of LRBP comes from studies inhibiting miR-122 expression, followed by determining the changes in hCG-induced LRBP expression. miR-122 expression was totally blocked by pretreating superovulated rats with an LNA-conjugated antagomir of miR-122 24 h prior to treatment with hCG. Importantly, abolishment of miR-122 expression by treatment with the antagomir abrogated the increase in LRBP mRNA and protein levels seen in response to hCG, further supporting the finding that the increased LRBP expression required for LHR mRNA downregulation is mediated by miR-122. Since miR-122 inhibition prevented hCG-induced LRBP upregulation, it would be expected that the decrease in LRBP might lead to a decrease in the association of LRBP with LHR mRNA. RNA electrophoretic gel mobility shift assay using cytosolic (S100) fractions of ovarian cell lysates and a ^{32}P-labeled LHR mRNA probe that is known to interact with LRBP showed a marked increase in the RNP complex formation between LRBP and labeled LHR mRNA probe following hCG treatment. This RNP complex formation was significantly reduced by miR-122 antagomir treatment.

Further studies showed that the inhibition observed for LRBP–LHR mRNP complex formation in the presence of miR-122 antagomir also led to abrogation of LHR mRNA downregulation. As shown in Fig. 6, the levels of LHR mRNA in the hCG-treated samples were reduced by

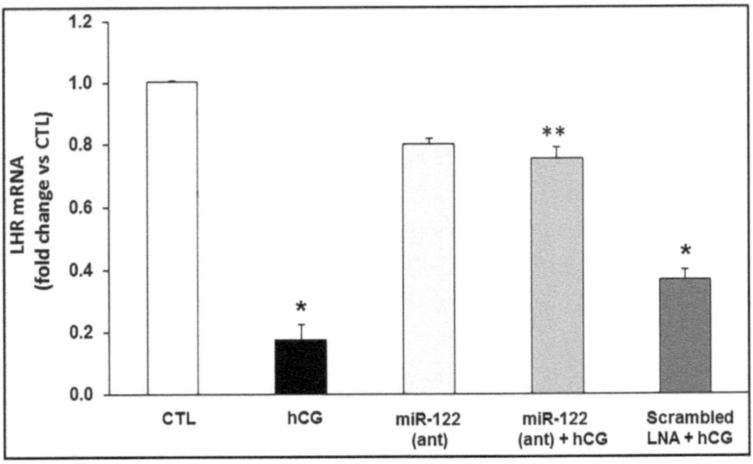

Fig. 6 Reversal of hCG-induced LHR mRNA downregulation by miR-122 antagomir. LNA-conjugated miR-122 antagomir or scrambled LNA control was injected into the bursa of the ovaries of superovulated rats on day 4, followed by hCG on day 5. Ovaries were collected 12 h later and were processed for total RNA isolation. Total RNAs were reverse transcribed, and the resulting cDNAs were subjected to real-time PCR quantitation using specific primers and probes for LHR. The *graph* represents changes in mRNA levels normalized to 18S rRNA and shown as fold change vs control (CTL) (Menon et al., 2015).

almost 80%. This was significantly reversed by pretreatment with an antagomir, which reduces miR-122 expression, bringing the LHR mRNA levels very close to the control levels. These results support the notion that miR-122-mediated upregulation of LRBP plays a crucial role in hCG-mediated LHR mRNA downregulation.

3.3 miR-122 Increases LRBP Expression Through the Activation of SREBP-1a/ SREBP-2

As stated earlier, mevalonate kinase (LRBP) is a member of the family of enzymes encoded by genes containing oxysterol responsive element. As shown by the elegant work of Brown and Goldstein (1997), oxysterols act by suppressing the proteolytic cleavage of the precursor of sterol response element-binding protein (SREBP) to its active form. The cleavage product, the active form of SREBP, binds to SRE, which causes transcriptional activation of genes regulated by SREBP. Conversely, when cellular concentrations of oxysterols are low, the cleavage of the precursor of SREBP is enhanced to produce an active form of SREBP, which then accelerates transcription of genes containing SRE elements in the promoter region. Therefore, it is likely that miR-122 might act at the level of SREBP processing to increase the expression of LRBP. Examination of the effect of miR-122 inhibition on SREBP levels showed that, as expected, hCG increased the activation of both SREBP-1a and SREBP-2, and that when miR-122 expression was inhibited using antagomir pretreatment, the activation of both SREBP-1a and SREBP-2 was completely blocked. This supports the notion that SREBP activation is an intermediary step in miR-122-mediated upregulation of LRBP in response to hCG treatment. This was further verified by demonstrating that hCG-mediated activation of both SREBP-1a and SREBP-2 was inhibited by pretreatment with U0126 or H-89, indicating that SREBP activation is downstream of cAMP/PKA-ERK1/2-miR-122 pathways and leads to LH/hCG-induced LHR downregulation.

The role of miR-122 in SREBP-mediated LRBP induction was further demonstrated by analyzing the binding of SREBP-1a to the promoter region of LRBP by ChIP assay, followed by real-time PCR to monitor LRBP expression. These studies showed that binding of SREBP-1a to the LRBP promoter was increased after hCG treatment. Most interestingly, the increase was almost completely abrogated by pretreatment with miR-122 antagomir prior to hCG treatment (Fig. 6). Thus, SREBP binds to the putative SRE element in the promoter region of the LRBP gene, which then increases the LRBP gene transcription. Inhibition of miR-122 by the antagomir treatment resulted in the inhibition of LRBP gene expression by decreasing the association of active SREBP to the LRBP promoter (Menon et al., 2015).

3.4 Downregulation of miR-122 During LHR Upregulation in Response to FSH Treatment

As stated earlier, it is now well established that during follicle growth in response to FSH, the growing follicles increase the expression of LHR. We now show that miR-122 plays a role in FSH-mediated upregulation of LHR expression by decreasing the expression of LRBP. The finding that rat granulosa cells treated with a dose of FSH (50 ng/mL) along with 17β-estradiol (E2) showed a time-dependent increase in LHR mRNA expression, and that this increase was accompanied by decreased levels of miR-122 levels, clearly shows that miR-122 levels remain suppressed when robust expression of LHR mRNA occurs. Analysis of miR-122 levels showed that FSH treatment produced a decline in miR-122 expression as predicted, prior to increasing LHR mRNA. However, this increase was not sustained throughout the FSH treatment period. After 24 h of FSH treatment, levels of miR-122 were comparable to that of the control. Thus, a decrease in miR-122 preceded the induction LHR mRNA in response to FSH treatment, suggesting that suppression of miR-122 expression triggers factors that increase LHR mRNA expression.

We have reported that an increase in miR-122 is always followed by a corresponding increase in LRBP expression, and that this increase in LRBP leads to an increase in its association with LHR mRNA, culminating in the accelerated degradation of LHR mRNA (Azhar & Menon, 1979; Menon et al., 2015; Menon, Gulappa, & Menon, 2017; Menon et al., 2013). Thus, it would be expected that a decrease in miR-122 levels would lead to a decreased association of LHR mRNA with LRBP. This in fact is the case, since analysis of the binding of LRBP with LHR mRNA by RNA electrophoretic mobility shift assay revealed a significant decrease in the association of LHR mRNA with LRBP during FSH-mediated induction of LHR mRNA. The intensity of the band representing the LRBP–LHR mRNP complex in the control was significantly higher compared to that seen after FSH treatment. This supports our hypothesis that the decrease in miR-122 is accompanied by a decrease in LRBP–LHR mRNP complex formation and inhibits LHR mRNA degradation, resulting in increased levels of LHR mRNA levels.

The suppressive effect of miR-122 on in LHR expression was further confirmed by overexpressing miR-122 in vitro by employing an adenoviral vector (Menon et al., 2017). Treatment of AdmiR-122-infected cells and cells infected with a GFP-tagged empty vector with FSH for another 24 h showed higher levels of LHR expression in the control group and complete suppression in the miR-122-infected group (Fig. 7). Furthermore, while FSH treatment decreased LRBP mRNA expression, overexpression of miR-122 reversed FSH-induced reduction in LRBP gene expression. Consistent with this finding, a decrease in the association of LHR mRNA with LRBP in the FSH-treated group was observed, but this decrease was abolished when miR-122 was overexpressed prior to FSH treatment. In summary, these findings suggest that overexpression of miR-122 initiated a series of events that culminated in accelerated degradation of LHR mRNA and prevented FSH-mediated induction of LHR mRNA. Thus, miR-122 regulates LHR levels by controlling LRBP levels (Fig. 8).

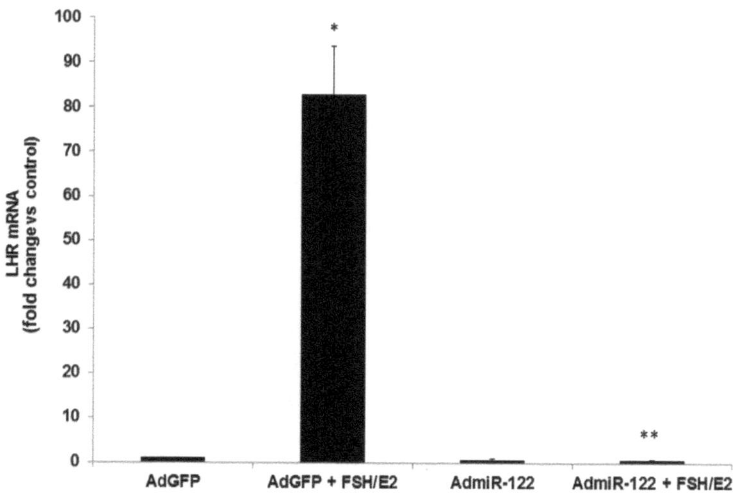

Fig. 7 Overexpression of miR-122 inhibited LHR upregulation. Cells were pretreated with AdmiR-122 (1×10^9 pfu) or AdGFP (1×10^9 pfu) for 24 h. Media were replaced after 24 h, and cells were treated with FSH (50 ng/mL) and E2 (1 nM) for 6 h. Total RNAs were isolated and reverse transcribed, and the resulting cDNAs were subjected to real-time PCR quantitation using specific primers and probes for LHR. The *graph* represents changes in mRNA levels normalized to 18S rRNA and shown as fold change vs control (Menon et al., 2017).

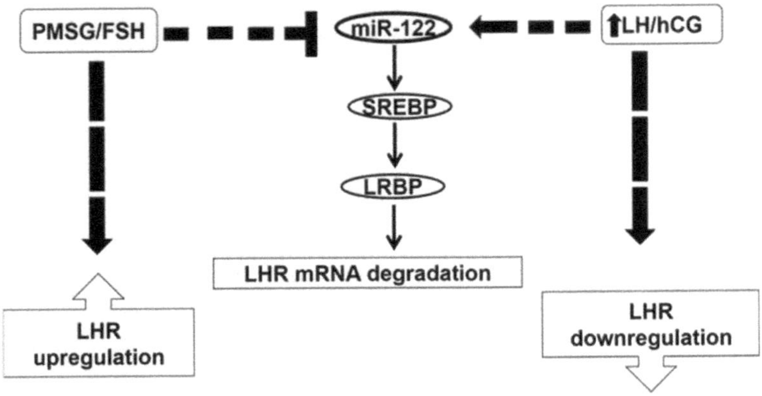

Fig. 8 miR-122/LRBP-mediated regulation of LHR levels in the ovary. miR-122 is expressed in response to a high dose of LH or hCG, which then activates SREBP and leads to increased expression of LRBP. Increase in LRBP causes a decrease in LHR mRNA. FSH causes decreased expression of miR-122 and LRBP during follicle growth. Decrease in LRBP causes an increase in LHR mRNA by preventing its degradation.

4. CONCLUSION AND FUTURE DIRECTIONS

miR-122 is an important molecule with therapeutic potential that has already been established in the liver, and studies are currently in progress on developing novel technologies for targeted delivery of miR-122 to liver and cancer cells. However, its function in the ovary and its role in fertility remain unexplored. On the basis of our studies, it can be speculated that miR-122 might serve as a gatekeeper for controlling LHR expression in the ovaries. Thus, approaches to alter endogenous miR-122 levels might potentially be beneficial in the treatment of infertility. We speculate that miR-122 could serve as a physiological mechanism to regulate LHR expression at the appropriate levels, thereby preventing ovarian hyperstimulation, a common complication encountered in assisted reproduction.

ACKNOWLEDGMENT

Supported by NIH Grant HD 06656.

REFERENCES

Azhar, S., & Menon, K. M. J. (1979). Receptor-mediated gonadotropin action in the ovary. Regulatory role of cyclic nucleotide phosphodiesterase(s) in intracellular adenosine 3′:5′-cyclic monophosphate turnover and gonadotropin-stimulated progesterone production by rat ovarian cells. *The Biochemical Journal*, *180*, 201–211.

Bjersing, L. (1968). On the morphology and endocrine function of granulosa cells in ovarian follicles and corpora lutea. Biochemical, histochemical, and ultrastructural studies on the porcine ovary with special reference to steroid hormone synthesis. *Acta Endocrinologica (Copenh), Suppl. 125*, 121–123.

Brown, M. S., & Goldstein, J. L. (1997). The SREBP pathway: Regulation of cholesterol metabolism by proteolysis of a membrane-bound transcription factor. *Cell, 89*, 331–340.

Chu, E., Schmitz, J., Ju, J., & Copur, S. (1999). An imunoprecipitation-RNA: rPCR method for the in vivo isolation of ribonucleoprotein complexes. In S. R. Haynes (Ed.), *Methods in Molecular Biology: Vol. 118. RNA-protein interaction protocols* (pp. 265–274). NJ: Humana Press Inc.

Filipowicz, W., & Grosshans, H. (2011). The liver-specific microRNA miR-122: Biology and therapeutic potential. *Progress in Drug Research, 67*, 221–238.

Fong, M. Y., Zhou, W., Liu, L., Alontaga, A. Y., Chandra, M., Ashby, J., et al. (2015). Breast-cancer-secreted miR-122 reprograms glucose metabolism in premetastatic niche to promote metastasis. *Nature Cell Biology, 17*, 183–194.

Ghosh, D. K., Peegel, H., Dunham, W. R., Sands, R. H., & Menon, K. M. J. (1988). Modulation of progesterone synthesis and cytochrome P450 levels in rat luteal cells during human chorionic gonadotropin-induced desensitized state. *Endocrinology, 123*, 514–522.

Hentze, M. W., & Kuhn, L. C. (1996). Molecular control of vertebrate iron metabolism: mRNA-based regulatory circuits operated by iron, nitric oxide, and oxidative stress. *Proceedings of the National Academy of Sciences of the United States of America, 93*, 8175–8182.

Hoffman, Y. M., Peegel, H., Sprock, M. J., Zhang, Q. Y., & Menon, K. M. J. (1991). Evidence that human chorionic gonadotropin/luteinizing hormone receptor down-regulation involves decreased levels of receptor messenger ribonucleic acid. *Endocrinology, 128*, 388–393.

Hunzicker-Dunn, M., Jungmann, R., Derda, D., & Birnbaumer, L. (1979). LH-induced desensitization of the adenylyl cyclase system in ovarian follicles. *Advances in Experimental Medicine and Biology, 112*, 27–44.

Ikeda, S., Nakamura, K., Kogure, K., Omori, Y., Yamashita, S., Kubota, K., et al. (2008). Effect of estrogen on the expression of luteinizing hormone-human chorionic gonadotropin receptor messenger ribonucleic acid in cultured rat granulosa cells. *Endocrinology, 149*, 1524–1533.

Jopling, C. (2012). Liver-specific microRNA-122: Biogenesis and function. *RNA Biology, 9*, 137–142.

Kash, J. C., & Menon, K. M. J. (1998). Identification of a hormonally regulated luteinizing hormone/human chorionic gonadotropin receptor mRNA binding protein. Increased mrna binding during receptor down-regulation. *The Journal of Biological Chemistry, 273*, 10658–10664.

Kash, J. C., & Menon, K. M. J. (1999). Sequence-specific binding of a hormonally regulated mRNA binding protein to cytidine-rich sequences in the lutropin receptor open reading frame. *Biochemistry, 38*, 16889–16897.

Kawate, N., Kletter, G. B., Wilson, B. E., Netzloff, M. L., & Menon, K. M. J. (1995). Identification of constitutively activating mutation of the luteinising hormone receptor in a family with male limited gonadotrophin independent precocious puberty (testotoxicosis). *Journal of Medical Genetics, 32*, 553–554.

Kawate, N., & Menon, K. M. J. (1994). Palmitoylation of luteinizing hormone/human choriogonadotropin receptors in transfected cells. Abolition of palmitoylation by mutation of Cys-621 and Cys-622 residues in the cytoplasmic tail increases ligand-induced internalization of the receptor. *The Journal of Biological Chemistry, 269*, 30651–30658.

Krutzfeldt, J., Rajewsky, N., Braich, R., Rajeev, K. G., Tuschl, T., Manoharan, M., et al. (2005). Silencing of microRNAs in vivo with 'antagomirs'. *Nature, 438*, 685–689.

Lagos-Quintana, M., Rauhut, R., Yalcin, A., Meyer, J., Lendeckel, W., & Tuschl, T. (2002). Identification of tissue-specific microRNAs from mouse. *Current Biology*, *12*, 735–739.

Landgraf, P., Rusu, M., Sheridan, R., Sewer, A., Iovino, N., Aravin, A., et al. (2007). A mammalian microRNA expression atlas based on small RNA library sequencing. *Cell*, *129*, 1401–1414.

LaPolt, P. S., Oikawa, M., Jia, X. C., Dargan, C., & Hsueh, A. J. (1990). Gonadotropin-induced up- and down-regulation of rat ovarian LH receptor message levels during follicular growth, ovulation and luteinization. *Endocrinology*, *126*, 3277–3279.

Li, Y., Fang, Y., Liu, Y., & Yang, X. (2015). MicroRNAs in ovarian function and disorders. *Journal of Ovarian Research*, *8*, 51.

Lu, D. L., Peegel, H., Mosier, S. M., & Menon, K. M. J. (1993). Loss of lutropin/human choriogonadotropin receptor messenger ribonucleic acid during ligand-induced down-regulation occurs post transcriptionally. *Endocrinology*, *132*, 235–240.

Menon, K. M. J., Clouser, C. L., & Nair, A. K. (2005). Gonadotropin receptors: Role of post-translational modifications and post-transcriptional regulation. *Endocrine*, *26*, 249–257.

Menon, B., Franzo-Romain, M., Damanpour, S., & Menon, K. M. J. (2011). Luteinizing hormone receptor mRNA down-regulation is mediated through ERK-dependent induction of RNA binding protein. *Molecular Endocrinology*, *25*, 282–290.

Menon, B., Gulappa, T., & Menon, K. M. J. (2015). miR-122 regulates LH receptor expression by activating sterol response element binding protein in rat ovaries. *Endocrinology*, *156*, 3370–3380.

Menon, B., Gulappa, T., & Menon, K. M. J. (2017). Molecular regulation of LHCGR expression by miR-122 during follicle growth in the rat ovary. *Molecular and Cellular Endocrinology*, *442*, 81–89.

Menon, B., Sinden, J., Franzo-Romain, M., Botta, R. B., & Menon, K. M. J. (2013). Regulation of LH receptor mRNA binding protein by miR-122 in rat ovaries. *Endocrinology*, *154*, 4826–4834.

Munshi, U. M., Clouser, C. L., Peegel, H., & Menon, K. M. J. (2005). Evidence that palmitoylation of carboxyl terminus cysteine residues of the human luteinizing hormone receptor regulates postendocytic processing. *Molecular Endocrinology*, *19*, 749–758.

Nagy, E., & Rigby, W. F. (1995). Glyceraldehyde-3-phosphate dehydrogenase selectively binds AU-rich RNA in the NAD(+)-binding region (Rossmann fold). *The Journal of Biological Chemistry*, *270*, 2755–2763.

Nair, A. K., Kash, J. C., Peegel, H., & Menon, K. M. J. (2002). Post-transcriptional regulation of luteinizing hormone receptor mRNA in the ovary by a novel mRNA-binding protein. *The Journal of Biological Chemistry*, *277*, 21468–21473.

Nair, A. K., & Menon, K. M. J. (2004). Isolation and characterization of a novel trans-factor for luteinizing hormone receptor mRNA from ovary. *The Journal of Biological Chemistry*, *279*, 14937–14944.

Peegel, H., Randolph, J., Jr., Midgley, A. R., & Menon, K. M. J. (1994). In situ hybridization of luteinizing hormone/human chorionic gonadotropin receptor messenger ribonucleic acid during hormone-induced down-regulation and the subsequent recovery in rat corpus luteum. *Endocrinology*, *135*, 1044–1051.

Pioli, P. A., Hamilton, B. J., Connolly, J. E., Brewer, G., & Rigby, W. F. (2002). Lactate dehydrogenase is an AU-rich element-binding protein that directly interacts with AUF1. *The Journal of Biological Chemistry*, *277*, 35738–35745.

Ross, J. (1993). Messenger RNA turnover in cell-free extracts from higher eukaryotes. In S. R. Haynes (Ed.), *RNA-protein interaction protocols Vol. 118.* (pp. 459–476). Totowa, NJ: Humana Press.

Ross, J. (1995). mRNA stability in mammalian cells. *Microbiological Reviews*, *59*, 423–450.

Salilew-Wondim, D., Ahmad, I., Gebremedhn, S., Sahadevan, S., Hossain, M. D., Rings, F., et al. (2014). The expression pattern of microRNAs in granulosa cells of subordinate and dominant follicles during the early luteal phase of the bovine estrous cycle. *PLoS One*, *9*, e106795.

Segaloff, D. L., Wang, H. Y., & Richards, J. S. (1990). Hormonal regulation of luteinizing hormone/chorionic gonadotropin receptor mRNA in rat ovarian cells during follicular development and luteinization. *Molecular Endocrinology*, *4*, 1856–1865.

Sen, K. K., Azhar, S., & Menon, K. M. J. (1979). Receptor-mediated gonadotropin action in the ovary. Desensitization of gonadotropin binding sites, activation of adenosine 3′:5′-cyclic monophosphate-dependent protein kinase(s), and regulation of steroidogenesis in rat ovary. *The Journal of Biological Chemistry*, *254*, 5664–5671.

Sirotkin, A. V., Ovcharenko, D., Grossmann, R., Laukova, M., & Mlyncek, M. (2009). Identification of microRNAs controlling human ovarian cell steroidogenesis via a genome-scale screen. *Journal of Cellular Physiology*, *219*, 415–420.

Thakral, S., & Ghoshal, K. (2015). miR-122 is a unique molecule with great potential in diagnosis, prognosis of liver disease, and therapy both as miRNA mimic and antimir. *Current Gene Therapy*, *15*, 142–150.

Wang, L., & Menon, K. M. J. (2005). Regulation of luteinizing hormone/chorionic gonadotropin receptor messenger ribonucleic acid expression in the rat ovary: Relationship to cholesterol metabolism. *Endocrinology*, *146*, 423–431.

Wang, L., Nair, A. K., & Menon, K. M. J. (2007). Ribonucleic acid binding protein-mediated regulation of luteinizing hormone receptor expression in granulosa cells: Relationship to sterol metabolism. *Molecular Endocrinology*, *21*, 2233–2241.

Zeleznik, A. J., Midgley, A. R., Jr., & Reichert, L. E., Jr. (1974). Granulosa cell maturation in the rat: Increased binding of human chorionic gonadotropin following treatment with follicle-stimulating hormone in vivo. *Endocrinology*, *95*, 818–825.

Zeleznik, A. J., Schuler, H. M., & Reichert, L. E., Jr. (1981). Gonadotropin-binding sites in the rhesus monkey ovary: Role of the vasculature in the selective distribution of human chorionic gonadotropin to the preovulatory follicle. *Endocrinology*, *109*, 356–362.

Zhu, H., Wang, H., & Ascoli, M. (1995). The lutropin/choriogonadotropin receptor is palmitoylated at intracellular cysteine residues. *Molecular Endocrinology*, *9*, 141–150.

CHAPTER FOUR

The Roles of Neurokinins and Endogenous Opioid Peptides in Control of Pulsatile LH Secretion

Peyton W. Weems*, Michael N. Lehman†, Lique M. Coolen†, Robert L. Goodman‡,[1]

*Graduate Program in Neuroscience, University of Mississippi Medical Center, Jackson, MS, United States
†University of Mississippi Medical Center, Jackson, MS, United States
‡West Virginia University, Morgantown, WV, United States
[1]Corresponding author: e-mail address: rgoodman@hsc.wvu.edu

Contents

Abstract

Work over the last 15 years on the control of pulsatile LH secretion has focused largely on a set of neurons in the arcuate nucleus (ARC) that contains two stimulatory neuropeptides, critical for fertility in humans (kisspeptin and neurokinin B (NKB)) and the inhibitory endogenous opioid peptide (EOP), dynorphin, and are now known as KNDy (kisspeptin–NKB–dynorphin) neurons. In this review, we consider the role of each of the KNDy peptides in the generation of GnRH pulses and the negative feedback actions of ovarian steroids, with an emphasis on NKB and dynorphin. With regard to negative

Vitamins and Hormones, Volume 107
ISSN 0083-6729
https://doi.org/10.1016/bs.vh.2018.01.011

feedback, there appear to be important species differences. In sheep, progesterone inhibits GnRH pulse frequency by stimulating dynorphin release, and estradiol inhibits pulse amplitude by suppressing kisspeptin. In rodents, the role of KNDy neurons in estrogen negative feedback remains controversial, progesterone may inhibit GnRH via dynorphin, but the physiological significance of this action is unclear. In primates, an EOP, probably dynorphin, mediates progesterone negative feedback, and estrogen inhibits kisspeptin expression. In contrast, there is now compelling evidence from several species that kisspeptin is the output signal from KNDy neurons that drives GnRH release during a pulse and may also act within the KNDy network to affect pulse frequency. NKB is thought to act within this network to initiate each pulse, although there is some redundancy in tachykinin signaling in rodents. In ruminants, dynorphin terminates GnRH secretion at the end of pulse, most likely acting on both KNDy and GnRH neurons, but the data on the role of this EOP in rodents are conflicting.

1. INTRODUCTION

In female mammals, the control of cyclic ovarian function is regulated via complex feedback actions among the brain, the pituitary gland, and the gonads, known as the hypothalamic–pituitary–gonadal (HPG) axis. The primary hierarchical component of the HPG axis are hypothalamic neurons which synthesize and release the decapeptide, gonadotropin-releasing hormone (GnRH) (Herbison, 2015). In most species, GnRH cell bodies are dispersed across the rostral–caudal extent of the preoptic area (POA) and hypothalamus (Silverman, Livne, & Witkin, 1994), with the majority sending their projections to the median eminence (ME) (King, Tobet, Snavely, & Arimura, 1982). At the ME, GnRH is secreted into the hypophysial portal blood vessels and carried to the anterior pituitary where it acts to stimulate the synthesis and release of the gonadotropins, luteinizing hormone (LH) and follicle-stimulating hormone (FSH). LH and FSH then act on somatic cells in the ovaries to stimulate follicular development, including maturation of ova and ovulation, maintenance of the corpus luteum, and synthesis and secretion of estradiol (E_2) and progesterone (Goodman & Inskeep, 2015; Zeleznik & Plant, 2015).

The complex sequence of events that comprises the ovarian cycle requires two different patterns of LH release: tonic LH secretion and the LH surge (Goodman & Inskeep, 2015; Zeleznik & Plant, 2015). Tonic GnRH and LH secretion occurs at low levels throughout the follicular and luteal phases, stimulates secretion of E_2 and progesterone, and, in turn, is controlled by the negative feedback actions of these steroids. In contrast, the GnRH and LH surge is a massive increase in secretion that occurs once/cycle at the end of the follicular phase; it induces ovulation and luteinization and is triggered by the positive feedback actions of E_2. Not surprisingly, these two modes of secretion are controlled by different neural systems, with those responsible for the GnRH surge varying markedly among species in which they have been examined (Goodman, 2015; Plant, 2012). A detailed consideration of these neural mechanisms is beyond the scope of this chapter, and interested readers are referred to several recent reviews (Christian & Moenter, 2010; Goodman, 2015; Goodman & Inskeep, 2015; Khan & Kauffman, 2012). In contrast, the basic mechanisms controlling tonic GnRH secretion, which are the primary focus of this chapter, show much less species variability (Goodman, 2015).

Tonic secretion of GnRH and LH occurs in an episodic pattern, an observation that was first made in ovariectomized (OVX) monkeys (Dierschke, Bhattacharya, Atkinson, & Knobil, 1970) and has since been shown in many mammalian species and endocrine conditions (Karsch, 1987). The pulsatile release of GnRH into the portal circulation is essential for normal reproductive function as exposure of gonadotropes to continuous GnRH inhibits gonadotropin secretion (Belchetz, Plant, Nakai, Keogh, & Knobil, 1978). It also plays a central role in control of reproductive function: changes in the pattern of GnRH, and thus LH, pulses largely determine key events of the follicular phase (Goodman & Inskeep, 2015; Zeleznik & Plant, 2015), are responsible for the onset of fertility at puberty (Goodman, 2015), and initiate regular transitions between fertility and infertility in seasonal breeders (Karsch et al., 1984). Studies using the sheep model, which allows for direct and frequent sampling of both hypophysial portal and jugular blood in unanesthetized animals, provide the best available description of pulsatile GnRH secretion. In OVX ewes, GnRH pulses occur in 5-min long (on average) square waves approximately once every hour, with no secretion occurring between each pulse (Moenter, Brand, Midgley, & Karsch, 1992). Based on this pattern, it is clear that the secretory activity of the GnRH neurons responsible for pulsatile release must be tightly synchronized, but the mechanisms responsible for the synchronization of the widely scattered population of GnRH neurons remain controversial.

A second major unresolved question relevant to the control of tonic GnRH and LH secretion is the neural mechanisms underlying steroid negative feedback. It has been recognized since 1980 that although both ovarian steroids inhibit episodic LH secretion, they alter different aspects of it: progesterone inhibits LH pulse frequency and E_2 inhibits pulse amplitude

(Goodman & Karsch, 1980), while actually increasing pulse frequency slightly (Karsch, Foster, Bittman, & Goodman, 1983). These actions are responsible for the changes in the pattern of episodic LH secretion seen during the ovarian cycle of many species (Goodman, 2015), including humans (Zeleznik & Plant, 2015) and reflect actions of progesterone (Skinner et al., 1998) and E_2 (Evans, Dahl, Mauger, & Karsch, 1995) on episodic GnRH secretion, at least in sheep. The effects of these two steroids on GnRH secretion must be mediated by interneurons because GnRH neurons do not contain progesterone receptors (PR) or the alpha form of estrogen receptors (ERα) that mediates estrogen negative feedback (Herbison, 2015). The phenotype of these steroid-responsive interneurons was heavily debated throughout the 20th century, with at least eight different neurotransmitters favored by different investigators (Herbison, 1998). Thus, the neural substrates underlying two critical characteristics of tonic GnRH secretion, how GnRH release is synchronized and inhibited by ovarian steroids, has only begun to be identified over the last decade.

A key observation in 2003 opened the door to the contemporary era of reproductive neuroendocrinology: rare mutations in the receptor for a specific neuropeptide, kisspeptin, caused infertility in humans (de Roux et al., 2003; Seminara et al., 2003). Numerous subsequent studies in a variety of species have led to the general view that kisspeptin neurons, which also contain PR and ERα, play a key role in synchronizing GnRH release during episodic secretion and may also mediate the negative feedback actions of E_2 and progesterone. In 2007, the report that a subpopulation of kisspeptin neurons in the arcuate nucleus (ARC) of the sheep also contained the tachykinin, neurokinin B (NKB), and the endogenous opioid peptide (EOP), dynorphin (Goodman et al., 2007), raised the possibility that these two neuropeptides were also involved. This observation was soon confirmed in several other species (Lehman, Hileman, & Goodman, 2013; Navarro et al., 2009; Overgaard, Ruiz-Pino, Castellano, Tena-Sempere, & Mikkelsen, 2014; Wakabayashi et al., 2010) which, together with the report that mutations in genes for NKB or its receptor (NK3R) caused infertility in humans (Topaloglu et al., 2009), rekindled interest in the role of neurokinins and EOP in the control of episodic LH secretion. Thus, these neurons, now known as KNDy cells (kisspeptin–NKB–dynorphin), contain two stimulatory peptides necessary for reproduction in humans and an inhibitory EOP. They will be the primary focus of this review; specifically, we will first briefly summarize the relevant historical literature and studies implicating kisspeptin in these processes, and then focus on the roles of NKB and dynorphin.

2. NEURAL SYSTEMS CRITICAL FOR EPISODIC LH SECRETION AND THE NEGATIVE FEEDBACK ACTIONS OF OVARIAN STEROIDS

Although the concept that different neural systems controlled tonic LH secretion and the LH surge was formally proposed in 1949 (Everett, Sawyer, & Markee, 1949), compelling evidence for the neuroanatomical regions containing each system did not occur until the deafferentation experiments of Halasz and colleagues in 1965. These demonstrated that knife cuts that partially or completely disrupted input from the POA into the medial basal hypothalamus (MBH) blocked ovulation and thus arrested the estrous cycle in ovary-intact rats (Halasz & Gorski, 1967), but did not prevent follicular development or E_2 secretion. Subsequent work confirmed that these knife cuts did not alter pulsatile LH secretion in intact or OVX rats (Blake & Sawyer, 1974; Ohkura, Tsukamura, & Maeda, 1991; Soper & Weick, 1980), sheep (Jackson, Kuehl, McDowell, & Zaleski, 1978), monkeys (Krey, Butler, & Knobil, 1975), and guinea pigs (King et al., 1998). The lack of effect of deafferentation in OVX animals led to the conclusion that the MBH contains the neural elements that produce episodic GnRH secretion. This conclusion is consistent with the high number of GnRH cell bodies in the primate MBH (Herbison, 2015) and evidence in sheep that only the population of GnRH neurons in the MBH express Fos (an index of neural activation) when episodic LH secretion is stimulated (Boukhliq, Goodman, Berriman, Adrian, & Lehman, 1999). On the other hand, few (mice) or no (rats) GnRH cell bodies are found in the rodent MBH (Herbison, 2015); there are two possible explanations for this apparent paradox: early work that demonstrated GnRH projections under the optic chiasm in rats (Hoffman & Gibbs, 1982), or the more recent discovery that GnRH projections to the ME in mice have characteristics of both axons and dendrites (Herde, Iremonger, Constantin, & Herbison, 2013). Thus, inputs to these "dendrons" within the MBH may maintain GnRH pulses in deafferented rodents. The lack of effect of deafferentation on follicular development and E_2 secretion in ovary-intact animals implied that the negative feedback actions of ovarian steroids also occur in this area. The latter has been confirmed by studies demonstrating that similar knife cuts had no effect on the ability of E_2 to inhibit LH pulses in rats (Blake, Norman, & Sawyer, 1974), sheep (Jackson et al., 1978), and monkeys (Krey et al., 1975) or on progesterone inhibition of LH secretion in rats (Blake, 1977) and pulse frequency in sheep (Whisnant & Goodman, 1994).

Two different approaches to identifying specific areas within the MBH responsible for episodic GnRH secretion have both pointed to the ARC: small lesions within this area and recordings of multiunit electrical activity (MUA). Thus, radiofrequency lesions of the ARC, but not other areas within the MBH, disrupted pulsatile LH secretion in OVX monkeys (Plant et al., 1978) and rats (Soper & Weick, 1980). Electrical recordings revealed regular bursts (or volleys) of increased MUA that always correlated with the onset of LH pulses (Kawakami, Uemura, & Hayashi, 1982; Mori et al., 1991; Thiery & Pelletier, 1981; Wilson et al., 1984), and the neural circuitry involved became known as "the GnRH pulse generator." These MUA volleys were recorded within the ARC of several species although they were sometime also seen in regions some distance from this nucleus (Silverman, Wilson, Kesner, & Knobil, 1986). Thus, these early studies pointed to the ARC as the likely site within the MBH containing the GnRH pulse generator. However, they did not provide information on sites of steroid negative feedback within the MBH because inhibitory effects of steroids could not be observed after disruption of pulses, and changes in MUA could reflect neural input from any area. Other approaches during this period included local administration of steroids and anatomical descriptions of ERα and PR expression within the MBH. But results from this work are difficult to interpret because diffusion of steroid from the site of administration would likely reach much of MBH and the pituitary in rodents, and populations of different neurons-containing ERα and/or PR are found throughout the MBH. Thus, workers fell back on pharmacological means to identify the type of neurons mediating negative feedback which, as noted earlier, failed to pinpoint the specific neural systems involved. It was only recently that the use of region-specific knock down of ERα in mice has provided strong evidence that the ARC is the site of E_2 negative feedback (Yeo & Herbison, 2014).

3. KISSPEPTIN AND EPISODIC LH SECRETION

The initial observation that mutations in the gene for the kisspeptin receptor (now known as *Kiss1r*) prevented the onset of puberty in boys and girls (de Roux et al., 2003; Seminara et al., 2003), implied that kisspeptin was critical for episodic GnRH secretion because increases in GnRH pulses were known to drive puberty. Subsequent work demonstrating that both knockout of *Kiss1* in rats (Uenoyama et al., 2015) and administration of a KISS1R antagonist to OVX ewes (Roseweir et al., 2009) blocked episodic

LH secretion supported this hypothesis. Anatomical studies soon identified two populations of kisspeptin neurons in rodents, one in a region adjacent to the POA, known as the rostral periventricular area of the third ventricle (RP3V), and the other in the ARC (Oakley, Clifton, & Steiner, 2009). In light of the classical work identifying these two areas as being responsible for the GnRH surge and tonic GnRH secretion, respectively, it was logical to suggest that the RP3V population was involved in the surge, while the ARC population was necessary for episodic GnRH release. Moreover, previous work had identified the RP3V as the site of E_2 positive feedback in rodents (Herbison, 2008) and it was soon demonstrated that E_2 increased kisspeptin expression in the RP3V, but decreased it in the ARC (Smith, Cunningham, Rissman, Clifton, & Steiner, 2005). Thus, the initial observations on these two kisspeptin populations quickly led to the hypothesis that the RP3V kisspeptin neurons mediate E_2 positive feedback, while inhibition of the ARC kisspeptin population is responsible of steroid negative feedback (Dungan, Clifton, & Steiner, 2006). There is now compelling evidence that the RP3V population is critical for the GnRH surge in rodents and considerable progress has been made identifying the neural circuitry involved (Christian & Moenter, 2010; Herbison, 2015; Khan & Kauffman, 2012). An analogous, but more scattered, kisspeptin population has been implicated in the GnRH surge in several other species, including sheep, guinea pigs, and primates (Lehman et al., 2013), but other neural systems and areas are equally important in these species (Goodman, 2015). For example, there is strong evidence that the ARC kisspeptin population plays a key role in E_2 positive feedback in sheep (Caraty et al., 1998; Merkley, Coolen, Goodman, & Lehman, 2015; Smith, Li, Pereira, & Clarke, 2009). On the other hand, the role of ARC kisspeptin neurons in steroid negative feedback is more complicated and controversial, as discussed later.

3.1 Role in Pulse Generation

It is now generally accepted that kisspeptin-containing neurons in the ARC play a central role in GnRH pulse generation. As noted earlier, this was inferred from the seminal observation on the role of kisspeptin in fertility, but the current hypothesis was not fully developed until it became clear that KNDy neurons have several characteristics important for GnRH pulse generation: (1) they form a reciprocal interconnected network that includes projections under the third ventricle connecting both sides of the ARC (Burke, Letts, Krajewski, & Rance, 2006; Foradori, Amstalden,

Goodman, & Lehman, 2006; Krajewski, Burke, Anderson, McMullen, & Rance, 2010; Wakabayashi, Yamamura, Sakamoto, Mori, & Okamura, 2013); (2) they have direct projections to GnRH cell bodies and/or terminals (Herbison, 2015; Kumar et al., 2015; Lehman et al., 2013; Merkley et al., 2015); (3) they contain NK3R (Amstalden et al., 2010; Burke et al., 2006; Krajewski et al., 2005), and (4) they are clustered close to the anatomical sites of MUA volleys that correlate with LH pulses (Ohkura et al., 2009). Based on these data, four groups (Lehman, Coolen, & Goodman, 2010; Navarro et al., 2009; Rance, Krajewski, Smith, Cholanian, & Dacks, 2010; Wakabayashi et al., 2010) working independently proposed that the synchronous activity of KNDy neurons drives episodic GnRH secretion, with kisspeptin being the output signal and NKB and Dyn acting within the KNDy network to initiate and terminate each GnRH pulse, respectively (Fig. 1). Thus, each GnRH pulse is

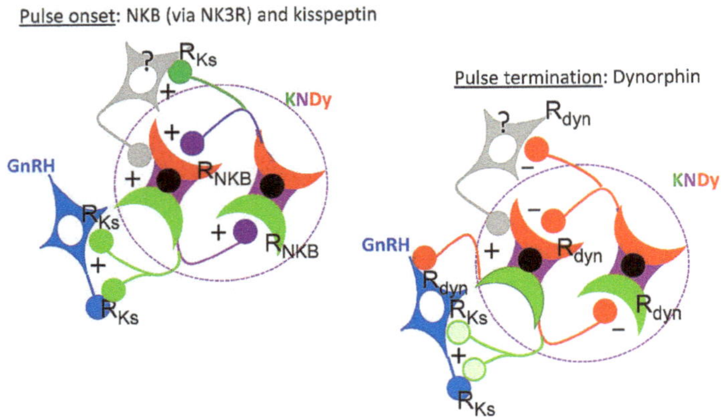

Fig. 1 Model for control of KNDy neural activity proposed to drive episodic GnRH secretion. Each GnRH pulse is initiated by NKB (*purple*) acting within the KNDy network (within *dashed oval*), which stimulates kisspeptin (*green*) release to drive GnRH (*blue*) secretion and possibly activate unidentified KISS1R-containing ARC neurons (*gray*) that reinforce the stimulatory actions of NKB on KNDy neurons. GnRH release is then terminated by dynorphin (*red*) release from KNDy neurons acting either directly on KNDy neurons and/or GnRH neurons and/or the unidentified Kiss1r-containing neurons. *Abbreviations*: R_{dyn}, kappa-opioid receptor; R_{Ks}, KISS1R; R_{NKB}, NK3R. Note that the *color* in each terminal indicates the biologically active transmitter (possibly due to selective expression of postsynaptic receptors) and does not reflect selective transport of that peptide to the terminal. *Revised from Lehman, M. N., Coolen, L. M., & Goodman, R. L. (2010). Minireview: Kisspeptin/neurokinin B/dynorphin (KNDy) cells of the arcuate nucleus: A central node in the control of gonadotropin-releasing hormone secretion.* Endocrinology, 151*(8), 3479–3489; reproduced by permission of the Endocrine Society.*

triggered by an initial increase in NKB from a few KNDy neurons, which stimulates further NKB release within this network. This positive feedback loop produces release of kisspeptin onto GnRH neurons and hence the extremely rapid increase in GnRH secretion that occurs at the onset of a pulse (Moenter et al., 1992). Stimulation of KNDy neurons by NKB is also predicted to stimulate release of Dyn, which initially holds kisspeptin release in check and then completely suppresses KNDy activity, terminating the GnRH pulse and preventing GnRH secretion between pulses. Dyn binding would also gradually suppress Dyn release from KNDy neurons, which would eventually reset the activity of the population allowing for NKB to trigger the next GnRH pulse. There is now compelling evidence, reviewed in detail elsewhere (Goodman, Coolen, & Lehman, 2014; Okamura et al., 2013), that KNDy neurons represent the GnRH pulse generator including the recent reports that optogenetic stimulation of these neurons can produce LH pulses in vivo (Han, McLennan, Czieselsky, & Herbison, 2015), and preliminary data that optogenetic inhibition of them suppresses episodic LH secretion (Clarkson et al., 2017).

There is now considerable evidence supporting the proposed role of kisspeptin as the output signal from KNDy neurons. First, almost all GnRH neurons contain *Kiss1r* (Irwig et al., 2004; Smith et al., 2009), while none express NK3R in sheep (Amstalden et al., 2010), and either few (Krajewski et al., 2005; Navarro et al., 2015) or none (Navarro, Gottsch, et al., 2011) contain this receptor in rats and mice. Second, kisspeptin pulses correlation with GnRH pulses in perfusates of the monkey MBH (Keen, Wegner, Bloom, Ghatei, & Terasawa, 2008). Third, kisspeptin potently stimulates GnRH electrical activity in slice preparation, while NK3R agonists have no effect (Navarro, Gottsch, et al., 2011). Finally, knockout of *Kiss1r* in GnRH neurons disrupts both ovarian cycles and the OVX-induced increase in LH secretion in mice (Novaira et al., 2014; Yeo, Clarkson, & Herbison, 2014). Thus, there is general agreement that kisspeptin release from KNDy neurons is the proximate stimulus for each GnRH pulse. Whether kisspeptin also plays some role within the KNDy network is less clear because exogenous kisspeptin did not affect MUA volleys associated with LH pulses (Kinsey-Jones, Li, Luckman, & O'Byrne, 2008; Ohkura et al., 2009), but intra-ARC administration of a KISS1R antagonist slowed LH pulse frequency (Goodman et al., 2013; Li et al., 2009). Any action of kisspeptin on KNDy neural activity is most likely indirect since these neurons do not express *Kiss1r* (Higo, Iijima, & Ozawa, 2017; Smith et al., 2011).

3.2 Role in Steroid Negative Feedback

The possible role of kisspeptin in mediating steroid negative feedback is complicated by the different actions of E_2 and progesterone on GnRH pulse amplitude and frequency. Moreover, most studies have focused on the inhibitory actions of E_2 largely because progesterone plays a minor physiological role during the short estrous cycle of rodents. A few studies have examined the effects of progesterone on kisspeptin expression in animals with 14-day luteal phases. There are reports that progesterone treatment decreases the number of ARC neurons-containing kisspeptin in chronically OVX sheep (Smith, Clay, Caraty, & Clarke, 2007) and monkeys (Alcin et al., 2013), but progesterone does not inhibit LH pulses in the long-term absence of E_2 in these species (Karsch, Legan, Hauger, & Foster, 1977; Yamaji, Dierschke, Bhattacharya, & Knobil, 1972), and kisspeptin expression is not lower in the luteal phase of the menstrual cycle (Eghlidi, Haley, Noriega, Kohama, & Urbanski, 2010; Smith, Shahab, Pereira, Pau, & Clarke, 2010). A more extensive study in sheep (Goodman & Inskeep, 2015) found no effect of progesterone on mean kisspeptin expression, and no correlation between kisspeptin expression and LH pulse frequency (Fig. 2). Thus, it appears unlikely that kisspeptin plays an important role in the negative feedback actions of progesterone. In contrast, several lines of evidence support the hypothesis that a decrease in ARC kisspeptin mediates the negative feedback actions of E_2. Thus, E_2 inhibits expression of kisspeptin in the ARC in every mammal studied to date (Goodman & Lehman, 2012), an action that is mediated by nonclassical ERα signaling mechanisms (Gottsch et al., 2009) similar to those mediating E_2 negative feedback (Glidewell-Kenney et al., 2007). There is also no evidence for inhibitory effects of E_2 in mice in which *Kiss1r* has been selectively knocked out of GnRH neurons (Novaira et al., 2014; Yeo et al., 2014). Similarly, lesioning of KNDy (and other NK3R containing) neurons using an NK3-saporin conjugate dramatically reduced the increase in LH secretion in OVX rats, although some inhibitory effects of E_2 (and a few KNDy) neurons remained (Mittelman-Smith et al., 2012). In ewes, kisspeptin expression correlated with LH pulse amplitude (Fig. 2) and Fos expression increased in KNDy (but not POA kisspeptin neurons) shortly after OVX (Merkley et al., 2012). Arguing against this hypothesis are reports that the firing rate of KNDy neurons is not higher in slices from gonadectomized mice (Cholanian, Krajewski-Hall, Levine, McMullen, & Rance, 2014; de Croft et al., 2012; Ruka, Burger, & Moenter, 2013) (but this was seen in

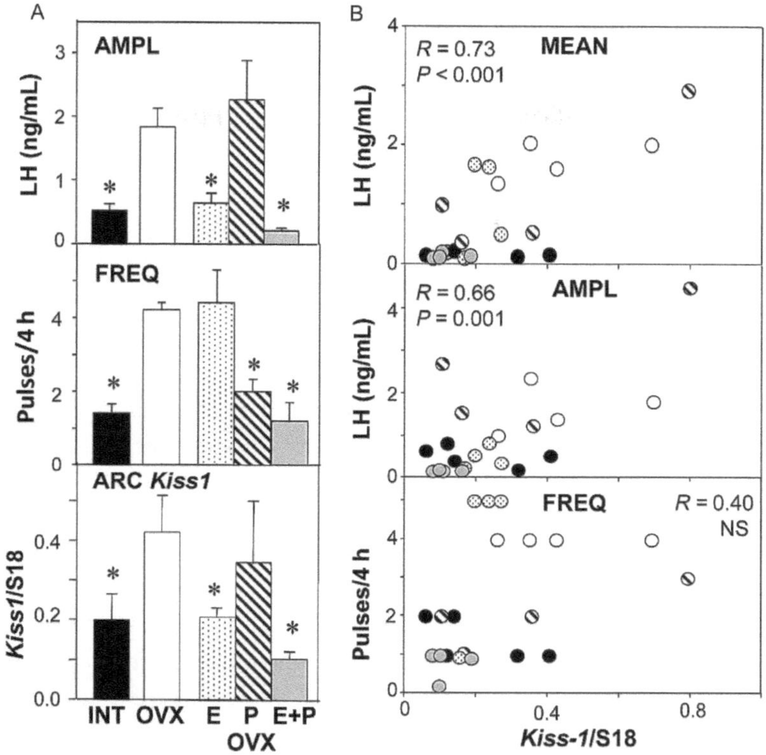

Fig. 2 Effects of ovarian steroids on episodic LH secretion and *Kiss1* mRNA expression in the caudal ARC of ewes. Ewes ($n=4$–5/group) were ovariectomized (OVX) and given either no implants (*clear symbols*) or implants that produced luteal phase levels of estradiol (*striped symbols*), progesterone (*stippled symbols*), or both (*shaded symbols*). Eight days later, LH pulse patterns were monitored for 4 h, and tissue was immediately collected and frozen. Samples from ovary-intact (INT) luteal phase ewes (*black symbols*) were collected at the same time. The caudal ARC was then microdissected, mRNA extracted, and *Kiss1* mRNA determined by qRT-PCR. Bars on the right depict mean ± SEM values for LH pulse amplitude (*top panel*), LH pulse frequency (*middle panel*), and *Kiss1* mRNA (*bottom panel*). *$P<0.05$ vs OVX. *Right panels* present correlations of mean LH (*top*), LH pulse amplitude (*middle*), and LH pulse frequency (*bottom*) with *Kiss1* mRNA expression in individual ewes. *Reproduced from Goodman, R. L., & Inskeep, E. K. (2015). Control of the ovarian cycle of the sheep. In T. M. Plant & A. J. Zeleznik (Eds.),* Knobil and Neill's physiology of reproduction, *4th ed., Vol. 2 (pp. 1259–1305). Amsterdam: Elsevier, with permission from Elsevier.*

one recent report; Ruka, Burger, & Moenter, 2016), Fos does not increase in these neurons in OVX rats (Adachi et al., 2007), and E_2 strongly inhibits LH secretion in mice in which ERα has been deleted from kisspeptin neurons (Dubois et al., 2015). The latter observation indicates that either KNDy neurons are not direct targets for E_2 negative feedback (but could still be inhibited by estrogen-responsive interneurons) or that compensatory changes occurred that allowed another system to assume this function. Thus, a decrease in kisspeptin release from KNDy neurons appears to mediate the negative feedback actions of E_2 on GnRH pulse amplitude in ewes, but its role in rodents remains unclear.

4. NEUROKININS AND EPISODIC LH SECRETION

Three members of the neurokinin (or tachykinin) family of peptides may play some role in the control of episodic LH secretion: NKB, substance P (SP), and neurokinin A (NKA). Members of this family all contain the sequence Phe-X-Gly-Leu-Met-NH_2 at their carboxyl terminus (Maggio, 1988). Two genes encode for these neurokinins, with *TAC3* (*TAC2* in rodents) encoding the precursor for NKB and *TAC1* containing sequences for both SP and NKA (Otsuka & Yoshioka, 1993). The three tachykinin receptors, NK1R, NK2R, and NK3R, are in the rhodopsin-like family of G protein-coupled receptors and act via G_q to increase intracellular calcium concentrations. These three receptors show specificity for SP, NKA, and NKB, respectively (Seabrook, Bowery, & Hill, 1995), although the degree of specificity varies among them (Table 1).

4.1 Hypothalamic Anatomy of the Tachykinins

Immunohistochemical (IHC) and in situ hybridization (ISH) techniques have provided similar information on the location of NKB-producing neurons in the diecephalon, with the majority of them in the ARC. This population has now been observed in rats, mice, sheep, goats, monkeys, and humans. In rodents, scattered NKB-containing cells are also found in the POA, anterior hypothalamus (AHA), dorsomedial hypothalamus, and diagonal band of Broca (Lucas, Hurley, Krause, & Harlan, 1992; Marksteiner, Sperk, & Krause, 1992; Merchenthaler, Maderdrut, O'Harte, & Conlon,

Table 1 Selectivity of Three Major Tachykinins at Tachykinin Receptors (nM)

	SP	NKA	NKB
NK1R	2	16	70
NK2R	2200	3	25
NK3R	18,000	1300	4

Based on response of rat ventral tegmental slices (Seabrook et al., 1995).

1992), while in sheep a few NKB-immunoreactive (ir) neurons are occasionally observed in the POA (Foradori et al., 2006; Goubillon et al., 2000). In humans, the number of cells with mRNA for NKB in the dBB, POA, and AHA were similar to those in the infundibular nucleus (analogous to the ARC in other species) (Chawla, Gutierrez, Young, McMullen, & Rance, 1997). In contrast, there was some divergence in location of SP-producing cells between IHC and ISH results. Using IHC, these were largely limited to the ARC in humans (Hrabovszky et al., 2013), monkeys (Kalil, Ramaswamy, & Plant, 2015; Ronnekleiv, Kelly, & Eskay, 1984), and sheep (Fergani et al., 2016) and found primarily in the ARC and ventromedial nucleus (VMN) of rats (Ljungdahl, Hokfelt, & Nilsson, 1978). In humans, many more SP-producing neurons were identified using ISH (Chawla et al., 1997), with heavy concentrations in the premammillary region (PMR), posterior hypothalamus, and mammillary body; in mice neurons-containing SP mRNA were located primarily in the ARC and VMN (Navarro et al., 2015), similar to the distribution in rats using IHC. The mismatched between these two techniques in humans could be due to neurons containing only NKA, but there have been no IHC studies of NKA expression in humans.

Both NK1R and NK3R are expressed more extensively than SP and NKB in the hypothalamus, but there is virtually no NK2R (Saffroy, Torrens, Glowinski, & Beaujouan, 2003) found in this area of the brain. In rats (Nakaya, Kaneko, Shigemoto, Nakanishi, & Mizuno, 1994), sheep (Fergani et al., 2016), and guinea pigs (Yip & Chahl, 2000), NK1R-expressing cells are found in many regions of the POA-hypothalamus, with relatively high concentrations in the lateral hypothalamus, the VMN, ARC, and PMR. Expression of NK3R is also extensively distributed throughout the hypothalamus with high concentrations in many nuclei, including the PVN, ARC, PMR, and lateral hypothalamus in rats (Krajewski et al., 2005), guinea pigs (Yip & Chahl, 2001), sheep (Amstalden et al., 2010), and humans (Mileusnic et al., 1999). There are some species differences (i.e., higher expression in POA and DMN in rats than sheep) but these differences probably have little relevance to the control of episodic LH secretion.

There are also species similarities and differences in the cellular expression of tachykinins and their receptors. Colocalization of kisspeptin and NKB in ARC neurons (Fig. 3, top panels) has been observed in all mammalian species studied to date, including humans (Hrabovszky et al., 2010), although the extent of this colocalization varies among species and

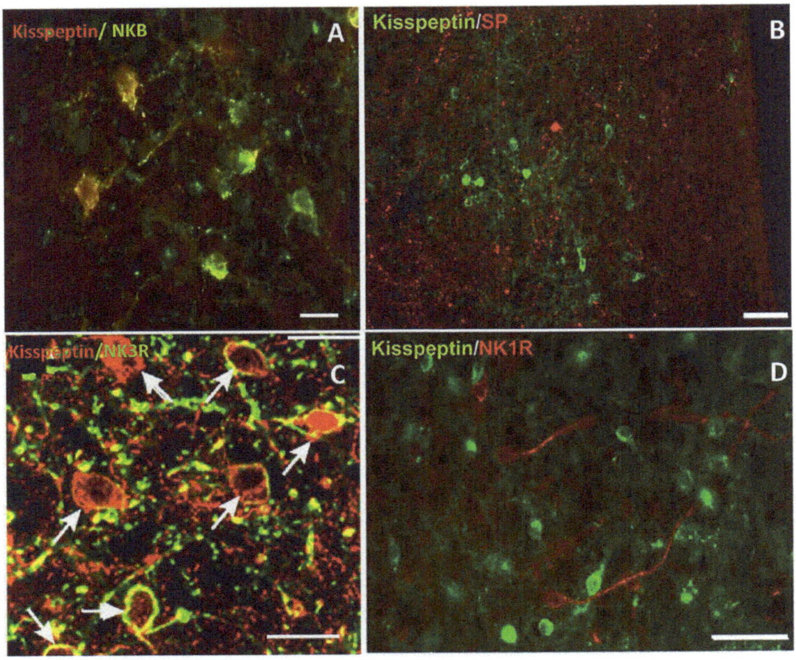

Fig. 3 Colocalization of NKB or NK3R (A and C), but not SP or NK1R (B and D) in ovine KNDy neurons. Each panel presents the overlay of two confocal images taken of the same section that has been stained immunohistochemically for kisspeptin (*red*) and NKB or NK3R (*green*) in Panels A and C, or kisspeptin (*green*) and SP or NK1R (*red*) in (B) and (D). *Arrows* in Panel C indicate examples of dual labeled cells. Magnification bars are 20 μm in (A and C), 100 μm in (B), and 50 μm in (D). *Photomicrographs from Goodman, R. L., Lehman, M. N., Smith, J. T., Coolen, L. M., de Oliveira, C. V., Jafarzadehshirazi, M. R., et al. (2007). Kisspeptin neurons in the arcuate nucleus of the ewe express both dynorphin A and neurokinin B.* Endocrinology, 148*(12), 5752–5760; Fergani, C., Mazzella, L., Coolen, L. M., McCosh, R. B., Hardy, S. L., Newcomb, N., et al. (2016). Do substance P and neurokinin A play important roles in the control of LH secretion in ewes?* Endocrinology, 157*(12), 4829–4841, with permission from the Endocrine Society; and Ahn et al. (2015). Journal of Neuroendocrinol, 27, 100–110, with permission from the British Society for Neuroendocrinology.*

is affected by sex and endocrine status (Lehman et al., 2013). It is important to point out that this colocalization is unique to the ARC; it is not seen in kisspeptin-ir or NKB-ir neurons in other regions of the brain. In contrast, little or no colocalization of SP and kisspeptin is seen in the ARC of mice (Simavli et al., 2015), sheep (Fig. 3, top panel; Fergani et al., 2016), monkeys (Kalil et al., 2015), or rats (based on indirect evidence Rance & Bruce, 1994), but has been observed directly (Hrabovszky et al., 2013) and indirectly (Rance & Young, 1991) in humans. Specifically, in tissue from postmenopausal women, 24.3% of the kisspeptin-ir neurons contained SP; SP was also found in NKB-containing neurons, 25% of which contained both NKB and kisspeptin, but the majority of kisspeptin–NKB dual labeled neurons did not containing SP (Hrabovszky et al., 2013). There are also species differences in the cellular localization of NK3R and NK1R. Specifically, NK3R was detected in 11%–16% of GnRH neurons in rats (Krajewski et al., 2005) and mice (Ruiz-Pino et al., 2015), using IHC and single-cell RT-PCR, respectively, but was not found in murine GnRH neurons with ISH (Navarro, Gottsch, et al., 2011) or in ovine GnRH neurons with IHC (Amstalden et al., 2010). On the other hand, virtually all KNDy neurons contain NK3R in OVX mice, based on ISH (Navarro et al., 2009; Navarro, Gottsch, et al., 2011) and single-cell RT-PCR (Navarro et al., 2015; Ruka et al., 2013), while IHC found that about 65% of them contained this receptor in ovary-intact rats (Krajewski et al., 2005) and sheep (Amstalden et al., 2010) (Fig. 3, bottom panels). In the latter studies, a significant percentage of NK3R-ir neurons in the ARC did not contain a KNDy peptide. The quantitative differences among these species may well reflect the different endocrine status because there is evidence that steroids inhibit NK3R expression in mice (Ruka et al., 2013). There is considerably less information on expression of NK1R in GnRH or KNDy neurons, but the two reports available indicated marked species differences. In OVX mice (Navarro et al., 2015), single-cell RT-PCR analysis found that 23% of GnRH and 49% of KNDy neurons contain *Tacr1*, while in OVX ewes (Fergani et al., 2016) IHC did not detect any NK1R in GnRH cells and only 6% of KNDy neurons contained this receptor (Fig. 3, bottom panel), despite numerous NK1R-ir cells and fibers in the ARC. Not surprisingly, no *Tacr2* was found in GnRH or KNDy neurons of mice (Navarro et al., 2015) since this receptor is not expressed in the hypothalamus. Based on the location of these receptors, one can infer that NKB is likely to be the only tachykinin affecting GnRH secretion in sheep and that it acts indirectly via KNDy neurons. In rodents both NKB and SP, acting on either GnRH or KNDy neurons can potentially affect GnRH secretion.

4.2 Role in Pulse Generation

Early work in the 1990s (reviewed by Fergani & Navarro, 2017) exploring the effects of tachykinins on LH secretion focused primarily on SP, with a few studies of NKA effects, and only one, which found no effect, of NKB (Sahu & Kalra, 1992). Because no consistent effects were observed, there was little subsequent functional work despite evidence that expression of both NKB and SP increased in the ARC of postmenopausal women (Rance & Young, 1991) that suggested a stimulatory role for these tachykinins in humans. Although these anatomical observations were confirmed in several other species, including rats (Rance & Bruce, 1994), sheep (Pillon, Caraty, Fabre-Nys, & Bruneau, 2003), and monkeys (Sandoval-Guzman, Stalcup, Krajewski, Voytko, & Rance, 2004), a 2004 review on tachykinins (Almeida et al., 2004) concluded that: "Nowadays, the physiological significance of one of the tachykinin receptors, the NK_3 type, and of one of the tachykinins, NKB, remains largely unknown." As noted earlier, this changed dramatically with the reports that ARC kisspeptin neurons also contained NKB (Goodman et al., 2007) and that mutations disrupting NKB signaling caused infertility (Gianetti et al., 2010; Topaloglu et al., 2009). Most subsequent work has focused on testing the hypothesis that NKB acts on KNDy neurons to initiate each GnRH pulse, although a few studies of NKA and SP have begun to appear.

There is now considerable evidence supporting the proposed role for NKB in pulse generation, although there are some apparent differences between ruminants and rodents. One of the most consistent observations across species is that NKB stimulates GnRH secretion by inducing kisspeptin release from KNDy neurons. Thus NKB, or senktide (a NK3R agonist), induces Fos in KNDy neurons in rats (Navarro, Castellano, et al., 2011) and sheep (Sakamoto et al., 2012), and stimulates electrical activity of KNDy (de Croft, Boehm, & Herbison, 2013; Navarro, Gottsch, et al., 2011; Ruka et al., 2013, 2016), but not GnRH (Navarro, Gottsch, et al., 2011) neurons, in slice preparations. Moreover, although some GnRH neurons in rodents contain NK3R and kisspeptin-independent effects of senktide have been observed (Gaskins, Glanowska, & Moenter, 2013), NK3R agonists are unable to stimulate LH secretion in the absence of *Kiss1r* (Garcia-Galiano et al., 2012; Grachev, Li, Lin, et al., 2012). These data from mice are supported by the report in monkeys that downregulation of KISS1R blocks the effects of senktide, while suppression of NK3R does not affect the stimulatory actions of kisspeptin (Ramaswamy, Seminara, & Plant, 2011), and the clinical observation that kisspeptin can restore episodic LH secretion in patients with mutations that disrupt NKB–NK3R signaling (Young et al., 2013). This hierarchical relationship may also explain the less severe reproductive phenotype in some of these patients than in those with *Kiss1r* deficiencies (Gianetti et al., 2010). In ruminants, there is also direct evidence that NKB signaling is critical for the activity of the pulse generator because icv administration of NKB increased the frequency of MUA volleys in OVX goats (Wakabayashi et al., 2010) and intra-ARC treatments of NKB stimulated episodic LH secretion in OVX ewes (Goodman et al., 2013). Although both these studies support a stimulatory role for NKB on pulse generation, a decrease in LH secretion was seen with the highest dose of NKB in goats that was not apparent in ewes. The simplest explanation for this discrepancy is that NKB acts in the ARC to stimulate the pulse generator and acts elsewhere in the hypothalamus to inhibit the response of GnRH neurons to the output of this generator. Further support for the stimulatory role of NKB comes from reports that specific NK3R antagonists when given icv (Fraser et al., 2015; Li, Millar, Clarke, & Smith, 2015) or into the ARC (Goodman et al., 2013) blocked or greatly slowed pulsatile LH secretion in OVX ewes. Similarly, the ability of male pheromones to induce a MUA volley and a LH pulse, which is mediated by kisspeptin release from KNDy neurons (De Bond, Li, Millar, Clarke, & Smith, 2013), can also be blocked by a NK3R antagonist (Sakamoto et al., 2013). Thus, endogenous release of NKB, most likely within the ARC, is essential for the pulsatile secretion of GnRH in these species.

Interpretation of studies on NKB signaling in rodents is complicated because the effects of NK3R agonists depend on the endocrine status of these animals, and there is considerable redundancy in tachykinin signaling in both rats and mice. Initial reports indicated that senktide inhibited LH secretion in rats (Sandoval-Guzman & Rance, 2004), but as the effects of NKB agonists were studied more extensively, it became evident that senktide was generally stimulatory when endogenous LH secretion was low but inhibitory when LH secretion was elevated (Fergani & Navarro, 2017), although one species difference was noted: senktide induced LH secretion in gonad-intact male mice (Navarro, Gottsch, et al., 2011), but not intact male rats (Ruiz-Pino et al., 2012). Interestingly, the inhibitory effects of senktide in OVX rats (given low E_2 treatment) are mediated by dynorphin (Grachev, Li, Kinsey-Jones, et al., 2012; Kinsey-Jones et al., 2012), as will be discussed in Section 5.3. Redundancy in tachykinin signaling was first indicated by evidence that all three tachykinin receptors had to be blocked to inhibit episodic LH secretion in OVX rats (Noritake et al., 2011). This was also evident in the effects of tachykinins on electrical activity of KNDy neurons in murine slices (de Croft et al., 2013): exogenous SP, NKA, and NKB each depolarized these cells, and all three tachykinin receptors had to be blocked to prevent the stimulatory actions of exogenous NKB. More recent studies examining the effects of specific NK1R and NK2R agonists in vivo have revealed potential differences between SP and NKA. In general, NKA–NK2R signaling parallels that of NKB–NK3R, with stimulatory effects when endogenous LH secretion is low and inhibition when it is high in both mice and rats (Navarro et al., 2015; Ruiz-Pino et al., 2015). Moreover, early work using the EOP antagonist, naloxone, is consistent with the inhibitory actions of NK2R signaling in castrated rats occurring via dynorphin (Kalra, Sahu, Bonavera, & Kalra, 1992), as is the case for NK3R signaling. In contrast, a different pattern was observed with the NKR1 agonist: it stimulated LH secretion in both intact and OVX mice (Navarro et al., 2015), but had no effect in gonad-intact adult male and female rats (Ruiz-Pino et al., 2015). Although SP–NK1R and NKA–NK2R signaling thus likely occurs via different pathways, the stimulatory action of each occurs via kisspeptin since it is not seen in *Kiss1r*-null mice (Navarro et al., 2015). Thus, NKB signaling appears to occur via all three tachykinin receptors in rodents, which is consistent with the relatively low EC_{50} values for NKB at each receptor (Table 1). The pathways by which SP and NKA affect GnRH secretion in rodents remain unclear; SP likely acts via NK1R (Table 1), which are found on GnRH, KNDy, and RP3V kisspeptin neurons (Navarro et al., 2015), but NKA is unlikely to act via NK3R despite its similar effects (Table 1) or NK1R (based on different effects of the specific agonists), and no NK2R are not found in the hypothalamus (Saffroy et al., 2003). The redundancy in tachykinin signaling may well account for the modest effects of *Tac2* (True, Nasrin Alam, Cox, Chan, & Seminara, 2015) and *Tacr3* (Yang, Caligioni, Chan, & Seminara, 2012) knockouts in mice, compared to the infertility seen in humans with analogous loss-of-function mutations (Gianetti et al., 2010; Topaloglu et al., 2009). It is interesting to note, however, that in spite of this redundancy, deficiencies in either *Tac2* (True et al., 2015) or *Tac1* (Simavli et al., 2015) alone delay puberty in female mice, so either gene cannot completely compensate for the loss of the other.

It is important to point out that the redundancy in tachykinin signaling evident in the control of GnRH pulse generation in rodents was not seen in ruminants when the effects of tachykinin agonists were examined. Thus, the minimal effective dose of NKA (2.0 nmol) and SP (10 nmol) needed to stimulate episodic LH secretion in ovary-intact anestrous ewes was 10–20 fold and 50–100 fold, respectively, that of NKB (0.1–0.2 nmol) (Fig. 4A and B) (Fergani et al., 2016). Similarly, when the ability of specific NK1R, NK2R, and NK3R agonists to stimulate MUA in OVX goats was examined, a much higher dose of the NK2R agonist, than the NK3R agonist, was needed to increase MUA (Fig. 4C), and the NKR1 agonist had no effect even

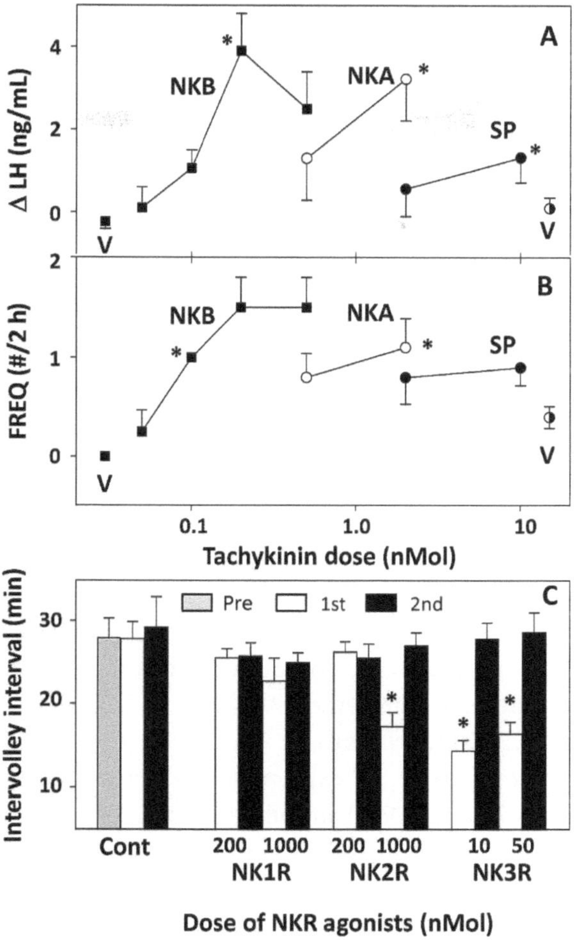

Fig. 4 Dose–response effects of injections of NKB (*solid squares*), NKA (*open circles*), and SP (*solid circles*) into the third ventricle of anestrous ewes on episodic LH secretion (Panels A and B). Panel A presents the mean (±SEM) change (average value in the 2 h after injection minus the average value before injection) in LH concentrations. Panel B depicts the mean (±SEM) LH pulse frequency during the 2 h postinjection. *Lowest dose that produces a significant effect ($P < 0.05$) compared to control values. Responses to vehicle (V) for NKB (*squares*) are shown on the *left*; those for NKA and SP (*circles*) are on the *right*. Panel C presents the mean (±SEM) time between MUA volleys for the first (*open bars*) and second (*black bars*) volley after a single injection of two doses of specific agonists to NK1R, NK2R, or NK3R in OVX goats. *Gray bar* is average time between volleys before the first control injection. *$P < 0.05$ vs control (second period) intervolley interval. *Data redrawn from Fergani, C., Mazzella, L., Coolen, L. M., McCosh, R. B., Hardy, S. L., Newcomb, N., et al. (2016). Do substance P and neurokinin A play important roles in the control of LH secretion in ewes?* Endocrinology, 157*(12), 4829–4841; Yamamura, T., Wakabayashi, Y., Ohkura, S., Navarro, V. M., & Okamura, H. (2015). Effects of intravenous administration of neurokinin receptor subtype-selective agonists on gonadotropin-releasing hormone pulse generator activity and luteinizing hormone secretion in goats. The Journal of Reproduction and Development,* 61*(1), 20–29.*

at the highest dose tested (Yamamura, Wakabayashi, Ohkura, Navarro, & Okamura, 2015). Moreover, as noted earlier, unlike in rats (Noritake et al., 2011), NK3R antagonists alone are able to inhibit episodic LH secretion in OVX ewes (Fraser et al., 2015; Goodman et al., 2013; Li et al., 2015). Although there is relatively little information on the possibility of redundancy in tachykinin signaling in humans, what is available suggests minimal duplication since a specific NK3R antagonist inhibits LH secretion in women (Fraser, Ramael, Hoveyda, Gheyle, & Combalbert, 2016) and infertility occurs with *TAC3* or *TAC3R* mutations (Topaloglu et al., 2009).

4.3 Role in Steroid Negative Feedback

Whether inhibition of tachykinins contributes to steroid negative feedback remains an open question. Studies of NKB and SP expression are consistent with this hypothesis. Expression of NKB in the ARC increases in postmenopausal women (Rance & Young, 1991) and following OVX in rats (Burke et al., 2006), sheep (Nestor et al., 2012), and monkeys (Sandoval-Guzman et al., 2004), and expression of this tachykinin is inhibited by E_2 in every species tested to date (Abel, Voytko, & Rance, 1999; Navarro, Castellano, et al., 2011; Navarro et al., 2009; Rance & Bruce, 1994; Weems, Smith, et al., 2017). SP expression also increases in postmenopausal women (Rance & Young, 1991) and after castration of monkeys (Kalil et al., 2015), and it is inhibited by E_2 in OVX mice (Navarro et al., 2015) and rats (Ruiz-Pino et al., 2015). However, it is impossible to draw any functional interpretations from these correlational studies because the parallel changes in kisspeptin expression seen under these different endocrine conditions may be more important. Similar issues confound interpretation of studies on Fos expression (Merkley et al., 2012) and genetic deletion of ERα in KNDy neurons (Dubois et al., 2015) discussed in Section 3.2, although inhibition of LH in the absence of ERα in KNDy cells might reflect an inhibition of ARC NKB neurons that do not contain kisspeptin (Burke et al., 2006). At this time, the only evidence implicating NKB in steroid negative feedback are reports that castration increases (Ruka et al., 2013, 2016) and steroids (E_2 or dihydrotestosterone) inhibit (Ruka et al., 2016) the stimulatory effects of senktide on the electrical activity of KNDy neurons. The changes in response to this NK3R agonist could reflect alterations in expression of this receptor because castration increased *Tacr3* expression in KNDy neurons, based on single-cell RT-PCR (Ruka et al., 2013). In contrast, testosterone treatment of castrated mice did not inhibit expression of NK3R in KNDy neurons when examined using ISH (Navarro, Gottsch, et al., 2011). We are not aware of any other studies that examined the effects of steroids on NK3R expression, but we observed no effect of E_2 on NK1R expression in sheep (Fergani et al., 2016). Thus, evaluation of the hypothesis that NKB mediates steroid negative feedback awaits studies in which kisspeptin or NKB expression in KNDy neurons is selectively inhibited.

Finally, it is interesting to note, parenthetically, that there is accumulating evidence that NKB participates in E_2 positive feedback in sheep, but that this action occurs in the retrochiasmatic area (RCh) rather than the ARC. Specifically, local administration of senktide to the RCh produced a surge-like (i.e., very high nonepisodic LH concentrations) increase in LH (Billings et al., 2010; Porter et al., 2014), while an NK3R antagonist acted in the RCh to blunt the estrogen-induced LH surge (Porter et al., 2014). The NKB-responsive neurons in the RCh act via kisspeptin release from KNDy neurons (Grachev et al., 2016), but the source of NKB input and the phenotype of these NK3R-containing cells remain to be determined. In summary, there is now compelling evidence that NKB acts on KNDy neurons to initiate each GnRH pulse, with considerable redundancy among tachykinins in rodents, but the role of NKB in steroid negative feedback remains to be determined.

5. EOP AND EPISODIC LH SECRETION

A potential role for EOP in control of episodic LH secretion was first proposed in 1980 based on the ability of an EOP receptor antagonist, naloxone, to increase LH pulses during the luteal phase of the menstrual cycle (Quigley & Yen, 1980). Numerous studies since then have explored the role of EOP in the negative feedback actions of both progesterone and E_2, but in terms of GnRH pulse generation the EOP, dynorphin, remains the least studied KNDy peptide. There are three families of EOP, each encoded by a different gene: the endorphins (primarily β-endorphin), the enkephalins (met-enkephin and leu-enkephalin), and the dynorphins (dynorphin A and dynorphin B). There are also three major EOP receptors: δ, μ, and κ (KOR) (Raynor et al., 1994). Dynorphin acts primarily via KOR (Table 2), while β-endorphin and the enkephalins have similar affinities for δ and μ receptors, and lower affinities for KOR. Although there is some evidence that each of these peptides can affect gonadotropin secretion, the vast majority of studies on EOP have focused on β-endorphin and dynorphin, so we will not consider the enkephalins further in this chapter.

Table 2 Selectivity of Endogenous Opioid Peptides at Opioid Receptors (nM)

	β-Endorphin	**Met-enkephelin**	**Leu-enkephalin**	**Dynorphin**
Delta	1.0	1.7	4.0	1000
Kappa	52	1000	1000	0.5
Mu	1.0	0.7	3.4	32

Binding potencies (K_i) of ligands for cloned EOP receptors (Raynor et al., 1994).

5.1 Hypothalamic Anatomy of β-Endorphin and Dynorphin

The precursor for β-endorphin, proopiomelanocortin (POMC), is found in both the hypothalamus and pituitary and can give rise to several important biologically active peptides including β-endorphin, α-melanocyte-stimulating hormone (α-MSH), and adrenocorticotropin. Thus, ISH for the mRNA-encoding POMC or IHC for β-endorphin or αMSH have all been used to identify neurons synthesizing these peptides, which are largely limited to the ARC in a number of species, including rats (Finley, Lindstrom, & Petrusz, 1981), sheep (Whisnant, Curto, & Goodman, 1992), and primates (Khachaturian, Lewis, Haber, Akil, & Watson, 1984). The immediate precursor for dynorphin, prodynorphin, has several possible products including Dyn A 1–17, Dyn A 1–13, Dyn A 1–8, and Dyn B, and ISH for preprodynorphin (PPD) and IHC for prodynorphin, dynorphin A, or dynorphin B have been used in neuroanatomical studies in rats (Burke et al., 2006; Fallon & Leslie, 1986; Watson et al., 1983), sheep (Foradori, Goodman, & Lehman, 2005), primates (Lewis, Khachaturian, & Watson, 1983), and humans (Abe et al., 1988). All these approaches demonstrated wide-spread distribution of dynorphin-producing cells throughout the POA and MBH, including magnocellular neurons in the supraoptic and paraventricular nuclei, and parvicellular neurons in the POA, periventricular region of the AHA, and ARC. In sheep (Foradori et al., 2005), PPD-containing neurons were observed in the VMN that did not contain the dynorphin A protein, indicating possible synthesis of other prodynorphin products in this nucleus. Both β-endorphin and dynorphin cell bodies send projections to several areas important for reproduction including the ovine ARC and POA (Dufourny, Caraty, Clarke, Robinson, & Skinner, 2005; Foradori et al., 2005). Synapses containing dynorphin were observed on 40% of POA, and 90% of MBH, GnRH neurons, while β-endorphin synapses were seen on approximately 70% of GnRH neurons in both areas (Goodman et al., 2004).

Although POMC and dynorphin neurons are both found in the ARC, these comprise separate neuronal populations as no colocalization of the two has been reported in any species. A number of studies have identified major differences in the colocalization of steroid receptors and other neurotransmitters in POMC and dynorphin neurons. Virtually, all ARC dynorphin neurons contain PR (Fig. 5A) (Foradori et al., 2002) and ERα (Lehman et al., 2013), and PR has also been found in the vast majority of ovine dynorphin neurons in the AHA and POA (Foradori et al., 2002) and dynorphin-containing neurons in the RP3V of rats (Simerly, Young, & Carr, 1996), but no magnocellular dynorphin neurons in sheep contained this receptor (Foradori et al., 2002). In contrast, far fewer POMC neurons contain PR and/or ERα, with percentages ranging from 20% to 33% in various species (Bethea & Widmann, 1996; Fox, Harlan, Shivers, & Pfaff, 1990; Lehman & Karsch, 1993; Olster & Blaustein, 1990). Not surprisingly, there

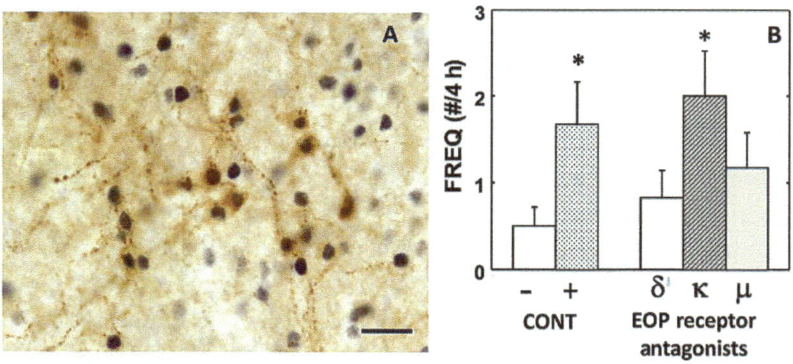

Fig. 5 Evidence that dynorphin from KNDy neurons mediates progesterone negative feedback in ewes. Panel A illustrates that virtually all dynorphin-ir neurons (*brown cytoplasm*) in the ARC contain PR (*black nucleus*). *Brown cytoplasm* indicates that staining reflects dynorphin, not cytoplasmic PR. Panel B depicts the effects of MBH microimplants containing antagonists to different EOP receptors on LH pulse frequency during the luteal phase of the ovine estrous cycle. Microimplants containing either naloxone (+ control) or norBNI (κ) significantly (*$P < 0.05$) increased pulse frequency compared to empty microimplants (− controls), but antagonists to either the δ EOP or the μ EOP receptor did not. *Figures reproduced from Foradori, C. D., Coolen, L. M., Fitzgerald, M. E., Skinner, D. C., Goodman, R. L., & Lehman, M. N. (2002). Colocalization of progesterone receptors in parvicellular dynorphin neurons of the ovine preoptic area and hypothalamus.* Endocrinology, 143*(11), 4366–4374; Goodman, R. L., Coolen, L. M., Anderson, G. M., Hardy, S. L., Valent, M., Connors, J. M., et al. (2004). Evidence that dynorphin plays a major role in mediating progesterone negative feedback on gonadotropin-releasing hormone neurons in sheep.* Endocrinology, 145*(6), 2959–2967, with permission from the Endocrine Society.*

are also major differences in the neuropeptides found in these two EOP populations. As noted previously, dynorphin is found in almost all KNDy neurons (using either kisspeptin or NKB as a marker) in all animals studied to date (Lehman et al., 2013). There is less information on whether there are dynorphin neurons in the ARC that do not contain either NKB or kisspeptin; in sheep essentially all dynorphin neurons contain NKB (Foradori et al., 2006) and kisspeptin (Goodman et al., 2007), while in rats 50%–60% of them (depending on location within the ARC) contain NKB (Burke et al., 2006). The degree of colocalization of dynorphin with kisspeptin and/or NKB in humans remains unclear; there is indirect evidence for this based on ISH data in postmenopausal women (Rometo & Rance, 2008), but low levels of colocalization were observed in men (Hrabovszky et al., 2012). Whether this reflects differences in endocrine status or species, or is caused by technical difficulties in analysis of peptide expression in postmortem human samples is hard to assess at this time. In addition to these neuropeptides, there is strong evidence that KNDy neurons are also glutamatergic in mice (Ciofi, Leroy, & Tramu, 2006; Cravo et al., 2011) and sheep (Merkley et al., 2015). The most common peptide also found in POMC neurons of rodents is CART (cocaine- and amphetamine-regulated transcript) (Elias et al., 1998), which like α-MSH, suppresses appetite (Elias et al., 1998) so that these neurons are thought to play a major role in the control of food intake. However, there appear to be major species differences in the colocalization of these peptides, because CART is not found in α-MSH-containing neurons in monkeys (Grayson et al., 2006) or humans (Menyhert et al., 2007), but is colocalized in neurons that contain orexigenic peptides (e.g., neuropeptide Y) (Menyhert et al., 2007). CART has also recently been colocalized with kisspeptin and NKB in human tissue (Skrapits et al., 2014, 2015). Using ISH, there is extensive colocalization of CART with POMC neurons of the sheep (Adam, Archer, Findlay, Thomas, & Marie, 2002), but whether it is also found in KNDy neurons has yet to be determined.

5.2 Role in Steroid Negative Feedback

Although best known for their function in nociception, EOPs also inhibit gonadotropin release. In humans, morphine and other opiates decrease serum LH (Delitala, Grossman, & Besser, 1983), and chronic opiate addiction has long been recognized to cause infertility in females (Morley, 1981). Furthermore, it is now generally accepted that the inhibition of GnRH/LH pulse frequency by progesterone is mediated by an inhibitory EOP (Lehman et al., 2010). Blockade of EOP action with naloxone, or similar EOP antagonists, increases LH pulse frequency during the luteal phase in several species (Brooks, Lamming, Lees, & Haynes, 1986; Ferin, Van Vugt, & Wardlaw, 1984; Whisnant & Goodman, 1988), including humans (Ropert, Quigley, & Yen, 1981) and GnRH secretion in sheep (Horton, Cummins, & Clarke, 1987). Moreover, this stimulatory action is also observed in progesterone-treated OVX, but not OVX, animals (Goodman, Gibson, Skinner, & Lehman, 2002; Whisnant, Havern, & Goodman, 1991) and women (Shoupe, Montz, & Lobo, 1985).

Although naloxone has similar affinities for KOR and μ opioid receptors (κ: 2.3 nM; μ: 0.9 nM; δ: 17 nM Raynor et al., 1994), most early work focused on the possible role of β-endorphin. Observations supporting a role for this EOP included evidence that progesterone increased concentrations of β-endorphin in the hypophysial portal blood of primates (Wardlaw, Wehrenberg, Ferin, Antunes, & Frantz, 1982) and POMC mRNA levels in the hypothalamus of rats (Wilcox & Roberts, 1985; Wise, Scarbrough, Weiland, & Larson, 1990) and sheep (Whisnant et al., 1992). However, more recent studies in sheep have shown that norbinaltorphimine (nor-BNI), a specific KOR antagonist, increased LH pulse frequency in luteal phase ewes when implanted in the MBH, whereas antagonists to the other two EOP receptors did not (Fig. 5B) (Goodman et al., 2004). These observations, led to the hypothesis that ARC dynorphin neurons, almost all of which contain PR (Foradori et al., 2002), mediate the negative feedback actions of progesterone in sheep. Several lines of evidence now support this hypothesis: (1) levels of mRNA for PPD decrease in the ARC of OVX ewes (Foradori, Goodman, Adams, Valent, & Lehman, 2005); (2) progesterone increases concentrations of dynorphin in CSF collected from the third ventricle (Foradori, Goodman, Adams, et al., 2005); and (3) local administration of the PR antagonist, RU486, to the ARC blocks the negative feedback actions of progesterone (Goodman et al., 2011). There is also indirect evidence that progesterone stimulates dynorphin in humans since ARC levels of mRNA for PPD are lower after menopause (Rometo & Rance, 2008), but there are contradictory data in rodents. In mice, prepro-DYN mRNA in the ARC is inhibited by ovarian steroids (Navarro et al., 2009), and knockout of KOR decreased LH secretion in OVX mice (Gottsch et al., 2009). Conversely, central administration of nor-BNI increased pulse frequency and mean LH levels in pregnant rats in which LH pulses are strongly suppressed by progesterone (Gallo, 1990). Thus, although Dyn appears to mediate progesterone negative feedback in ewes and primates, its role in rodents is less clear, possibly because of the abbreviated luteal phase in cycling rats and mice (Lehman et al., 2010).

In comparison to their role in progesterone negative feedback, the role of EOP in general, and dynorphin specifically, in estrogen negative feedback is unclear. As was the case for progesterone, early studies using naloxone, or similar EOP antagonists, reported increased LH secretion during the follicular phase of the ovine estrous cycle (Goodman et al., 2002) and human menstrual cycle (Quigley & Yen, 1980) and when estrogen is given to OVX rats (Van Vugt, Sylvester, Aylsworth, & Meites, 1982), sheep (Goodman et al., 2002; Whisnant & Goodman, 1988), and women (Shoupe et al., 1985). Release of β-endorphin in the ME is higher during the ovine follicular phase (Conover, Kuljis, Rabii, & Advis, 1993) and lower in OVX, than in ovary-intact, rats (Sarkar & Yen, 1985), but there appear to be species differences in the control of POMC mRNA in the ARC. Based on ISH, POMC expression is lower in OVX, than in E_2-treated OVX, ewes (Whisnant et al., 1992) and after menopause (Abel & Rance, 1999), but E_2 inhibits POMC expression in rats (Wilcox & Roberts, 1985). Perhaps more importantly, in rats the effects of naloxone on GnRH release were not diminished by OVX (Karahalios & Levine, 1988), and in sheep administration of naloxone significantly increased GnRH pulse amplitude in OVX ewes, regardless of estradiol replacement (Goodman, Parfitt, Evans, Dahl, & Karsch, 1995). The results of work focused specifically on the possible role of dynorphin in mediating E_2 negative feedback have been conflicting. In sheep, E_2 increased the number of dynorphin-containing neurons in the ARC (Weems, Smith, et al., 2017), but in mice dynorphin expression in the ARC is inhibited by estrogen (Gottsch et al., 2009; Navarro et al., 2009). A recent study in OVX rats found that central nor-BNI administration significantly increased LH release only in the presence of estrogen (Mostari et al., 2013), providing functional support for the hypothesis that dynorphin–KOR signaling is involved in the estrogen negative feedback. Moreover, E_2 increases the inhibitory effects of dynorphin on the electrical activity of KNDy neurons (Ruka et al., 2016). Nevertheless, it is unlikely that ARC dynorphin plays a critical role in estrogen negative feedback in mice because the effects of this steroid on PPD expression occur via the classical estrogen response element (Gottsch et al., 2009), which is not required for inhibition of LH secretion by estrogen in this species (Glidewell-Kenney et al., 2007). Thus, although some reports support a role for dynorphin in mediating estrogen negative feedback, the majority of studies are not consistent with this hypothesis.

It is interesting to note that recent data raise the possibility that dynorphin release from KNDy neurons inhibits GnRH secretion during the preovulatory LH surge in rats. Thus, lesioning of these neurons with an NK3R-saporin conjugate increased the amplitude of the steroid-induced LH surge (Helena et al., 2015; Mittelman-Smith, Krajewski-Hall, McMullen, & Rance, 2016), and this effect could be reversed with microinjections of dynorphin into the RP3V (Helena et al., 2015). These data are consistent with earlier reports that PPD mRNA is reduced in the medial POA during the surge on proestrus (Zhang, McCoy, & Gallo, 2002), that local administration of dynorphin to this area blocks the LH surge (Zhang & Gallo, 2002), and that local blockade of KOR advances the surge (Smith & Gallo, 1997). In contrast, an agonist for μ opioid receptors delayed the LH surge in sheep, while a KOR agonist did not (Walsh & Clarke, 1996). However, since opioid antagonists do not advance the timing of the LH surge in the ewe (Currie, Joseph, & Rawlings, 1991; Horton & Clarke, 1988), the physiological relevance of these observations is unclear.

5.3 Role in Pulse Generation

Previous studies have also provided strong evidence for the proposed role of dynorphin on the GnRH pulse generator, with consistent effects of KOR agonists observed across species. Specifically, icv administration of dynorphin blocks volleys of MUA and LH pulses in OVX goats (Wakabayashi et al., 2010) and direct intra-ARC administration of a dynorphin agonist decreased LH pulse frequency in estrogen-replaced OVX rats (Grachev, Li, Kinsey-Jones, et al., 2012). In vitro, dynorphin inhibits spontaneous electrical activity in KNDy neurons (de Croft et al., 2013; Ruka et al., 2013, 2016) and a KOR agonist blocked the prolonged volley of action potentials in these neurons induced by brief (10 s), high frequency (20 Hz), optogenetic stimulation (Qiu et al., 2016). Moreover, dynorphin release from ARC neurons mediates the inhibitory effects of NK3R agonists seen when LH secretion is elevated because this inhibition can be blocked with the KOR antagonist, nor-BNI (Grachev, Li, Kinsey-Jones, et al., 2012; Kinsey-Jones et al., 2012). The mechanisms underlying the increase in dynorphin effectiveness in OVX rats remain a mystery, although it could reflect dynorphin release from non-KNDy ARC neurons that contain NK3R (Burke et al., 2006). In contrast to the consistent effects of KOR agonists, some differences between ruminants and rodents were observed with nor-BNI. When given icv in OVX goats (Wakabayashi et al., 2010) or into the ARC of OVX ewes (Goodman et al., 2013), nor-BNI increased LH pulse frequency (in both) and the occurrence of MUA in goats. Moreover, the ability of naloxone to prolong the plateau and falling phases (pulse termination) of GnRH release with no effect on the rising phase (pulse onset) in OVX sheep indicates that dynorphin may play an important role in terminating each pulse (Goodman et al., 1995). Similarly, the ability of this antagonist to increase GnRH secretion between pulses indicates that dynorphin release may ensure that episodic GnRH secretion remains tightly coupled with little or no secretion between pulses (Goodman et al., 1995). However, several lines of evidence in rodents are not consistent with the proposed role of dynorphin in pulse termination. Specifically, nor-BNI had no effect on LH pulse frequency when given either into the ARC (Grachev, Li, Kinsey-Jones, et al., 2012) or into the third ventricle (Mostari et al., 2013). Similarly, this KOR antagonist did not affect spontaneous single cell electrical activity in mice (de Croft et al., 2013; Ruka et al., 2013), but did block the inhibitory actions of exogenous dynorphin (Ruka et al., 2013). On the other hand, chronic peripheral infusion of nor-BNI advanced pubertal development in female rats (Mostari et al., 2013), presumably due to an increase in episodic GnRH secretion, and nor-BNI prolonged the volley of action potentials induced by optogenetic stimulation in vitro (Qiu et al., 2016). Thus, there is strong evidence in ruminants for the proposed role of dynorphin in GnRH pulse termination, but the role of endogenous release of this EOP in episodic GnRH secretion in rodents remains to be resolved.

Two important aspects of dynorphin action in GnRH pulse generation are just beginning to be addressed: (1) where dynorphin acts to inhibit GnRH secretion and (2) what accounts for the lag time between pulse initiation by NKB and pulse termination by dynorphin. With regard to the first question, previous rodent studies found very low levels of KOR in KNDy cells in mice and no detectable mRNA for any of the three EOP receptors in GnRH neurons of the rat (Mitchell et al., 1997; Sannella & Petersen, 1997). However, using IHC we recently observed a high degree of KOR colocalization in KNDy cells (Fig. 6), with greater than 90% of these neurons containing this EOP receptor (Weems, Witty, et al., 2016); in contrast only 11% of the kisspeptin neurons in the POA expressed KOR. Surprisingly, GnRH cells also showed high levels of colocalization in sheep,

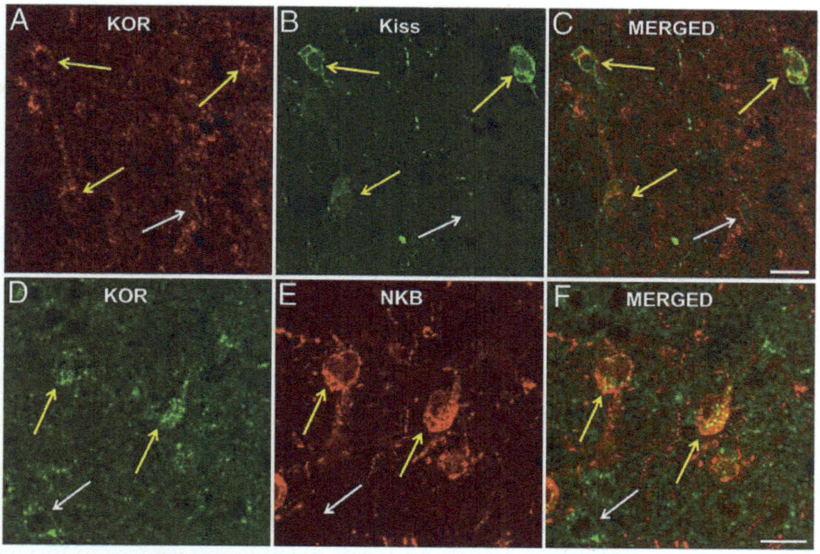

Fig. 6 Representative images showing KOR colocalization in KNDy cells. *Upper panels* (A–C) are confocal images of 1-μm-thick optical sections showing KOR-ir (*red*) and kisspeptin-ir (*green*) cells in the ARC. *Lower panels* are confocal (D–F) images of KOR-ir (*green*) and NKB-ir (*red*) in the ARC. Cells colocalizing KOR and either kisspeptin or NKB are denoted by *yellow arrows*. Single-labeled KOR cells are denoted by *white arrows*. Scale bars: 50 μm. *Kiss*, kisspeptin. *Reproduced from Weems, P. W., Coolen, L. M., Hileman, S. M., Hardy, S. L., McCosh, R. B., Goodman, R. L., et al. (2016). Kappa opioid receptors are internalized in arcuate KNDy cells during GnRH pulse termination in the ewe.* Annual Meeting of Society for Neuroscience, San Diego CA, *Abtsr 60.04; Weems, P. W., Witty, C. F., Amstalden, M., Coolen, L. M., Goodman, R. L., & Lehman, M. N. (2016). Kappa-opioid receptor is colocalized in GnRH and KNDy cells in the female ovine and rat brain.* Endocrinology, 157*(6), 2367–2379, with permission from the Endocrine Society.*

ranging from 74.4% to 95.4% for GnRH cells in the POA and MBH, respectively (Weems, Witty, et al., 2016). Because of previous rodent data suggesting that GnRH neurons did not contain mRNA for KOR, we also performed dual IHC for GnRH and KOR in the POA of rats and found that virtually all of them expressed the KOR protein. Differences between rodents and sheep could be due to sensitivity of the techniques, or to species differences in the relative importance of Dyn signaling for modulation of pulse frequency. Nonetheless, these findings thus allow one to infer that, at least in sheep, dynorphin terminates GnRH pulses by two possible pathways: (1) indirectly via the KNDy cell network and/or (2) directly at the level of GnRH cell bodies.

A key unexplained characteristic of the KNDy model for pulse generation is the mechanism responsible for the delay between pulse activation by NKB and pulse termination by dynorphin. Given the effect of naloxone on GnRH pulse shape in sheep (Goodman et al., 1995), inhibition of GnRH secretion by dynorphin likely begins relatively rapidly (within a few minutes of pulse onset). Therefore, the most likely possibilities meditating this time delay are (1) a lag between the release of NKB and dynorphin, (2) differences in receptor turnover and/or trafficking to the plasma membrane, or (3) intracellular signaling mechanisms in KNDy cells that are faster for NKB than dynorphin. A recent report that KNDy peptides are segregated in separate vesicles in rats (Murakawa, Iwata, Takeshita, & Ozawa, 2016) is consistent with independent release mechanisms (Vaaga, Borisovska, & Westbrook, 2014) that would allow for a delay in release. However, it remains unknown if NKB and dynorphin are coreleased or if their release is independently modulated.

There are no published data that address possible differences in intracellular signaling, but evidence from other systems suggests receptor turnover and trafficking may play a key role in this delay. KOR is a G_i protein-coupled receptor, with KOR activation causing increased potassium channel conductance and decreased calcium conductance via the G-protein β-γ subunit (Dhawan et al., 1996). These ion channel changes hyperpolarize KOR-containing neurons, resulting in decreased cell firing and neurotransmitter release (Bruchas & Chavkin, 2010). Like most GPCRs, KORs are desensitized by receptor phosphorylation and endocytosed after repeated or sustained exposure to agonists (Chavkin, James, & Goldstein, 1982). Thus, internalization of KOR could be an important step in activity-dependent resetting of the KNDy network. Alterations in KOR trafficking have been shown in magnocellular neurons-containing arginine vasopressin (AVP). Dynorphin is colocalized with AVP in secretory vesicles (Shuster, Riedl, Li, Vulchanova, & Elde, 2000; Whitnall, Gainer, Cox, & Molineaux, 1983), and KOR is expressed within the membrane of these vesicles (Shuster, Riedl, Li, Vulchanova, & Elde, 1999). Thus when AVP neurons fire in a phasic pattern, AVP and dynorphin are released during the burst of high frequency action potentials and KOR is also inserted into the plasma membrane (Shuster et al., 1999), allowing dynorphin to inhibit the presynaptic neuron and terminate the burst of action potentials (Brown, 2004; Brown, Leng, Ludwig, & Bourque, 2006). We have begun to determine whether KOR is internalized, or inserted into the KNDy cell membrane, during a pulse by examining the intracellular location of KOR

(Fig. 7) at times that approximate pulse onset and termination (4 min and 12 min, respectively, after injection of NKB into the third ventricle to induce a LH pulse). Preliminary data indicate that KOR begins to be internalized 4 min after NKB injection and this internalization increases at 12 min after injection (Weems, Coolen, et al., 2016). Since GPCR internalization can be used as an index of endogenous ligand release, these data indicate that dynorphin release begins shortly after the start of a pulse and continues for at least 8 min. Interestingly, KOR internalization was seen in MBH GnRH neurons only at the 12 min time point (Weems, Coolen, et al., 2017), so that dynorphin release on to these neurons only occurs near the end of a pulse. If supported by additional data, these observations indicate that there is little or

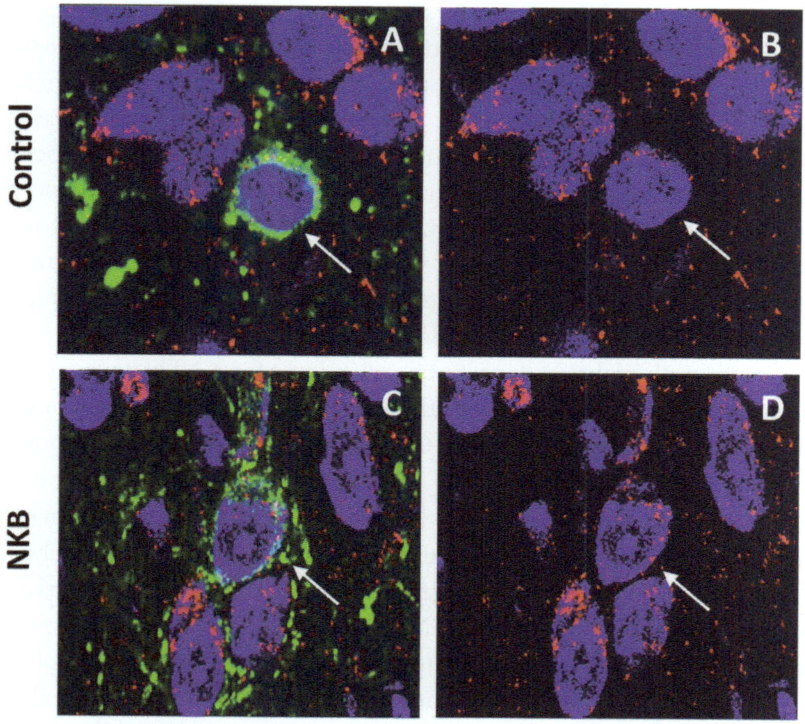

Fig. 7 Internalization of KOR during a NKB-induced LH pulse. Confocal images of ARC sections from ewes that received either control (A and B) or NKB (C and D) injections into the third ventricle 12 min before tissue collection. Sections have been stained for dynorphin (*green*), KOR (*red*), and with Fluoro-Nissl (*blue*), which labels all neurons in the section. *Left panels* (A and C) are overlays of all three channels, while *right panels* (B and D) present only KOR and Fluoro-Nissl. *Arrows* indicate a KNDy neuron with KOR internalization after NKB (A and B) and absence of internalization in a KNDy neuron after saline (control) injection.

no delay between NKB and dynorphin release at the start of a pulse, and that GnRH pulse termination likely occurs in response to dynorphin's inhibitory actions on both KNDy and GnRH neurons.

6. CONCLUSION

In summary, there is now compelling evidence for the KNDy model of GnRH pulse generation in ruminants and, based on our preliminary data, we propose that shortly following NKB stimulation of KNDy cells, dynorphin acts via KOR to restrain kisspeptin, and thus GnRH secretion, during a pulse and that this EOP then acts indirectly within the KNDy network and directly at the level of the GnRH neuron to terminating each pulse. The support for this hypothesis in rodents remains somewhat controversial. Recent optogenetic data provide strong support that KNDy neurons play a key role in GnRH pulse generation, but the underlying mechanisms may differ from those in ruminants. It is clear that other tachykinins can substitute for NKB in rats and mice, and the effect of *Tac1* and *Tac2* knockouts suggests an important role for at least two of them (NKB and SP), and possibly NKA as well. The effects of KOR agonists demonstrate that dynorphin can inhibit KNDy cell activity, but whether endogenous dynorphin release, or some other neurotransmitter, normally plays a role in GnRH pulse termination in rodents remains an open question.

The strength of the evidence supporting a role for KNDy neurons in steroid negative feedback varies with the steroid involved. There is strong evidence that progesterone inhibits GnRH pulse frequency by stimulating release of dynorphin from KNDy neurons in sheep and probably during pregnancy in rats. The evidence in other species, including humans, is largely consistent with this hypothesis, but is more circumstantial. The role of KNDy neurons in estrogen negative feedback is more controversial and the relative contributions of each KNDy peptide to any inhibitory actions of this steroid are largely speculative. It seems unlikely that dynorphin plays an important role based on GnRH measurements in ewes and the inhibitory effects of E_2 on dynorphin expression in rodents. Since E_2 inhibits expression of both kisspeptin and NKB in KNDy neurons, either one (or both) could play a role in the negative feedback actions of this steroid. However, strong tests of their physiological roles are difficult, in part, because experiments using antagonists or lesions cannot distinguish between permissive and stimulatory actions of these neurotransmitters. Consequently, most data supporting a role for either kisspeptin or NKB in estrogen negative feedback remain correlational in nature. All the data in sheep are consistent with a role for either kisspeptin or NKB, but the data in rodents are conflicting, particularly since knock out of ERα in kisspeptin neurons did not affect the negative feedback actions of E_2. Despite these unresolved questions, the field of reproductive neuroendocrinology has made great strides in the last 15 years in unraveling the neural mechanisms controlling episodic GnRH secretion. The focus on KNDy neurons as a key nodal point in the generation and regulation of GnRH/LH pulses will likely expand in the future to include other sets of interconnected neurons as important participants in this functional circuitry.

REFERENCES

Abe, J., Okamura, H., Kitamura, T., Ibata, Y., Minamino, N., Matsuo, H., et al. (1988). Immunocytochemical demonstration of dynorphin(PH-8P)-like immunoreactive elements in the human hypothalamus. *The Journal of Comparative Neurology*, *276*(4), 508–513.

Abel, T. W., & Rance, N. E. (1999). Proopiomelanocortin gene expression is decreased in the infundibular nucleus of postmenopausal women. *Brain Research. Molecular Brain Research*, *69*(2), 202–208.

Abel, T. W., Voytko, M. L., & Rance, N. E. (1999). The effects of hormone replacement therapy on hypothalamic neuropeptide gene expression in a primate model of menopause. *The Journal of Clinical Endocrinology and Metabolism*, *84*(6), 2111–2118.

Adachi, S., Yamada, S., Takatsu, Y., Matsui, H., Kinoshita, M., Takase, K., et al. (2007). Involvement of anteroventral periventricular metastin/kisspeptin neurons in estrogen positive feedback action on luteinizing hormone release in female rats. *The Journal of Reproduction and Development*, *53*(2), 367–378.

Adam, C. L., Archer, Z. A., Findlay, P. A., Thomas, L., & Marie, M. (2002). Hypothalamic gene expression in sheep for cocaine- and amphetamine-regulated transcript, proopiomelanocortin, neuropeptide Y, agouti-related peptide and leptin receptor and responses to negative energy balance. *Neuroendocrinology*, *75*(4), 250–256.

Alcin, E., Sahu, A., Ramaswamy, S., Hutz, E. D., Keen, K. L., Terasawa, E., et al. (2013). Ovarian regulation of kisspeptin neurones in the arcuate nucleus of the rhesus monkey (macaca mulatta). *Journal of Neuroendocrinology*, *25*(5), 488–496.

Almeida, T. A., Rojo, J., Nieto, P. M., Pinto, F. M., Hernandez, M., Martin, J. D., et al. (2004). Tachykinins and tachykinin receptors: Structure and activity relationships. *Current Medicinal Chemistry*, *11*(15), 2045–2081.

Amstalden, M., Coolen, L. M., Hemmerle, A. M., Billings, H. J., Connors, J. M., Goodman, R. L., et al. (2010). Neurokinin 3 receptor immunoreactivity in the septal region, preoptic area and hypothalamus of the female sheep: Colocalisation in neurokinin B cells of the arcuate nucleus but not in gonadotrophin-releasing hormone neurones. *Journal of Neuroendocrinology*, *22*(1), 1–12.

Belchetz, P. E., Plant, T. M., Nakai, Y., Keogh, E. J., & Knobil, E. (1978). Hypophysial responses to continuous and intermittent delivery of hypopthalamic gonadotropin-releasing hormone. *Science*, *202*(4368), 631–633.

Bethea, C. L., & Widmann, A. A. (1996). Immunohistochemical detection of progestin receptors in hypothalamic beta-endorphin and substance P neurons of steroid-treated monkeys. *Neuroendocrinology*, *63*(2), 132–141.

Billings, H. J., Connors, J. M., Altman, S. N., Hileman, S. M., Holaskova, I., Lehman, M. N., et al. (2010). Neurokinin B acts via the neurokinin-3 receptor in the retrochiasmatic area to stimulate luteinizing hormone secretion in sheep. *Endocrinology*, *151*(8), 3836–3846.

Blake, C. A. (1977). A medial basal hypothalamic site of synergistic action of estrogen and progesterone on the inhibition of pituitary luteinizing hormone release. *Endocrinology*, *101*(4), 1130–1134.

Blake, C. A., Norman, R. L., & Sawyer, C. H. (1974). Localization of inhibitory actions of estrogen and nicotine on release of luteinizing hormone in rats. *Neuroendocrinology*, *16*(1), 22–35.

Blake, C. A., & Sawyer, C. H. (1974). Effects of hypothalamic deafferentation on the pulsatile rhythm in plasma concentrations of luteinizing hormone in ovariectomized rats. *Endocrinology*, *94*(3), 730–736.

Boukhliq, R., Goodman, R. L., Berriman, S. J., Adrian, B., & Lehman, M. N. (1999). A subset of gonadotropin-releasing hormone neurons in the ovine medial basal hypothalamus is activated during increased pulsatile luteinizing hormone secretion1. *Endocrinology*, *140*(12), 5929–5936.

Brooks, A. N., Lamming, G. E., Lees, P. D., & Haynes, N. B. (1986). Opioid modulation of LH secretion in the ewe. *Journal of Reproduction and Fertility*, *76*(2), 693–708.

Brown, C. H. (2004). Rhythmogenesis in vasopressin cells. *Journal of Neuroendocrinology*, *16*(9), 727–739.

Brown, C. H., Leng, G., Ludwig, M., & Bourque, C. W. (2006). Endogenous activation of supraoptic nucleus kappa-opioid receptors terminates spontaneous phasic bursts in rat magnocellular neurosecretory cells. *Journal of Neurophysiology*, *95*(5), 3235–3244.

Bruchas, M. R., & Chavkin, C. (2010). Kinase cascades and ligand-directed signaling at the kappa opioid receptor. *Psychopharmacology*, *210*(2), 137–147.

Burke, M. C., Letts, P. A., Krajewski, S. J., & Rance, N. E. (2006). Coexpression of dynorphin and neurokinin B immunoreactivity in the rat hypothalamus: Morphologic evidence of interrelated function within the arcuate nucleus. *The Journal of Comparative Neurology*, *498*(5), 712–726.

Caraty, A., Fabre-Nys, C., Delaleu, B., Locatelli, A., Bruneau, G., Karsch, F. J., et al. (1998). Evidence that the mediobasal hypothalamus is the primary site of action of estradiol in inducing the preovulatory gonadotropin releasing hormone surge in the ewe. *Endocrinology*, *139*(4), 1752–1760.

Chavkin, C., James, I. F., & Goldstein, A. (1982). Dynorphin is a specific endogenous ligand of the kappa opioid receptor. *Science*, *215*(4531), 413–415.

Chawla, M. K., Gutierrez, G. M., Young, W. S., 3rd, McMullen, N. T., & Rance, N. E. (1997). Localization of neurons expressing substance P and neurokinin B gene transcripts in the human hypothalamus and basal forebrain. *The Journal of Comparative Neurology*, *384*(3), 429–442.

Cholanian, M., Krajewski-Hall, S. J., Levine, R. B., McMullen, N. T., & Rance, N. E. (2014). Electrophysiology of arcuate neurokinin B neurons in female Tac2-EGFP transgenic mice. *Endocrinology*, *155*(7), 2555–2565.

Christian, C. A., & Moenter, S. M. (2010). The neurobiology of preovulatory and estradiol-induced gonadotropin-releasing hormone surges. *Endocrine Reviews*, *31*(4), 544–577.

Ciofi, P., Leroy, D., & Tramu, G. (2006). Sexual dimorphism in the organization of the rat hypothalamic infundibular area. *Neuroscience*, *141*(4), 1731–1745.

Clarkson, J., Han, S. Y., Piet, R., Porteous, R., McLennan, T., Colledge, W. H., et al. (2017). In *The role of arcuate kisspeptin neurons in the generation of luteinizing hormone pulses ENDO2017 abstracts*. OR16-11.

Conover, C. D., Kuljis, R. O., Rabii, J., & Advis, J. P. (1993). Beta-endorphin regulation of luteinizing hormone-releasing hormone release at the median eminence in ewes: Immunocytochemical and physiological evidence. *Neuroendocrinology, 57*(6), 1182–1195.

Cravo, R. M., Margatho, L. O., Osborne-Lawrence, S., Donato, J., Jr., Atkin, S., Bookout, A. L., et al. (2011). Characterization of Kiss1 neurons using transgenic mouse models. *Neuroscience, 173*, 37–56.

Currie, W. D., Joseph, I. B., & Rawlings, N. C. (1991). Morphine, naloxone and the gonadotrophin surge in ewes. *Journal of Reproduction and Fertility, 92*(2), 407–414.

De Bond, J. A., Li, Q., Millar, R. P., Clarke, I. J., & Smith, J. T. (2013). Kisspeptin signaling is required for the luteinizing hormone response in anestrous ewes following the introduction of males. *PLoS One, 8*(2), e57972.

de Croft, S., Boehm, U., & Herbison, A. E. (2013). Neurokinin B activates arcuate kisspeptin neurons through multiple tachykinin receptors in the male mouse. *Endocrinology, 154*(8), 2750–2760.

de Croft, S., Piet, R., Mayer, C., Mai, O., Boehm, U., & Herbison, A. E. (2012). Spontaneous kisspeptin neuron firing in the adult mouse reveals marked sex and brain region differences but no support for a direct role in negative feedback. *Endocrinology, 153*(11), 5384–5393.

de Roux, N., Genin, E., Carel, J. C., Matsuda, F., Chaussain, J. L., & Milgrom, E. (2003). Hypogonadotropic hypogonadism due to loss of function of the KiSS1-derived peptide receptor GPR54. *Proceedings of the National Academy of Sciences of the United States of America, 100*(19), 10972–10976.

Delitala, G., Grossman, A., & Besser, M. (1983). Differential effects of opiate peptides and alkaloids on anterior pituitary hormone secretion. *Neuroendocrinology, 37*(4), 275–279.

Dhawan, B. N., Cesselin, F., Raghubir, R., Reisine, T., Bradley, P. B., Portoghese, P. S., et al. (1996). International Union of Pharmacology. XII. Classification of opioid receptors. *Pharmacological Reviews, 48*(4), 567–592.

Dierschke, D. J., Bhattacharya, A. N., Atkinson, L. E., & Knobil, E. (1970). Circhoral oscillations of plasma LH levels in the ovariectomized rhesus monkey. *Endocrinology, 87*(5), 850–853.

Dubois, S. L., Acosta-Martinez, M., DeJoseph, M. R., Wolfe, A., Radovick, S., Boehm, U., et al. (2015). Positive, but not negative feedback actions of estradiol in adult female mice require estrogen receptor alpha in kisspeptin neurons. *Endocrinology, 156*(3), 1111–1120.

Dufourny, L., Caraty, A., Clarke, I. J., Robinson, J. E., & Skinner, D. C. (2005). Progesterone-receptive beta-endorphin and dynorphin B neurons in the arcuate nucleus project to regions of high gonadotropin-releasing hormone neuron density in the ovine preoptic area. *Neuroendocrinology, 81*(3), 139–149.

Dungan, H. M., Clifton, D. K., & Steiner, R. A. (2006). Minireview: Kisspeptin neurons as central processors in the regulation of gonadotropin-releasing hormone secretion. *Endocrinology, 147*(3), 1154–1158.

Eghlidi, D. H., Haley, G. E., Noriega, N. C., Kohama, S. G., & Urbanski, H. F. (2010). Influence of age and 17beta-estradiol on kisspeptin, neurokinin B, and prodynorphin gene expression in the arcuate-median eminence of female rhesus macaques. *Endocrinology, 151*(8), 3783–3794.

Elias, C. F., Lee, C., Kelly, J., Aschkenasi, C., Ahima, R. S., Couceyro, P. R., et al. (1998). Leptin activates hypothalamic CART neurons projecting to the spinal cord. *Neuron, 21*(6), 1375–1385.

Evans, N. P., Dahl, G. E., Mauger, D., & Karsch, F. J. (1995). Estradiol induces both qualitative and quantitative changes in the pattern of gonadotropin-releasing hormone secretion during the presurge period in the ewe. *Endocrinology, 136*(4), 1603–1609.

Everett, J. W., Sawyer, C. H., & Markee, J. E. (1949). A neurogenic timing factor in control of the ovulatory discharge of luteinizing hormone in the cyclic rat. *Endocrinology*, *44*(3), 234–250.

Fallon, J. H., & Leslie, F. M. (1986). Distribution of dynorphin and enkephalin peptides in the rat brain. *The Journal of Comparative Neurology*, *249*(3), 293–336.

Fergani, C., Mazzella, L., Coolen, L. M., McCosh, R. B., Hardy, S. L., Newcomb, N., et al. (2016). Do substance P and Neurokinin a play important roles in the control of LH secretion in ewes? *Endocrinology*, *157*(12), 4829–4841.

Fergani, C., & Navarro, V. (2017). Expanding the role of Tachykinins in the neuroendocrine control of reproduction. *Reproduction*, *153*, R1–R14.

Ferin, M., Van Vugt, D., & Wardlaw, S. (1984). The hypothalamic control of the menstrual cycle and the role of endogenous opioid peptides. *Recent Progress in Hormone Research*, *40*, 441–485.

Finley, J. C., Lindstrom, P., & Petrusz, P. (1981). Immunocytochemical localization of beta-endorphin-containing neurons in the rat brain. *Neuroendocrinology*, *33*(1), 28–42.

Foradori, C. D., Amstalden, M., Goodman, R. L., & Lehman, M. N. (2006). Colocalisation of dynorphin a and neurokinin B immunoreactivity in the arcuate nucleus and median eminence of the sheep. *Journal of Neuroendocrinology*, *18*(7), 534–541.

Foradori, C. D., Coolen, L. M., Fitzgerald, M. E., Skinner, D. C., Goodman, R. L., & Lehman, M. N. (2002). Colocalization of progesterone receptors in parvicellular dynorphin neurons of the ovine preoptic area and hypothalamus. *Endocrinology*, *143*(11), 4366–4374.

Foradori, C. D., Goodman, R. L., Adams, V. L., Valent, M., & Lehman, M. N. (2005). Progesterone increases dynorphin a concentrations in cerebrospinal fluid and preprodynorphin messenger ribonucleic acid levels in a subset of dynorphin neurons in the sheep. *Endocrinology*, *146*(4), 1835–1842.

Foradori, C. D., Goodman, R. L., & Lehman, M. N. (2005). Distribution of preprodynorphin mRNA and dynorphin-a immunoreactivity in the sheep preoptic area and hypothalamus. *Neuroscience*, *130*(2), 409–418.

Fox, S. R., Harlan, R. E., Shivers, B. D., & Pfaff, D. W. (1990). Chemical characterization of neuroendocrine targets for progesterone in the female rat brain and pituitary. *Neuroendocrinology*, *51*(3), 276–283.

Fraser, G. L., Hoveyda, H. R., Clarke, I. J., Ramaswamy, S., Plant, T. M., Rose, C., et al. (2015). The NK3 receptor antagonist ESN364 interrupts pulsatile LH secretion and moderates levels of ovarian hormones throughout the menstrual cycle. *Endocrinology*, *156*(11), 4214–4225.

Fraser, G. L., Ramael, S., Hoveyda, H. R., Gheyle, L., & Combalbert, J. (2016). The NK3 receptor antagonist ESN364 suppresses sex hormones in men and women. *The Journal of Clinical Endocrinology and Metabolism*, *101*, 417–426.

Gallo, R. V. (1990). Kappa-opioid receptor involvement in the regulation of pulsatile luteinizing hormone release during early pregnancy in the rat. *Journal of Neuroendocrinology*, *2*(5), 685–691.

Garcia-Galiano, D., van Ingen Schenau, D., Leon, S., Krajnc-Franken, M. A., Manfredi-Lozano, M., Romero-Ruiz, A., et al. (2012). Kisspeptin signaling is indispensable for neurokinin B, but not glutamate, stimulation of gonadotropin secretion in mice. *Endocrinology*, *153*(1), 316–328.

Gaskins, G. T., Glanowska, K. M., & Moenter, S. M. (2013). Activation of neurokinin 3 receptors stimulates GnRH release in a location-dependent but kisspeptin-independent manner in adult mice. *Endocrinology*, *154*(11), 3984–3989.

Gianetti, E., Tusset, C., Noel, S. D., Au, M. G., Dwyer, A. A., Hughes, V. A., et al. (2010). TAC3/TACR3 mutations reveal preferential activation of gonadotropin-releasing hormone release by neurokinin B in neonatal life followed by reversal in adulthood. *The Journal of Clinical Endocrinology and Metabolism*, *95*(6), 2857–2867.

Glidewell-Kenney, C., Hurley, L. A., Pfaff, L., Weiss, J., Levine, J. E., & Jameson, J. L. (2007). Nonclassical estrogen receptor alpha signaling mediates negative feedback in the female mouse reproductive axis. *Proceedings of the National Academy of Sciences of the United States of America*, *104*(19), 8173–8177.

Goodman, R. L. (2015). Neuroendocrine control of gonadotropin secretion: Comparative aspects. In T. M. Plant & A. J. Zeleznik (Eds.), *Knobil and Neill's physiology of reproduction: Vol. 2* (4th ed., pp. 1537–1574). Amsterdam: Elsevier.

Goodman, R. L., Coolen, L. M., Anderson, G. M., Hardy, S. L., Valent, M., Connors, J. M., et al. (2004). Evidence that dynorphin plays a major role in mediating progesterone negative feedback on gonadotropin-releasing hormone neurons in sheep. *Endocrinology*, *145*(6), 2959–2967.

Goodman, R. L., Coolen, L. M., & Lehman, M. N. (2014). A role for neurokinin B in pulsatile GnRH secretion in the ewe. *Neuroendocrinology*, *99*(1), 18–32.

Goodman, R. L., Gibson, M., Skinner, D. C., & Lehman, M. N. (2002). Neuroendocrine control of pulsatile GnRH secretion during the ovarian cycle: Evidence from the ewe. *Reproduction Supplement*, *59*, 41–56.

Goodman, R. L., Hileman, S. M., Nestor, C. C., Porter, K. L., Connors, J. M., Hardy, S. L., et al. (2013). Kisspeptin, neurokinin B, and dynorphin act in the arcuate nucleus to control activity of the GnRH pulse generator in ewes. *Endocrinology*, *154*(11), 4259–4269.

Goodman, R. L., Holaskova, I., Nestor, C. C., Connors, J. M., Billings, H. J., Valent, M., et al. (2011). Evidence that the arcuate nucleus is an important site of progesterone negative feedback in the ewe. *Endocrinology*, *152*(9), 3451–3460.

Goodman, R. L., & Inskeep, E. K. (2015). Control of the ovarian cycle of the sheep. In T. M. Plant & A. J. Zeleznik (Eds.), *Knobil and Neill's physiology of reproduction: Vol. 2* (4th ed., pp. 1259–1305). Amsterdam: Elsevier.

Goodman, R. L., & Karsch, F. J. (1980). Pulsatile secretion of luteinizing hormone: Differential suppression by ovarian steroids. *Endocrinology*, *107*(5), 1286–1290.

Goodman, R. L., & Lehman, M. N. (2012). Kisspeptin neurons from mice to men: Similarities and differences. *Endocrinology*, *153*(11), 5105–5118.

Goodman, R. L., Lehman, M. N., Smith, J. T., Coolen, L. M., de Oliveira, C. V., Jafarzadehshirazi, M. R., et al. (2007). Kisspeptin neurons in the arcuate nucleus of the ewe express both dynorphin A and neurokinin B. *Endocrinology*, *148*(12), 5752–5760.

Goodman, R. L., Parfitt, D. B., Evans, N. P., Dahl, G. E., & Karsch, F. J. (1995). Endogenous opioid peptides control the amplitude and shape of gonadotropin-releasing hormone pulses in the ewe. *Endocrinology*, *136*(6), 2412–2420.

Gottsch, M. L., Navarro, V. M., Zhao, Z., Glidewell-Kenney, C., Weiss, J., Jameson, J. L., et al. (2009). Regulation of Kiss1 and dynorphin gene expression in the murine brain by classical and nonclassical estrogen receptor pathways. *The Journal of Neuroscience*, *29*(29), 9390–9395.

Goubillon, M. L., Forsdike, R. A., Robinson, J. E., Ciofi, P., Caraty, A., & Herbison, A. E. (2000). Identification of neurokinin B-expressing neurons as an highly estrogen-receptive, sexually dimorphic cell group in the ovine arcuate nucleus. *Endocrinology*, *141*(11), 4218–4225.

Grachev, P., Li, X. F., Kinsey-Jones, J. S., di Domenico, A. L., Millar, R. P., Lightman, S. L., et al. (2012). Suppression of the GnRH pulse generator by neurokinin B involves a kappa-opioid receptor-dependent mechanism. *Endocrinology*, *153*(10), 4894–4904.

Grachev, P., Li, X. F., Lin, Y. S., Hu, M. H., Elsamani, L., Paterson, S. J., et al. (2012). GPR54-dependent stimulation of luteinizing hormone secretion by neurokinin B in prepubertal rats. *PLoS One*, 7(9), e44344.

Grachev, P., Porter, K. L., Coolen, L. M., McCosh, R. B., Connors, J. M., Hileman, S. M., et al. (2016). Surge-like Luteinising hormone secretion induced by Retrochiasmatic area NK3R activation is mediated primarily by arcuate Kisspeptin Neurones in the ewe. *Journal of Neuroendocrinology*, *28*(6), 1–12.

Grayson, B. E., Allen, S. E., Billes, S. K., Williams, S. M., Smith, M. S., & Grove, K. L. (2006). Prenatal development of hypothalamic neuropeptide systems in the nonhuman primate. *Neuroscience*, *143*(4), 975–986.

Halasz, B., & Gorski, R. A. (1967). Gonadotrophic hormone secretion in female rats after partial or total interruption of neural afferents to the medial basal hypothalamus. *Endocrinology*, *80*(4), 608–622.

Han, S. Y., McLennan, T., Czieselsky, K., & Herbison, A. E. (2015). Selective optogenetic activation of arcuate kisspeptin neurons generates pulsatile luteinizing hormone secretion. *Proceedings of the National Academy of Sciences of the United States of America*, *112*(42), 13109–13114.

Helena, C. V., Toporikova, N., Kalil, B., Stathopoulos, A. M., Pogrebna, V. V., Carolino, R. O., et al. (2015). KNDy neurons modulate the magnitude of the steroid-induced luteinizing hormone surges in ovariectomized rats. *Endocrinology*, *156*(11), 4200–4213.

Herbison, A. E. (1998). Multimodal influence of estrogen upon gonadotropin-releasing hormone neurons. *Endocrine Reviews*, *19*(3), 302–330.

Herbison, A. E. (2008). Estrogen positive feedback to gonadotropin-releasing hormone (GnRH) neurons in the rodent: The case for the rostral periventricular area of the third ventricle (RP3V). *Brain Research Reviews*, *57*(2), 277–287.

Herbison, A. E. (2015). Physiology of the adult gonadotropin-releasing hormone neuronal network. In T. M. Plant & A. J. Zeleznik (Eds.), *Knobil and Neill's physiology of reproduction*: *Vol. 1* (4th ed., pp. 388–467). Amsterdam: Elsivier.

Herde, M. K., Iremonger, K. J., Constantin, S., & Herbison, A. E. (2013). GnRH neurons elaborate a long-range projection with shared axonal and dendritic functions. *The Journal of Neuroscience*, *33*(31), 12689–12697.

Higo, S., Iijima, N., & Ozawa, H. (2017). Characterisation of Kiss1r (Gpr54)-expressing neurones in the arcuate nucleus of the female rat hypothalamus. *Journal of Neuroendocrinology*, *29*(2), 1–8.

Hoffman, G. E., & Gibbs, F. P. (1982). LHRH pathways in rat brain: 'Deafferentation' spares a sub-chiasmatic LHRH projection to the median eminence. *Neuroscience*, 7(8), 1979–1993.

Horton, R. J., & Clarke, I. J. (1988). Lack of an effect of morphine or naloxone on the oestrogen-induced LH surge in anoestrous ewes. *The Journal of Endocrinology*, *119*(1), 89–93.

Horton, R. J., Cummins, J. T., & Clarke, I. J. (1987). Naloxone evokes large-amplitude GnRH pulses in luteal-phase ewes. *Journal of Reproduction and Fertility*, *81*(1), 277–286.

Hrabovszky, E., Borsay, B. A., Racz, K., Herczeg, L., Ciofi, P., Bloom, S. R., et al. (2013). Substance P immunoreactivity exhibits frequent colocalization with kisspeptin and neurokinin B in the human infundibular region. *PLoS One*, *8*(8), e72369.

Hrabovszky, E., Ciofi, P., Vida, B., Horvath, M. C., Keller, E., Caraty, A., et al. (2010). The kisspeptin system of the human hypothalamus: Sexual dimorphism and relationship with gonadotropin-releasing hormone and neurokinin B neurons. *The European Journal of Neuroscience*, *31*(11), 1984–1998.

Hrabovszky, E., Sipos, M. T., Molnar, C. S., Ciofi, P., Borsay, B. A., Gergely, P., et al. (2012). Low degree of overlap between kisspeptin, neurokinin B, and dynorphin immunoreactivities in the infundibular nucleus of young male human subjects challenges the KNDy neuron concept. *Endocrinology*, *153*(10), 4978–4989.

Irwig, M. S., Fraley, G. S., Smith, J. T., Acohido, B. V., Popa, S. M., Cunningham, M. J., et al. (2004). Kisspeptin activation of gonadotropin releasing hormone neurons and regulation of KiSS-1 mRNA in the male rat. *Neuroendocrinology*, *80*(4), 264–272.

Jackson, G. L., Kuehl, D., McDowell, K., & Zaleski, A. (1978). Effect of hypothalamic deafferentation on secretion of luteinizing hormone in the ewe. *Biology of Reproduction*, *18*(5), 808–819.

Kalil, B., Ramaswamy, S., & Plant, T. M. (2015). The distribution of substance P and kisspeptin in the mediobasal hypothalamus of the male rhesus monkey and a comparison of intravenous administration of these Peptides to release GnRH as reflected by LH secretion. *Neuroendocrinology*, *103*, 711–723.

Kalra, P. S., Sahu, A., Bonavera, J. J., & Kalra, S. P. (1992). Diverse effects of tachykinins on luteinizing hormone release in male rats: Mechanism of action. *Endocrinology*, *131*(3), 1195–1201.

Karahalios, D. G., & Levine, J. E. (1988). Naloxone stimulation of in vivo LHRH release is not diminished following ovariectomy. *Neuroendocrinology*, *47*(6), 504–510.

Karsch, F. J. (1987). Central actions of ovarian steroids in the feedback regulation of pulsatile secretion of luteinizing hormone. *Annual Review of Physiology*, *49*, 365–382.

Karsch, F. J., Bittman, E. L., Foster, D. L., Goodman, R. L., Legan, S. J., & Robinson, J. E. (1984). Neuroendocrine basis of seasonal reproduction. *Recent Progress in Hormone Research*, *40*, 185–232.

Karsch, F. J., Foster, D. L., Bittman, E. L., & Goodman, R. L. (1983). A role for estradiol in enhancing luteinizing hormone pulse frequency during the follicular phase of the estrous cycle of sheep. *Endocrinology*, *113*(4), 1333–1339.

Karsch, F. J., Legan, S. J., Hauger, R. L., & Foster, D. L. (1977). Negative feedback action of progesterone on tonic luteinizing hormone secretion in the ewe: Dependence on the ovaries. *Endocrinology*, *101*(3), 800–806.

Kawakami, M., Uemura, T., & Hayashi, R. (1982). Electrophysiological correlates of pulsatile gonadotropin release in rats. *Neuroendocrinology*, *35*(1), 63–67.

Keen, K. L., Wegner, F. H., Bloom, S. R., Ghatei, M. A., & Terasawa, E. (2008). An increase in kisspeptin-54 release occurs with the pubertal increase in luteinizing hormone-releasing hormone-1 release in the stalk-median eminence of female rhesus monkeys in vivo. *Endocrinology*, *149*(8), 4151–4157.

Khachaturian, H., Lewis, M. E., Haber, S. N., Akil, H., & Watson, S. J. (1984). Pro-opiomelanocortin peptide immunocytochemistry in rhesus monkey brain. *Brain Research Bulletin*, *13*(6), 785–800.

Khan, A. R., & Kauffman, A. S. (2012). The role of kisspeptin and RFamide-related peptide-3 neurones in the circadian-timed preovulatory luteinising hormone surge. *Journal of Neuroendocrinology*, *24*(1), 131–143.

King, J. C., Ronsheim, P., Liu, E., Powers, L., Slonimski, M., & Rubin, B. S. (1998). Fos expression in luteinizing hormone-releasing hormone neurons of guinea pigs, with knife cuts separating the preoptic area and the hypothalamus, demonstrating luteinizing hormone surges. *Biology of Reproduction*, *58*(2), 323–329.

King, J. C., Tobet, S. A., Snavely, F. L., & Arimura, A. A. (1982). LHRH immunopositive cells and their projections to the median eminence and organum vasculosum of the lamina terminalis. *The Journal of Comparative Neurology*, *209*(3), 287–300.

Kinsey-Jones, J. S., Grachev, P., Li, X. F., Lin, Y. S., Milligan, S. R., Lightman, S. L., et al. (2012). The inhibitory effects of neurokinin B on GnRH pulse generator frequency in the female rat. *Endocrinology*, *153*(1), 307–315.

Kinsey-Jones, J. S., Li, X. F., Luckman, S. M., & O'Byrne, K. T. (2008). Effects of kisspeptin-10 on the electrophysiological manifestation of gonadotropin-releasing hormone pulse generator activity in the female rat. *Endocrinology*, *149*(3), 1004–1008.

Krajewski, S. J., Anderson, M. J., Iles-Shih, L., Chen, K. J., Urbanski, H. F., & Rance, N. E. (2005). Morphologic evidence that neurokinin B modulates gonadotropin-releasing hormone secretion via neurokinin 3 receptors in the rat median eminence. *The Journal of Comparative Neurology*, *489*(3), 372–386.
Krajewski, S. J., Burke, M. C., Anderson, M. J., McMullen, N. T., & Rance, N. E. (2010). Forebrain projections of arcuate neurokinin B neurons demonstrated by anterograde tract-tracing and monosodium glutamate lesions in the rat. *Neuroscience*, *166*(2), 680–697.
Krey, L. C., Butler, W. R., & Knobil, E. (1975). Surgical disconnection of the medial basal hypothalamus and pituitary function in the rhesus monkey. I. Gonadotropin secretion. *Endocrinology*, *96*(5), 1073–1087.
Kumar, D., Candlish, M., Periasamy, V., Avcu, N., Mayer, C., & Boehm, U. (2015). Specialized subpopulations of kisspeptin neurons communicate with GnRH neurons in female mice. *Endocrinology*, *156*(1), 32–38.
Lehman, M. N., Coolen, L. M., & Goodman, R. L. (2010). Minireview: Kisspeptin/neurokinin B/dynorphin (KNDy) cells of the arcuate nucleus: A central node in the control of gonadotropin-releasing hormone secretion. *Endocrinology*, *151*(8), 3479–3489.
Lehman, M. N., Hileman, S. M., & Goodman, R. L. (2013). Neuroanatomy of the kisspeptin signaling system in mammals: Comparative and developmental aspects. *Advances in Experimental Medicine and Biology*, *784*, 27–62.
Lehman, M. N., & Karsch, F. J. (1993). Do gonadotropin-releasing hormone, tyrosine hydroxylase-, and beta-endorphin-immunoreactive neurons contain estrogen receptors? A double-label immunocytochemical study in the Suffolk ewe. *Endocrinology*, *133*(2), 887–895.
Lewis, M. E., Khachaturian, H., & Watson, S. J. (1983). Comparative distribution of opiate receptors and three opioid peptide neuronal systems in rhesus monkey central nervous system. *Life Sciences*, *33*(Suppl. 1), 239–242.
Li, X. F., Kinsey-Jones, J. S., Cheng, Y., Knox, A. M., Lin, Y., Petrou, N. A., et al. (2009). Kisspeptin signalling in the hypothalamic arcuate nucleus regulates GnRH pulse generator frequency in the rat. *PLoS One*, *4*(12), e8334.
Li, Q., Millar, R. P., Clarke, I. J., & Smith, J. T. (2015). Evidence that Neurokinin B controls basal gonadotropin-releasing hormone secretion but is not critical for estrogen-positive feedback in sheep. *Neuroendocrinology*, *101*(2), 161–174.
Ljungdahl, A., Hokfelt, T., & Nilsson, G. (1978). Distribution of substance P-like immunoreactivity in the central nervous system of the rat—I. Cell bodies and nerve terminals. *Neuroscience*, *3*(10), 861–943.
Lucas, L. R., Hurley, D. L., Krause, J. E., & Harlan, R. E. (1992). Localization of the tachykinin neurokinin B precursor peptide in rat brain by immunocytochemistry and in situ hybridization. *Neuroscience*, *51*(2), 317–345.
Maggio, J. E. (1988). Tachykinins. *Annual Review of Neuroscience*, *11*, 13–28.
Marksteiner, J., Sperk, G., & Krause, J. E. (1992). Distribution of neurons expressing neurokinin B in the rat brain: Immunohistochemistry and in situ hybridization. *The Journal of Comparative Neurology*, *317*(4), 341–356.
Menyhert, J., Wittmann, G., Lechan, R. M., Keller, E., Liposits, Z., & Fekete, C. (2007). Cocaine- and amphetamine-regulated transcript (CART) is colocalized with the orexigenic neuropeptide Y and agouti-related protein and absent from the anorexigenic alpha-melanocyte-stimulating hormone neurons in the infundibular nucleus of the human hypothalamus. *Endocrinology*, *148*(9), 4276–4281.
Merchenthaler, I., Maderdrut, J. L., O'Harte, F., & Conlon, J. M. (1992). Localization of neurokinin B in the central nervous system of the rat. *Peptides*, *13*(4), 815–829.

Merkley, C. M., Coolen, L. M., Goodman, R. L., & Lehman, M. N. (2015). Evidence for changes in numbers of synaptic inputs onto KNDy and GnRH Neurones during the preovulatory LH surge in the ewe. *Journal of Neuroendocrinology*, *27*(7), 624–635.

Merkley, C. M., Porter, K. L., Coolen, L. M., Hileman, S. M., Billings, H. J., Drews, S., et al. (2012). KNDy (kisspeptin/neurokinin B/dynorphin) neurons are activated during both pulsatile and surge secretion of LH in the ewe. *Endocrinology*, *153*(11), 5406–5414.

Mileusnic, D., Lee, J. M., Magnuson, D. J., Hejna, M. J., Krause, J. E., Lorens, J. B., et al. (1999). Neurokinin-3 receptor distribution in rat and human brain: An immunohistochemical study. *Neuroscience*, *89*(4), 1269–1290.

Mitchell, V., Prevot, V., Jennes, L., Aubert, J. P., Croix, D., & Beauvillain, J. C. (1997). Presence of mu and kappa opioid receptor mRNAs in galanin but not in GnRH neurons in the female rat. *Neuroreport*, *8*(14), 3167–3172.

Mittelman-Smith, M. A., Krajewski-Hall, S. J., McMullen, N. T., & Rance, N. E. (2016). Ablation of KNDy neurons results in hypogonadotropic hypogonadism and amplifies the steroid-induced LH surge in female rats. *Endocrinology*, *157*(5), 2015–2027.

Mittelman-Smith, M. A., Williams, H., Krajewski-Hall, S. J., Lai, J., Ciofi, P., McMullen, N. T., et al. (2012). Arcuate kisspeptin/neurokinin B/dynorphin (KNDy) neurons mediate the estrogen suppression of gonadotropin secretion and body weight. *Endocrinology*, *153*(6), 2800–2812.

Moenter, S. M., Brand, R. M., Midgley, A. R., & Karsch, F. J. (1992). Dynamics of gonadotropin-releasing hormone release during a pulse. *Endocrinology*, *130*(1), 503–510.

Mori, Y., Nishihara, M., Tanaka, T., Shimizu, T., Yamaguchi, M., Takeuchi, Y., et al. (1991). Chronic recording of electrophysiological manifestation of the hypothalamic gonadotropin-releasing hormone pulse generator activity in the goat. *Neuroendocrinology*, *53*(4), 392–395.

Morley, J. E. (1981). The endocrinology of the opiates and opioid peptides. *Metabolism*, *30*(2), 195–209.

Mostari, P., Ieda, N., Deura, C., Minabe, S., Yamada, S., Uenoyama, Y., et al. (2013). Dynorphin-kappa opioid receptor signaling partly mediates estrogen negative feedback effect on LH pulses in female rats. *The Journal of Reproduction and Development*, *59*(3), 266–272.

Murakawa, H., Iwata, K., Takeshita, T., & Ozawa, H. (2016). Immunoelectron microscopic observation of the subcellular localization of kisspeptin, neurokinin B and dynorphin A in KNDy neurons in the arcuate nucleus of the female rat. *Neuroscience Letters*, *612*, 161–166.

Nakaya, Y., Kaneko, T., Shigemoto, R., Nakanishi, S., & Mizuno, N. (1994). Immunohistochemical localization of substance P receptor in the central nervous system of the adult rat. *The Journal of Comparative Neurology*, *347*(2), 249–274.

Navarro, V. M., Bosch, M. A., Leon, S., Simavli, S., True, C., Pinilla, L., et al. (2015). The integrated hypothalamic tachykinin-kisspeptin system as a central coordinator for reproduction. *Endocrinology*, *156*(2), 627–637.

Navarro, V. M., Castellano, J. M., McConkey, S. M., Pineda, R., Ruiz-Pino, F., Pinilla, L., et al. (2011). Interactions between kisspeptin and neurokinin B in the control of GnRH secretion in the female rat. *American Journal of Physiology. Endocrinology and Metabolism*, *300*(1), E202–E210.

Navarro, V. M., Gottsch, M. L., Chavkin, C., Okamura, H., Clifton, D. K., & Steiner, R. A. (2009). Regulation of gonadotropin-releasing hormone secretion by kisspeptin/dynorphin/neurokinin B neurons in the arcuate nucleus of the mouse. *The Journal of Neuroscience*, *29*(38), 11859–11866.

Navarro, V. M., Gottsch, M. L., Wu, M., Garcia-Galiano, D., Hobbs, S. J., Bosch, M. A., et al. (2011). Regulation of NKB pathways and their roles in the control of Kiss1 neurons in the arcuate nucleus of the male mouse. *Endocrinology, 152*(11), 4265–4275.
Nestor, C. C., Briscoe, A. M., Davis, S. M., Valent, M., Goodman, R. L., & Hileman, S. M. (2012). Evidence of a role for kisspeptin and neurokinin B in puberty of female sheep. *Endocrinology, 153*(6), 2756–2765.
Noritake, K., Matsuoka, T., Ohsawa, T., Shimomura, K., Sanbuissho, A., Uenoyama, Y., et al. (2011). Involvement of neurokinin receptors in the control of pulsatile luteinizing hormone secretion in rats. *The Journal of Reproduction and Development, 57*(3), 409–415.
Novaira, H. J., Sonko, M. L., Hoffman, G., Koo, Y., Ko, C., Wolfe, A., et al. (2014). Disrupted kisspeptin signaling in GnRH neurons leads to hypogonadotrophic hypogonadism. *Molecular Endocrinology, 28*(2), 225–238.
Oakley, A. E., Clifton, D. K., & Steiner, R. A. (2009). Kisspeptin signaling in the brain. *Endocrine Reviews, 30*(6), 713–743.
Ohkura, S., Takase, K., Matsuyama, S., Mogi, K., Ichimaru, T., Wakabayashi, Y., et al. (2009). Gonadotrophin-releasing hormone pulse generator activity in the hypothalamus of the goat. *Journal of Neuroendocrinology, 21*(10), 813–821.
Ohkura, S., Tsukamura, H., & Maeda, K. (1991). Effects of various types of hypothalamic deafferentation on luteinizing hormone pulses in ovariectomized rats. *Journal of Neuroendocrinology, 3*(5), 503–508.
Okamura, H., Tsukamura, H., Ohkura, S., Uenoyama, Y., Wakabayashi, Y., & Maeda, K. (2013). Kisspeptin and GnRH pulse generation. *Advances in Experimental Medicine and Biology, 784*, 297–323.
Olster, D. H., & Blaustein, J. D. (1990). Immunocytochemical colocalization of progestin receptors and beta-endorphin or enkephalin in the hypothalamus of female guinea pigs. *Journal of Neurobiology, 21*(5), 768–780.
Otsuka, M., & Yoshioka, K. (1993). Neurotransmitter functions of mammalian tachykinins. *Physiological Reviews, 73*(2), 229–308.
Overgaard, A., Ruiz-Pino, F., Castellano, J. M., Tena-Sempere, M., & Mikkelsen, J. D. (2014). Disparate changes in kisspeptin and neurokinin B expression in the arcuate nucleus after sex steroid manipulation reveal differential regulation of the two KNDy peptides in rats. *Endocrinology, 155*(10), 3945–3955.
Pillon, D., Caraty, A., Fabre-Nys, C., & Bruneau, G. (2003). Short-term effect of oestradiol on neurokinin B mRNA expression in the infundibular nucleus of ewes. *Journal of Neuroendocrinology, 15*(8), 749–753.
Plant, T. M. (2012). A comparison of the neuroendocrine mechanisms underlying the initiation of the preovulatory LH surge in the human, old world monkey and rodent. *Frontiers in Neuroendocrinology, 33*(2), 160–168.
Plant, T. M., Krey, L. C., Moossy, J., McCormack, J. T., Hess, D. L., & Knobil, E. (1978). The arcuate nucleus and the control of gonadotropin and prolactin secretion in the female rhesus monkey (Macaca Mulatta). *Endocrinology, 102*(1), 52–62.
Porter, K. L., Hileman, S. M., Hardy, S. L., Nestor, C. C., Lehman, M. N., & Goodman, R. L. (2014). Neurokinin-3 receptor activation in the retrochiasmatic area is essential for the full pre-ovulatory luteinising hormone surge in ewes. *Journal of Neuroendocrinology, 26*(11), 776–784.
Qiu, J., Nestor, C. C., Zhang, C., Padilla, S. L., Palmiter, R. D., Kelly, M. J., et al. (2016). High-frequency stimulation-induced peptide release synchronizes arcuate kisspeptin neurons and excites GnRH neurons. *eLife, 5*, 1–24.
Quigley, M. E., & Yen, S. S. (1980). The role of endogenous opiates in LH secretion during the menstrual cycle. *The Journal of Clinical Endocrinology and Metabolism, 51*(1), 179–181.
Ramaswamy, S., Seminara, S. B., & Plant, T. M. (2011). Evidence from the agonadal juvenile male rhesus monkey (Macaca mulatta) for the view that the action of neurokinin B to

trigger gonadotropin-releasing hormone release is upstream from the kisspeptin receptor. *Neuroendocrinology*, *94*(3), 237–245.

Rance, N. E., & Bruce, T. R. (1994). Neurokinin B gene expression is increased in the arcuate nucleus of ovariectomized rats. *Neuroendocrinology*, *60*(4), 337–345.

Rance, N. E., Krajewski, S. J., Smith, M. A., Cholanian, M., & Dacks, P. A. (2010). Neurokinin B and the hypothalamic regulation of reproduction. *Brain Research*, *1364*, 116–128.

Rance, N. E., & Young, W. S., 3rd. (1991). Hypertrophy and increased gene expression of neurons containing neurokinin-B and substance-P messenger ribonucleic acids in the hypothalami of postmenopausal women. *Endocrinology*, *128*(5), 2239–2247.

Raynor, K., Kong, H., Chen, Y., Yasuda, K., Yu, L., Bell, G. I., et al. (1994). Pharmacological characterization of the cloned kappa-, delta-, and mu-opioid receptors. *Molecular Pharmacology*, *45*(2), 330–334.

Rometo, A. M., & Rance, N. E. (2008). Changes in prodynorphin gene expression and neuronal morphology in the hypothalamus of postmenopausal women. *Journal of Neuroendocrinology*, *20*(12), 1376–1381.

Ronnekleiv, O. K., Kelly, M. J., & Eskay, R. L. (1984). Distribution of immunoreactive substance P neurons in the hypothalamus and pituitary of the rhesus monkey. *The Journal of Comparative Neurology*, *224*(1), 51–59.

Ropert, J. F., Quigley, M. E., & Yen, S. S. (1981). Endogenous opiates modulate pulsatile luteinizing hormone release in humans. *The Journal of Clinical Endocrinology and Metabolism*, *52*(3), 583–585.

Roseweir, A. K., Kauffman, A. S., Smith, J. T., Guerriero, K. A., Morgan, K., Pielecka-Fortuna, J., et al. (2009). Discovery of potent kisspeptin antagonists delineate physiological mechanisms of gonadotropin regulation. *The Journal of Neuroscience*, *29*(12), 3920–3929.

Ruiz-Pino, F., Garcia-Galiano, D., Manfredi-Lozano, M., Leon, S., Sanchez-Garrido, M. A., Roa, J., et al. (2015). Effects and interactions of tachykinins and dynorphin on FSH and LH secretion in developing and adult rats. *Endocrinology*, *156*(2), 576–588.

Ruiz-Pino, F., Navarro, V. M., Bentsen, A. H., Garcia-Galiano, D., Sanchez-Garrido, M. A., Ciofi, P., et al. (2012). Neurokinin B and the control of the gonadotropic axis in the rat: Developmental changes, sexual dimorphism, and regulation by gonadal steroids. *Endocrinology*, *153*(10), 4818–4829.

Ruka, K. A., Burger, L. L., & Moenter, S. M. (2013). Regulation of arcuate neurons coexpressing kisspeptin, neurokinin B, and dynorphin by modulators of neurokinin 3 and kappa-opioid receptors in adult male mice. *Endocrinology*, *154*(8), 2761–2771.

Ruka, K. A., Burger, L. L., & Moenter, S. M. (2016). Both estrogen and androgen modify the response to activation of neurokinin-3 and kappa-opioid receptors in arcuate kisspeptin neurons from male mice. *Endocrinology*, *157*(2), 752–763.

Saffroy, M., Torrens, Y., Glowinski, J., & Beaujouan, J. C. (2003). Autoradiographic distribution of tachykinin NK2 binding sites in the rat brain: Comparison with NK1 and NK3 binding sites. *Neuroscience*, *116*(3), 761–773.

Sahu, A., & Kalra, S. P. (1992). Effects of tachykinins on luteinizing hormone release in female rats: Potent inhibitory action of neuropeptide K. *Endocrinology*, *130*(3), 1571–1577.

Sakamoto, K., Murata, K., Wakabayashi, Y., Yayou, K., Ohkura, S., Takeuchi, Y., et al. (2012). Central administration of neurokinin B activates kisspeptin/NKB neurons in the arcuate nucleus and stimulates luteinizing hormone secretion in ewes during the non-breeding season. *The Journal of Reproduction and Development*, *58*(6), 700–706.

Sakamoto, K., Wakabayashi, Y., Yamamura, T., Tanaka, T., Takeuchi, Y., Mori, Y., et al. (2013). A population of kisspeptin/neurokinin B neurons in the arcuate nucleus may be the central target of the male effect phenomenon in goats. *PLoS One*, *8*(11), e81017.

Sandoval-Guzman, T., & Rance, N. E. (2004). Central injection of senktide, an NK3 receptor agonist, or neuropeptide Y inhibits LH secretion and induces different patterns of Fos expression in the rat hypothalamus. *Brain Research*, *1026*(2), 307–312.
Sandoval-Guzman, T., Stalcup, S. T., Krajewski, S. J., Voytko, M. L., & Rance, N. E. (2004). Effects of ovariectomy on the neuroendocrine axes regulating reproduction and energy balance in young cynomolgus macaques. *Journal of Neuroendocrinology*, *16*(2), 146–153.
Sannella, M. I., & Petersen, S. L. (1997). Dual label in situ hybridization studies provide evidence that luteinizing hormone-releasing hormone neurons do not synthesize messenger ribonucleic acid for mu, kappa, or delta opiate receptors. *Endocrinology*, *138*(4), 1667–1672.
Sarkar, D. K., & Yen, S. S. (1985). Changes in beta-endorphin-like immunoreactivity in pituitary portal blood during the estrous cycle and after ovariectomy in rats. *Endocrinology*, *116*(5), 2075–2079.
Seabrook, G. R., Bowery, B. J., & Hill, R. G. (1995). Pharmacology of tachykinin receptors on neurones in the ventral tegmental area of rat brain slices. *European Journal of Pharmacology*, *273*(1–2), 113–119.
Seminara, S. B., Messager, S., Chatzidaki, E. E., Thresher, R. R., Acierno, J. S., Jr., Shagoury, J. K., et al. (2003). The GPR54 gene as a regulator of puberty. *The New England Journal of Medicine*, *349*(17), 1614–1627.
Shoupe, D., Montz, F. J., & Lobo, R. A. (1985). The effects of estrogen and progestin on endogenous opioid activity in oophorectomized women. *The Journal of Clinical Endocrinology and Metabolism*, *60*(1), 178–183.
Shuster, S. J., Riedl, M., Li, X., Vulchanova, L., & Elde, R. (1999). Stimulus-dependent translocation of kappa opioid receptors to the plasma membrane. *The Journal of Neuroscience*, *19*(7), 2658–2664.
Shuster, S. J., Riedl, M., Li, X., Vulchanova, L., & Elde, R. (2000). The kappa opioid receptor and dynorphin co-localize in vasopressin magnocellular neurosecretory neurons in guinea-pig hypothalamus. *Neuroscience*, *96*(2), 373–383.
Silverman, A. J., Livne, I., & Witkin, J. W. (1994). The gonadotropin-releasing hormone (GnRH) neuronal systems: Immunocytochemistry and in situ hybridization. In E. Knobil & J. D. Neill (Eds.), *The physiology of reproduction: Vol. 1* (2nd ed., pp. 1683–1709). New York NY: Raven Press.
Silverman, A. J., Wilson, R., Kesner, J. S., & Knobil, E. (1986). Hypothalamic localization of multiunit electrical activity associated with pulsatile LH release in the rhesus monkey. *Neuroendocrinology*, *44*(2), 168–171.
Simavli, S., Thompson, I. R., Maguire, C. A., Gill, J. C., Carroll, R. S., Wolfe, A., et al. (2015). Substance p regulates puberty onset and fertility in the female mouse. *Endocrinology*, *156*(6), 2313–2322.
Simerly, R. B., Young, B. J., & Carr, A. M. (1996). Co-expression of steroid hormone receptors in opioid peptide-containing neurons correlates with patterns of gene expression during the estrous cycle. *Brain Research. Molecular Brain Research*, *40*(2), 275–284.
Skinner, D. C., Evans, N. P., Delaleu, B., Goodman, R. L., Bouchard, P., & Caraty, A. (1998). The negative feedback actions of progesterone on gonadotropin-releasing hormone secretion are transduced by the classical progesterone receptor. *Proceedings of the National Academy of Sciences of the United States of America*, *95*(18), 10978–10983.
Skrapits, K., Borsay, B. A., Herczeg, L., Ciofi, P., Bloom, S. R., Ghatei, M. A., et al. (2014). Colocalization of cocaine- and amphetamine-regulated transcript with kisspeptin and neurokinin B in the human infundibular region. *PLoS One*, *9*(8), e103977.
Skrapits, K., Borsay, B. A., Herczeg, L., Ciofi, P., Liposits, Z., & Hrabovszky, E. (2015). Neuropeptide co-expression in hypothalamic kisspeptin neurons of laboratory animals and the human. *Frontiers in Neuroscience*, *9*, 29.

Smith, J. T., Clay, C. M., Caraty, A., & Clarke, I. J. (2007). KiSS-1 messenger ribonucleic acid expression in the hypothalamus of the ewe is regulated by sex steroids and season. *Endocrinology*, *148*(3), 1150–1157.

Smith, J. T., Cunningham, M. J., Rissman, E. F., Clifton, D. K., & Steiner, R. A. (2005). Regulation of Kiss1 gene expression in the brain of the female mouse. *Endocrinology*, *146*(9), 3686–3692.

Smith, M. J., & Gallo, R. V. (1997). The effect of blockade of kappa-opioid receptors in the medial preoptic area on the luteinizing hormone surge in the proestrous rat. *Brain Research*, *768*(1–2), 111–119.

Smith, J. T., Li, Q., Pereira, A., & Clarke, I. J. (2009). Kisspeptin neurons in the ovine arcuate nucleus and preoptic area are involved in the preovulatory luteinizing hormone surge. *Endocrinology*, *150*(12), 5530–5538.

Smith, J. T., Li, Q., Yap, K. S., Shahab, M., Roseweir, A. K., Millar, R. P., et al. (2011). Kisspeptin is essential for the full preovulatory LH surge and stimulates GnRH release from the isolated ovine median eminence. *Endocrinology*, *152*(3), 1001–1012.

Smith, J. T., Shahab, M., Pereira, A., Pau, K. Y., & Clarke, I. J. (2010). Hypothalamic expression of KISS1 and gonadotropin inhibitory hormone genes during the menstrual cycle of a non-human primate. *Biology of Reproduction*, *83*(4), 568–577.

Soper, B. D., & Weick, R. F. (1980). Hypothalamic and extrahypothalamic mediation of pulsatile discharges of luteinizing hormone in the ovariectomized rat. *Endocrinology*, *106*(1), 348–355.

Thiery, J. C., & Pelletier, J. (1981). Multiunit activity in the anterior median eminence and adjacent areas of the hypothalamus of the ewe in relation to LH secretion. *Neuroendocrinology*, *32*(4), 217–224.

Topaloglu, A. K., Reimann, F., Guclu, M., Yalin, A. S., Kotan, L. D., Porter, K. M., et al. (2009). TAC3 and TACR3 mutations in familial hypogonadotropic hypogonadism reveal a key role for neurokinin B in the central control of reproduction. *Nature Genetics*, *41*(3), 354–358.

True, C., Nasrin Alam, S., Cox, K., Chan, Y. M., & Seminara, S. B. (2015). Neurokinin B is critical for normal timing of sexual maturation but dispensable for adult reproductive function in female mice. *Endocrinology*, *156*(4), 1386–1397.

Uenoyama, Y., Nakamura, S., Hayakawa, Y., Ikegami, K., Watanabe, Y., Deura, C., et al. (2015). Lack of pulse and surge modes and glutamatergic stimulation of luteinising hormone release in Kiss1 knockout rats. *Journal of Neuroendocrinology*, *27*(3), 187–197.

Vaaga, C. E., Borisovska, M., & Westbrook, G. L. (2014). Dual-transmitter neurons: Functional implications of co-release and co-transmission. *Current Opinion in Neurobiology*, *29*, 25–32.

Van Vugt, D. A., Sylvester, P. W., Aylsworth, C. F., & Meites, J. (1982). Counteraction of gonadal steroid inhibition of luteinizing hormone release by naloxone. *Neuroendocrinology*, *34*(4), 274–278.

Wakabayashi, Y., Nakada, T., Murata, K., Ohkura, S., Mogi, K., Navarro, V. M., et al. (2010). Neurokinin B and dynorphin A in kisspeptin neurons of the arcuate nucleus participate in generation of periodic oscillation of neural activity driving pulsatile gonadotropin-releasing hormone secretion in the goat. *The Journal of Neuroscience*, *30*(8), 3124–3132.

Wakabayashi, Y., Yamamura, T., Sakamoto, K., Mori, Y., & Okamura, H. (2013). Electrophysiological and morphological evidence for synchronized GnRH pulse generator activity among kisspeptin/neurokinin B/dynorphin A (KNDy) neurons in goats. *The Journal of Reproduction and Development*, *59*(1), 40–48.

Walsh, J. P., & Clarke, I. J. (1996). Effects of central administration of highly selective opioid mu-, delta- and kappa-receptor agonists on plasma luteinizing hormone (LH), prolactin,

and the estrogen-induced LH surge in ovariectomized ewes. *Endocrinology*, *137*(9), 3640–3648.
Wardlaw, S. L., Wehrenberg, W. B., Ferin, M., Antunes, J. L., & Frantz, A. G. (1982). Effect of sex steroids on beta-endorphin in hypophyseal portal blood. *The Journal of Clinical Endocrinology and Metabolism*, *55*(5), 877–881.
Watson, S. J., Khachaturian, H., Taylor, L., Fischli, W., Goldstein, A., & Akil, H. (1983). Pro-dynorphin peptides are found in the same neurons throughout rat brain: Immunocytochemical study. *Proceedings of the National Academy of Sciences of the United States of America*, *80*(3), 891–894.
Weems, P. W., Coolen, L. M., Hileman, S. M., Hardy, S. L., McCosh, R. B., Goodman, R. L., et al. (2016). In *Kappa opioid receptors are internalized in arcuate KNDy cells during GnRH pulse termination in the ewe Annual Meeting of Society for Neuroscience, San Diego CA*. Abtsr. 60.04.
Weems, P., Coolen, L. M., Hileman, S. M., Hardy, S. L., McCosh, R. B., Goodman, R. L., et al. (2017). In *Kappa opioid receptors are internalized in MBH GnRH cells during GnRH pulse termination* ENDO2017 Abstracts. SUN 472.
Weems, P., Smith, J., Clarke, I. J., Coolen, L. M., Goodman, R. L., & Lehman, M. N. (2017). Effects of season and estradiol on KNDy neuron peptides, colocalization with D2 dopamine receptors, and dopaminergic inputs in the ewe. *Endocrinology*, *158*(4), 831–841.
Weems, P. W., Witty, C. F., Amstalden, M., Coolen, L. M., Goodman, R. L., & Lehman, M. N. (2016). Kappa-opioid receptor is colocalized in GnRH and KNDy cells in the female ovine and rat brain. *Endocrinology*, *157*(6), 2367–2379.
Whisnant, C. S., Curto, K., & Goodman, R. L. (1992). Immunocytochemical localization of beta endorphin and gonadal steroid regulation of proopiomelanocortin messenger ribonucleic acid in the ewe. *Neuroendocrinology*, *56*(6), 812–821.
Whisnant, C. S., & Goodman, R. L. (1988). Effects of an opioid antagonist on pulsatile luteinizing hormone secretion in the ewe vary with changes in steroid negative feedback. *Biology of Reproduction*, *39*(5), 1032–1038.
Whisnant, C. S., & Goodman, R. L. (1994). Effect of anterior hypothalamic deafferentation on the negative feedback of gonadal steroids on luteinizing hormone pulse frequency in the ewe. *Domestic Animal Endocrinology*, *11*(2), 151–159.
Whisnant, S. C., Havern, R. L., & Goodman, R. L. (1991). Endogenous opioid suppression of luteinizing hormone pulse frequency and amplitude in the ewe: Hypothalamic sites of action. *Neuroendocrinology*, *54*(6), 587–593.
Whitnall, M. H., Gainer, H., Cox, B. M., & Molineaux, C. J. (1983). Dynorphin-A-(1-8) is contained within vasopressin neurosecretory vesicles in rat pituitary. *Science*, *222*(4628), 1137–1139.
Wilcox, J. N., & Roberts, J. L. (1985). Estrogen decreases rat hypothalamic proopiomelanocortin messenger ribonucleic acid levels. *Endocrinology*, *117*(6), 2392–2396.
Wilson, R. C., Kesner, J. S., Kaufman, J. M., Uemura, T., Akema, T., & Knobil, E. (1984). Central electrophysiologic correlates of pulsatile luteinizing hormone secretion in the rhesus monkey. *Neuroendocrinology*, *39*(3), 256–260.
Wise, P. M., Scarbrough, K., Weiland, N. G., & Larson, G. H. (1990). Diurnal pattern of proopiomelanocortin gene expression in the arcuate nucleus of proestrous, ovariectomized, and steroid-treated rats: A possible role in cyclic luteinizing hormone secretion. *Molecular Endocrinology*, *4*(6), 886–892.
Yamaji, T., Dierschke, D. J., Bhattacharya, A. N., & Knobil, E. (1972). The negative feedback control by estradiol and progesterone of LH secretion in the ovariectomized rhesus monkey. *Endocrinology*, *90*(3), 771–777.
Yamamura, T., Wakabayashi, Y., Ohkura, S., Navarro, V. M., & Okamura, H. (2015). Effects of intravenous administration of neurokinin receptor subtype-selective agonists

on gonadotropin-releasing hormone pulse generator activity and luteinizing hormone secretion in goats. *The Journal of Reproduction and Development*, *61*(1), 20–29.
Yang, J. J., Caligioni, C. S., Chan, Y. M., & Seminara, S. B. (2012). Uncovering novel reproductive defects in neurokinin B receptor null mice: Closing the gap between mice and men. *Endocrinology*, *153*(3), 1498–1508.
Yeo, S. H., Clarkson, J., & Herbison, A. E. (2014). Kisspeptin-gpr54 signaling at the GnRH neuron is necessary for negative feedback regulation of luteinizing hormone secretion in female mice. *Neuroendocrinology*, *100*(2–3), 191–197.
Yeo, S. H., & Herbison, A. E. (2014). Estrogen-negative feedback and estrous cyclicity are critically dependent upon estrogen receptor-alpha expression in the arcuate nucleus of adult female mice. *Endocrinology*, *155*(8), 2986–2995.
Yip, J., & Chahl, L. A. (2000). Localization of tachykinin receptors and Fos-like immunoreactivity induced by substance P in guinea-pig brain. *Clinical and Experimental Pharmacology & Physiology*, *27*(11), 943–946.
Yip, J., & Chahl, L. A. (2001). Localization of NK1 and NK3 receptors in guinea-pig brain. *Regulatory Peptides*, *98*(1–2), 55–62.
Young, J., George, J. T., Tello, J. A., Francou, B., Bouligand, J., Guiochon-Mantel, A., et al. (2013). Kisspeptin restores pulsatile LH secretion in patients with neurokinin B signaling deficiencies: Physiological, pathophysiological and therapeutic implications. *Neuroendocrinology*, *97*(2), 193–202.
Zeleznik, A. J., & Plant, T. M. (2015). Control of the menstrual cycle. In T. M. Plant & A. J. Zeleznik (Eds.), *Knobil and Neill's physiology of reproduction*: *Vol. 2* (4th ed., pp. 1307–1361). Amsterdam: Elsevier.
Zhang, Q., & Gallo, R. V. (2002). Effect of prodynorphin-derived opioid peptides on the ovulatory luteinizing hormone surge in the proestrous rat. *Endocrine*, *18*(1), 27–32.
Zhang, Q., McCoy, J. M., & Gallo, R. V. (2002). Further studies on possible dynorphin involvement in the ovulatory luteinizing hormone surge in the proestrous rat. *Endocrine*, *18*(3), 231–238.

CHAPTER FIVE

Interaction of Melatonin and BMP-6 in Ovarian Steroidogenesis

Fumio Otsuka[1]
Okayama University Graduate School of Medicine, Dentistry and Pharmaceutical Sciences, Okayama, Japan
[1]Corresponding author: e-mail address: fumiotsu@md.okayama-u.ac.jp

Contents

Abstract

The bone morphogenetic protein (BMP) system in the ovary plays a physiological role as a luteinization inhibitor in growing follicles. BMP-6 secreted from oocytes and granulosa cells can exert an inhibitory effect on follicle-stimulating hormone (FSH) actions by suppressing adenylate cyclase activity downstream of the FSH receptor. The inhibition of FSH-induced progesterone production by BMP-6 is impaired by melatonin treatment in granulosa cells. Intracellular Smad signaling induced by BMP-6 is suppressed by melatonin, suggesting that melatonin has a regulatory role in BMP receptor signaling in granulosa cells. Since the expression of BMP-6 in granulosa cells is increased in patients with polycystic ovary syndrome, melatonin may play an important role in the maintenance of progesterone production by suppressing BMP-6 signaling, leading to the preservation of ovarian function.

1. INTRODUCTION

Normal follicle growth occurs as a result of complex interactions between gonadotropins secreted from the anterior pituitary and autocrine/paracrine factors expressed in the ovary. The physiological process

Vitamins and Hormones, Volume 107
ISSN 0083-6729
https://doi.org/10.1016/bs.vh.2018.01.012

of normal folliculogenesis is integrated by the actions of oocytes, in which oocyte-derived growth factors command neighboring follicular cells to integrate various functions of granulosa and theca cells (Eppig, 2001; Gilchrist et al., 2006; Hsueh, Kawamura, Cheng, & Fauser, 2015). Growth factors locally expressed in the ovary including bone morphogenetic proteins (BMPs) exhibit various activities for regulating steroidogenesis as well as mitogenesis in granulosa cells (Findlay et al., 2002; Shimasaki, Moore, Erickson, & Otsuka, 2003; Shimasaki, Moore, Otsuka, & Erickson, 2004).

Expression of mRNAs and/or proteins for the ovarian BMP system has been demonstrated in cellular components of growing follicles (Erickson & Shimasaki, 2003), wherein the biofunctional activity of BMPs in the regulation of ovarian function has been gradually uncovered (Otsuka & Inagaki, 2011; Otsuka, McTavish, & Shimasaki, 2011; Shimasaki et al., 2004) (Fig. 1). Control of follicle-stimulating hormone (FSH) responsiveness in

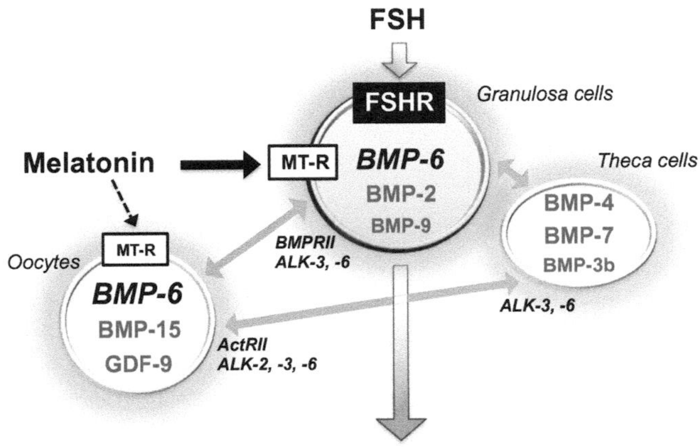

Fig. 1 BMP system in the ovarian follicle and effects of melatonin. BMP-6 and -15 and GDF-9 are expressed by oocytes and exhibit biological activities in granulosa cells expressing BMPRII, ALK-3 and -6, and theca cells expressing ALK-3 and -6. Theca cells express BMP-3b, -4, and -7, which exhibit biological activities in granulosa cells and oocytes. Granulosa cells express BMP-2 and -6 and possibly BMP-9, which act on oocytes and theca cells. In addition to these paracrine actions, BMPs also have autocrine effects in the ovary. The BMP system acts to regulate cell proliferation as well as differentiation during folliculogenesis under the influence of FSH. On the other hand, melatonin has effects on functions in ovarian follicles, especially in granulosa cells and oocytes. The actions of melatonin on these cells are mediated mainly by the membrane receptors MT1 and MT2 and also by other binding sites in the nucleus or cytosol. Acetylserotonin *O*-methyltransferase (ASMT), in addition to MT1, is also expressed in the cumulus–oocyte complex, while the MT2 gene is expressed only in oocytes.

growing follicles is important for the establishment of dominant follicles and subsequent normal ovulation. Recent studies have provided evidence that the BMP system plays important roles in the maintenance of female fertility in mammals in an autocrine/paracrine manner (Shimasaki et al., 2004). The activity of the BMP system regulates FSH action and FSH receptor (FSHR) sensitivity during the normal process of folliculogenesis (Otsuka & Inagaki, 2011; Otsuka et al., 2011; Shimasaki et al., 2004).

On the other hand, melatonin, an indolamine secreted mainly from the pineal gland, is functionally linked to the physiologic regulation of circadian and seasonal rhythms and also to the activities of various hormones and cytokines (Borjigin, Zhang, & Calinescu, 2012; Pandi-Perumal et al., 2008; Reiter, Rosales-Corral, Manchester, & Tan, 2013; Reiter, Tamura, Tan, & Xu, 2014). In this chapter, a new viewpoint of melatonin actions is introduced by focusing on ovarian steroidogenesis (Nakamura et al., 2014). The finding of a functional interaction between melatonin activity and BMP-6 signaling may be linked to the establishment of therapeutic approaches for various ovarian disorders represented in polycystic ovary syndrome (PCOS).

2. BMP SYSTEM IN THE OVARY

BMPs, which belong to the transforming growth factor (TGF)-β superfamily, were originally identified to be inducers of ectopic bone growth and cartilage formation. Substantial progress has been made in clarifying the wide-ranging roles of BMPs from bone to other systemic actions (Wagner et al., 2010). In various organs and tissues, BMPs regulate cell growth, differentiation, and apoptosis, and they also play functional roles as autocrine/paracrine factors in the control of steroidogenesis (Otsuka, 2010, 2013).

BMP ligands and receptors, including BMP-2, -3, -3b, -4, -6, -7, -15 and growth differentiation factor (GDF)-9 and BMP type-I and type-II receptors, are expressed in cell-dependent patterns in ovarian cells that undergo dynamic changes during follicular development and corpora luteal morphogenesis (Erickson & Shimasaki, 2003; Shimasaki et al., 2004) (Fig. 1). Circulating BMP-9 is also involved in the regulation of ovarian steroidogenesis (Hosoya et al., 2015). Seven type-I receptors (ALK-1, -2, -3, -4, -5, -6, and -7) and five type-II receptors (ActRII, ActRIIB, AMHRII, BMPRII, and TβRII) for TGF-β superfamily members have been characterized in mammals (Mueller, 2015). Binding of BMP ligands to the corresponding receptors results in phosphorylation of type-I receptors, leading to activation of downstream signaling molecules, Smads (Mueller, 2015). Pathway-restricted Smad1/5/8 are phosphorylated by type-I receptors and then interact with a common mediator, Smad4, to form a hetero-oligomeric complex with Smad1/5/8. The complex is then translocated into the nucleus, where it binds directly or indirectly to target DNA, being followed by specific gene transcription.

The finding that defects of the oocyte-specific growth factors GDF-9 and BMP-15 result in female infertility in mammals was an important breakthrough in the reproduction field (Dong et al., 1996; Galloway et al., 2000). BMP-15 is predominantly expressed in oocytes (Otsuka et al., 2000) and has been shown to induce granulosa cell mitosis and inhibit the actions of FSH by suppressing FSHR expression (Otsuka & Shimasaki, 2002; Otsuka, Yamamoto, Erickson, & Shimasaki, 2001). The lack of bioactive BMP-15 and its mitotic effects lead to arrest of follicle development as shown in homozygotes of *BMP15*-mutant sheep called Inverdale strain (Otsuka et al., 2000). The reproductive phenotypes caused by aberrant BMP activities combined with data from in vitro studies have established roles of the endogenous BMP system including BMP-15 signaling in the regulation of follicular development and also shown the biological importance of oocyte-secreted factors in the field of mammalian reproduction (Findlay et al., 2002; Monniaux, 2016; Otsuka et al., 2005; Pangas, 2012; Shimasaki et al., 2003, 2004).

3. FUNCTIONAL ROLES OF BMP-6 IN OVARIAN FOLLICULOGENESIS

The expression of BMP ligands as well as BMP receptor type-IA (BMPRIA; also called ALK-3), type-IB (BMPRIB/ALK-6), and type-II (ActRII, BMPRII) within different compartments of the antral follicle has been demonstrated in ovaries from various species including rodents, humans, and ruminants, and it has been shown that the expression undergoes dynamic changes during follicular development and luteal morphogenesis (Erickson & Shimasaki, 2003; Shimasaki et al., 2004) (Fig. 1).

Regulation of FSH responsiveness in granulosa cells is critical for the steps of follicular selection, dominant follicle formation, and subsequent ovulation in mammals. This process elicited by FSHR signaling is further modulated by many autocrine/paracrine factors (Findlay et al., 2002). BMPs play a key role in female fertility by regulating steroidogenesis and mitosis in granulosa cells. BMP ligands commonly suppress FSH-induced progesterone synthesis (Inagaki et al., 2009; Lee, Otsuka, Moore, & Shimasaki, 2001; Miyoshi et al., 2007; Otsuka & Inagaki, 2011; Otsuka et al., 2011; Shimasaki et al., 2004). Oocyte-derived BMP-15 attenuates FSH actions by suppressing FSHR expression in granulosa cells (Otsuka, Yamamoto, et al., 2001). BMP-2 from granulosa cells, BMP-4 and -7 from theca cells, oocyte- and granulosa-derived BMP-6, and BMP-9 in serum and granulosa cells also suppress FSH-induced production of progesterone (Hosoya et al., 2015; Inagaki et al., 2009; Lee et al., 2001; Miyoshi et al., 2007; Otsuka, Moore, & Shimasaki, 2001).

BMP-6 is specifically expressed in oocytes and granulosa cells of healthy Graafian follicles (Erickson & Shimasaki, 2003) (Fig. 1). BMP-6 inhibits FSH actions by controlling cAMP synthesis through suppression of adenylate cyclase activity in granulosa cells (Otsuka, Moore, & Shimasaki, 2001) (Fig. 2). It has been shown that BMP-6 expression in granulosa cells, but not that in oocytes, rapidly decreases at the time when the dominant follicle is selected (Erickson & Shimasaki, 2003), suggesting that BMP-6 contributes to the selection process of dominant follicles. BMP-4 and -7 increase FSH-induced production of estradiol and suppress production of progesterone (Miyoshi et al., 2007; Shimasaki et al., 1999). BMP-7 also contributes to the recruitment process of primordial follicles into the growing follicle pool (Lee et al., 2001). The actions of BMP-7 on FSH-induced estradiol production occur through suppression of ERK1/2-MAPK downstream of FSHR signaling (Miyoshi et al., 2007). BMP-2 and -4 stimulate FSH-induced p38-MAPK phosphorylation, leading to an increase of FSH-induced estradiol production (Inagaki et al., 2009). In addition, some of the BMP actions are extracellularly regulated by the binding protein follistatin and noggin (Mulloy & Rider, 2015; Otsuka, Moore, Iemura, Ueno, & Shimasaki, 2001), while BMP–Smad signal activities are also regulated by FSHR signaling, leading to fine-tuning of the mutual sensitivities of BMPs and FSH (Miyoshi et al., 2006). Nevertheless, circulating levels of BMP-4 and -7 in serum (<400 pg/mL) have been reported to be lower than their biologically active levels (David et al., 2008; Sugimoto et al., 2007), whereas BMP-9 is likely to exist in human serum and plasma at relatively high concentrations of 2–12 ng/mL (David et al., 2008). Thus, each BMP ligand derived from separate sources differentially regulates FSH-induced steroidogenesis in granulosa cells by a ligand-dependent mechanism. The finding that FSH-induced progesterone synthesis is commonly inhibited by ovarian BMPs suggests a key role of the BMP system as luteinization inhibitors (Shimasaki et al., 2003, 2004) (Fig. 2).

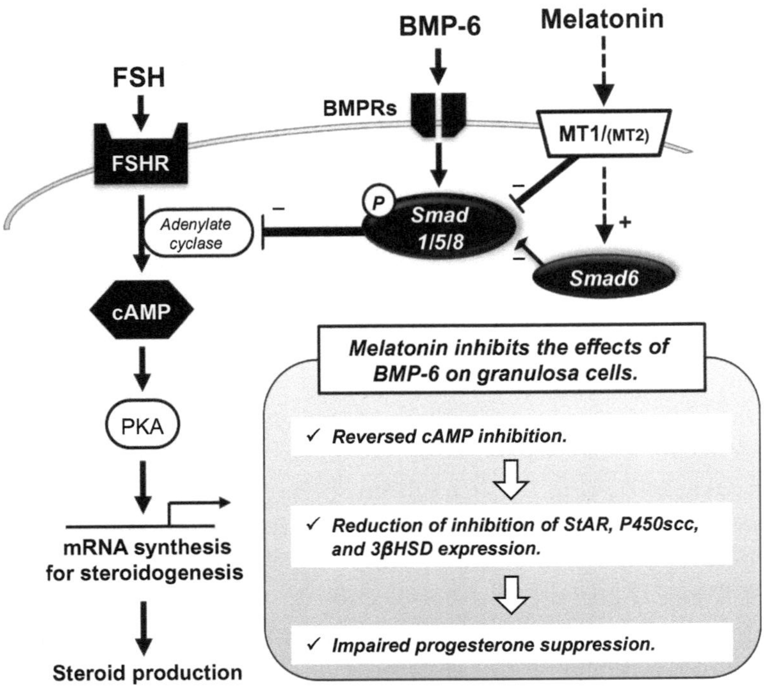

Fig. 2 Interaction of BMP-6 signaling and melatonin action in ovarian granulosa cells. BMP-6 expressed in oocytes and granulosa cells inhibits FSH actions by suppressing adenylate cyclase activity. The inhibitory effect of BMP-6 on progesterone production induced by FSH is impaired by melatonin treatment. BMPR signaling, including Smad1/5/8 phosphorylation and Id-1 transcription, is suppressed by melatonin, suggesting that melatonin plays an inhibitory role in BMPR signaling in granulosa cells. As a possible mechanism of the effects of melatonin, it was found that melatonin upregulates inhibitory Smad6 expression. Thus, melatonin plays a role in the maintenance of progesterone production by suppressing BMP-6 signaling, which involves suppression of cAMP inhibition and impairment of reduced steroidogenetic enzyme expression.

Analysis of human ovary tissues from women with PCOS has shown that GDF-9 mRNA expression was delayed and reduced during the growth and differentiation phase compared with BMP-15 mRNA expression (Teixeira Filho et al., 2002). It was also reported that protein expression of GDF-9 and BMP-15 was not different in oocytes from PCOS cases and control cases, whereas GDF-9 protein expression in cumulus granulosa cells was decreased (Zhao et al., 2010). Recent clinical studies have also shown enhanced expression of BMP-6 in granulosa cells isolated from PCOS patients (Khalaf et al., 2013; Kim, Kang, Yoon, Shim, & Lee, 2014; Schmidt et al., 2014).

4. MELATONIN RECEPTOR EXPRESSION IN THE OVARY

Melatonin is a biological rhythm regulator and has a variety of other essential functions. Melatonin plays regulatory roles in the physiologic circadian rhythm, seasonal formation, and the activities of hormones and cytokines expressed in various tissues and organs (Borjigin et al., 2012; Pandi-Perumal et al., 2008; Reiter et al., 2013, 2014). In the periphery, melatonin transduces seasonal information that is involved in defining the length of the night (Borjigin et al., 2012). Biological activities of melatonin are elicited by two types of G protein-coupled receptors, called MT1 and MT2, which are expressed in the brain as well as various peripheral tissues (Dubocovich, 2007). Of note, melatonin is functionally involved in the formation of reproductive rhythm for seasonal animals by acting on the pars tuberalis of the pituitary expressing a high density of melatonin receptors (Hazlerigg, Morgan, & Messager, 2001; Pevet et al., 2006). Both the receptors of MT1 and MT2 are expressed in various endocrine tissues including hypothalamic gonadotropin-releasing hormone neurons and ovarian follicles, suggesting that melatonin action is functionally linked to the reproductive system comprised of the hypothalamo–pituitary–ovary (HPO) system.

Melatonin has wide-ranging effects, at the levels of the gonads, on various component cells in ovarian follicles, particularly granulosa cells and oocytes (Reiter et al., 2013). Bioactivities of melatonin are mainly mediated by the membrane receptors MT1 and MT2 and also by other binding sites localized in the nucleus and/or the cytosol. Expression of MT1 and MT2 in the rat ovary has been observed (Soares, Masana, Ersahin, & Dubocovich, 2003), and the binding capacity to melatonin was shown to be increased in the stages of the proestrus phase compared with that in metestrus. This finding indicates the possibility of expressional regulation of melatonin receptors by estrogen action in the follicles. It has also been reported that the ovarian MT1 and MT2 receptors were differentially located at the cell membrane, cytoplasm, and nuclear membrane in bovine granulosa cells and also regulated by melatonin action (Wang et al., 2012). The finding that a key synthetic enzyme of melatonin, acetylserotonin *O*-methyltransferase (ASMT), is also expressed in the bovine cumulus–oocyte complex, in addition to MT1, indicates the possibility of local melatonin productivity in the ovary, while the MT2 gene is expressed only in oocytes (El-Raey et al., 2011).

Melatonin also functions as a direct and powerful free-radical scavenger to reduce oxidative stress at the level of the ovary without interacting with the proper receptors (Tamura et al., 2009, 2013). Because of its amphiphilic property, melatonin effectively passes through various physiological barriers and acts as a highly effective antioxidant against oxidative stress caused by reactive species (Cruz, Leal, Cruz, Tan, & Reiter, 2014). The direct activities of melatonin and its metabolites involve the scavenging of both reactive oxygen species (ROS) and reactive nitrogen species (Cruz et al., 2014). ROS have been shown to suppress progesterone production through inhibition of the activities of steroidogenic enzymes and intracellular carrier proteins involved in transport of cholesterol to mitochondria in luteal cells in the ovary. It has been shown that melatonin abolished the suppression of progesterone synthesis elicited by ROS in luteinized granulosa cells (Tamura et al., 2009, 2013). Thus, melatonin is likely to protect granulosa cells from ROS in growing follicles during the ovulation process and also may support normal luteinization of granulosa cells and enhancement of subsequent fertilization.

5. INTERACTION OF MELATONIN AND BMP-6 IN GRANULOSA CELLS

As mentioned earlier, both melatonin and BMP-6 are involved in progesterone synthesis and luteinization process in the ovary, implying the existence of a mutual regulatory pathway between melatonin and BMP receptor signaling in granulosa cells. In order to clarify the functional interrelationship between melatonin and BMP-6 and its molecular mechanism, we used rat primary granulosa cells obtained from immature female rat ovaries (Nakamura et al., 2014). In vitro treatment with melatonin (0–1000 nM) did not affect basal steroidogenetic activities of estradiol or progesterone production by granulosa cells in a serum-free condition. Treatment with melatonin also had no direct effect on FSH-induced estradiol or progesterone production by granulosa cells, and the results were not affected by the presence of cocultured oocytes.

Of note, the inhibitory effect of BMP-6 on FSH-induced progesterone production was reversed by cotreatment with melatonin in a concentration-dependent manner (100–1000 nM) (Nakamura et al., 2014) (Fig. 3A). The same effect on progesterone production was also observed in granulosa cells cocultured with oocytes by treatment with melatonin. In accordance with

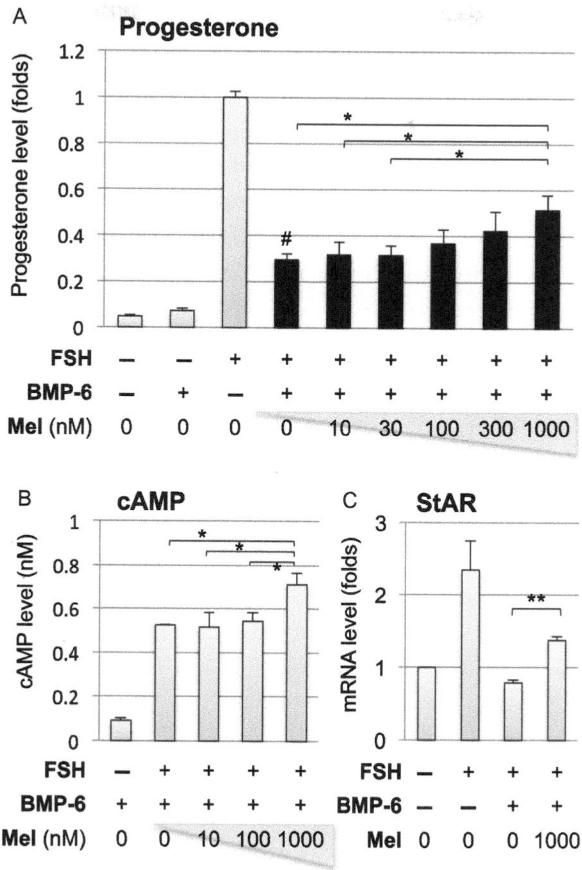

Fig. 3 Effects of melatonin on BMP-6 activity in rat granulosa cells. Key data that show functional interaction of melatonin and BMP-6 in granulosa cells were extracted from our previous report (Nakamura et al., 2014). (A) Progesterone production: Rat granulosa cells were cultured with melatonin (Mel) and/or BMP-6 (100 ng/mL) in the presence or absence of FSH (30 ng/mL). After 48-h culture, the levels of progesterone in the medium were determined, and the fold changes normalized by the levels in the medium treated with FSH alone are shown. (B) cAMP synthesis: Granulosa cells were cultured in a serum-free medium containing IBMX (0.1 mM), FSH in combination with melatonin and/or BMP-6. After 48-h culture, the levels of cAMP in the medium were determined. (C) StAR mRNA level: Total cellular RNA was extracted from granulosa cells treated with FSH, BMP-6, and melatonin for 48 h, and StAR mRNA level was determined by quantitative PCR. The expression levels of target gene mRNA were standardized by RPL19 level in each sample, and then levels of mRNA of each target gene were expressed as fold changes. Results are shown as means ± SEM. The results were analyzed by ANOVA and the unpaired *t*-test. #, $P < 0.05$ vs the group treated with FSH alone; *, $P < 0.05$ and **, $P < 0.01$ between the indicated groups vs control groups.

the changes in progesterone production, FSH-induced cAMP levels in the medium were not affected by melatonin treatment, whereas the inhibitory effect of BMP-6 on cAMP was reversed by melatonin in granulosa cells (Fig. 3B). Cotreatment with melatonin also reversed the inhibitory effects of BMP-6 on the mRNA expression of StAR, P450scc, and 3βHSD2 induced by FSH in granulosa cells (Nakamura et al., 2014) (Fig. 3C).

Since the inhibitory effect of BMP-6 on FSH-induced progesterone production was impaired by cotreatment with melatonin, we further investigated the interaction of melatonin and BMPR signaling in granulosa cells. It was found that BMP-6-induced Smad1/5/8 phosphorylation was suppressed by melatonin in the presence or absence of FSH (Nakamura et al., 2014). Induction of mRNA expression of the BMP target gene Id-1 by BMP-6 with or without FSH was also suppressed by treatment with melatonin, although melatonin alone did not affect the Id-1 mRNA level. The expression levels of BMP-6 receptors, including type-I receptors (ALK-2 and ALK-6) and type-II receptors (ActRII and BMPRII), were not changed by melatonin. The expression level of MT1, a predominant melatonin receptor in granulosa cells, was not affected by BMP-6 and melatonin treatment. It was found that melatonin treatment upregulated the expression of Smad6, but not that of Smad7, in granulosa cells (Nakamura et al., 2014).

Namely, BMPR–Smad1/5/8 signaling induced by BMP-6 was suppressed by melatonin (Fig. 2). MT1 expression in granulosa cells was not affected by BMP-6, and BMPR expression remained stable under the condition of melatonin treatment. However, the expression of inhibitory Smad6 was found to be upregulated by melatonin, suggesting a new regulatory mechanism of melatonin in BMP-6 signal intensity in granulosa cells (Fig. 2).

6. OTHER ENDOCRINE REGULATION BY MELATONIN AND BMPs

In ovarian steroidogenesis, the potent inhibitory effects of BMP-6 on FSH-induced progesterone production and cAMP synthesis were reversed by cotreatment with melatonin, indicating that melatonin has an antagonistic effect on BMP-6 in steroidogenesis by granulosa cells. Similar regulatory functions of melatonin and BMPs are also seen in steroidogenesis within the adrenal cortex (Fig. 4).

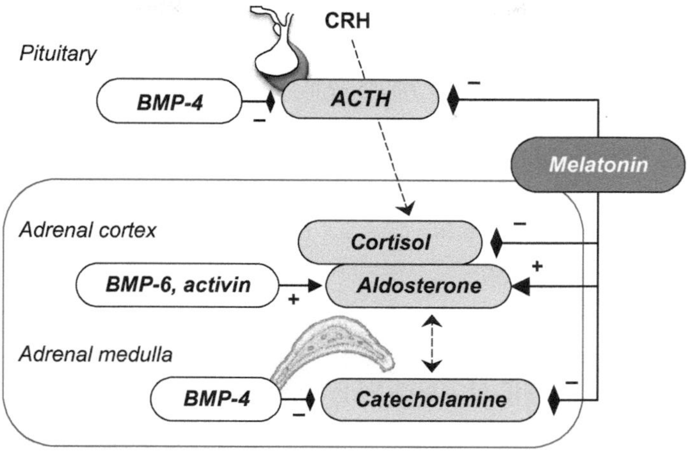

Fig. 4 HPA axis regulated by the interaction of melatonin and BMPs. Melatonin is involved in the HPA axis regulation and cortex–medullar interaction in the adrenal as follows: (1) melatonin suppresses ACTH production in cooperation with BMP-4 action in pituitary corticotrope cells, (2) melatonin suppresses adrenal cortisol production, (3) melatonin activates aldosterone production induced by ACTH and activin in adrenocortical cells, and (4) melatonin suppresses catecholamine production by cooperating with BMP-4 in adrenomedullar cells.

Adrenocortical and medullar tissues functionally interact in a paracrine manner through their close anatomical structures (Schinner & Bornstein, 2005). Endogenous glucocorticoids promote catecholamine synthesis by inducing catecholamine-synthesizing enzymes via the corticomedullary portal system in the adrenal. We earlier reported the presence of the BMP system in adrenomedullary cells (Kano et al., 2005) as well as in adrenocortical cells (Otani et al., 2008; Suzuki et al., 2004). For instance, BMP-6 is involved in the maintenance of angiotensin II-induced aldosterone production by upregulating MAPK signaling (Otani et al., 2008), while the activin system is linked to the adrenocorticotropin-induced cAMP–PKA pathway in adrenocortical cells (Hara et al., 2015). Endogenous BMP-4 and its signaling machinery are expressed in the adrenal medulla and they play a regulatory role in catecholamine synthesis induced by adrenocortical steroids (Komatsubara et al., 2017).

Melatonin receptors, mainly the MT1 receptor, have been detected in adrenal tissues and cells in mammals (Torres-Farfan et al., 2003). As for the effects of melatonin on adrenocortical hormones, it has been reported that melatonin, directly on the adrenal gland, inhibits glucocorticoid synthesis in response to adrenocorticotropin (ACTH) in various species (Richter et al., 2008; Torres-Farfan et al., 2003). Melatonin also suppresses ACTH secretion via BMP-4 action in corticotrope cells (Tsukamoto et al., 2013) (Fig. 4). Under the influence of ACTH, melatonin facilitates aldosterone production by cooperating with activin in adrenocortical cells (Hara et al., 2015). It was further revealed that melatonin plays an inhibitory role in catecholamine synthesis by cooperating with BMP-4 and glucocorticoid in adrenomedullar cells (Komatsubara et al., 2017). Hence, melatonin is involved not only in HPO regulation but also in the HPA axis and cortico–medullar interaction by regulating adrenal steroidogenesis.

7. CONCLUSION AND PERSPECTIVES

BMP-6 suppresses FSH-induced expression of steroidogenic enzymes and FSH-induced progesterone synthesis by inhibiting FSH action at the level of adenylate cyclase (Otsuka, Moore, & Shimasaki, 2001; Shimasaki et al., 2004) (Fig. 2). Cellular signaling of BMP-6, Smad1/5/8 phosphorylation, and Id-1 transcription was impaired by melatonin via upregulation of inhibitory Smad6 expression, suggesting that melatonin has a counteracting role in the BMP–Smad pathway and in the regulation of BMP-6 signal intensity for maintaining progesterone production in the ovary (Nakamura et al., 2014) (Fig. 2).

Recently, the possibility of effectiveness of melatonin for female infertility caused by PCOS has been proposed (Saha, Kaur, & Saha, 2012). Decreased qualities of oocytes and embryos, in addition to anovulation, are likely to be involved in the cause of infertility by PCOS. Oxidative stress induced by ROS may be responsible for the poor qualities of oocytes and granulosa cells. Considering that oxidative stress induced by ROS is likely to be responsible for determination of oocyte quality and apoptotic granulosa cells in PCOS, preservation of the melatonin level in follicular fluid is important for follicle growth, ovulation, and maintenance of oocyte quality (Reiter et al., 2013, 2014; Saha et al., 2012; Tamura et al., 2009, 2013). Melatonin may be an effective supplement for ameliorating ovarian function and oocyte quality in women with PCOS (Saha et al., 2012; Tamura et al., 2009).

Recent studies have also shown that BMP-6 mRNA level was increased in granulosa cells isolated from PCOS patients compared with that in granulosa cells from control patients undergoing in vitro fertilization (Khalaf et al., 2013; Kim et al., 2014; Schmidt et al., 2014). BMP-2, -4, and -6 were undetectable in serum of PCOS patients, but BMP-7 was detected in serum from 3 of 20 patients, albeit at the lower limit of the standard curve (van Houten et al., 2013). In the human ovary, BMP-6 is predominantly expressed in granulosa cells of healthy tertiary follicles but not in atretic follicles (Shi et al., 2009). Increased expression of BMP-6 in granulosa cells of PCOS patients may indicate the existence of a certain disturbance of follicle development. The effect of melatonin that counteracts BMP-6 activity in granulosa cells may be important for compensating the reduced progesterone level and recovering the possible arrest of follicular growth in PCOS patients.

Collectively, it was shown that melatonin has antagonistic effects on BMP-6 actions in granulosa cells. Since ovarian BMP-6 acts as a luteinization inhibitor for normal folliculogenesis, melatonin supplementation may contribute to balancing the biological activity of endogenous BMP-6 for maintaining progesterone production and luteinization in growing follicles.

ACKNOWLEDGMENTS

This work was supported in part by Grants-in-Aid for Scientific Research (No. 24591364 and 15K09434), Foundation for Growth Science, Astellas Foundation for Research on Metabolic Disorders, Japan Foundation for Applied Enzymology (Japan), and The Uehara Memorial Foundation. I am grateful to Drs. Eri Nakamura, Takeshi Hosoya, Takayuki Hara, Motoshi Komatsubara, and Nahoko Iwata for their efforts in experiments.

DISCLOSURE STATEMENT

The authors have nothing to disclose.

REFERENCES

Borjigin, J., Zhang, L. S., & Calinescu, A. A. (2012). Circadian regulation of pineal gland rhythmicity. *Molecular and Cellular Endocrinology*, *349*, 13–19.

Cruz, M. H., Leal, C. L., Cruz, J. F., Tan, D. X., & Reiter, R. J. (2014). Essential actions of melatonin in protecting the ovary from oxidative damage. *Theriogenology*, *82*, 925–932.

David, L., Mallet, C., Keramidas, M., Lamande, N., Gasc, J. M., Dupuis-Girod, S., et al. (2008). Bone morphogenetic protein-9 is a circulating vascular quiescence factor. *Circulation Research*, *102*, 914–922.

Dong, J., Albertini, D. F., Nishimori, K., Kumar, T. R., Lu, N., & Matzuk, M. M. (1996). Growth differentiation factor-9 is required during early ovarian folliculogenesis. *Nature*, *383*, 531–535.

Dubocovich, M. L. (2007). Melatonin receptors: Role on sleep and circadian rhythm regulation. *Sleep Medicine*, *8*(Suppl. 3), 34–42.

El-Raey, M., Geshi, M., Somfai, T., Kaneda, M., Hirako, M., Abdel-Ghaffar, A. E., et al. (2011). Evidence of melatonin synthesis in the cumulus oocyte complexes and its role in enhancing oocyte maturation in vitro in cattle. *Molecular Reproduction and Development*, *78*, 250–262.

Eppig, J. J. (2001). Oocyte control of ovarian follicular development and function in mammals. *Reproduction*, *122*, 829–838.

Erickson, G. F., & Shimasaki, S. (2003). The spatiotemporal expression pattern of the bone morphogenetic protein family in rat ovary cell types during the estrous cycle. *Reproductive Biology and Endocrinology, 1*, 9.

Findlay, J. K., Drummond, A. E., Dyson, M. L., Baillie, A. J., Robertson, D. M., & Ethier, J. F. (2002). Recruitment and development of the follicle: The roles of the transforming growth factor-beta superfamily. *Molecular and Cellular Endocrinology, 191*, 35–43.

Galloway, S. M., McNatty, K. P., Cambridge, L. M., Laitinen, M. P., Juengel, J. L., Jokiranta, T. S., et al. (2000). Mutations in an oocyte-derived growth factor gene (BMP15) cause increased ovulation rate and infertility in a dosage-sensitive manner. *Nature Genetics, 25*, 279–283.

Gilchrist, R. B., Ritter, L. J., Myllymaa, S., Kaivo-Oja, N., Dragovic, R. A., Hickey, T. E., et al. (2006). Molecular basis of oocyte-paracrine signalling that promotes granulosa cell proliferation. *Journal of Cell Science, 119*, 3811–3821.

Hara, T., Otsuka, F., Tsukamoto-Yamauchi, N., Inagaki, K., Hosoya, T., Nakamura, E., et al. (2015). Mutual effects of melatonin and activin on induction of aldosterone production by human adrenocortical cells. *The Journal of Steroid Biochemistry and Molecular Biology, 152*, 8–15.

Hazlerigg, D. G., Morgan, P. J., & Messager, S. (2001). Decoding photoperiodic time and melatonin in mammals: What can we learn from the pars tuberalis? *Journal of Biological Rhythms, 16*, 326–335.

Hosoya, T., Otsuka, F., Nakamura, E., Terasaka, T., Inagaki, K., Tsukamoto-Yamauchi, N., et al. (2015). Regulatory role of BMP-9 in steroidogenesis by rat ovarian granulosa cells. *The Journal of Steroid Biochemistry and Molecular Biology, 147*, 85–91.

Hsueh, A. J., Kawamura, K., Cheng, Y., & Fauser, B. C. (2015). Intraovarian control of early folliculogenesis. *Endocrine Reviews, 36*, 1–24.

Inagaki, K., Otsuka, F., Miyoshi, T., Yamashita, M., Takahashi, M., Goto, J., et al. (2009). p38-Mitogen-activated protein kinase stimulated steroidogenesis in granulosa cell-oocyte cocultures: Role of bone morphogenetic proteins 2 and 4. *Endocrinology, 150*, 1921–1930.

Kano, Y., Otsuka, F., Takeda, M., Suzuki, J., Inagaki, K., Miyoshi, T., et al. (2005). Regulatory roles of bone morphogenetic proteins and glucocorticoids in catecholamine production by rat pheochromocytoma cells. *Endocrinology, 146*, 5332–5340.

Khalaf, M., Morera, J., Bourret, A., Reznik, Y., Denoual, C., Herlicoviez, M., et al. (2013). BMP system expression in GCs from polycystic ovary syndrome women and the in vitro effects of BMP4, BMP6, and BMP7 on GC steroidogenesis. *European Journal of Endocrinology, 168*, 437–444.

Kim, J. W., Kang, K. M., Yoon, T. K., Shim, S. H., & Lee, W. S. (2014). Study of circulating hepcidin in association with iron excess, metabolic syndrome, and BMP-6 expression in granulosa cells in women with polycystic ovary syndrome. *Fertility and Sterility, 102*, 548–554 e2.

Komatsubara, M., Hara, T., Hosoya, T., Toma, K., Tsukamoto-Yamauchi, N., Iwata, N., et al. (2017). Melatonin regulates catecholamine biosynthesis by modulating bone morphogenetic protein and glucocorticoid actions. *Journal of Steroid Biochemistry and Molecular Biology, 165*, 182–189.

Lee, W., Otsuka, F., Moore, R. K., & Shimasaki, S. (2001). Effect of bone morphogenetic protein-7 on folliculogenesis and ovulation in the rat. *Biology of Reproduction, 65*, 994–999.

Miyoshi, T., Otsuka, F., Inagaki, K., Otani, H., Takeda, M., Suzuki, J., et al. (2007). Differential regulation of steroidogenesis by bone morphogenetic proteins in granulosa cells: Involvement of extracellularly regulated kinase signaling and oocyte actions in follicle-stimulating hormone-induced estrogen production. *Endocrinology, 148*, 337–345.

Miyoshi, T., Otsuka, F., Suzuki, J., Takeda, M., Inagaki, K., Kano, Y., et al. (2006). Mutual regulation of follicle-stimulating hormone signaling and bone morphogenetic protein system in human granulosa cells. *Biology of Reproduction, 74,* 1073–1082.

Monniaux, D. (2016). Driving folliculogenesis by the oocyte-somatic cell dialog: Lessons from genetic models. *Theriogenology, 86,* 41–53.

Mueller, T. D. (2015). Mechanisms of BMP-receptor interaction and activation. *Vitamins and Hormones, 99,* 1–61.

Mulloy, B., & Rider, C. C. (2015). The bone morphogenetic proteins and their antagonists. *Vitamins and Hormones, 99,* 63–90.

Nakamura, E., Otsuka, F., Terasaka, T., Inagaki, K., Hosoya, T., Tsukamoto-Yamauchi, N., et al. (2014). Melatonin counteracts BMP-6 regulation of steroidogenesis by rat granulosa cells. *The Journal of Steroid Biochemistry and Molecular Biology, 143,* 233–239.

Otani, H., Otsuka, F., Inagaki, K., Suzuki, J., Miyoshi, T., Kano, Y., et al. (2008). Aldosterone breakthrough caused by chronic blockage of angiotensin II type 1 receptors in human adrenocortical cells: Possible involvement of bone morphogenetic protein-6 actions. *Endocrinology, 149,* 2816–2825.

Otsuka, F. (2010). Multiple endocrine regulation by bone morphogenetic protein system. *Endocrine Journal, 57,* 3–14.

Otsuka, F. (2013). Multifunctional bone morphogenetic protein system in endocrinology. *Acta Medica Okayama, 67,* 75–86.

Otsuka, F., & Inagaki, K. (2011). Unique bioactivities of bone morphogenetic proteins in regulation of reproductive endocrine functions. *Reproductive Medicine and Biology, 10,* 131–142.

Otsuka, F., McTavish, K. J., & Shimasaki, S. (2011). Integral role of GDF-9 and BMP-15 in ovarian function. *Molecular Reproduction and Development, 78,* 9–21.

Otsuka, F., Moore, R. K., Iemura, S.-I., Ueno, N., & Shimasaki, S. (2001). Follistatin inhibits the function of the oocyte-derived factor BMP-15. *Biochemical and Biophysical Research Communications, 289,* 961–966.

Otsuka, F., Moore, R. K., & Shimasaki, S. (2001). Biological function and cellular mechanism of bone morphogenetic protein-6 in the ovary. *The Journal of Biological Chemistry, 276,* 32889–32895.

Otsuka, F., Moore, R. K., Wang, X., Sharma, S., Miyoshi, T., & Shimasaki, S. (2005). Essential role of the oocyte in estrogen amplification of follicle-stimulating hormone signaling in granulosa cells. *Endocrinology, 146,* 3362–3367.

Otsuka, F., & Shimasaki, S. (2002). A negative feedback system between oocyte bone morphogenetic protein 15 and granulosa cell kit ligand: Its role in regulating granulosa cell mitosis. *Proceedings of the National Academy of Sciences of the United States of America, 99,* 8060–8065.

Otsuka, F., Yamamoto, S., Erickson, G. F., & Shimasaki, S. (2001). Bone morphogenetic protein-15 inhibits follicle-stimulating hormone (FSH) action by suppressing FSH receptor expression. *The Journal of Biological Chemistry, 276,* 11387–11392.

Otsuka, F., Yao, Z., Lee, T. H., Yamamoto, S., Erickson, G. F., & Shimasaki, S. (2000). Bone morphogenetic protein-15: Identification of target cells and biological functions. *The Journal of Biological Chemistry, 275,* 39523–39528.

Pandi-Perumal, S. R., Trakht, I., Srinivasan, V., Spence, D. W., Maestroni, G. J., Zisapel, N., et al. (2008). Physiological effects of melatonin: Role of melatonin receptors and signal transduction pathways. *Progress in Neurobiology, 85,* 335–353.

Pangas, S. A. (2012). Bone morphogenetic protein signaling transcription factor (SMAD) function in granulosa cells. *Molecular and Cellular Endocrinology, 356,* 40–47.

Pevet, P., Agez, L., Bothorel, B., Saboureau, M., Gauer, F., Laurent, V., et al. (2006). Melatonin in the multi-oscillatory mammalian circadian world. *Chronobiology International, 23,* 39–51.

Reiter, R. J., Rosales-Corral, S. A., Manchester, L. C., & Tan, D. X. (2013). Peripheral reproductive organ health and melatonin: Ready for prime time. *International Journal of Molecular Sciences*, *14*, 7231–7272.
Reiter, R. J., Tamura, H., Tan, D. X., & Xu, X. Y. (2014). Melatonin and the circadian system: Contributions to successful female reproduction. *Fertility and Sterility*, *102*, 321–328.
Richter, H. G., Torres-Farfan, C., Garcia-Sesnich, J., Abarzua-Catalan, L., Henriquez, M. G., Alvarez-Felmer, M., et al. (2008). Rhythmic expression of functional MT1 melatonin receptors in the rat adrenal gland. *Endocrinology*, *149*, 995–1003.
Saha, L., Kaur, S., & Saha, P. K. (2012). Pharmacotherapy of polycystic ovary syndrome—An update. *Fundamental & Clinical Pharmacology*, *26*, 54–62.
Schinner, S., & Bornstein, S. R. (2005). Cortical-chromaffin cell interactions in the adrenal gland. *Endocrine Pathology*, *16*, 91–98.
Schmidt, J., Weijdegard, B., Mikkelsen, A. L., Lindenberg, S., Nilsson, L., & Brannstrom, M. (2014). Differential expression of inflammation-related genes in the ovarian stroma and granulosa cells of PCOS women. *Molecular Human Reproduction*, *20*, 49–58.
Shi, J., Yoshino, O., Osuga, Y., Koga, K., Hirota, Y., Hirata, T., et al. (2009). Bone morphogenetic protein-6 stimulates gene expression of follicle-stimulating hormone receptor, inhibin/activin beta subunits, and anti-Mullerian hormone in human granulosa cells. *Fertility and Sterility*, *92*, 1794–1798.
Shimasaki, S., Moore, R. K., Erickson, G. F., & Otsuka, F. (2003). The role of bone morphogenetic proteins in ovarian function. *Reproduction Supplement*, *61*, 323–337.
Shimasaki, S., Moore, R. K., Otsuka, F., & Erickson, G. F. (2004). The bone morphogenetic protein system in mammalian reproduction. *Endocrine Reviews*, *25*, 72–101.
Shimasaki, S., Zachow, R. J., Li, D., Kim, H., Iemura, S., Ueno, N., et al. (1999). A functional bone morphogenetic protein system in the ovary. *Proceedings of the National Academy of Sciences of the United States of America*, *96*, 7282–7287.
Soares, J. M., Jr., Masana, M. I., Ersahin, C., & Dubocovich, M. L. (2003). Functional melatonin receptors in rat ovaries at various stages of the estrous cycle. *The Journal of Pharmacology and Experimental Therapeutics*, *306*, 694–702.
Sugimoto, H., Yang, C., LeBleu, V. S., Soubasakos, M. A., Giraldo, M., Zeisberg, M., et al. (2007). BMP-7 functions as a novel hormone to facilitate liver regeneration. *The FASEB Journal*, *21*, 256–264.
Suzuki, J., Otsuka, F., Inagaki, K., Takeda, M., Ogura, T., & Makino, H. (2004). Novel action of activin and bone morphogenetic protein in regulating aldosterone production by human adrenocortical cells. *Endocrinology*, *145*, 639–649.
Tamura, H., Nakamura, Y., Korkmaz, A., Manchester, L. C., Tan, D. X., Sugino, N., et al. (2009). Melatonin and the ovary: Physiological and pathophysiological implications. *Fertility and Sterility*, *92*, 328–343.
Tamura, H., Takasaki, A., Taketani, T., Tanabe, M., Kizuka, F., Lee, L., et al. (2013). Melatonin as a free radical scavenger in the ovarian follicle. *Endocrine Journal*, *60*, 1–13.
Teixeira Filho, F. L., Baracat, E. C., Lee, T. H., Suh, C. S., Matsui, M., Chang, R. J., et al. (2002). Aberrant expression of growth differentiation factor-9 in oocytes of women with polycystic ovary syndrome. *The Journal of Clinical Endocrinology and Metabolism*, *87*, 1337–1344.
Torres-Farfan, C., Richter, H. G., Rojas-Garcia, P., Vergara, M., Forcelledo, M. L., Valladares, L. E., et al. (2003). mt1 melatonin receptor in the primate adrenal gland: Inhibition of adrenocorticotropin-stimulated cortisol production by melatonin. *The Journal of Clinical Endocrinology and Metabolism*, *88*, 450–458.
Tsukamoto, N., Otsuka, F., Ogura-Ochi, K., Inagaki, K., Nakamura, E., Toma, K., et al. (2013). Melatonin receptor activation suppresses adrenocorticotropin production via BMP-4 action by pituitary AtT20 cells. *Molecular and Cellular Endocrinology*, *375*, 1–9.

van Houten, E. L., Laven, J. S., Louwers, Y. V., McLuskey, A., Themmen, A. P., & Visser, J. A. (2013). Bone morphogenetic proteins and the polycystic ovary syndrome. *Journal of Ovarian Research*, *6*, 32.

Wagner, D. O., Sieber, C., Bhushan, R., Borgermann, J. H., Graf, D., & Knaus, P. (2010). BMPs: From bone to body morphogenetic proteins. *Science Signaling*, *3*, mr1.

Wang, S. J., Liu, W. J., Wu, C. J., Ma, F. H., Ahmad, S., Liu, B. R., et al. (2012). Melatonin suppresses apoptosis and stimulates progesterone production by bovine granulosa cells via its receptors (MT1 and MT2). *Theriogenology*, *78*, 1517–1526.

Zhao, S. Y., Qiao, J., Chen, Y. J., Liu, P., Li, J., & Yan, J. (2010). Expression of growth differentiation factor-9 and bone morphogenetic protein-15 in oocytes and cumulus granulosa cells of patients with polycystic ovary syndrome. *Fertility and Sterility*, *94*, 261–267.

CHAPTER SIX

The Effect of Steroid Hormones on Ovarian Follicle Development

Chia-Hong Chou*, Mei-Jou Chen†,1

*Department of Obstetrics and Gynecology, National Taiwan University Hospital, Taipei, Taiwan
†National Taiwan University Livia Shangyu Wan Scholar, Department of Obstetrics and Gynecology, National Taiwan University Hospital, Taipei, Taiwan
[1]Corresponding author: e-mail address: mjchen04@ntu.edu.tw

Contents

Abstract

Ovarian follicular cell undergoes extensive proliferation and differentiation during the period that the follicle evolved from the primordial state to its ovulatory phase and then corpus luteum status. During this process, different and various levels of steroid hormones in follicular fluid, or obtained from circulation and adjacent cells as granulosa and theca cells, or from environment and diet will all affect the process of follicular growth and development. Differential steroid hormones might have differential effects

Vitamins and Hormones, Volume 107
ISSN 0083-6729
https://doi.org/10.1016/bs.vh.2018.01.013

on ovarian folliculogenesis via the effect on granulosa cell growth and follicular fluid formation that involves cell proliferation, apoptosis, and angiogenesis within the follicle. In this chapter we further discuss the role of various steroid hormones such as estrogens, progesterone, and androgens on ovarian follicular growth and development. Various stages of follicle development that might be disturbed by the steroid hormones are also discussed in this chapter.

1. INTRODUCTION

The progression of a primordial follicle to the ovulatory stage follicle requires nearly 1 year in female. During the course of folliculogenesis, growth is achieved by cell proliferation of granulosa cells and formation of follicular fluid, whereas development involves cell proliferation, apoptosis, and angiogenesis within the follicle. In this chapter we discuss the role of various steroid hormones such as estrogens, progesterone, and androgens on follicular growth and development.

2. ESTROGEN EFFECT ON OVARIAN FOLLICLE DEVELOPMENT

2.1 Estrogen Does Not Have a Direct Effect on Granulosa Cell Proliferation

The levels of steroid hormones in follicular fluid obtained from women with normal ovarian function are closely related to the follicular size and health of the aspirated follicles, and also correlated to the phase of the menstrual cycle in which the follicular fluid sample obtained. A significant correlation exists between the granulosa cell DNA contents and concentration of estradiol (E2) in follicular fluid. Such correlation is restricted to nonpreovulatory follicles with a diameter >6 mm. An abundance of evidence supports the notion that E2 acts as a local mitogen in follicles (Westergaard, Christensen, & McNatty, 1986).

Ovarian follicular cell undergoes extensive proliferation and differentiation from the time that the follicle escapes from the primordial state to its ovulatory capacity. A study examined the dynamic modification of the phosphorylation state of the histone H3 N-terminal tail in granulosa cells (GCs) during follicular development. Estradiol of the rat ovarian follicles induces the activity of the mitotic kinase and thus provides evidence for mitotic regulation in follicle development (Aurora et al., 2005).

Estrogen functions via activation of two estrogen receptors (estrogen receptor-alpha [ER-α] and estrogen receptor-beta [ER-β]). ER-β is widely distributed in tissues throughout the body, including tissues previously considered to be estrogen-insensitive. The epithelium of ovarian granulosa cells is ER-α-negative, but ER-β-positive and ER-β appear to be involved in important physiologic processes, such as differentiation, extracellular matrix organization, and stromal–epithelial communication. In tissues in which both ERs are expressed, the two receptors counteract each other. In the uterus, mammary gland, and immune system, ER-α promotes proliferation, whereas ER-β has proapoptotic and prodifferentiating functions (Morani, Warner, & Gustafsson, 2008).

Estrogen plays an important role in the aging of estrogen-responsive tissues. Disruption of the aromatase gene caused estrogen deficiency in mice results in a significant inhibition of telomerase maintenance of telomeres in mouse ovaries in a tissue-specific manner. Such inhibition entails a significant shortening of telomeres and compromised proliferation in the ovarian follicular granulosa cells. Decreased levels of protooncogene c-Myc and telomerase reverse transcriptase were noted in response to estrogen deficiency by gene expression analysis. Estrogen replacement therapy leads to an increase in telomerase reverse transcriptase gene expression, telomerase activity, telomere length, and ovarian tissue growth, thereby restore ovary development to normal (Bayne et al., 2011).

Primordial follicles are generated early in life and remain dormant for prolonged periods in the mammal ovaries. The growth of primordial follicles resumes via primordial follicle activation, and growth continues until the preovulatory stage under regulation of estrogen and other growth factors. The phosphatidylinositol-3 kinase (PI3K)/acute transforming retrovirus thymoma protein kinase (Akt) signaling pathway is activated in GCs through the process of ovarian follicular growth. A p110δ isoform of the PI3K catalytic subunit encoded by the Pik3cd gene knockout strategy was used to identify the role of PI3K in the development of follicle (Li et al., 2013). Pik3cd-null female mice are subfertile, exhibit fewer growing follicles and more atretic antral follicles in the ovary, and respond poorly to exogenous gonadotropins as follicle-stimulating hormone (FSH) stimulation in vitro compared with controls. p110δ isoform of the PI3K catalytic subunit is crucial for the PI3K pathway in both FSH- and E2-stimulated follicle growth in ovarian granulosa cells; however, the p110δ isoform is not required for primordial follicle activation and oocyte development (Li et al., 2013). Therefore, estrogen does not have a direct effect on GC proliferation, which may due to the lack of ER-α; however, estrogen increases telomerase reverse transcriptase gene expression, telomerase activity, telomere length, mitotic kinase activation, and PI3K signaling, all of which are critical for follicle development.

2.2 Antiapoptosis Effect of Estrogen on Granulosa Cells

Apoptotic cell death is a mechanism underlying ovarian follicle atresia. In a study that using hypophysectomized and diethylstilbestrol (DES) capsules implanted immature rates to study the regulation of follicle cell apoptosis by sex steroids (Billig, Furuta, & Hsueh, 1993). In granulosa cells, an increase in apoptotic DNA fragmentation was seen 12h after withdrawal of DES implants, followed by a 25-fold increase at 48h. In situ analysis of DNA fragmentation on histologic sections of ovaries demonstrated that apoptosis induced by DES withdrawal is confined to the granulosa cells in early antral and preantral follicles. No increase in DNA breakdown was detected in thecal cells and interstitial tissues or GCs of primordial and primary follicles. In contrast, replacement with DES or estradiol benzoate completely prevented the observed ovarian weight loss and suppressed the increase in GC apoptosis. Treatment with estradiol benzoate suppressed apoptosis 2 days after removal of DES implants in a dose-dependent fashion. Furthermore, the antiatretogenic effect of estrogen was blocked by treatment with testosterone (0.5mg twice daily), which increased ovarian apoptotic DNA fragmentation and decreased ovarian weight in DES-treated animals in a time-dependent manner. Therefore, sex steroids play an important role in the regulation of apoptotic cell death of ovary, with estrogens preventing apoptosis and androgens antagonizing the effect of estrogens (Billig et al., 1993).

Several para/autocrine effects of estrogen within the ovary include increased ovarian weight, stimulation of GC growth, augmentation of FSH action, and attenuation of apoptosis through regulation of gonadotropin synthesis and secretion from the hypothalamic–pituitary axis.

ERα is present in all three components of the murine hypothalamic–pituitary–ovarian axis. In contrast, ERβ is easily detectable in ovarian granulosa cells but is low-to-absent in the pituitary of the adult mouse. The different expression pattern for the two ERs suggests the presence of separate roles for each ER in the regulation of ovarian function. The lack of ER-α in the hypothalamic–pituitary axis of ER-α-knockout (α-ERKO) mice results in chronic elevation of serum luteinizing hormone (LH) and is the primary basis for the ovarian phenotype with features of polycystic follicles and anovulation. Prolonged treatment with a gonadotropin releasing hormone (GnRH) antagonist reduces serum LH levels and prevents the α-ERKO cystic ovarian phenotype. To investigate a direct role for ER-α within the ovary, immature α-ERKO females were stimulated to ovulate with exogenous gonadotropins. The ovulatory capacity in immature α-ERKO females is reduced compared with age-matched wild-type; however, oocytes collected from α-ERKO mice were able to undergo successful in vitro fertilization. A similar discrepancy in oocyte yield was observed after superovulation of peripubertal wild-type and α-ERKO females. In addition, ovaries from immature superovulated α-ERKO females had several ovulatory, but unruptured follicles.

These studies indicated that treatment of α-ERKO females with a GnRH antagonist decreased the serum LH levels to within the wild-type range and also prevented development of the characteristic cystic and hemorrhagic follicle phenotype. A lack of functional ER-α within the ovary had no effect on the regulation of ER-β, LH-receptor, cyclin-D2, P450-side chain cleavage enzyme, prostaglandin synthase (PGS)-2, and progesterone receptor (PGR) which are required for follicular maturation and ovulation. Reduced number of ovulations following the administration of exogenous gonadotropins in the α-ERKO suggests an intraovarian role for ER-α in follicular development and ovulation (Couse, Bunch, Lindzey, Schomberg, & Korach, 1999).

NF-kappaB has also been implicated as an important regulator of genes controlling apoptosis, and estradiol may modulate NF-kappaB activation. After assay the NF-kappaB activity in relation to the levels of E2 and follicle growth on days 4, 6, and 8 of the first follicular wave following ovulation, the concentration of E2 in follicular fluid decreased and the activity of NF-kappaB in GCs lowered in accompany with the follicle growth. NF-kappaB activity in vitro and a minimal incidence of apoptosis ($<4\%$), as measured by the Annexin V and TUNEL assays, are associated with production of E2; however, inhibition of NF-kappaB in vitro leads to suppression of apoptosis. These results indicate that follicular NF-kappaB activation is associated with E2 production (Valdez & Turzillo, 2005). Experiments were also conducted to test whether or not E2 protects GCs from Fas ligand (FasL)-induced apoptosis and whether or not protection involves modulation of the cell cycle of proliferation. Treatment of cultured bovine GCs with E2 decreases susceptibility to FasL-induced apoptosis. The effects of E2 are mediated through the estrogen receptor and are not mediated by stimulation of IGF production. E2 also increases the percentage of cells progressing from the G1-to-S phase of the cell cycle and increases expression of cyclin D2 protein and the cell proliferation marker, Ki67. Progression from the G1-to-S phase of the cell cycle was necessary for the protective effect of E2. Blocking progression from the G1-to-S phase with the cdk2 inhibitor, roscovitine, or blocking cells in the S phase with hydroxyurea, prevents protection by E2. E2 stimulates transition from the G1-to-S phase and protects against apoptosis in ovarian granulosa cells only when cell cycle progression is unperturbed (Quirk, Cowan, & Harman, 2006).

Complementary DNA array was used to study the pattern of ovarian mRNA expression from day-2 and -4 Swiss mice to study the genes involved in ovarian primordial-to-primary follicle transition. The expression profiles of 1176 genes in neonatal mouse ovaries on days 2 and 4, which contain primordial and primary follicles, respectively, were compared. The results revealed that 26% of the genes were differentially expressed between day-2 and -4 ovaries, with 19% upregulated and 7% downregulated on day 4. Analysis of differentially expressed genes revealed that the primordial-to-primary follicle stage transition is associated with induction in the expression of mainly growth factors, immune-related factors, hormone and hormone receptors, and signal transducers. The transition is also associated with proliferation of granulosa cells and absence of apoptosis. The primary follicles express ER-β and are responsive to estrogen actions in vitro with respect to the increase in the number of primary follicles and GC proliferation (Dharma, Modi, & Nandedkar, 2009).

The antifolliculogenic effect of oral dienogest was evaluated by using female cynomolgus monkeys started since day 7 of the menstrual cycle. In this study, the plasma E2 level declined within 24 h after dinogest treatment, while dienogest did not decrease the FSH level prior to the E2 decline. After the decline in the E2 level, the low level of E2 was sustained for >11 days. It is thought that a single oral dose of dienogest induces atresia of the dominant follicle. On histologic examination, three animals that showed a decline in the E2 level also revealed apoptotic changes in GCs with scattered aromatase expression in the ovarian dominant follicles. Such finding suggested that dienogest inhibits the plasma E2 level by inducing the atresia of the ovarian dominant follicle directly (Sasagawa et al., 2008).

During mouse embryonic development, oocytes develop in germline cysts, formed by several rounds of cell division and incomplete cytokinesis. Shortly after birth, cysts break down and individual oocytes are enclosed by

GCs to form primordial follicles. At the same time, two-thirds of the oocytes die by apoptosis, with only one-third surviving. The steroid hormones, E2 and P4, as well as the phytoestrogen, genistein, inhibit cyst breakdown and primordial follicle assembly. E2 could exert its effect through the effects of membrane-bound estrogen receptors to inhibit the cyst breakdown and therefore involved in estrogen signaling in neonatal oocyte development (Chen, Breen, & Pepling, 2009).

The cytotoxic drug, cis-platinum, could induce premature ovarian failure by reducing the viability of human granulosa cells. Steroid hormones as E2 and P4 decreased in human granulosa-luteal cells which incubated with cis-platinum. The E2 production was more pronounced than P4 with cis-platinum treatment. Such findings demonstrated that cis-platinum induces apoptosis of human granulosa-luteal cells in culture with impaired steroidogenesis, which may be one mechanism by which a cis-platinum-containing regimen induces premature ovarian failure (Chatterjee, Helal, Mobberley, Ryder, & Bajoria, 2014).

CCAAT enhancer binding proteins (CEBP)-β has been identified in the ovary and is critical for follicular growth, ovulation, and luteinization in mice. In the CEBP-β gene knocked down porcine GCs, knockdown of CEBP-β significantly increased the expression of p-ERK1/2 and arrested the GCs at the S phase of the cell cycle, but had no effects on cell apoptosis. More importantly, knockdown of CEBP-β markedly downregulated the expression of E2 and P4 in the culture medium. Messenger RNA expression of bcl-2 (antiapoptosis), StAR, and Runx2 (steroid hormone synthesis) was upregulated, while genes related to apoptosis (caspase-3 and p53), hormonal synthesis (CYP11A1), and cell cycle (cyclinA1, cyclinB1, and cyclinD1) were downregulated, suggesting that knockdown of CEBP-β may inhibit apoptosis, and regulate the cell cycle and hormone secretions at the transcriptional level in porcine GCs (Zhen et al., 2014).

2.3 Effect of Estrogen on Follicular Angiogenesis

The rapid, controlled, and cyclic nature of angiogenesis in the ovarian follicle suggests the potential for sex steroids to influence neovascularization. Angiogenesis is regulated by a local balance between the levels of endogenous stimulators and inhibitors. There is limited data regarding the direct effect of estrogen on angiogenesis of ovarian GCs; however, existing evidence suggests that estrogens stimulate angiogenesis via effects on endothelial cells. Estradiol-17beta [E(2)β] and its metabolites, which are sequentially synthesized by cytochrome P450s and catechol-O-methyltransferase (COMT) to form 2- and 4-hydroxyestradiol [OHE(2)] and 2- and 4-methoxestradiol [ME(2)], are elevated during pregnancy. A study investigated whether or not cytochrome P450s and COMT are expressed in uterine artery endothelial cells, and whether or not E(2)β and its metabolites modulate cell proliferation via ER-α and/or ER-β and play roles in physiologic uterine angiogenesis during pregnancy by using the 5-bromodeoxyuridine proliferation assay (Jobe et al., 2010). Using confocal microscopy and Western blot analyses to determine enzyme locations and levels, we observed CYP1A1, CYP1A2, CYP1B1, CYP3A4, and COMT expression in uterine artery endothelial cells; however, all of these expressions were similar between nonpregnant and pregnant uterine artery endothelial cells. E(2)β, 2-OHE(2), 4-OHE(2), and 4-ME(2) treatments stimulated proliferation in pregnant, but not nonpregnant uterine artery endothelial cells in a concentration-dependent fashion; 2-ME(2) did not stimulate proliferation in either cell type. Proliferative responses of pregnant uterine artery endothelial cells to E(2)β were solely mediated by ER-β, whereas responses to E(2)β metabolites were neither ER-α- nor ER-β-mediated. This study demonstrated an important vascular role for E(2)β, its cytochrome P450- and COMT-derived metabolites, and ER-β in uterine angiogenesis regulation during pregnancy that may be dysfunctional in preeclampsia and other cardiovascular disorders (Jobe et al., 2010).

Vascular endothelial growth factor (VEGF), another factor that plays a pivotal role in the regulation of microvascular permeability and angiogenesis, processes essential for normal endometrial growth and implantation. Estrogen could stimulate VEGF expression in the uterus at the transcriptional level; however, the VEGF gene promoter lacks a consensus estrogen response element. Chromatin immunoprecipitation showed that both Sp1 and Sp3 were associated with a proximal, GC-rich region of the promoter of VEGF gene before E2 treatment. Both the α- and β-subunits of the transcription factor, hypoxia-inducible factor 1 (HIF-1), which regulates VEGF expression in response to hypoxia and several hormones and growth factors, were present in the uterus. E2 can induce recruitment of the α- and β-subunits of HIF-1 to the VEGF promoter, which contains the hypoxia response element to which HIF-1 binds. This binding was transient, matching the pattern of E2-induced VEGF expression. These results indicate that HIF-1 is an important mediator of E2-induced VEGF expression in the uterus (Kazi, Jones, & Koos, 2005).

2.4 Antiangiogenesis Effect of Estrogen Metabolites

2-Methoxyestradiol (2-ME) is an estradiol metabolite with antiangiogenic and antitumor activity. 2-ME is formed by GC COMT activity and is present in the normal follicle at high concentrations. In this unique microenvironment, 2-ME may regulate selected cell types via autocrine and/or paracrine action. By in vitro culture studies using primary cultures of hormone- and/or growth factor-stimulated porcine GCs, as well as two types of endothelial cells, primary cultures of porcine endothelial cells, and a spontaneously transformed rabbit endothelial vascular cell line, 2-ME or E2 exert differential mitotic and/or apoptotic effects on endothelial cells and GC (Shang, Konidari, & Schomberg, 2001). The 2-ME, but not E2, suppressed epidermal growth factor-stimulated rabbit endothelial vascular cells and epidermal growth factor/insulin-stimulated porcine endothelial cell proliferation. In contrast, 2-ME did not attenuate FSH/INS-stimulated porcine GC proliferation. E2 was not antimitotic to porcine GCs and was nonapoptotic in either cell type. 2-ME significantly increased apoptosis of rabbit endothelial vascular cells and attenuated the cell movement. These results show that 2-ME may have autocrine and/or paracrine action at its site of production and as a candidate ovarian angiogenesis inhibitor (Shang et al., 2001).

Cultured GCs are able to produce 2-ME under hypoxic conditions. 2-ME could be a local effector in determining the fine tuning responsible for follicle angiogenesis (Basini et al., 2011). A study was conducted to quantify 2-ME in human corpus lutea (CL) of different ages, determine the expression of cytochrome-P450-1A1 (CYP1A1) and COMT in CL, and clarify the action of 2-ME on VEGF secretion and luteal angiogenesis. CL were collected from 15 women via minilaparotomy for tubal sterilization. Granulosa lutein cells were harvested 36 h after hCG administration in patients undergoing IVF. The results demonstrated that plasma levels of E2 decreased in the late luteal phase in association with an increase in luteal tissue 2-ME concentrations. There was a significant reduction of angiogenic activity in late CL. There was no significant variation in CYP1A1 and COMT expression in all CL. At physiologic doses, 2-ME inhibited basal VEGF by granulosa lutein cells and diminished the angiogenic activity in conditioned media, but did not prevent VEGF production stimulated by hCG. The study demonstrated the role of 2-ME in physiologic luteolysis by reducing angiogenesis through inhibition of new vessel growth (Basini et al., 2008; Kohen et al., 2013).

3. ANDROGEN EFFECT ON OVARIAN FOLLICLE DEVELOPMENT

3.1 Apoptotic Phenomenon in Follicles and Corpora Lutea

Follicles and CL were evaluated for the presence of DNA fragmentation and characteristics of apoptosis in a clinical study that enrolled 116 human ovarian follicles (primordial to dominant) and 5 CL from a total of 27 premenopausal women (Yuan & Giudice, 1997). Follicle functional status was evaluated by determining follicle sizes and follicular fluid androgen-to-estrogen (A:E) ratios. No apoptosis was observed in 67 primordial, primary, or secondary follicles. Positive staining for DNA fragmentation was found in a limited number of GCs in 0.1–2 mm follicles, whereas abundant staining in GCs was detected in 2.1–9.9 mm follicles. In contrast, no DNA fragmentation was detected in dominant follicles (10–16 mm). The frequency of apoptosis in follicles was calculated to be 37% in 0.1–2 mm follicles, 50% in 2.1–5 mm follicles, and 27% in 5.1–9.9 mm follicles. Abundant low-molecular weight DNA laddering was only found in androgen-dominant follicles and not in estrogen-dominant follicles. Positive staining for DNA fragmentation and low-molecular weight DNA laddering was observed in degenerating, but not healthy appearing CL. In the former, DNA fragmentation was found primarily in large luteal cells. These data suggest that follicular atresia in human ovaries results from normal programmed cell death and primarily occurs in the GC layers of the early antral and antral follicles <10 mm in size. Furthermore, because apoptosis occurs as early as the 200-mm stage, follicle selection may begin as early as the initial formation of the antrum. The results also suggest that degeneration of the CL occurs by apoptotic mechanisms (Yuan & Giudice, 1997).

3.2 Nuclear Progesterone Receptor Mediates the Susceptibility to Apoptosis in Luteinizing GCs

A study investigated if PGR-mediated effects are involved in regulating the susceptibility to apoptosis in LH receptor-stimulated human luteinizing GCs (Svensson, Markström, Shao, Andersson, & Billig, 2001). Luteinizing GCs were isolated from follicular aspirates after oocyte removal. The cells were treated with or without PGR antagonist, progesterone, androgens, or steroid to evaluate their effects.

Addition of the PR antagonists (RU 486 or Org 31710) in vitro to human luteinizing GCs caused an increase in caspase-3 activity and an increase in internucleosomal DNA fragmentation. No effect on DNA fragmentation was noted after addition of dexamethasone, dihydrotestosterone, or picrotoxin. Such findings revealed that nuclear PGR-mediated effects are involved in regulating the susceptibility to apoptosis in LH receptor-stimulated human luteinizing GCs (Svensson et al., 2001).

3.3 Granulosa Cells Significantly Decreased With Age

The incidence of apoptotic GCs associated with the women's age on ovarian fecundity. In a study that enrolled 21 normoovulatory women underwent ovulation induction for standard IVF and were divided into four groups according to their ages. Granulosa cells were isolated by follicle aspiration after the ovarian hyperstimulation with gonadotrophins and tested for the incidence of apoptosis by using fluorescence microscopy (Sadraie, Saito, Kaneko, Saito, & Hiroi, 2000). In this study, GCs in the older women revealed a significant increase in the number of apoptotic cells. Though the endometrial thickness and follicular E2, P4, and free testosterone levels were not significantly different among the four different age groups, the number of total oocytes and the number of mature oocytes obtained significantly decreased with age (Sadraie et al., 2000).

3.4 Androgens Stimulate Early Primate Follicle Development

Excess androgens are associated with a characteristic morphology of polyfollicular ovaries. However, if such feature is due to direct androgen action on follicular development or due to interference with gonadotropin release at the level of the pituitary or hypothalamus is not clear. To elucidate potential androgen effects on the ovary, using in situ hybridization, the cellular localization of androgen receptor (AR) mRNA was studied in rhesus monkey. Researchers compared the relative abundance of AR transcripts in monkeys during follicular and luteal phases of the menstrual cycle and in monkeys treated with testosterone to investigate the regulation of ovarian AR gene expression (Weil et al., 1998). AR mRNA expression was most abundant in GCs of healthy preantral and antral follicles in the primate ovary. Theca interna and stromal cells also expressed AR mRNA, but to a lesser degree than GCs. No significant cycle stage effects were noted in AR mRNA levels. The AR mRNA level was significantly increased in GCs and decreased in theca interna and stromal cells of testosterone-treated monkeys. GC AR mRNA abundance was positively correlated with expression of the proliferation-specific antigen Ki-67 and negatively correlated with GC apoptosis. These data show that primate ovary AR gene expression is most abundant in GCs of healthy growing follicles, where its expression is upregulated by testosterone. The positive correlation between granulosa AR gene expression and cell proliferation and the negative correlation with programmed cell death suggest that androgens stimulate early primate follicle development (Weil et al., 1998).

In sheep, excess prenatal testosterone leads to increased ovarian follicular recruitment and persistence, and finally leads to reproductive dysfunction. In a study which injected twice weekly with testosterone propionate or dihydrotestosterone (DHT) propionate from days 30 to 90 of gestation in pregnant sheep to investigate the follicular disruption and the developmental ontogeny of ovarian proliferation and apoptotic factors (Salvetti, Ortega, Veiga-Lopez, & Padmanabhan, 2012). Prenatal testosterone treatment induced changes in expression of proliferative and apoptotic markers by assessing the changes in ovarian expression of proliferating cell nuclear antigen (PCNA), BCL2, BAX, activated CASP3, and FAS/FASLG at prenatal, postnatal, prepubertal, and after puberty stages. Prenatal testosterone, but not DHT, increased PCNA and decreased BCL2 in granulosa/theca cells of antral follicles postnatally at 10- and 21-month age, but decreased CASP3 in granulosa/theca cells of antral follicles at prepubertal and after puberty. Both treatments decreased BAX immunostaining in GCs of fetal primordial/primary follicles. No differential effect of FAS expression at any developmental time point in any follicular compartment. The effects on BAX appear to be programmed by androgenic actions and PCNA, BCL2, and CASP3 by estrogenic actions of testosterone. Overall, the findings demonstrate that fetal exposure to excess testosterone disrupts the ovarian proliferation and apoptosis balance, thus providing a basis for the follicular disruptions (Salvetti et al., 2012).

3.5 Androgen Factors Involved in Follicular Arrest

The effects of testosterone, DHT, and 17β-estradiol on human ovarian stromal tissue survival in culture and their effects on cell death were investigated by tissue cultures (Otala et al., 2004). The study revealed that ARs were detected in the ovarian stroma and GCs of the primordial follicles, although they were more clearly seen in primary follicles and more advanced stage follicles. Testosterone only marginally suppressed ovarian tissue apoptosis in vitro. DHT was more effective than testosterone, whereas 17β-estradiol had no notable effect on the viability of the tissue. The effects of androgens on the ovarian tissue may be mediated through ARs because blocking the receptors with an AR antagonist reversed the suppressive effect of DHT (Otala et al., 2004; Salvetti et al., 2012).

Our previous study investigated the effects of androgens on metabolism, as well as factors involved in follicular arrest and the reduced number of estrus cycles (Chen et al., 2015). The DHT-treated rats had fewer estrus cycles, a higher number of large arrested follicles, and an increase in body weight gain compared with dehydroepiandrostenedione (DHEA)- and placebo-treated rats. In cultured rat GCs, DHT suppressed FSH-induced GC proliferation and increased the accumulation of cells in the G2/M phase. DHT decreased phosphorylated Akt (p-Akt) and cyclin D1 levels by increasing PTEN. DHT-promoted PTEN expression was regulated by peroxisome proliferator-activated receptor-gamma (PPAR-γ) in GCs. In the large follicles of the DHT-treated rats, the expression of PPAR-γ and PTEN was higher, but the expression of p-Akt and PCNA was lower. Thus, DHT and DHEA produced differential effects on metabolism in prepubertal female rats, such as the clinical manifestations of women with PCOS. DHT treatment may affect ovarian follicular maturation by altering GC proliferation via the regulation of enhancing PPAR-γ-dependent PTEN/p-Akt expression in the GCs (Chen et al., 2015). Basic studies have implied that DHT may be useful for enhancing human ovarian tissue survival in vitro.

4. PROGESTERONE EFFECT ON OVULATION

4.1 Clinical Evidence of Progesterone Effect on Ovulation

The relationships and interindividual variations in urinary and serum reproductive hormone levels relative to ultrasound-observed ovulation in menstrual cycles of normally menstruating women were studied previously (Roos et al., 2015). This study enrolled 40 normally menstruating and aged between 18- and 49-year-old women without known subfertility. The enrolled subjected collected daily urine and bloody samples for hormone profiles and transvaginal ultrasound for a complete menstrual cycle. Serum LH, P4, E2, urinary LH, pregnanediol-3-glucuronide, and estrone-3-glucuronide were measured. As shown in this close observational study (Roos et al., 2015), menstrual cycle length varied from 22 to 37 days (median, 27 days) and ovulation that was evidenced by ultrasound ranged from day 8 to 26 (median, day 15). Serum and urinary hormone profiles showed excellent agreement. Estrogen and LH peaks in urine and serum showed a range of signal characteristics across the study group before and after ovulation. The rise in estrogen and LH always occurred before ovulation; the rise in P4 from baseline always occurred after ovulation. The study demonstrated that the beginning of the surge in serum and urinary LH was an excellent predictor of ovulation. The rise in P4 and pregnanediol-3-glucuronide above baseline was a consistent marker of luteinization, thus confirming ovulation. Both LH and P4 surges delivered clear, sharp signals in all volunteers, allowing reliable detection and confirmation of ovulation (Roos et al., 2015).

In a clinical study that enrolled 302 women aged between 20 and 40 years old and classified into two groups according to ovulatory status (Sanchez et al., 2016). This study was conducted to assess the correlation between low levels of P4 and ovulation by ultrasound monitoring in infertile patients with regular menstrual cycles. All related factors as age, body mass index, duration of infertility, FSH, thyroid-stimulating hormone, LH, and E2 were comparable between two groups. The study results revealed that there was a significant association between the percentage of ovulation by ultrasound monitoring and the percentages of patients who presented with low levels of P4. The study suggested that low serum levels of P4 are associated with a low percentage of ovulation in infertile women with regular menstrual cycles and women with unexplained infertility (Sanchez et al., 2016).

Another clinical study was conducted to identify an accurate P4 threshold to confirm ovulation in the assessment of a woman's fertility (Leiva, Bouchard, Boehringer, Abulla, & Ecochard, 2015). The study included 107 women over 326 menstrual cycles and tracked daily first morning urine, changes in observed cervical mucus discharge, serum P4, and ultrasonography to identify the day of ovulation. The study revealed that a random serum P4 level >5 ng/mL confirmed ovulation (Leiva et al., 2015).

4.2 Molecular Mechanism of Progesterone Effect on Ovulation

Ovulation is a precisely timed process by which a mature oocyte is released from an ovarian follicle. This process is initiated by the pituitary surge of LH, temporally associated with transcriptional regulation of numerous genes, and is presumed to involve the synthesis and/or activation of specific proteases that degrade the follicle wall. The PGR, a nuclear receptor transcription factor, is induced in GCs of preovulatory follicles in response to the LH surge and has been shown to be essential for ovulation. P4 is activated through the PGR. The study conducted on nonluteinized GCs collected from monkeys undergoing controlled ovarian stimulation protocols demonstrated that the nuclear PGR appears to mediate P4 action in the dominant follicle promoting ovulation in primates. The in vitro and in vivo effects of PGR knockdown in luteinized granulosa cells also support the hypothesis that P4 enhances its own synthesis in the primate CL by promoting luteinization (Bishop, Hennebold, Kahl, & Stouffer, 2016).

Mice lacking PGR fail to ovulate and are infertile. In a study that using mice model to elucidate PGR-regulated genes in the ovulation process, matrix metalloproteinases (MMP-2 and MMP-9) were found not being targets of PGR during ovulation, but the other two proteases (ADAMTS-1 [a disintegrin and metalloproteinase with thrombospondin-like motifs] and cathepsin L [a lysosomal cysteine protease]) were noted to be involved in the PGR action as transcriptional targets (Robker et al., 2000). ADAMTS-1 is induced after LH stimulation in GCs of preovulatory FSH, but the highest levels of cathepsin L mRNA occur in preovulatory follicles in response to LH in a PGR-dependent manner. The identification of two regulated proteases in the ovary, together with abnormal expression in anovulatory PGR knockout mice, suggests that each plays a critical role in follicular rupture and represents a major advance in our understanding of the proteolytic events that control ovulation (Robker et al., 2000).

Progesterone acting through its nuclear receptor plays an essential role in ovulation by mediating the expression of genes involved in ovulation and/or luteal formation (Mishra, Park, Wilson, & Jo, 2015). Microarray analysis using rat GCs treated with hCG with and without PGR antagonist (RU486) showed that the transcript of Xlr5c-like mRNA was downregulated by RU486 (Mishra et al., 2015). The Xlr5c-like mRNA level is transiently upregulated in GCs of periovulatory follicles after hCG stimulation in PMSG-primed rat ovaries. The transient induction of Xlr5c-like mRNA was mimicked by hCG treatment in cultured GCs from preovulatory ovaries. LH-activated PKA, MEK, PI3K, and p38 signaling is involved in the increase in Xlr5c-like mRNA, while the increase in Xlr5c-like mRNA was abolished by RU486. The inhibitory effect of RU486 was reversed by synthetic progestin (medroxyprogesterone), but not by dexamethasone. Furthermore, mutation of SP1/SP3 and PGR response element sites in the promoter region of Xlr5c-like decreased Xlr5c-like reporter activity. RU486 also inhibited Xlr5c-like reporter activity. A ChIP assay verified binding of PGR and SP3 to the Xlr5c-like promoter in periovulatory GCs. Xlr5c-like knockdown in GC cultures resulted in reduced levels of mRNA for Snap25, Cxcr4, and Adamts1. Progesterone/PGR mediates the LH-induced increase in Xlr5c-like mRNA.

In turn, Xlr5c-like is involved in regulating the expression of specific ovulatory genes, such as Snap25, Cxcr4, and Adamts1, possibly acting in the nucleus of periovulatory GCs (Mishra et al., 2015).

There are three types of local factors induced by the mid-cycle LH surge and play a critical role in ovulation and/or luteinization of the primate follicle (Stouffer, Xu, & Duffy, 2007). The ovulatory gonadotropin surge increases prostaglandin (PG) levels in follicles prior to rupture. LH stimulation lead to the synthesis of the "inducible" form of PG G/H synthase (PGS2), other aspects of PG synthesis (notably a phospholipase A2 [cPLA2], and a PGE synthase [PGES]) and metabolism (15-hydroxy PG dehydrogenase [HPGD]) also appear to be LH-regulated and may control the timing of the PG rise in the ovulatory follicle. Intrafollicular ablation and replacement of PGs suggest that PGE2 is essential for release of the oocyte, but not necessary for follicle rupture or luteinization. Novel PGE-regulated genes have been identified in macaque GCs, including adipose differentiation-related protein. Similar studies indicate that the rise in P4 synthesis, as well as the induction of the genomic PGR in GCs, is essential for both ovulation and luteinization of the primate follicle. Progesterone could control cell cycle activity (via cyclin B1 and cyclin-dependent kinase inhibitor p27), lipid metabolism, proteases, and protease inhibitors (matrix metalloproteinase or MMP1; tissue inhibitor of MMP or TIMP1) and cell health in the GC layer. Two classes of angiogenic factors, originally proposed as important for embryonic and pathologic (tumorigenic) vasculogenesis, appear induced in the granulosa layer of the preovulatory follicle, i.e., VEGF and angiopoietin (ANGPT). Local injection of antagonists to VEGF (soluble VEGF receptor) and ANGPT (the natural antagonist ANGPT2) into the preovulatory follicle suppressed ovulation and luteinization in monkeys, possibly by disrupting the structure–function of existing vessels or preventing angiogenesis in the avascular granulosa layer (Stouffer et al., 2007).

4.3 Progesterone Increased Immune Protection When Fertilization and Implantation

To evaluate the leukocyte count on the day of ovulation, blood samples were obtained from healthy and regularly cycling women at the following three times: during menstruation, ovulation, and in the mid-luteal phase to investigate the levels of white blood cells (WBCs), neutrophils, lymphocytes, mixed cells, P4, and E2. The results demonstrated that compared to menstruation, WBCs and neutrophils increased around ovulation and remained stable in the mid-luteal phase, whereas lymphocyte and mixed cell counts did not change throughout the menstrual cycle. As for the association between sex hormone variation with leukocyte changes and menstruation and ovulation, it was found a positive correlation existed between E2 and WBC, and negative correlations existed between P4 and WBCs and between P4 and the neutrophil count, but not between ovulation and the mid-luteal phase. The results showed that peripheral leukocyte changes taking place in the second half of the cycle are already observable on the day of ovulation and are associated with sex hormone variation. For such reason, it could be speculated that these changes accompanied with elevated progesterone levels may lead to increased immune protection against pathogens at a time when fertilization and implantation occur (Nowak, Borkowska, & Pawlowski, 2016).

A study was conducted to characterize the immune cell types within the primate corpus luteum (CL, 36). Luteal tissue was collected from rhesus females at discrete intervals during the luteal phase of the natural menstrual cycle and incubated with fluorescently labeled antibodies specific for immune cell surface proteins (CD11b [neutrophils and monocytes/macrophages], CD14 [monocytes/macrophages], CD16 [natural killer {NK} cells], CD20 [B-lymphocytes], and CD3epsilon [T-lymphocytes]) for analysis by flow cytometry. The results revealed that the number of CD11b(+) and CD14(+) cells increased significantly 3–4 days after serum P4 concentrations declined to <0.3 ng/mL. CD16(+) cells were the most abundant immune cell type in the CL during the mid- and mid–late luteal phases and were threefold increased 3–4 days after serum P4 decreased to baseline levels. CD3epsilon(+) cells tended to increase 3–4 days after P4 decline. The number of CD11b(+) and CD14(+) cells of CL increased after GnRH antagonist treatment, whereas progestin, but not estrogen, replacement suppressed the number of CD11b(+), CD14(+), and CD16(+) cells. The number of CD3epsilon(+) cells of CL was not changed with either GnRH antagonist or steroid treatment. These data suggest that an increased number of innate immune cells in primate CL after P4 synthesis declines play a role in onset of structural regression of the primate CL (Bishop, Xu, Molskness, Stouffer, & Hennebold, 2015).

4.4 Effects of Estradiol, Progesterone, and Corpus Luteum on the Dominant Follicle

To determine the effects of E2, P4, presence of a CL, and size of a dominant follicle on the characteristics and patterns of GnRH-induced LH release and subsequent ovulation, there was a study using lactating dairy cows with a GnRH-induced LH release and hourly collected blood samples for hormone assay were analyzed (Stevenson & Pulley, 2016). Injections of GnRH induced LH release during 6 h after each injection in all study subjects. The concentrations of LH induced by GnRH were inhibited at greater concentrations of P4 and in the presence of a CL. In contrast, GnRH-induced LH concentrations were increased when E2 was $\geq$4.0 pg/mL, but relatively unaffected by the size of the dominant follicle. Furthermore, the resulting incidences of ovulation were decreased at greater P4 concentrations and the presence of a CL, and increased at greater E2 concentrations and presence of a dominant follicle. In cows with or without a CL, the presence a dominant follicle did not increase the mean LH concentration or incidence of ovulation. While when premature P4 concentration was elevated at the time of GnRH treatment, subsequent LH concentrations and ovulation were suppressed. At a lower concentration of P4 or at higher concentration of E2, the pregnancy outcomes after timed artificial insemination were improved with similar incidence of ovulation. Such findings suggest a greater P4 or lesser at the time of artificial insemination may suppress pregnancy chance by other mechanisms.

Ovarian steroid hormones are major regulators of the physiology of the oviduct and reproductive events occurring within the oviduct. A study was conducted to establish an entire steroid profiling of the bovine oviductal fluid during the estrous cycle (Lamy et al., 2016). Contralateral and ipsilateral (to the CL or preovulatory follicle) oviducts of cows were classified into four stages of the estrous cycle: postovulatory, mid-luteal, late luteal, and preovulatory on the basis of the ovarian morphology and intrafollicular steroid concentrations. The concentrations of P4 in ipsilateral oviductal fluid increased from postovulatory to mid-luteal, then decreased from late-luteal to preovulatory, and were 4–16 times higher than the contralateral oviductal fluid. Most P4 metabolites followed similar patterns of variation. The concentrations of preovulatory estradiol-17β were significantly higher compared with all other stages, with no difference regarding the side of ovulation. Concentrations of androstenedione displayed a pattern similar to that of E2, whereas other androgens, estrone, and corticoids did not vary between stages or sides. The results demonstrated a highly concentrated and fluctuating hormonal environment in the bovine oviductal fluid (Lamy et al., 2016).

5. CONCLUSION

Degeneration of the CL occurs by apoptotic mechanisms. The nuclear PGR-mediated effects are involved in regulating the susceptibility to apoptosis in LH receptor-stimulated human luteinizing GCs. Apoptotic GCs increased while the number of total oocytes and the number of mature oocytes significantly decreased with age. However, E2, P4, and free testosterone levels were not significantly changed with age. The positive correlation between granulosa AR gene expression and cell proliferation and negative correlation with apoptosis suggest that androgens stimulate early primate follicle development; however, fetal exposure to excess testosterone disrupts the ovarian proliferation/apoptosis balance. DHT may be useful for enhancing human ovarian tissue survival in vitro.

The surge of serum P4 is a signal for confirmation of ovulation. The level of P4 or E2 is critical for the time of artificial insemination. The expression of the PGR is critically involved in proteolytic events that control ovulation. Progesterone/PGR mediate the LH-induced increase in the expression of specific ovulatory genes. The effect of P4 mediates immune protection against pathogens at a time when fertilization and implantation typically occur. Understanding the cascade of P4 leading to ovulation and luteinization of the primate follicle may help to consider novel ovary-selective approaches to human reproduction.

REFERENCES

Aurora, B., Ruiz-Cortés, Z. T., Kimmins, S., Monaco, L., Burns, K. H., Sassone-Corsi, P., et al. (2005). Estrogen mediates phosphorylation of histone H3 in ovarian follicle and mammary epithelial tumor cells via the mitotic kinase. *Molecular Endocrinology*, *19*(12), 2991–3000.

Basini, G., Bussolati, S., Santini, S. E., Bianchi, F., Careri, M., Mangia, A., et al. (2008). Hydroxyestrogens inhibit angiogenesis in swine ovarian follicles. *The Journal of Endocrinology*, *199*(1), 127–135.

Basini, G., Grasselli, F., Bussolati, S., Baioni, L., Bianchi, F., Musci, M., et al. (2011). Hypoxia stimulates the production of the angiogenesis inhibitor 2-methoxyestradiol by swine granulosa cells. *Steroids*, *76*(13), 1433–1436.

Bayne, S., Li, H., Jones, M. E., Pinto, A. R., van Sinderen, M., Drummond, A., et al. (2011). Estrogen deficiency reversibly induces telomere shortening in mouse granulosa cells and ovarian aging in vivo. *Protein & Cell*, *2*(4), 333–346.

Billig, H., Furuta, I., & Hsueh, A. J. (1993). Estrogens inhibit and androgens enhance ovarian granulose cell apoptosis. *Endocrinology*, *133*(5), 2204–2212.

Bishop, C. V., Hennebold, J. D., Kahl, C. A., & Stouffer, R. L. (2016). Knockdown of progesterone receptor (PGR) in macaque granulosa cells disrupts ovulation and progesterone production. *Biology of Reproduction*, *94*(5), 109.

Bishop, C. V., Xu, F., Molskness, T. A., Stouffer, R. L., & Hennebold, J. D. (2015). Dynamics of immune cell types within the macaque corpus luteum during the menstrual cycle: Role of progesterone. *Biology of Reproduction*, *93*(5), 112.

Chatterjee, R., Helal, M., Mobberley, M., Ryder, T., & Bajoria, R. (2014). Impaired steroidogenesis and apoptosis of granulosa-luteal cells in primary culture induced by cisplatinum. *American Journal of Obstetrics and Gynecology*, *210*(3), 252.e1–252.e7.

Chen, Y., Breen, K., & Pepling, M. E. (2009). Estrogen can signal through multiple pathways to regulate oocyte cyst breakdown and primordial follicle assembly in the neonatal mouse ovary. *The Journal of Endocrinology*, *202*(3), 407–417.

Chen, M. J., Chou, C. H., Chen, S. U., Yang, W. S., Yang, Y. S., & Ho, H. N. (2015). The effect of androgens on ovarian follicle maturation: Dihydrotestosterone suppress

FSH-stimulated granulosa cell proliferation by upregulating PPARγ-dependent PTEN expression. *Scientific Reports*, *5*, 18319.

Couse, J. F., Bunch, D. O., Lindzey, J., Schomberg, D. W., & Korach, K. S. (1999). Prevention of the polycystic ovarian phenotype and characterization of ovulatory capacity in the estrogen receptor-alpha knockout mouse. *Endocrinology*, *140*(12), 5855–5865.

Dharma, S. J., Modi, D. N., & Nandedkar, T. D. (2009). Gene expression profiling during early folliculogenesis in the mouse ovary. *Fertility and Sterility*, *91*(5 Suppl), 2025–2036.

Jobe, S. O., Ramadoss, J., Koch, J. M., Jiang, Y., Zheng, J., & Magness, R. R. (2010). Estradiol-17beta and its cytochrome P450- and catechol-O-methyltransferase-derived metabolites stimulate proliferation in uterine artery endothelial cells: Role of estrogen receptor-alpha versus estrogen receptor-beta. *Hypertension*, *55*(4), 1005–1011.

Kazi, A. A., Jones, J. M., & Koos, R. D. (2005). Chromatin immunoprecipitation analysis of gene expression in the rat uterus in vivo: Estrogen-induced recruitment of both estrogen receptor alpha and hypoxia-inducible factor 1 to the vascular endothelial growth factor promoter. *Molecular Endocrinology*, *19*(8), 2006–2019.

Kohen, P., Henríquez, S., Rojas, C., Gerk, P. M., Palomino, W. A., Strauss, J. F., 3rd, et al. (2013). 2-Methoxyestradiol in the human corpus luteum throughout the luteal phase and its influence on lutein cell steroidogenesis and angiogenic activity. *Fertility and Sterility*, *100*(5), 1397–1404.

Lamy, J., Liere, P., Pianos, A., Aprahamian, F., Mermillod, P., & Saint-Dizier, M. (2016). Steroid hormones in bovine oviductal fluid during the estrous cycle. *Theriogenology*, *86*(6), 1409–1420.

Leiva, R., Bouchard, T., Boehringer, H., Abulla, S., & Ecochard, R. (2015). Random serum progesterone threshold to confirm ovulation. *Steroids*, *101*, 125–129.

Li, Q., He, H., Zhang, Y. L., Li, X. M., Guo, X., Huo, R., et al. (2013). Phosphoinositide 3-kinase p110δ mediates estrogen- and FSH-stimulated ovarian follicle growth. *Molecular Endocrinology*, *27*(9), 1468–1482.

Mishra, B., Park, J. Y., Wilson, K., & Jo, M. (2015). X-linked lymphocyte regulated gene 5c-like (Xlr 5c-like) is a novel target of progesterone action in granulosa cells of periovulatory rat ovaries. *Molecular and Cellular Endocrinology*, *412*, 226–238.

Morani, A., Warner, M., & Gustafsson, J. A. (2008). Biological functions and clinical implications of oestrogen receptors alfa and beta in epithelial tissues. *Journal of Internal Medicine*, *264*(2), 128–142.

Nowak, J., Borkowska, B., & Pawlowski, B. (2016). Leukocyte changes across menstruation, ovulation, and mid-luteal phase and association with sex hormone variation. *American Journal of Human Biology*, *28*(5), 721–728.

Otala, M., Mäkinen, S., Tuuri, T., Sjöberg, J., Pentikäinen, V., Matikainen, T., et al. (2004). Effects of testosterone, dihydrotestosterone, and 17beta-estradiol on human ovarian tissue survival in culture. *Fertility and Sterility*, *82*(Suppl. 3), 1077–1085.

Quirk, S. M., Cowan, R. G., & Harman, R. M. (2006). The susceptibility of granulosa cells to apoptosis is influenced by oestradiol and the cell cycle. *The Journal of Endocrinology*, *189*(3), 441–453.

Robker, R. L., Russell, D. L., Espey, L. L., Lydon, J. P., O'Malley, B. W., & Richards, J. S. (2000). Progesterone-regulated genes in the ovulation process: ADAMTS-1 and cathepsin L proteases. *Proceedings of the National Academy of Sciences of the United States of America*, *97*(9), 4689–4694.

Roos, J., Johnson, S., Weddell, S., Godehardt, E., Schiffner, J., Freundl, G., et al. (2015). Monitoring the menstrual cycle: Comparison of urinary and serum reproductive hormones referenced to true ovulation. *The European Journal of Contraception & Reproductive Health Care*, *20*(6), 438–450.

Sadraie, S. H., Saito, H., Kaneko, T., Saito, T., & Hiroi, M. (2000). Effects of aging on ovarian fecundity in terms of the incidence of apoptotic granulosa cells. *Journal of Assisted Reproduction and Genetics, 17*(3), 168–173.

Salvetti, N. R., Ortega, H. H., Veiga-Lopez, A., & Padmanabhan, V. (2012). Developmental programming: Impact of prenatal testosterone excess on ovarian cell proliferation and apoptotic factors in sheep. *Biology of Reproduction, 87*(1): 22, 1–10.

Sanchez, E. G., Giviziez, C. R., Sanchez, H. M., Agostinho, P. L., Barros, P. S., & Approbato, M. S. (2016). Low progesterone levels and ovulation by ultrasound assessment in infertile patients. *JBRA Assisted Reproduction, 20*(1), 13–16.

Sasagawa, S., Shimizu, Y., Nagaoka, T., Tokado, H., Imada, K., & Mizuguchi, K. (2008). Dienogest, a selective progestin, reduces plasma estradiol level through induction of apoptosis of granulosa cells in the ovarian dominant follicle without follicle-stimulating hormone suppression in monkeys. *Journal of Endocrinological Investigation, 31*(7), 636–641.

Shang, W., Konidari, I., & Schomberg, D. W. (2001). 2-Methoxyestradiol, an endogenous estradiol metabolite, differentially inhibits granulosa and endothelial cell mitosis: A potential follicular antiangiogenic regulator. *Biology of Reproduction, 65*(2), 622–627.

Stevenson, J. S., & Pulley, S. L. (2016). Feedback effects of estradiol and progesterone on ovulation and fertility of dairy cows after gonadotropin-releasing hormone-induced release of luteinizing hormone. *Journal of Dairy Science, 99*(4), 3003–3015.

Stouffer, R. L., Xu, F., & Duffy, D. M. (2007). Molecular control of ovulation and luteinization in the primate follicle. *Frontiers in Bioscience, 12*, 297–307 [Review].

Svensson, E. C., Markström, E., Shao, R., Andersson, M., & Billig, H. (2001). Progesterone receptor antagonists Org 31710 and RU 486 increase apoptosis in human periovulatory granulosa cells. *Fertility and Sterility, 76*(6), 1225–1231.

Valdez, K. E., & Turzillo, A. M. (2005). Regulation of nuclear factor-kappa B (NF-kappaB) activity and apoptosis by estradiol in bovine granulosa cells. *Molecular and Cellular Endocrinology, 243*(1–2), 66–73.

Weil, S. J., Vendola, K., Zhou, J., Adesanya, O. O., Wang, J., Okafor, J., et al. (1998). Androgen receptor gene expression in the primate ovary: Cellular localization, regulation, and functional correlations. *The Journal of Clinical Endocrinology and Metabolism, 83*(7), 2479–2485.

Westergaard, L., Christensen, I. J., & McNatty, K. P. (1986). Steroid levels in ovarian follicular fluid related to follicle size and health status during the normal menstrual cycle in women. *Human Reproduction*, (4), 227–232.

Yuan, W., & Giudice, L. C. (1997). Programmed cell death in human ovary is a function of follicle and corpus luteum status. *The Journal of Clinical Endocrinology and Metabolism, 82*(9), 3148–3155.

Zhen, Y. H., Wang, L., Riaz, H., Wu, J. B., Yuan, Y. F., Han, L., et al. (2014). Knockdown of CEBPβ by RNAi in porcine granulosa cells resulted in S phase cell cycle arrest and decreased progesterone and estradiol synthesis. *The Journal of Steroid Biochemistry and Molecular Biology, 143*, 90–98.

CHAPTER SEVEN

GABA-A Receptors Mediate Tonic Inhibition and Neurosteroid Sensitivity in the Brain

Doodipala Samba Reddy[1]
College of Medicine, Texas A&M University Health Science Center, Bryan, TX, United States
[1]Corresponding author: e-mail address: reddy@medicine.tamhsc.edu

Contents

Abstract

Neurosteroids like allopregnanolone (AP) are positive allosteric modulators of synaptic and extrasynaptic GABA-A receptors. AP and related neurosteroids exhibit a greater potency for δ-containing extrasynaptic receptors. The δGABA-A receptors, which are expressed extrasynaptically in the dentate gyrus and other regions, contribute to tonic inhibition, promoting network shunting as well as reducing seizure susceptibility. Levels of endogenous neurosteroids fluctuate with ovarian cycle. Natural and synthetic neurosteroids maximally potentiate tonic inhibition in the hippocampus and provide robust protection against a variety of limbic seizures and status epilepticus. Recently, a consensus neurosteroid pharmacophore model has been proposed at extrasynaptic δGABA-A receptors based on structure–activity relationship for functional activation of tonic currents and seizure protection. Aside from anticonvulsant actions, neurosteroids have been found to be powerful anxiolytic and anesthetic agents. Neurosteroids and Zn^{2+} have preferential affinity for δ-containing receptors. Thus, Zn^{2+} can prevent neurosteroid activation of extrasynaptic δGABA-A receptor-mediated tonic inhibition. Recently, we demonstrated that Zn^{2+} selectively inhibits extrasynaptic δGABA-A receptors and thereby fully prevents AP activation of tonic inhibition and seizure protection. We confirmed that neurosteroids exhibit greater sensitivity at extrasynaptic δGABA-A receptors. Overall, extrasynaptic GABA-A receptors are primary mediators of tonic inhibition in the brain and play a key role in the pathophysiology of epilepsy and other neurological disorders.

Vitamins and Hormones, Volume 107
ISSN 0083-6729
https://doi.org/10.1016/bs.vh.2017.12.001

1. INTRODUCTION

Neurosteroids, also referred to as neuroactive steroids, are endogenous steroids that rapidly alter neuronal excitability through interaction with neuronal membrane receptors. Neuroactive steroids refer to steroids that are synthesized by an endocrine gland that reach the brain through the bloodstream and have effects on brain function. The term "neurosteroid" was originally coined in 1981 by the French endocrinologist Étienne-Émile Baulieu to refer to steroids that are synthesized de novo in the nervous system from cholesterol, independently of the peripheral steroidogenic endocrine glands. The terms "neuroactive steroids" or "neurosteroids" interchangeably refer to steroids that rapidly alter the excitability of neurons by virtue of their pharmacological actions as positive allosteric modulators of GABA-A receptors, which are protein receptors for GABA, the primary inhibitory neurotransmitter in the brain.

Neurosteroids are endogenous steroids that are synthesized in the brain, the adrenal glands, and the gonads. The most widely studied neurosteroids include allopregnanolone (AP), allotetrahydrodeoxycorticosterone (THDOC), and androstanediol (AD) (Table 1). They have potent and selective effects on the GABA-A receptors. Neurosteroids do not directly interact with classical steroid hormone receptors, but their metabolites can bind to intracellular steroid receptors. GABA-A receptors are responsible for the majority of inhibitory currents in the brain. Structurally, they are pentameric channels made from various subunits (α1–6, β1–4, γ1–3, δ, ε, θ, ρ1–3). In adult neurons, GABA-A channels conduct chloride ions into the cell and hyperpolarize the neuronal membrane. Based on their location on neuronal sites, GABA-A receptors are stratified into synaptic (primarily γ-containing) and extrasynaptic (primarily δ-containing) receptors (Akk et al., 2007, 2005). Synaptic receptors, which are present ubiquitously within the brain, produce phasic currents in response to vesicular release of GABA. This current is often assessed by recording mIPSCs. Extrasynaptic receptors, which are expressed in specific brain regions including hippocampus, thalamus, and cerebellum (Whissell, Lecker, Wang, Yu, & Orser, 2015), generate nondesensitizing tonic currents that are continuously activated by extracellular GABA. Tonic current is often assessed by recording basal tone shift. Tonic current sets the baseline and shunting inhibition via continuous chloride conductance in specific neurons in the hippocampus. The hippocampus is a seahorse-shaped structure considered to be an important part of the

Table 1 Chemical Names of Natural and Synthetic Neurosteroids

Neurosteroid	Chemical Name	Key Structural Feature
Allopregnanolone (AP)	3α-Hydroxy-5α-pregnan-20-one	3α-Hydroxyl, 5α-reduction, 17-acetyl
Ganaxolone	3α-Hydroxy-3β-methyl-5α-pregnan-20-one	3β-Methyl
Pregnanolone	3α-Hydroxy-5β-pregnan-20-one	5β-Reduction
Tetrahydrodeoxycorticosterone (THDOC)	3α,21-Dihydroxy-5α-pregnan-20-one	21-Hydroxyl
Alfaxolone	3α-Hydroxy-5α-pregnan-11,20-dione	11-Carbonyl
Alphadolone	3α,21-Dihydroxy-5α-pregnane-11,20-dione	
Androstanediol	5α-Androstan-3α,17β-diol	17-Hydroxyl
Androsterone	5α-Androstan-3α-ol-17-one	17-Carbonyl
Etiocholanolone	3α-Hydroxy-5β-androstan-17-one	17-Carbonyl, 5β-reduction
ORG-20599	21-Chloro-3α-hydroxy-2β-(4-morpholinyl)-5α-pregnan-20-one	2β-(4-Morpholinyl), 17-acyl chloride

limbic system. It is one of the most extensively studied areas of the brain and has important functions in learning, memory, and emotion. The dentate gyrus, a subregion of hippocampus, is a critical region for controlling inputs from afferent neuronal pathways (Coulter & Carlson, 2007). Dentate gyrus granule cells (DGGCs) highly express δ-subunit GABA-A receptors that modulate network inhibition within the hippocampus (Carver & Reddy, 2013; Glykys, Mann, & Mody, 2008). The tonic current of DGGCs is highly (~95%) dependent on δ-containing receptors (Carver, Wu, Gangisetty, & Reddy, 2014; Glykys et al., 2008; Stell, Brickley, Tang, Farrant, & Mody, 2003). Neurosteroids exhibit preferential sensitivity at extrasynaptic δGABA-A receptors (Brown, Kerby, Bonnert, Whiting, & Wafford, 2002; Meera, Wallner, & Otis, 2011; Reddy, 2004, 2006, 2010; Reddy & Kulkarni, 2000; Reddy & Rogawski, 2001, 2009).

Research on the characterization of neurosteroid structure–activity relationship (SAR) for extrasynaptic δGABA-A receptors is essential for the development of novel therapeutic agents for epilepsy and related brain disorders. Emerging GABAergic mechanisms of neurosteroids are providing innovative therapies for epilepsy, anxiety disorders, Fragile X syndrome, and postpartum depression (Reddy, 2016). In this chapter, we review our current knowledge on the biosynthesis of neurosteroids, as well as their interactions with synaptic and extrasynaptic GABA-A receptors to further understand the role of neurosteroid-sensitive tonic inhibition in the brain.

2. BIOSYNTHESIS OF NEUROSTEROIDS

It has been known since the 1940s, from the pioneering work of Hans Selye, that steroid hormones progesterone and deoxycorticosterone can exert anesthetic and anticonvulsant actions. Research during the past two decades has revealed that neurosteroids mediate these neurological effects of steroid hormones. A variety of neurosteroids are present in the brain, including pregnane, androstane, and estrogen classes. The pathways for the production of the three prototype neurosteroids are widely studied, including AP, THDOC, and AD. Neurosteroids are synthesized in the brain from cholesterol or steroid hormone precursors via progressive A-ring reductions. The steroid precursors of neurosteroids, such as progesterone and deoxycorticosterone, are mainly synthesized in the gonads, adrenal gland, and fetoplacental unit. These conversion steps occur in peripheral tissues such as reproductive endocrine tissues, liver, and skin that are rich in the two reducing activities. Since neurosteroids are highly lipophilic and can readily cross the blood–brain barrier, neurosteroids synthesized in peripheral tissues accumulate in the brain and can influence brain function (Akk et al., 2009).

Neurosteroids are also synthesized within the brain from cholesterol. All necessary enzymes for the biosynthesis of neurosteroids are present in the brain. Neurosteroid synthesis occurs in glia and neurons in many brain regions, including the hippocampus and neocortex. Several neurosteroids including AP and pregnanolone are generated by enzymatic pathways. Biosynthetic pathways of neurosteroid production are derived from specific enzyme inhibitors. The enzyme steroid 5α-reductase (NADPH-Δ4-3-oxosteroid-5α-oxidoreductase; EC 1.3.99.5) is an NADPH-dependent enzyme that was originally characterized to catalyze the reduction of testosterone to the more potent androgen dihydrotestosterone. Progesterone and deoxycorticosterone and other 3-keto-pragnane steroids are also substrates for the 5α-reductase enzyme. Two isozymes of 5α-reductase have been identified in humans. Type 1 5α-reductase is found in the brain, skin, gut, liver, and other tissues. The type 2 5α-reductase, found primarily in the peripheral reproductive tissues (prostate, genitalia testis), is effectively inhibited by finasteride. Dutasteride inhibits both type 1 and type 2 isozymes. The 5α-reductase enzymes are present in the human brain. 5α-Reductase activity has been identified in both neurons and glial cells in the brain. In humans, the 5α-reductase and 3α-HSOR enzymes have been found in neocortex and subcortical white matter as well as in hippocampal tissues. Thus, it is likely that neurosteroids can be formed from their parent hormonal steroids directly in the target brain region. Steroid precursors readily enter the brain so that pools of peripherally synthesized precursors are readily available for local neurosteroid biosynthesis. Since the activity of the 3α-HSOR is far greater than that of the 5α-reductase, steroid 5α-reduction is the rate-limiting step in the biosynthesis of neurosteroids. In addition, neurosteroids may be produced locally in brain not only from their steroid hormone precursors but also from cholesterol. The initial step in steroidogenesis is the conversion of cholesterol to pregnanolone by the mitochondrial enzyme P450scc. Mechanism-based, irreversible inhibition of 5α-reductase by finasteride and dutasteride can shut down the neurosteroid biosynthesis in the brain and thereby affects neuronal excitability and behavior (Ellsworth, Azzolina, Cimis, Bull, & Harris, 1998).

Neurosteroids play a pathophysiological role in catamenial epilepsy, a menstrual cycle-related seizure clustering in women with epilepsy. Neurosteroid levels in the brain fluctuate with the ovarian cycle, with peak levels during the luteal phase and lowest levels during perimenstrual phase. Progesterone is a major precursor of pregnane neurosteroids, such as AP and pregnanolone in the brain (Reddy, Castaneda, O'Malley, & Rogawski, 2004; Tuveri et al., 2008). Seizures often decrease in the midluteal phase when serum progesterone levels are high and increase premenstrually when serum progesterone levels drop, igniting the perimenstrual-type catamenial epilepsy (Reddy, 2016). Thus, perimenstrual catamenial seizures are ascribed to withdrawal of the protective effects of neurosteroids (Reddy, 2016).

3. NEUROSTEROID ACTIVATION OF SYNAPTIC AND EXTRASYNAPTIC RECEPTORS

There is abundant evidence that neurosteroids rapidly alter neuronal excitability through direct interaction with GABA-A receptors (Fig. 1).

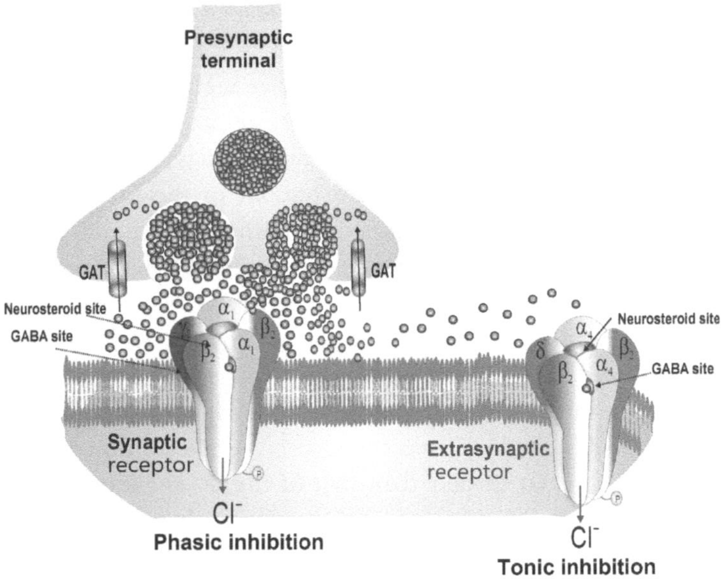

Fig. 1 Schematic illustration of neurosteroid-sensitive synaptic and extrasynaptic δGABA-A receptors in the brain.

Unlike benzodiazepines and barbiturates, neurosteroids are a special group of compounds present within the brain for modulating GABA-A receptor function. Hence, neurosteroids are often referred to as endogenous modulators of GABA-A receptors in the brain (Majewska, Harrison, Schwartz, Barker, & Paul, 1986; Reddy, 2003). GABA-A receptors are ligand-gated chloride channels which, when activated by GABA, hyperpolarize the neurons through influx of chloride ions. Based on location, GABA-A receptors can be classified into synaptic and extrasynaptic receptors (Fig. 1). Synaptic and extrasynaptic receptors differ in their GABA affinity (Mortensen, Patel, & Smart, 2011), desensitization rate, agonist efficacy (Bianchi & Macdonald, 2002, 2003), and neurosteroid sensitivity (Brown et al., 2002; Wohlfarth, Bianchi, & Macdonald, 2002).

A sophisticated technique called "patch-clamp electrophysiology" is used to study neurosteroid activity on electrical currents in neurons dissociated from brain or brain slices. Since neurons communicate by electrical signals, any changes in such electrical currents are thought to impact the function of the neuron. Electrophysiological studies show that neurosteroids are positive allosteric agonists of synaptic and extrasynaptic GABA-A receptors (Carver & Reddy, 2013; Reddy & Estes, 2016) (Fig. 1). The mode of action for neurosteroids depends on the concentration: at high concentrations (>1 μM), neurosteroids directly activate receptors, whereas at low concentrations ($<1\ \mu M$), they allosterically potentiate GABA-A receptor currents. This is achieved by binding to "neurosteroid binding sites" on the receptor channel (Harrison & Simmonds, 1984; Hosie, Clarke, da Silva, & Smart, 2009; Hosie, Wilkins, da Silva, & Smart, 2006). Although the exact interaction site is unclear, there are two discrete binding sites for neurosteroids including an allosteric site within the α-subunit transmembrane domain and a site of direct activation at the α–β subunit interface (Hosie et al., 2009, 2006; Hosie, Wilkins, & Smart, 2007). The neurosteroid binding sites are quite different from binding sites for barbiturates, benzodiazepines, or cage convulsants. The efficacy of neurosteroids on GABA-A receptor function is stereospecific and adheres to specific structural features. SAR studies at synaptic γGABA-A receptors show that the C3α-OH steroid structure is essential for the binding affinity and the receptor-enhancing function by neurosteroids (Harrison, Majewska, Harrington, & Barker, 1987; Mitchell, Herd, Gunn, Lambert, & Belelli, 2008). Apart from 5α-H stereoselectivity, the C17- or C20-ketone group is important for high-potency positive allosteric modulation (Covey et al., 2000; Kokate, Svensson, & Rogawski, 1994; Qian et al., 2014; Reddy & Jian, 2010; Upasani et al., 1997). In addition, the lipophilic properties of neurosteroids contribute to the potency and efficacy of receptor binding (Chisari et al., 2009). So far, over 500 analogs have been tested for CNS activity and over a dozen have been tested in clinical trials (Reddy & Estes, 2016).

Unlike benzodiazepines, neurosteroids can exert robust modulatory effects at GABA-A receptors. At higher concentrations, neurosteroids directly activate GABA-A receptors (Kokate et al., 1994; Reddy & Rogawski, 2002). This is a unique mechanistic feature of neurosteroids. The concentration of neurosteroid which is able to directly gate GABA-A receptors ($\geq 1\ \mu M$) is greater than the observed physiological range of endogenous neurosteroids in the brain (0.1–$0.3\ \mu M$) (Carver & Reddy, 2013; Gunn et al., 2015; Reddy, 2003). Nevertheless, exogenous administration of neurosteroids can reach sufficiently high concentrations within the brain to achieve direct activation of GABA-A receptors (Reddy & Estes, 2016; Reddy & Jian, 2010). Hence, the mechanism of action of neurosteroids is complex. At nanomolar levels, they follow allosteric agonism, meaning that the resulting maximal efficacy of GABA potentiation reaches a limiting value when the allosteric site is fully occupied. At higher (micromolar) levels that exceed allosteric-saturating levels, neurosteroids follow direct agonism, defined when the resulting maximal efficacy of chloride conductance attains a maximal value per available receptor sites within the neurons. Thus, it is thought that the overall net inhibition during neurosteroid therapy occurs via allosteric binding and direct activation mechanisms. These can ultimately result in maximal inhibition at the neuronal networks and thereby regulate the neuronal network excitability in the brain.

4. NEUROSTEROID SENSITIVITY ON GABAergic TONIC INHIBITION

Do neurosteroids show receptor subtype specificity? Numerous studies have been conducted over the past two decades using recombinant GABA-A receptors and native neurons with defined subtype composition. There is strong evidence that neurosteroids bind to all GABA-A receptor isoforms, but δ-containing receptors exhibit preferential sensitivity at extracellular GABA concentrations (Brown et al., 2002; Carver & Reddy, 2016; Meera et al., 2011; Stell et al., 2003) (Table 2). Unlike benzodiazepines, neurosteroids are able to modulate all isoforms of GABA-A receptors, including

Table 2 Correlation of Neurosteroid Activation of Tonic Currents and Anticonvulsant Activity

	Tonic Currents		Seizure Protection
Neurosteroid	**$E_{1\mu M}$ (pA)[a]**	**$EF_{(twofold\ GABA)}$ (nM)[b]**	**ED_{50} (mg/kg)[c]**
3α5αAP	100.6	80	4.2 (2.7–5.8)
3βMe-AP (GX)	64.0	290	1.5 (1.3–1.7)
3α5β-AP	44.4	780	7.7 (6.6–8.8)
3β5α-AP	15.3	>10,000	>100
THDOC	66.6	410	5.0 (2.6–7.4)
ALFX	40.9	990	8.8 (6.1–11.4)
ORG-20599	86.4	120	18.6 (16.6–20.6)
AD	33.2	1710	44.0 (30.2–58.8)

[a]$E_{1\mu M}$ values represent the mean normalized tonic current responses of drug at 1 μM concentration coapplied with 1 μM GABA. GABA 1 μM tonic current: 0.66 ± 0.22 pA/pF, 19.6 pA.
[b]EF values represent the effective functional concentration of drug (nM) required to double or triple the 1 μM GABA response.
[c]ED_{50} values represent the dose in milligrams per kilogram that protected 50% of animals in the 6-Hz seizure stimulation test. 95% confidence intervals are listed in *parenthesis*, according to a normal distribution.

those that contain benzodiazepine-insensitive α4 and α6 subunits or those that lack the obligatory γ2 subunit required for benzodiazepine sensitivity. The allosteric binding of neurosteroids to low-efficacy δGABA-A receptors induces a pronounced conformational change, greater channel opening, and nondesensitizing tonic inhibition (Bianchi & Macdonald, 2003). Increased δGABA-A receptor expression enhances neurosteroid sensitivity through greater potentiation of tonic current (Carver et al., 2014; Sanna et al., 2009; Wu, Gangisetty, Carver, & Reddy, 2013). Conversely, deficient δGABA-A receptor expression reduces the sensitivity to neurosteroids (Carver & Reddy, 2016; Mihalek et al., 1999; Peng et al., 2002; Spigelman et al., 2002). In the dentate gyrus, ~95% tonic currents are generated by extrasynaptic δGABA-A receptors (Glykys et al., 2008). Mutations to the *Gabrd* gene result in dysfunctional GABA-A receptor currents, and these mutations have been linked with clinical cases of generalized epilepsies (Dibbens et al., 2004; Feng et al., 2006). Thus, diminished δGABA-A receptor-mediated tonic inhibition exerts a significant impact on network excitability, behavior, and seizure susceptibility.

Despite the emergent translational potential of the extrasynaptic GABA-A receptors, the neurosteroid structures required for functional activation of extrasynaptic receptors is poorly understood. Recently, we identified a consensus neurosteroid pharmacophore model at extrasynaptic δGABA-A receptors for activation of tonic currents (Carver & Reddy, 2016). We recorded neurosteroid allosteric potentiation of GABA-A receptors as well as direct activation of tonic currents using a wide array of natural and synthetic neurosteroids. Using an extensive library of neurosteroids, we found that, for all neurosteroids, the C3α-OH group remains obligatory for extrasynaptic receptor functional activity, as C3β-OH epimers were inactive in activating tonic currents (Table 2). AP and related pregnane analogs exhibited the highest potency and maximal efficacy in promoting tonic currents. Alterations at the C17 or C20 region of the neurosteroid molecule drastically altered the transduction kinetics of tonic current activation. The androstane analogs had the weakest modulatory response among the analogs tested. Neurosteroid potentiation of tonic currents was completely (~95%) diminished in granule cells from δ-knockout mice, suggesting that δ-subunit receptors are essential for neurosteroid activity. Therefore, the low-to-high efficacy transduction produced by neurosteroids is δ-subunit mediated under physiological conditions. These findings are consistent with the significance of δGABA-A receptors to tonic inhibition and neurosteroid sensitivity (Carver & Reddy, 2016; Carver et al., 2014; Wu et al., 2013).

The effects of δ-subunit-specific mutations on the transduction of the neurosteroid interaction have yet to be determined. The Q241 and N407/Y410 site-specific residues of the α1-subunit were determined to be crucial for neurosteroid binding (Hosie et al., 2007). More recently, T235W and Q240W mutations in the α_4-subunit rendered δ-containing receptors (α4βδ) insensitive to positive allosteric modulation by 3α5α-AP (Jensen et al., 2013). Therefore, the higher affinity for the α_4-subunit binding interface and δ-dependent channel transduction greatly contribute to neurosteroid binding.

5. NEUROSTEROID AND ZINC INTERACTIONS IN MODULATING TONIC INHIBITION

Zn^{2+} is the most abundant in the synaptic vesicles of hippocampal mossy fibers. Zn^{2+} release occurs with neuronal excitation including seizure events and exerts powerful excitability effects in the hippocampal circuits. Zn^{2+} inhibits synaptic GABA-A receptors, but its interaction is less well appreciated at the extrasynaptic receptors, which respond sensitively to endogenous neurosteroids. Three discrete binding sites contribute to Zn^{2+} inhibition, including one site within the channel pore and two at the external amino-terminus of the α–β interfaces; the inclusion of the γ-subunit disrupts two of the sites, resulting in reduction in sensitivity (Hosie, Dunne, Harvey, & Smart, 2003). Although the δ-containing receptors are more sensitive to a blockage by Zn^{2+} than γ-containing receptors (Mangan et al., 2005), the ability of Zn^{2+} to modify extrasynaptic δGABA-A receptors is poorly understood. Recently, we described a selective functional blockade by Zn^{2+} of neurosteroid-sensitive, extrasynaptic GABA-A receptors in the mouse hippocampus dentate gyrus (Carver, Chuang, & Reddy, 2016). Zn^{2+} displayed a concentration-dependent, reversible, noncompetitive blockade of AP-sensitive tonic current in DGGCs (IC_{50}, 16 μM). Tonic current was fully blocked by Zn^{2+}, akin to the GABA-A receptor antagonist gabazine. Zn^{2+} inhibition of tonic current was lacking in DGGCs from δ-subunit knockout mice. Moreover, AP-activated synaptic receptor-mediated phasic currents were not affected by Zn^{2+}. Finally, intrahippocampal infusion of Zn^{2+} elicited rapid epileptiform activity and significantly blocked the antiseizure activity of AP in the kindling model of epilepsy. Due to their high affinity for δ-containing receptors, Zn^{2+} and neurosteroids influence inhibition through opposing action, albeit at different allosteric sites

(Hosie et al., 2003). Our data suggest that Zn^{2+} is a δ-selective noncompetitive antagonist and therefore hinders neurosteroid transduction of extrasynaptic GABA-A channels. By demonstrating that extracellular Zn^{2+} prevents neurosteroid augmentation of tonic current and protection against limbic seizures, our findings provide novel implications of this potential antagonistic interaction in a variety of neurological conditions like seizures, epileptogenesis, and epilepsy and conditions with compromised neurovascular unit or Zn^{2+} transport pathways in the brain. Therefore, extracellular Zn^{2+} can prevent neurosteroid activation of extrasynaptic δGABA-A receptor-mediated tonic inhibition and thereby impair their ability to promote neuroprotection and seizure suppression in the brain.

6. CONCLUSIONS AND FUTURE DIRECTIONS

Synaptic GABA-A receptors are primary mediators of rapid inhibition in the brain and play a key role in the pathophysiology of epilepsy and other neurological disorders. The δ-subunit GABA-A receptors are expressed extrasynaptically in the dentate gyrus and contribute to tonic inhibition, promoting network shunting as well as reducing seizure susceptibility. Recently, we identified an SAR for neurosteroid modulation of extrasynaptic GABA-A receptor-mediated tonic inhibition in the murine DGGCs. A consensus neurosteroid pharmacophore model at extrasynaptic δGABA-A receptors has emerged for activation of tonic current and seizure protection. Our recent studies provide strong evidence that neurosteroids exhibit greater sensitivity at extrasynaptic δGABA-A receptors. In addition, there is growing evidence on the pharmacodynamic interactions between neurosteroids and zinc, especially at extrasynaptic GABA-A receptors. Zn^{2+} is a noncompetitive inhibitor of extrasynaptic GABA-A receptors and can block neurosteroid modulation of tonic inhibition. Therefore, Zn^{2+} inhibition of neurosteroid-sensitive, extrasynaptic GABA-A receptors in the hippocampus has direct implications in many brain hyperexcitability conditions like seizures, epileptogenesis, and epilepsy. Zn^{2+} interactions may aid to further understand the physiology of extrasynaptic GABA-A receptors. Additional research is warranted to understand the contributory role of tonic inhibition in pathophysiology of epilepsy, sleep, and neuropsychiatric conditions. Alterations in ovarian hormones and modification of tonic inhibition are proposed as potential pathophysiological mechanisms underlying catamenial epilepsy.

ACKNOWLEDGMENTS

This work was partly supported by National Institutes of Health Grant R01NS051398 (to D.S.R.). This work was also partly supported by the CounterACT Program, National Institutes of Health, Office of the Director, and the National Institute of Neurologic Disorders and Stroke (Grant U01 NS083460) (to D.S.R.). The author declares no competing financial interests.

REFERENCES

Akk, G., Covey, D. F., Evers, A. S., Steinbach, J. H., Zorumski, C. F., & Mennerick, S. (2007). Mechanisms of neurosteroid interactions with GABA-A receptors. *Pharmacology & Therapeutics*, *116*, 35–57.

Akk, G., Covey, D. F., Evers, A. S., Steinbach, J. H., Zorumski, C. F., & Mennerick, S. (2009). The influence of the membrane on neurosteroid actions at GABA-A receptors. *Psychoneuroendocrinology*, *34*(Suppl. 1), S59–66.

Akk, G., Shu, H. J., Wang, C., Steinbach, J. H., Zorumski, C. F., Covey, D. F., et al. (2005). Neurosteroid access to the GABA-A receptor. *The Journal of Neuroscience*, *25*, 11605–11613.

Bianchi, M. T., & Macdonald, R. L. (2002). Slow phases of GABA-A receptor desensitization: Structural determinants and possible relevance for synaptic function. *The Journal of Physiology*, *544*, 3–18.

Bianchi, M. T., & Macdonald, R. L. (2003). Neurosteroids shift partial agonist activation of GABA-A receptor channels from low- to high-efficacy gating patterns. *The Journal of Neuroscience*, *23*, 10934–10943.

Brown, N., Kerby, J., Bonnert, T. P., Whiting, P. J., & Wafford, K. A. (2002). Pharmacological characterization of a novel cell line expressing human alpha(4)beta(3)delta GABA-A receptors. *British Journal of Pharmacology*, *136*, 965–974.

Carver, C. M., Chuang, S. H., & Reddy, D. S. (2016). Zinc selectively blocks neurosteroid-sensitive extrasynaptic deltaGABAA receptors in the hippocampus. *The Journal of Neuroscience*, *36*, 8070–8077.

Carver, C. M., & Reddy, D. S. (2013). Neurosteroid interactions with synaptic and extrasynaptic GABA-A receptors: Regulation of subunit plasticity, phasic and tonic inhibition, and neuronal network excitability. *Psychopharmacology*, *230*, 151–188.

Carver, C. M., & Reddy, D. S. (2016). Neurosteroid structure-activity relationships for functional activation of extrasynaptic deltaGABA-A receptors. *The Journal of Pharmacology and Experimental Therapeutics*, *357*, 188–204.

Carver, C. M., Wu, X., Gangisetty, O., & Reddy, D. S. (2014). Perimenstrual-like hormonal regulation of extrasynaptic delta-containing GABA-A receptors mediating tonic inhibition and neurosteroid sensitivity. *The Journal of Neuroscience*, *34*, 14181–14197.

Chisari, M., Eisenman, L. N., Krishnan, K., Bandyopadhyaya, A. K., Wang, C., Taylor, A., et al. (2009). The influence of neuroactive steroid lipophilicity on GABAA receptor modulation: Evidence for a low-affinity interaction. *Journal of Neurophysiology*, *102*, 1254–1264.

Coulter, D. A., & Carlson, G. C. (2007). Functional regulation of the dentate gyrus by GABA-mediated inhibition. *Progress in Brain Research*, *163*, 235–243.

Covey, D. F., Nathan, D., Kalkbrenner, M., Nilsson, K. R., Hu, Y., Zorumski, C. F., et al. (2000). Enantioselectivity of pregnanolone-induced gamma-aminobutyric acid(A) receptor modulation and anesthesia. *The Journal of Pharmacology and Experimental Therapeutics*, *293*, 1009–1016.

Dibbens, L. M., Feng, H. J., Richards, M. C., Harkin, L. A., Hodgson, B. L., Scott, D., et al. (2004). GABRD encoding a protein for extra- or peri-synaptic GABA-A receptors is a susceptibility locus for generalized epilepsies. *Human Molecular Genetics*, *13*, 1315–1319.

Ellsworth, K. P., Azzolina, B. A., Cimis, G., Bull, H. G., & Harris, G. S. (1998). Cloning, expression and characterization of rhesus macaque types 1 and 2 5alpha-reductase: Evidence for mechanism-based inhibition by finasteride. *The Journal of Steroid Biochemistry and Molecular Biology*, *66*, 271–279.

Feng, H. J., Kang, J. Q., Song, L., Dibbens, L., Mulley, J., & Macdonald, R. L. (2006). Delta subunit susceptibility variants E177A and R220H associated with complex epilepsy alter channel gating and surface expression of alpha4beta2delta GABA-A receptors. *The Journal of Neuroscience*, *26*, 1499–1506.

Glykys, J., Mann, E. O., & Mody, I. (2008). Which GABA-A receptor subunits are necessary for tonic inhibition in the hippocampus? *The Journal of Neuroscience*, *28*, 1421–1426.

Gunn, B. G., Cunningham, L., Mitchell, S. G., Swinny, J. D., Lambert, J. J., & Belelli, D. (2015). GABA-A receptor-acting neurosteroids: A role in the development and regulation of the stress response. *Frontiers in Neuroendocrinology*, *36*, 28–48.

Harrison, N. L., Majewska, M. D., Harrington, J. W., & Barker, J. L. (1987). Structure-activity relationships for steroid interaction with the gamma-aminobutyric acidA receptor complex. *The Journal of Pharmacology and Experimental Therapeutics*, *241*, 346–353.

Harrison, N. L., & Simmonds, M. A. (1984). Modulation of the GABA receptor complex by a steroid anaesthetic. *Brain Research*, *323*, 287–292.

Hosie, A. M., Clarke, L., da Silva, H., & Smart, T. G. (2009). Conserved site for neurosteroid modulation of GABA-A receptors. *Neuropharmacology*, *56*, 149–154.

Hosie, A. M., Dunne, E. L., Harvey, R. J., & Smart, T. G. (2003). Zinc-mediated inhibition of GABA-A receptors: Discrete binding sites underlie subtype specificity. *Nature Neuroscience*, *6*, 362–369.

Hosie, A. M., Wilkins, M. E., da Silva, H. M., & Smart, T. G. (2006). Endogenous neurosteroids regulate GABA-A receptors through two discrete transmembrane sites. *Nature*, *444*, 486–489.

Hosie, A. M., Wilkins, M. E., & Smart, T. G. (2007). Neurosteroid binding sites on GABA-A receptors. *Pharmacology & Therapeutics*, *116*, 7–19.

Jensen, M. L., Wafford, K. A., Brown, A. R., Belelli, D., Lambert, J. J., & Mirza, N. R. (2013). A study of subunit selectivity, mechanism and site of action of the delta selective compound 2 (DS2) at human recombinant and rodent native GABA-A receptors. *British Journal of Pharmacology*, *168*, 1118–1132.

Kokate, T. G., Svensson, B. E., & Rogawski, M. A. (1994). Anticonvulsant activity of neurosteroids: Correlation with gamma-aminobutyric acid-evoked chloride current potentiation. *The Journal of Pharmacology and Experimental Therapeutics*, *270*, 1223–1229.

Majewska, M. D., Harrison, N. L., Schwartz, R. D., Barker, J. L., & Paul, S. M. (1986). Steroid hormone metabolites are barbiturate-like modulators of the GABA receptor. *Science*, *232*, 1004–1007.

Mangan, P. S., Sun, C., Carpenter, M., Goodkin, H. P., Sieghart, W., & Kapur, J. (2005). Cultured hippocampal pyramidal neurons express two kinds of GABAA receptors. *Molecular Pharmacology*, *67*, 775–788.

Meera, P., Wallner, M., & Otis, T. S. (2011). Molecular basis for the high THIP/gaboxadol sensitivity of extrasynaptic GABA-A receptors. *Journal of Neurophysiology*, *106*, 2057–2064.

Mihalek, R. M., Banerjee, P. K., Korpi, E. R., Quinlan, J. J., Firestone, L. L., Mi, Z. P., et al. (1999). Attenuated sensitivity to neuroactive steroids in gamma-aminobutyrate type A receptor delta subunit knockout mice. *Proceedings of the National Academy of Sciences of the United States of America*, *96*, 12905–12910.

Mitchell, E. A., Herd, M. B., Gunn, B. G., Lambert, J. J., & Belelli, D. (2008). Neurosteroid modulation of GABA-A receptors: Molecular determinants and significance in health and disease. *Neurochemistry International*, *52*, 588–595.

Mortensen, M., Patel, B., & Smart, T. G. (2011). GABA potency at GABA-A receptors found in synaptic and extrasynaptic zones. *Frontiers in Cellular Neuroscience*, *6*, 1.

Peng, Z., Hauer, B., Mihalek, R. M., Homanics, G. E., Sieghart, W., Olsen, R. W., et al. (2002). GABA-A receptor changes in delta subunit-deficient mice: Altered expression of alpha4 and gamma2 subunits in the forebrain. *The Journal of Comparative Neurology*, *446*, 179–197.

Qian, M., Krishnan, K., Kudova, E., Li, P., Manion, B. D., Taylor, A., et al. (2014). Neurosteroid analogues. 18. Structure-activity studies of ent-steroid potentiators of gamma-aminobutyric acid type A receptors and comparison of their activities with those of alphaxalone and allopregnanolone. *Journal of Medicinal Chemistry*, *57*, 171–190.

Reddy, D. S. (2003). Pharmacology of endogenous neuroactive steroids. *Critical Reviews in Neurobiology*, *15*, 197–234.

Reddy, D. S. (2004). Testosterone modulation of seizure susceptibility is mediated by neurosteroids 3alpha-androstanediol and 17beta-estradiol. *Neuroscience*, *129*, 195–207.

Reddy, D. S. (2006). Physiological role of adrenal deoxycorticosterone-derived neuroactive steroids in stress-sensitive conditions. *Neuroscience*, *138*, 911–920.

Reddy, D. S. (2010). Neurosteroids: Endogenous role in the human brain and therapeutic potentials. *Progress in Brain Research*, *186*, 113–137.

Reddy, D. S. (2016). Catamenial epilepsy: Discovery of an extrasynaptic molecular mechanism for targeted therapy. *Frontiers in Cellular Neuroscience*, *10*(101), 1–14.

Reddy, D. S., Castaneda, D. C., O'Malley, B. W., & Rogawski, M. A. (2004). Anticonvulsant activity of progesterone and neurosteroids in progesterone receptor knockout mice. *The Journal of Pharmacology and Experimental Therapeutics*, *310*, 230–239.

Reddy, D. S., & Estes, W. A. (2016). Clinical potential of neurosteroids for CNS disorders. *Trends in Pharmacological Sciences*, *37*, 543–561.

Reddy, D. S., & Jian, K. (2010). The testosterone-derived neurosteroid androstanediol is a positive allosteric modulator of GABAA receptors. *The Journal of Pharmacology and Experimental Therapeutics*, *334*, 1031–1041.

Reddy, D. S., & Kulkarni, S. K. (2000). Development of neurosteroid-based novel psychotropic drugs. *Progress in Medicinal Chemistry*, *37*, 135–175.

Reddy, D. S., & Rogawski, M. A. (2001). Enhanced anticonvulsant activity of neuroactive steroids in a rat model of catamenial epilepsy. *Epilepsia*, *42*, 337–344.

Reddy, D. S., & Rogawski, M. A. (2002). Stress-induced deoxycorticosterone-derived neurosteroids modulate GABA-A receptor function and seizure susceptibility. *The Journal of Neuroscience*, *22*, 3795–3805.

Reddy, D. S., & Rogawski, M. A. (2009). Neurosteroid replacement therapy for catamenial epilepsy. *Neurotherapeutics*, *6*, 392–401.

Sanna, E., Mostallino, M. C., Murru, L., Carta, M., Talani, G., Zucca, S., et al. (2009). Changes in expression and function of extrasynaptic GABA-A receptors in the rat hippocampus during pregnancy and after delivery. *The Journal of Neuroscience*, *29*, 1755–1765.

Spigelman, I., Li, Z., Banerjee, P. K., Mihalek, R. M., Homanics, G. E., & Olsen, R. W. (2002). Behavior and physiology of mice lacking the GABAA-receptor delta subunit. *Epilepsia*, *43*(Suppl. 5), 3–8.

Stell, B. M., Brickley, S. G., Tang, C. Y., Farrant, M., & Mody, I. (2003). Neuroactive steroids reduce neuronal excitability by selectively enhancing tonic inhibition mediated by delta subunit-containing GABA-A receptors. *Proceedings of the National Academy of Sciences of the United States of America*, *100*, 14439–14444.

Tuveri, A., Paoletti, A. M., Orru, M., Melis, G. B., Marotto, M. F., Zedda, P., et al. (2008). Reduced serum level of THDOC, an anticonvulsant steroid, in women with perimenstrual catamenial epilepsy. *Epilepsia*, *49*, 1221–1229.

Upasani, R. B., Yang, K. C., Acosta-Burruel, M., Konkoy, C. S., McLellan, J. A., Woodward, R. M., et al. (1997). 3alpha-Hydroxy-3beta-(phenylethynyl)-5beta-pregnan-20-ones: Synthesis and pharmacological activity of neuroactive steroids with high affinity for GABA-A receptors. *Journal of Medicinal Chemistry*, *40*, 73–84.

Whissell, P. D., Lecker, I., Wang, D. S., Yu, J., & Orser, B. A. (2015). Altered expression of delta-GABA-A receptors in health and disease. *Neuropharmacology*, *88*, 24–35.

Wohlfarth, K. M., Bianchi, M. T., & Macdonald, R. L. (2002). Enhanced neurosteroid potentiation of ternary GABA-A receptors containing the delta subunit. *The Journal of Neuroscience*, *22*, 1541–1549.

Wu, X., Gangisetty, O., Carver, C. M., & Reddy, D. S. (2013). Estrous cycle regulation of extrasynaptic delta-containing GABA-A receptor-mediated tonic inhibition and limbic epileptogenesis. *The Journal of Pharmacology and Experimental Therapeutics*, *346*, 146–160.

CHAPTER EIGHT

GATA Regulation and Function During the Ovarian Life Cycle

Jill Bennett-Toomey*, Carlos Stocco†,1
*Ohio Northern University, Ada, OH, United States
†College of Medicine, University of Illinois at Chicago, Chicago, IL, United States
1Corresponding author: e-mail address: costocco@uic.edu

Contents

Abstract

GATA4 and GATA6 are the sole GATA factors expressed in the ovary during embryonic development and adulthood. Up today, GATA4 and GATA6 are the only transcription factors that have been conditionally deleted during ovarian development and at each major stage of follicle maturation. The evidence from these transgenic mice revealed that GATA4 and GATA6 are crucial for follicles assembly, granulosa cell differentiation, postnatal follicle growth, and luteinization. Thus, conditional knockdown of both factors in the granulosa cells at any stage of development leads to female infertility. GATA targets impacting female reproduction include genes involved in steroidogenesis, hormone signaling, ovarian hormones, extracellular matrix organization, and apoptosis/cell division.

Vitamins and Hormones, Volume 107
ISSN 0083-6729
https://doi.org/10.1016/bs.vh.2018.01.014

1. INTRODUCTION

The formation, maintenance, and maturation of ovarian follicles require the timely activation of genes involved in differentiation, cell survival, steroidogenesis, and hormone production. Several of the genes included in these various processes contain within their regulatory elements the motif, WGATAR. This DNA motif is recognized by a family of transcription factors called GATA. (Molkentin, Lin, Duncan, & Olson, 1997; Viger, Guittot, Anttonen, Wilson, & Heikinheimo, 2008).

The GATA family consists of six transcription factors, GATA1 to GATA6, which was formed by two duplications early in vertebrate evolution (Tang et al., 2014). The first duplication led to the emergence of GATA1, GATA2, and GATA3 and the second resulted in GATA4, GATA5, and GATA6 (Tang et al., 2014). GATA1, GATA2, and GATA3 are expressed in hematopoietic cell lineages and play critical roles in the differentiation and proliferation of hematopoietic stem cells and immune cells (Weiss & Orkin, 1995). GATA4, GATA5, and GATA6 are mainly associated with organ development in tissues of mesodermal and endodermal origin such as the heart, gut, and gonads (Molkentin, 2000). The expression and function of GATA4 and GATA6 in the gonads are the focus of this review.

GATA4 was discovered as a tissue-specific, retinoic acid-inducible, and developmentally regulated transcription factor (Arceci, King, Simon, Orkin, & Wilson, 1993). This initial finding also provided the first evidence for the high expression of GATA4 in the ovary of adult female mice. Since then, GATA factors have been associated with ovarian function in an ever increasing number of species, including, mammals, fish, and birds (reviewed in Lowry & Atchley, 2000). Moreover, a GATA-binding protein was found in Drosophila identified as dGATAb, which is expressed in ovarian follicular cells where it binds and activates the yolk protein genes Yp1 and Yp2 (Lossky & Wensink, 1995). Also, the *C. elegans* ELT-1 protein, a homolog of the vertebrate GATA family, is highly expressed in the germ line of both sexes (Shim, 1999). Thus, the presence of GATA-binding proteins in reproductive tissues throughout evolution underscores the role of these factors in reproduction.

All members of the GATA family can be divided into three distinct functional domains (Fig. 1). (A) The amino-terminal region contains two transcriptional activation domains. In mouse GATA4, the activation domain 1 is located between aa 1 and 74 and activation domain 2 between aa 130 and

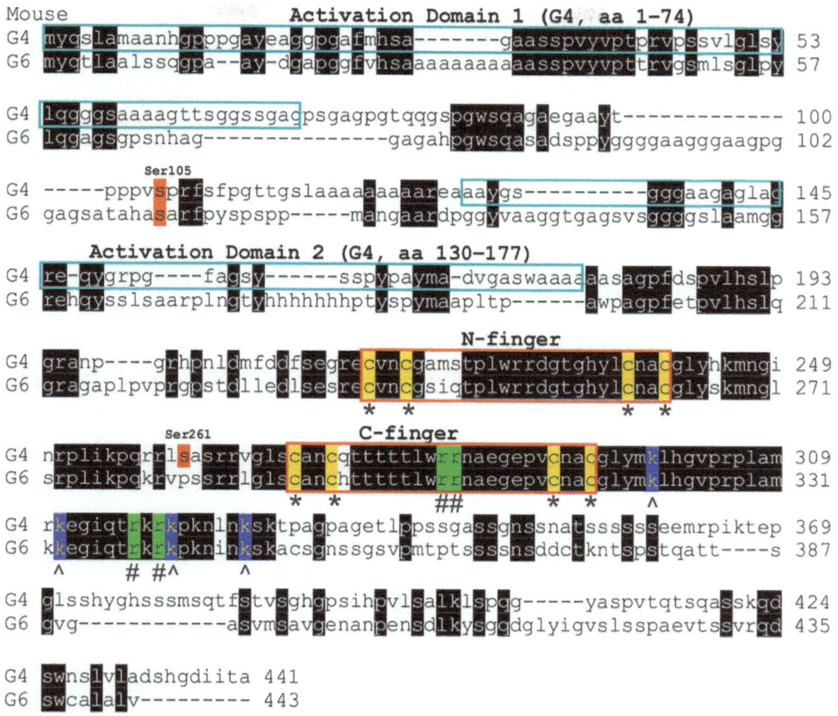

Fig. 1 Structure–function analysis of GATA4. *Numbers* indicate amino acid residues in the mouse GATA4 protein (G4) and GATA6 protein (G6). Residues in *red* indicate phosphorylation sites, Ser105 and Ser261. Residues in *yellow* (*) are the cysteines involved in the DNA binding that form the zinc fingers. In *green* (#) are the residues involved in the nuclear translocation domain. In *purple* (^) are the lysine residues conserved between GATA1 and GATA4 that may be targeted for acetylation.

177 (Morrisey, Ip, Tang, & Parmacek, 1997). (B) The central core contains two zinc finger DNA-binding domains: a C-zinc finger and an N-zinc finger. These domains have been well characterized for GATA1 in which the C-zinc finger is involved in DNA binding and recognizes the (A/T)GATA (A/G) DNA motif. The N-zinc finger however does not bind DNA directly, although it appears to modulate DNA binding through interaction with other transcription factors (Mackay et al., 1998). GATA4 and GATA6 share these characteristics with GATA1 because mouse GATA4 is 70% identical to mouse GATA1 within the DNA-binding region and GATA4 and GATA6 have 86% homology in their DNA-binding region (Molkentin, 2000). The central region also contains a nuclear localization region, which for GATA4 consists of two clusters of Arginine amino acids, Arg282, Arg283, Arg317, and Arg319 separated by a 33-amino acid spacer, which

forms a nonclassical nuclear localization region (Philips, Kwok, & Chong, 2007). (C) Finally, the carboxy-terminal domain also appears to be involved on transcriptional activation as amino acids 362–400 in GATA4 are essential for mediating cardiogenesis in *Xenopus* embryos via physical and functional interaction with CDK4 (Gallagher et al., 2014).

2. GATA EXPRESSION IN THE OVARY

2.1 Development

The fetal ovary expresses GATA2, GATA4, and GATA6. However, GATA2 is expressed only transiently, accumulating rapidly from embryonic day (E) 11.5 and ceases by E15.5. This short expression of GATA2 is restricted to the germ cell lineage as its localization coincides with the expression of Oct4, a germ cell marker (Siggers, Smith, & Greenfield, 2002). The role of GATA2 in germ cell development has not been elucidated.

The gonads initially develop as sexually bipotential structures called genital ridges, which subsequently differentiate to become testes or ovaries. The genital ridges originate from the proliferation of the coelomic epithelium, a process that starts approximately at E10.5 in the mouse. GATA4 protein is already present in the pregenital ridge region as early as E10 and is required for the proliferation of the coelomic epithelium, leading to the formation of the genital ridge (Hu, Okumura, & Page, 2013). GATA4 remains expressed at E11.5 in the primitive gonads of both XX and XY mouse embryos. By E13.5, GATA4 is found in most ovarian somatic cells where it is expressed at high levels over the entire fetal period (Anttonen, Ketola, Parviainen, Pusa, & Heikinheimo, 2003). These somatic cells eventually develop into pregranulosa cells. Germ cells, however, do not express GATA4 (Lavoie, McCoy, & Blake, 2004). The expression of GATA4 decreases in mouse fetal ovaries shortly after ovarian differentiation, but it is maintained at high levels in fetal testes (Heikinheimo et al., 1997; Vaskivuo et al., 2001; Viger, Mertineit, Trasler, & Nemer, 1998). Similarly, GATA6 expression has been clearly documented during ovarian development and is detectable in the ovaries at E13.5 (Padua, Fox, Jiang, Morse, & Tevosian, 2014) and in the somatic cells of the prenatal ovary (Lavoie, McCoy, et al., 2004).

In humans, GATA4 is present in granulosa cells at 23 weeks of gestation; however, the number of GATA4-positive cells decreases by week 33 (Vaskivuo et al., 2001). GATA4 is also present in the ovaries of pigs during the entire fetal period (Anttonen et al., 2003; McCoard, Wise, Fahrenkrug, & Ford, 2001). The abundant expression of GATA4 and

GATA6 in the fetal ovary suggests that they play a vital role in the early steps of differentiation (see below). However, GATA4-deficient and GATA6-deficient mice die in utero before the differentiation of the gonad (Kuo et al., 1997; Molkentin et al., 1997; Morrisey et al., 1998), precluding studies of the role of these factors in testicular or ovarian formation. Interestingly, testis cords do not form in animals carrying a null mutation of FOG2 (friend of GATA2), a partner of GATA4, or a mutation in GATA4 ($GATA4^{ki}$) that abrogates its interaction with FOG2 (Tevosian et al., 2002). Similarly, the GATA4–FOG2 complex is indirectly required for the expression of forkhead transcription factor, FOXL2 (Manuylov, Smagulova, Leach, & Tevosian, 2008), which is one of the earliest expressed female-specific genes and it is essential for female reproductive development (Uhlenhaut et al., 2009). These reports demonstrate that GATA4 and its ability to interact with FOG2 are essential for the determination and differentiation of testes and ovaries. The role of GATA factors in gonad development has been recently reviewed (Tevosian, 2014).

2.2 Adult Gonads

GATA4 and GATA6 are readily detectable in the ovary of several species including rodents (Anttonen et al., 2003; Heikinheimo et al., 1997; Viger et al., 1998) and humans (Laitinen et al., 2000; Vaskivuo et al., 2001). Ovarian expression of GATA4 mRNA increases around the time of birth. Thus, GATA factors are detectable in the granulosa cells of primary follicles indicating that GATA4 becomes highly expressed during recruitment of primordial follicles. Moderate GATA4 expression can be found in the theca cells, the germinal epithelium, and the interstitial cells of adult ovaries. GATA4 is also strongly expressed in the nucleus of ovarian surface epithelial cells (Capo-chichi et al., 2003; Lavoie, McCoy, et al., 2004). As the follicle develops, GATA4 becomes uniformly expressed in healthy follicles and is expressed in granulosa and theca cells of secondary and antral follicles and to a lesser degree in stromal cells (Anttonen et al., 2003; Heikinheimo et al., 1997; Laitinen et al., 2000; Vaskivuo et al., 2001; Viger et al., 1998). As with GATA4, GATA6 is clearly detectable in granulosa cells of adult animals. GATA6 mRNA is also found in the germinal epithelium and the corpus luteum but not in oocytes, theca cells, or interstitial cells (Heikinheimo et al., 1997; Lavoie, McCoy, et al., 2004).

GATA4 and GATA6 expression increases in granulosa cells during follicle growth from the preantral to the preovulatory stage (Heikinheimo et al., 1997). Interestingly, in preovulatory follicles, cumulus granulosa cells, which are the cells adjacent to the oocyte, express less GATA4 mRNA than mural granulosa cells, which are located near the basement membrane of follicles (Bennett, Wu, Gossen, Zhou, & Stocco, 2012). This suggests a differential regulation of GATA within the follicle. After ovulation, GATA4 mRNA expression is significantly reduced in the corpus luteum, while increasing GATA6 expression can be found in the luteal tissue (Convissar et al., 2015; Heikinheimo et al., 1997). This is clearly observed in equine chorionic gonadotropin (eCG)-primed immature mice treated with human chorionic gonadotropin (hCG), which induces a steady increase in GATA6 expression in the luteinizing granulosa cells (Convissar et al., 2015). GATA4 binding to the *Cyp19a1* promoter is also downregulated during luteinization (Kwintkiewicz, Cai, & Stocco, 2007), indicating a decrease in GATA4 activity. Thus, GATA6 is more abundant and active than GATA4 in luteinizing granulosa cells.

Luteal cells from rodents, farm animals, and humans express GATA4 and GATA6 (Gillio-Meina, Hui, & LaVoie, 2003; Heikinheimo et al., 1997; Laitinen et al., 2000; Lavoie, McCoy, et al., 2004; Taniguchi, Komiyama, Viger, & Okuda, 2009). As expected, based on the changes observed during luteinization, GATA6 expression in the corpus luteum is higher than the expression of GATA4 (Anttonen et al., 2003; Gillio-Meina et al., 2003; Heikinheimo et al., 1997; Lavoie, McCoy, et al., 2004; Silverman, Eimerl, & Orly, 1999). However, although some reports showed that the corpora lutea of eCG- and hCG-treated mice are virtually devoid of GATA4, both GATA4 and GATA6 remain expressed in the corpus luteum (Anttonen et al., 2003; Bennett et al., 2012; Gillio-Meina et al., 2003; Heikinheimo et al., 1997).

2.3 Follicular Atresia

The vast majority of follicles undergo programmed cell death during maturation. In mice, granulosa cells undergoing apoptosis lack GATA4 message but continue to express significant levels of GATA6 (Heikinheimo et al., 1997). Additional evidence obtained in rats indicates that during the early stages of atresia, the granulosa cells closest to the antrum lose GATA4 expression, but retain GATA6 expression. As atresia advances, GATA4, but not GATA6, is lost from all granulosa cells (Lavoie, McCoy, et al., 2004). This suggests that GATA4 protects granulosa cells from undergoing apoptosis. This role for GATA4 is supported by findings in human granulosa cells (Laitinen et al., 2000; Vaskivuo et al., 2001). Moreover, follicles are disorganized and show morphological characteristics of atresia (pyknotic nuclei and increased staining for cleaved caspase 3 in mice lacking GATA4 and GATA6 expression within their granulosa cells (Bennett et al., 2012).

3. REGULATION OF GATA EXPRESSION AND ACTIVITY IN ADULT OVARIES

In immature mice, administration of eCG stimulates follicular expression of GATA4 and GATA6 messages (Heikinheimo et al., 1997). In pigs, GATA4 and GATA6 mRNA expression is low in small follicles and reaches the highest level of expression after follicles have developed to the preovulatory state (Gillio-Meina et al., 2003). In mouse sertoli tumor cells (MSC-1) and mouse granulosa tumor cells (NT-1), follicle-stimulating hormone (FSH) increases GATA4 and GATA6 mRNA expression (Heikinheimo et al., 1997). We have demonstrated that primary rat granulosa cells respond to FSH with an increase in GATA4 mRNA and protein expression (Kwintkiewicz et al., 2007). Those results, taken together with the observation that women carrying an inactivating mutation in the FSH receptor show little or negligible GATA4 expression in the ovary (Vaskivuo et al., 2002), provide a compelling argument for a major role of FSH in the regulation of GATA4 in the ovary. However, this would apply only to later stages of folliculogenesis as GATA4 is expressed in primary and secondary follicles, which are known to be independent of FSH action.

Posttranslational modifications may also be an important mechanism to stimulate GATA4 transcriptional activity and DNA-binding capacity (Fig. 2). In particular, FSH stimulates GATA4 phosphorylation at Ser105 in primary ovarian granulosa cells (Kwintkiewicz et al., 2007) and follicles (Fig. 2B). Phosphorylation of Ser105 enhances the transcriptional potency of GATA4 (Liang et al., 2001). Accordingly, GATA4 overexpression together with FSH stimulation synergistically activates the expression and promoter activity of aromatase, whereas mutation of Ser105 prevents these effects (Kwintkiewicz et al., 2007). Phosphorylation at Ser105 seems to stimulate the translocation of GATA4 to the nucleus and to increase the DNA-binding capacity of GATA4 (Kwintkiewicz et al., 2007). As shown in Fig. 2B, in follicles treated with FSH, pSer105-GATA4 is located exclusively in the nucleus, whereas after staining with an antibody against total GATA4, signal can be detected in the cytoplasm and the nucleus.

FSH effects on GATA4 activity in the ovary appear to be mostly mediated by extracellular signal-regulated kinases 1 and 2 (ERK1/2).

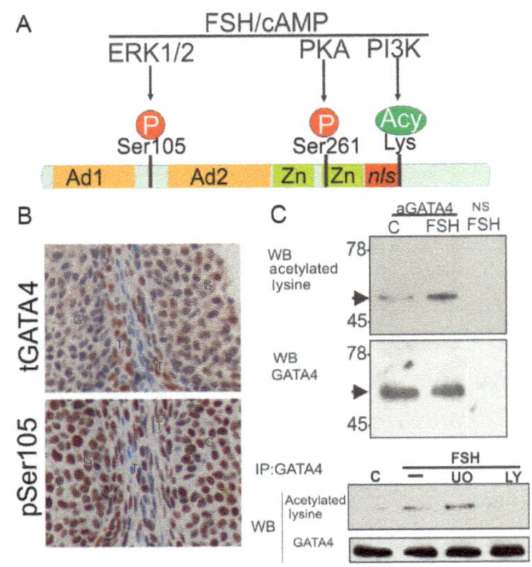

Fig. 2 (A) Scheme of the possible effect of FSH signaling on GATA4 posttranscriptional modifications. Activation of the FSH receptor increases cAMP which activates the protein kinase A, PKA. cAMP also participates on PI3K and ERK1/2 activation. *Acy*, acetylation; *AD*, transcription activation domain; *NLS*, nuclear localization signal; *P*, phosphorylation; *Zn*, zinc finger. (B) Immunohistochemical localization of GATA4[pS105] and total GATA4 in the ovaries of immature rats treated with FSH. Sections were probed with an anti-GATA4 (*top panel*) or an anti-[pS105]-GATA4 (*bottom*) antibody. Notice the predominant nuclear localization of pSer105-GATA4 in granulosa cells (G); *T*, theca cells. (C) Acetyl-GATA4 increases after FSH treatment. (*Top*) Primary granulosa cells were treated with FSH for 24 h. Nuclear extracts (50 μg of protein) were immunoprecipitated with an anti-GATA4 antibody or normal serum (NS), followed by sequential Western blotting with antiacetylated lysine or anti-GATA4 antibodies. (*Bottom*) Cells were treated with MAPK/ERK inhibitor, UO126 (UO) or PI3K inhibitor, LY294002 (LY) for 1 h before FSH treatment. Total GATA4 and acetyl-GATA4 were determined 24 h after the addition of FSH to the media.

Thus, in granulosa cells, we observed that FSH-induced GATA4 phosphorylation at Ser105 is only partially prevented by inhibitors of protein kinase A (PKA) and phosphoinositide 3-kinase (PI3K), but fully blocked by ERK1/2 inhibitors. Moreover, ERK1/2 inhibitors decrease GATA4 phosphorylation to undetectable levels even in untreated cells (Kwintkiewicz et al., 2007). In addition, based on the facts that the cAMP/PKA signaling pathway is used by gonadotropins and that GATA4 transactivates several PKA-responsive promoters (Leung & Steele, 1992), it has been suggested that phosphorylation of GATA4 on Ser261 by PKA may play a vital role in the activation of GATA4 in the ovary (Tremblay & Viger, 2003).

Acetylation of GATA4 may also regulate stability, DNA-binding capacity, and transcriptional activity (Kawamura et al., 2005; Yanazume et al.,

2003). In primary granulosa cells, we have observed that FSH increases not only GATA4 phosphorylation (Fig. 2B; Kwintkiewicz et al., 2007) but also acetylation (Fig. 2C, unpublished). Cotreatment with an inhibitor of PI3K (LY294002) blocks GATA4 acetylation (Fig. 2C). However, the acetylated residues and the role that this modification has on the regulation of GATA4 function in granulosa cells remain to be investigated.

Phosphorylation and acetylation may regulate the interaction of GATA factors with a wide variety of factors and cofactors. These interactions are crucial to determine the tissue-specific actions of GATA4 (Molkentin, 2000). For instance, GATA4 regulates cardiac gene expression by interacting with factors uniquely expressed in the myocardium (Durocher, Charron, Warren, Schwartz, & Nemer, 1997; Molkentin et al., 1998; Morin, Charron, Robitaille, & Nemer, 2000). Moreover, GATA4's interaction with hepatic nuclear factor-1α contributes to the regulation of fatty acid-binding proteins in the liver and the gastrointestinal tract (Divine et al., 2004). The interaction of GATA4 with transcription factors highly expressed in the ovary such as steroidogenic factor 1 (SF1:NR5A1) and liver receptor homolog 1 (LRH1:NR5A2) has been inferred from cotransfection experiments. GATA4 interacts with LRH1 and SF1 to synergistically activate the promoter of 3βHSD2 (Martin et al., 2005), AMH (Tremblay & Viger, 1999), inhibin α-subunit (Robert, Miyamoto, Taniguchi, & Viger, 2006), and Cyp19a1 (Bouchard, Taniguchi, & Viger, 2005). GATA4 modulates tissue-specific transcription of the *P450c17* gene by direct interaction with Sp1 (Fluck & Miller, 2004). This evidence indicates that several proteins can act as partners of GATA4 to regulate ovarian gene expression.

Taken together, these findings indicate that GATA function is tightly controlled by gonadotropins at both the levels of expression and at the post-transcriptional levels. In addition, although no all the ovarian partners of GATA have been elucidated yet, GATA4 interaction with SF1 and LRH1 in the granulosa cells provides another mechanism to closely regulate GATA function in the ovary.

4. CONDITIONAL KNOCKOUT MODELS OF GATA IN THE OVARY

Due to the embryonic lethality of knockout mice for GATA4 and GATA6 (Kuo et al., 1997; Molkentin et al., 1997; Morrisey et al., 1998), conditional knockouts have been utilized to gain perspective on the roles of these factors in the ovary. Floxed GATA4 mice, either by targeting exon

2 (Pu, Ishiwata, Juraszek, Ma, & Izumo, 2004), which removes the translational start site and the N-terminal activation domain of GATA4, or exons 3–5 (Watt, Battle, Li, & Duncan, 2004), which codes for the zinc finger-binding domains and nuclear localization signal, have been generated. Floxed GATA6 mice, targeting exon 2 (Sodhi, Li, & Duncan, 2006), which encodes the majority of the GATA6 protein, have also been generated. To date, at least five conditional knockouts have been developed to study the role of GATA4 and GATA6 at specific times during ovarian development, folliculogenesis, and luteinization. We will briefly describe these models following the progress from early development to corpus luteum formation.

4.1 Conditional Knockdown During Embryo Development

As mentioned earlier, GATA4 is expressed in the genital ridge across many organisms, including mammals (Anttonen et al., 2003; McCoard et al., 2001; Viger et al., 1998), chickens (Oreal, Mazaud, Picard, Magre, & Carre-Eusebe, 2002), and fish (Li et al., 2012). The expression of GATA4 in the genital ridge has been mostly linked to testis development since GATA4 participates in the induction of the SRY gene (Miyamoto, Taniguchi, Hamel, Silversides, & Viger, 2008). The first indication that GATA4 was involved in the formation of the gonads was observed in animals carrying a null mutation for friend of GATA (FOG2), a partner of GATA4, or a mutation in GATA4 ($GATA4^{ki}$) that abrogates its interaction with FOG2 but maintains GATA4 expression. In the FOG2 null or $GATA4^{ki/ki}$ mice, testis cords do not form at E12.5 (Crispino et al., 2001; Tevosian et al., 2002). Loss of GATA4–FOG2 interaction also arrests ovarian differentiation and disrupts the expression of SPRR2D (small proline-rich 2d) and FOXL2, which are implicated in ovarian development (Manuylov et al., 2008). These findings demonstrate that GATA4 plays a role in both ovary and testis development.

Moreover, inducible knockdown of GATA4 demonstrated that this factor is also required for the development of the genital ridge prior to its role in gonadal sex determination (Hu et al., 2013). Using a tamoxifen-inducible Cre/Lox system to delete GATA4 in mouse embryos after E8.5, it was shown that GATA4 is essential for the thickening of the coelomic epithelium and for the expression of SF1, which is needed for the differentiation of the gonadal–adrenal primordium (Hu et al., 2013).

In an effort to determine the consequences of GATA4 loss in ovarian development, Efimenko et al. (2013) generated mice carrying $GATA4^{flox/flox}$ and Cre recombinase under the control of the promoter of SF1. In SF1-Cre mice, Cre recombinase is active on E10.5 in the somatic cells of the gonadal ridge (Bingham, Verma-Kurvari, Parada, & Parker, 2006). Ovarian development was not affected in SF1-Cre:GATA4$^{flox/flox}$ mice (Efimenko et al., 2013), despite that GATA4 protein is absent in the somatic cells of the gonads at E12.5. However, these animals have decreased expression of genes crucial for ovarian development such as follistatin and FOXL2. Since the initiation of ovarian differentiation occurs at E11.5, this finding demonstrates that deletion of GATA4 at E12.5 has no effect on embryonic ovarian development once this checkpoint is cleared.

Interestingly, by postnatal day (PND) 7, SF1-Cre:GATA4$^{flox/flox}$ mice show a significant reduction in ovarian size as consequence of the reduced number of growing follicles. However, at PND3, control and SF1-Cre: GATA4$^{flox/flox}$ mice have a comparable number of primordial follicles. By PND7, knockdown mice have more primordial follicles than control mice, suggesting that follicle activation is impaired (Efimenko et al., 2013). This is supported by a decrease in proliferation observed in the ovaries of conditional mutants. Markers of follicle growth, such as AMH and 3βHSD, are also low in the granulosa cells of SF1-Cre:GATA4$^{flox/flox}$ mice. Consequently, the ovaries of adult conditional mutants have few antral follicles accompanied by increased apoptosis of the granulosa cells. Moreover, inclusion bodies, hemorrhagic follicles, and ovarian cysts are present in ovaries of adult SF1-Cre:GATA4$^{flox/flox}$ mice. Not surprisingly, SF1-Cre: GATA4$^{flox/flox}$ female mice are infertile. Finally, follicular development cannot be rescued by exogenous gonadotropins in SF1-Cre:GATA4$^{flox/flox}$ mice (Efimenko et al., 2013).

SF1-Cre:GATA6$^{flox/flox}$ mice were also generated primarily to determine the role of GATA6 within the adrenal cortex (Pihlajoki et al., 2013). These conditional mutant mice have no obvious reproductive phenotype, as ovarian weight, estrogen levels, and fertility are not affected (Pihlajoki et al., 2013).

The double conditional knockouts of GATA4 and GATA6 using SF1-Cre were later generated to determine the roles of both factors during embryonic ovarian development (Padua et al., 2014). Deletion occurs by E13.5 and by E15.5 GATA4 and GATA6 are not detectable in the ovarian somatic cells of SF1-Cre:GATA4/6$^{flox/flox}$ embryos. At E15.5 the expression of FOXL2 was drastically reduced in the double conditional mutants. However, the development of germ cells proceeds normally. Interestingly, by E18.5 the oogonias do not localize to the ovarian cortex and germ cell nests remain centrally localized in the double knockout animals. By PND4,

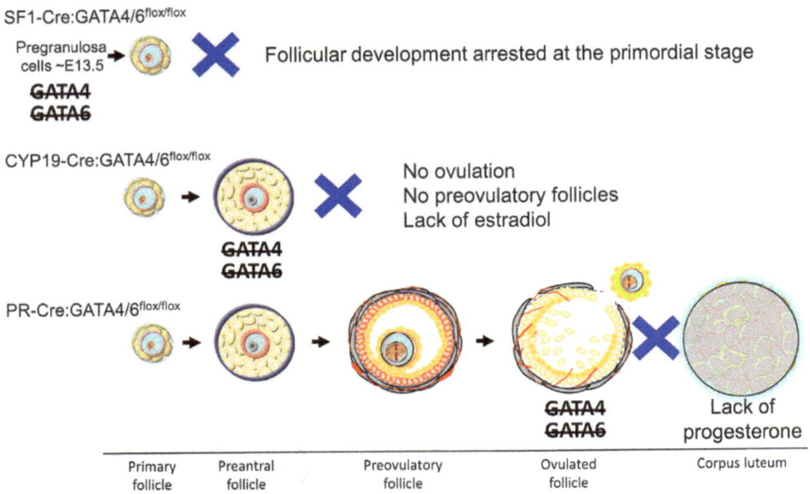

Fig. 3 Scheme of the main characteristics observed in double GATA4 and GATA6 ovarian conditional knockout mice. The schemes represent finding observed in SF1-Cre: GATA4/6$^{f/f}$ (Padua et al., 2014), CYP19-Cre; GATA4/6$^{f/f}$ (Bennett et al., 2012), and PR-Cre:GATA4/6$^{f/f}$ (Convissar et al., 2015) mice. Strikethrough GATA4 and GATA6 suggest approximately the stage of follicle development at which the conditional knockdown in the granulosa cells takes place.

the ovaries of double knockout mice contain no growing follicles as they remain arrested in a primordial state in germ cell clusters. Similar to the single GATA4 knockout mice, the ovaries of SF1-Cre:GATA4/6$^{flox/flox}$ are smaller in size, have decreased cell proliferation, and by PND6 became apoptotic. Oocyte growth was also absent, and the majority of follicles disappear by PND9. Further fertility analyses are not possible as females do not survive after 2 weeks, likely due to adrenal defects (Padua et al., 2014).

In summary, knockdown of GATA4 or GATA6 using SF1-Cre causes loss of ovarian function after birth or no effects at all, respectively. The SF1-Cre-driven deletion of both factors blocks early follicular development (Fig. 3). These findings suggest overlapping and cooperative roles of GATA4 and GATA6 at the beginning of ovarian development and that GATA activity is required for proper follicular assembly.

4.2 Conditional Knockdown in Adult Animals

(A) GATA4$^{flox/flox}$ mice targeting exons 3–5 were crossed with AMHR2-Cre to generate AMHR2-Cre;GATA4$^{flox/flox}$ mice (Kyronlahti et al., 2011). AMHR2-Cre drives the expression of Cre recombinase in proliferating granulosa cells and in the mesenchyme of developing Müllerian ducts

(E11.5), adult oviducts, and uterus (Jamin, Arango, Mishina, Hanks, & Behringer, 2002). AMHR2-Cre:GATA4$^{flox/flox}$ females are subfertile compared to controls. By 6 months, the AMHR2-Cre:GATA4$^{flox/flox}$ females had a distinct ovarian phenotype characterized by large nonhemorrhagic cysts and low serum AMH levels. These animals had abnormal responses to gonadotropin and treatment with eCG and hCG resulted in the formation of hemorrhagic cysts and lack of oocyte release. Also, CYP19a1 expression is lower in these conditional knockouts mice when compared to controls. Despite these abnormalities, AMHR2-Cre;GATA4$^{flox/flox}$ female mice are only subfertile and produce litters with a reduced number of pups per litter. AMHR2-Cre;GATA6$^{flox/flox}$ females were embryonic lethal due to leaky expression of yolk sac endoderm (Pihlajoki, Farkkila, Soini, Heikinheimo, & Wilson, 2016).

(B) Single (GATA4 or GATA6) and double (GATA4/GATA6) conditional knockout mice were also generated using Cre recombinase under the control of the ovarian promoter of CYP19a1 (Bennett et al., 2012). To maximize the recombination of the floxed genes, animals carrying one floxed allele (exon 2) and one null allele (GATA4$^{flox/-}$ and GATA6$^{flox/-}$) were used in this report. CYP19a1 is active in the granulosa cells of growing early antral follicles. Thus, in CYP19-Cre mice, knockdown is achieved at the antral stage (Fan, Liu, Cahill, & Richards, 2008).

CYP19-Cre:GATA4$^{flox/-}$ animals are subfertile. Gonadotropin stimulation results in the formation of hemorrhagic follicles, whereas eCG/hCG treatment revealed impaired ovulation, decreased expression of steroidogenic genes, and low serum estradiol levels (Bennett, Baumgarten, & Stocco, 2013; Bennett et al., 2012). In contrast, CYP19-Cre:GATA6$^{flox/-}$ mice have normal ovarian function (Bennett et al., 2013, 2012).

The overlapping expression of GATA4 and GATA6 in the granulosa cells led to the simultaneous conditional deletion of GATA4 and GATA6. CYP19-Cre:GATA4/6$^{flox/-}$ mice are infertile and spend less time in proestrus than control females. Double conditional knockout animals had significantly lower serum estradiol levels and higher serum FSH and LH levels. Morphologically, the double knockouts lack preovulatory follicles and corpora lutea and have increased follicular atresia. After exogenous gonadotropin treatment, double knockout females fail to ovulate, and their ovaries are completely devoid of antral follicles. As expected, these animals also have reduced expression of steroidogenic genes, CYP19a1 in particular, leading to diminished serum estradiol levels (Bennett et al., 2012). Interestingly, the decrease in serum estradiol correlated with an increase in serum FSH levels in double knockout mutants. This observation led to the idea that the granulosa cells of these animals may not respond to FSH. In fact, it was shown that exogenous gonadotropins stimulate FSH receptor expression in control mice but not in mice lacking GATA4 and GATA6. This finding suggests that a failure to express adequate FSH receptor levels may significantly contribute to the phenotype observed in double knockouts.

Summing up, knockdown of GATA6 in granulosa cells of growing follicles has no effect on folliculogenesis; however, depletion of both GATA4 and GATA6 stops the formation of preovulatory follicles leading to the accumulation of secondary follicles (Fig. 3). This suggests that GATA6 partially compensates by the loss of GATA4 in growing follicles and that GATA4 plays a more significant role in follicle growth.

4.3 Conditional Knockdown During Luteinization

Conditional knockdown of GATA4 and GATA6 in granulosa cells disrupts follicle development and ovulation, which also prevents the formation of corpora lutea (Bennett et al., 2013, 2012). Therefore, studies to examine the roles of GATA4 and GATA6 in luteal function require the knockdown of GATA after or during luteinization. For this purpose, Convissar et al. (2015) targeted GATA4 and GATA6 using a Cre/lox system in which Cre recombinase is fused to the progesterone receptor (PR-Cre mice). In PR-Cre mice, Cre recombinase is expressed only in granulosa cells during the luteinization process and leads to the targeted deletion of floxed genes in the corpus luteum (Soyal et al., 2005). PR-Cre:GATA4$^{flox/flox}$, PR-Cre:GATA6$^{flox/flox}$ and PR-Cre: GATA4/6$^{flox/flox}$ animals were generated (Convissar et al., 2015). Unique to this model is that PR-Cre:GATA4$^{flox/flox}$ and PR-Cre:GATA6$^{flox/flox}$ females are subfertile, suggesting that GATA4 and GATA6 only partially compensate for each other in the corpus luteum. This clearly differs from findings obtained during development or during postnatal folliculogenesis where knockdown of GATA6 has no effects on ovarian function. In addition, the estrous cycle and ovulation rates were normal in double conditional mutants carrying PR-Cre. However, similar to previous findings, double knockout animals are infertile (Convissar et al., 2015). Thus, this report provides the first demonstration that the deletion of GATA6 in ovarian cells leads to a notable reproductive phenotype and suggests a crucial role of GATA6 on the regulation of luteal function.

PR-Cre:GATA4/6$^{flox/flox}$ animals show no differences in size or corpora lutea number when compared to controls (Fig. 3). However, serum progesterone levels were significantly lower in conditional mutant females mated with fertile males as well as after treatment with eCG and hCG. This decrease in serum progesterone appears to be the main cause of infertility since administration of progesterone to double knockout animals recovers implantation. In gonadotropin-stimulated mutant animals, the reduction in serum progesterone is accompanied by an abrupt decrease in the expression of steroidogenic genes (see below). Interestingly, lack of GATA4 and GATA6 expression in luteal cells also decreases the expression of the receptors for luteinizing hormone and prolactin, which are important luteotropic hormones (Stocco, Telleria, & Gibori, 2007).

Noteworthy, knockdown of GATA6 using SF1-Cre or CYP19-Cre has no effect on luteal function. However, luteal function was significantly affected in the PR-Cre:GATA6$^{flox/flox}$ mice. This is intriguing as SF1-Cre and CYP19-Cre would induce the knockdown of GATA6 before luteal cell differentiation. It is possible that early knockdown of GATA6 triggers compensatory mechanisms by GATA4 that fully rescue luteal function, but when the knockdown of GATA6 is done during luteinization, these mechanisms cannot be activated on time. This idea is supported by the decrease in GATA4 expression that takes place during luteinization, whereas GATA6 levels steadily increase during the same period (Convissar et al., 2015).

5. MOLECULAR TARGETS OF GATA4 AND GATA6 IN THE OVARY

5.1 Steroidogenic Genes

Evidence suggests that GATA4 and GATA6 may regulate the expression of every steroidogenic enzyme involved in the synthesis of progesterone, androgens, and estrogens. However, most insights into the role of GATA in steroidogenesis were provided by in vitro transactivation experiments, showing that these factors can target the promoters of steroidogenic genes. However, because transactivation experiments do not necessarily recapitulate the in vivo gene regulation, the development of conditional knockout mice of GATA4 and GATA6 has provided crucial data to understand the effects of these factors in ovarian function.

The steroidogenic acute regulator (StAR) was the first gene identified as a possible target of GATA4 in the ovary. StAR mediates the delivery of cholesterol to the mitochondria (Fig. 4). The promoter of StAR in several species contains a GATA site located within 100 bp of the transcription start site (Lavoie & King, 2009). Mutation of this binding site ablates both basal and

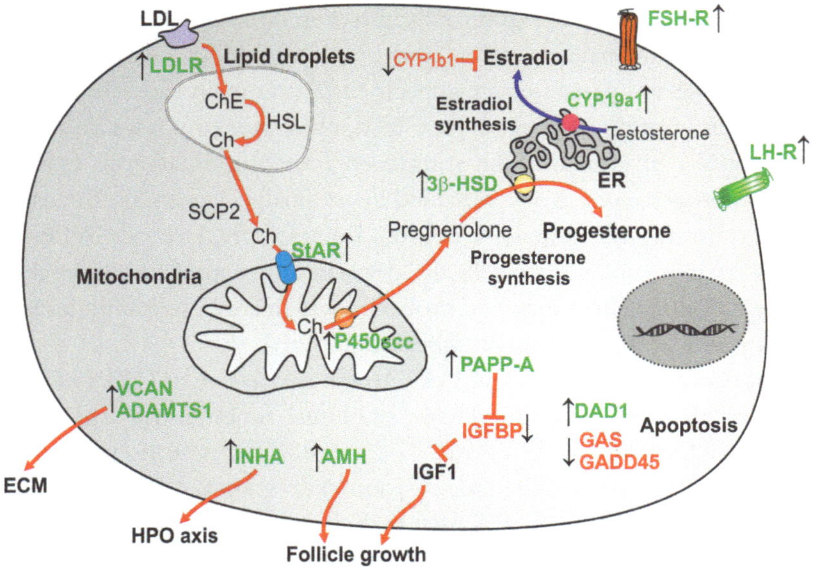

Fig. 4 Ovarian genes controlled by GATA4 and GATA6. In *green* are genes shown to be downregulated in granulosa cells lacking GATA4 or GATA6, suggesting that these factors are needed to stimulate their expression. Genes shown in *red* were increase in the absence of GATAs, indicating that these factors may repress their expression in granulosa cells. *Ch*, cholesterol; *ChE*, Ch esters; *HPO*, hypothalamus pituitary ovary axis; *HSL*, hormone-sensitive lipase; *LDL*, low-density lipoprotein; *LDLR*, LDL receptor; *SCP2*, sterol carrier protein 2. *Arrows* indicate the possible effect that GATA factors may have on the regulation of the targeted gene: *upward arrow*, upregulation; *downward arrow*, downregulation. *Lines with flat heads* indicate inhibition.

FSH-driven StAR expression in rat granulosa cells (Silverman et al., 1999). Whereas, forced expression of GATA4 in porcine granulosa cells increases StAR promoter activity (Gillio-Meina et al., 2003). The GATA site on the StAR promoter binds recombinant GATA4 and GATA6 (LaVoie, Singh, & Hui, 2004). Subsequently, using ChIP, it was also shown that GATA4 binds to the StAR promoter in mouse primary granulosa cells and that cAMP stimulates this binding (Hiroi, Christenson, Chang, et al., 2004; Hiroi, Christenson, & Strauss, 2004).

In vitro experiments using small interference RNA in porcine luteal cells showed that GATA6 knockdown reduces *Star* mRNA and progesterone levels by 60%, whereas knockdown of GATA4 has no effect on basal and cAMP-induced StAR expression (Hui & Lavoie, 2008). These findings suggest that GATA6 plays a more dominant role in the regulation of StAR in luteal cells. Likewise, our in vivo and in vitro results revealed that in the corpus luteum, StAR is highly sensitive to GATA6 depletion (Convissar et al., 2015). For instance, using PR-Cre:GATA4/6$^{flox/flox}$ mice, we found a 50% reduction in StAR mRNA levels 36 h after hCG, which seems to be triggered by only a slight decrease in GATA6 expression (Convissar et al., 2015). Moreover, the basal expression of StAR is reduced to almost undetectable levels in cells that express minimal levels of GATA4 and GATA6. In cells lacking GATA factors, cAMP treatment is still able to stimulate StAR; however, StAR expression is 18 times lower than in cAMP-treated control cells (Convissar et al., 2015). These findings confirm a crucial role of GATA factors in StAR expression in vivo and provide evidence suggesting a prominent role of GATA6 in the regulation of StAR in luteal cells.

Once cholesterol reaches the inner mitochondrial membrane, its transformation into steroid hormones begins. The synthesis of progesterone is the simplest of all the steroidogenic pathways, including only two enzymes (Fig. 4). The first modification to cholesterol occurs in the inner mitochondrial membrane where it is converted to pregnenolone by the cytochrome P450 side-chain cleavage (CYP11A1) enzyme. One of the first pieces of evidence, suggesting that GATA may participate in the regulation of CYP11a1, was the colocalization of GATA4/6 and CYP11a1 in the corpus luteum of rats (Lavoie, McCoy, et al., 2004). Later a GATA-binding site at −475/−470 was described in the CYP11a1 promoter. This site is conserved in the CYP11a1 promoter of mice, rats, and humans (Sher, Yivgi-Ohana, & Orly, 2007). Mutation of this GATA element reduces CYP11a1 promoter activity in granulosa cells. Gel shift data suggest that GATA4 is the predominant occupant of this element throughout folliculogenesis and during granulosa cell luteinization (Sher et al., 2007). Strikingly, GATA4 and GATA6 seem to regulate CYP11a1 in luteal cells indirectly. For instance, luteal cells lacking GATA4 and GATA6 respond to cAMP by expressing CYP11a1 at levels equivalent to control cells (Convissar et al., 2015). Thus, the decrease in CYP11a1 found in the corpus luteum of PR-Cre: GATA4/6$^{flox/flox}$ mice appears to be due to the inability of luteal cells to respond to LH (see below).

However, GATA may indirectly play a role in maintaining optimal conditions to sustain CYP11a1 activity. Thus, ferredoxin 1 (Fdx1), which shuttles electrons from ferredoxin reductase to CYP11a1, decreases in CYP19-Cre:GATA4/6$^{flox/flox}$ mice. Additionally, FSH stimulation of low-density lipoprotein receptor (Ldlr), which is required for the uptake of cholesterol, is also decreased in these conditional mutant animals.

Conversion of pregnenolone into progesterone occurs in the endoplasmic reticulum and is catalyzed by 3β-hydroxysteroid dehydrogenase/$\Delta^5\Delta^4$ isomerase (3β-HSD; Fig. 4). GATA has been proposed to regulate the expression of both 3β-HSD2 (Martin et al., 2005) and 3β-HSD1 (Peng, Huang, Jin, Jiang, & Payne, 2004). However, most data have been obtained in nonovarian steroidogenic cells. A site located at −196 in the human 3β-HSD2 promoter confers responsiveness to GATA factors (Martin et al., 2005). Inhibition of GATA4 expression blunts 3β-HSD2 promoter activity in MA-10 (testis cell line) and CV-1 (kidney cell line) cells, whereas overexpressing GATA4 or GATA6 has the opposite effect (Martin et al., 2005). Transient knockdown of GATA4 using siRNA in luteinizing porcine granulosa cells reduces basal 3β-HSD1 mRNA, whereas transient knockdown of GATA4 or GATA6 has no effect on the response of 3β-HSD1 to cAMP (Hui & Lavoie, 2008).

Findings in PR-Cre:GATA4/$6^{flox/flox}$ mice indicate that 3β-HSD1 is highly sensitive to GATA6 depletion. For instance, in vivo time-course experiments demonstrated that 3β-HSD1 and GATA6 expression decreases simultaneously even before GATA4 expression is reduced. In granulosa cells of PR-Cre:GATA4/$6^{flox/flox}$ mice, in which GATA expression was reduced more than 90% with respect to controls, basal expression of 3β-HSD1 was reduced to almost undetectable levels (Convissar et al., 2015). In contrast to porcine granulosa cells, 3β-HSD1 expression is not stimulated by cAMP in mouse GATA4- and GATA6-deficient cells (Convissar et al., 2015). It is probable that this difference is due to a more efficient knockdown of GATA factors in cells isolated from PR-Cre:GATA4/$6^{flox/flox}$ mice. However, together these findings provide strong evidence for a role of GATA in the expression of 3β-HSD1 in ovarian cells.

The next gene in the steroidogenic pathway that appears to be regulated by GATA transcription factors is P450c17. It has been shown that GATA4 or GATA6 and Sp1 together regulate the expression of P450c17 in human adrenal NCI-H295A cells (Fluck & Miller, 2004). In the ovary, P450c17 expression in theca cells facilitates the production of androgens from progestins. Patients with polycystic ovary syndrome (PCOS) typically have hyperandrogenism of both ovarian and adrenal origin (Rosenfield, 1999). Ovarian theca cells from PCOS women have increased P450c17 and GATA6 expression (Wood et al., 2003). These reports suggest that, in rodents and humans, GATA6 may play a role in the production of androgens.

Aromatase (CYP19a1) is essential for estradiol production and a hallmark of granulosa cell differentiation in preovulatory follicles (Stocco, 2008).

CYP19a1 is abundantly expressed in the granulosa cells of preovulatory follicles due to the potent stimulatory effect of FSH. We demonstrated that FSH stimulation of CYP19a1 is mediated by GATA4 (Kwintkiewicz et al., 2007). Thus, in rat granulosa cells, FSH promotes GATA4 binding to the CYP19a1 promoter. In addition, basal and FSH-induced CYP19a1 expression is enhanced by GATA4 overexpression, whereas depletion of GATA4 via small interference RNA blunts the induction of CYP19a1 expression by FSH (Kwintkiewicz et al., 2007). These results along with the decrease in serum estradiol levels observed in CYP19-Cre:GATA4/$6^{flox/-}$ mice illustrate clearly the importance of GATA4 on the maintenance of estradiol biosynthesis in follicles (Fig. 4).

GATA may also indirectly contribute to maintaining estrogen synthesis in granulosa cells. For instance, the expression of cytochrome P4501b1 (CYP1b1), which inactivates estradiol (Han, Kim, Hwang, Song, & Jeong, 2010), increases fivefold in CYP19-Cre:GATA4/$6^{flox/-}$ animals. These findings suggest that in the absence of GATA4 and GATA6 expression, not only the synthesis of estradiol is significantly impaired, but also its inactivation is stimulated. Moreover, an increase in apolipoprotein-E (ApoE) was observed in CYP19-Cre:GATA4/$6^{flox/-}$ animals. In the ovary, ApoE may limit androgen production, thereby limiting follicular estrogen synthesis. In fact, in humans, the levels of ApoE in follicular fluid decrease as serum estrogen levels increase during the menstrual cycle (Brown, Hay, & Schreiber, 1989). In addition, ApoE has been shown to increase in atretic follicles (Nicosia, Moger, Dyer, Prack, & Williams, 1992), which corresponds to the increase in apoptosis observed in follicles lacking GATA factors (see below). Taken together these findings suggest that both the increase in CYP1b1 and ApoE along with the decrease in aromatase contribute to the reduction in estradiol observed in CYP19-Cre:GATA4/$6^{flox/-}$ animals (Fig. 4).

A decrease in the mRNA expression of the membrane receptor plexin C1 (PlxnC1) was also observed in CYP19-Cre:GATA4/$6^{flox/-}$ mice (Bennett et al., 2013). Plexins are receptors for semaphorins (Sema). PlexinB1 and its ligand Sema4D are expressed in ovarian follicles of FSH-stimulated mice (Regev, Goldman, & Shalev, 2005, 2007). Although the function of these proteins in the ovary is unknown, PlxnC1 and Sema7A are involved in the regulation of cytoskeleton components including actin and cofilin (Messina et al., 2011; Scott, McClelland, Fricke, & Fender, 2009). Cofilin inactivation plays a major role in ovarian steroidogenesis (Karlsson et al., 2010), suggesting that GATA also contributes to maintaining optimal steroid synthesis by regulating plexins.

5.2 Hormones

Anti-Müllerian hormone (AMH) is highly expressed in granulosa cells of preantral and early antral follicles (Salmon, Handyside, & Joyce, 2004) and progressively decreases toward the preovulatory stage (Baarends et al., 1995; Gruijters, Visser, Durlinger, & Themmen, 2003; Weenen et al., 2004). AMH inhibits the recruitment of primordial follicles into the pool of growing follicles and reduces the responsiveness of growing follicles to FSH (Knight & Glister, 2006). GATA4 and a GATA-binding site are essential for full AMH promoter activity in rat Sertoli cells (Tremblay, Robert, & Viger, 2001; Watanabe, Clarke, Lane, Wang, & Donahoe, 2000). Coexpression of GATA4 and SF1 in CV-1 cells, a kidney cell line, enhances the activity of the AMH promoter in part by a direct interaction between these proteins (Tremblay & Viger, 1999).

In agreement with the lack of follicle maturation observed in the absence of GATA factors (Bennett et al., 2012), AMH mRNA expression is higher in the ovary of CYP19-Cre:GATA4/6$^{flox/-}$ animals when compared with controls. These are intriguing findings since the expression of AMH has been shown to be stimulated by GATA4 (Viger et al., 2008). Thus, the mechanism that leads to the sustained expression of AMH in CYP19-Cre:GATA4/6$^{flox/-}$ mice needs further analysis. It is possible that because follicles do not progress beyond the secondary state in conditional mutants, secondary follicle accumulation leads to a relative increase in AMH. Alternatively, since FSH inhibits AMH (Grynberg et al., 2012), the lack of FSH action due to diminished FSH receptor (see below) may also contribute to the sustained levels of AMH in CYP19-Cre:GATA4/6$^{flox/-}$ animals.

In SF1-Cre:GATA4/6$^{flox/flox}$ mice, AMH staining was found only in a small number of follicles (Efimenko et al., 2013). This finding suggests that during early folliculogenesis, GATA4 and GATA6 are needed to maintain AMH expression. Further in vivo studies are required to determine whether GATA factors regulate AMH, especially in animals with normal expression of GATA factors. Thus, ChIP could be used to determine the interaction of GATA4 or GATA6 with the AMH promoter in granulosa cells of secondary follicles isolated from wild-type animals.

GATA response elements have also been found in the promoter region of inhibin alpha (INHA) (Tremblay & Viger, 2001). Inhibins are heterodimers of the common α subunit with either βa or βb subunits (inhibins A and B, respectively). The balance between activins (β homodimers) and inhibins shifts throughout follicle development, with primary to preantral follicles mainly expressing β subunits to form activins, and larger preovulatory follicles expressing more α subunit, to form inhibin A or B. This shift is controlled by FSH (Knight, Satchell, & Glister, 2012). Inhibin provides a negative feedback mechanism from the gonads to the pituitary for inhibiting gonadotropin production, whereas locally it regulates oocyte maturation, granulosa cell function, and steroid hormone production (Cook, Thompson, Jardetzky, & Woodruff, 2004).

The expression of the subunit α is regulated by GATA factors in vitro (Anttonen et al., 2006; Robert et al., 2006). Thus, overexpression studies demonstrated that GATA4 can activate the inhibin α-subunit promoter (Tremblay & Viger, 2001) and that both GATA4 and GATA6 synergize with LRH1 in the regulation of this promoter (Robert et al., 2006). In addition, a twofold decrease in the expression of the subunit α was observed in granulosa cells of CYP19-Cre:GATA4/6$^{flox/-}$ mice when compared with wild-type animals. In eCG-treated CYP19-Cre:GATA4/6$^{flox/-}$ mice, a decrease in the expression of the βa and βb subunits was also observed. GATA4 also activates the promoter of the βb subunit gene in testicular cell lines (Feng, Wu, Zhang, & Chen, 2000).

AMH is expressed in the granulosa cells of growing follicles and acts mostly locally to modulate follicle growth by inhibiting FSH actions (Knight & Glister, 2006). In contrast, ovarian inhibin acts as an endocrine hormone, as part of a negative feedback loop, to regulate FSH synthesis in the anterior pituitary (Findlay et al., 2001). This evidence suggests that GATA4 and GATA6 indirectly participate in the regulation of folliculogenesis and of the hypothalamus–pituitary–ovary axis by regulating AMH and INHA, respectively (Fig. 4).

5.3 Gonadotropin Receptors

5.3.1 FSH Receptor

FSH is crucial for follicle growth and granulosa cell differentiation. Thus, FSH receptor and FSH-deficient female mice are infertile due to a block in folliculogenesis before antral follicle formation (Dierich et al., 1998; Kumar, Wang, Lu, & Matzuk, 1997). Interestingly, this block in folliculogenesis also occurs in CYP19-Cre:GATA4/6$^{flox/-}$ mice (Bennett et al., 2012). We demonstrated that granulosa cells of CYP19-Cre:GATA4/6$^{flox/-}$ mice express extremely low levels of FSH receptor protein. In addition, we showed that GATA4 interacts with a GATA response element found in the FSH receptor promoter and that overexpression of GATA4 stimulates the activity of this promoter. Finally, the mutation of this GATA-binding site diminishes FSH receptor promoter activity. These results demonstrate that GATA4 is required to sustain FSH receptor expression in granulosa cells. Thus, the downregulation of the FSH receptor in the granulosa cells of CYP19-Cre:GATA4/6$^{flox/-}$ mice provides a molecular mechanism to explain the abnormal follicle growth and infertility of these conditional knockout animals.

Reciprocally, FSH seems to be necessary to maintain GATA expression in the gonads of several species. Women carrying an inactivating mutation in the FSH receptor gene express low levels of GATA4 in their ovaries (Vaskivuo et al., 2002). In mice, FSH enhances the expression of GATA4 and GATA6 transcripts (Heikinheimo et al., 1997; Kwintkiewicz et al., 2007). In pigs, GATA4 mRNA levels are low in small follicles and reach the highest level of expression in preovulatory follicles (Gillio-Meina et al., 2003), suggesting that the expression of this transcription factor increases during FSH-induced follicle maturation. GATA4 expression is also low in granulosa cells of FSH receptor knockout animals (Balla, Danilovich, Yang, & Sairam, 2003). In addition, we have demonstrated that FSH signaling leads to the phosphorylation of GATA4 on serine 105 (Kwintkiewicz et al., 2007), a modification that is known to increase the transcriptional activity of GATA4 (Liang et al., 2001). This evidence, along with the reduction in FSH receptor expression after the loss of GATA function, suggests the presence of a positive feedback system between FSH and GATA in ovarian granulosa cells. We propose that this positive feedback is critical for the differentiation of granulosa cells and the rapid growth of preantral and early antral follicle to the preovulatory stage.

5.3.2 LH Receptor

LH receptor is crucial for luteal formation and function (Duncan, McNeilly, Fraser, & Illingworth, 1996; Stouffer, 2003). We have observed that in luteal cells of PR-Cre:GATA4/6$^{flox/flox}$ mice, the expression of the LH receptor is extremely low, suggesting that GATA factors play a fundamental role in the upregulation of this receptor in the corpus luteum. It remains to be determined whether GATA inhibits the mechanisms involved in LH receptor mRNA degradation or directly stimulates the activity of the promoter of this gene. This latter possibility is supported by evidence indicating that GATA4 expression induces LH receptor promoter activity in adrenocortical tumors (Rahman et al., 2004).

5.4 Ovulation-Related and Ovarian/Fertility Genes

Microarray analysis also revealed that the lack of GATA4 and GATA6 affects the mRNA expression of Grem1 and Grem2 in granulosa cells.

Grem1 expression was significantly low in CYP19-Cre:GATA4$^{flox/-}$ mice animals. In contrast, Grem2 expression decreased by twofold in CYP19-Cre:GATA6$^{flox/-}$ mice, suggesting that GATA4 and GATA6 may specifically target Grem1 and Grem2, respectively. Grem1 expression in granulosa cells is stimulated by eCG, GDF9, and BMP4 (Pangas, Jorgez, & Matzuk, 2004; Sudo, Avsian-Kretchmer, Wang, & Hsueh, 2004). Both Grem1 and Grem2 prevent the inhibitory effect of BMP4 on granulosa cell steroidogenesis (Sudo et al., 2004). Our findings demonstrate that the absence of GATA factors abolished the increase of Grem1 and Grem2 expression induced by eCG. Therefore, GATA4 and GATA6 may contribute to the healthy development of follicles by maintaining Grem1/2 expression, which in turn inhibits BMP4. Noteworthy, Grem2 is one of the few genes affected by in the absence of GATA6, but whether the Grem2 gene is directly regulated by GATA6 remains to be determined.

5.5 Extracellular Structure and Organization

Antrum formation does not occur in CYP19-Cre:GATA4/6$^{flox/-}$ mice, suggesting that GATA4 and GATA6 are needed for the build up of the extracellular components involved in this process. In fact, in these mutant mice, the expression of genes involved in the reorganization of the extracellular matrix (ECM) was impaired. Within the genes affected are ADAMTS1 and versican (VCAN). ADAMTS (a disintegrin and metalloproteinase with thrombo spondin motifs) are proteinases that cleave proteoglycans present in the ECM and are required for normal ovulation (Brown et al., 2010), whereas VCAN is known to be produced by granulosa cells of small growing and antral follicles (Doyle, Russell, Sriraman, & Richards, 2004; Russell, Ochsner, Hsieh, Mulders, & Richards, 2003). In granulosa cells, the expression of ADAMTS1 and VCAN increases after treatment of FSH, but this increase is abolished by the deletion of both GATA4 and GATA6 (Bennett et al., 2013). Since versican is a large proteoglycan that crosslinks with hyaluronan (Rahmani et al., 2006; Zimmermann & Dours-Zimmermann, 2008), it may contribute substantially to the osmotic potential of the follicular fluid. These findings suggest that in the absence of GATA factors, the formation and composition of the follicular ECM is significantly compromised (Fig. 4). Therefore, lack of versican expression may explain, at least in part, the lack of antral follicle formation observed in GATA conditional knockout animals.

5.6 Insulin-Like Growth Factor (IGF) System

IGF1 is required for the differentiation of granulosa cells to the preovulatory stage (Zhou et al., 2013). A slight reduction in IGF1 mRNA expression occurs in the granulosa cells of CYP19-Cre:GATA4/6$^{flox/-}$ mice (Bennett et al., 2013). However, IGF-binding protein (IGFBP) 2, IGFBP4, and IGFBP5 are elevated in these mice. In particular, IGFBP4 is among the most upregulated genes in CYP19-Cre:GATA4/6$^{flox/-}$ mice. IGFBPs inhibit the interaction of IGF1 with the IGF1 receptor and are known to prevent follicular growth and maturation (Monget, Monniaux, Pisselet, & Durand, 1993; Silva, Figueiredo, & van den Hurk, 2009). Thus, the down-regulation of IGFBPs during granulosa cell differentiation to the preovulatory stage seems to be required for normal folliculogenesis (Funston, Seidel, Klindt, & Roberts, 1996; Monget et al., 2002; Rees, Clemmons, Horvitz, Clarke, & Busby, 1998). For instance, IGFBP5-overexpressing female mice are subfertile (Salih et al., 2004), whereas IGFBP4 inhibits LH-induced progesterone and FSH-induced estradiol production in human granulosa cells (Wright, Holly, Galea, Brincat, & Mason, 2002). IGFBP levels are mainly regulated via proteolytic degradation by specific proteinases such as pregnancy-associated plasma protein-A (PAPP-A), which is highly expressed in healthy antral follicles and positively correlates with dominant follicle development (Hourvitz et al., 2002; Mazerbourg et al., 2001). Interestingly, PAPP-A expression in granulosa cells is significantly decreased in CYP19-Cre:GATA4/6$^{flox/-}$ mice. These findings suggest that in the absence of GATA factors, IGF1 biological activity declines due to a reduction of IGF1, an increase in IGFBPs, and a decrease in PAPP-A expression (Fig. 4). These effects may significantly impair follicle growth.

5.7 Apoptosis/Cell Division

Genes involved in apoptosis and proliferation are also affected by the deletion of GATA factors in granulosa cells (Bennett et al., 2013). Within these genes, defender against apoptotic cell death (DAD1), a negative regulator of programmed cell death (Nakashima et al., 1993), is decreased in CYP19-Cre:GATA4/6$^{flox/-}$ mice (Bennett et al., 2013). In contrast, growth arrest-specific and growth arrest and DNA damage-inducible proteins (GADD45), which are inhibitors of proliferation (Hutchison, Bassett, & White, 2010; Perugini et al., 2009; Sacilotto, Espert, Castillo, Franco, & Lopez-Rodas, 2011) were increased in CYP19-Cre:GATA4/6$^{flox/-}$ mice (Bennett et al., 2013). FSH-induced proliferation is significantly reduced in granulosa cells lacking GATA4 and GATA6 (Bennett et al., 2013). These changes are supported by a decrease in proliferating cell nuclear antigen, a marker of proliferation, in the follicles of conditional mutant animals. Moreover, staining for cleaved caspase 3, a marker of apoptosis, increases in the ovaries of CYP19-Cre:GATA4/6$^{flox/-}$ mice (Bennett et al., 2013). These findings suggest that the lack of GATA factors in granulosa cells not only increases apoptosis but also halts proliferation (Fig. 4).

6. CONCLUSIONS

We have focused this review on the role of GATA4 and GATA6 on steroidogenesis and ovarian function in adult mice. However, conditional knockout models have shown that these factors are crucial to sustain ovarian function and that their roles span from embryonic granulosa development to granulosa cells proliferation, maturation, and differentiation in the adult ovary. As expected, from early in vitro studies, the knockdown of GATA in the ovary significantly affected steroidogenesis. However, microarray analysis of conditional mutants revealed that these factors might be involved in several other aspects of granulosa cell function. Thus, both GATA4 and GATA6 are needed to maintain granulosa cell proliferation, to prevent apoptosis, for the formation of the follicular antrum, to preserve the activity of the IGF system, and to maintain the follicular hormonal milieu. However, it remains to be determined whether GATA4 and GATA6 directly target the regulatory regions of genes involved in this wide variety of processes. For instance, studies to determine the genome-wide interaction of GATA4 and GATA6 with regulatory elements (ChIP-seq), preferably using in vivo nontransgenic systems, are needed to elucidate direct and indirect targets of these factors in the ovary.

Finally, the novel finding that GATA4 and GATA6 participate in the regulation of the receptors for FSH and LH provides evidence for a leading role of GATA on the maintenance of ovarian function in reproductive-age animals. Thus, because FSH and LH signaling is crucial for normal folliculogenesis and luteal function, respectively, GATA factors can be placed upstream of almost all pathways and systems necessary to maintain normal ovarian function and consequently to sustain female fertility.

ACKNOWLEDGMENTS

This work was supported by NIH Grant number R56HD086054 (C.S.).

Disclosure Summary: The authors have nothing to disclose.

This chapter does not require any permissions.

REFERENCES

Anttonen, M., Ketola, I., Parviainen, H., Pusa, A. K., & Heikinheimo, M. (2003). FOG-2 and GATA-4 are coexpressed in the mouse ovary and can modulate mullerian-inhibiting substance expression. *Biology of Reproduction*, *68*, 1333–1340.

Anttonen, M., Parviainen, H., Kyronlahti, A., Bielinska, M., Wilson, D. B., Ritvos, O., et al. (2006). GATA-4 is a granulosa cell factor employed in inhibin-alpha activation by the TGF-beta pathway. *Journal of Molecular Endocrinology*, *36*, 557–568.

Arceci, R. J., King, A. A., Simon, M. C., Orkin, S. H., & Wilson, D. B. (1993). Mouse GATA-4: A retinoic acid-inducible GATA-binding transcription factor expressed in endodermally derived tissues and heart. *Molecular and Cellular Biology*, *13*, 2235–2246.

Baarends, W. M., Uilenbroek, J. T., Kramer, P., Hoogerbrugge, J. W., van Leeuwen, E. C., Themmen, A. P., et al. (1995). Anti-mullerian hormone and anti-mullerian hormone type II receptor messenger ribonucleic acid expression in rat ovaries during postnatal development, the estrous cycle, and gonadotropin-induced follicle growth. *Endocrinology*, *136*, 4951–4962.

Balla, A., Danilovich, N., Yang, Y., & Sairam, M. R. (2003). Dynamics of ovarian development in the FORKO immature mouse: Structural and functional implications for ovarian reserve. *Biology of Reproduction*, *69*, 1281–1293.

Bennett, J., Baumgarten, S. C., & Stocco, C. (2013). GATA4 and GATA6 silencing in ovarian granulosa cells affects levels of mRNAs involved in steroidogenesis, extracellular structure organization, IGF-I activity, and apoptosis. *Endocrinology*, *154*, 4845–4858.

Bennett, J., Wu, Y. G., Gossen, J., Zhou, P., & Stocco, C. (2012). Loss of GATA-6 and GATA-4 in granulosa cells blocks folliculogenesis, ovulation, and follicle stimulating hormone receptor expression leading to female infertility. *Endocrinology*, *153*, 2474–2485.

Bingham, N. C., Verma-Kurvari, S., Parada, L. F., & Parker, K. L. (2006). Development of a steroidogenic factor 1/Cre transgenic mouse line. *Genesis*, *44*, 419–424.

Bouchard, M. F., Taniguchi, H., & Viger, R. S. (2005). Protein kinase A-dependent synergism between GATA factors and the nuclear receptor, liver receptor homolog-1, regulates human aromatase (CYP19) PII promoter activity in breast cancer cells. *Endocrinology*, *146*, 4905–4916.

Brown, H. M., Dunning, K. R., Robker, R. L., Boerboom, D., Pritchard, M., Lane, M., et al. (2010). ADAMTS1 cleavage of versican mediates essential structural remodeling of the ovarian follicle and cumulus-oocyte matrix during ovulation in mice. *Biology of Reproduction*, *83*, 549–557.

Brown, S. A., Hay, R. V., & Schreiber, J. R. (1989). Relationship between serum estrogen and level of apolipoprotein E in human ovarian follicular fluid. *Fertility and Sterility*, *51*, 639–643.

Capo-chichi, C. D., Roland, I. H., Vanderveer, L., Bao, R., Yamagata, T., Hirai, H., et al. (2003). Anomalous expression of epithelial differentiation-determining GATA factors in ovarian tumorigenesis. *Cancer Research*, *63*, 4967–4977.

Convissar, S. M., Bennett, J., Baumgarten, S. C., Lydon, J. P., DeMayo, F. J., & Stocco, C. (2015). GATA4 and GATA6 knockdown during luteinization inhibits progesterone production and gonadotropin responsiveness in the corpus luteum of female mice. *Biology of Reproduction*, *93*, 133.

Cook, R. W., Thompson, T. B., Jardetzky, T. S., & Woodruff, T. K. (2004). Molecular biology of inhibin action. *Seminars in Reproductive Medicine*, *22*, 269–276.

Crispino, J. D., Lodish, M. B., Thurberg, B. L., Litovsky, S. H., Collins, T., Molkentin, J. D., et al. (2001). Proper coronary vascular development and heart morphogenesis depend on interaction of GATA-4 with FOG cofactors. *Genes & Development*, *15*, 839–844.

Dierich, A., Sairam, M. R., Monaco, L., Fimia, G. M., Gansmuller, A., LeMeur, M., et al. (1998). Impairing follicle-stimulating hormone (FSH) signaling in vivo: Targeted disruption of the FSH receptor leads to aberrant gametogenesis and hormonal imbalance. *Proceedings of the National Academy of Sciences of the United States of America*, *95*, 13612–13617.

Divine, J. K., Staloch, L. J., Haveri, H., Jacobsen, C. M., Wilson, D. B., Heikinheimo, M., et al. (2004). GATA-4, GATA-5, and GATA-6 activate the rat liver fatty acid binding protein gene in concert with HNF-1alpha. *American Journal of Physiology. Gastrointestinal and Liver Physiology*, *287*, G1086–1099.

Doyle, K. M., Russell, D. L., Sriraman, V., & Richards, J. S. (2004). Coordinate transcription of the ADAMTS-1 gene by luteinizing hormone and progesterone receptor. *Molecular Endocrinology*, *18*, 2463–2478.

Duncan, W. C., McNeilly, A. S., Fraser, H. M., & Illingworth, P. J. (1996). Luteinizing hormone receptor in the human corpus luteum: Lack of down-regulation during maternal recognition of pregnancy. *Human Reproduction*, *11*, 2291–2297.

Durocher, D., Charron, F., Warren, R., Schwartz, R. J., & Nemer, M. (1997). The cardiac transcription factors Nkx2-5 and GATA-4 are mutual cofactors. *The EMBO Journal*, *16*, 5687–5696.

Efimenko, E., Padua, M. B., Manuylov, N. L., Fox, S. C., Morse, D. A., & Tevosian, S. G. (2013). The transcription factor GATA4 is required for follicular development and normal ovarian function. *Developmental Biology*, *381*, 144–158.

Fan, H. Y., Liu, Z., Cahill, N., & Richards, J. S. (2008). Targeted disruption of Pten in ovarian granulosa cells enhances ovulation and extends the life span of luteal cells. *Molecular Endocrinology*, *22*, 2128–2140.

Feng, Z. M., Wu, A. Z., Zhang, Z., & Chen, C. L. (2000). GATA-1 and GATA-4 transactivate inhibin/activin beta-B-subunit gene transcription in testicular cells. *Molecular Endocrinology*, *14*, 1820–1835.

Findlay, J. K., Drummond, A. E., Dyson, M., Baillie, A. J., Robertson, D. M., & Ethier, J. F. (2001). Production and actions of inhibin and activin during folliculogenesis in the rat. *Molecular and Cellular Endocrinology*, *180*, 139–144.

Fluck, C. E., & Miller, W. L. (2004). GATA-4 and GATA-6 modulate tissue-specific transcription of the human gene for P450c17 by direct interaction with Sp1. *Molecular Endocrinology*, *18*, 1144–1157.

Funston, R. N., Seidel, G. E., Jr., Klindt, J., & Roberts, A. J. (1996). Insulin-like growth factor I and insulin-like growth factor-binding proteins in bovine serum and follicular fluid before and after the preovulatory surge of luteinizing hormone. *Biology of Reproduction*, *55*, 1390–1396.

Gallagher, J. M., Yamak, A., Kirilenko, P., Black, S., Bochtler, M., Lefebvre, C., et al. (2014). Carboxy terminus of GATA4 transcription factor is required for its cardiogenic activity and interaction with CDK4. *Mechanisms of Development*, *134*, 31–41.

Gillio-Meina, C., Hui, Y. Y., & LaVoie, H. A. (2003). GATA-4 and GATA-6 transcription factors: Expression, immunohistochemical localization, and possible function in the porcine ovary. *Biology of Reproduction*, *68*, 412–422.

Gruijters, M. J., Visser, J. A., Durlinger, A. L., & Themmen, A. P. (2003). Anti-Mullerian hormone and its role in ovarian function. *Molecular and Cellular Endocrinology*, *211*, 85–90.

Grynberg, M., Pierre, A., Rey, R., Leclerc, A., Arouche, N., Hesters, L., et al. (2012). Differential regulation of ovarian anti-Müllerian hormone (AMH) by estradiol through α- and β-estrogen receptors. *The Journal of Clinical Endocrinology & Metabolism*, *97*, E1649–E1657.

Han, E. H., Kim, H. G., Hwang, Y. P., Song, G. Y., & Jeong, H. G. (2010). Prostaglandin E2 induces CYP1B1 expression via ligand-independent activation of the ERalpha pathway in human breast cancer cells. *Toxicological Sciences*, *114*, 204–216.

Heikinheimo, M., Ermolaeva, M., Bielinska, M., Rahman, N. A., Narita, N., Huhtaniemi, I. T., et al. (1997). Expression and hormonal regulation of transcription factors GATA-4 and GATA-6 in the mouse ovary. *Endocrinology, 138*, 3505–3514.

Hiroi, H., Christenson, L. K., Chang, L., Sammel, M. D., Berger, S. L., & Strauss, J. F., 3rd. (2004). Temporal and spatial changes in transcription factor binding and histone modifications at the steroidogenic acute regulatory protein (stAR) locus associated with stAR transcription. *Molecular Endocrinology, 18*, 791–806.

Hiroi, H., Christenson, L. K., & Strauss, J. F., 3rd. (2004). Regulation of transcription of the steroidogenic acute regulatory protein (StAR) gene: Temporal and spatial changes in transcription factor binding and histone modification. *Molecular and Cellular Endocrinology, 215*, 119–126.

Hourvitz, A., Kuwahara, A., Hennebold, J. D., Tavares, A. B., Negishi, H., Lee, T. H., et al. (2002). The regulated expression of the pregnancy-associated plasma protein-A in the rodent ovary: A proposed role in the development of dominant follicles and of corpora lutea. *Endocrinology, 143*, 1833–1844.

Hu, Y. C., Okumura, L. M., & Page, D. C. (2013). Gata4 is required for formation of the genital ridge in mice. *PLoS Genetics, 9*, e1003629.

Hui, Y. Y., & Lavoie, H. A. (2008). GATA4 reduction enhances 3′,5′-cyclic adenosine 5′-monophosphate-stimulated steroidogenic acute regulatory protein messenger ribonucleic acid and progesterone production in luteinized porcine granulosa cells. *Endocrinology, 149*, 5557–5567.

Hutchison, M. R., Bassett, M. H., & White, P. C. (2010). SCF, BDNF, and Gas6 are regulators of growth plate chondrocyte proliferation and differentiation. *Molecular Endocrinology, 24*, 193–203.

Jamin, S. P., Arango, N. A., Mishina, Y., Hanks, M. C., & Behringer, R. R. (2002). Requirement of Bmpr1a for Mullerian duct regression during male sexual development. *Nature Genetics, 32*, 408–410.

Karlsson, A. B., Maizels, E. T., Flynn, M. P., Jones, J. C., Shelden, E. A., Bamburg, J. R., et al. (2010). Luteinizing hormone receptor-stimulated progesterone production by preovulatory granulosa cells requires protein kinase A-dependent activation/dephosphorylation of the actin dynamizing protein cofilin. *Molecular Endocrinology, 24*, 1765–1781.

Kawamura, T., Ono, K., Morimoto, T., Wada, H., Hirai, M., Hidaka, K., et al. (2005). Acetylation of GATA-4 is involved in the differentiation of embryonic stem cells into cardiac myocytes. *The Journal of Biological Chemistry, 280*, 19682–19688.

Knight, P. G., & Glister, C. (2006). TGF-beta superfamily members and ovarian follicle development. *Reproduction, 132*, 191–206.

Knight, P. G., Satchell, L., & Glister, C. (2012). Intra-ovarian roles of activins and inhibins. *Molecular and Cellular Endocrinology, 359*, 53–65.

Kumar, T. R., Wang, Y., Lu, N., & Matzuk, M. M. (1997). Follicle stimulating hormone is required for ovarian follicle maturation but not male fertility. *Nature Genetics, 15*, 201–204.

Kuo, C. T., Morrisey, E. E., Anandappa, R., Sigrist, K., Lu, M. M., Parmacek, M. S., et al. (1997). GATA4 transcription factor is required for ventral morphogenesis and heart tube formation. *Genes & Development, 11*, 1048–1060.

Kwintkiewicz, J., Cai, Z., & Stocco, C. (2007). Follicle-stimulating hormone-induced activation of Gata4 contributes in the up-regulation of Cyp19 expression in rat granulosa cells. *Molecular Endocrinology, 21*, 933–947.

Kyronlahti, A., Vetter, M., Euler, R., Bielinska, M., Jay, P. Y., Anttonen, M., et al. (2011). GATA4 deficiency impairs ovarian function in adult mice. *Biology of Reproduction, 84*, 1033–1044.

Laitinen, M. P., Anttonen, M., Ketola, I., Wilson, D. B., Ritvos, O., Butzow, R., et al. (2000). Transcription factors GATA-4 and GATA-6 and a GATA family cofactor, FOG-2, are expressed in human ovary and sex cord-derived ovarian tumors. *The Journal of Clinical Endocrinology and Metabolism, 85*, 3476–3483.

Lavoie, H. A., & King, S. R. (2009). Transcriptional regulation of steroidogenic genes: STARD1, CYP11A1 and HSD3B. *Experimental Biology and Medicine (Maywood, NJ)*, *234*, 880–907.

Lavoie, H. A., McCoy, G. L., & Blake, C. A. (2004). Expression of the GATA-4 and GATA-6 transcription factors in the fetal rat gonad and in the ovary during postnatal development and pregnancy. *Molecular and Cellular Endocrinology*, *227*, 31–40.

LaVoie, H. A., Singh, D., & Hui, Y. Y. (2004). Concerted regulation of the porcine steroidogenic acute regulatory protein gene promoter activity by follicle-stimulating hormone and insulin-like growth factor I in granulosa cells involves GATA-4 and CCAAT/enhancer binding protein beta. *Endocrinology*, *145*, 3122–3134.

Leung, P. C., & Steele, G. L. (1992). Intracellular signaling in the gonads. *Endocrine Reviews*, *13*, 476–498.

Li, J., Chen, W., Wang, D., Zhou, L., Sakai, F., Guan, G., et al. (2012). GATA4 is involved in the gonadal development and maturation of the teleost fish tilapia, Oreochromis niloticus. *The Journal of Reproduction and Development*, *58*, 237–242.

Liang, Q., Wiese, R. J., Bueno, O. F., Dai, Y. S., Markham, B. E., & Molkentin, J. D. (2001). The transcription factor GATA4 is activated by extracellular signal-regulated kinase 1- and 2-mediated phosphorylation of serine 105 in cardiomyocytes. *Molecular and Cellular Biology*, *21*, 7460–7469.

Lossky, M., & Wensink, P. C. (1995). Regulation of Drosophila yolk protein genes by an ovary-specific GATA factor. *Molecular and Cellular Biology*, *15*, 6943–6952.

Lowry, J. A., & Atchley, W. R. (2000). Molecular evolution of the GATA family of transcription factors: Conservation within the DNA-binding domain. *Journal of Molecular Evolution*, *50*, 103–115.

Mackay, J. P., Kowalski, K., Fox, A. H., Czolij, R., King, G. F., & Crossley, M. (1998). Involvement of the N-finger in the self-association of GATA-1. *The Journal of Biological Chemistry*, *273*, 30560–30567.

Manuylov, N. L., Smagulova, F. O., Leach, L., & Tevosian, S. G. (2008). Ovarian development in mice requires the GATA4-FOG2 transcription complex. *Development*, *135*, 3731–3743.

Martin, L. J., Taniguchi, H., Robert, N. M., Simard, J., Tremblay, J. J., & Viger, R. S. (2005). GATA factors and the nuclear receptors, steroidogenic factor 1/liver receptor homolog 1, are key mutual partners in the regulation of the human 3beta-hydroxysteroid dehydrogenase type 2 promoter. *Molecular Endocrinology*, *19*, 2358–2370.

Mazerbourg, S., Overgaard, M. T., Oxvig, C., Christiansen, M., Conover, C. A., Laurendeau, I., et al. (2001). Pregnancy-associated plasma protein-A (PAPP-A) in ovine, bovine, porcine, and equine ovarian follicles: Involvement in IGF binding protein-4 proteolytic degradation and mRNA expression during follicular development. *Endocrinology*, *142*, 5243–5253.

McCoard, S. A., Wise, T. H., Fahrenkrug, S. C., & Ford, J. J. (2001). Temporal and spatial localization patterns of Gata4 during porcine gonadogenesis. *Biology of Reproduction*, *65*, 366–374.

Messina, A., Ferraris, N., Wray, S., Cagnoni, G., Donohue, D. E., Casoni, F., et al. (2011). Dysregulation of Semaphorin7A/beta1-integrin signaling leads to defective GnRH-1 cell migration, abnormal gonadal development and altered fertility. *Human Molecular Genetics*, *20*, 4759–4774.

Miyamoto, Y., Taniguchi, H., Hamel, F., Silversides, D. W., & Viger, R. S. (2008). A GATA4/WT1 cooperation regulates transcription of genes required for mammalian sex determination and differentiation. *BMC Molecular Biology*, *9*, 44.

Molkentin, J. D. (2000). The zinc finger-containing transcription factors GATA-4, -5, and -6. Ubiquitously expressed regulators of tissue-specific gene expression. *The Journal of Biological Chemistry*, *275*, 38949–38952.

Molkentin, J. D., Lin, Q., Duncan, S. A., & Olson, E. N. (1997). Requirement of the transcription factor GATA4 for heart tube formation and ventral morphogenesis. *Genes & Development, 11*, 1061–1072.

Molkentin, J. D., Lu, J. R., Antos, C. L., Markham, B., Richardson, J., Robbins, J., et al. (1998). A calcineurin-dependent transcriptional pathway for cardiac hypertrophy. *Cell, 93*, 215–228.

Monget, P., Fabre, S., Mulsant, P., Lecerf, F., Elsen, J. M., Mazerbourg, S., et al. (2002). Regulation of ovarian folliculogenesis by IGF and BMP system in domestic animals. *Domestic Animal Endocrinology, 23*, 139–154.

Monget, P., Monniaux, D., Pisselet, C., & Durand, P. (1993). Changes in insulin-like growth factor-I (IGF-I), IGF-II, and their binding proteins during growth and atresia of ovine ovarian follicles. *Endocrinology, 132*, 1438–1446.

Morin, S., Charron, F., Robitaille, L., & Nemer, M. (2000). GATA-dependent recruitment of MEF2 proteins to target promoters. *The EMBO Journal, 19*, 2046–2055.

Morrisey, E. E., Ip, H. S., Tang, Z., & Parmacek, M. S. (1997). GATA-4 activates transcription via two novel domains that are conserved within the GATA-4/5/6 subfamily. *The Journal of Biological Chemistry, 272*, 8515–8524.

Morrisey, E. E., Tang, Z., Sigrist, K., Lu, M. M., Jiang, F., Ip, H. S., et al. (1998). GATA6 regulates HNF4 and is required for differentiation of visceral endoderm in the mouse embryo. *Genes & Development, 12*, 3579–3590.

Nakashima, T., Sekiguchi, T., Kuraoka, A., Fukushima, K., Shibata, Y., Komiyama, S., et al. (1993). Molecular cloning of a human cDNA encoding a novel protein, DAD1, whose defect causes apoptotic cell death in hamster BHK21 cells. *Molecular and Cellular Biology, 13*, 6367–6374.

Nicosia, M., Moger, W. H., Dyer, C. A., Prack, M. M., & Williams, D. L. (1992). Apolipoprotein-E messenger RNA in rat ovary is expressed in theca and interstitial cells and presumptive macrophage, but not in granulosa cells. *Molecular Endocrinology, 6*, 978–988.

Oreal, E., Mazaud, S., Picard, J. Y., Magre, S., & Carre-Eusebe, D. (2002). Different patterns of anti-Mullerian hormone expression, as related to DMRT1, SF-1, WT1, GATA-4, Wnt-4, and Lhx9 expression, in the chick differentiating gonads. *Developmental Dynamics, 225*, 221–232.

Padua, M. B., Fox, S. C., Jiang, T., Morse, D. A., & Tevosian, S. G. (2014). Simultaneous gene deletion of gata4 and gata6 leads to early disruption of follicular development and germ cell loss in the murine ovary. *Biology of Reproduction, 91*, 24.

Pangas, S. A., Jorgez, C. J., & Matzuk, M. M. (2004). Growth differentiation factor 9 regulates expression of the bone morphogenetic protein antagonist gremlin. *The Journal of Biological Chemistry, 279*, 32281–32286.

Peng, L., Huang, Y., Jin, F., Jiang, S. W., & Payne, A. H. (2004). Transcription enhancer factor-5 and a GATA-like protein determine placental-specific expression of the type I human 3beta-hydroxysteroid dehydrogenase gene, HSD3B1. *Molecular Endocrinology, 18*, 2049–2060.

Perugini, M., Kok, C. H., Brown, A. L., Wilkinson, C. R., Salerno, D. G., Young, S. M., et al. (2009). Repression of Gadd45alpha by activated FLT3 and GM-CSF receptor mutants contributes to growth, survival and blocked differentiation. *Leukemia, 23*, 729–738.

Philips, A. S., Kwok, J. C., & Chong, B. H. (2007). Analysis of the signals and mechanisms mediating nuclear trafficking of GATA-4. Loss of DNA binding is associated with localization in intranuclear speckles. *The Journal of Biological Chemistry, 282*, 24915–24927.

Pihlajoki, M., Farkkila, A., Soini, T., Heikinheimo, M., & Wilson, D. B. (2016). GATA factors in endocrine neoplasia. *Molecular and Cellular Endocrinology, 421*, 2–17.

Pihlajoki, M., Gretzinger, E., Cochran, R., Kyronlahti, A., Schrade, A., Hiller, T., et al. (2013). Conditional mutagenesis of Gata6 in SF1-positive cells causes gonadal-like differentiation in the adrenal cortex of mice. *Endocrinology, 154*, 1754–1767.

Pu, W. T., Ishiwata, T., Juraszek, A. L., Ma, Q., & Izumo, S. (2004). GATA4 is a dosage-sensitive regulator of cardiac morphogenesis. *Developmental Biology, 275*, 235–244.

Rahman, N. A., Kiiveri, S., Rivero-Muller, A., Levallet, J., Vierre, S., Kero, J., et al. (2004). Adrenocortical tumorigenesis in transgenic mice expressing the inhibin alpha-subunit promoter/simian virus 40 T-antigen transgene: Relationship between ectopic expression of luteinizing hormone receptor and transcription factor GATA-4. *Molecular Endocrinology, 18*, 2553–2569.

Rahmani, M., Wong, B. W., Ang, L., Cheung, C. C., Carthy, J. M., Walinski, H., et al. (2006). Versican: Signaling to transcriptional control pathways. *Canadian Journal of Physiology and Pharmacology, 84*, 77–92.

Rees, C., Clemmons, D. R., Horvitz, G. D., Clarke, J. B., & Busby, W. H. (1998). A protease-resistant form of insulin-like growth factor (IGF) binding protein 4 inhibits IGF-1 actions. *Endocrinology, 139*, 4182–4188.

Regev, A., Goldman, S., & Shalev, E. (2005). Expression of plexin-B1 in the mouse ovary and its possible role in follicular development. *Fertility and Sterility, 84*(Suppl. 2), 1210–1219.

Regev, A., Goldman, S., & Shalev, E. (2007). Semaphorin-4D (Sema-4D), the Plexin-B1 ligand, is involved in mouse ovary follicular development. *Reproductive Biology and Endocrinology, 5*, 12.

Robert, N. M., Miyamoto, Y., Taniguchi, H., & Viger, R. S. (2006). LRH-1/NR5A2 cooperates with GATA factors to regulate inhibin alpha-subunit promoter activity. *Molecular and Cellular Endocrinology, 257–258*, 65–74.

Rosenfield, R. L. (1999). Ovarian and adrenal function in polycystic ovary syndrome. *Endocrinology and Metabolism Clinics of North America, 28*, 265–293.

Russell, D. L., Ochsner, S. A., Hsieh, M., Mulders, S., & Richards, J. S. (2003). Hormone-regulated expression and localization of versican in the rodent ovary. *Endocrinology, 144*, 1020–1031.

Sacilotto, N., Espert, A., Castillo, J., Franco, L., & Lopez-Rodas, G. (2011). Epigenetic transcriptional regulation of the growth arrest-specific gene 1 (Gas1) in hepatic cell proliferation at mononucleosomal resolution. *PLoS One, 6*, e23318.

Salih, D. A., Tripathi, G., Holding, C., Szestak, T. A., Gonzalez, M. I., Carter, E. J., et al. (2004). Insulin-like growth factor-binding protein 5 (Igfbp5) compromises survival, growth, muscle development, and fertility in mice. *Proceedings of the National Academy of Sciences of the United States of America, 101*, 4314–4319.

Salmon, N. A., Handyside, A. H., & Joyce, I. M. (2004). Oocyte regulation of anti-Mullerian hormone expression in granulosa cells during ovarian follicle development in mice. *Developmental Biology, 266*, 201–208.

Scott, G. A., McClelland, L. A., Fricke, A. F., & Fender, A. (2009). Plexin C1, a receptor for semaphorin 7a, inactivates cofilin and is a potential tumor suppressor for melanoma progression. *The Journal of Investigative Dermatology, 129*, 954–963.

Sher, N., Yivgi-Ohana, N., & Orly, J. (2007). Transcriptional regulation of the cholesterol side chain cleavage cytochrome P450 gene (CYP11A1) revisited: Binding of GATA, cyclic adenosine 3′,5′-monophosphate response element-binding protein and activating protein (AP)-1 proteins to a distal novel cluster of cis-regulatory elements potentiates AP-2 and steroidogenic factor-1-dependent gene expression in the rodent placenta and ovary. *Molecular Endocrinology, 21*, 948–962.

Shim, Y. H. (1999). elt-1, a gene encoding a Caenorhabditis elegans GATA transcription factor, is highly expressed in the germ lines with msp genes as the potential targets. *Molecules and Cells, 9*, 535–541.

Siggers, P., Smith, L., & Greenfield, A. (2002). Sexually dimorphic expression of Gata-2 during mouse gonad development. *Mechanisms of Development, 111*, 159–162.

Silva, J. R., Figueiredo, J. R., & van den Hurk, R. (2009). Involvement of growth hormone (GH) and insulin-like growth factor (IGF) system in ovarian folliculogenesis. *Theriogenology, 71*, 1193–1208.

Silverman, E., Eimerl, S., & Orly, J. (1999). CCAAT enhancer-binding protein beta and GATA-4 binding regions within the promoter of the steroidogenic acute regulatory protein (StAR) gene are required for transcription in rat ovarian cells. *The Journal of Biological Chemistry, 274*, 17987–17996.

Sodhi, C. P., Li, J., & Duncan, S. A. (2006). Generation of mice harbouring a conditional loss-of-function allele of Gata6. *BMC Developmental Biology, 6*, 19.

Soyal, S. M., Mukherjee, A., Lee, K. Y., Li, J., Li, H., DeMayo, F. J., et al. (2005). Cre-mediated recombination in cell lineages that express the progesterone receptor. *Genesis, 41*, 58–66.

Stocco, C. (2008). Aromatase expression in the ovary: Hormonal and molecular regulation. *Steroids, 73*, 473–487.

Stocco, C., Telleria, C., & Gibori, G. (2007). The molecular control of corpus luteum formation, function, and regression. *Endocrine Reviews, 28*, 117–149.

Stouffer, R. L. (2003). Progesterone as a mediator of gonadotrophin action in the corpus luteum: Beyond steroidogenesis. *Human Reproduction Update, 9*, 99–117.

Sudo, S., Avsian-Kretchmer, O., Wang, L. S., & Hsueh, A. J. (2004). Protein related to DAN and cerberus is a bone morphogenetic protein antagonist that participates in ovarian paracrine regulation. *The Journal of Biological Chemistry, 279*, 23134–23141.

Tang, Y., Wei, Y., He, W., Wang, Y., Zhong, J., & Qin, C. (2014). GATA transcription factors in vertebrates: Evolutionary, structural and functional interplay. *Molecular Genetics and Genomics, 289*, 203–214.

Taniguchi, H., Komiyama, J., Viger, R. S., & Okuda, K. (2009). The expression of the nuclear receptors NR5A1 and NR5A2 and transcription factor GATA6 correlates with steroidogenic gene expression in the bovine corpus luteum. *Molecular Reproduction and Development, 76*, 873–880.

Tevosian, S. G. (2014). Transgenic mouse models in the study of reproduction: Insights into GATA protein function. *Reproduction, 148*, R1–R14.

Tevosian, S. G., Albrecht, K. H., Crispino, J. D., Fujiwara, Y., Eicher, E. M., & Orkin, S. H. (2002). Gonadal differentiation, sex determination and normal Sry expression in mice require direct interaction between transcription partners GATA4 and FOG2. *Development, 129*, 4627–4634.

Tremblay, J. J., Robert, N. M., & Viger, R. S. (2001). Modulation of endogenous GATA-4 activity reveals its dual contribution to Mullerian inhibiting substance gene transcription in sertoli cells. *Molecular Endocrinology, 15*, 1636–1650.

Tremblay, J. J., & Viger, R. S. (1999). Transcription factor GATA-4 enhances Mullerian inhibiting substance gene transcription through a direct interaction with the nuclear receptor SF-1. *Molecular Endocrinology, 13*, 1388–1401.

Tremblay, J. J., & Viger, R. S. (2001). GATA factors differentially activate multiple gonadal promoters through conserved GATA regulatory elements. *Endocrinology, 142*, 977–986.

Tremblay, J. J., & Viger, R. S. (2003). Transcription factor GATA-4 is activated by phosphorylation of serine 261 via the cAMP/protein kinase a signaling pathway in gonadal cells. *The Journal of Biological Chemistry, 278*, 22128–22135.

Uhlenhaut, N. H., Jakob, S., Anlag, K., Eisenberger, T., Sekido, R., Kress, J., et al. (2009). Somatic sex reprogramming of adult ovaries to testes by FOXL2 ablation. *Cell, 139*, 1130–1142.

Vaskivuo, T. E., Aittomaki, K., Anttonen, M., Ruokonen, A., Herva, R., Osawa, Y., et al. (2002). Effects of follicle-stimulating hormone (FSH) and human chorionic gonadotropin in individuals with an inactivating mutation of the FSH receptor. *Fertility and Sterility, 78*, 108–113.

Vaskivuo, T. E., Anttonen, M., Herva, R., Billig, H., Dorland, M., te Velde, E. R., et al. (2001). Survival of human ovarian follicles from fetal to adult life: Apoptosis, apoptosis-related proteins, and transcription factor GATA-4. *The Journal of Clinical Endocrinology and Metabolism, 86,* 3421–3429.

Viger, R. S., Guittot, S. M., Anttonen, M., Wilson, D. B., & Heikinheimo, M. (2008). Role of the GATA family of transcription factors in endocrine development, function, and disease. *Molecular Endocrinology, 22,* 781–798.

Viger, R. S., Mertineit, C., Trasler, J. M., & Nemer, M. (1998). Transcription factor GATA-4 is expressed in a sexually dimorphic pattern during mouse gonadal development and is a potent activator of the Mullerian inhibiting substance promoter. *Development, 125,* 2665–2675.

Watanabe, K., Clarke, T. R., Lane, A. H., Wang, X., & Donahoe, P. K. (2000). Endogenous expression of Mullerian inhibiting substance in early postnatal rat sertoli cells requires multiple steroidogenic factor-1 and GATA-4-binding sites. *Proceedings of the National Academy of Sciences of the United States of America, 97,* 1624–1629.

Watt, A. J., Battle, M. A., Li, J., & Duncan, S. A. (2004). GATA4 is essential for formation of the proepicardium and regulates cardiogenesis. *Proceedings of the National Academy of Sciences of the United States of America, 101,* 12573–12578.

Weenen, C., Laven, J. S., Von Bergh, A. R., Cranfield, M., Groome, N. P., Visser, J. A., et al. (2004). Anti-mullerian hormone expression pattern in the human ovary: Potential implications for initial and cyclic follicle recruitment. *Molecular Human Reproduction, 10,* 77–83.

Weiss, M. J., & Orkin, S. H. (1995). GATA transcription factors: Key regulators of hematopoiesis. *Experimental Hematology, 23,* 99–107.

Wood, J. R., Nelson, V. L., Ho, C., Jansen, E., Wang, C. Y., Urbanek, M., et al. (2003). The molecular phenotype of polycystic ovary syndrome (PCOS) theca cells and new candidate PCOS genes defined by microarray analysis. *The Journal of Biological Chemistry, 278,* 26380–26390.

Wright, R. J., Holly, J. M., Galea, R., Brincat, M., & Mason, H. D. (2002). Insulin-like growth factor (IGF)-independent effects of IGF binding protein-4 on human granulosa cell steroidogenesis. *Biology of Reproduction, 67,* 776–781.

Yanazume, T., Hasegawa, K., Morimoto, T., Kawamura, T., Wada, H., Matsumori, A., et al. (2003). Cardiac p300 is involved in myocyte growth with decompensated heart failure. *Molecular and Cellular Biology, 23,* 3593–3606.

Zhou, P., Baumgarten, S. C., Wu, Y., Bennett, J., Winston, N., Hirshfeld-Cytron, J., et al. (2013). IGF-I signaling is essential for FSH stimulation of AKT and steroidogenic genes in granulosa cells. *Molecular Endocrinology, 27,* 511–523.

Zimmermann, D. R., & Dours-Zimmermann, M. T. (2008). Extracellular matrix of the central nervous system: From neglect to challenge. *Histochemistry and Cell Biology, 130,* 635–653.

CHAPTER NINE

Involvement of Bone Morphogenetic Proteins (BMP) in the Regulation of Ovarian Function

Sheena L.P. Regan*[,1], Phil G. Knight[†], John L. Yovich*[,‡], Yee Leung[§], Frank Arfuso*, Arun Dharmarajan*

*Stem Cell and Cancer Biology Laboratory, School of Pharmacy and Biomedical Sciences, Curtin Health Innovation Research Institute, Curtin University, Perth, WA, Australia
[†]School of Biological Sciences, Hopkins Building, University of Reading, Reading, United Kingdom
[‡]PIVET Medical Centre, Perth, WA, Australia
[§]Western Australian Gynaecologic Cancer Service, King Edward Memorial Hospital for Women, Perth, WA, Australia
[1]Corresponding author: e-mail address: sheenaregan@aapt.net.au

Contents

Vitamins and Hormones, Volume 107
ISSN 0083-6729
https://doi.org/10.1016/bs.vh.2018.01.015

Abstract

Primordial germ cells migrate to the fetal gonads and proliferate during gestation to generate a fixed complement of primordial follicles, the so-called ovarian reserve. Primordial follicles comprise an oocyte arrested at the diplotene stage of meiosis, surrounded by a layer of pregranulosa cells. Activation of primordial follicles to grow beyond this arrested stage is of particular interest because, once activated, they are subjected to regulatory mechanisms involved in growth, selection, maturation, and ultimately, ovulation or atresia. The vast majority of follicles succumb to atresia and are permanently lost from the quiescent or growing pool of follicles. The bone morphogenetic proteins (BMPs), together with other intraovarian growth factors, are intimately involved in regulation of follicle recruitment, dominant follicle selection, ovulation, and atresia.

Activation of primordial follicles appears to be a continuous process, and the number of small antral follicles at the beginning of the menstrual cycle provides an indirect indication of ovarian reserve. Continued antral follicle development during the follicular phase of the menstrual cycle is driven by follicle stimulating hormone (FSH) and luteinizing hormone (LH) in conjunction with many intraovarian growth factors and inhibitors interrelated in a complex web of regulatory balance.

The BMP signaling system has a major intraovarian role in many species, including the human, in the generation of transcription factors that influence proliferation, steroidogenesis, cell differentiation, and maturation prior to ovulation, as well as formation of corpora lutea after ovulation. At the anterior pituitary level, BMPs also contribute to the regulation of gonadotrophin production.

1. OVERVIEW OF FOLLICULOGENESIS

The underlying physiological processes of reproduction in females and males are similar in humans and other mammals. In the female gonad (ovary) the oocyte is encapsulated by layers of follicular somatic cells that proliferate, and later differentiate and mature to form preovulatory follicles. At ovulation, the mature follicle wall ruptures and the oocyte is expelled from the follicle and is thereafter potentially destined for fertilization. The recruitment of follicles, their growth, and the expulsion of the oocyte are dependent on complex signaling mechanisms involving the hypothalamic–pituitary–gonadal axis. Neurotransmitters and neuropeptides from the hypothalamus stimulate the release of gonadotrophin releasing hormone (GnRH) into the hypophyseal portal system, which, in turn, stimulates the anterior pituitary to release gonadotrophic hormones that act on the ovary to promote follicular growth. Several predominant growth factors that regulate the transcription of genes and control the recruitment and selection of the

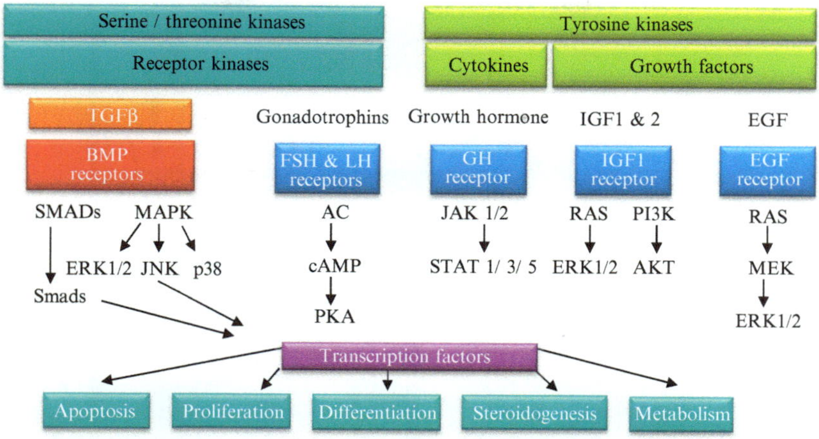

Fig. 1 Overview of the TGFβ and the growth hormone kinase signaling interaction. Major serine and tyrosine kinases and receptors, and signaling pathways involved in ovarian regulation (Amsterdam et al., 2003; Fan et al., 2009; Manna et al., 2002; Miyazono, Kamiya, & Morikawa, 2010; Moore et al., 2001; Rice, Ojha, Whitehead, & Mason, 2007; Tajima et al., 2003).

dominant follicles belong to the transforming growth factor beta (TGFβ) superfamily including the subfamilies of bone morphogenetic proteins (BMPs), inhibins, and activins, growth differentiation factors (GDFs), and anti-Müllerian hormone (AMH) (Fig. 1) (Edson, 2009; Eppig, 2001; Erickson & Shimasaki, 2001; Fabre et al., 2006; Gilchrist, Ritter, & Armstrong, 2004; Knight & Glister, 2006; McNatty et al., 2004; Otsuka, 2013).

At approximately 26 weeks of gestation in humans, the reproductive potential of the fetus is established (Childs et al., 2010). By this stage the primordial follicles are fully formed and begin a process of initial activation followed by eventual demise or ovulation and potential fertilization, over the reproductive life span of the individual (Pangas, 2012). Activated primordial follicles grow and differentiate into preantral follicles (Fig. 2). With further development, preantral follicles mature into antral follicles with the formation of a fluid-filled central compartment (Rodgers & Irving-Rodgers, 2010). At the onset of puberty, cyclic increases in gonadotrophin secretion from the anterior pituitary raise follicle-stimulating hormone (FSH) to a threshold point sufficient to rescue a growing cohort of small antral follicles and initiate cyclic recruitment (Fig. 2) (Gougeon, 1986; Richards, 1994).

Antral follicles contain an oocyte surrounded by cumulus granulosa cells that form a continuum with mural granulosa cells lining the antrum. The follicle wall is composed of granulosa and theca cells separated by the

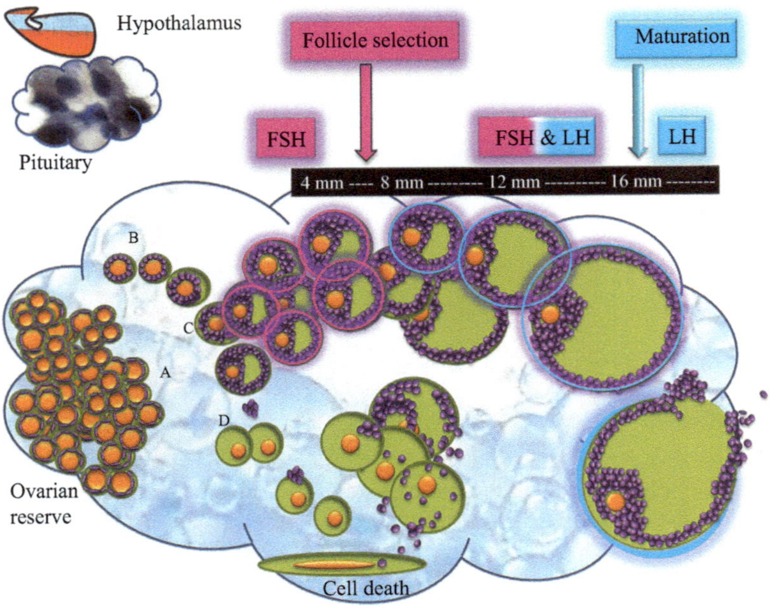

Fig. 2 Folliculogenesis: Activation of the primordial follicle, dominant follicle selection, growth and maturation before ovulation. Ovarian reserve of primordial follicle with squamous pregranulosa cells (A), activation and initial recruitment of primary follicle with cuboidal granulosa cells (B), secondary follicle with multiple layers of granulosa and no antral cavity (C), and cell death of preantral follicles (D). FSHR expression (pink rings) and FSH secretion promote antral follicle formation, followed by dominant follicle selection (pink arrow) based on LHR expression (blue rings). Proliferation and growth increase the size of the follicle. Maturation of the follicle and the LH surge differentiate the follicle cells in a complex process of luteinization. Diameter of the follicles at the respective stages of folliculogenesis is indicated in mm scale.

basal lamina. Stromal cells within a connective tissue matrix are encapsulated by a layer of epithelial cells at the ovarian surface (Erickson & Shimasaki, 2003; Rodgers & Irving-Rodgers, 2010).

Folliculogenesis involves the stage-dependent expression of intraovarian growth factors and their receptors that regulate proliferation and differentiation of granulosa and thecal cells (Erickson & Shimasaki, 2003; Gougeon, 1986). The mature follicle or preovulatory follicle completes differentiation and commences luteinization (morphological and steroidogenic capacity changes) prior to ovulation, and then ruptures, releasing the oocyte in the proximity of the opening of the fallopian tube (Ainsworth, Tsang, Downey, Marcus, & Armstrong, 1980; Rodgers & Irving-Rodgers, 2010).

The number of preovulatory follicles selected for dominance and ovulation varies according to species and is dependent on the regulation by the gonadotrophins (FSH and LH (luteinizing hormone)) and the interaction with intraovarian growth factors (Ginther et al., 2005; Gougeon, 1986). TGFβ family members, including BMPs, have been shown to play a major role in the recruitment and growth of the ovarian follicle (Edson, 2009; Erickson & Shimasaki, 2001; Fabre et al., 2006; Knight & Glister, 2006; Otsuka, 2013). There has been considerable interest in the type 1 BMP receptor (BMPR1B) which binds to the BMP ligands 2, 4, 6, 7, and 15 culminating in altered gene transcription (Miyazono, Maeda, & Imamura, 2005). A naturally occurring point mutation of the BMPR1B gene in the Booroola Merino (BB) sheep results in partial attenuation of receptor function and increases ovulation rate (Mulsant et al., 2001; Souza, MacDougall, Campbell, McNeilly, & Baird, 2001; Wilson et al., 2001). In the human clinical context, ovulation rate is increased during in vitro fertilization (IVF) treatment by administration of FSH to stimulate the growth of multiple follicles (Edwards, Lobo, & Bouchard, 1996; Edwards & Steptoe, 1983). BMPR1B-mediated signaling and its interaction with the signaling of FSH receptor (FSHR) and LH receptor (LHR) appear to have a central role in this process.

2. THE OVARIAN RESERVE

Oogonia proliferate in the ovary before commencing meiosis from approximately weeks 9 to 11 of gestation in humans (Fig. 3). Germ cell cysts (egg nests) containing multiple oogonia are infiltrated by somatic cells, forming individual primordial follicles, each with a single layer of somatic cells surrounding the oocyte (Pangas, 2012). The somatic cells differentiate into granulosa cells, and the oocyte resides in the dictyate-stage of meiotic prophase 1 until the mid-cycle LH surge triggers meiotic progression in the follicles selected for ovulation (Edwards et al., 1996). The progressive decline of the ovarian reserve is well documented and is related to chronological age (Almog, Shehata, Shalom-Paz, Tan, & Tulandi, 2011; Hansen, Hodnett, Knowlton, & Craig, 2011).

The total number of germ cells peaks at over six million at ~26 weeks gestation. At birth the number of germ cells (oocytes) has already declined by ~80%, and this decline continues inexorably throughout the reproductive life span of the individual (Fig. 3) (Monniaux et al., 2014). At puberty, the levels of gonadotrophins increase sufficiently to promote tertiary follicles to

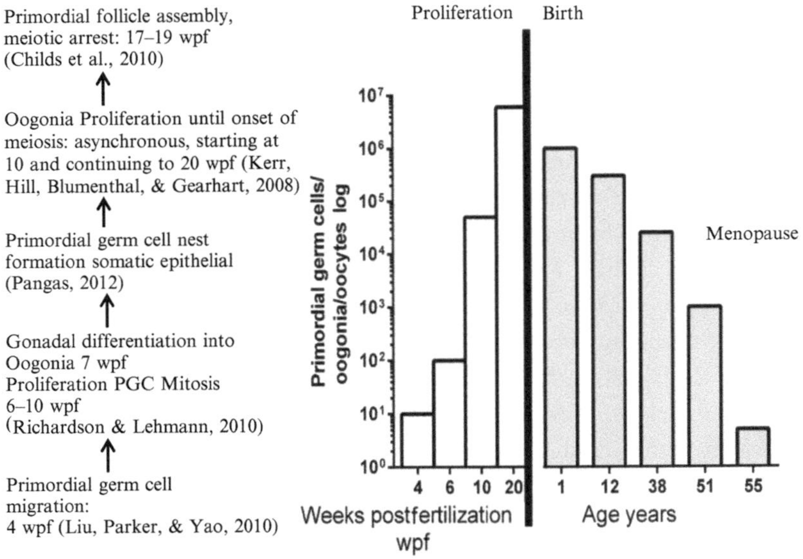

Fig. 3 Primordial germ cell proliferation and oogenesis before birth, and the loss of primordial follicles from birth to menopause. *Based on Baerwald, A., Adams, G., & Pierson, R. (2012). Ovarian antral folliculogenesis during the human menstrual cycle: A review. Human Reproduction Update, 18, 73–91; Fabre, S., Pierre, A., Mulsant, P., Bodin, L., Di Pasquale, E., Persani, L., et al. (2006). Regulation of ovulation rate in mammals: Contribution of sheep genetic models. Reproductive Biology and Endocrinology, 4, 20; Knight, P.G., & Glister, C. (2006). TGF-β superfamily members and ovarian follicle development. Reproduction, 132, 191–206; Matsuda, F., Inoue, N., Manabe, N., & Ohkura, S. (2012). Follicular growth and atresia in mammalian ovaries: Regulation by survival and death of granulosa cells. Journal of Reproduction and Development, 58, 44–50; Skinner, M.K. (2005). Regulation of primordial follicle assembly and development. Human Reproduction Update, 11, 461–471; Webb, R., & Campbell, B. K. (2007). Development of the dominant follicle: Mechanisms of selection and maintenance of oocyte quality. Society of Reproduction and Fertility Supplement, 64, 141–163.*

continue growth, and to resist apoptosis (Matsuda, Inoue, Manabe, & Ohkura, 2012). Ultrasonographic estimates of the number of small antral follicles growing (AFC) or serum levels of AMH (secretion by the small antral follicles) are strongly correlated to the ovarian reserve (Hansen et al., 2011; van Rooij et al., 2005).

3. INTRAOVARIAN REGULATORS OF FOLLICULOGENESIS

Inducement of FSHR and LHR expression and modulation of responsiveness to gonadotrophins appear to be under the control of various intraovarian growth and development regulators (Fig. 1) (Baerwald,

Adams, & Pierson, 2012; Erickson, Wang, & Hsueh, 1979; Fan et al., 2009). BMPs, GDF9, AMH, inhibins, activins, and BMP/activin binding proteins have been implicated directly or indirectly, in vivo, by experiments that involve treatments, such as ligand infusion and active or passive immunization (Al-Samerria, Al-Ali, McFarlane, & Almahbobi, 2015; Campbell, Kendall, & Baird, 2009; Juengel, Hudson, Whiting, & McNatty, 2004; Knight, Satchell, & Glister, 2012), and by evidence from natural mutations and knockout gene models in several species (Araújo et al., 2010; Di Pasquale et al., 2006; Feary et al., 2007; McNatty et al., 2007). In vitro culture of isolated granulosa and theca cells and ovarian tissue explants has provided substantive data on the influence of these growth factors on steroidogenesis and cell proliferation (Brankin, Quinn, Webb, & Hunter, 2005; Campbell, Souza, Skinner, Webb, & Baird, 2006; Glister, Kemp, & Knight, 2004; McNatty et al., 2009; Nilsson & Skinner, 2003).

The early acquisition of granulosal FSHR and LHR facilitates dominant follicle growth in the face of declining FSH levels during the follicular phase of the cycle (Fig. 2) (LaPolt, Tilly, Aihara, Nishimori, & Hsueh, 1992; Sen et al., 2014). Acquisition of granulosal LH responsiveness supplements the FSHR-mediated conversion of androstenedione to estradiol by P450 aromatase (CYP19A1), maintaining a positive estrogen to androgen ratio in the follicle. As the antral follicle increases in size, more estrogen and anti-apoptotic factors are produced to ensure the survival of the dominant ovulatory follicle (Amsterdam et al., 2003). With reduced FSHR and LHR density, the granulosa cells of subordinate follicles have a reduced capacity to convert theca-derived androgens to estrogens and are destined for atresia (Hillier, Whitelaw, & Smyth, 1994; Xu et al., 1995).

4. ROLE OF BMPS IN OVARIAN REGULATION

The body of work investigating the role of BMPs as ovarian regulators ranges from studies on primordial germ cell migration through to inducement of ovulation and corpus luteum formation (Erickson & Shimasaki, 2001, 2003; Knight & Glister, 2006; Miyazono et al., 2010; Otsuka, 2013; Pangas, 2012; Shimasaki, Moore, Otsuka, & Erickson, 2004). Various mammalian species have been used as in vivo and in vitro research models including polyovulatory rodents and pigs, monoovulatory species (sheep, cattle) and humans (Edson et al., 2010; Raz, 2003; Regan et al., 2017, 2015). The ability to create global and conditional gene knockout models in mice and to use natural and created mutations or specific cell lines to study the effects of perturbing specific BMP pathway components has facilitated research in this area. Furthermore, treatment with BMPs as well as blocking receptors and signaling pathways has been used extensively to examine the roles of BMP signaling in ovarian function.

In human studies, in vivo and in vitro research has progressed substantially with the rise of IVF centers, providing an accessible source of follicular material from gonadotrophin-stimulated patients undergoing oocyte retrieval. However, the availability of nonpathogenic human ovarian tissue, free from exogenous gonadotrophin stimulation, is very limited and is therefore infrequently used in research (Bomsel Helmreich et al., 1979; Fowler, Sorsa, Harris, Knight, & Mason, 2001; Garcia, Jones, Acosta, & Wright, 1981; Gougeon, 1986; Klein et al., 2000; MacNaughton, Banah, McCloud, Hee, & Burger, 1992). Given the stage-specific nature of ovarian regulation and variation among species, caution should be used when interpreting results (Erickson & Shimasaki, 2003; Otsuka, 2010). In addition, the complex interactions and feedback loops between locally produced growth factors and other components of the hypothalamic–pituitary–ovarian axis complicate the interpretation of in vivo experiments exploring the intraovarian roles of specific growth factors (Zeleznik, 2001).

5. BMPs: MEMBERS OF THE TGFβ SUPERFAMILY

The TGF superfamily consists of over 40 different ligands and can be divided into several subfamilies including the BMP subfamily that is the focus of this review. As with other TGFβ superfamily members, BMP signaling pathways are operational in numerous tissues and organs across the life-course, where they exert complex inhibitory and stimulatory control over cell proliferation, apoptosis, and cell differentiation (Massagué, 2008).

There are seven TGFβ type 1 receptors, commonly referred to as ALK1 to ALK7, and six type 2 TGFβ receptors. The BMP ligands, 2, 4, 6, 7, and 15, form a receptor–ligand complex with the type 1 TGFβ receptor BMPR1B (ALK6) and a type 2 TGFβ receptor BMPR2 (Fig. 4) (Miyazono et al., 2010). The heterotetrameric receptor complex initiates phosphorylation of the intracellular substrate molecules, receptor-regulated Smads (Smads 1, 5, and/or 8 in the case of BMP signaling). The Smad forms a complex with a common mediator, Smad 4, and translocates to the nucleus. In the nucleus, allocated specific cofactors for each BMP ligand initiates transcription of genes required by the cell (Mitsui et al., 2015; Moore, Otsuka, & Shimasaki, 2003). BMP signaling is modulated at different levels by specific repressor and activator molecules in the nucleus, cytoplasm, extracellular

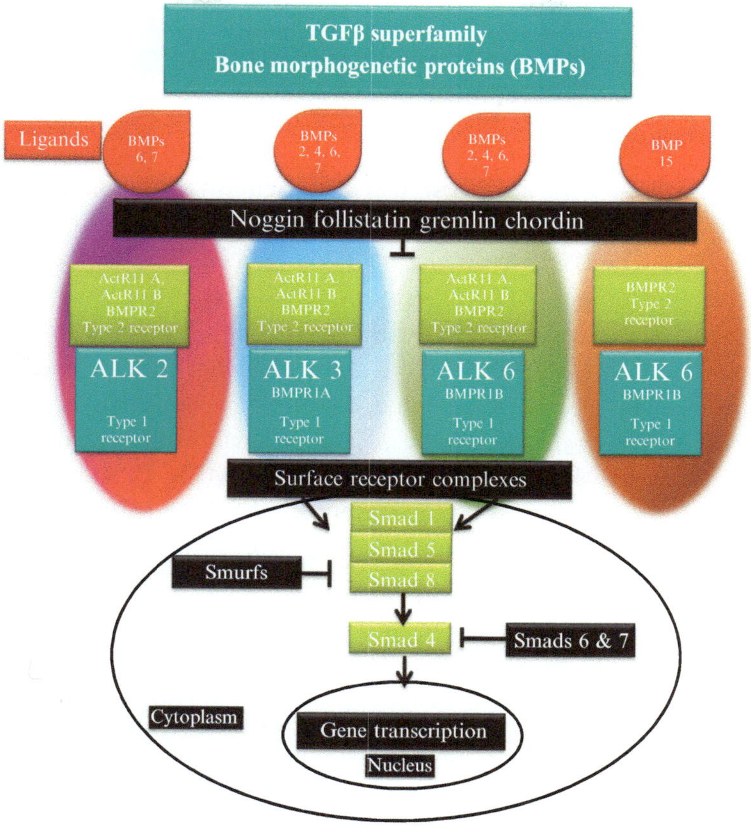

Fig. 4 BMP signaling pathway. BMP ligand signaling and the formation of hetero-tetrameric receptor complex with a type 1 and type 2 TGFβ receptor (four oval shapes). Phosphorylation of downstream signaling molecules via the receptor regulated Smads (R-Smads) via the common universal Smad 4 molecule to initiate translocation to the nucleus for gene transcription involved in cell proliferation, steroidogenesis, cell differentiation, and cell death.

fluid, and extracellular matrix. Intracellular modulators that attenuate signaling include inhibitory Smads 6 and 7 and extracellular BMP inhibitory binding proteins include follistatin, noggin, chordin, and gremlin. These nonsignaling secreted proteins sequester BMP ligands and modulate their binding to signaling receptors, generally inhibiting their actions (Miyazono et al., 2010).

In addition to activation of canonical Smad-mediated signaling pathway, BMPs can activate mitogen-activated protein kinases (MAPK) such as extracellular signal-regulated kinases 1 and 2 (ERK 1/2) (Inagaki et al., 2009). The cross talk between the Smads and the ERK 1 and ERK 2 signaling adds another dimension to the complex signaling that regulates folliculogenesis (Tajima et al., 2003).

6. OBSERVATIONS ON INTRAOVARIAN ROLES OF DIFFERENT BMPs

6.1 BMP2

Evidence suggests BMP2 is involved in oocyte endowment, primordial pool assembly, and activation of primordial follicles (Lawson et al., 1999; Ying & Zhao, 2001). The BMPs including BMP2 have been shown to be involved with the regulation of FSH synthesis through a competitive binding mechanism whereby activin increases synthesis and inhibin suppressed synthesis (Lee et al., 2007). BMP2 also increased granulosa cell estrogen and inhibin B production and FSHR expression in culture; however, it had no effect on proliferation (Shi et al., 2011; Souza, Campbell, McNeilly, & Baird, 2002). BMP2 has a relatively low affinity to bind with BMPR1B, and binds preferentially with another type 1 TGFβ receptor, BMPR1A (ALK3); however, species differences exist (Miyazono et al., 2010).

6.2 BMP4

BMP4 is produced by both theca and granulosa cells in the bovine and human model (Glister et al., 2004; Khalaf et al., 2013). BMP4 has been shown to bind precociously with BMPR1A and BMPR1B in sheep (Fabre et al., 2006), signaling via the Smad intermediate molecules to modulate ovarian function. BMP4 produced by granulosa and theca cells signals via BMPR1B to activate Smad 1 that inhibits steroidogenic acute regulatory protein (StAR) and P450 Side Chain Cleavage (CYP11A1) gene expression in the granulosa cells to progesterone synthesis during the proliferative phase of follicle development and the early onset of the LH surge and ovulation (Pierre et al., 2004). In addition, BMP4 modulates cell function via alternative signaling pathways to the TGFβ which involves the MAPK family, in particular ERK 1/2 (Fan et al., 2009; Moore et al., 2001).

In particular, BMP4 inhibits progesterone production of small antral follicles by negatively influencing cyclic adenosine monophosphate (cAMP) levels and expression of FSHR, StAR, and CYP11A1 of the granulosa cells (Fabre et al., 2003; Mulsant et al., 2001). In contrast, large antral follicles are not responsive to BMP4-induced inhibition of granulosal progesterone secretion or mitogenic growth, which indicates that BMP4 may be more involved in regulation up to dominant follicle selection (Fabre et al., 2006; Tanwar & McFarlane, 2011). While FSH-induced progesterone biosynthesis was inhibited by BMP4, it was shown to enhance FSH-induced estrogen production in rat (Shimasaki et al., 1999) and sheep (Fabre et al., 2003; Mulsant et al., 2001). BMP4 suppressed IGF-induced progesterone production by bovine granulosa cells while enhancing estrogen, inhibin, and activin production (Glister et al., 2004).

BMP4 is also involved in oocyte endowment, primordial follicle assembly, and primordial follicle activation (Lawson et al., 1999; Lee, Otsuka, Moore, & Shimasaki, 2001; Lee et al., 2004; Nilsson & Skinner, 2003; Ying & Zhao, 2001). At the pituitary level, BMP reduces steroidogenic factor (SF-1) transcriptional activity on the LHβ promoter (Pierre et al., 2004). SF-1 is a key activator of steroidogenic endocrine function, and BMP4 and SF-1 are found in gonadotrope cells that produce LH in the anterior pituitary (Ingraham et al., 1994; Val, Lefrançois-Martinez, Veyssière, & Martinez, 2003). The direct link between BMP4 and LH synthesis via BMPR1B-induced phosphorylation of Smad 1 provides an explanation for the increase in LHRs as the BMP ligands and receptors decline during folliculogenesis, which releases their inhibitory effect (Nicol et al., 2008; Regan et al., 2017). In addition, BMP4 inhibits ovine pituitary FSHβ expression and reduces the concentration of FSH (Faure et al., 2005).

6.3 BMP7

At the ovarian level, BMP7 is mostly expressed by theca cells, but with some expression by granulosa cells (Glister et al., 2004; Khalaf et al., 2013). In the rat, BMP7 has a similar biological effect as BMP4, 6, and 15 in stimulating granulosal proliferation, suppressing FSH-induced progesterone biosynthesis and increasing FSH-induced CYP19A1 expression (Lee et al., 2001; Shimasaki et al., 1999). In monoovulatory species, the effects on granulosa cells appear inconsistent in the bovine. FSH-induced hormone secretion was not altered by BMP7, whereas in the goat, BMP7 increased FSHR and decreased LHR mRNA expression. In humans, BMP7 increased FSHR expression (Shi et al., 2010; Zhu et al., 2013). Differences in response are likely due to granulosa cells either being luteinized (caused by the addition of serum during culture) or attributed to being collected from gonadotrophin-stimulated ovaries in humans.

6.4 BMP6

BMP6 is expressed by the oocyte in follicles from the primordial stage, whereas granulosa cells also express BMP6 in antral follicles in humans and other species (Glister et al., 2004; Khalaf et al., 2013; Wu et al., 2007).

BMP6 is involved in proliferation, steroidogenesis, and cytodifferentiation of granulosa cells in a species-specific manner. In the goat, BMP6 did not increase granulosal FSHR, yet it increased LHR expression (Zhu et al., 2013). In humans, BMP6 increased FSHR mRNA, whereas in the rat, its effect was exerted downstream of the FSHR (Ogura Nose et al., 2012; Otsuka, Moore, Shimasaki, et al., 2001). BMP6 has been shown to increase proliferation of granulosa cells, in contrast with BMP2 or BMP4, in cultured sheep granulosa cells. However, in the rat, BMP6 had no effect on granulosa cell proliferation (Campbell et al., 2006; Otsuka, Moore, et al., 2001). BMP6 mRNA levels in granulosa cells were significantly increased in women with polycystic ovarian syndrome (Khalaf et al., 2013). At the pituitary level, a sheep study revealed that BMP6 inhibits expression of FSHβ, which reduces the synthesis of FSH by gonadotrophs (Faure et al., 2005).

6.5 BMP15

BMP15 is exclusively secreted by the oocyte and has a strong affinity for BMPR1B (ALK6) in sheep and humans; yet it binds precociously to ALK6, ALK2, and ALK3 in mice, indicating that substantial species differences exist (Inagaki & Shimasaki, 2010; Otsuka & Shimasaki, 2002; Pulkki et al., 2012). From the primary follicle stage onward, expression of BMP15 mRNA progressively increased, peaking in sheep early antral follicles, followed by a sequential decrease in larger antral follicle (Feary et al., 2007). It is likely that a BMP15 concentration gradient is established from the oocyte via the cumulus cells to the granulosa cells. BMP15 has been shown to suppress cumulus apoptosis (Hussein, Froiland, Amato, Thompson, & Gilchrist, 2005). Oocytes surrounded by cumulus cells with greater levels of BMP15 mRNA were shown to yield an increased pregnancy rate after IVF (Li et al., 2014). Moreover, an association between high levels of BMP15 in the follicular fluid and oocyte quality has been reported (Li et al., 2014; Wu et al., 2007). Specifically, BMP15 increases proliferation of granulosa cells and cytodifferentiation, independently of FSH, yet BMP15 suppresses FSH-regulated progesterone synthesis by reducing StAR and CYP11A1 in the rat (Otsuka, Moore, et al., 2001).

It has also been shown that BMP15 decreased FSH-induced progesterone biosynthesis by decreasing FSHR mRNA, ultimately inhibiting luteinization (Moore et al., 2003; Otsuka & Shimasaki, 2002). In parallel, BMP15 promoted granulosa cell mitosis independent of the FSHR-Smad 1,5,8 pathway via activation of the ERK 1/2 pathway (Lee et al., 2001; Moore et al., 2003; Otsuka, Yamamoto, Erickson, & Shimasaki, 2001; Shimasaki et al., 1999). Moreover, in the pituitary, BMP15 has been associated with increasing FSHβ expression (Otsuka & Shimasaki, 2002).

7. OTHER TGFβ SUPERFAMILY MEMBERS (GDF9, GDF3, AMH)

7.1 GDF9

GDF9 and GDF3 are TGFβ superfamily members with close homology to the BMPs, and signal via ALK4 and 5, and via Smad 2 and 3. GDF9 is also produced by the oocyte from primordial follicles onward and is known to suppress LHR expression along with BMP6 and 7. GDF9 gene knockout mice have arrested follicle growth at the primary stage, which indicates a very early essential role in reproduction (Kaivo-Oja et al., 2005).

7.2 GDF3

GDF3 has been detected in the cytoplasm of the oocyte of the resting primordial and primary follicles in humans, and not in the granulosa cells until antral cavity formation (Shi et al., 2012). GDF3 is found in antral granulosa cells in the human ovary, and increases LHR expression, yet it reduces the inhibitory effect of BMP6 and 7 on LHR expression (Shi et al., 2012). Around the time of dominant follicle selection, expression of BMP6 and 7 decreases and GDF3 expression increases.

7.3 AMH

Another TGFβ superfamily member that has attracted considerable interest with regard to its intraovarian role is AMH. While several BMPs have been shown to promote primordial follicle activation and increase the pool of growing follicles in cultured neonatal mouse ovaries, AMH exerts an inhibitory effect on follicle growth (Gouedard et al., 2000; Josso, di Clemente, & Gouédard, 2001). While AMH inhibits FSH-dependent follicle growth and estrogen production at a later follicle stage, several BMPs have the opposite effect. AMH signals via its own type 2 receptor (AMHR2) forming a signaling complex with either BMPR1B (ALK6) or BMPR1A (ALK3) (Josso & di Clemente, 2003). It remains unclear how AMH promotes primordial follicle activation since AMHR2 is evidently not expressed by the primordial follicle (Weenen et al., 2004).

AMH protein first appears in the activated primordial follicle, and its concentration peaks in the small antral follicles, followed by a steady decline at the time of dominant follicle selection, along with activin and BMP6 (Rice et al., 2007). In sheep and humans, AMH is not present in mural granulosa cells from large antral follicles; however, it has been reported in the cumulus granulosa cells (Campbell, Clinton, & Webb, 2012; Weenen et al., 2004). BMP6 has been shown to increase the secretion of AMH in humans which in turn has an inhibitory effect on primary follicle formation, thus preserving the ovarian reserve (Rice et al., 2007; Shi et al., 2009).

When AMH is blocked directly by immunizing sheep against AMH, their ovulation rate increases, whereas the mitogenic activity of granulosa cells remains the same (Campbell et al., 2012). AMH attenuation has, therefore, been identified as a possible contributor to the observed increase in ovulation rate of the Booroola Merino sheep carrying a mutation in BMPR1B.

8. BMPs AND PRIMORDIAL FOLLICLE ACTIVATION

Once the resting primordial follicle is assembled, it is only a matter of time before activation occurs. However, this activation process remains poorly understood. Several theories are proposed as to the activation of the primordial follicle (Fig. 5). To begin with, morphological data exist

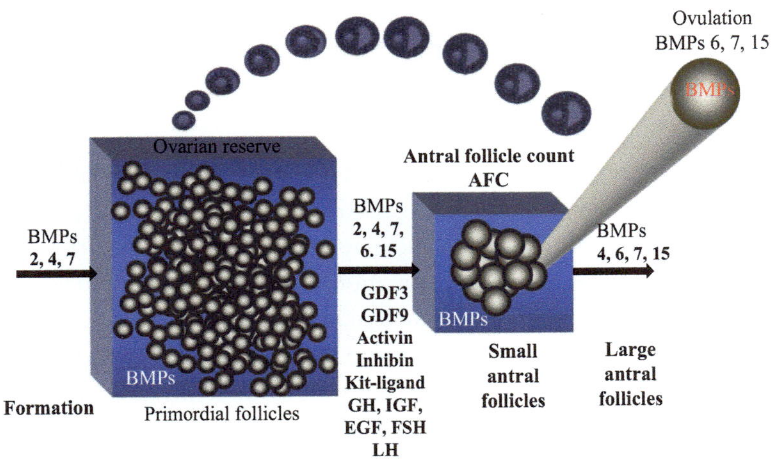

Fig. 5 BMP signaling and follicle development. The bone morphogenetic proteins (BMP) ligands 2, 4, 6, 7, and 15 via the receptor (R) BMPR1B are involved in embryonic ovarian formation of primordial follicles; activation of the primordial to primary follicle; antral follicle formation and recruitment into cyclic folliculogenesis to ovulation.

showing that the mesenchymal pericyte migrates toward the primordial follicle and aligns itself adjacent to the primordial follicle (Bukovsky, 2016). Resting primordial follicles may be in the vicinity but remain dormant, which indicates that local signaling factors initiate the primordial follicle to grow. The migration of the pericyte may be under the control of either neural or cytokine factors such as platelet-derived growth factor beta, or an immune response. However, the inhibitory gradient theory is compelling, because when resting primordial follicles are removed from the ovary, activation occurs spontaneously (Hussein et al., 2005; Suzuki et al., 2015). Furthermore, the rate of activation of the primordial follicle is proportional to the ovarian reserve (Anzalone, Hong, & LaPolt, 2001).

As the ovary ages, there is a reduction in angiogenesis, vascular endothelial growth factor, and the number of pericytes (Liu et al., 2009; Mattioli et al., 2001; Robinson et al., 2009; Taylor, Wilson, Hillier, Wiegand, & Fraser, 2007). The TGFβ super family is involved with the proliferation of pericytes (Sweeney, Ayyadurai, & Zlokovic, 2016). The pericyte has a number of known receptors, one of which is the BMPR1B (ALK 6) and ALK 5, and therefore would have the ability to respond to BMP and GDF ligands present at that time, with an affinity for the receptor.

BMP15 expression in the oocyte is not evident until the primary follicle stage in humans and sheep (Galloway et al., 2002; Li et al., 2014). Sheep with a BMP15 or BMP6 gene knockout show a much later primary to secondary follicle stage arrest in growth, resulting in infertility (Galloway et al., 2002; McNatty et al., 2007). GDF9, also secreted by the oocyte, is reported to be expressed in the primordial follicles and has been shown to increase primordial to primary follicle conversion (Vitt, Hayashi, Klein, & Hsueh, 2000). Whereas, GDF9 knockout mice are infertile due to arrested primordial to primary growth transition (Dong, Albertini, Nishimori, Kumar, & Matzuk, 1996). AMH has an inhibitory effect on primordial follicle activation in the mouse (Durlinger et al., 2002). The AMHR2 is essential for AMH responsiveness, forming a signaling complex with BMPR1B. However, in humans, AMHR2 is evidently not expressed in the primordial follicle only appearing during the primary to secondary transition (Rice et al., 2007).

Other factors implicated in primordial follicle activation include death receptor (VASA or DEAD-box4) and leukemia inhibitory factor (LIF), forkhead box O3 (Foxo 3), growth hormone, and the phosphatidylinositol 3-kinase (PI3K)-AKT signaling axis pathway (Albamonte, Albamonte, Stella, Zuccardi, & Vitullo, 2013; Castrillon, Miao, Kollipara, Horner, &

DePinho, 2003; John, Shirley, Gallardo, & Castrillon, 2007; Reddy et al., 2008; Slot et al., 2006). Phosphatase and tensin homolog (PTEN) inhibitors or AKT stimulants (including the BMP ligands) appear to influence proliferation, migration, and activation of the primordial follicle, and continued growth (Hsueh, Kawamura, Cheng, & Fauser, 2015; Kawamura et al., 2013).

In rodents, BMP4 and 7 have been reported to enhance primordial activation (Skinner, 2005) and to enhance primary to preantral growth (Lee et al., 2004). Immunization to inhibit BMP4 and BMPR1B signaling reduced the conversion of primordial follicles to primary follicles in mice, which conserved the ovarian pool of primordial follicles over time (Al-Samerria et al., 2015; Nilsson & Skinner, 2003). Similarly, Booroola sheep, with a partially attenuating mutation to the BMPR1B, retained more primordial follicles over time compared to the wild-type sheep (Ruoss, Tadros, O'Shea, McFarlane, & Almahbobi, 2009).

9. BMP RECEPTOR ACTIVITY IN THE OVARY

The BMP ligands strongly activate the BMPR1B, and their role in the regulation of gonadotrophin receptor expression has been previously reported (Miyazono et al., 2010; Shi et al., 2012, 2011, 2010; Zhu et al., 2013). The BMPR1B is first expressed in primordial follicles on the oocyte and the granulosa cells of primary follicles throughout folliculogenesis (mural and cumulus) (Abir et al., 2008). Androgen receptors are first expressed in the transitional follicle between the primordial and primary stage, and are early regulators of ovarian development, particularly the inducement of FSHR on the granulosa cell (Fig. 6) (Erickson et al., 1979; Nielsen et al., 2011; Rice et al., 2007; Sen & Hammes, 2010; Sen et al., 2014).

Dominant follicle selection takes place when the androgen receptor expression reduces and estrogen production increases. BMPR1B and FSHR expression has also been shown to decrease at this time, followed by an increase in LHR expression (Fig. 7) (Regan et al., 2016, 2017). At the time of maturation of the follicle, downregulation of BMPR1B, FSHR, and LHR expression is associated with reduced proliferation and a shift from estrogen to progesterone synthesis in the ovulatory follicles in both human and animal models (Figs. 6 and 7) (Regan et al., 2016, 2017, 2015; Rice et al., 2007). This shift in steroidogenesis requires the progesterone-suppressive BMP signaling to be downregulated in the largest follicles (Regan et al., 2016, 2017). In addition, medium to large antral follicles require substantial androgen substrate to generate estrogen, and a reduction

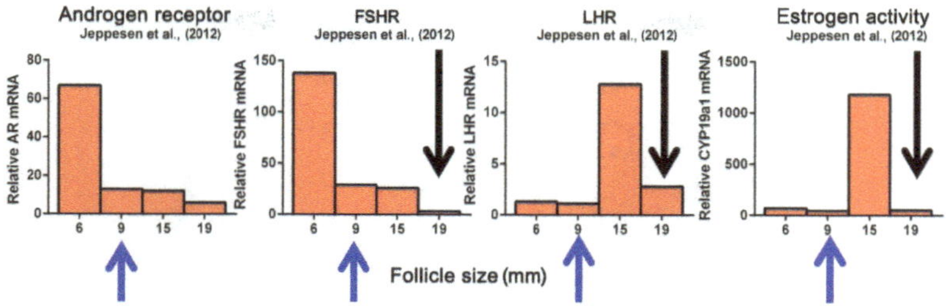

Fig. 6 The stage-specific relationship between granulosal receptor expression and estrogen activity during folliculogenesis in a natural cycle. Dominant follicle selection took place when the androgen receptor and FSHR expression decreased, and LHR expression increased (indicated by the *blue upwards-arrow*). Downregulation of FSHRs, LHRs, and the cessation of proliferation occurs in the preovulatory follicles in humans and animals (indicated by the *black downwards-arrow*). *CYP19a1* is the gene that encodes aromatase, essential for the production of estrogen. *Based on Gasperin, B.G., Ferreira, R., Rovani, M.T., Bordignon, V., Duggavathi, R., Buratini, J., et al. (2014). Expression of receptors for BMP15 is differentially regulated in dominant and subordinate follicles during follicle deviation in cattle.* Animal Reproduction Science, 144, *72–78; Jeppesen, J.V., Kristensen, S.G., Nielsen, M.E., Humaidan, P., Dal Canto, M., Fadini, R., et al. (2012). Lh-receptor gene expression in human granulosa and cumulus cells from antral and preovulatory follicles,* The Journal of Clinical Endocrinology and Metabolism, 97, *E1524–E1531.*

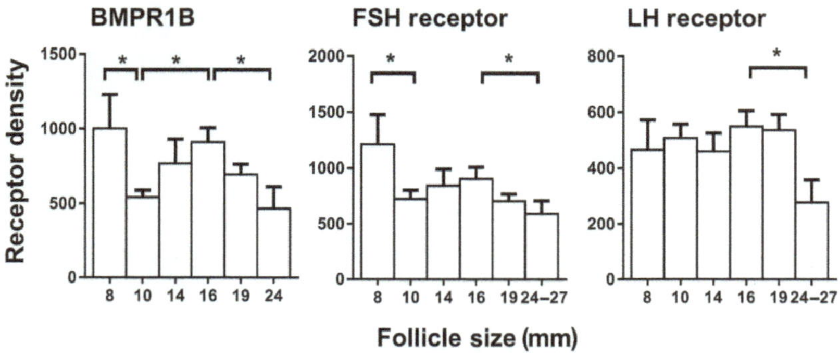

Fig. 7 The stage-specific relationship between granulosal receptor expression during folliculogenesis in human IVF cycles. Granulosal BMPR1B, FSHR, and LHR protein density and follicle size profile of young patients with a typical ovarian reserve for the age group. The patients were 23–30 years old and stimulated with gonadotrophins during an IVF cycle. Values in graphs are means ± SEM, and differences were considered significant if $*P < 0.05$ (Regan et al., 2016, 2017).

in BMP signaling reflects the ability of BMPs to regulate thecal androgen production (Glister, Richards, & Knight, 2005).

Granulosa cells are unique in the ovary because they express FSHRs, which are required for the synthesis of estrogen expression (Fig. 8) (Miller & Auchus, 2011). Theca cells express LHRs and synthesize androgens, which are used by the granulosa cells as substrate for estrogen synthesis.

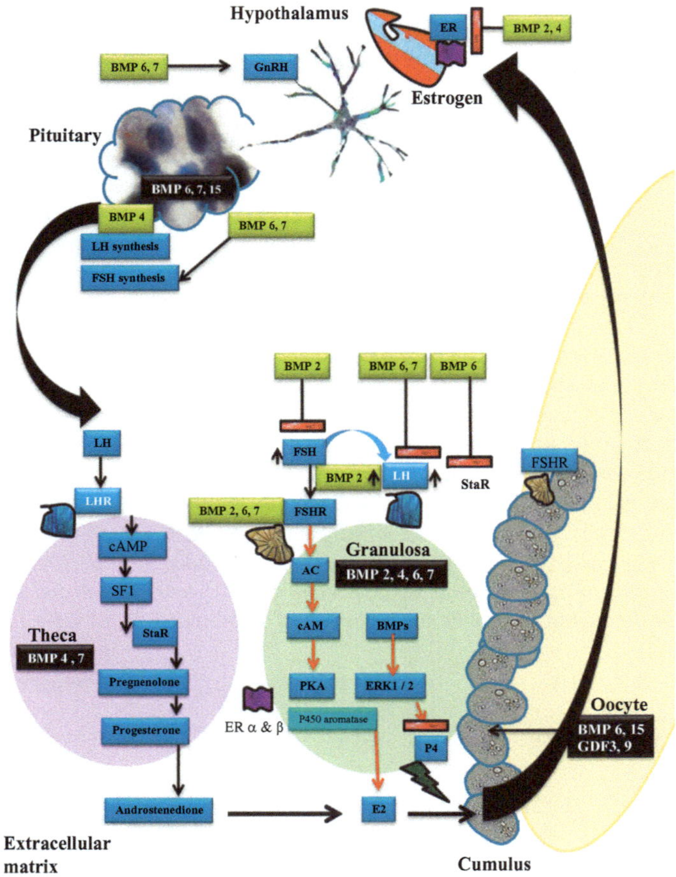

Fig. 8 Hypothalamic–pituitary–ovarian axis of regulation and the granulosa and theca cell interaction. Hypothalamic, pituitary control of ovarian growth and differentiation of the ovarian follicle comprised of the theca (*circle on left*), granulosa (*circle on center*), cumulus granulosa cells (attached to oocyte), and oocyte (*yellow oval on right*) surrounded by extracellular matrix. BMPs production by cell type is indicated in *black*. BMP signaling activity (*green*) and inhibition (*red bar*). Delayed expression of LHR (*white text*) on granulosa compared to the theca cells; estroen receptor (ER). *Based on Bao, B., Garverick, H. A., Smith, G. W., Smith, M. F., Salfen, B. E., & Youngquist, R. S. (1997). Changes in messenger ribonucleic acid encoding luteinizing hormone receptor, cytochrome P450-side chain cleavage, and aromatase are associated with recruitment and selection of bovine ovarian follicles. Biology of Reproduction, 56, 1158–1168; Chen, A.Q., Yu, S., Wang, Z., Xu, Z., & Yang, Z. (2008). Stage-specific expression of bone morphogenetic protein type I and type II receptor genes: Effects of follicle-stimulating hormone on ovine antral follicles,* Animal Reproduction Science, 111, *391–399; Dijke, P., Korchynskyi, O., Valdimarsdottir, G., & Goumans, M.-J. (2003). Controlling cell fate by bone morphogenetic protein receptors. Molecular and Cellular Endocrinology, 211, 105–113; Feary, E., Juengel, J., Smith, P., French, M., O'Connell, A., Lawrence, S., et al. (2007). Patterns of expression of messenger RNAs encoding GDF9, BMP15, TGFBR1, BMPR1B, and BMPR2 during follicular development*

and characterization of ovarian follicular populations in ewes carrying the woodlands FecX2W mutation. Biology of Reproduction, 77, *990–998; Fitzpatrick, S. L., Carlone, D. L., Robker, R. L., & Richards, J. S. (1997). Expression of aromatase in the ovary: Down-regulation of mRNA by the ovulatory luteinizing hormone surge. Steroids, 62, 197–206; Hillier, S., Whitelaw, P., & Smyth, C. (1994). Follicular oestrogen synthesis: The 'two-cell, two-gonadotrophin' model revisited.* Molecular and Cellular Endocrinology, 100, *51–54; Hussein, T., Froiland, D., Amato, F., Thompson, J., and Gilchrist, R. (2005). Oocytes prevent cumulus cell apoptosis by maintaining a morphogenic paracrine gradient of bone morphogenetic proteins.* Journal of Cell Science, 118, *5257–5268; Kayani, A. R., Glister, C., & Knight, P. G. (2009). Evidence for an inhibitory role of bone morphogenetic protein(s) in the follicular–luteal transition in cattle. Reproduction, 137, 67–78; Miller, W., & Bose, H. (2011). Early steps in steroidogenesis: Intracellular cholesterol trafficking.* Journal of Lipid Research, 52, *2111–2135; Miyazono, K., Maeda, S., & Imamura, T. (2005). BMP receptor signaling: Transcriptional targets, regulation of signals, and signaling cross-talk.* Cytokine & Growth Factor Reviews, 6, *251–263; Miyoshi, T., Otsuka, F., Inagaki, K., Otani, H., Takeda, M., Suzuki, J., et al. (2007). Differential regulation of steroidogenesis by bonemorphogenetic proteins in granulosa cells: Involvement of extracellularly regulated kinase signaling and oocyte actions in follicle-stimulating hormone-induced estrogen production. Endocrinology, 148, 337–345; Moore, R.K., Otsuka, F., & Shimasaki, S. (2001). Role of ERK1/2 in the differential synthesis of progesterone and estradiol by granulosa cells.* Biochemical and Biophysical Research Communications, 289, *796–800; Nicol, L., Faure, M.O., McNeilly, J.R., Fontaine, J., Taragnat, C., and McNeilly A.S. (2008). Bone morphogenetic protein-4 interacts with activin and GnRH to modulate gonadotrophin secretion in LbetaT2 gonadotrophs.* Journal of Endocrinology, 196, *497–507; Pierre, A., Pisselet, C., Dupont, J., Mandon-Pepin, B., Monniaux, D., Monget P., et al. (2004). Molecular basis of bone morphogenetic protein-4 inhibitory action on progesterone secretion by ovine granulosa cells.* Journal of Molecular Endocrinology, 33, *805–814; Rice, S., Ojha, K., Whitehead, S., & Mason H. (2007). Stage-specific expression of androgen receptor, follicle-stimulating hormone receptor, and anti-Müllerian hormone type II receptor in single, isolated, human preantral follicles: Relevance to polycystic ovaries.* The Journal of Clinical Endocrinology and Metabolism, 92, *1034–1040; Seger, R., Hanoch, T., Rosenberg, R., Dantes, A., Merz, W. E., Strauss, J. F., et al. (2001). The ERK signaling cascade inhibits gonadotropin-stimulated steroidogenesis. Journal of Biological Chemistry, 276, 13957–13964; Sugawara, T., Saito, M., & Fujimoto, S. (2000). Sp1 and SF-1 interact and cooperate in the regulation of human steroidogenic acute regulatory protein gene expression. Endocrinology, 141, 2895–2903; Sullivan, M. W., Stewart Akers, A., Krasnow, J. S., Berga, S. L., & Zeleznik, A. J. (1999). Ovarian responses in women to recombinant follicle-stimulating hormone and luteinizing hormone (LH): A role for LH in the final stages of follicular maturation. The Journal of Clinical Endocrinology and Metabolism, 84, 228–232; Takeda, M., Otsuka, F., Takahashi, H., Inagaki, K., Miyoshi, T., Tsukamoto, N., et al. (2012). Interaction between gonadotropin-releasing hormone and bone morphogenetic protein-6 and -7 signaling in LβT2 gonadotrope cells. Molecular and Cellular Endocrinology, 348, 147–154; ten Dijke, P., Korchynskyi, O., Valdimarsdottir, G., & Goumans, M. (2003). Controlling cell fate by bone morphogenetic protein receptors. Molecular and Cellular Endocrinology, 211, 105–113; Yamamoto, T., Saatcioglu, F., & Matsuda, T. (2002). Cross-talk between bone morphogenic proteins and estrogen receptor signaling. Endocrinology, 143, 2635–2642; Yuan, W., Bao, B., Garverick, H. A., Youngquist, R. S., & Lucy, M. C. (1998). Follicular dominance in cattle is associated with divergent patterns of ovarian gene expression for insulin-like growth factor (IGF)-I, IGF-II, and IGF binding protein-2 in dominant and subordinate follicles. Domestic Animal Endocrinology, 15(1), 55–63.*

The receptor density of BMPRIB on granulosal cells fluctuates biphasically during menstrual cycle in unison with the FSHR in young women and Merino sheep (Regan et al., 2016, 2015).

As pituitary FSH secretion is reduced, the follicles with granulosal LHRs have the capacity to supplement the FSH-dependent synthesis of estrogen. The follicles with granulosal-expressed LHR continue to grow and become the selected dominant follicles (Fig. 2). The extent of androstenedione conversion to estrogen continues to increase, which creates a positive estrogen feedback loop to the hypothalamic–pituitary complex, leading to further GnRH release (Fig. 8) (Faure et al., 2005).

Proliferation of the granulosa and theca cells continues as the rise in estrogen level promotes proliferation, until a threshold level is reached, which culminates in the generation of the LH surge (Austin et al., 2001; Ginther et al., 2005). In the event that reduced conversion of androstenedione to estrogen occurs, androgen levels rise, creating an androgen dominant follicle. Greater androstenedione-to-estrogen ratios have been shown to result in an elevated level of granulosa cell apoptosis and follicle demise (Yuan & Giudice, 1997).

10. BMPs AND DOMINANT FOLLICLE SELECTION

In monovular species such as humans and cattle, follicle divergence (i.e., selection of a dominant follicle from a pool of growing 4–8 mm antral follicles) occurs at a stage of the cycle when pituitary FSH secretion reduces and LH secretion increases (Fig. 2) (Austin et al., 2001; Edwards et al., 1996). During a natural cycle, the follicles with a higher density of gonadotrophin receptors are presumed to be more responsive to the gonadotrophins and continue to increase in size (Bächler, Menshykau, De Geyter, & Iber, 2014; Gougeon, 1986; LaPolt et al., 1992).

FSHRs are expressed exclusively by granulosa cells that respond to pituitary-derived FSH by proliferating and increasing estrogen output. In turn, this promotes expression of LHR by granulosa cells of the selected dominant follicle, enabling them to respond to LH pulses and survive the fall in FSH (Fig. 2).

Downregulation of granulosal BMPR1B and FSHR expression has been observed at the stage of cyclic dominant follicle selection, occurring between ~8–10 mm in the human and ~1–1.7 mm in the Merino sheep (Regan et al., 2016, 2015). As mentioned earlier, follicles are selected as a consequence of the decline in pituitary FSH and only follicles with the newly acquired LHR can sustain estrogen production during the preovulatory phase. Suppression of granulosal progesterone synthesis, in favor of FSH-dependent estrogen production, appears to be governed by the action of the BMPs (Knight & Glister, 2006; Moore et al., 2001).

At this stage in follicle development, BMP15 and inhibins increase to stimulate proliferation and further growth of the follicle and oocyte (Feary et al., 2007; Yding Andersen, 2017). The ability of the follicle to reach the FSH–estrogen threshold before ovulation with sufficient granulosal LHR appears to be of paramount importance to the survival of the selected dominant follicle. Alternatively, follicular regression proceeds followed by atresia (Campbell, Dobson, Baird, & Scaramuzzi, 1999; Ginther et al., 2012; Luo, Gumen, Haughian, & Wiltbank, 2011; Picton & McNeilly, 1991).

11. BMPs AND OVULATION RATE

The number of preovulatory follicles that develop can be artificially enhanced by exogenous rFSH stimulation, such as that used in IVF treatment cycles, or by a naturally occurring mutation-induced increase in responsiveness, such as that seen in the Booroola Merino sheep (Fig. 9) (Mulsant et al., 2001; Souza et al., 2001; Wilson et al., 2001).

Fig. 9 Booroola Merino and wild-type Merino sheep, Armidale NSW. The Booroola (red number on sheep backs) sheep have a naturally occurring gene mutation that partially attenuates the BMPR1B receptor signaling and increases the ovulation rate to ~5. Wild-type Merino (blue numbers). University of New England, breeding program, 1964 to 2010, Dr Tim O'Shea (deceased).

The TGFβ type 1 receptor BMPR1B has been localized on sheep granulosa cells from the primordial follicle stage onward (Al-Samerria & Almahbobi, 2014; Anthony et al., 2015; Chen, Yu, Wang, Xu, & Yang, 2008; Erickson & Shimasaki, 2003; Gasperin et al., 2014). The level of expression increased sequentially from primordial to antral follicles in sheep. BMPR1B is expressed mainly on granulosa cells and the oocyte in the bovine and human model, and to a lesser degree in their theca cells (Abir et al., 2008; Glister et al., 2004). It has also been demonstrated that BMP ligands are produced in a stage-specific manner by follicular cells in animals and humans (Gasperin et al., 2014; Glister et al., 2004; Regan et al., 2016, 2017, 2015).

The Booroola Merino, with a naturally occurring point mutation of the BMPR1B gene, has an increased ovulation rate (Fabre et al., 2006; Mulsant et al., 2001; Souza et al., 2001; Wilson et al., 2001). This increase is likely due to the follicles being more sensitive to FSH at an earlier follicle size (Baird & Campbell, 1998; McNatty et al., 1985).

The Booroola sheep follicles contained significantly fewer granulosa cells than the normal wild-type (Campbell et al., 2006; McNatty et al., 1985). Studies conducted on the granulosa cells show that, when stimulated in vitro by LH or FSH, they produced more cAMP, estrogen, and androstenedione from the same number of cells from the large antral follicle (Campbell et al., 2006). An increased cellular capacity to produce estrogen would compensate for the reduced number of granulosa cells. Taken together, it is apparent that the Booroola sheep produce multiple follicles because of the greater density of receptors for FSH and LH due to the attenuated BMPR1B signal (Regan et al., 2015).

Recently, compelling data show that the expression of mature surface granulosa receptors for FSHR, LHR, and BMPR1B is significantly elevated in the Booroola compared to the young wild-type Merino sheep (Regan et al., 2015). In another study, conflicting results were observed; however, the Booroola sheep were much older (6–10 years compared to 4 years). In addition, mRNA expression was measured rather than the mature expressed protein itself. It is plausible that the Booroola mutation of the BMPR1B signal may partially eliminate the BMP ligand-induced suppression of FSHR and LHR expression. Partial attenuation of BMP action may thus lead to an upregulation of FSHR, earlier acquisition of LHR, and an increase of BMPR1B itself in the Booroola sheep. The increased signaling resulted in multiple ovulations and an increased birth rate of ~5 (Otsuka,

Yamamoto, et al., 2001; Regan et al., 2015). The findings clearly show the effect of the repression exerted by the BMPs in regulating ovulation rate.

Evidence indicates a strong connection between the role of BMPs, AMH, and the gonadotropin-dependent regulation of ovulation rate. BMP4, 6, and 15 increase the transcriptional activity of the AMH promotor activity via SF-1 (Anthony et al., 2015). Yet it still remains unclear as to why immunization against AMH increased the ovulation rate but did not reduce proliferation of granulosa cells, as reported in the Booroola mutation (Campbell et al., 2012). A possible explanation may be related to the stage-specific downregulation of AMH after dominant follicle selection. The greatest mitogenic activity of granulosa cells occurs after this divergence when AMH is low, whereas dominant follicle selection occurs when AMH is high (Austin et al., 2001). Immunization against AMH alone would, therefore, increase ovulation rate but not granulosa cell proliferation governed by other BMPs. Whereas immunization against BMPR1B or an attenuating mutation in BMPR1B, such as that in Booroola sheep, would affect both proliferation and ovulation rate.

Identification of the BMP ligand responsible for the increased ovulation rate or decreased proliferation of the granulosa cells has not been achieved. BMP15 has been associated with an increase in ovulation rate in sheep with specific mutations in the BMP15 gene (Hanrahan et al., 2004; McNatty et al., 2009). Heterozygous Inverdale sheep with an inactivation mutation for BMP15 exhibit an increase in ovulation rate, whereas in homozygous carriers, follicle development did not progress past the primary follicle stage (Braw-Tal et al., 1993; McNatty et al., 2009). In sheep, short-term immunization against BMP15 increased the ovulation rate from 1–2 to $\geq$3 without affecting plasma progesterone concentration (Juengel et al., 2004, 2011).

In another study, complete neutralization of BMP15 prevented exogenous FSH-induced follicle rescue resulting in anovulation, which indicates that BMP15 is required for FSHR transcription (McNatty et al., 2009). Sheep with ovarian infusion of BMP6 showed a reduced cycle length and size of the preovulatory follicles (Campbell et al., 2009). Although the effect of the infusion was short-lived, the estrogen and androstenedione increased with no change to the ovulation rate. Findings from the Booroola sheep indicate that a combination of BMP15 and BMP6 signal attenuation (via BMPR1B) may be responsible for the reduced mitogenic activity of granulosa cells and the increased ovulation rate, with BMP2 and 4 influencing primordial to preantral follicle development.

12. BMPs AND THE TERMINAL STAGE OF FOLLICULOGENESIS

Granulosa cell proliferation continues, and the ovarian follicle increases in size from 10 to 20 + mm in the human, producing large quantities of estrogen that reach a critical level, triggering the release of LH from the pituitary and the onset of the LH surge. The release of LH surge initiates numerous events, changing the granulosa cells and the oocyte in preparation for the expulsion of the oocyte from the follicle, and corpus luteum formation.

The cells of the follicle differentiate morphologically, inducing cytoskeletal reorganization, expansion of the granulosa cell, and cessation of mitogenic proliferation. Resumption of meiosis and oocyte maturation takes place. The cumulus cells expand away from the oocyte-cumulus complex, severing the morphological cumulus gap junctions where cross talk linkages and concentration gradient of BMP15 and 6 radiating from the oocyte are disrupted (Hussein et al., 2005). Angiogenic cells infiltrate the degenerating basal lamina in preparation for blood vessel formation in the developing corpus luteum.

The BMP-induced suppression of progesterone synthesis by the granulosa cells is released. The granulosa cell acquires the ability to synthesize large amounts of progesterone (Westergaard, Christensen, & McNatty, 1986). The BMPR1B expression density on the cell surface is reduced in the largest follicles either by the degradation of receptors or by reduced BMPR1B mRNA production (Menon & Menon, 2014; Regan et al., 2016, 2015; Zhang & Roy, 2004).

This process, collectively referred to as luteinization, appears to be associated with BMPR1B, FSHR, and LHR downregulation in the leading dominant follicles (Fan et al., 2009; Izadyar, Zeinstra, & Bevers, 1998; Regan et al., 2017, 2015). Furthermore, LHR density peaks in the preovulatory follicle in the wild type and the Booroola followed by a significant reduction in the leading dominant follicle during the LH surge coincident with a reduced mitogenic index and reduced estrogen levels (Jeppesen et al., 2012; LaPolt et al., 1992; Ophir et al., 2014; Regan et al., 2017, 2015).

In women reaching the end of their reproductive life span, dysregulation of granulosal BMPR1B (Fig. 10) and FSHR occurs (Regan et al., 2016, 2017). As the ovarian pool of primordial follicles depletes in women, receptor density continues to increase with follicle growth.

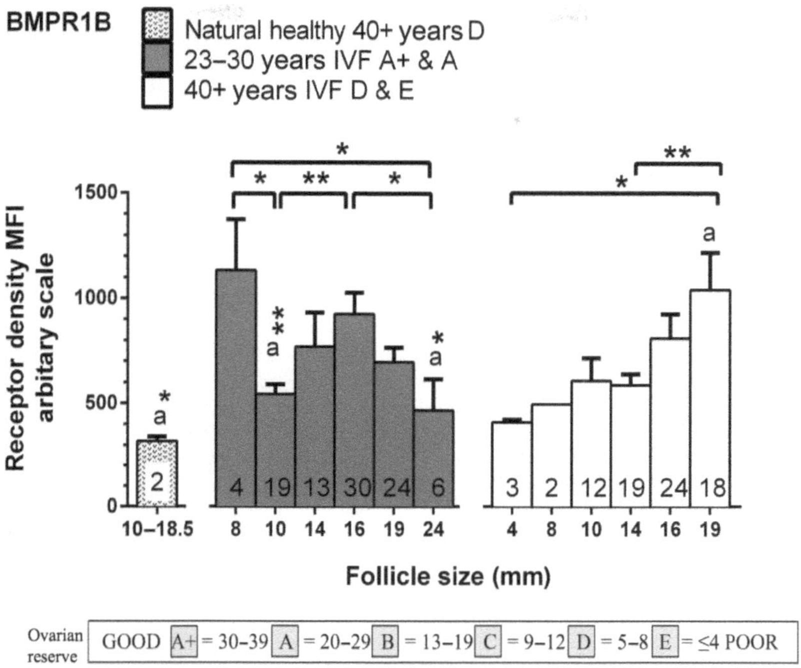

Fig. 10 Granulosal BMPR1B density from follicles of different sizes collected from young and older IVF patients compared to an unstimulated natural healthy cycle. Granulosal BMPR1B protein density and follicle size profile of a natural healthy unstimulated patient of 41 years with an AFC of D, before the LH surge (*patterned bar*). Patients, 23–30 years stimulated, IVF cycle with an AFC of A+ and A (*gray bar*). Patients, 40+ years stimulated IVF cycle with an AFC of D and E (*white bar*). IVF patients were grouped according to ovarian reserve measured indirectly by the antral follicle count (AFC). The antral follicle count, is defined as the number of follicles between 2 and 10 mm in diameter, combining the number collected from both ovaries; that were present on ~day 5 of a preliminary assessment cycle, without rFSH. Group A+ = 30–39 small follicles; group A = 20–29 small follicles; group B = 13–19 small follicles; group C = 9–12 small follicles, group D = 5–8 small follicles; group E = ≤4 small follicles. Mean fluorescent intensity (MFI) was obtained using an average of ~ 8000 granulosa cells per follicle for the direct measurement of receptor protein expression. The data were subjected to statistical verification using one-way ANOVA with an uncorrected Fisher's LSD. Values in graphs are means ± SEM, and differences were considered significant if $*P < 0.05$ and $**P < 0.01$. The letter, "a" signifies a statistical difference to the matching letter with an attached asterisk(s) (a*, a**). The number within the column represents the number of follicles analyzed for that group (Regan et al., 2016).

It is proposed that the lack of receptor downregulation results in a reduced response to the LH surge and reduced output of progesterone, and is associated with poor oocyte quality and pregnancy rates (Regan et al., 2016, 2017).

REFERENCES

Abir, R., Ben-Haroush, A., Melamed, N., Felz, C., Krissi, H., & Fisch, B. (2008). Expression of bone morphogenetic proteins 4 and 7 and their receptors IA, IB, and II in human ovaries from fetuses and adults. *Fertility and Sterility*, *89*, 1430–1440.

Ainsworth, L., Tsang, B. K., Downey, B. R., Marcus, G. J., & Armstrong, D. T. (1980). Interrelationships between follicular fluid steroid levels, Gonadotropic stimuli, and oocyte maturation during preovulatory development of porcine follicles. *Biology of Reproduction*, *23*, 621–627.

Albamonte, M. I., Albamonte, M. S., Stella, I., Zuccardi, L., & Vitullo, A. D. (2013). The infant and pubertal human ovary: Balbiani's body-associated VASA expression, immunohistochemical detection of apoptosis-related BCL2 and BAX proteins, and DNA fragmentation. *Human Reproduction*, *28*, 698–706.

Almog, B., Shehata, F., Shalom-Paz, E., Tan, S. L., & Tulandi, T. (2011). Age-related normogram for antral follicle count: McGill reference guide. *Fertility and Sterility*, *95*, 663–666.

Al-Samerria, S., Al-Ali, I., McFarlane, J. R., & Almahbobi, G. (2015). The impact of passive immunisation against BMPRIB and BMP4 on follicle development and ovulation in mice. *Reproduction*, *149*, 403–411.

Al-Samerria, S., & Almahbobi, G. (2014). Three-dimensional image analysis to quantify the temproro-spacial expression of cellular receptors. *Journal of Medical and Bioengineering*, *3*, 179–182.

Amsterdam, A., Sasson, R., Keren Tal, I., Aharoni, D., Dantes, A., Rimon, E., et al. (2003). Alternative pathways of ovarian apoptosis: Death for life. *Biochemical Pharmacology*, *66*, 1355–1362.

Anthony, E., Pierre, A., Clemente, N. d., Picard, J.-Y., Jarrier, P., Mansanet, C., et al. (2015). Anti-Müllerian hormone regulation by the bone morphogenetic proteins in the sheep ovary: Deciphering a direct regulatory pathway. *Endocrinology*, *156*, 301–313.

Anzalone, C. R., Hong, L.-S., Lu, J. K. H., & LaPolt, P. S. (2001). Influences of age and ovarian follicular reserve on estrous cycle patterns, ovulation, and hormone secretion in the Long-Evans rat. *Biology of Reproduction*, *64*, 1056–1062.

Araújo, V. R., Silva, C. M. G., Magalhães, D. M., Silva, G. M., Báo, S. N., Silva, J. R. V., et al. (2010). Effect of bone morphogenetic protein-7 (BMP-7) on in vitro survival of caprine preantral follicles. *Pesquisa Veterinária Brasileira*, *30*, 305–310.

Austin, E. J., Mihm, M., Evans, A. C. O., Knight, P. G., Ireland, J. L. H., Ireland, J. J., et al. (2001). Alterations in intrafollicular regulatory factors and apoptosis during selection of follicles in the first follicular wave of the bovine estrous cycle. *Biology of Reproduction*, *64*, 839–848.

Bächler, M., Menshykau, D., De Geyter, C., & Iber, D. (2014). Species-specific differences in follicular antral sizes result from diffusion-based limitations on the thickness of the granulosa cell layer. *Molecular Human Reproduction*, *20*, 208–221.

Baerwald, A., Adams, G., & Pierson, R. (2012). Ovarian antral folliculogenesis during the human menstrual cycle: A review. *Human Reproduction Update*, *18*, 73–91.

Baird, D., & Campbell, B. (1998). Follicle selection in sheep with breed differences in ovulation rate. *Molecular and Cellular Endocrinology*, *145*, 89–95.

Bomsel Helmreich, O., Gougeon, A., Thebault, A., Saltarelli, D., Milgrom, E., Frydman, R., et al. (1979). Healthy and atretic human follicles in the preovulatory phase: Differences in evolution of follicular morphology and steroid content of follicular fluid. *The Journal of Clinical Endocrinology and Metabolism*, *48*, 686–694.

Brankin, V., Quinn, R. L., Webb, R., & Hunter, M. G. (2005). BMP-2 and -6 modulate porcine theca cell function alone and co-cultured with granulosa cells. *Domestic Animal Endocrinology*, *29*, 593–604.

Braw-Tal, R., McNatty, K., Smith, P., Heath, D., Hudson, N., Phillips, D., et al. (1993). Ovaries of ewes homozygous for the X-linked Inverdale gene (FecXI) are devoid of secondary and tertiary follicles but contain many abnormal structures. *Biology of Reproduction, 49*, 895–907.

Bukovsky, A. (2016). Involvement of blood mononuclear cells in the infertility, age-associated diseases and cancer treatment. *World Journal of Stem Cells, 8*, 399–427.

Campbell, B., Clinton, M., & Webb, R. (2012). The role of anti-mullerian hormone (AMH) during follicle development in a monovulatory species (sheep). *Endocrinology, 153*, 4533–4543.

Campbell, B. K., Dobson, H., Baird, D. T., & Scaramuzzi, R. J. (1999). Examination of the relative role of FSH and LH in the mechanism of ovulatory follicle selection in sheep. *Journal of Reproduction and Fertility, 117*, 355–367.

Campbell, B. K., Kendall, N. R., & Baird, D. T. (2009). Effect of direct ovarian infusion of bone morphogenetic protein 6 (BMP6) on ovarian function in sheep. *Biology of Reproduction, 81*, 1016–1023.

Campbell, B. K., Souza, C. J. H., Skinner, A. J., Webb, R., & Baird, D. T. (2006). Enhanced response of granulosa and theca cells from sheep carriers of the FecB mutation in vitro to gonadotropins and bone morphogenic protein-2, -4, and -6. *Endocrinology, 147*, 1608–1620.

Castrillon, D. H., Miao, L., Kollipara, R., Horner, J. W., & DePinho, R. A. (2003). Suppression of ovarian follicle activation in mice by the transcription factor Foxo3a. *Science, 301*, 215–218.

Chen, A. Q., Yu, S., Wang, Z., Xu, Z., & Yang, Z. (2008). Stage-specific expression of bone morphogenetic protein type I and type II receptor genes: Effects of follicle-stimulating hormone on ovine antral follicles. *Animal Reproduction Science, 111*, 391–399.

Childs, A. J., Bayne, R. A. L., Murray, A. A., Silva, S. J. M. D., Collins, C. S., Spears, N., et al. (2010). Differential expression and regulation by activin of the neurotrophins BDNF and NT4 during human and mouse ovarian development. *Developmental Dynamics, 239*, 1211–1219.

Di Pasquale, E., Rossetti, R., Marozzi, A., Bodega, B., Borgato, S., Cavallo, L., et al. (2006). Identification of new variants of human BMP15 gene in a large cohort of women with premature ovarian failure. *The Journal of Clinical Endocrinology and Metabolism, 91*, 1976–1979.

Dong, J., Albertini, D., Nishimori, K., Kumar, T., Lu, N., & Matzuk, M. (1996). Growth differentiation factor-9 is required during early ovarian folliculogenesis. *Nature, 383*, 531–535.

Durlinger, A. L. L., Gruijters, M. J. G., Kramer, P., Karels, B., Ingraham, H. A., Nachtigal, M. W., et al. (2002). Anti-Mullerian hormone inhibits initiation of primordial follicle growth in the mouse ovary. *Endocrinology, 143*, 1076–1084.

Edson, M. A. (2009). The mammalian ovary from genesis to revelation. *Endocrine Reviews, 30*, 624–712.

Edson, M. A., Nalam, R. L., Clementi, C., Franco, H. L., DeMayo, F. J., Lyons, K. M., et al. (2010). Granulosa cell-expressed BMPR1A and BMPR1B have unique functions in regulating fertility but act redundantly to suppress ovarian tumor development. *Molecular Endocrinology, 24*, 1251–1266.

Edwards, R. G., Lobo, R., & Bouchard, P. (1996). Time to revolutionize ovarian stimulation. *Human Reproduction, 11*, 917–919.

Edwards, R. G., & Steptoe, P. C. (1983). Current status of in-vitro fertilisation and implantation of human embryos. *Lancet (London, England), 2*, 1265–1269.

Eppig, J. (2001). Oocyte control of ovarian follicular development and function in mammals. *Reproduction, 122*, 829–838.

Erickson, G., & Shimasaki, S. (2001). The physiology of folliculogenesis: The role of novel growth factors. *Fertility and Sterility, 76*, 943–949.

Erickson, G., & Shimasaki, S. (2003). The spatiotemporal expression pattern of the bone morphogenetic protein family in rat ovary cell types during the estrous cycle. *Reproductive Biology and Endocrinology*, *1*(9), 1–20.

Erickson, G. F., Wang, C., & Hsueh, A. J. W. (1979). FSH induction of functional LH receptors in granulosa cells cultured in a chemically defined medium. *Nature*, *279*, 336–338.

Fabre, S., Pierre, A., Mulsant, P., Bodin, L., Di Pasquale, E., Persani, L., et al. (2006). Regulation of ovulation rate in mammals: Contribution of sheep genetic models. *Reproductive Biology and Endocrinology*, *4*, 20.

Fabre, S., Pierre, A., Pisselet, C., Mulsant, P., Lecerf, F., Pohl, J., et al. (2003). The Booroola mutation in sheep is associated with an alteration of the bone morphogenetic protein receptor-IB functionality. *The Journal of Endocrinology*, *177*, 435–444.

Fan, H., Liu, Z., Shimada, M., Sterneck, E., Johnson, P. F., Hedrick, S. M., et al. (2009). MAPK3/1 (ERK1/2) in ovarian granulosa cells are essential for female fertility. *Science*, *324*, 938–941.

Faure, M. O., Nicol, L., Fabre, S., Fontaine, J., Mohoric, N., McNeilly, A., et al. (2005). BMP-4 inhibits follicle-stimulating hormone secretion in ewe pituitary. *The Journal of Endocrinology*, *186*, 109–121.

Feary, E., Juengel, J., Smith, P., French, M., O'Connell, A., Lawrence, S., et al. (2007). Patterns of expression of messenger RNAs encoding GDF9, BMP15, TGFBR1, BMPR1B, and BMPR2 during follicular development and characterization of ovarian follicular populations in ewes carrying the woodlands FecX2W mutation. *Biology of Reproduction*, *77*, 990–998.

Fowler, P. A., Sorsa, T., Harris, W. J., Knight, P. G., & Mason, H. D. (2001). Relationship between follicle size and gonadotrophin surge attenuating factor (GnSAF) bioactivity during spontaneous cycles in women. *Human Reproduction*, *16*, 1353–1358.

Galloway, S., Gregan, S., Wilson, T., McNatty, K., Juengel, J., Ritvos, O., et al. (2002). Bmp15 mutations and ovarian function. *Molecular and Cellular Endocrinology*, *191*, 15–18.

Garcia, J., Jones, G. S., Acosta, A. A., & Wright, G. L. (1981). Corpus Luteum function after follicle aspiration for oocyte retrieval. *Fertility and Sterility*, *36*, 565–572.

Gasperin, B. G., Ferreira, R., Rovani, M. T., Bordignon, V., Duggavathi, R., Buratini, J., et al. (2014). Expression of receptors for BMP15 is differentially regulated in dominant and subordinate follicles during follicle deviation in cattle. *Animal Reproduction Science*, *144*, 72–78.

Gilchrist, R., Ritter, L., & Armstrong, D. (2004). Oocyte-somatic cell interactions during follicle development in mammals. *Animal Reproduction Science*, *82–83*, 431–446.

Ginther, O. J., Beg, M. A., Gastal, E. L., Gastal, M. O., Baerwald, A. R., & Pierson, R. A. (2005). Systemic concentrations of hormones during the development of follicular waves in mares and women: A comparative study. *Reproduction*, *130*, 379–388.

Ginther, O. J., Khan, F. A., Hannan, M. A., Rodriguez, M. B., Pugliesi, G., & Beg, M. A. (2012). Role of LH in luteolysis and growth of the ovulatory follicle and estradiol regulation of LH secretion in heifers. *Theriogenology*, *77*, 1442–1452.

Glister, C., Kemp, C., & Knight, P. (2004). Bone morphogenetic protein (BMP) ligands and receptors in bovine ovarian follicle cells: Actions of BMP-4, -6 and 7 on granulosa cells and differential modulation of Smad-1 phosphorylation by follistatin. *Reproduction*, *127*, 239–254.

Glister, C., Richards, S., & Knight, P. (2005). Bone morphogenetic proteins (BMP) -4, -6, and -7 potently suppress basal and luteinizing hormone-induced androgen production by bovine theca interna cells in primary culture: Could ovarian hyperandrogenic dysfunction be caused by a defect in thecal BMP signaling? *Endocrinology*, *146*, 1883–1892.

Gouedard, L., Chen, Y. G., Thevenet, L., Racine, C., Borie, S., Lamarre, I., et al. (2000). Engagement of bone morphogenetic protein type IB receptor and Smad1 signaling by anti-Mullerian hormone and its type II receptor. *The Journal of Biological Chemistry*, *275*.

Gougeon, A. (1986). Dynamics of follicular growth in the human: A model from preliminary results. *Human Reproduction, 1*, 81–87.

Hanrahan, J., Gregan, S., Mulsant, P., Mullen, M., Davis, G., Powell, R., et al. (2004). Mutations in the genes for oocyte-derived growth factors GDF9 and BMP15 are associated with both increased ovulation rate and sterility in Cambridge and Belclare sheep (Ovis aries). *Biology of Reproduction, 70*, 900–909.

Hansen, K. R., Hodnett, G. M., Knowlton, N., & Craig, L. B. (2011). Correlation of ovarian reserve tests with histologically determined primordial follicle number. *Fertility and Sterility, 95*, 170–175.

Hillier, S., Whitelaw, P., & Smyth, C. (1994). Follicular oestrogen synthesis: The 'two-cell, two-gonadotrophin' model revisited. *Molecular and Cellular Endocrinology, 100*, 51–54.

Hsueh, A. J. W., Kawamura, K., Cheng, Y., & Fauser, B. C. J. M. (2015). Intraovarian control of early folliculogenesis. *Endocrine Reviews, 36*, 1–24.

Hussein, T., Froiland, D., Amato, F., Thompson, J., & Gilchrist, R. (2005). Oocytes prevent cumulus cell apoptosis by maintaining a morphogenic paracrine gradient of bone morphogenetic proteins. *Journal of Cell Science, 118*, 5257–5268.

Inagaki, K., Otsuka, F., Miyoshi, T., Yamashita, M., Takahashi, M., Goto, J., et al. (2009). p38-mitogen-activated protein kinase stimulated steroidogenesis in granulosa cell-oocyte cocultures: Role of bone morphogenetic proteins 2 and 4. *Endocrinology, 150*, 1921–1930.

Inagaki, K., & Shimasaki, S. (2010). Impaired production of BMP-15 and GDF-9 mature proteins derived from proproteins WITH mutations in the proregion. *Molecular and Cellular Endocrinology, 328*, 1–7.

Ingraham, H. A., Lala, D. S., Ikeda, Y., Luo, X., Shen, W. H., Nachtigal, M. W., et al. (1994). The nuclear receptor steroidogenic factor 1 acts at multiple levels of the reproductive axis. *Genes & Development, 8*, 2302–2312.

Izadyar, F., Zeinstra, E., & Bevers, M. M. (1998). Follicle-stimulating hormone and growth hormone act differently on nuclear maturation while both enhance developmental competence of in vitro matured bovine oocytes. *Molecular Reproduction and Development, 51*, 339–345.

Jeppesen, J. V., Kristensen, S. G., Nielsen, M. E., Humaidan, P., Dal Canto, M., Fadini, R., et al. (2012). Lh-receptor gene expression in human granulosa and cumulus cells from antral and preovulatory follicles. *The Journal of Clinical Endocrinology and Metabolism, 97*, E1524–E1531.

John, G. B., Shirley, L. J., Gallardo, T. D., & Castrillon, D. H. (2007). Specificity of the requirement for Foxo3 in primordial follicle activation. *Reproduction, 133*, 855–863.

Josso, N., & Clemente, N. d. (2003). Transduction pathway of anti-Müllerian hormone, a sex-specific member of the TGF-β family. *Trends in Endocrinology and Metabolism, 14*, 91–97.

Josso, N., di Clemente, N., & Gouédard, L. (2001). Anti-Müllerian hormone and its receptors. *Molecular and Cellular Endocrinology, 179*, 25–32.

Juengel, J., Hudson, N., Whiting, L., & McNatty, K. (2004). Effects of immunization against bone morphogenetic protein 15 and growth differentiation factor 9 on ovulation rate, fertilization, and pregnancy in ewes. *Biology of Reproduction, 70*, 557–561.

Juengel, J., Quirke, L., Lun, S., Heath, D., Johnstone, P., & McNatty, K. (2011). Effects of immunizing ewes against bone morphogenetic protein 15 on their responses to exogenous gonadotrophins to induce multiple ovulations. *Reproduction, 142*, 565–572.

Kaivo-Oja, N., Mottershead, D. G., Mazerbourg, S., Myllymaa, S., Duprat, S., Gilchrist, R. B., et al. (2005). Adenoviral gene transfer allows Smad-responsive gene promoter analyses and delineation of type I receptor usage of transforming growth factor-β family ligands in cultured human granulosa luteal cells. *The Journal of Clinical Endocrinology and Metabolism, 90*, 271–278.

Kawamura, K., Cheng, Y., Suzuki, N., Deguchi, M., Sato, Y., Takae, S., et al. (2013). Hippo signaling disruption and Akt stimulation of ovarian follicles for infertility treatment. *Proceedings of the National Academy of Sciences of the United States of America, 110*, 17474–17479.

Kerr, C. L., Hill, C. M., Blumenthal, P. D., & Gearhart, J. D. (2008). Expression of pluripotent stem cell markers in the human fetal ovary. *Human Reproduction, 23*, 589–599.

Khalaf, M., Morera, J., Bourret, A., Reznik, Y., Denoual, C., Herlicoviez, M., et al. (2013). BMP system expression in GCs from polycystic ovary syndrome women and the in vitro effects of BMP4, BMP6, and BMP7 on GC steroidogenesis. *European Journal of Endocrinology, 168*, 437–444.

Klein, N. A., Battaglia, D. E., Woodruff, T. K., Padmanabhan, V., Giudice, L. C., Bremner, W. J., et al. (2000). Ovarian follicular concentrations of Activin, Follistatin, Inhibin, insulin-like growth factor I (IGF-I), IGF-II, IGF-binding Protein-2 (IGFBP-2), IGFBP-3, and vascular endothelial growth factor in spontaneous menstrual cycles of normal women of advanced reproductive age. *The Journal of Clinical Endocrinology and Metabolism, 85*, 4520–4525.

Knight, P. G., & Glister, C. (2006). TGF-β superfamily members and ovarian follicle development. *Reproduction, 132*, 191–206.

Knight, P. G., Satchell, L., & Glister, C. (2012). Intra-ovarian roles of activins and inhibins. *Molecular and Cellular Endocrinology, 359*, 53–65.

LaPolt, P. S., Tilly, J. L., Aihara, T., Nishimori, K., & Hsueh, A. J. (1992). Gonadotropin-induced up- and down-regulation of ovarian follicle-stimulating hormone (FSH) receptor gene expression in immature rats: Effects of pregnant mare's serum gonadotropin, human chorionic gonadotropin, and recombinant FSH. *Endocrinology, 130*, 1289–1295.

Lawson, K. A., Dunn, N. R., Roelen, B. A., Zeinstra, L. M., Davis, A. M., Wright, C. V., et al. (1999). Bmp4 is required for the generation of primordial germ cells in the mouse embryo. *Genes & Development, 13*, 424–436.

Lee, K., Khivansara, V., Santos, M., Lamba, P., Yuen, T., Sealfon, S., et al. (2007). Bone morphogenetic protein 2 and activin A synergistically stimulate follicle-stimulating hormone beta subunit transcription. *Journal of Molecular Endocrinology, 38*, 315–330.

Lee, W., Otsuka, F., Moore, R., & Shimasaki, S. (2001). Effect of bone morphogenetic protein-7 on folliculogenesis and ovulation in the rat. *Biology of Reproduction, 65*, 994–999.

Lee, W.-S., Yoon, S.-J., Yoon, T.-K., Cha, K.-Y., Lee, S.-H., Shimasaki, S., et al. (2004). Effects of bone morphogenetic protein-7 (BMP-7) on primordial follicular growth in the mouse ovary. *Molecular Reproduction and Development, 69*, 159–163.

Li, Y., Li, R.-Q., Ou, S.-B., Zhang, N.-F., Reng, L., Wei, L.-N., et al. (2014). Increased GDF9 and BMP15 mRNA levels in cumulus granulosa cells correlate with oocyte maturation, fertilization, and embryo quality in humans. *Reproductive Biology and Endocrinology, 12*, 81.

Liu, Z., Kobayashi, K., van Dinther, M., van Heiningen, S. H., Valdimarsdottir, G., van Laar, T., et al. (2009). VEGF and inhibitors of TGF{beta} type-I receptor kinase synergistically promote blood-vessel formation by inducing {alpha}5-integrin expression. *Journal of Cell Science, 122*, 3294–3302.

Liu, C.-F., Parker, K., & Yao, H. H. C. (2010). WNT4/β-catenin pathway maintains female germ cell survival by inhibiting activin βB in the mouse fetal ovary. *PLoS ONE, 5*, e10382.

Luo, W., Gumen, A., Haughian, J., & Wiltbank, M. (2011). The role of luteinizing hormone in regulating gene expression during selection of a dominant follicle in cattle. *Biology of Reproduction, 84*, 369–378.

MacNaughton, J., Banah, M., McCloud, P., Hee, J., & Burger, H. (1992). Age related changes in follicle stimulating hormone, luteinizing hormone, oestradiol and immunoreactive inhibin in women of reproductive age. *Clinical Endocrinology, 36*, 339–345.

Manna, P., Dyson, M., Eubank, D., Clark, B., Lalli, E., Sassone Corsi, P., et al. (2002). Regulation of steroidogenesis and the steroidogenic acute regulatory protein by a member of the cAMP response-element binding protein family. *Molecular Endocrinology, 16*, 184–199.

Massagué, J. (2008). A very private TGF-[beta] receptor embrace. *Molecular Cell, 29*, 149–150.

Matsuda, F., Inoue, N., Manabe, N., & Ohkura, S. (2012). Follicular growth and atresia in mammalian ovaries: Regulation by survival and death of granulosa cells. *Journal of Reproduction and Development, 58*, 44–50.

Mattioli, M., Barboni, B., Turriani, M., Galeati, G., Zannoni, A., Castellani, G., et al. (2001). Follicle activation involves vascular endothelial growth factor production and increased blood vessel extension. *Biology of Reproduction, 65*, 1014–1019.

McNatty, K., Heath, D., Hudson, N., Lun, S., Juengel, J., & Moore, L. (2009). Gonadotrophin-responsiveness of granulosa cells from bone morphogenetic protein 15 heterozygous mutant sheep. *Reproduction, 138*, 545–551.

McNatty, K. P., Hudson, N. L., Whiting, L., Reader, K. L., Lun, S., Western, A., et al. (2007). The effects of immunizing sheep with different BMP15 or GDF9 peptide sequences on ovarian follicular activity and ovulation rate. *Biology of Reproduction, 76*, 552–560.

McNatty, K. P., Henderson, K. M., Lun, S., Heath, D. A., Ball, K., Hudson, N. L., et al. (1985). Ovarian activity in Booroola X Romney ewes which have a major gene influencing their ovulation rate. *Journal of Reproduction and Fertility, 73*, 109–120.

McNatty, K., Moore, L., Hudson, N., Quirke, L., Lawrence, S., Reader, K., et al. (2004). The oocyte and its role in regulating ovulation rate: A new paradigm in reproductive biology. *Reproduction, 128*, 379–386.

Menon, K. M. J., & Menon, B. (2014). Regulation of luteinizing hormone receptor expression by an RNA binding protein: Role of ERK signaling. *The Indian Journal of Medical Research, 140*, 112–119.

Miller, W. L., & Auchus, R. J. (2011). The molecular biology, biochemistry, and physiology of human steroidogenesis and its disorders. *Endocrine Reviews, 32*(1), 81–151.

Mitsui, Y., Hirata, H., Arichi, N., Hiraki, M., Yasumoto, H., Chang, I., et al. (2015). Inactivation of bone morphogenetic protein 2 may predict clinical outcome and poor overall survival for renal cell carcinoma through epigenetic pathways. *Oncotarget, 6*, 9577–9591.

Miyazono, K., Kamiya, Y., & Morikawa, M. (2010). Bone morphogenetic protein receptors and signal transduction. *Journal of Biochemistry, 147*, 35–51.

Miyazono, K., Maeda, S., & Imamura, T. (2005). BMP receptor signaling: Transcriptional targets, regulation of signals, and signaling cross-talk. *Cytokine & Growth Factor Reviews, 6*, 251–263.

Monniaux, D., Clément, F., Dalbiès-Tran, R., Estienne, A., Fabre, S., Mansanet, C., et al. (2014). The ovarian reserve of primordial follicles and the dynamic reserve of antral growing follicles: What is the link? *Biology of Reproduction, 90*, 85. 81–11.

Moore, R. K., Otsuka, F., & Shimasaki, S. (2001). Role of ERK1/2 in the differential synthesis of progesterone and estradiol by granulosa cells. *Biochemical and Biophysical Research Communications, 289*, 796–800.

Moore, R. K., Otsuka, F., & Shimasaki, S. (2003). Molecular basis of bone morphogenetic protein-15 signaling in granulosa cells. *The Journal of Biological Chemistry, 278*, 304–310.

Mulsant, P., Lecerf, F., Fabre, S., Schibler, L., Monget, P., Lanneluc, I., et al. (2001). Mutation in bone morphogenetic protein receptor-IB is associated with increased ovulation rate in Booroola Mérino ewes. *Proceedings of the National Academy of Sciences of the United States of America*, *98*, 5104–5109.

Nicol, L., Faure, M. O., McNeilly, J. R., Fontaine, J., Taragnat, C., & McNeilly, A. S. (2008). Bone morphogenetic protein-4 interacts with activin and GnRH to modulate gonadotrophin secretion in LbetaT2 gonadotrophs. *Journal of Endocrinology*, *196*, 497–507.

Nielsen, M. E., Rasmussen, I. A., Kristensen, S. G., Christensen, S. T., Møllgård, K., Wreford Andersen, E., et al. (2011). In human granulosa cells from small antral follicles, androgen receptor mRNA and androgen levels in follicular fluid correlate with FSH receptor mRNA. *Molecular Human Reproduction*, *17*, 63–70.

Nilsson, E. E., & Skinner, M. K. (2003). Bone morphogenetic protein-4 acts as an ovarian follicle survival factor and promotes primordial follicle development. *Biology of Reproduction*, *69*, 1265–1272.

Ogura Nose, S., Yoshino, O., Osuga, Y., Shi, J., Hiroi, H., Yano, T., et al. (2012). Anti-Mullerian hormone (AMH) is induced by bone morphogenetic protein (BMP) cytokines in human granulosa cells. *European Journal of Obstetrics, Gynecology, and Reproductive Biology*, *164*, 44–47.

Ophir, L., Yung, Y., Maman, E., Rubinstein, N., Yerushalmi, G. M., Haas, J., et al. (2014). Establishment and validation of a model for non-luteinized human mural granulosa cell culture. *Molecular and Cellular Endocrinology*, *384*, 165–174.

Otsuka, F. (2010). Multiple endocrine regulation by bone morphogenetic protein system. *Endocrine Journal*, *57*, 3–14.

Otsuka, F. (2013). Multifunctional bone morphogenetic protein system in endocrinology. *Acta Medica Okayama*, *67*, 75–86.

Otsuka, F., Moore, R. K., & Shimasaki, S. (2001). Biological function and cellular mechanism of bone morphogenetic Protein-6 in the ovary. *The Journal of Biological Chemistry*, *276*, 32889–32895.

Otsuka, F., & Shimasaki, S. (2002). A novel function of bone morphogenetic protein-15 in the pituitary: Selective synthesis and secretion of FSH by gonadotropes. *Endocrinology*, *143*, 4938–4941.

Otsuka, F., Yamamoto, S., Erickson, G., & Shimasaki, S. (2001). Bone morphogenetic protein-15 inhibits follicle-stimulating hormone (FSH) action by suppressing FSH receptor expression. *The Journal of Biological Chemistry*, *276*, 11387–11392.

Pangas, S. A. (2012). Regulation of the ovarian reserve by members of the transforming growth factor beta family. *Molecular Reproduction and Development*, *79*, 666–679.

Picton, H. M., & McNeilly, A. S. (1991). Evidence to support a follicle-stimulating hormone threshold theory for follicle selection in ewes chronically treated with gonadotrophin-releasing hormone agonist. *Journal of Reproduction and Fertility*, *93*, 43–51.

Pierre, A., Pisselet, C., Dupont, J., Mandon-Pepin, B., Monniaux, D., Monget, P., et al. (2004). Molecular basis of bone morphogenetic protein-4 inhibitory action on progesterone secretion by ovine granulosa cells. *Journal of Molecular Endocrinology*, *33*, 805–814.

Pulkki, M., Mottershead, D., Pasternack, A., Muggalla, P., Ludlow, H., van Dinther, M., et al. (2012). A covalently dimerized recombinant human bone morphogenetic protein-15 variant identifies bone morphogenetic protein receptor type 1B as a key cell surface receptor on ovarian granulosa cells. *Endocrinology*, *153*, 1509–1518.

Raz, E. (2003). Primordial germ-cell development: The zebrafish perspective. *Nature Reviews. Genetics*, *4*, 690–700.

Reddy, P., Liu, L., Adhikari, D., Jagarlamudi, K., Rajareddy, S., Shen, Y., et al. (2008). Oocyte-specific deletion of Pten causes premature activation of the primordial follicle pool. *Science*, *319*, 611–613.

Regan, S. L. P., Knight, P. G., Yovich, J., Stanger, J., Leung, Y., Arfuso, F., et al. (2016). Dysregulation of granulosal bone morphogenetic protein receptor 1B density is associated with reduced ovarian reserve and the age-related decline in human fertility. *Molecular and Cellular Endocrinology*, *425*, 84–93.

Regan, S. L. P., Knight, P. G., Yovich, J. L., Stanger, J. D., Leung, Y., Arfuso, F., et al. (2017). Infertility and ovarian follicle reserve depletion are associated with dysregulation of the FSH and LH receptor density in human antral follicles. *Molecular and Cellular Endocrinology*, *446*, 40–51.

Regan, S. L. P., McFarlane, J. R., O'Shea, T., Andronicos, N., Arfuso, F., Dharmarajan, A., et al. (2015). Flow cytometric analysis of FSHR, BMPR1B, LHR and apoptosis in granulosa cells and ovulation rate in merino sheep. *Reproduction*, *150*, 151–163.

Rice, S., Ojha, K., Whitehead, S., & Mason, H. (2007). Stage-specific expression of androgen receptor, follicle-stimulating hormone receptor, and anti-müllerian hormone type II receptor in single, isolated, human preantral follicles: Relevance to polycystic ovaries. *The Journal of Clinical Endocrinology and Metabolism*, *92*, 1034–1040.

Richards, J. S. (1994). Hormonal control of gene expression in the ovary. *Endocrine Reviews*, *15*, 725–751.

Richardson, B. E., & Lehmann, R. (2010). Mechanisms guiding primordial germ cell migration: Strategies from different organisms. *Nature Reviews. Molecular Cell Biology*, *11*, 37–49.

Robinson, R. S., Woad, K. J., Hammond, A. J., Laird, M., Hunter, M. G., & Mann, G. E. (2009). Angiogenesis and vascular function in the ovary. *Reproduction*, *138*, 869–881.

Rodgers, R. J., & Irving-Rodgers, H. F. (2010). Formation of the ovarian follicular antrum and follicular fluid. *Biology of Reproduction*, *82*, 1021–1029.

Ruoss, C., Tadros, A., O'Shea, T., McFarlane, J., & Almahbobi, G. (2009). Ovarian follicle development in Booroola sheep exhibiting impaired bone morphogenetic protein signalling pathway. *Reproduction*, *138*, 689–696.

Sen, A., & Hammes, S. (2010). Granulosa cell-specific androgen receptors are critical regulators of ovarian development and function. *Molecular Endocrinology*, *24*, 1393–1403.

Sen, A., Prizant, H., Light, A., Biswas, A., Hayes, E., Lee, H.-J., et al. (2014). Androgens regulate ovarian follicular development by increasing follicle stimulating hormone receptor and microRNA-125b expression. *Proceedings of the National Academy of Sciences of the United States of America*, *111*(8), 3008–3013.

Shi, J., Yoshino, O., Osuga, Y., Akiyama, I., Harada, M., Koga, K., et al. (2012). Growth differentiation factor 3 is induced by bone morphogenetic protein 6 (BMP-6) and BMP-7 and increases luteinizing hormone receptor messenger RNA expression in human granulosa cells. *Fertility and Sterility*, *97*, 979–983.

Shi, J., Yoshino, O., Osuga, Y., Koga, K., Hirota, Y., Hirata, T., et al. (2009). Bone morphogenetic protein-6 stimulates gene expression of follicle-stimulating hormone receptor, inhibin/activin beta subunits, and anti-Müllerian hormone in human granulosa cells. *Fertility and Sterility*, *92*, 1794–1798.

Shi, J., Yoshino, O., Osuga, Y., Koga, K., Hirota, Y., Nose, E., et al. (2011). Bone morphogenetic protein-2 (BMP-2) increases gene expression of FSH receptor and aromatase and decreases gene expression of LH receptor and StAR in human granulosa cells. *American Journal of Reproductive Immunology*, *1989*(65), 421–427.

Shi, J., Yoshino, O., Osuga, Y., Nishii, O., Yano, T., & Taketani, Y. (2010). Bone morphogenetic protein 7 (BMP-7) increases the expression of follicle-stimulating hormone (FSH) receptor in human granulosa cells. *Fertility and Sterility*, *93*, 1273–1279.

Shimasaki, S., Moore, R., Otsuka, F., & Erickson, G. (2004). The bone morphogenetic protein system in mammalian reproduction. *Endocrine Reviews*, *25*, 72–101.

Shimasaki, S., Zachow, R. J., Li, D., Kim, H., Iemura, S.-I., Ueno, N., et al. (1999). A functional bone morphogenetic protein system in the ovary. *Proceedings of the National Academy of Sciences of the United States of America, 96*, 7282–7287.

Skinner, M. K. (2005). Regulation of primordial follicle assembly and development. *Human Reproduction Update, 11*, 461–471.

Slot, K. A., Kastelijn, J., Bachelot, A., Kelly, P. A., Binart, N., & Teerds, K. J. (2006). Reduced recruitment and survival of primordial and growing follicles in GH receptor-deficient mice. *Reproduction, 131*, 525–532.

Souza, C., Campbell, B., McNeilly, A., & Baird, D. (2002). Effect of bone morphogenetic protein 2 (BMP2) on oestradiol and inhibin A production by sheep granulosa cells, and localization of BMP receptors in the ovary by immunohistochemistry. *Reproduction, 123*, 363–369.

Souza, C., MacDougall, C., Campbell, B., McNeilly, A., & Baird, D. (2001). The Booroola (FecB) phenotype is associated with a mutation in the bone morphogenetic receptor type 1 B (BMPR1B) gene. *The Journal of Endocrinology, 169*, R1–R6.

Suzuki, N., Yoshioka, N., Takae, S., Sugishita, Y., Tamura, M., Hashimoto, S., et al. (2015). Successful fertility preservation following ovarian tissue vitrification in patients with primary ovarian insufficiency. *Human Reproduction, 30*, 608–615.

Sweeney, M. D., Ayyadurai, S., & Zlokovic, B. V. (2016). Pericytes of the neurovascular unit: Key functions and signaling pathways. *Nature Neuroscience, 19*, 771–783.

Tajima, K., Dantes, A., Yao, Z., Sorokina, K., Kotsuji, F., Seger, R., et al. (2003). Down-regulation of steroidogenic response to gonadotropins in human and rat preovulatory granulosa cells involves mitogen-activated protein kinase activation and modulation of DAX-1 and steroidogenic factor-1. *The Journal of Clinical Endocrinology and Metabolism, 88*, 2288–2299.

Tanwar, P., & McFarlane, J. (2011). Dynamic expression of bone morphogenetic protein 4 in reproductive organs of female mice. *Reproduction, 142*, 573–579.

Taylor, P. D., Wilson, H., Hillier, S. G., Wiegand, S. J., & Fraser, H. M. (2007). Effects of inhibition of vascular endothelial growth factor at time of selection on follicular angiogenesis, expansion, development and atresia in the marmoset. *Molecular Human Reproduction, 13*, 729–736.

Val, P., Lefrançois-Martinez, A. M., Veyssière, G., & Martinez, A. (2003). SF-1 a key player in the development and differentiation of steroidogenic tissues. *Nuclear Receptor, 1*, 8.

van Rooij, I. A. J., Broekmans, F. J. M., Scheffer, G. J., Looman, C. W. N., Habbema, J. D. F., de Jong, F. H., et al. (2005). Serum antimüllerian hormone levels best reflect the reproductive decline with age in normal women with proven fertility: A longitudinal study. *Fertility and Sterility, 83*, 979–987.

Vitt, U., Hayashi, M., Klein, C., & Hsueh, A. (2000). Growth differentiation factor-9 stimulates proliferation but suppresses the follicle-stimulating hormone-induced differentiation of cultured granulosa cells from small antral and preovulatory rat follicles. *Biology of Reproduction, 62*, 370–377.

Weenen, C., Laven, J. S. E., von Bergh, A. R. M., Cranfield, M., Groome, N. P., Visser, J. A., et al. (2004). Anti-Mullerian hormone expression pattern in the human ovary: Potential implications for initial and cyclic follicle recruitment. *Molecular Human Reproduction, 10*, 77–83.

Westergaard, L., Christensen, I. J., & McNatty, K. P. (1986). Steroid levels in ovarian follicular fluid related to follicle size and health status during the normal menstrual cycle in women. *Human Reproduction, 1*, 227–232.

Wilson, T., Wu, X.-Y., Juengel, J. L., Ross, I. K., Lumsden, J. M., Lord, E. A., et al. (2001). Highly prolific Booroola sheep have a mutation in the intracellular kinase domain of bone morphogenetic protein IB receptor (ALK-6) that is expressed in both oocytes and granulosa cells. *Biology of Reproduction, 64*, 1225–1235.

Wu, Y.-T., Tang, L., Cai, J., X-E, L., Xu, J., Zhu, X.-M., et al. (2007). High bone morphogenetic protein-15 level in follicular fluid is associated with high quality oocyte and subsequent embryonic development. *Human Reproduction, 22*, 1526–1531.

Xu, Z., Garverick, H. A., Smith, G. W., Smith, M. F., Hamilton, S. A., & Youngquist, R. S. (1995). Expression of follicle-stimulating hormone and luteinizing hormone receptor messenger ribonucleic acids in bovine follicles during the first follicular wave. *Biology of Reproduction, 53*, 951–957.

Yding Andersen, C. (2017). Inhibin-B secretion and FSH isoform distribution may play an integral part of follicular selection in the natural menstrual cycle. *MHR: Basic Science of Reproductive Medicine, 23*, 16–24.

Ying, Y., & Zhao, G. Q. (2001). Cooperation of endoderm-derived BMP2 and extraembryonic ectoderm-derived BMP4 in primordial germ cell generation in the mouse. *Developmental Biology, 232*, 484–492.

Yuan, W., & Giudice, L. (1997). Programmed cell death in human ovary is a function of follicle and corpus luteum status. *The Journal of Clinical Endocrinology and Metabolism, 82*, 3148–3155.

Zeleznik, A. J. (2001). Follicle selection in primates: "many are called but few are chosen". *Biology of Reproduction, 65*, 655–659.

Zhang, Y.-M., & Roy, S. K. (2004). Downregulation of follicle-stimulating hormone (fsh)-receptor messenger rna levels in the hamster ovary: Effect of the endogenous and exogenous FSH. *Biology of Reproduction, 70*, 1580–1588.

Zhu, G., Cui, Y., Qinglin-Wang, Yonggang-Kang, Yanzhi-Lv, Wang, G., et al. (2013). Bone morphogenetic proteins (BMP) 2, 4, 6 and 7 affect ovarian follicular development through regulation of follicle-stimulating hormone receptor (FSHR) and luteinizing hormone receptor (LHR) expression in goat granulosa cells. *Journal of Cell Biology and Genetics, 3*(1), 14–21.

CHAPTER TEN

Gonadotropin Surge-Attenuating Factor: A Nonsteroidal Ovarian Hormone Controlling GnRH-Induced LH Secretion in the Normal Menstrual Cycle

Ioannis E. Messinis[1], Christina I. Messini, George Anifandis, Antonios Garas, Alexandros Daponte
Faculty of Medicine, School of Health Sciences, University of Thessaly, Larissa, Greece
[1]Corresponding author: e-mail address: messinis@med.uth.gr

Contents

Abstract

Gonadotropin surge-attenuating factor (GnSAF) is a nonsteroidal ovarian substance, which attenuates the endogenous LH surge in superovulated women. Different molecular sequences have been found, but only one of them has shown substantial homology to a known substance of the human genome. A molecular mass of 12.5 kDa showing identity to the carboxyl-terminal fragment of human serum albumin and expressing GnSAF bioactivity in vitro has been identified. It has been suggested that in the normal menstrual cycle the in vivo bioactivity of GnSAF increases under the influence of the intercycle rise of FSH. GnSAF is considered the "missing link" between the ovaries and the hypothalamo-pituitary system, maintaining the pituitary in a state of low responsiveness to GnRH in the early- to midfollicular phase of the cycle. A marked decline in GnSAF bioactivity in the late follicular phase facilitates the onset and the full expression of the midcycle LH surge.

Vitamins and Hormones, Volume 107
ISSN 0083-6729
https://doi.org/10.1016/bs.vh.2018.01.002

1. INTRODUCTION

Gonadotropin surge-attenuating factor (GnSAF) was defined for the first time in 1989 as a nonsteroidal substance produced by the human ovaries, which under conditions of multiple follicular development attenuates the endogenous LH surge (Messinis & Templeton, 1989). The evidence for the existence of GnSAF was based on data obtained from studies in superovulated women undergoing in vitro fertilization (IVF) treatment. In the first two published studies, clomiphene citrate plus FSH or human menopausal gonadotropin (HMG) were used, but subsequently similar information was also obtained from studies using FSH alone (Messinis & Templeton, 1986; Messinis, Templeton, & Baird, 1985, 1986). In all these studies, it was shown that the endogenous LH surge was markedly attenuated both in amplitude and in duration, although on several occasions the LH surge was completely abolished (Messinis et al., 1985, 1986; Messinis & Templeton, 1986). Additionally, a marked suppression of the LH response to GnRH was found in superovulated women and this is possibly part of the mechanism of the LH surge attenuation (Messinis & Templeton, 1989). From these studies, it became apparent for the first time that a novel ovarian mechanism controlling gonadotropin secretion was present, the activity of which was enhanced by the superovulation induction process.

Human studies were preceded by studies in animals. In the late 1970s, there were a few case reports in monkeys on the blockage of the LH surge under various conditions; however, observations were first made during superovulation induction (Goodman & Hodgen, 1978). In these animals, multiple follicular development resulted in the blockage of the endogenous LH surge with parallel marked reduction of the LH response to GnRH. Especially, although administration of HMG to cynomolgus monkeys increased markedly serum estradiol levels, in none of the animals a LH surge occurred (Littman & Hodgen, 1984). It was subsequently shown that the GnRH-induced LH secretion was suppressed in oophorectomized monkeys by the administration of porcine follicular fluid, indicating that this inhibitory activity was derived from the ovaries (Sopelak & Hodgen, 1984). It was proposed then that the ovaries of these animals produced a substance, named gonadotropin surge-inhibiting factor, which blocked the positive feedback mechanism (GnSIF) (Schenken & Hodgen, 1983). However, it is still unclear if GnSIF in monkeys is the same with GnSAF in women. This distinction is important because, in contrast to monkeys, in women with superovulated ovaries the endogenous LH surge is not always blocked. It has been reported that in about 20% of superovulated cycles an endogenous LH surge takes place (Janssens et al., 2000; Messinis & Templeton, 1987a; Wang et al., 2016), but it is markedly attenuated and therefore the term GnSAF as opposed to GnSIF is more appropriate for women. However, due to the similarities in the action of these two putative factors on the secretion of LH from the pituitary, some of the researches, who have attempted to purify the candidate molecule, have used a combined term GnSAF/IF.

As it was mentioned earlier, in the first two studies in women, superovulation was induced with the administration of clomiphene plus HMG or FSH (Messinis et al., 1985, 1986). The dose of clomiphene was 150 mg/day for 5 days followed by daily injections of 225 IU FSH or HMG. Measurement of LH was performed in 6-hourly blood samples collected from the time point at which the leading follicle reached the size of 14 mm in diameter by ultrasound. In these cycles, an endogenous LH surge was detected invariably when the leading follicle reached the diameter of 18 mm. Although the LH surge was markedly attenuated, luteinization with normal increase in serum progesterone levels was noted in all cases. However, it was subsequently understood that the occurrence of the endogenous LH surge was dependent on the treatment protocol used for superovulation induction, as it will be discussed later.

2. OVARIAN STIMULATION

Nowadays, ovarian stimulation with the use of FSH is applied to the majority of the women undergoing IVF treatment. With this approach, multiple follicular development takes place, which leads to profound changes in the secretion of reproductive hormones. The most prominent hormonal change is the rapid increase in serum estradiol concentrations to supraphysiological levels (Messinis & Templeton, 1986). Additionally, inhibin bioactivity is significantly increased (Tsonis, Messinis, Templeton, McNeilly, & Baird, 1988). These changes lead to a potentiation of the ovarian-negative feedback mechanism, and as a consequence the secretion of the two gonadotropins from the pituitary is markedly suppressed (Messinis, 2006). This reduction in LH secretion has been clearly shown in several studies, but the suppression of FSH is not evident, because of the exogenous administration of this hormone. Serum LH levels start to decline from the beginning of the FSH treatment, and the suppression persists through to midcycle and during the luteal phase, where the negative feedback mechanism is further enhanced by the hypersecretion of progesterone from the multiple corpora lutea (Messinis et al., 1998; Messinis & Templeton, 1987b). The rapid increase in serum estradiol concentrations during FSH treatment would be also expected to result in the potentiation of the positive feedback mechanism, as the threshold level of estradiol is exceeded soon after the onset of ovarian stimulation. Nevertheless, as opposed to the negative, the positive feedback mechanism in FSH superovulated cycles is weakened leading to the blockage or the marked attenuation of the endogenous LH surge (Messinis, 2006).

It has become evident that the occurrence of an endogenous LH surge in superovulated cycles, in which neither agonists nor antagonists are used, depends on the protocol applied for ovarian stimulation. Specifically, when clomiphene citrate is used either alone or sequentially with gonadotropins, the LH surge occurs invariably, but it is always markedly attenuated (Messinis et al., 1985, 1986). In the same cycles, the attenuated LH surge is timed by the ovulatory follicle, i.e., the surge starts when the leading follicle becomes on average 18 mm in diameter. However, when FSH is given alone without clomiphene, the LH surge may be blocked or may occur prematurely, i.e., before full follicle maturation to the ovulatory stage and it is again markedly attenuated (Glasier, Thatcher, Wickings, Hillier, & Baird, 1988; Messinis & Templeton, 1986, 1987a). The difference in the time of onset of the LH surge between a regimen containing FSH alone and a regimen containing clomiphene plus FSH can be explained by the antiestrogenic properties of clomiphene. This drug is usually administered for 5 days, but due to its long half-life it occupies pituitary estrogen receptors for a longer period of time, so that estradiol becomes unable to trigger the positive feedback mechanism for several days. It is estimated that by the time the estrogen receptors are cleared from clomiphene the leading follicle has become ovulatory in size and this coincides with the onset of the endogenous LH surge elicited by estradiol. When clomiphene is given for a longer than 5 days period, a LH surge is not displayed despite the marked increase in estradiol levels (Messinis & Templeton, 1988). In that case, the blockage of the LH surge is due to the persistent clomiphene action, although GnSAF may also contribute. Such protocols have been used successfully for ovarian stimulation without the need for the use of GnRH analogs (Kato et al., 2012).

An attenuated LH surge has specific characteristics with markedly reduced amplitude and duration (~24 h). Its clinical importance has not been fully studied. Nevertheless, luteinization with increased serum progesterone levels occurs invariably, while conception has taken place after fertilization in vitro of the collected oocytes (Messinis et al., 1985). This approach, however, i.e., oocyte recovery based on the onset of an attenuated LH surge, is inconvenient and although it was used at the beginning of IVF development it was soon abandoned.

To prevent a premature LH surge, in the early 1980s, the GnRH agonists were introduced in the ovarian stimulation protocols and a little later the antagonists. Although systematic reviews and meta-analyses have not shown any significant difference in the clinical outcome of IVF/intracytoplasmic sperm injection (ICSI) treatment between agonists and antagonists cycles (Al-Inany et al., 2011; Xiao, Su, & Zeng, 2014), there are differences in the efficacy of these two types of drugs to prevent a premature LH surge during stimulation with FSH. Regarding the agonists, they have been proven quite effective in preventing a premature LH surge even when they are given in lower than the standard doses (Dal Prato, Borini, Coticchio, Cattoli, & Flamigni, 2004; Janssens et al., 2000). In contrast, with the antagonists several LH peaks have been detected during treatment with these drugs (Albano et al., 2000; Borm & Mannaerts, 2000; Messinis et al., 2005), which are possibly due to the inability of the GnRH antagonists to block the positive feedback effect of estradiol (Messinis et al., 2010). This in conjunction with the limited efficacy, in terms of duration of action, of the single daily dose of 0.25 mg of the antagonists to protect the pituitary from the stimulating action of the rising estradiol levels (Griesinger et al., 2006) indicates that during ovarian stimulation there are short periods of time between two injections of an antagonist at which the pituitary remains unprotected from the stimulating action of estradiol. This can activate the positive feedback mechanism leading to the occurrence of a LH surge. Since, however, only isolated LH peaks and no real LH surges take place in GnRH antagonists cycles, it is suggested that GnSAF produced in high amounts acts synergistically with the GnRH antagonist on the pituitary and therefore an abortive rather than a clearly defined LH surge is displayed (Messinis, Messini, Dimitraki, Anifandis, & Dafopoulos, 2015).

3. ISOLATION OF GNSAF

Although the first indication for the existence of GnSAF was presented several years ago, up to now this factor has been only partially characterized, while there is no consensus regarding the structure of this molecule. Several attempts have been made to isolate GnSAF from various biological materials. In the first attempt, human follicular fluid obtained from superovulated women was used (Fowler, Templeton, & Messinis, 1993). By applying serial ultrafiltration, a molecule of 10–30 kDa was found. Subsequently, a sequence demonstrating GnSAF/IF bioactivity in vitro was purified from Sertoli cell-conditioned medium as a monomeric protein with a molecular mass of 37 kDa (Tio, Koppenaal, Bardin, & Cheng, 1994). However, the isolated molecule in these experiments also showed inhibin activity by inhibiting not only GnRH-induced LH but also basal FSH secretion. A little later, a 69-kDa monomeric protein demonstrating GnSAF bioactivity in vitro was purified from porcine follicular fluid (Danforth & Cheng, 1995). An antibody raised against that protein identified in human follicular fluid two GnSAF/IF-like homologues of 63 and 59 kDa, respectively (Mroueh et al., 1996).

A subsequent study, using several biochemical steps, reported the purification of GnSAF from human follicular fluid as a molecule of 12.5 kDa (Pappa et al., 1999). That study demonstrated identity of GnSAF with the carboxyl (C) terminal fragment of human serum albumin (HSA), and it was the first that showed homology to a known protein of the human genome. The bioactivity of GnSAF was tested in an in vitro bioassay, using rat pituitary cells in culture, and was found to be different from that of inhibin. That the candidate molecule showed identity to the C-terminal fragment of HSA is not exceptional, as different substances, which are fragments of larger molecules, have shown activities that are not expressed by the mother molecules (O'Reilly et al., 1994, 1997). Following these findings, experiments were performed with the use of human luteinized granulosa cells obtained from women undergoing IVF treatment and the results supported the theory of the C-terminal fragment of HSA as a GnSAF candidate. Especially, in these experiments, the expression of the mRNA of HSA was demonstrated in the granulosa cells (Karligiotou, Kollia, Kallitsaris, & Messinis, 2006). With the use of reverse transcription polymerase chain reaction, the expression of mRNA transcripts corresponding to the NH2- and the C-terminal fragments of HSA was found in the cytoplasm, while full-length mRNA transcripts of HSA were not detected in the granulosa cells. More recently, with the use of in situ hybridization and immunocytochemistry, mRNA transcripts corresponding to the C-terminal fragment of HSA were detected in human luteinized granulosa cells in high intensity that after the addition of FSH became similar to that in hepatic HepG2 cells, which were used as a positive control (Karligiotou et al., 2011). FSH acting on the ovaries is considered the principal stimulator of GnSAF production during superovulation induction. Furthermore, recombinant polypeptides corresponding to the subdomain IIIB of HSA, which were produced with the use of the pichia pastoris expression and secretion system, demonstrated bioactivity of GnSAF in an in vitro bioassay system (Tavoulari, Frillingos, Karatza, Messinis, & Seferiadis, 2004). Such an activity was not shown by the other domains or the full length of HSA. These results provided further support to the initial findings that GnSAF corresponds to the C-terminal fragment of HSA.

Two subsequent studies on the purification of GnSAF have contributed further to the existing confusion in terms of the nature of GnSAF. One of them reported two new sequences of 60–66 kDa from human granulosa cell-conditioned medium (Fowler et al., 2002). The other and more recent study reported two different bioactive molecules one heteromeric of 160 kDa and one monomeric of 40 kDa, both showing GnSAF bioactivity in vitro (Hendriks, Lambalk, Helder, & de Koning, 2011). In that study, the mitogen-activated protein kinases (MAPK) phosphorylation pathway was attenuated by the candidate molecules. Of these two molecules, the NH2-terminus of the high molecular weight showed high homology with human β-chain of the complement C3, while that of the low molecular weight with other proteins, members of the DING family. Based on these findings, a synthetic molecule of 26 amino acids was produced, which exerted a significant inhibitory action on GnRH-induced LH secretion in vitro without affecting basal FSH levels (Hendriks et al., 2011). No further research has been reported with this synthetic molecule.

It is evident from the above discussion that further experiments are needed for the purification of GnSAF to homogeneity. Although there are great differences in the molecules proposed so far as putative GnSAF candidates, clinical research studies have invariably demonstrated the increased in vivo bioactivity of this factor during superovulation induction with FSH (Messinis, 2006). It is evident that GnSAF does not affect basal secretion of LH, but it attenuates LH secretion regulated by the pulsatile action of GnRH.

4. PHYSIOLOGICAL ROLE OF GnSAF

Accumulating evidence has indicated that GnSAF is an important component of the feedback loop, playing a physiological role during the normal menstrual cycle. Although in this context GnSAF exerts a negative effect on LH secretion, this factor's action seems to be more important during the positive feedback mechanism.

4.1 Negative Feedback

It has been established that the negative effect of the ovaries on the hypothalamo-pituitary system is mediated by the two steroids, estradiol, and progesterone, although nonsteroidal substances, such as inhibin, also participate. The principal mediator of the negative feedback effect during the follicular phase of the cycle is estradiol and during the luteal phase progesterone (Messinis, Messini, & Dafopoulos, 2014). Several clinical research studies have investigated the role of estradiol either after the exogenous administration of this substance in normally cycling or postmenopausal women or after the removal of endogenous estrogen, such as following ovariectomy (Alexandris et al., 1997; Tsai & Yen, 1971; Yen & Tsai, 1971). An additional way to study the role of estradiol is to increase the endogenous production of this steroid, as it happens during ovarian stimulation with FSH for IVF treatment, where a marked suppression in serum LH levels occurs (Messinis et al., 1998). Information obtained from in vivo experiments in women has demonstrated that circulating progesterone in the normal follicular phase also contributes to the control of LH secretion particularly in the context of the negative feedback (Dafopoulos et al., 2004). This has been also shown in experiments performed in ovariectomized premenopausal women treated immediately after the operation with estradiol plus progesterone or in normally menstruating women treated with the antiprogestagen compound, mifepristone, in the follicular phase of the cycle (Messinis, Krishnan, Kazem, Khadilkar, & Templeton, 1997; Messinis et al., 2002). The amount of progesterone produced by the follicles is minimal and this explains the low circulating levels of this steroid in the follicular phase. Although part of this amount is derived from the adrenals, the contribution of the ovaries is also significant (Alexandris et al., 1997).

In the luteal phase of the cycle, progesterone is produced in high amounts by the corpus luteum. When normally menstruating women were ovariectomized in the midluteal phase, the immediate increase in FSH and LH levels after the operation was not prevented by estradiol replacement alone, but only when estradiol was combined with progesterone (Messinis et al., 2002). This suggests that, although estradiol plays an endocrine role in the luteal phase, progesterone is the principal mediator of the ovarian negative effect on the pituitary at that stage of the cycle.

Experiments in animals have shown the role of inhibin in the control of FSH secretion (Molskness, Woodruff, Hess, Dahl, & Stouffer, 1996; Rivier, Rivier, & Vale, 1986). Inhibin B produced by the follicles, which are recruited under the intercycle rise of FSH or "FSH window," participates at the beginning of the cycle in the negative feedback mechanism. On the other hand, inhibin A, produced mainly by the corpus luteum, participates in the control of FSH secretion during the luteal phase. It has been suggested, however, that the two inhibins are important regulators of the intercycle rise of FSH, with inhibin A playing a role in the opening and inhibin B in the closure of the "FSH window" (Messinis, 2006). Other nonsteroidal ovarian substances, such as activin and follistatin, do not seem to have an endocrine role. Follistatin is a carrier protein for activin, while the latter is released from the complex with follistatin in different tissues exerting local paracrine actions (de Kretser, Hedger, Loveland, & Phillips, 2002).

Regarding GnSAF, this factor is produced during the early follicular phase of the cycle. Nevertheless, as it will be discussed later, the role of GnSAF is evident in the context of the positive rather than the negative feedback mechanism by interacting with the estradiol action on the hypothalamo-pituitary system. All in vitro experiments performed so far in the different studies have not shown any effect of GnSAF on the basal secretion of LH, suggesting that this factor is not a component of the negative feedback mechanism.

4.2 Positive Feedback

It has been established that estradiol mediates the positive action of the ovaries on the hypothalamo-pituitary system. The positive feedback is a unique characteristic of the normal menstrual cycle that is responsible for the occurrence of the midcycle LH surge, maintaining thus the integrity of the cycle. This pivotal role of estradiol has been demonstrated in in vivo experiments in women following the exogenous administration of this hormone (Liu & Yen, 1983). This means that by increasing serum estradiol to levels similar to those at the preovulatory stage within at least 48 h an endogenous LH surge is displayed, the characteristics of which resemble those of the midcycle surge, although it may be smaller in amplitude and shorter in duration. During the normal menstrual cycle, estradiol levels increase gradually from the early- to the late follicular phase, while this steroid sensitizes the pituitary to GnRH during the same period of time. The latter information has been derived mainly from experiments involving the iv administration of two submaximal pulses of GnRH, 10 μg each, 2 h apart, during the follicular phase of the cycle (Lasley, Wang, & Yen, 1975). The response to the second as compared to the first pulse is enhanced only when the environment is highly estrogenic, as it happens in the late follicular phase or in postmenopausal women after the administration of exogenous estrogen. This phenomenon is called "self-priming" effect of GnRH on the pituitary and is crucial for the full expression of the positive feedback effect and the occurrence of the midcycle LH surge. In such an experimental model, the response to the first GnRH pulse is maximal at 30 min and this value represents the pituitary sensitivity to GnRH with LH being secreted from the so-called "first pool" of the gonadotrophs, while the whole area under the curve following the two injections represents the pituitary reserve with LH being secreted from the "second pool" or "reserve pool" (Wang, Lasley, Lein, & Yen, 1976). Estradiol during the follicular phase of the cycle builds up the pituitary reserves and increases the duration of LH secretory events and consequently the secreted LH mass (Quyyumi, Pinkerton, Evans, & Veldhuis, 1993). This enhancement of LH secretion by estradiol is very important for the occurrence of the endogenous LH surge.

Part of the mechanism via which estradiol sensitizes the pituitary to GnRH is the increase in the number of GnRH receptors on the pituitary gonadotrophs (Laws, Webster, & Miller, 1990). Another possibility is an estradiol-dependent reduced metabolism of GnRH, which increases the availability of this hormone to the gonadotrophs (Danforth, Elkind-Hirsch, & Hodgen, 1990). Estradiol may also reduce the threshold of GnRH, which is required for the pituitary to become sensitive to this hypothalamic hormone (Ramey, Highsmith, Wilfinger, & Baldwin, 1987).

The low circulating levels of progesterone during the follicular phase also seem to participate in the process of pituitary sensitization to GnRH and thus in the mechanism of the occurrence of the midcycle LH surge. Relevant information has been obtained from experiments in women after the administration of the progesterone antagonist, mifepristone, during the follicular phase of the cycle (Kazem et al., 1996). In these experiments, the response of LH and FSH to two small pulses of GnRH injected iv 2 h apart was reduced significantly in the cycles in which the women received mifepristone. Data from studies in animals have indicated that progesterone is also produced at the central nervous system under the influence of estradiol (Soma, Sinchak, Lakhter, Schlinger, & Micevych, 2005). It has been demonstrated from studies in knockout mice for the progesterone receptor that this neuroprogesterone produced under these conditions in the hypothalamus as well as its receptors are obligatory for the generation of the positive feedback mechanism (Chappell et al., 1999; Micevych et al., 2003). Whether a similar mechanism is also operating in women's hypothalamus is not known. Apart from a possible role of progesterone for the onset of the LH surge, the consequent increase in the levels of this steroid is critical for the amplification, but also for the termination of the LH surge (Dafopoulos et al., 2006; Messinis & Templeton, 1990).

4.3 GnSAF

It has been established that activation of the pituitary gonadotrophs by the GnRH pulses is crucial for the secretion of LH during the normal menstrual cycle. In cycles superovulated with the exogenous administration of FSH, the LH response to GnRH is markedly reduced (Messinis & Templeton, 1989). In the already described experimental model of the two GnRH pulses, which differentiate between the two functionally related pools of the pituitary gonadotrophs, it has been shown that in superovulated with FSH cycles, the response of LH to the two GnRH pulses is markedly suppressed despite the supraphysiological concentrations of estradiol (Messinis & Templeton, 1991). Although it would be expected in the presence of the very high estradiol levels the pituitary gonadotrophs to be highly responsive to GnRH, the opposite was found. This suggests that GnSAF affects both the pituitary sensitivity and the pituitary reserve reducing markedly the secretion of LH. Based on this information, the reduced response to the first GnRH pulse (10 μg) iv at 30 min (ΔLH) reflects the increased bioactivity of GnSAF, and this has been used as an in vivo bioassay for GnSAF in women (Messinis, 2006).

With the use of this method, the in vivo bioactivity of GnSAF can be estimated during the normal menstrual cycle providing important information on the physiological role of GnSAF. Experiments in the early follicular phase have shown a gradual reduction in the ΔLH response to GnRH at a time when serum estradiol levels increased but not significantly (Messinis, Hirsch, & Templeton, 1991). This decrease in ΔLH, therefore, can be attributed to the increased production of GnSAF from small antral follicles. Since FSH is the hormone that stimulates the increased bioactivity of GnSAF in superovulated cycles, it is possible that in the early follicular phase the intercycle rise of FSH stimulates the production of GnSAF (Fig. 1). Further clinical research, during the rest of the normal menstrual cycle, has demonstrated that the ΔLH response to GnRH remains low with no significant changes in the midfollicular phase showing a marked increase only in the

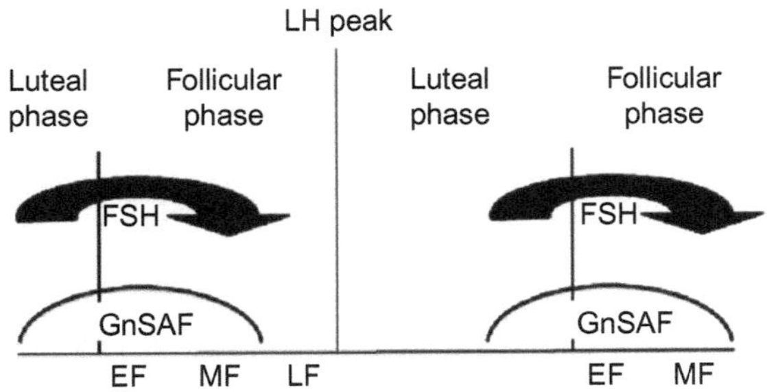

Fig. 1 GnSAF bioactivity increases during the luteal–follicular transition under the stimulating action of the intercycle rise of FSH. *EF*, early follicular; *MF*, midfollicular; *LF*, late follicular phase.

late follicular phase (Messinis et al., 1994). This pattern of ΔLH changes, i.e., low response in the early follicular phase, remaining stable in the midfollicular and increasing in the late follicular phase is different from the pattern of estradiol changes, which show a continuous rise from the early to mid- and late follicular phase until the onset of the midcycle LH surge (Dafopoulos et al., 2004). Since estradiol is the principal ovarian hormone that sensitizes the pituitary gonadotrophs to GnRH, it would be expected this sensitization, expressed as an increase in ΔLH, to occur not only in the late but also from the early follicular phase up to the midcycle LH surge.

In fact, in contrast to the described data in the follicular phase of the normal menstrual cycle, different results have been obtained in postmenopausal women during treatment with exogenous estrogen in a "simulated" cycle (Dafopoulos et al., 2004). In these experiments, a "simulated" follicular phase of 14 days duration was created, in which serum estradiol levels increased steadily from the beginning to the end of the experimental period mimicking the pattern of estradiol changes in the follicular phase of the normal menstrual cycle. At the same time, the pituitary sensitivity to GnRH (ΔLH) increased in parallel with the estradiol rise (Fig. 2). It is clear, therefore, that the secretion of LH as a response to GnRH (pituitary sensitivity) is differentially controlled by estradiol according to the level of the ovarian function. When the ovaries are functioning normally, as it happens during the follicular phase of the menstrual cycle, GnSAF is produced, which antagonizes the sensitizing action of estradiol on the pituitary response to GnRH. In contrast, when the ovarian function is terminated, as after menopause, there is no production of GnSAF and this allows exogenous estradiol to exert its sensitizing effect on the pituitary from the beginning till the end of its administration.

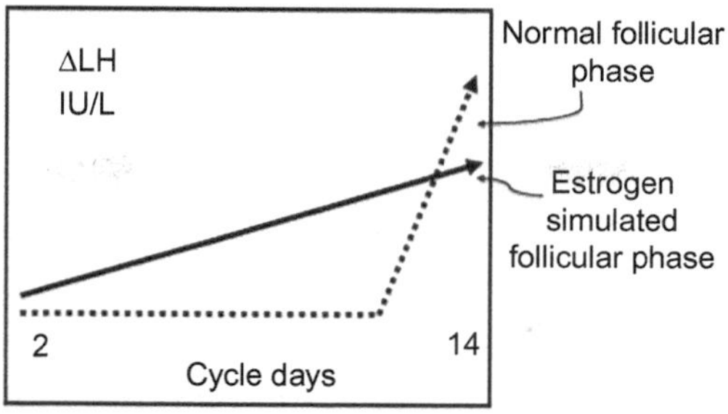

Fig. 2 Pattern of changes in pituitary sensitivity (ΔLH) during the follicular phase of the normal menstrual cycle and a "simulated" follicular phase created in postmenopausal women after the exogenous administration of estrogen. *The diagram was based on data presented in Dafopoulos, K., Kotsovassilis, C. G., Milingos, S., Kallitsaris, A., Galazios, G., Zintzaras, E., et al. (2004). Changes in pituitary sensitivity to GnRH in estrogen-treated post-menopausal women: Evidence that gonadotrophin surge attenuating factor plays a physiological role.* Human Reproduction, 19, *1985–1992.*

Clinical research in the luteal phase of the normal menstrual cycle has demonstrated a gradual reduction in the pituitary sensitivity to GnRH (ΔLH) from the day of ovulation to the end of the cycle (Messinis, Koutsoyiannis, et al., 1993; Messinis, MacTavish, & Templeton, 1993). Whether this suggests increased production of GnSAF in the second half of the cycle is not known, although under conditions of superovulation induction the activity of GnSAF is also high in the luteal phase (Messinis, MacTavish, et al., 1993). Further research in women following ovariectomy, performed in the midluteal phase of the normal cycle, has suggested that in unstimulated spontaneous cycles the bioactivity of GnSAF is higher in the late as compared to the early luteal phase (Messinis, Koutsoyiannis, et al., 1993; Messinis, MacTavish, et al., 1993; Messinis et al., 2002). As it was already mentioned, increased bioactivity of GnSAF has been also demonstrated in the early- to midfollicular phase of the normal cycle (Messinis et al., 1991, 1994). All these data taken together indicate that during the normal menstrual cycle there are clear changes in the bioactivity of GnSAF. Especially, increased bioactivity has been shown during the luteal–follicular transition, i.e., during the intercycle rise of FSH. Since FSH is considered the main stimulator of GnSAF production, it is likely that the intercycle rise of FSH apart from stimulating the production of estradiol and inhibin B from the recruited antral follicles also increases the bioactivity of GnSAF (Fig. 1). It is evident that GnSAF is derived from a cohort of small antral follicles from which the dominant follicle is selected under the action of FSH. This is supported by in vitro and in vivo studies showing that the contribution of the selected large ovulatory follicle is negligible (Dimitraki et al., 2014; Fowler, Sorsa, Harris, Knight, & Mason, 2001).

It has become evident from the above discussion that GnSAF is a constituent of the ovarian hormonal milieu, which is secreted from the growing follicles together with other hormones under the stimulating action of FSH. As such, GnSAF must play an important role in the physiology of the menstrual cycle. The increased bioactivity of GnSAF in the early- to the midfollicular phase indicates that the role of this ovarian hormone is notably apparent at that stage of the cycle. The most likely action of GnSAF is to antagonize the sensitizing effect of estradiol on the pituitary retaining the response of LH to GnRH during the early- and midfollicular phase of the cycle at a low level (Fig. 3). It is probable, therefore, that GnSAF is the "missing link" between the ovaries and the central nervous system, which maintains the pituitary in a state of low responsiveness to GnRH during the greater part of the follicular phase, keeping thus low levels of LH until the late follicular phase. From a physiological point of view, this is very important, because if LH levels were increased above a certain "ceiling" before full follicle maturation, this would be detrimental for the growth of the recruited by FSH antral follicles, affecting also the growth of the dominant follicle leading to premature luteinization and a subsequent

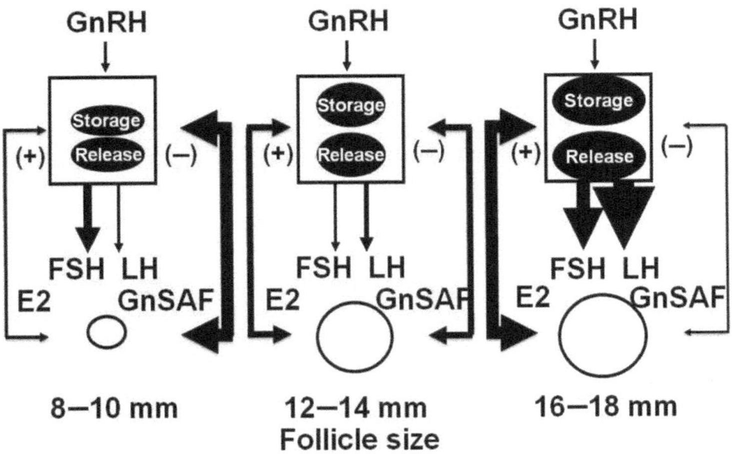

Fig. 3 Interaction between estradiol and GnSAF during the follicular phase of the normal menstrual cycle. Estradiol sensitizes the pituitary to GnRH, while GnSAF antagonizes the estradiol effect maintaining the pituitary in a state of low responsiveness to GnRH during the early- and midfollicular phase of the cycle. In the late follicular phase, the decreased bioactivity of GnSAF facilitates the positive feedback effect of estradiol leading to the midcycle LH surge.

negative impact on the endometrium (Loumaye, Engrand, Shoham, Hillier, & Baird, 2003).

Although GnSAF has an inhibitory action on the pituitary, it does not affect the basal (tonic) secretion of LH, which is controlled by ovarian steroids. GnSAF seems to interfere with the secretion of LH that is regulated by GnRH pulses. Despite the negative action of GnSAF, the final result is positive. In fact, it is the gradual decrease in GnSAF bioactivity during the follicular phase of the cycle particularly from the mid- to the late follicular phase that facilitates the positive feedback effect of estradiol and the occurrence of the midcycle LH surge. Since in superovulated women with constantly high levels of GnSAF bioactivity the endogenous LH surge is markedly attenuated, it is evident that under conditions of lower GnSAF bioactivity, as during the normal menstrual cycle, the dimensions of the LH surge are augmented. This indicates that GnSAF participates in the control of the amplitude of the midcycle LH surge via its decreasing almost eliminating activity toward the late follicular phase. It is, therefore, the "withdrawal" of GnSAF bioactivity in the late follicular phase of the menstrual cycle, which amplifies the midcycle LH surge. Whether GnSAF is also involved in the mechanism, which controls the time of the onset of the LH surge in the normal menstrual cycle, is not clear. The fact that in superovulated cycles a premature LH surge may occur indicates that GnSAF is less important than estradiol for timing the onset of the surge. Additionally, experiments in women during the early- and midfollicular phase of the cycle have shown that when preovulatory concentrations of estradiol are attained in serum after the exogenous administration of this steroid, an endogenous LH surge is induced invariably, suggesting that it is estradiol that is the main triggering of the LH surge at midcycle (Messinis, 2006). In this context, progesterone seems also to play a role, despite that there are conflicting data regarding an increase in serum levels of this steroid before the onset of the LH surge. Nevertheless, experiments in women have demonstrated that exogenous progesterone given after pretreatment with increasing doses of estradiol exerts a positive feedback effect before estradiol has reached the appropriate threshold level (Leyendecker, Wildt, Gips, Nocke, & Plotz, 1976). Furthermore, GnSAF in high amounts may block the positive feedback mechanism, since in about 80% of the cycles with FSH-induced multiple follicular development the LH surge is abolished (Janssens et al., 2000; Messinis & Templeton, 1987a; Wang et al., 2016). Therefore, the role of GnSAF in timing the onset of the midcycle LH surge in the normal menstrual cycle needs further investigation.

5. MECHANISM OF ACTION

The mechanism via which GnSAF attenuates the endogenous LH surge is not clear. Since GnRH is the hypothalamic gonadotropin-releasing hormone, at least part of this mechanism is the marked suppressesion of GnRH-induced LH secretion (Messinis & Templeton, 1989). The latter has been found in superovulated cycles and indicates that the action of GnSAF is exerted at least at the pituitary level. Results from experiments in women with the use of the above-described model of the two submaximal GnRH pulses have suggested that GnSAF reduces the size of the two functionally related pools of the pituitary gonadotrophs from which gonadotropins are released. Estradiol is the factor that sensitizes the pituitary to GnRH, increasing thus both the pituitary sensitivity and the gonadotropin reserves in the gonadotrophs. This interaction between GnRH and estradiol is crucial for triggering the positive feedback mechanism and the occurrence of the endogenous LH surge. The reduced pituitary sensitivity therefore during superovulation induction suggests that GnSAF antagonizes the action of estradiol on this gland rendering the gonadotrophs less sensitive to GnRH (Fig. 4). As a result and depending on the amount of GnSAF produced by the hyperstimulated ovaries, the endogenous LH surge is either blocked or markedly attenuated. Whether GnSAF also exerts an action at the hypothalamic level is not known.

Studies in animals have clearly demonstrated that the estradiol positive feedback also involves the hypothalamus with the production or activation

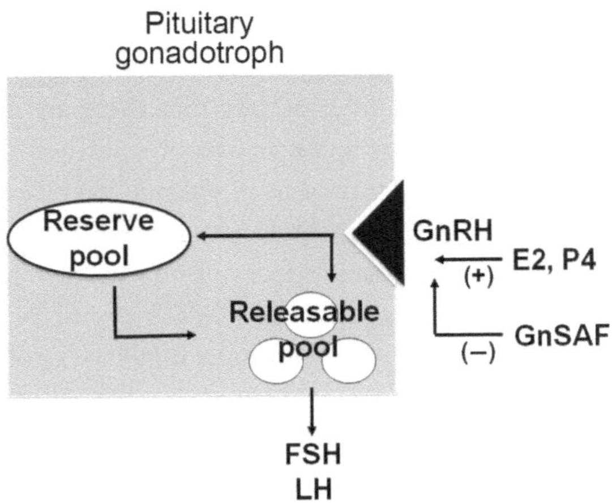

Fig. 4 The two functionally related pools of the pituitary gonadotroph. GnSAF acts on the pituitary to antagonize the sensitizing effect of estradiol on the LH response to GnRH. Progesterone also exerts a sensitizing action.

of various pathways. In particular, progesterone synthesized in the hypothalamus as well as its receptors play an important role for the expression of the positive feedback effect and the generation of the LH surge (Chappell et al., 1999; Micevych et al., 2003). Based on the fact that in animals progesterone produced in the hypothalamus under the stimulation by estradiol is obligatory for the expression of the positive feedback mechanism (Chappell et al., 1999), recent studies have implicated hypothalamic progesterone receptors in the action of GnSAF, via reduced phosphorylation/activation of these receptors in the pituitary gonadotrophs (Gordon et al., 2010; Gordon, Garrido-Gracia, Aguilar, & Sánchez-Criado, 2009; Gordon, Garrido-Gracia, Sánchez-Criado, & Aguilar, 2011). It is evident that further studies are needed to clarify the mechanism of action of GnSAF at the hypothalamo-pituitary system.

In the context of a purification procedure, a synthetic molecule of 26 amino acids, which exerted a significant inhibitory action on GnRH-induced LH secretion in vitro without affecting basal FSH levels, has been produced (Hendriks et al., 2011). These experiments showed that the MAPK phosphorylation pathway was attenuated. Nevertheless, no further research has been reported with this synthetic molecule. Certainly, the full characterization of GnSAF will provide the tool for the study of the mechanism of the cellular and molecular action of this substance at the hypothalamo-pituitary system.

Since the year of 2000, a RF-amide peptide has been isolated from quails, which reduced LH and FSH secretion from anterior pituitary cells in culture and has been named gonadotropin inhibitory hormone (GnIH) as opposed to gonadotropin-releasing hormone (Tsutsui et al., 2000). Subsequently, it was shown that the RF-amide peptide injected into chickens reduced LH secretion (Osugi et al., 2004). It has been suggested that the RF-amide-related protein 3 (RFRP-3), found in the hypothalamus and possibly in the ovaries, acts as a comodulator of GnSAF or even as a GnSAF candidate. However, RFRP-3 has not been widely accepted as an established inhibitory mechanism of gonadotropin secretion from the pituitary and further research is required.

6. CLINICAL IMPLICATIONS

Ovarian stimulation leading to superovulation induction is an established approach for infertility treatment using IVF/ICSI. Exogenous FSH is the basis of the different treatment regimens, but there is no consensus regarding the best protocol. Nevertheless, as estradiol concentrations rise rapidly and the threshold for the positive feedback effect is exceeded before full follicle maturation, a premature LH surge may occur (Messinis & Templeton, 1986). In the majority of FSH protocols either a GnRH agonist or a GnRH antagonist is included in order to block the positive feedback mechanism and prevent a premature LH surge. Since the endogenous LH surge is blocked in almost 80% of the cycles without the use of GnRH analogs (Janssens et al., 2000; Messinis & Templeton, 1987a; Wang et al., 2016) due to the overproduction of GnSAF, it is likely that the GnRH analogs are needed in only a small number of cycles (Wang et al., 2016). In these cases, serum LH should be measured on a daily basis, which may not be very convenient.

A premature LH surge leads to luteinization of the granulosa cells and increased serum progesterone concentrations. This steroid acts on the estrogen-primed endometrium and induces an advanced morphology in the glandular and stromal epithelium (Bourgain & Devroey, 2003). It has been shown that serum progesterone levels up to 3 ng/mL affect adversely the pregnancy rate in fresh but not in frozen–thawed cycles (Venetis, Kolibianakis, Bosdou, & Tarlatzis, 2013). This means that even without the use of an agonist or an antagonist in FSH-treated cycles the outcome of treatment is not affected if the "freeze-all" method is chosen. Evidence has been provided showing that in fresh cycles the GnRH antagonists provide similar clinical outcome with the agonists (Al-Inany et al., 2011). Although the antagonists at the daily dose of 0.25 mg do not block the positive feedback mechanism, they are effective with the help of GnSAF, which is produced by the superovulated ovaries (Messinis et al., 2010).

It is evident, therefore, that if GnSAF is characterized to homogeneity and after its synthetic production, it could be used in ovarian stimulation protocols instead of the GnRH analogs to prevent the premature LH surge by acting as a supplement to the endogenously produced substance. Another potential use of a synthetic GnSAF might be for contraceptive purposes by blocking the endogenous LH surge and therefore inhibiting ovulation. It is not unlikely exogenous GnSAF also to be used in patients with PCOS in order to decrease the high basal LH levels, although in such women there is no reduced production of GnSAF but rather a defect in the interaction of this factor with different ovarian substances on the hypothalamo-pituitary system (Dafopoulos, Venetis, Pournaras, Kallitsaris, & Messinis, 2009). It is evident that the full characterization of GnSAF is of outmost importance not only for the explanation of the physiological hormonal events during the normal menstrual cycle but also for its use in various clinical conditions.

7. CONCLUSIONS

Accumulated evidence derived from in vivo and in vitro experiments indicates that GnSAF is a nonsteroidal ovarian substance that plays an important role during the normal menstrual cycle. The bioactivity of this factor is high in the early- to midfollicular phase, decreasing gradually thereafter until the midcycle. This factor antagonizes the sensitizing effect of estradiol on the pituitary response to GnRH during the greater part of the follicular phase. With its decreasing activity in the late follicular phase, GnSAF facilitates the full expression of the positive feedback mechanism and the occurrence of a normal midcycle LH surge. It is suggested that GnSAF is the "missing link" between the ovaries and the hypothalamo-pituitary system, which maintains the pituitary in a state of low responsiveness to GnRH pulses in the early- to midfollicular phase, limiting thus the release of LH to low but adequate amounts for normal folliculogenesis. This factor appears to affect not the tonic but the pulsatile aspect of LH release that is controlled centrally by GnRH and modulated by the ovarian steroids. GnSAF therefore provides a negative action in the context of the positive feedback loop.

REFERENCES

Albano, C., Felberbaum, R. E., Smitz, J., Riethmüller-Winzen, H., Engel, J., Diedrich, K., et al. (2000). Ovarian stimulation with HMG: Results of a prospective randomized phase III European study comparing the luteinizing hormone-releasing hormone (LHRH)-antagonist cetrorelix and the LHRH-agonist buserelin. European Cetrorelix study group. *Human Reproduction*, *15*, 526–531.

Alexandris, E., Milingos, S., Kollios, G., Seferiadis, K., Lolis, D., & Messinis, I. E. (1997). Changes in gonadotrophin response to gonadotrophin releasing hormone in normal women following bilateral ovariectomy. *Clinical Endocrinology*, *47*, 721–726.

Al-Inany, H. G., Youssef, M. A., Aboulghar, M., Broekmans, F., Sterrenburg, M., Smit, J., et al. (2011). Gonadotrophin-releasing hormone antagonists for assisted reproductive technology. *Cochrane Database Systematic Reviews*, *5*, CD001750. https://doi.org/10.1002/14651858.CD001750.pub3 PMID: 21563131.

Borm, G., & Mannaerts, B. (2000). Treatment with the gonadotrophin-releasing hormone antagonist ganirelix in women undergoing ovarian stimulation with recombinant follicle stimulating hormone is effective, safe and convenient: Results of a controlled, randomized, multicentre trial. The European Orgalutran study group. *Human Reproduction*, *15*, 1490–1498.

Bourgain, C., & Devroey, P. (2003). The endometrium in stimulated cycles for IVF. *Human Reproduction Update*, *9*, 515–522.

Chappell, P. E., Schneider, J. S., Kim, P., Xu, M., Lydon, J. P., O'Malley, B. W., et al. (1999). Absence of gonadotropin surges and gonadotropin-releasing hormone self-priming in ovariectomized (OVX), estrogen (E2)-treated, progesterone receptor knockout (PRKO) mice. *Endocrinology*, *140*, 3653–3658.

Dafopoulos, K., Kotsovassilis, C. G., Milingos, S., Kallitsaris, A., Galazios, G., Zintzaras, E., et al. (2004). Changes in pituitary sensitivity to GnRH in estrogen-treated post-menopausal women: Evidence that gonadotrophin surge attenuating factor plays a physiological role. *Human Reproduction, 19*, 1985–1992.

Dafopoulos, K., Mademtzis, I., Vanakara, P., Kallitsaris, A., Stamatiou, G., Kotsovassilis, C., et al. (2006). Evidence that termination of the estradiol-induced luteinizing hormone surge in women is regulated by ovarian factors. *Journal of Clinical Endocrinology and Metabolism, 91*, 641–645.

Dafopoulos, K., Venetis, C., Pournaras, S., Kallitsaris, A., & Messinis, I. E. (2009). Ovarian control of pituitary sensitivity of luteinizing hormone secretion to gonadotropin-releasing hormone in women with the polycystic ovary syndrome. *Fertility and Sterility, 92*, 1378–1380.

Dal Prato, L., Borini, A., Coticchio, G., Cattoli, M., & Flamigni, C. (2004). Half-dose depot triptorelin in pituitary suppression for multiple ovarian stimulation in assisted reproduction technology: A randomized study. *Human Reproduction, 19*, 2200–2205.

Danforth, D. R., & Cheng, C. Y. (1995). Purification of a candidate gonadotropin surge inhibiting factor from porcine follicular fluid. *Endocrinology, 136*, 1658–1665.

Danforth, D. R., Elkind-Hirsch, K., & Hodgen, G. D. (1990). In vivo and in vitro modulation of gonadotropin-releasing hormone metabolism by estradiol and progesterone. *Endocrinology, 127*, 319–324.

de Kretser, D. M., Hedger, M. P., Loveland, K. L., & Phillips, D. J. (2002). Inhibins, activins and follistatin in reproduction. *Human Reproduction Update, 8*, 529–541.

Dimitraki, M., Messini, C. I., Dafopoulos, K., Gioka, T., Koutlaki, N., Garas, A., et al. (2014). Attenuating activity of the ovary on LH response to GnRH during the follicular phase of the cycle. *Clinical Endocrinology, 80*, 439–443.

Fowler, P. A., Sorsa, T., Harris, W. J., Knight, P. G., & Mason, H. D. (2001). Relationship between follicle size and gonadotrophin surge attenuating factor (GnSAF) bioactivity during spontaneous cycles in women. *Human Reproduction, 16*, 1353–1358.

Fowler, P. A., Sorsa-Leslie, T., Cash, P., Dunbar, B., Melvin, W., Wilson, Y., et al. (2002). A 60–66 kDa protein with gonadotrophin surge attenuating factor bioactivity is produced by human ovarian granulosa cells. *Molecular Human Reproduction, 8*, 823–832.

Fowler, P. A., Templeton, A., & Messinis, I. E. (1993). The ovarian modulation of gonadotrophin releasing hormone-induced luteinizing hormone secretion in women. *Human Reproduction, 8*(Suppl. 2), 112–116.

Glasier, A., Thatcher, S. S., Wickings, E. J., Hillier, S. G., & Baird, D. T. (1988). Superovulation with exogenous gonadotropins does not inhibit the luteinizing hormone surge. *Fertility and Sterility, 49*, 81–85.

Goodman, A. L., & Hodgen, G. D. (1978). Post partum patterns of circulating FSH, LH, prolactin, estradiol, and progesterone in nonsuckling cynomolgus monkeys. *Steroids, 31*, 731–744.

Gordon, A., Aguilar, R., Garrido-Gracia, J. C., Bellido, C., Millán, Y., Guil-Luna, S., et al. (2010). Human follicular fluid from superovulated women inhibits progesterone receptor-dependent gonadotropin-releasing hormone self-priming in an estrous cycle-dependent manner in the rat. *Journal of Endocrinological Investigation, 33*, 564–570.

Gordon, A., Garrido-Gracia, J. C., Aguilar, R., & Sánchez-Criado, J. E. (2009). Ovarian stimulation with FSH in the rat reduces proestrous GnRH-dependent LH secretion through a dual mechanism: Inhibition of LH synthesis and release. *Neuroscience Letters, 460*, 219–222.

Gordon, A., Garrido-Gracia, J. C., Sánchez-Criado, J. E., & Aguilar, R. (2011). Involvement of rat gonadotrope progesterone receptor in the ovary-mediated inhibitory action of FSH on LH synthesis. *Journal of Physiology and Biochemistry, 67*, 145–151.

Griesinger, G., Dawson, A., Schultze-Mosgau, A., Finas, D., Diedrich, K., & Felberbaum, R. (2006). Assessment of luteinizing hormone level in the gonadotropin-releasing hormone antagonist protocol. *Fertility and Sterility, 85*, 791–793.

Hendriks, M. L., Lambalk, C. B., Helder, M. N., & de Koning, J. (2011). Purification of a candidate gonadotrophin, surge-inhibiting/attenuating factor (GnSIF/AF) showing MAPK as a possible target. *Molecular Reproduction and Development, 78*, 292–304.

Janssens, R. M., Lambalk, C. B., Vermeiden, J. P., Schats, R., Bernards, J. M., Rekers-Mombarg, L. T., et al. (2000). Dose-finding study of triptorelin acetate for prevention of a premature LH surge in IVF: A prospective, randomized, double-blind, placebo-controlled study. *Human Reproduction, 15*, 2333–2340.

Karligiotou, E., Kollia, P., Kallitsaris, A., & Messinis, I. E. (2006). Expression of human serum albumin (HSA) mRNA in human granulosa cells: Potential correlation of the 95 amino acid long carboxyl terminal of HSA to gonadotrophin surge-attenuating factor. *Human Reproduction, 21*, 645–650.

Karligiotou, E., Kollia, P., Papaggeli, P., Samara, S., Vagena, A., Dafopoulos, K., et al. (2011). FSH modulatory effect on human granulosa cells: A gene-protein candidate for gonadotrophin surge-attenuating factor. *Reproductive Biomedicine Online, 23*, 440–488.

Kato, K., Takehara, Y., Segawa, T., Kawachiya, S., Okuno, T., Kobayashi, T., et al. (2012). Minimal ovarian stimulation combined with elective single embryo transfer policy: Age-specific results of a large, single-centre, Japanese cohort. *Reproductive Biology Endocrinology, 10*, 35.

Kazem, R., Messinis, I. E., Fowler, P., Groome, N. P., Knight, P. G., & Templeton, A. A. (1996). Effect of mifepristone (RU486) on the pituitary response to gonadotrophin releasing hormone in women. *Human Reproduction, 11*, 2585–2590.

Lasley, B. L., Wang, C. F., & Yen, S. S. (1975). The effects of estrogen and progesterone on the functional capacity of the gonadotrophs. *Journal of Clinical Endocrinology and Metabolism, 41*, 820–826.

Laws, S. C., Webster, J. C., & Miller, W. L. (1990). Estradiol alters the effectiveness of gonadotropin-releasing hormone (GnRH) in ovine pituitary cultures: GnRH receptors versus responsiveness to GnRH. *Endocrinology, 127*, 381–386.

Leyendecker, G., Wildt, L., Gips, H., Nocke, W., & Plotz, E. J. (1976). Experimental studies on the positive feedback effect of progesterone, 17 alpha-hydroxyprogesterone and 20 alpha-dihydroprogesterone on the pituitary release of LH and FSH in the human female. The estrogen priming of the progesterone feedback on pituitary gonadotropins in the eugonadal woman. *Archiv für Gynäkologie, 221*, 29–45.

Littman, B. A., & Hodgen, G. D. (1984). Human menopausal gonadotropin stimulation in monkeys: Blockade of the luteinizing hormone surge by a highly transient ovarian factor. *Fertility and Sterility, 41*, 440–447.

Liu, J. H., & Yen, S. S. (1983). Induction of midcycle gonadotropin surge by ovarian steroids in women: A critical evaluation. *Journal of Clinical Endocrinology and Metabolism, 57*, 797–802.

Loumaye, E., Engrand, P., Shoham, Z., Hillier, S. G., & Baird, D. T. (2003). Clinical evidence for an LH 'ceiling' effect induced by administration of recombinant human LH during the late follicular phase of stimulated cycles in World Health Organization type I and type II anovulation. *Human Reproduction, 18*, 314–322.

Messinis, I. E. (2006). Ovarian feedback, mechanism of action and possible clinical implications. *Human Reproduction Update, 12*, 557–571.

Messinis, I. E., Hirsch, P., & Templeton, A. A. (1991). Follicle stimulating hormone stimulates the production of gonadotrophin surge attenuating factor (GnSAF) in vivo. *Clinical Endocrinology, 35*, 403–407.

Messinis, I. E., Koutsoyiannis, D., Milingos, S., Tsahalina, E., Seferiadis, K., Lolis, D., et al. (1993). Changes in pituitary response to GnRH during the luteal-follicular transition of the human menstrual cycle. *Clinical Endocrinology, 38*, 159–163.

Messinis, I. E., Krishnan, M., Kazem, R., Khadilkar, S., & Templeton, A. A. (1997). Effect of mifepristone on folliculogenesis in women treated with recombinant FSH. *Clinical Endocrinology (Oxf), 46*, 309–314.

Messinis, I. E., Lolis, D., Zikopoulos, K., Tsahalina, E., Seferiadis, K., & Templeton, A. A. (1994). Effect of an increase in FSH on the production of gonadotrophin-surge-attenuating factor in women. *Journal of Reproduction and Fertility, 101*, 689–695.

Messinis, I. E., Loutradis, D., Domali, E., Kotsovassilis, C. P., Papastergiopoulou, L., Kallitsaris, A., et al. (2005). Alternate day and daily administration of GnRH antagonist may prevent premature luteinization to a similar extent during FSH treatment. *Human Reproduction, 20*, 3192–3197.

Messinis, I. E., MacTavish, A., & Templeton, A. A. (1993). Activity of gonadotrophin surge-attenuating factor during the luteal phase in superovulated women. *Journal of Reproduction and Fertility, 97*, 271–275.

Messinis, I. E., Messini, C. I., & Dafopoulos, K. (2014). Novel aspects of the endocrinology of the menstrual cycle. *Reproductive Biomedicine Online, 28*, 714–722.

Messinis, I. E., Messini, C. I., Dimitraki, M., Anifandis, G., & Dafopoulos, K. (2015). Gonadotropin surge attenuating factor: A physiological regulator. *Hellenic Journal of Obstetrics and Gynecology, 14*, 5–18.

Messinis, I. E., Milingos, S., Alexandris, E., Mademtzis, I., Kollios, G., & Seferiadis, K. (2002). Evidence of differential control of FSH and LH responses to GnRH by ovarian steroids in the luteal phase of the cycle. *Human Reproduction, 17*, 299–303.

Messinis, I. E., Milingos, S., Zikopoulos, K., Hasiotis, G., Seferiadis, K., & Lolis, D. (1998). Luteinizing hormone response to gonadotrophin-releasing hormone in normal women undergoing ovulation induction with urinary or recombinant follicle stimulating hormone. *Human Reproduction, 13*, 2415–2420.

Messinis, I. E., & Templeton, A. (1986). The effect of pulsatile follicle stimulating hormone on the endogenous luteinizing hormone surge in women. *Clinical Endocrinology, 25*, 633–640.

Messinis, I. E., & Templeton, A. A. (1987a). Endocrine and follicle characteristics of cycles with and without endogenous luteinizing hormone surges during superovulation induction with pulsatile follicle-stimulating hormone. *Human Reproduction, 2*, 11–16.

Messinis, I. E., & Templeton, A. A. (1987b). Disparate effects of endogenous and exogenous oestradiol on luteal phase function in women. *Journal of Reproduction and Fertility, 79*, 549–554.

Messinis, I. E., & Templeton, A. (1988). Blockage of the positive feedback effect of oestradiol during prolonged administration of clomiphene citrate to normal women. *Clinical Endocrinology (Oxf), 29*, 509–516.

Messinis, I. E., & Templeton, A. A. (1989). Pituitary response to exogenous LHRH in superovulated women. *Journal of Reproduction and Fertility, 87*, 633–639.

Messinis, I. E., & Templeton, A. A. (1990). Effects of supraphysiological concentrations of progesterone on the characteristics of the oestradiol-induced gonadotrophin surge in women. *Journal of Reproduction and Fertility, 88*, 513–519.

Messinis, I. E., & Templeton, A. (1991). Attenuation of gonadotrophin release and reserve in superovulated women by gonadotrophin surge attenuating factor (GnSAF). *Clinical Endocrinology, 34*, 259–263.

Messinis, I. E., Templeton, A., & Baird, D. T. (1985). Endogenous luteinizing hormone surge during superovulation induction with sequential use of clomiphene citrate and pulsatile human menopausal gonadotropin. *Journal of Clinical Endocrinology and Metabolism, 61*, 1076–1080.

Messinis, I. E., Templeton, A., & Baird, D. T. (1986). Endogenous luteinizing hormone surge in women during induction of multiple follicular development with pulsatile follicle stimulating hormone. *Clinical Endocrinology, 24*, 193–201.

Messinis, I. E., Vanakara, P., Zavos, A., Verikouki, C., Georgoulias, P., & Dafopoulos, K. (2010). Failure of the GnRH antagonist ganirelix to block the positive feedback effect of exogenous estrogen in normal women. *Fertility and Sterility, 94*, 1554–1556.

Micevych, P., Sinchak, K., Mills, R. H., Tao, L., LaPolt, P., & Lu, J. K. (2003). The luteinizing hormone surge is preceded by an estrogen-induced increase of hypothalamic progesterone in ovariectomized and adrenalectomized rats. *Neuroendocrinology, 78*, 29–35.

Molskness, T. A., Woodruff, T. K., Hess, D. L., Dahl, K. D., & Stouffer, R. L. (1996). Recombinant human inhibin-A administered early in the menstrual cycle alters concurrent pituitary and follicular, plus subsequent luteal, function in rhesus monkeys. *Journal Clinical Endocrinology Metabolism, 81*, 4002–4006.

Mroueh, J. M., Arbogast, L. K., Fowler, P., Templeton, A. A., Friedman, C. I., & Danforth, D. R. (1996). Identification of gonadotrophin surge-inhibiting factor (GnSIF)/attenuin in human follicular fluid. *Human Reproduction, 11*, 490–496.

O'Reilly, M. S., Boehm, T., Shing, Y., Fukai, N., Vasios, G., Lane, W. S., et al. (1997). Endostatin: An endogenous inhibitor of angiogenesis and tumor growth. *Cell, 88*, 277–285.

O'Reilly, M. S., Holmgren, L., Shing, Y., Chen, C., Rosenthal, R. A., Moses, M., et al. (1994). Angiostatin: A novel angiogenesis inhibitor that mediates the suppression of metastases by a Lewis lung carcinoma. *Cell, 79*, 315–328.

Osugi, T., Ukena, K., Bentley, G. E., O'Brien, S., Moore, I. T., Wingfield, J. C., et al. (2004). Gonadotropin-inhibitory hormone in Gambel's white-crowned sparrow (Zonotrichia Leucophrys Gambelii): cDNA identification, transcript localization and functional effects in laboratory and field experiments. *Journal of Endocrinology, 182*, 33–42.

Pappa, A., Seferiadis, K., Fotsis, T., Shevchenko, A., Marselos, M., Tsolas, O., et al. (1999). Purification of a candidate gonadotrophin surge attenuating factor from human follicular fluid. *Human Reproduction, 14*, 1449–1456.

Quyyumi, S. A., Pinkerton, J. V., Evans, W. S., & Veldhuis, J. D. (1993). Estradiol amplifies the amount of luteinizing hormone (LH) secreted in response to increasing doses of gonadotropin-releasing hormone by specifically augmenting the duration of evoked LH secretory events and hence their mass. *Journal of Clinical Endocrinology and Metabolism, 76*, 594–600.

Ramey, J. W., Highsmith, R. F., Wilfinger, W. W., & Baldwin, D. M. (1987). The effects of gonadotropin-releasing hormone and estradiol on luteinizing hormone biosynthesis in cultured rat anterior pituitary cells. *Endocrinology, 120*, 1503–1513.

Rivier, C., Rivier, J., & Vale, W. (1986). Inhibin-mediated feedback control of follicle-stimulating hormone secretion in the female rat. *Science, 234*, 205–208.

Schenken, R. S., & Hodgen, G. D. (1983). Follicle-stimulating hormone induced ovarian hyperstimulation in monkeys: Blockade of the luteinizing hormone surge. *Journal of Clinical Endocrinol and Metabolism, 57*, 50–55.

Soma, K. K., Sinchak, K., Lakhter, A., Schlinger, B. A., & Micevych, P. E. (2005). Neurosteroids and female reproduction: Estrogen increases 3beta-HSD mRNA and activity in rat hypothalamus. *Endocrinology, 146*, 4386–4390.

Sopelak, V. M., & Hodgen, G. D. (1984). Blockade of the estrogen-induced luteinizing hormone surge in monkeys: A nonsteroidal, antigenic factor in porcine follicular fluid. *Fertility and Sterility, 41*, 108–113.

Tavoulari, S., Frillingos, S., Karatza, P., Messinis, I. E., & Seferiadis, K. (2004). The recombinant subdomain IIIB of human serum albumin displays activity of gonadotrophin surge-attenuating factor. *Human Reproduction, 19*, 849–858.

Tio, S., Koppenaal, D., Bardin, C. W., & Cheng, C. Y. (1994). Purification of gonadotropin surge-inhibiting factor from Sertoli cell-enriched culture medium. *Biochemical and Biophysical Research Communications, 199*, 1229–1236.

Tsai, C. C., & Yen, S. S. (1971). The effect of ethinyl estradiol administration during early follicular phase of the cycle on the gonadotropin levels and ovarian function. *Journal of Clinical Endocrinology and Metabolism, 33*, 917–923.

Tsonis, C. G., Messinis, I. E., Templeton, A. A., McNeilly, A. S., & Baird, D. T. (1988). Gonadotropic stimulation of inhibin secretion by the human ovary during the follicular and early luteal phase of the cycle. *Journal of Clinical Endocrinology and Metabolism, 66*, 915–921.

Tsutsui, K., Saigoh, E., Ukena, K., Teranishi, H., Fujisawa, Y., Kikuchi, M., et al. (2000). A novel avian hypothalamic peptide inhibiting gonadotropin release. *Biochemical and Biophysical Research Communications, 275*, 661–667.

Venetis, C. A., Kolibianakis, E. M., Bosdou, J. K., & Tarlatzis, B. C. (2013). Progesterone elevation and probability of pregnancy after IVF: A systematic review and meta-analysis of over 60 000 cycles. *Human Reproduction Update, 19*, 433–457.

Wang, H. L., Lai, H. H., Chuang, T. H., Shih, Y. W., Huang, S. C., Lee, M. J., et al. (2016). A patient friendly corifollitropin alfa protocol without routine pituitary suppression in normal responders. *PLoS One, 11*, e0154123. https://doi.org/10.1371/journal.pone.0154123.

Wang, C. F., Lasley, B. L., Lein, A., & Yen, S. S. (1976). The functional changes of the pituitary gonadotrophs during the menstrual cycle. *Journal of Clinical Endocrinology and Metabolism, 42*, 718–728.

Xiao, J. S., Su, C. M., & Zeng, X. T. (2014). Comparisons of GnRH antagonist versus GnRH agonist protocol in supposed normal ovarian responders undergoing IVF: A systematic review and meta-analysis. *PLoS One, 9*, e106854. https://doi.org/10.1371/journal.pone.0106854.

Yen, S. S., & Tsai, C. C. (1971). The effect of ovariectomy on gonadotropin release. *Journal of Clinical Investigation, 50*, 1149–1153.

CHAPTER ELEVEN

Luteolysis and the Auto-, Paracrine Role of Cytokines From Tumor Necrosis Factor α and Transforming Growth Factor β Superfamilies

António M. Galvão*,†,1, Dariusz Skarzynski†, Graça Ferreira-Dias*
*C.I.I.S.A., Faculty of Veterinary Medicine, University of Lisbon, Lisbon, Portugal
†Institute of Animal Reproduction and Food Research of PAS, Olsztyn, Poland
[1]Corresponding author: e-mail address: a.galvao@pan.olsztyn.pl

Contents

Abstract

Successful pregnancy establishment demands optimal luteal function in mammals. Nonetheless, regression of the corpus luteum (CL) is absolutely necessary for normal female cyclicity. This dichotomy relies on intricate molecular signals and rapidly activated biological responses, such as angiogenesis, extracellular matrix (ECM) remodeling, or programmed cell death. The CL establishment and growth after ovulation depend not only on the luteinizing hormone-mediated endocrine signal but also on a number of auto-, paracrine interactions promoted by cytokines and growth factors like fibroblast growth factor 2, vascular endothelial growth factor A, and tumor necrosis factor α (TNF), which coordinate vascularigenesis and ECM reorganization as well as steroidogenesis. With the organ fully developed, the release of the uterine prostaglandin F2α activates luteolysis, an intricate process supported by intraluteal interactions that ensure the loss

Vitamins and Hormones, Volume 107
ISSN 0083-6729
https://doi.org/10.1016/bs.vh.2018.01.001

of steroidogenic function (functional luteolysis) and the involution of the organ (structural luteolysis). This chapter provides an overview of the local action of cytokines during luteal function, with particular emphasis on the role of TNF and transforming growth factor β superfamilies during luteolysis.

1. INTRODUCTION

The corpus luteum (CL) is responsible for progesterone (P4) secretion, the hormone which prepares the endometrium for pregnancy establishment. However, different mechanisms among mammalian species control the functional lifespan of the CL in both cyclicity and pregnancy. Its transient nature relies on the fast regulation of biological processes such as cell growth and differentiation, angiogenesis, or cell death and collagen deposition and extracellular matrix (ECM) remodeling. Thus, soon after ovulation the remaining cells and vessels from the follicle differentiate into the CL tissue, mainly under the regulatory action of luteinizing hormone (LH). Importantly, a large number of local auto-, paracrine interactions prompt steroidogenesis and as a result P4 increases rapidly (Richards, Russell, Robker, Dajee, & Alliston, 1998), as well as angiogenesis is promoted (Redmer & Reynolds, 1996) until the organ is fully developed. Conversely, in the absence of pregnancy the resumption of a new estrous cycle demands the regression of the CL. Likewise, following the main luteolytic signal from the uterine PGF2α (McCracken, Custer, & Lamsa, 1999), a plethora of local mediators interact to ensure the functional demise and the structural regression of the CL (Davis & Rueda, 2002). This chapter describes the importance of auto-, paracrine interactions for luteal regulation, with particular emphasis on the luteolytic action of the cytokines from the tumor necrosis factor α (TNF) and transforming growth factor β (TGF) superfamilies.

2. THE CORPUS LUTEUM: GENERAL VIEW AND HISTORICAL STUDIES

After ovulation, granulosa cells from the recently ovulated follicle undergo luteinization, contributing for the formation of the CL. This transient endocrine gland was first associated with pregnancy by the end of the 19th century (Wintersteiner & Allen, 1934). In 1895, the anatomist Johannes Sobotta (1869–1945) published the first drawings of the mouse CL, mainly describing an adenoid and ductless structure (described in Frobenius, 1999). It was in 1934 that the isolation of P4 was originally reported. Different groups contributed for the discovery. While in Europe Heinrich Slotta (1895–1987) and Ludwig Fraenkel (1870–1951) conducted their research in Breslau, nowadays Wroclaw (Poland), in the United States, Oskar Wintersteiner (1898–1971) was also a pioneer on CL research. Independent studies from these scientists resulted in the purification and publication of the formula of P4 (Frobenius, 1999). Furthermore, experimentation by Fraenkel in 1901 demonstrated that bilateral ovariectomy in pregnant rabbits led to the interruption of gestation (Frobenius, 1999).

The importance of the uterus for luteolysis initiation was first shown by Loeb (1923) after studies in guinea pig, in which hysterectomized animals stopped showing ovarian cyclicity and started presenting abnormal luteal persistence. Subsequently, similar results were observed in cyclic sheep, cow, pig, mare, rabbit, and rat (Anderson, Bland, & Melampy, 1969). The sheep has long been used as a model to study luteolysis. Various studies on uterine and ovarian transplantation carried out in 1969 in small ruminants confirmed the interaction between these two organs during luteolysis (Anderson et al., 1969). In the same year, Pharriss and Wyngarden (1969) demonstrated that PGF2α shortened pseudopregnancy in rats, with the consequent decrease of P4 luteal content. Two years later, the contracurrent mechanism that allows for the direct transfer of PGF2α from the uterine vein to the ovarian artery was demonstrated in sheep (McCracken et al., 1972). Studies in the mare confirmed that at the end of luteal phase PGF2α action determines P4 decrease in plasma, defined as functional luteolysis (Douglas & Ginther, 1972; Henderson & McNatty, 1975). But, in contrast to what was found in ruminants, no differences between intrauterine and intramuscular routes of PGF2α administration were observed in the equidae (Douglas & Ginther, 1975). Anatomical variations between the mare and ruminants account for the absence of an utero-ovarian transfer system, once the ovarian artery does not have a major contact with the utero-ovarian vein (Ginther, 1976).

3. LUTEAL ESTABLISHMENT AND FUNCTION

3.1 Molecular Control of Steroidogenesis

The main targets of P4 action in the reproductive tract are the mammary gland, the hypothalamo-pituitary axis, the uterus, and the ovary itself (Niswender, Juengel, Silva, Rollyson, & McIntush, 2000). Noteworthy, the local action of P4 on auto-, paracrine signaling mediators ensures luteal

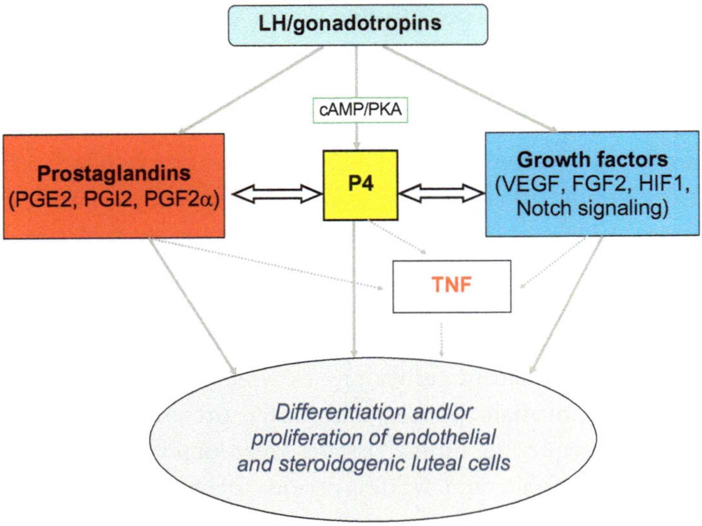

Fig. 1 Hypothetical representation of the local signaling promoting corpus luteum (CL) establishment. Luteinizing hormone (LH) mediates progesterone (P4) secretion through cyclic adenosine monophosphate (cAMP)/protein kinase A (PKA) activation. Prostaglandins (PG), growth factors like vascular endothelial growth factor (VEGF) A, hypoxia-inducible factor-1 (HIF1), and Notch signaling or the cytokine tumor necrosis factor α (TNF) also support P4 secretion and promote CL growth and development.

establishment as an endocrine organ. In the last decades, a great deal of research has been characterizing the complexity of local interactions, leading to CL establishment. Primarily led by LH, the amplification of the luteotrophic signal supports mainly vascularization, ECM remodeling, and steroidogenesis (Fig. 1) (Juengel, Meberg, Turzillo, Nett, & Niswender, 1995; Niswender & Nett, 1994; Skarzynski, Ferreira-Dias, & Okuda, 2008). The establishment of the CL as an endocrine organ can be seen as a bidimensional paradigm. First, a profound tissue reorganization is promoted by a shift in cellular processes, in which the intense cell division mainly driven by estradiol (E2) gives place to cell differentiation, sustained by the drop in circulating follicle-stimulating hormone and the rise in LH (Crisp & Channing, 1972; Ginther, Utt, Bergfelt, & Beg, 2006; Jablonka-Shariff, Grazul-Bilska, Redmer, & Reynolds, 1993). Second, P4 secretion and its circulating levels must ensure implantation and pregnancy progression (Graham & Clarke, 1997). The luteal cells are originated from the differentiation and proliferation of postovulatory granulosa and theca cells (Murphy, 2000). Simultaneously, extensive restructuring and microvascularization promote the development of the follicular tissue into the fully developed CL (Ferreira-Dias, Bravo, Mateus, Redmer, & Medeiros, 2006; Robinson & Woad, 2017). A full description of the factors contributing for luteal establishment goes beyond the scope of this review. Nonetheless, we highlight the main events responsible for the molecular control of steroidogenesis during luteal establishment.

First and foremost, molecular regulation of CL formation is strongly dependent on the pituitary gland (Hutchison & Zeleznik, 1984). Studies in ewes in which the pituitary gland was removed evidenced a drop in the main luteal steroidogenic enzymes like steroidogenic acute regulatory protein (StAR), 3β-hydroxysteroid dehydrogenase (3βHSD), and cholesterol side-chain cleavage enzyme (CYP11A1) (Juengel, Nett, et al., 1995). Both LH and growth hormone were shown to be determinant for CL establishment and maintenance not only in ruminants (Niswender & Nett, 1994) but also in the mare (Watson, Pedersen, Thomson, & Fraser, 2000). Furthermore, in the mare the neutralization of LH with antibodies confirmed the dependence of equine CL on pituitary hormones (Pineda & Ginther, 1972). The LH luteotrophic role was similarly evidenced when the administration of GnRH to diestrus mares caused an increase in P4 production (Johnson, Becker, & Roma, 1988). Moreover, in vivo and in vitro studies confirmed that P4 secretion is stimulated after exposition to LH or hCG (Kelly, Hoyer, & Wise, 1988). Indeed, Ginther and others reported a low LH concentration in mid diestrus, but the occurrence of occasional LH pulses should ensure the required levels for CL maintenance in the mare (Ginther, 1992). In rhesus monkey, LH was shown to mediate the structural remodeling necessary during luteinization of granulosa cells, once the collection of these cells before LH surge and their culture in vitro produced substantial less amounts of P4 (Channing, 1970). Several studies reported that LH promotes the transient expression of luteotrophic factors in the CL, such as P4 receptor (Natraj & Richards, 1993), cyclooxygenase-2 (COX-2) (Lim et al., 1999), early growth response protein-1 (Egr-1) (Espey et al., 2000), and nerve growth factor IB (Nur77) (Park et al., 2001). The Egr-1 activates cyclic adenosine monophosphate (cAMP)/protein kinase A (PKA) and MAPK pathway and *Egr-1* knockout mice lack ovulation and CL formation (Russell, Doyle, Gonzales-Robayna, Pipaon, & Richards, 2003). Concerning Nurr77, it was shown to regulate transcriptional activity of steroidogenic cells (Chang & Chung, 1995) and it has been associated with apoptosis in nonovarian cells (Liu & Baker, 1994). Nonetheless, studies in *Nurr77* knockout mice did not show any aberrant reproductive phenotype, suggesting the existence of alternative pathways to Nur77 (Crawford, Sadovsky, Woodson, Lee, & Milbrandt, 1995). Besides these transiently activated pathways, factors permanently triggered by LH were clearly reviewed elsewhere (Stocco, Telleria, & Gibori, 2007).

3.2 Structural Organization and Vascularization

A biological process tightly linked with ovarian function is the angiogenesis, equally important during both luteal establishment and regression. Vessel proliferation is crucial for follicular growth and in preantral follicle around 40% of proliferative cells in the theca are of endothelial origin (Martelli et al., 2009). Indeed, the follicular vascular bed provides the foundation of luteal vasculature (Robinson et al., 2009). It was evidenced in the developing follicle that granulosa and theca cells produce angiogenic factors (Redmer, Rone, & Goodman, 1985). The LH surge induces several cellular and biochemical changes leading to ovulation (Reynolds & Redmer, 1999). A number of these processes, like break down of the basement membrane and immune-like response, support angiogenesis during luteal establishment (Dickson & Fraser, 2000). A great deal of research has recently characterized numerous local mediators with proangiogenic activity. Over the last decade the vascular endothelial growth factor A (VEGF) has been described as the main angiogenic factor produced by luteal cells (Redmer & Reynolds, 1996; Reynolds, Grazul-Bilska, & Redmer, 2002). We have also shown in the equine CL the proangiogenic cross talk between TNF and VEGF (Galvão, Ferreira-Dias, & Skarzynski, 2013). In vitro studies with equine luteal cells revealed that TNF increased VEGF protein expression (Galvão, Henriques, et al., 2012). In contrast, when cells were treated with VEGF, we found that the concentration of TNF in culture media was also augmented (Galvão, Ferreira-Dias, et al., 2013) what might suggest the involvement of both pathways in luteal vascularization in the mare. Another important target of TNF during vascularigenesis is the nitric oxide (NO) system, a very potent vascular mediator (Kimura et al., 2001). The NO and TNF axis has been shown to be determinant during CL establishment, promoting not only angiogenesis (Ferreira-Dias et al., 2011) but also the secretion of the luteo-supportive P4 (Ferreira-Dias et al., 2011) and PGE2 (Galvão, Skarzynski, & Ferreira-Dias, 2016; Harris, Squires, Michael, Bernal, & Abayasekara, 2001). Acting majorly on its type 2 receptor (TNFR2), TNF was able to stimulate the in vitro secretion of P4 and PGE2 and to inhibit PGF2α from early luteal phase CL (early-CL) isolated cells (Galvão, Skarzynski, et al., 2012). This suggests the existence of a feed-forward loop between VEGF and TNF on P4 and PGE2 secretion, which promotes either angiogenesis or secretory function.

Most recently, factors like fibroblast growth factor 2 (FGF2) (Robinson & Woad, 2017) or the delta-Notch signaling pathway have been implicated in the determination of cell fate and patterning associated with endothelial cell sprouting and vascular development (Benedito & Hellström, 2013; Murta et al., 2015; Vorontchikhina, Zimmermann, Shawber, Tang, & Kitajewski, 2005). Importantly, FGF2 was shown to be a key proangiogenic factor that responds to PGF2α in early CL in cows (Zalman et al., 2012). Indeed, PGF2α regulates angiogenic factors in the CL in a stage-dependent manner (Zalman et al., 2012). Thus, as one of the most vascularized organs in the body (Reynolds, Grazul-Bilska, Killilea, & Redmer, 1994), the CL function relies on a tight regulation of angiogenesis. Eventually, the microvasculature characterizing more than 50% of the cell population of a developed CL (O'Shea, Rodgers, & D'Occhio, 1989) is necessary to drain the P4 from the local microvasculature into the main stream, in order to act systemically.

4. LUTEOLYSIS

4.1 Luteolysis and the Uterine PGF2α

In the absence of pregnancy, luteolysis ensures the decrease in P4 production, which is followed by structural regression of the CL. The complexity of the luteolytic process has drawn the attention of the scientific community during the last decades, with a vast number of local signaling factors being described as luteolysis mediators in several species. With the CL fully developed and endocrinologically active, luteolysis appears as a default program ready to be triggered in the mid stage of the luteal phase primarily by the uterine PGF2α. The exception is preconized by signals for maternal recognition of pregnancy, which block the luteolytic process. In this chapter, we revisit the activation of the luteolytic process by the uterine PGF2α and the amplification of the signal by 2 cytokine superfamilies with significance for both functional and structural luteolysis: (i) the TNF superfamily; and (ii) transforming growth factor β (TGF) superfamily.

Alongside with the major luteolysin, the eicosanoid PGF2α (McCracken, Glew, & Scaramuzzi, 1970), cytokines such as TNF (Friedman, Weiss,

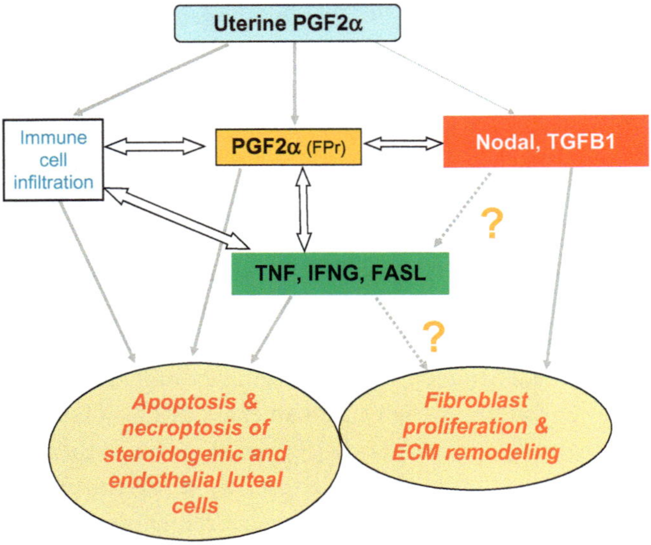

Fig. 2 Hypothetical representation of luteolytic interactions, highlighting the involvement Nodal and transforming growth factor β1 (TGFB1) in intraluteal prostaglandin (PG) F2α amplification. Their synchronous action with other cytokines such as tumor necrosis factor α (TNF), FAS-Ligand (FASL), and interferon γ (IFNG) hampers steroidogenesis and promotes luteal cell death and extracellular matrix (ECM) remodeling. Prostaglandin (PG) F2α receptor (FPr).

Levy, & Meidan, 2000; Galvão, Skarzynski, et al., 2012), Fas ligand, interferon γ (IFNG) (Davis & Rueda, 2002; Galvão, Henriques, et al., 2012), NO (Vega et al., 2000), Nodal (Galvão et al., 2016), and leukotrienes (Skarzynski, Jaroszewski, & Okuda, 2005) have been reported to contribute for luteolysis (Fig. 2). The collective action of these local mediators triggers (i) functional impairment and (ii) activation of programmed cell death mechanisms, which ultimately are responsible for the deep remodeling and elimination of the majority of the cellular content of the organ.

4.2 Intraluteal Activation of the Luteolytic Signal

The molecular regulation of luteolysis diverges among species. Nonetheless, it has been assumed that the main steps for luteal regression include (1) decrease in intracellular cholesterol transport and demise in P4 secretion (functional luteolysis); and (2) vasoconstriction and ECM remodeling, which is supported by leukocytes influx. Generally, neutrophils and macrophages invade the CL and secrete cytokines, capable of triggering apoptosis of luteal cells, and complete the involution of the organ (structural luteolysis) (Auletta & Flint, 1988; Hansel & Convey, 1983).

The luteolytic action of PGF2α is mediated by its specific plasma membrane receptor (FPr) (Powell, Hammarström, & Samuelsson, 1975). Importantly, PGF2α can upregulate the expression of its receptor, as shown in rat (Olofsson et al., 1996), cow (Rao, Estergreen, Carman, & Moss, 1979), or pig (Gadsby, Balapure, Britt, & Fitz, 1990) during late luteal phase. In the mare, besides being reported in follicular cells before and after ovulation (Sayasith, Bouchard, Doré, & Sirois, 2006), FPr transcription in the CL was shown to be unresponsive to PGF2α (Beg et al., 2005). Nonetheless, luteolysis onset seems to rely not exclusively on FPr transcription, once porcine luteal cells express the receptor from day 4 postovulation (Gadsby et al., 1990), but the responsiveness to PGF2α is achieved only after day 13 postovulation (Guthrie & Polge, 1976). Also in bovine CL, PGF2α was shown to be ineffective on luteolysis mediation before day 5 of luteal phase (Henricks, Long, Hill, & Dickey, 1974). Similar to what was seen for pigs, in cows, PGF2α receptors were already expressed on day 2 (Wiltbank, Shiao, Bergfelt, & Ginther, 1995). Additional auto-, paracrine mechanisms must determine the refractoriness of the CL to uterine, ovarian, or exogenous PGF2α. A differential affinity of FPr to PGF2α throughout the estrous cycle may contribute for the refractoriness seen in the early stages of the CL. As demonstrated for ruminants, unresponsiveness of the CL to PGF2α could be due to the existence of high- and low-affinity states of the FPr (McCracken et al., 1999). Recent studies highlighted that the early CL resistance to PGF2α in cows might rely on the activation of survival pathways, such as the FGF2, which enables the CL to evade luteolysis (Mondal et al., 2011). In the mare, the sensitivity of the CL to the luteolytic PGF2α should be also different, mainly because the systemic route for PG transport accounts for the metabolization of 90% of the PGs in lungs (Ferreira & Vane, 1967). As reported by Ginther, the sensitivity of equine CL is 18 times higher than in sheep (Ginther, 1992) and the affinity of the equine CL for PGF2α is 10 times greater than cow corpora lutea (Kimball & Wyngarden, 1977). Another characteristic of the mare is the responsiveness of the CL to the luteolytic eicosanoid, since a very rapid drop in plasma P4 is seen after a single PGF2α administration in mid luteal phase (Johnson et al., 1988). Particularly for the mare, studies from the past decade contradicted previous reports evidencing the absence of the luteolytic effect of PGF2α seen in the early luteal phase until day 5 postovulation (reviewed in Aurich, 2011). More recently, early administration of PGF2α on day 2 postovulation did impair luteal function and shortened interovulatory interval (Coffman et al., 2014; Nie, Johnson, Wenzel, & Braden, 2003; Troedsson et al., 2001). Nevertheless, a reliable comparison between studies is difficult to make, due to the diversity in nature and concentration of the lutelytic agents used, alongside with different responsiveness between species subjected to the pharmacological treatment. Ultimately, the determination of dosage per kg of body mass had to be provided (Handler, Wüstenhagen, Schams, Kindahl, & Aurich, 2004). Thus, it is arguable that the equine CL highly depends on auto-, paracrine activations for the amplification of the luteolytic signal. In conclusion, different molecular mechanisms must be temporarily activated during CL life span, in order to render the organ responsive to the uterine PGF2α and promote luteolysis.

4.3 Luteolytic Signal Amplification

In ruminants, one of the first responses of the CL to uterine PGF2α is the secretion of luteal oxytocin (OXT) (Jones & Flint, 1989; Theodosis, Wooding, Sheldrick, & Flint, 1986). Nonetheless, this episode is not seen, for instance, in the mare (Stevenson, Parkinson, & Wathes, 1991). As a consequence, intraluteal PGF2α is augmented, amplifying this way the luteolytic signal (Tsai & Wiltbank, 1997). In this regard, production of PGF2α by luteal tissue has been reported also in sheep (Rexroad & Guthrie, 1979), pigs (Guthrie & Polge, 1976), and horses (Watson & Sertich, 1990). A previous study in equine CL evidenced a threefold increase in the mRNA level of the rate-limiting enzyme involved in PG synthetic pathway, the COX2, after exogenous treatment with PGF2α (Beg et al., 2005). Using an in vitro model with equine luteal cells, we have also confirmed an increase in the protein expression of both COX2 and PGF2α synthase (PGFS) following PGF2α treatment (Galvão et al., 2016). Overall, these findings support the existence of an intraluteal autocrine loop to augment the luteolytic signal, which might present a species-specific regulation.

4.3.1 TNF Superfamily and Luteolysis

TNF superfamily comprises an increasing number of structurally related ligand–receptor pairs, as TNF, its receptors TNFR1 and TNFR2, Fas ligand, and its receptor FAS (Gravestein & Borst, 1998; Tansey & Szymkowski, 2009). We will discuss the luteolytic actions of two conserved cytokines from TNF superfamily: (i) TNF and (ii) FASL and their receptors, as well as the interferon family member IFNG (De Andrea, Ravera, Gioia, Gariglio, &

Landolfo, 2002), luteolytic action of which has been broadly studied in association with TNF and FASL.

A major event during luteolysis is the infiltration of the CL by immune cells (Bauer, Reibiger, & Spanel-Borowski, 2001; Cannon, Petroff, & Pate, 2003). Two particular leukocytes, macrophages and T lymphocytes, increase in number after initiation of the luteolytic process by PGF2α in cows (Bauer et al., 2001), humans (Best, Pudney, Welch, Burger, & Hill, 1996), mares (Al-Zi'abi, Fraser, & Watson, 2002), or mice (Petrovská, Dimitrov, & Michael, 1996). Ultimately, these cells represent the main source of inflammatory cytokines such as TNF, IFNG, or interleukin 1β (Okuda & Sakumoto, 2003; Pate & Landis Keyes, 2001). Numerous studies have demonstrated the role of TNF/TNFR1 system in luteolysis in cows (Petroff, Petroff, & Pate, 1999; Sakumoto, Berisha, Kawate, Schams, & Okuda, 2000; Skarzynski, Bah, et al., 2003), pigs (Wuttke, Pitzel, Knoke, Theiling, & Jarry, 1997), mare (Galvão, Skarzynski, et al., 2012), mice (Henkes et al., 2008), or humans (Vaskivuo et al., 2002). For instance, TNF has been shown to regulate StAR in rodents (Abdo, Hisheh, Arfuso, & Dharmarajan, 2008; Chen, Feng, & Liu, 1999), or to decrease P4 secretion and induce IFNG and FAS-mediated apoptosis of luteal steroidogenic and endothelial cells in bovine CL (Pru, Lynch, Davis, & Rueda, 2003; Taniguchi, Yokomizo, & Okuda, 2002). In the mare, acting on TNFR1, this cytokine not only promoted the in vitro secretion of PGF2α but also decreased P4 and PGE2 concentrations in the culture media of luteal cells isolated from the late luteal phase CL (Galvão, Skarzynski, et al., 2012). Interestingly, structural luteolysis, here characterized by decreased cell viability of cultured mid-luteal phase (mid-CL) cells and increased activity of the apoptosis marker PARP1, was seen only when the cytokines TNF and IFNG were used in association (Galvao et al., 2010). Nonetheless, in bovine CL, TNF itself promoted structural luteolysis, first inducing cell death of luteal endothelial cells in a dose- and time-dependent manner (Friedman et al., 2000). Another difference found between these two species concerns steroidogenesis modulation by either FASL or IFNG. In vitro studies with bovine luteal cells reported that IFNG was able to decrease P4 secretion (Fairchild & Pate, 1991; Nothnick & Pate, 1990). However, in the mare this cytokine mainly regulated structural luteolysis events, always in association with TNF or FASL (Galvao et al., 2010; Galvão, Skarzynski, et al., 2012). Furthermore, FASL has been recurrently associated with structural regression in cattle (Taniguchi et al., 2002), mice (Quirk, Harman, Huber, & Cowan, 2000), or rats (Roughton, Lareu, Bittles, & Dharmarajan, 1999).

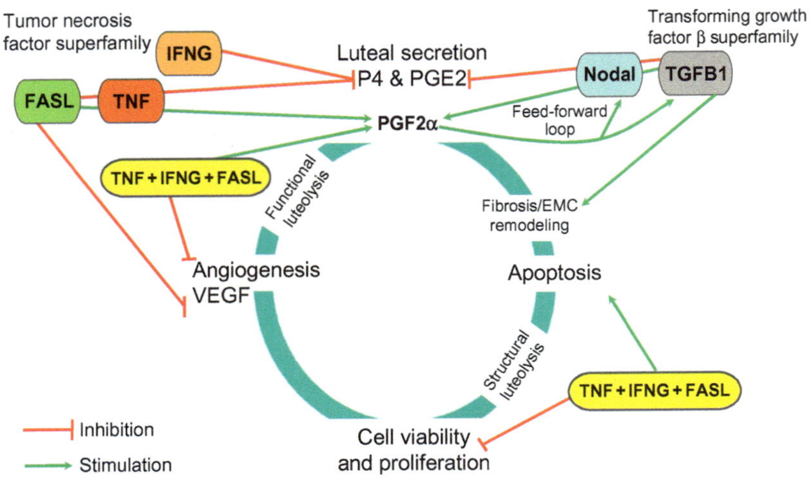

Fig. 3 Diagram depicting the involvement tumor necrosis factor and transforming growth factor β superfamilies in luteolysis. Tumor necrosis factor α (TNF), Fas ligand (FASL), interferon γ (IFNG), and Nodal and transforming growth factor β1 (TGFB1) mediate functional luteolysis, through downregulation of progesterone (P4) and prostaglandin (PG) E2 and upregulation of PGF2α. Structural luteolysis, here characterized by apoptosis and decreased cell viability, extracellular matrix (ECM) remodeling, and angioregression, is mainly supported by the cytokine synergistic action (TNF + IFNG + FASL).

Nonetheless, we have shown that FASL was able to modulate secretory activity of in vitro cultured equine luteal cells (Galvao et al., 2010). It augmented the concentration of PGF2α in culture media and diminished P4 and PGE2 (Galvao et al., 2010). More strikingly, in the mare the treatment with FASL alone did not affect cellular viability and apoptosis of luteal cells, except when used in the cytokine association TNF + IFNG + FASL (Galvao et al., 2010; Fig. 3). Accordingly, the synergism between these cytokines during structural regression was seen also in other species. As previously reported, TNF and IFNG mediate cytotoxic effects in luteal cells in the cow, through nitric oxide (NO) production (Benyo & Pate, 1992; Skarzynski et al., 2007). In fact, reactive oxygen species, mainly originated from nonsteroidogenic cells (Aten, Kolodecik, Rossi, Debusscher, & Behrman, 1998), were linked with either functional or structural luteolysis (Kato, Sugino, Takiguchi, Kashida, & Nakamura, 1997; Korzekwa et al., 2004; Korzekwa, Okuda, Woclawek-Potocka, Murakami, & Skarzynski, 2006). Their harmful properties were shown to be suppressed by P4, possibly through the inhibition of cell death mechanisms activation usually seen before the decline of steroidogenic capacity (Sugino et al., 1996). Conversely, blocking NO synthase (NOS) activity prevented the P4 downregulation seen during luteolysis activation and prolonged luteal phase in cows (Jaroszewski & Hansel, 2000; Skarzynski, Jaroszewski, et al., 2003). In the mare, it was also observed an increase in endothelial NOS (eNOS) expression during late CL stage (Ferreira-Dias et al., 2011). Furthermore, in vitro studies revealed that the cytokine association TNF + IFNG + FASL promoted NO production by mid-CL cells, through upregulation of mRNA transcription and protein expression of eNOS (Galvão, Szóstek, Skarzynski, & Ferreira-Dias, 2013). However, when those cytokines were used alone, no response was produced in equine CL (Galvão, Szóstek, et al., 2013). This highlights the need for a synchronous action of these cytokines during structural regression in the mare (Fig. 3).

Apoptosis or programmed cell death is another important component of the luteolytic process and can be activated through two main pathways: (i) intrinsic and (ii) extrinsic. The intrinsic pathway relies on stimuli originated within the cell, usually in response to drugs, radiation, or growth factors withdrawal. First, changes in the mitochondrial permeability are triggered through the unbalanced ratio between proapoptotic and anti-apoptotic Bcl-2 family members (Adams & Cory, 1998; Czabotar, Lessene, Strasser, & Adams, 2014). Conversely, cytokines such as TNF and FASL can mediate apoptosis through the extrinsic activation of death receptors TNFR1 and FAS (Nagata, 1997). Ultimately, both pathways lead to the recruitment of intracellular proteases—the caspases (CASP), which after cleavage can induce DNA fragmentation and chromatin condensation (Kitazumi & Tsukahara, 2011). In this regard, CASP3 is a pivotal mediator of apoptosis, so far characterized in the luteolytic process of rats (Yang et al., 2015), mare (Ferreira-Dias et al., 2007), ewes (Lee, Banu, McCracken, & Arosh, 2016; Rueda, Hendry, Tilly, & Hamernik, 1999), and cattle (Skarzynski et al., 2005). The cytokines TNF, IFNG, and FASL have been intimately connected with apoptosis in the CL (Benyo & Pate, 1992; Galvao et al., 2010; Hojo, Oda, Lee, Acosta, & Okuda, 2010; Taniguchi et al., 2002). These extrinsic apoptotic mediators were markedly upregulated in bovine CL, shortly after PGF2α treatment (Davis & Rueda, 2002; Okuda & Sakumoto, 2003). Nonetheless, recent studies in bovine CL revealed the involvement of TNF and IFN in CASP-independent programmed form of necrosis or inflammatory cell death named necroptosis (Hojo et al., 2016). Indeed, death ligands TNF and

FASL were shown to induce not only apoptosis but also necroptosis in a number of different tissues (Vanlangenakker, Vanden Berghe, & Vandenabeele, 2012). Necroptosis is regulated by the receptor-interacting protein kinase (RIPK) 1 and RIPK3 (Cho et al., 2009), which are both known as cellular stress sensors (Declercq, Vanden Berghe, & Vandenabeele, 2009). Not only RIPK1 and RIPK3 were shown to be upregulated in bovine CL after PGFα treatment in vivo, but also the in vitro treatment with the cytokine association TNF + IFNG supported mRNA and protein expression of RIPK1 and RIPK3 (Hojo et al., 2016). However, single treatments with TNF and IFNG alone did not affect RIPKs, what underlines once more the multifactorial nature of the luteolytic process. Thus, the synergism between TNF and IFNG appears to be a hallmark for structural luteolysis, mediating not only programmed cell death apoptosis but also necroptosis as recently characterized in the bovine CL.

Another important role of the cytokines from TNF superfamily during luteolysis is the support of vascular regression. The immune–endothelial interface is preponderant for the regulation of cellular processes throughout luteal phase, in particular during luteolysis. Endothelial cells present immunological functions, being important mediators of innate and adaptive immunity (Pober & Sessa, 2007). The close dialogue between the two cell types contributes for the amplification of the inflammatory signal, under which the activated endothelium secretes cytokines and chemokines that attract more immune cells and mediate their extravasation (Mai, Virtue, Shen, Wang, & Yang, 2013). Here one should notice the importance of the basal membrane degradation during luteolysis, which facilitates the invasion of the CL by immune cells. The basal membrane provides support to the blood vessels and is degraded by matrix metalloproteases (MMP) 2 and 9 (Klein & Bischoff, 2011). In bovine luteal cells, PGF2α in association with IFNG increased *MMP9* transcription (Abe, Sakumoto, & Okuda, 2015). In the same study the authors concluded that PGF2α and IFNG during luteolysis may be involved in EMC remodeling and detachment of luteal cells from the CL tissue by increasing the production of MMPs (Abe et al., 2015). Still in cows, the cytokines TNF and IFNG were shown to regulate endothelial cell function (Korzekwa, Bodek, Bukowska, Blitek, & Skarzynski, 2011). Noteworthy, these cytokines can enter a cross talk with endothelin-1 (ET-1; mainly secreted by endothelial cells), promoting death of microvascular endothelial cells of the bovine CL (Pru et al., 2003). In the mare, TNF showed a luteal stage-dependent effect on angiogenesis, acting as a proangiogenic factor in early- and mid-CL isolated cells, or as antiangiogenic in late-CL cells (Galvão, Ferreira-Dias, et al., 2013). In the late luteal stage, not only TNF but also IFNG and the cytokine association TNF + IFNG + FASL presented antiangiogenic properties. We have shown that the in vitro treatment with TNF + IFNG + FASL upregulated the mRNA transcription of the antiangiogenic mediator thrombospondin (THBS) 1 and its receptor CD36 (Galvão, Ferreira-Dias, et al., 2013; Galvão, Henriques, et al., 2012). Another interesting finding in the mare was the negative effect of FASL on VEGF. In mid-CL isolated cells, FASL severely decreased VEGF protein expression, a similar response seen for the cytokine association TNF + IFNG + FASL (Galvão, Henriques, et al., 2012). Indeed, FASL involvement in equine CL regression goes beyond its proapoptotic role demonstrated for bovine CL (Taniguchi et al., 2002). Besides hampering secretory activity of luteal cells (Galvao et al., 2010), FASL also supported angioregression in mid-CL (Galvão, Ferreira-Dias, et al., 2013; Galvão, Henriques, et al., 2012).

In conclusion, the cytokines TNF, IFNG, and FASL play a crucial role in the amplification of luteolytic signal, downregulating P4 and PGE2, upregulating PGF2α, mediating vascular regression, and promoting apoptosis and necroptosis of luteal cells.

4.3.2 Transforming Growth Factor β Superfamily and Luteolysis

TGF superfamily represents a large superfamily of cytokines, including two families: the TGF/activin/Nodal family and the bone morphogenetic protein/growth and differentiation factor/Mullerian-inhibiting substance family (Chang, Brown, & Matzuk, 2002). The ligands TGF, Nodal, and the coreceptor Cripto-1 exert their biological action through cell surface receptors (Activin receptor-like kinase—ALK), which phosphorylate intracellular signaling molecules (Smad), culminating on gene transcription regulation (Wang & Tsang, 2007). They are involved in different cellular processes in the adult, like cell growth, cell differentiation, and apoptosis, among others (Jones, Stoikos, Findlay, & Salamonsen, 2006; Wang & Tsang, 2007). Particularly in the CL, the ligand TGFβ1 (TGFB1) has been shown to be expressed in cattle (Farberov & Meidan, 2014; Maroni & Davis, 2012; Miyamoto, Okuda, Schweigert, & Schams, 1992), monkey (Chen et al., 2003), rat (Teerds & Dorrington, 1992), hamster (Roy & Kole, 1995), marmoset (Wehrenberg, Giebel, & Rune, 1998), woman (Chegini & Williams, 1992), and mare (Galvao et al., 2017). Nonetheless, the involvement of Nodal in luteal regression was shown exclusively for the equine CL (Galvão et al.,

2016). Like most of the ligands from TGFβ family, TGFB1 and Nodal activate serine/threonine kinase receptors that phosphorylate Smad proteins to regulate gene expression (Schier, 2003, 2009). TGFB1 role on luteal function was first reported in the woman CL (Chegini & Williams, 1992). Few years later, studies in marmosets linked the costaining of 3βHSD and TGFB1 to the luteal establishment (Wehrenberg et al., 1998). But, Miyamoto and coworkers have shown an opposing role for this cytokine, once the in vivo treatment of bovine CL with TGFB1 decreased the LH-induced P4 secretion in a dose-dependent manner (Miyamoto et al., 1992). More recently, TGFB1 downregulated StAR expression and P4 secretion, through both canonical (SMAD3) and noncanonical (ERK1/2) signaling pathways (Fang, Chang, Cheng, Leung, & Sun, 2014), and TGFB1 itself was amplified by PGF2α (Hou et al., 2008). The suppressive role of TGFB1 on P4 demise and functional luteolysis was confirmed also in the mare. Galvao and coworkers have shown that TGFB1 downregulated P4 and PGE2 in culture media of mid-CL cells, through inhibition of protein expression of StAR, CYP11A1, and 3βHSD and microsomal PGES-1 and cytosolic PGES (Galvao et al., 2017). But, PGF2α was upregulated via COX2 and PGFS translation (Galvao et al., 2017). In line with our findings for TGFB1, Nodal also decreased the expression of both P4 and PGE2 synthetic enzymes and supported PGF2α expression (Galvão et al., 2016). Evidences that PGF2α promoted phosphorylation of Smad3, a marker of TGFB1 and Nodal signaling activation, confirmed the existence of a feed-forward loop between TGFB1/Nodal system and the luteolytic eicosanoid (Fig. 3) (Galvão et al., 2016, 2017). Importantly, the pharmacological inhibition of the type I receptors for both pathways—the ALK5, ALK4, and ALK7, efficiently reverted the luteolytic role of PGF2α on P4 and PGE2 synthetic machinery downregulation and also the upregulation of its own enzymes COX2 and PGFS (unpublished data). This emphasizes the importance of TGB1 and Nodal signaling for intraluteal amplification of the luteolytic signal.

Despite its well-established involvement in functional luteolysis, TGFB1 has also been shown to mediate structural luteolysis in several species. In bovine CL, TGFB1 mediated stromal vascular reorganization, inhibiting the growth of microvascular endothelial cells (CLENDO cells) and disrupting capillary morphogenesis and endothelial barrier function (Maroni & Davis, 2011). Furthermore, Farberov and Meidan unraveled the relationship between TGFB1 and the antiangiogenic factor THBS1, in which the vasoactive compound was able to amplify TGFB1 and promote the phosphorylation of SMAD2 in luteal bovine endothelial cells (Farberov & Meidan, 2014). Thus, TGFB1 was shown to directly regulate luteal endothelium, participating in the vascular regression necessary for luteolysis accomplishment.

Another good evidence of TGFB1 luteolytic role is its profibrotic action previously shown. Maroni and Davis have evidenced that the treatment of luteal fibroblasts with TGFB1 induced the expression of laminin, collagen (COLL1), and MMP1 (Maroni & Davis, 2012). Importantly, luteal fibroblasts do not express PGF2α receptors as the steroidogenic cells, suggesting that paracrine interactions are fundamental for the luteolytic signal propagation. Thus, PGF2α-induced TGFB1 produced in steroidogenic cells regulated the expression of profibrotic mediators from luteal fibroblasts (Maroni & Davis, 2012). Using an in vitro system with mid-CL isolated cells, we verified that TGFB1 promoted the transcription of the profibrotic genes *COLL1A2* and *COLL3A2* in a dose-dependent manner (Fig. 4A and B). This suggests that in regressing CL, TGFB1 expression might trigger the transcription of profibrotic mediators in the mare. Also, we have assessed the transcription of MMPs after TGFB1 treatment and verified that the mRNA level of *MMP1* and *MMP9* was drastically increased by TGFB1 at 0.1 and 1 ng/mL (Fig. 4C and D), whereas *MMP2* was slightly increased exclusively by TGFB1 at 5 ng/mL (Fig. 4E). As seen for women CL, MMP2 expression peaked in late CL (Duncan, McNeilly, & Illingworth, 1998), denoting a role in luteolysis. However, MMP9 has been associated with strong ECM remodeling after ovulation and during CL establishment (Duncan et al., 1998). The abovementioned study from Maroni and Davis linked TGFB1 with MMP1 during luteolysis (Maroni & Davis, 2012). Nonetheless, we verified a very strong response of MMP9 and MMP1 to TGFB1. Indeed, the present analysis considers exclusively the mRNA transcription of MMPs, whereas a more reliable measurement would be the assessment of their activity by zymography. Finally, another component of ECM, *fibronectin* (*FN*) *1* mRNA was also increased by TGFB1 at 1 and 5 ng/mL (Fig. 4F). Overall, we suggest that TGFB1 might be involved in structural regression, playing a profibrotic role through collagen deposition and ECM remodeling.

Nodal and TGFB1 seem to play important roles in the luteolytic process. So far, Nodal has been exclusively linked to functional regression in in vitro studies in equine CL, downregulating P4 and PGE_2 and upregulating

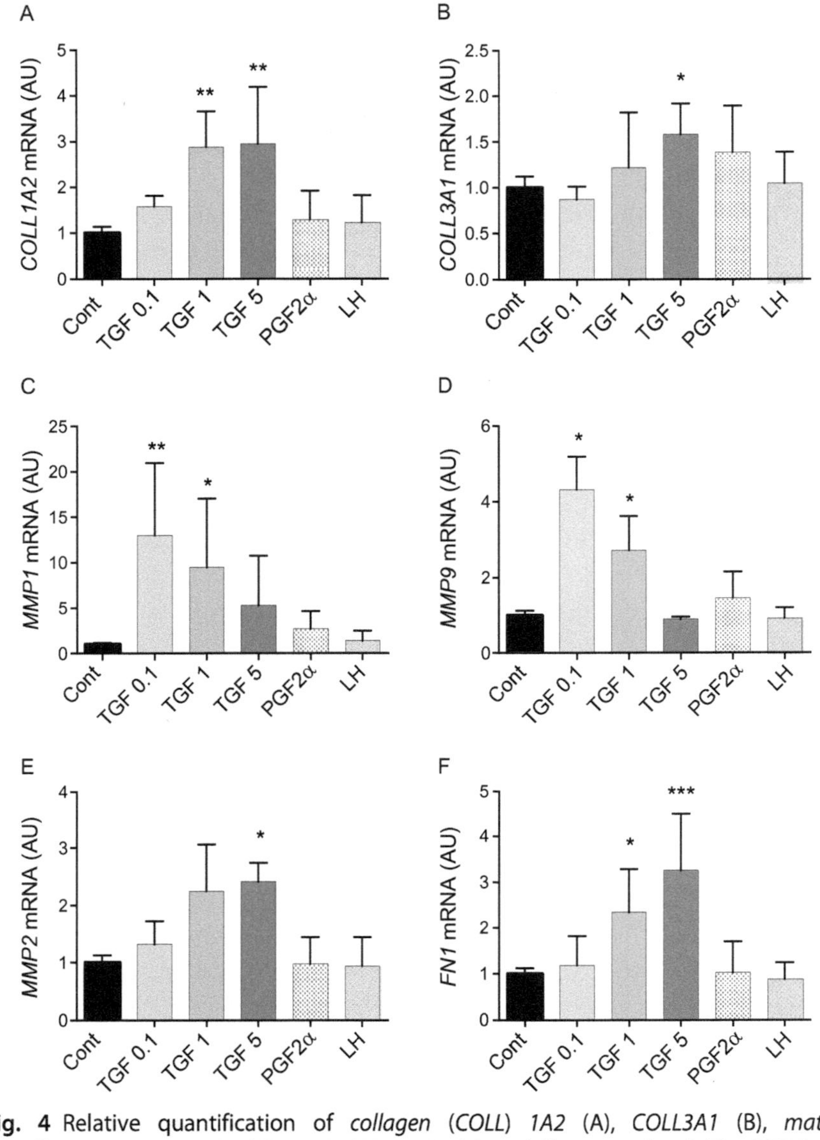

Fig. 4 Relative quantification of *collagen* (*COLL*) *1A2* (A), *COLL3A1* (B), *matrix metalloproteinase* (*MMP*) *1* (C), *MMP9* (D), *MMP2* (E), and *fibronectin* (*FN1*) (F) mRNA level by real time PCR ($n=6$), in mid-CL cells after in vitro treatment with: (i) saline; (ii) TGFB1 (TGF) in different concentrations (0.1 to 5 ng/mL); (iii) PGF2α (10^{-7}M); and (iv) LH (10 ng/mL). Transcription normalized with the housekeeping gene *β-2 microglobulin* (*B2MG*). Bars represent mean ± SEM. Luteal tissue and venous blood from jugular vein were collected postmortem at the local abattoir from randomly designated cyclic Lusitano mares. Luteal structures were classified in different luteal stages (early, mid, and late CL) as previously described (Galvão, Henriques, et al., 2012; Galvao et al., 2010). All methodologies for luteal cells' isolation and culture, real-time PCR, and statistical analysis were recently described (Galvão et al., 2016). $*P < 0.05$; $**P < 0.01$; $***P < 0.001$ (all primers in Table 1).

Table 1 Primer Sequences Used in Real-Time PCR Analysis of Equine CL

Gene (Accession Number)	Sequence 5′–3	Amplicon (bp)
COL1A2 (XM_001492939.3)	Forward: CAAGGGCATTAGGGGACACA	196
	Reverse: ACCCACACTTCCATCGCTTC	
COL3A1 (AF117954.1)	Forward: CAAAGGAGAGCCAGGAGCAC	98
	Reverse: GACCGCAGCACTCTTAGCACA	
FN1 (XM_001489104.5)	Forward: CTCCCTGGTGGTGCTCATCT	143
	Reverse: CGGCAGAAGAGGCATTTGAT	
MMP2 (XM_001493281.2)	Forward: TCCCACTTTGATGACGACGA	115
	Reverse: TTGCCGTTGAAGAGGAAAGG	
MMP9 (NM_001111302.19)	Forward: GCGGTAAGGTGCTGCTGTTC	177
	Reverse: GAAGCGGTCCTGGGAGAAGT	
MMP1 (XM_014860019.1)	Forward: TCTCACAGCTTCCCAACAGT	230
	Reverse: CATCTGGGCTGCTTCATCAC	
B2MG (X69083)	Forward: CGGGCTACTCTCCCTGACTG	92
	Reverse: TTGGCTTTCCATTCTCTGCTG	

PGF2α. Conversely, TGFB1 has been vastly investigated in this context, with in vivo and in vitro reports describing its involvement in functional and structural luteolysis. Noteworthy, a feed-forward loop between Nodal and TGFB1 and PGF2α seems to be determinant for the local amplification of the luteolytic signal.

5. CONCLUSIONS

In mammalian species, optimal luteal function is determinant for reproductive success. Progesterone secreted by steroidogenic luteal cells signals not only at the endometrium level but also locally, activating auto-, paracrine mediators such as TNF, cAMP/protein kinase A, FGF2, and NO that ensure major cellular processes such as neovascularization and ECM remodeling, steroidogenesis, and cell survival or apoptosis inhibition. This very fast sequence of events enables the rapid elevation of circulating levels of P4, crucial for pregnancy establishment. In the absence of fertilization, luteolysis is triggered by the uterine PGF2α. Consequently, a myriad of intraluteal interactions is started and culminate with: (i) steroidogenesis impairment—function luteolysis; and (ii) angioregression, death of luteal and endothelial cells, and involution of the organ—structural luteolysis. In this regard, auto-, paracrine factors like the cytokines TNF, FASL, IFNG, Nodal, and TGFB1 represent potential tools to manipulate the luteolytic process, once they were shown to downregulate steroidogenesis, block angiogenesis and support different forms of cell death, like apoptosis and necroptosis. New factors, such as the TGFβ superfamily member Nodal, are being characterized shedding new light on the complexity of the luteolytic process. A better understanding of their mechanisms of action will potentially reveal new therapies to tackle luteal insufficiency or new strategies to manipulate the estrous cycle.

ACKNOWLEDGMENTS

Work supported by Program "Juventus Plus" (IP2014011373)—Polish Ministry of Science and Higher Education. A.G. was supported by the FCT fellowship (SFRH/BPD/79001/2011).

REFERENCES

Abdo, M., Hisheh, S., Arfuso, F., & Dharmarajan, A. (2008). The expression of tumor necrosis factor-alpha, its receptors and steroidogenic acute regulatory protein during corpus luteum regression. *Reproductive Biology and Endocrinology*, *6*, 50.

Abe, H., Sakumoto, R., & Okuda, K. (2015). Expression of matrix metalloproteinases in bovine luteal cells induced by prostaglandin F2α, interferon γ and tumor necrosis factor α. *The Journal of Reproduction and Development*, *61*, 277–286.

Adams, J. M., & Cory, S. (1998). The Bcl-2 protein family: Arbiters of cell survival. *Science*, *281*, 1322–1326.

Al-Zi'abi, M. O., Fraser, H. M., & Watson, E. D. (2002). Cell death during natural and induced luteal regression in mares. *Reproduction*, *123*, 67–77.

Anderson, L. L., Bland, K. P., & Melampy, R. M. (1969). Comparative aspects of uterine-luteal relationships. *Recent Progress in Hormone Research*, *25*, 57–104.

Aten, R. F., Kolodecik, T. R., Rossi, M. J., Debusscher, C., & Behrman, H. R. (1998). Prostaglandin f2alpha treatment in vivo, but not in vitro, stimulates protein kinase C-activated superoxide production by nonsteroidogenic cells of the rat corpus luteum. *Biology of Reproduction*, *59*, 1069–1076.

Auletta, F. J., & Flint, A. P. (1988). Mechanisms controlling corpus luteum function in sheep, cows, nonhuman primates, and women especially in relation to the time of luteolysis. *Endocrine Reviews*, *9*, 88–105.

Aurich, C. (2011). Reproductive cycles of horses. *Animal Reproduction Science*, *124*, 220–228.

Bauer, M., Reibiger, I., & Spanel-Borowski, K. (2001). Leucocyte proliferation in the bovine corpus luteum. *Reproduction*, *121*, 297–305.

Beg, M. A., Gastal, E. L., Gastal, M. O., Ji, S., Wiltbank, M. C., & Ginther, O. J. (2005). Changes in steady-state concentrations of messenger ribonucleic acids in luteal tissue during prostaglandin F2alpha induced luteolysis in mares. *Animal Reproduction Science*, *90*, 273–285.

Benedito, R., & Hellström, M. (2013). Notch as a hub for signaling in angiogenesis. *Experimental Cell Research*, *319*, 1281–1288.

Benyo, D. F., & Pate, J. L. (1992). Tumor necrosis factor-alpha alters bovine luteal cell synthetic capacity and viability. *Endocrinology*, *130*, 854–860.

Best, C. L., Pudney, J., Welch, W. R., Burger, N., & Hill, J. A. (1996). Localization and characterization of white blood cell populations within the human ovary throughout the menstrual cycle and menopause. *Human Reproduction*, *11*, 790–797.

Cannon, M. J., Petroff, M. G., & Pate, J. L. (2003). Effects of prostaglandin F2alpha and progesterone on the ability of bovine luteal cells to stimulate T lymphocyte proliferation. *Biology of Reproduction*, *69*, 695–700.

Chang, H., Brown, C. W., & Matzuk, M. M. (2002). Genetic analysis of the mammalian transforming growth factor-β superfamily. *Endocrine Reviews*, *23*, 787–823.

Chang, S. F., & Chung, B. C. (1995). Difference in transcriptional activity of two homologous CYP21A genes. *Molecular Endocrinology*, *9*, 1330–1336.

Channing, C. P. (1970). Influences of the in vivo and in vitro hormonal environment upon luteinization of granulosa cells in tissue culture. *Recent Progress in Hormone Research*, *26*, 589–622.

Chegini, N., & Williams, R. S. (1992). Immunocytochemical localization of transforming growth factors (TGFs) TGF-alpha and TGF-beta in human ovarian tissues. *The Journal of Clinical Endocrinology and Metabolism*, *74*, 973–980.

Chen, Y. J., Feng, Q., & Liu, Y. X. (1999). Expression of the steroidogenic acute regulatory protein and luteinizing hormone receptor and their regulation by tumor necrosis factor alpha in rat corpora lutea. *Biology of Reproduction*, *60*, 419–427.

Chen, X.-L., Gao, H.-J., Wei, P., Song, X.-X., Hu, Z.-Y., & Liu, Y.-X. (2003). Regulatory effect of IFN-gamma on expression of TGF-beta 1, T beta R-II, and StAR in corpus luteum of pregnant rhesus monkey. *Acta Pharmacologica Sinica*, *24*, 435–441.

Cho, Y. S., Challa, S., Moquin, D., Genga, R., Ray, T. D., Guildford, M., et al. (2009). Phosphorylation-driven assembly of the RIP1-RIP3 complex regulates programmed necrosis and virus-induced inflammation. *Cell*, *137*, 1112–1123.

Coffman, E. A., Pinto, C. R. F., Snyder, H. K., Leisinger, C. A., Cole, K., & Whisnant, C. S. (2014). Antiluteogenic effects of serial prostaglandin F2α administration in cycling mares. *Theriogenology*, *82*, 1241–1245.

Crawford, P. A., Sadovsky, Y., Woodson, K., Lee, S. L., & Milbrandt, J. (1995). Adrenocortical function and regulation of the steroid 21-hydroxylase gene in NGFI-B-deficient mice. *Molecular and Cellular Biology*, *15*, 4331–4336.

Crisp, T. M., & Channing, C. P. (1972). Fine structural events correlated with progestin secretion during luteinization of rhesus monkey granulosa cells in culture. *Biology of Reproduction, 7*, 55–72.

Czabotar, P. E., Lessene, G., Strasser, A., & Adams, J. M. (2014). Control of apoptosis by the BCL-2 protein family: Implications for physiology and therapy. *Nature Reviews. Molecular Cell Biology, 15*, 49–63.

Davis, J. S., & Rueda, B. R. (2002). The corpus luteum: An ovarian structure with maternal instincts and suicidal tendencies. *Frontiers in Bioscience, 7*, d1949–1978.

De Andrea, M., Ravera, R., Gioia, D., Gariglio, M., & Landolfo, S. (2002). The interferon system: An overview. *European Journal of Paediatric Neurology, 6*(Suppl. A), A41-46-58.

Declercq, W., Vanden Berghe, T., & Vandenabeele, P. (2009). RIP kinases at the crossroads of cell death and survival. *Cell, 138*, 229–232.

Dickson, S. E., & Fraser, H. M. (2000). Inhibition of early luteal angiogenesis by gonadotropin-releasing hormone antagonist treatment in the primate. *The Journal of Clinical Endocrinology and Metabolism, 85*, 2339–2344.

Douglas, R. H., & Ginther, O. J. (1972). Effect of prostaglandin F2a on length of diestrus in mares. *Prostaglandins, 2*, 265–268.

Douglas, R. H., & Ginther, O. J. (1975). Route of prostaglandin F2alpha injection and luteolysis in mares (38519). *Proceedings of the Society for Experimental Biology and Medicine, 148*, 263–269.

Duncan, W. C., McNeilly, A. S., & Illingworth, P. J. (1998). The effect of luteal "rescue" on the expression and localization of matrix metalloproteinases and their tissue inhibitors in the human corpus Luteum. *The Journal of Clinical Endocrinology and Metabolism, 83*, 2470–2478.

Espey, L. L., Ujioka, T., Russell, D. L., Skelsey, M., Vladu, B., Robker, R. L., et al. (2000). Induction of early growth response protein-1 gene expression in the rat ovary in response to an ovulatory dose of human chorionic gonadotropin. *Endocrinology, 141*, 2385–2391.

Fairchild, D. L., & Pate, J. L. (1991). Modulation of bovine luteal cell synthetic capacity by interferon-gamma. *Biology of Reproduction, 44*, 357–363.

Fang, L., Chang, H.-M., Cheng, J.-C., Leung, P. C. K., & Sun, Y.-P. (2014). TGF-β1 downregulates StAR expression and decreases progesterone production through Smad3 and ERK1/2 signaling pathways in human granulosa cells. *The Journal of Clinical Endocrinology and Metabolism, 99*, E2234–2243.

Farberov, S., & Meidan, R. (2014). Functions and transcriptional regulation of thrombospondins and their interrelationship with fibroblast growth factor-2 in bovine luteal cells. *Biology of Reproduction, 91*, 58.

Ferreira, S. H., & Vane, J. R. (1967). Prostaglandins: Their disappearance from and release into the circulation. *Nature, 216*, 868–873.

Ferreira-Dias, G., Bravo, P. P., Mateus, L., Redmer, D. A., & Medeiros, J. A. (2006). Microvascularization and angiogenic activity of equine corpora lutea throughout the estrous cycle. *Domestic Animal Endocrinology, 30*, 247–259.

Ferreira-Dias, G., Costa, A. S., Mateus, L., Korzekwa, A. J., Galvão, A., Redmer, D. A., et al. (2011). Nitric oxide stimulates progesterone and prostaglandin E2 secretion as well as angiogenic activity in the equine corpus luteum. *Domestic Animal Endocrinology, 40*, 1–9.

Ferreira-Dias, G., Mateus, L., Costa, A. S., Solá, S., Ramalho, R. M., Castro, R. E., et al. (2007). Progesterone and caspase-3 activation in equine cyclic corpora lutea. *Reproduction in Domestic Animals, 42*, 380–386.

Friedman, A., Weiss, S., Levy, N., & Meidan, R. (2000). Role of tumor necrosis factor alpha and its type I receptor in luteal regression: Induction of programmed cell death in bovine corpus luteum-derived endothelial cells. *Biology of Reproduction, 63*, 1905–1912.

Frobenius, W. (1999). Ludwig Fraenkel: 'Spiritus rector' of the early progesterone research. *European Journal of Obstetrics and Gynecology and Reproductive Biology, 83*, 115–119.

Gadsby, J. E., Balapure, A. K., Britt, J. H., & Fitz, T. A. (1990). Prostaglandin F2 alpha receptors on enzyme-dissociated pig luteal cells throughout the estrous cycle. *Endocrinology, 126*, 787–795.

Galvão, A. M., Ferreira-Dias, G., & Skarzynski, D. J. (2013). Cytokines and angiogenesis in the corpus luteum. *Mediators of Inflammation, 2013*, 420186.

Galvão, A., Henriques, S., Pestka, D., Lukasik, K., Skarzynski, D., Mateus, L. M., et al. (2012). Equine luteal function regulation may depend on the interaction between cytokines and vascular endothelial growth factor: An in vitro study. *Biology of Reproduction, 86*, 187.

Galvao, A. M., Ramilo, D. W., Skarzynski, D. J., Lukasik, K., Tramontano, A., Mollo, A., et al. (2010). Is FAS/Fas ligand system involved in equine corpus luteum functional regression? *Biology of Reproduction, 83*, 901–908.

Galvão, A., Skarzynski, D., & Ferreira-Dias, G. (2016). Nodal promotes functional luteolysis via down-regulation of progesterone and prostaglandins E2 and promotion of PGF2α synthetic pathways in mare corpus luteum. *Endocrinology, 157*, 858–871.

Galvão, A., Skarzynski, D. J., Szóstek, A., Silva, E., Tramontano, A., Mollo, A., et al. (2012). Cytokines tumor necrosis factor-α and interferon-γ participate in modulation of the equine corpus luteum as autocrine and paracrine factors. *Journal of Reproductive Immunology, 93*, 28–37.

Galvão, A. M., Szóstek, A. Z., Skarzynski, D. J., & Ferreira-Dias, G. M. (2013). Role of tumor necrosis factor-α, interferon-γ and Fas-ligand on in vitro nitric oxide activity in the corpus luteum. *Cytokine, 64*, 18–21.

Galvao, A., Wolodko, K., Adamowski, M., Rebordao, M.R., Skarzynski, D.J., and Ferreira-Dias, G. (2017). The involvement of transforming growth factor beta 1 in equine functional luteolysis. (Edinburgh: Fertility 2017), p. 138.

Ginther, O. J. (1976). *Comparative anatomy of uteroovarian vasculature.* Veterinary Scope.

Ginther, O. J. (1992). *Reproductive biology of the mare: Basic and applied aspects* (2nd ed.). Cross Plains, WI: Equiservices.

Ginther, O. J., Utt, M. D., Bergfelt, D. R., & Beg, M. A. (2006). Controlling interrelationships of progesterone/LH and estradiol/LH in mares. *Animal Reproduction Science, 95*, 144–150.

Graham, J. D., & Clarke, C. L. (1997). Physiological action of progesterone in target tissues. *Endocrine Reviews, 18*, 502–519.

Gravestein, L. A., & Borst, J. (1998). Tumor necrosis factor receptor family members in the immune system. *Seminars in Immunology, 10*, 423–434.

Guthrie, H. D., & Polge, C. (1976). Luteal function and oestrus in gilts treated with a synthetic analogue of prostaglandin F-2alpha (ICI 79,939) at various times during the oestrous cycle. *Journal of Reproduction and Fertility, 48*, 423–425.

Handler, J., Wüstenhagen, A., Schams, D., Kindahl, H., & Aurich, C. (2004). Estrous cycle characteristics, luteal function, secretion of oxytocin (OT) and plasma concentrations of 15-keto-13,14-dihydro-PGF2alpha (PGF2alpha-metabolite) after administration of low doses of prostaglandin F2alpha (PGF2alpha) in pony mares. *Theriogenology, 61*, 1573–1582.

Hansel, W., & Convey, E. M. (1983). Physiology of the estrous cycle. *Journal of Animal Science, 57*(Suppl. 2), 404–424.

Harris, T. E., Squires, P. E., Michael, A. E., Bernal, A. L., & Abayasekara, D. R. (2001). Human granulosa-lutein cells express functional EP1 and EP2 prostaglandin receptors. *Biochemical and Biophysical Research Communications, 285*, 1089–1094.

Henderson, K. M., & McNatty, K. P. (1975). A biochemical hypothesis to explain the mechanism of luteal regression. *Prostaglandins, 9*, 779–797.

Henkes, L. E., Sullivan, B. T., Lynch, M. P., Kolesnick, R., Arsenault, D., Puder, M., et al. (2008). Acid sphingomyelinase involvement in tumor necrosis factor α-regulated vascular and steroid disruption during luteolysis in vivo. *Proceedings of the National Academy of Sciences of the United States of America*, *105*, 7670–7675.

Henricks, D. M., Long, J. T., Hill, J. R., & Dickey, J. F. (1974). The effect of prostaglandin F2 alpha during various stages of the oestrous cycle of beef heifers. *Journal of Reproduction and Fertility*, *41*, 113–120.

Hojo, T., Oda, A., Lee, S.-H., Acosta, T. J., & Okuda, K. (2010). Effects of tumor necrosis factor α and interferon γ on the viability and mRNA expression of TNF receptor type I in endothelial cells from the bovine corpus luteum. *The Journal of Reproduction and Development*, *56*, 515–519.

Hojo, T., Siemieniuch, M. J., Lukasik, K., Piotrowska-Tomala, K. K., Jonczyk, A. W., Okuda, K., et al. (2016). Programmed necrosis—A new mechanism of steroidogenic luteal cell death and elimination during luteolysis in cows. *Scientific Reports*, *6*, 38211.

Hou, X., Arvisais, E. W., Jiang, C., Chen, D., Roy, S. K., Pate, J. L., et al. (2008). Prostaglandin F2alpha stimulates the expression and secretion of transforming growth factor B1 via induction of the early growth response 1 gene (EGR1) in the bovine corpus luteum. *Molecular Endocrinology*, *22*, 403–414.

Hutchison, J. S., & Zeleznik, A. J. (1984). The rhesus monkey corpus luteum is dependent on pituitary gonadotropin secretion throughout the luteal phase of the menstrual cycle. *Endocrinology*, *115*, 1780–1786.

Jablonka-Shariff, A., Grazul-Bilska, A. T., Redmer, D. A., & Reynolds, L. P. (1993). Growth and cellular proliferation of ovine corpora lutea throughout the estrous cycle. *Endocrinology*, *133*, 1871–1879.

Jaroszewski, J. J., & Hansel, W. (2000). Intraluteal administration of a nitric oxide synthase blocker stimulates progesterone and oxytocin secretion and prolongs the life span of the bovine corpus luteum. *Proceedings of the Society for Experimental Biology and Medicine*, *224*, 50–55.

Johnson, A. L., Becker, S. E., & Roma, M. L. (1988). Effects of gonadotrophin-releasing hormone and prostaglandin F-2 alpha on corpus luteum function and timing of the subsequent ovulation in the mare. *Journal of Reproduction and Fertility*, *83*, 545–551.

Jones, D. S., & Flint, A. P. (1989). Nucleotide sequence of a full length cDNA clone encoding the oxytocin-neurophysin I precursor isolated from the ovine corpus luteum. *Nucleic Acids Research*, *17*, 7990.

Jones, R. L., Stoikos, C., Findlay, J. K., & Salamonsen, L. A. (2006). TGF-beta superfamily expression and actions in the endometrium and placenta. *Reproduction*, *132*, 217–232.

Juengel, J. L., Meberg, B. M., Turzillo, A. M., Nett, T. M., & Niswender, G. D. (1995). Hormonal regulation of messenger ribonucleic acid encoding steroidogenic acute regulatory protein in ovine corpora lutea. *Endocrinology*, *136*, 5423–5429.

Juengel, J. L., Nett, T. M., Tandeski, T. R., Eckery, D. C., Sawyer, H. R., & Niswender, G. D. (1995). Effects of luteinizing hormone and growth hormone on luteal development in hypophysectomized ewes. *Endocrine*, *3*, 323–326.

Kato, H., Sugino, N., Takiguchi, S., Kashida, S., & Nakamura, Y. (1997). Roles of reactive oxygen species in the regulation of luteal function. *Reviews of Reproduction*, *2*, 81–83.

Kelly, C. M., Hoyer, P. B., & Wise, M. E. (1988). In-vitro and in-vivo responsiveness of the corpus luteum of the mare to gonadotrophin stimulation. *Journal of Reproduction and Fertility*, *84*, 593–600.

Kimball, F. A., & Wyngarden, L. J. (1977). Prostaglandin F2alpha specific binding in equine corpora lutea. *Prostaglandins*, *13*, 553–564.

Kimura, H., Weisz, A., Ogura, T., Hitomi, Y., Kurashima, Y., Hashimoto, K., et al. (2001). Identification of hypoxia-inducible factor 1 ancillary sequence and its function in vascular endothelial growth factor gene induction by hypoxia and nitric oxide. *The Journal of Biological Chemistry*, *276*, 2292–2298.

Kitazumi, I., & Tsukahara, M. (2011). Regulation of DNA fragmentation: The role of caspases and phosphorylation. *The FEBS Journal*, *278*, 427–441.
Klein, T., & Bischoff, R. (2011). Physiology and pathophysiology of matrix metalloproteases. *Amino Acids*, *41*, 271–290.
Korzekwa, A. J., Bodek, G., Bukowska, J., Blitek, A., & Skarzynski, D. J. (2011). Characterization of bovine immortalized luteal endothelial cells: Action of cytokines on production and content of arachidonic acid metabolites. *Reproductive Biology and Endocrinology*, *9*, 27.
Korzekwa, A., Jaroszewski, J. J., Bogacki, M., Deptula, K. M., Maslanka, T. S., Acosta, T. J., et al. (2004). Effects of prostaglandin F(2alpha) and nitric oxide on the secretory function of bovine luteal cells. *The Journal of Reproduction and Development*, *50*, 411–417.
Korzekwa, A. J., Okuda, K., Woclawek-Potocka, I., Murakami, S., & Skarzynski, D. J. (2006). Nitric oxide induces apoptosis in bovine luteal cells. *The Journal of Reproduction and Development*, *52*, 353–361.
Lee, J., Banu, S. K., McCracken, J. A., & Arosh, J. A. (2016). Early pregnancy modulates survival and apoptosis pathways in the corpus luteum in sheep. *Reproduction*, *151*, 187–202.
Lim, C. S., Baumann, C. T., Htun, H., Xian, W., Irie, M., Smith, C. L., et al. (1999). Differential localization and activity of the A- and B-forms of the human progesterone receptor using green fluorescent protein chimeras. *Molecular Endocrinology*, *13*, 366–375.
Liu, D. Y., & Baker, H. W. G. (1994). Andrology: Disordered acrosome reaction of spermatozoa bound to the zona pellucida: A newly discovered sperm defect causing infertility with reduced sperm-zona pellucida penetration and reduced fertilization in vitro. *Human Reproduction*, *9*, 1694–1700.
Mai, J., Virtue, A., Shen, J., Wang, H., & Yang, X.-F. (2013). An evolving new paradigm: Endothelial cells—Conditional innate immune cells. *Journal of Hematology & Oncology*, *6*, 61.
Maroni, D., & Davis, J. S. (2011). TGFB1 disrupts the angiogenic potential of microvascular endothelial cells of the corpus luteum. *Journal of Cell Science*, *124*, 2501–2510.
Maroni, D., & Davis, J. S. (2012). Transforming growth factor Beta 1 stimulates profibrotic activities of luteal fibroblasts in cows. *Biology of Reproduction*, *87*, 127.
Martelli, A., Palmerini, M. G., Russo, V., Rinaldi, C., Bernabò, N., Di Giacinto, O., et al. (2009). Blood vessel remodeling in pig ovarian follicles during the periovulatory period: An immunohistochemistry and SEM-corrosion casting study. *Reproductive Biology and Endocrinology*, *7*, 72.
McCracken, J. A., Carlson, J. C., Glew, M. E., Goding, J. R., Baird, D. T., Grëen, K., et al. (1972). Prostaglandin F 2 identified as a luteolytic hormone in sheep. *Nature: New Biology*, *238*, 129–134.
McCracken, J. A., Custer, E. E., & Lamsa, J. C. (1999). Luteolysis: A neuroendocrine-mediated event. *Physiological Reviews*, *79*, 263–323.
McCracken, J. A., Glew, M. E., & Scaramuzzi, R. J. (1970). Corpus luteum regression induced by prostagland in F2-alpha. *The Journal of Clinical Endocrinology and Metabolism*, *30*, 544–546.
Miyamoto, A., Okuda, K., Schweigert, F. J., & Schams, D. (1992). Effects of basic fibroblast growth factor, transforming growth factor-β and nerve growth factor on the secretory function of the bovine corpus luteum in vitro. *The Journal of Endocrinology*, *135*, 103–114.
Mondal, M., Schilling, B., Folger, J., Steibel, J. P., Buchnick, H., Zalman, Y., et al. (2011). Deciphering the luteal transcriptome: Potential mechanisms mediating stage-specific luteolytic response of the corpus luteum to prostaglandin $F_{2\alpha}$. *Physiological Genomics*, *43*, 447–456.
Murphy, B. D. (2000). Models of luteinization. *Biology of Reproduction*, *63*, 2–11.

Murta, D., Batista, M., Silva, E., Trindade, A., Mateus, L., Duarte, A., et al. (2015). Differential expression of Notch component and effector genes during ovarian follicle and corpus luteum development during the oestrous cycle. *Reproduction, Fertility, and Development, 27*, 1038–1048.

Nagata, S. (1997). Apoptosis by death factor. *Cell, 88*, 355–365.

Natraj, U., & Richards, J. S. (1993). Hormonal regulation, localization, and functional activity of the progesterone receptor in granulosa cells of rat preovulatory follicles. *Endocrinology, 133*, 761–769.

Nie, G. J., Johnson, K. E., Wenzel, J. G. W., & Braden, T. D. (2003). Luteal function in mares following administration of oxytocin, cloprostenol or saline on day 0, 1 or 2 post-ovulation. *Theriogenology, 60*, 1119–1125.

Niswender, G. D., Juengel, J. L., Silva, P. J., Rollyson, M. K., & McIntush, E. W. (2000). Mechanisms controlling the function and life span of the corpus luteum. *Physiological Reviews, 80*, 1–29.

Niswender, G. D., & Nett, T. M. (1994). Corpus luteum and its control in infraprimate species. In E. Knobil (Ed.), *Knobil and Neill's physiology of reproduction* (pp. 781–816). Elsevier Academic Press Publications.

Nothnick, W. B., & Pate, J. L. (1990). Interleukin-1 beta is a potent stimulator of prostaglandin synthesis in bovine luteal cells. *Biology of Reproduction, 43*, 898–903.

Okuda, K., & Sakumoto, R. (2003). Multiple roles of TNF super family members in corpus luteum function. *Reproductive Biology and Endocrinology, 1*, 95.

Olofsson, J. I., Leung, C. H., Bjurulf, E., Ohno, T., Selstam, G., Peng, C., et al. (1996). Characterization and regulation of a mRNA encoding the prostaglandin F2alpha receptor in the rat ovary. *Molecular and Cellular Endocrinology, 123*, 45–52.

O'Shea, J. D., Rodgers, R. J., & D'Occhio, M. J. (1989). Cellular composition of the cyclic corpus luteum of the cow. *Journal of Reproduction and Fertility, 85*, 483–487.

Park, J. I., Park, H. J., Choi, H. S., Lee, K., Lee, W. K., & Chun, S. Y. (2001). Gonadotropin regulation of NGFI-B messenger ribonucleic acid expression during ovarian follicle development in the rat. *Endocrinology, 142*, 3051–3059.

Pate, J. L., & Landis Keyes, P. (2001). Immune cells in the corpus luteum: Friends or foes? *Reproduction, 122*, 665–676.

Petroff, M. G., Petroff, B. K., & Pate, J. L. (1999). Expression of cytokine messenger ribonucleic acids in the bovine corpus luteum. *Endocrinology, 140*, 1018–1021.

Petrovská, M., Dimitrov, D. G., & Michael, S. D. (1996). Quantitative changes in macrophage distribution in normal mouse ovary over the course of the estrous cycle examined with an image analysis system. *American Journal of Reproductive Immunology, 36*, 175–183.

Pharriss, B. B., & Wyngarden, L. J. (1969). The effect of prostaglandin F2α on the progestogen content of ovaries from pseudopregnant rats. *Proceedings of the Society for Experimental Biology and Medicine, 130*, 92–94.

Pineda, M. H., & Ginther, O. J. (1972). Inhibition of estrus and ovulation in mares treated with an antiserum against an equine pituitary fraction. *American Journal of Veterinary Research, 33*, 1775–1780.

Pober, J. S., & Sessa, W. C. (2007). Evolving functions of endothelial cells in inflammation. *Nature Reviews. Immunology, 7*, 803–815.

Powell, W. S., Hammarström, S., & Samuelsson, B. (1975). Occurrence and properties of a prostaglandin F2alpha receptor in bovine corpora lutea. *European Journal of Biochemistry, 56*, 73–77.

Pru, J. K., Lynch, M. P., Davis, J. S., & Rueda, B. R. (2003). Signaling mechanisms in tumor necrosis factor alpha-induced death of microvascular endothelial cells of the corpus luteum. *Reproductive Biology and Endocrinology, 1*, 17.

Quirk, S. M., Harman, R. M., Huber, S. C., & Cowan, R. G. (2000). Responsiveness of mouse corpora luteal cells to Fas antigen (CD95)-mediated apoptosis. *Biology of Reproduction, 63*, 49–56.

Rao, C. V., Estergreen, V. L., Carman, F. R., & Moss, G. E. (1979). Receptors for gonadotrophin and prostaglandin F2 alpha in bovine corpora lutea of early, mid and late luteal phase. *Acta Endocrinologica, 91*, 529–537.

Redmer, D. A., & Reynolds, L. P. (1996). Angiogenesis in the ovary. *Reviews of Reproduction, 1*, 182–192.

Redmer, D. A., Rone, J. D., & Goodman, A. L. (1985). Evidence for a non-steroidal angiotropic factor from the primate corpus luteum: Stimulation of endothelial cell migration in vitro. *Proceedings of the Society for Experimental Biology and Medicine, 179*, 136–140.

Rexroad, C. E., & Guthrie, H. D. (1979). Prostaglandin F2 alpha and progesterone release in vitro by ovine luteal tissue during induced luteolysis. *Advances in Experimental Medicine and Biology, 112*, 639–644.

Reynolds, L. P., Grazul-Bilska, A. T., Killilea, S. D., & Redmer, D. A. (1994). Mitogenic factors of corpora lutea. *Progress in Growth Factor Research, 5*, 159–175.

Reynolds, L. P., Grazul-Bilska, A. T., & Redmer, D. A. (2002). Angiogenesis in the female reproductive organs: Pathological implications. *International Journal of Experimental Pathology, 83*, 151–163.

Reynolds, L. P., & Redmer, D. A. (1999). Growth and development of the corpus luteum. *Journal of Reproduction and Fertility. Supplement, 54*, 181–191.

Richards, J. S., Russell, D. L., Robker, R. L., Dajee, M., & Alliston, T. N. (1998). Molecular mechanisms of ovulation and luteinization. *Molecular and Cellular Endocrinology, 145*, 47–54.

Robinson, R. S., & Woad, K. J. (2017). Luteal angiogenesis. In R. Meidan (Ed.), *The life cycle of the corpus luteum* (pp. 1–21). Springer International Publishing.

Robinson, R. S., Woad, K. J., Hammond, A. J., Laird, M., Hunter, M. G., & Mann, G. E. (2009). Angiogenesis and vascular function in the ovary. *Reproduction, 138*, 869–881.

Roughton, S. A., Lareu, R. R., Bittles, A. H., & Dharmarajan, A. M. (1999). Fas and Fas ligand messenger ribonucleic acid and protein expression in the rat corpus luteum during apoptosis-mediated luteolysis. *Biology of Reproduction, 60*, 797–804.

Roy, S. K., & Kole, A. R. (1995). Transforming growth factor-beta receptor type II expression in the hamster ovary: Cellular site(s), biochemical properties, and hormonal regulation. *Endocrinology, 136*, 4610–4620.

Rueda, B. R., Hendry, I. R., Tilly, J. L., & Hamernik, D. L. (1999). Accumulation of caspase-3 messenger ribonucleic acid and induction of caspase activity in the ovine corpus luteum following prostaglandin F2alpha treatment in vivo. *Biology of Reproduction, 60*, 1087–1092.

Russell, D. L., Doyle, K. M. H., Gonzales-Robayna, I., Pipaon, C., & Richards, J. S. (2003). Egr-1 induction in rat granulosa cells by follicle-stimulating hormone and luteinizing hormone: Combinatorial regulation by transcription factors cyclic adenosine 3′,5′-monophosphate regulatory element binding protein, serum response factor, sp1, and early growth response factor-1. *Molecular Endocrinology, 17*, 520–533.

Sakumoto, R., Berisha, B., Kawate, N., Schams, D., & Okuda, K. (2000). Tumor necrosis factor-alpha and its receptor in bovine corpus luteum throughout the estrous cycle. *Biology of Reproduction, 62*, 192–199.

Sayasith, K., Bouchard, N., Doré, M., & Sirois, J. (2006). Molecular cloning and gonadotropin-dependent regulation of equine prostaglandin F2alpha receptor in ovarian follicles during the ovulatory process in vivo. *Prostaglandins & Other Lipid Mediators, 80*, 81–92.

Schier, A. F. (2003). Nodal signaling in vertebrate development. *Annual Review of Cell and Developmental Biology, 19*, 589–621.

Schier, A. F. (2009). Nodal morphogens. *Cold Spring Harbor Perspectives in Biology, 1*, a003459.

Skarzynski, D. J., Bah, M. M., Deptula, K. M., Woclawek-Potocka, I., Korzekwa, A., Shibaya, M., et al. (2003). Roles of tumor necrosis factor-alpha of the estrous cycle in cattle: An in vivo study. *Biology of Reproduction, 69*, 1907–1913.

Skarzynski, D. J., Ferreira-Dias, G., & Okuda, K. (2008). Regulation of luteal function and corpus luteum regression in cows: Hormonal control, immune mechanisms and intercellular communication. *Reproduction in Domestic Animals, 43*(Suppl. 2), 57–65.

Skarzynski, D. J., Jaroszewski, J. J., Bah, M. M., Deptula, K. M., Barszczewska, B., Gawronska, B., et al. (2003). Administration of a nitric oxide synthase inhibitor counteracts prostaglandin F2-induced luteolysis in cattle. *Biology of Reproduction, 68*, 1674–1681.

Skarzynski, D. J., Jaroszewski, J. J., & Okuda, K. (2005). Role of tumor necrosis factor-alpha and nitric oxide in luteolysis in cattle. *Domestic Animal Endocrinology, 29*, 340–346.

Skarzynski, D. J., Woclawek-Potocka, I., Korzekwa, A., Bah, M. M., Piotrowska, K., Barszczewska, B., et al. (2007). Infusion of exogenous tumor necrosis factor dose dependently alters the length of the luteal phase in cattle: Differential responses to treatment with indomethacin and L-NAME, a nitric oxide synthase inhibitor. *Biology of Reproduction, 76*, 619–627.

Stevenson, K. R., Parkinson, T. J., & Wathes, D. C. (1991). Measurement of oxytocin concentrations in plasma and ovarian extracts during the oestrous cycle of mares. *Journal of Reproduction and Fertility, 93*, 437–441.

Stocco, C., Telleria, C., & Gibori, G. (2007). The molecular control of corpus luteum formation, function, and regression. *Endocrine Reviews, 28*, 117–149.

Sugino, N., Shimamura, K., Tamura, H., Ono, M., Nakamura, Y., Ogino, K., et al. (1996). Progesterone inhibits superoxide radical production by mononuclear phagocytes in pseudopregnant rats. *Endocrinology, 137*, 749–754.

Taniguchi, H., Yokomizo, Y., & Okuda, K. (2002). Fas-Fas ligand system mediates luteal cell death in bovine corpus luteum. *Biology of Reproduction, 66*, 754–759.

Tansey, M. G., & Szymkowski, D. E. (2009). The TNF superfamily in 2009: New pathways, new indications, and new drugs. *Drug Discovery Today, 14*, 1082–1088.

Teerds, K. J., & Dorrington, J. H. (1992). Immunohistochemical localization of transforming growth factor-beta 1 and -beta 2 during follicular development in the adult rat ovary. *Molecular and Cellular Endocrinology, 84*, R7–13.

Theodosis, D. T., Wooding, F. B., Sheldrick, E. L., & Flint, A. P. (1986). Ultrastructural localisation of oxytocin and neurophysin in the ovine corpus luteum. *Cell and Tissue Research, 243*, 129–135.

Troedsson, M. H., Ababneh, M. M., Ohlgren, A. F., Madill, S., Vetscher, N., & Gregas, M. (2001). Effect of periovulatory prostaglandin F2alpha on pregnancy rates and luteal function in the mare. *Theriogenology, 55*, 1891–1899.

Tsai, S. J., & Wiltbank, M. C. (1997). Prostaglandin F2alpha induces expression of prostaglandin G/H synthase-2 in the ovine corpus luteum: A potential positive feedback loop during luteolysis. *Biology of Reproduction, 57*, 1016–1022.

Vanlangenakker, N., Vanden Berghe, T., & Vandenabeele, P. (2012). Many stimuli pull the necrotic trigger, an overview. *Cell Death and Differentiation, 19*, 75–86.

Vaskivuo, T. E., Ottander, U., Oduwole, O., Isomaa, V., Vihko, P., Olofsson, J. I., et al. (2002). Role of apoptosis, apoptosis-related factors and 17β-hydroxysteroid dehydrogenases in human corpus luteum regression. *Molecular and Cellular Endocrinology, 194*, 191–200.

Vega, M., Urrutia, L., Iñiguez, G., Gabler, F., Devoto, L., & Johnson, M. C. (2000). Nitric oxide induces apoptosis in the human corpus luteum in vitro. *Molecular Human Reproduction, 6*, 681–687.

Vorontchikhina, M. A., Zimmermann, R. C., Shawber, C. J., Tang, H., & Kitajewski, J. (2005). Unique patterns of Notch1, Notch4 and Jagged1 expression in ovarian vessels during folliculogenesis and corpus luteum formation. *Gene Expression Patterns*, *5*, 701–709.

Wang, H., & Tsang, B. K. (2007). Nodal signalling and apoptosis. *Reproduction*, *133*, 847–853.

Watson, E. D., Pedersen, H. G., Thomson, S. R. M., & Fraser, H. M. (2000). Control of follicular development and luteal function in the mare: Effects of a GnRH antagonist. *Theriogenology*, *54*, 599–609.

Watson, E. D., & Sertich, P. L. (1990). Secretion of prostaglandins and progesterone by cells from corpora lutea of mares. *Journal of Reproduction and Fertility*, *88*, 223–229.

Wehrenberg, U., Giebel, J., & Rune, G. M. (1998). Possible involvement of transforming growth factor-beta 1 and transforming growth factor-beta receptor type II during luteinization in the marmoset ovary. *Tissue & Cell*, *30*, 360–367.

Wiltbank, M. C., Shiao, T. F., Bergfelt, D. R., & Ginther, O. J. (1995). Prostaglandin F2 alpha receptors in the early bovine corpus luteum. *Biology of Reproduction*, *52*, 74–78.

Wintersteiner, O., & Allen, W. M. (1934). Crystalline progestin. *The Journal of Biological Chemistry*, *107*, 321–336.

Wuttke, W., Pitzel, L., Knoke, I., Theiling, K., & Jarry, H. (1997). Immune-endocrine interactions affecting luteal function in pigs. *Journal of Reproduction and Fertility. Supplement*, *52*, 19–29.

Yang, Y., Sun, M., Shan, Y., Zheng, X., Ma, H., Ma, W., et al. (2015). Endoplasmic reticulum stress-mediated apoptotic pathway is involved in corpus luteum regression in rats. *Reproductive Sciences*, *22*, 572–584.

Zalman, Y., Klipper, E., Farberov, S., Mondal, M., Wee, G., Folger, J. K., et al. (2012). Regulation of angiogenesis-related prostaglandin f2alpha-induced genes in the bovine corpus luteum. *Biology of Reproduction*, *86*, 92.

CHAPTER TWELVE

Molecular Aspects and Clinical Relevance of GDF9 and BMP15 in Ovarian Function

Martina Belli, Shunichi Shimasaki[1]
University of California San Diego, School of Medicine, La Jolla, CA, United States
[1]Corresponding author: e-mail address: sshimasaki@ucsd.edu

Contents

Abstract

Growth and differentiation factor 9 (GDF9) and bone morphogenetic protein 15 (BMP15) are oocyte-secreted factors with a leading role in the control of ovarian function in female reproduction, modulating both the cell fate of the somatic granulosa cells and the quality and developmental competence of the egg. This short review aims to consolidate the molecular aspects of GDF9 and BMP15 and their integral actions in female fertility to understand particularly their effects on oocyte quality and fetal growth. The significant consequences of mutations in the *GDF9* and *BMP15* genes in women with dizygotic twins as well as the clinical relevance of these oocyte factors in the pathogenesis of primary ovarian insufficiency and polycystic ovary syndrome are also addressed.

Vitamins and Hormones, Volume 107
ISSN 0083-6729
https://doi.org/10.1016/bs.vh.2017.12.003

1. INTRODUCTION

The bidirectional communications between the oocyte and GCs in the ovary consist in the exchange of nutrients and signals that influence the growth and differentiation of both somatic cells and the egg by a complex regulatory network of autocrine, juxtacrine, and paracrine factors (Eppig, 2001; Gilchrist, Ritter, & Armstrong, 2004; Matzuk, Burns, Viveiros, & Eppig, 2002). Earlier observations on the importance of this communication throughout folliculogenesis provided evidence of the regulatory and nourishing roles of GCs and cumulus cells (CCs) on the oocyte (Anderson & Albertini, 1976; Gilula, Epstein, & Beers, 1978; Goodenough, Simon, & Paul, 1999), while more recent studies led to the conclusion that the oocyte exerts a key role in influencing somatic cells' proliferation, differentiation, and function through the production of soluble factors (Gilchrist, Lane, & Thompson, 2008). The oocyte is also able to influence its own developmental competence by inducing the production of positive regulatory factors in CCs that, in turn, act on it. For instance, factors produced by denuded oocytes, or exogenous recombinant GDF9 or BMP15, were able to increase the rates of blastocyst formation after in vitro maturation (IVM) and in vitro fertilization (IVF) of cumulus-oocyte complexes (COCs) (Dey et al., 2012; Gomez et al., 2012; Hussein, Sutton-McDowall, Gilchrist, & Thompson, 2011; Hussein, Thompson, & Gilchrist, 2006; Romaguera et al., 2010; Su, Hu, et al., 2014; Su, Wang, et al., 2014; Sudiman et al., 2014; Yeo, Gilchrist, Thompson, & Lane, 2008). Thus the oocyte, in addition to regulate the CC phenotype in its surrounding cells, can indirectly control and influence its own developmental competence.

2. DISCOVERY OF GDF9 AND BMP15

GDF9 and BMP15 were first described in 1993 (McPherron & Lee, 1993) and 1998 (Dube et al., 1998; Laitinen et al., 1998), respectively, as novel members of the TGF-β superfamily and oocyte-specific factors in the mouse ovary (Dube et al., 1998; Laitinen et al., 1998; McGrath, Esquela, & Lee, 1995; McPherron & Lee, 1993). The mRNAs of both proteins are expressed in oocytes of the postnatal mouse ovary throughout folliculogenesis with a very low or no expression in oocytes of primordial follicles and a strong expression in oocytes at all stages of developing follicles (Laitinen et al., 1998; McGrath et al., 1995). GDF9 mRNA is also detectable in ovulated oocytes (McGrath et al., 1995), and indeed, its EST cDNA was originally identified from a mouse two-cell embryo library (Laitinen et al., 1998), although its expression is rapidly lost thereafter (McGrath et al., 1995). These findings excited and rekindled many researchers who have been studying over a long period the role of OSFs in folliculogenesis and ovulation. It is now well recognized that these factors play pivotal roles in the normal proliferation and differentiation of GCs and for the acquisition of developmental competence by the oocyte.

3. GDF9 AND BMP15 STRUCTURES

All TGF-β superfamily members including GDF9 and BMP15 are synthetized as large preproproteins composed of a signal peptide, a prodomain that directs their dimerization and a mature domain (Shimasaki, Moore, Otsuka, & Erickson, 2004). After removal of the signal peptide, the dimeric proproteins are cleaved by furin-like proteases, and the dimeric mature proteins are secreted (Al-Musawi, Walton, Heath, Simpson, & Harrison, 2013; Shimasaki et al., 2004; Simpson et al., 2012). GDF9 and BMP15 are the most homologous proteins in amino acid sequence, and a peculiarity of these proteins is that the fourth cysteine residue is substituted by a serine. The fourth cysteine is the only cysteine that is involved in the formation of intersubunit disulfide bond and present in all TGF-β superfamily members except not only GDF9 and BMP15 but also GDF3, lefty-1, and lefty-2 (McPherron & Lee, 1993; Meno et al., 1997). Thus, a question was whether GDF9 and BMP15 proteins exist as monomers or noncovalently linked homodimers. They are now known to be noncovalently linked homodimers (Liao, Moore, Otsuka, & Shimasaki, 2003).

The proteolytic cleavage of the recombinant BMP15 and GDF9 proproteins expressed in vitro is not as efficient as expected, unless specific proteolytic enzymes are simultaneously overexpressed (Hayashi et al., 1999; Li, Rajanahally, Edson, & Matzuk, 2009; Liao et al., 2003; Liao, Moore, & Shimasaki, 2004; McNatty et al., 2005a; Mottershead et al., 2008; Otsuka, Moore, & Shimasaki, 2001; Pulkki et al., 2012; Reader et al., 2011; Saito, Yano, Sharma, McMahon, & Shimasaki, 2008). Similarly, in monkey follicular fluid, a strong and a faint band of GDF9 proprotein and mature protein, respectively, was observed, while BMP15 proprotein but not mature protein was detected (Duffy, 2003). Proproteins and mature proteins were also detected in human (Requena, Cruz, Agudo, Pacheco, & Garcia-Velasco, 2016; Wu et al., 2007) and dairy cow (Behrouzi, Colazo, & Ambrose, 2016) follicular fluid with different ratios. Thus, proproteins of GDF9 and BMP15 together with the respective mature proteins are secreted in vitro in cell culture and in vivo from the oocyte. Moreover, the secretion of mature and proproteins of these factors by the oocyte depends on the species analyzed, and it can be different if produced in vivo or during in vitro culture (Gilchrist et al., 2004; Lin et al., 2012; McNatty et al., 2006). For instance, it was demonstrated that in vitro, but not in vivo, mouse COCs are deficient of the mature protein of BMP15 (Mester et al., 2015; Yoshino, McMahon, Sharma, & Shimasaki, 2006), or that sheep oocytes produce only unprocessed GDF9 and BMP15 in follicular fluid (McNatty et al., 2006), while during IVM they secrete mature proteins of the two factors (Lin et al., 2012). Further, in mouse, IVM oocytes produce a mixture of pro-GDF9 and mature GDF9 (Gilchrist et al., 2004), while in rat, IVM oocytes produce only mature GDF9 (Lin et al., 2012).

Some TGF-β superfamily members are secreted as a complex consisting of the mature dimer noncovalently bound to the prodomain (Israel, Nove, Kerns, Moutsatsos, & Kaufman, 1992; Jones et al., 1994; Thies et al., 2000; Wang et al., 1990). Saito et al. demonstrated that purified and bioactive recombinant human BMP15 is composed of the dimeric mature and proregion proteins (Saito et al., 2008). In contrast, Simpson et al. reported that recombinant human, but not mouse, GDF9 is inactive due to a latent complex with proregion proteins (Simpson et al., 2012). However, it is not always the case since purified bioactive recombinant human GDF9 produced from HEK293 (Liao et al., 2004) and HEK293T cells (Mottershead et al., 2008) was also reported. Further, both recombinant human BMP15 and GDF9 are phosphorylated and the phosphorylation is necessary for their bioactivity (McMahon, Sharma, & Shimasaki, 2008; Saito et al., 2008). The kinase for BMP15 was later identified to be Fam20C (Tagliabracci et al., 2012). Thus, the phosphorylation status of BMP15 and GDF9 should also be addressed with regard to their biological activity. The necessity of standardization in preparation and biochemical analysis for recombinant GDF9 was previously proposed by Pangas and Matzuk (2005), but unfortunately without substantial clues until now.

4. GDF9/BMP15 HETERODIMERS

As stated earlier, GDF9 and BMP15 share most homologous amino acid sequences, lack the fourth cysteine necessary for the covalent linkage between two subunits, and are expressed in the same cell (the oocyte) at the same time from the primary stage throughout folliculogenesis. Based on these information and our earlier experience in discovering activin (now called activin AB) that is composed of β_A and β_B subunits of inhibin A and B, respectively (Ling et al., 1986), we speculated that GDF9 and BMP15 may form heterodimers and demonstrated the first direct physical evidence of the heterodimeric GDF9/BMP15 molecule (Liao et al., 2003). Thereafter over a decade the integrated bioactivity of GDF9 and BMP15, observed in different species and functional studies, let many researchers approach on the formation of a functional heterodimer (Hanrahan et al., 2004; McIntosh et al., 2008; McNatty et al., 2005a, 2005b, 2004; Mottershead, Ritter, & Gilchrist, 2012; Peng et al., 2013). However, only recent studies demonstrated the functional evidence of the purified recombinant human GDF9/BMP15 heterodimer, which was named cumulin (Mottershead et al., 2015), able to exhibit highly potent bioactivity on GCs and to improve oocyte quality. Specifically, cumulin stimulates thymidine incorporation in primary mouse mural GCs (ED_{50}, 0.6 ng/mL) and both SMAD2/3 and SMAD1/5/8 luciferase reporter activities in COV434 human GC line. Cumulin also regulates in vitro GC differentiation by increasing mRNA expression of *Ptx3*, *Has2*, *Tnfaip6*, and *Ptgs2*. Covalent cumulin, in which the fourth cysteine is recovered, is more potent than noncovalent procumulin in these assays. Noncovalent procumulin, however, is more effective in the improvement of oocyte quality than the covalent cumulin, proving the important functional activity of the prodomain in these factors.

Mottershead et al. also demonstrated interesting data, suggesting that mature GDF9 and BMP15 as well as pro-GDF9 and pro-BMP15 form heterodimers extracellularly. They further proposed that the latent GDF9 is activated following the heterodimerization with BMP15 to form cumulin, either within the oocyte or in the extracellular matrix (Mottershead et al., 2015). This could explain the potent synergistic action observed between BMP15 and GDF9 in terms of mitogenic and differentiating activity on GCs as well as on oocyte quality, raising cumulin to the position of main regulator of fertility in monoovulatory mammals. Therefore, the differential or synergistic effect of GDF9 and BMP15 may occur depending on whether they exert their bioactivity separately or cooperatively by forming cumulin, respectively.

BMP15 binds to the complex of ALK6 and BMPRII on GC surface and activates SMAD1/5/8 intracellular pathway (Moore, Otsuka, & Shimasaki, 2003; Pulkki et al., 2012; Shimasaki et al., 2004), whereas GDF9 activates ALK5 and BMPRII receptors and triggers SMAD2/3 pathway downstream (Kaivo-Oja et al., 2003; Mazerbourg et al., 2004; Roh et al., 2003; Vitt, Mazerbourg, Klein, & Hsueh, 2002), although it is controversial whether ALK5 is the only type I receptor for GDF9 as *Alk5* conditional knockout mice have no defects in follicular development and cumulus expansion (Li et al., 2011), which is a distinct phenotype from *Gdf9*-null mice (Dong et al., 1996). Interestingly, it is suggested that cumulin binds to a receptor complex formed by two BMPRII, one ALK6, and one ALK5/4 molecule, which can activate the two distinct SMAD pathways (i.e., SMAD1/5/8 and SMAD2/3). This last evidence is in contrast to the work of Peng et al., which showed that the GDF9/BMP15 heterodimer signals exclusively through SMAD2/3 pathway, since it can bind to but not activate the ALK6 receptor (Peng et al., 2013).

5. SPECIES-SPECIFIC ROLES OF GDF9 AND BMP15

The significance of GDF9 and BMP15 factors in the regulation of ovarian function emerged with the studies of deficient mouse models of GDF9 (Dong et al., 1996) and BMP15 (Yan et al., 2001). *Gdf9*-deficient female mice are infertile and show a block of folliculogenesis at the stage of primary follicles, when one layer of cuboidal GCs surrounds the oocyte (Dong et al., 1996). Ovaries from *Gdf9*-deficient female mice appear smaller in size, the number of primordial and primary follicles is higher compared with control, no corpora lutea are observed, and atretic follicles are missing. Without the support of highly differentiated GCs, oocytes are prevented to acquire meiotic competence, thus altering the reproductive function. Moreover, the hypogonadal phenotype related to the absence of GDF9 induces higher serum levels of FSH and LH as compared with control. Deficient males are not affected in the reproductive phenotype as they proved to be fertile (Carabatsos, Elvin, Matzuk, & Albertini, 1998; Dong et al., 1996).

In contrast, *Bmp15*-deficient female mice are subfertile, with a lower litter size and a lower number of litters per month (Pangas & Matzuk, 2004;

Yan et al., 2001), demonstrating a less essential role for BMP15 in mouse reproductive function compared with GDF9. *Bmp15*-null male mice are fertile with normal testis size (Yan et al., 2001), highlighting the importance of this factor restricted to the female reproductive function. Also, mice lacking BMP15 and GDF9 showed lower expression of epidermal growth factor receptor (EGFR) compared with wild-type mice, and the activation of SMAD2/3 and extracellular signal-regulated kinases 1 and 2 signaling in CCs triggered by the oocyte was poor (Su et al., 2010). Thus, the oocyte, via paracrine signaling of OSFs, guides the acquisition of responsiveness of CCs to EGF signals, promotes a correct cumulus expansion, and improves its own meiotic maturation and developmental competence in the follicle (Ritter, Sugimura, & Gilchrist, 2015; Su et al., 2010; Sugimura et al., 2015). Moreover, even though BMP15 was shown to have a role in the preservation of gap junction communication between the oocyte and CCs (Sugimura et al., 2014), the action of OSFs on cumulus expansion is species specific, as for some species (e.g., in the mouse) the activation of SMAD2/3 and/or SMAD1/5/8 is required (Buccione, Schroeder, & Eppig, 1990; Dragovic et al., 2007), whereas in the cow the presence of OSFs is not critical (Sugimura et al., 2014).

Of interest in species-specific differences in the role of GDF9 and BMP15 is the dissimilarity between mono- vs polyovulatory species (Dong et al., 1996; Galloway et al., 2000; Moore, Erickson, & Shimasaki, 2004). For instance, BMP15 action is more important in sheep (monoovulatory) rather than mice (polyovulatory) in the first stage of follicle development (Eppig, 2001). In fact, while *Bmp15*-null female mice show impaired fertility but an overall normal follicular morphology (Yan et al., 2001), sheep with spontaneous homozygous mutations in the *Bmp15* gene show no further follicle development beyond the primary stage, which lead to sterility (Eppig, 2001; Galloway et al., 2000; Hanrahan et al., 2004).

Why is the role of BMP15 in female mice during folliculogenesis less critical than that in ewes? One of the relevant findings is that the functional mature BMP15 in the mouse ovary is not detectable before the ovulatory stage (Yoshino et al., 2006). Three days before ovulation, however, the mature BMP15 is produced and induces cumulus expansion in mouse COCs along with increased expression of essential EGF-like growth factors (Park et al., 2004; Yoshino et al., 2006). These findings are consistent with the phenotype of *Bmp15*-null mice, which exhibit normal folliculogenesis but have defects in the ovulation process (Yan et al., 2001). The species-specific differences in the phenotypes caused by *Bmp15* mutations may thus be attributed to the temporal variations in the production of the mature form of BMP15. Additionally, Crawford et al. reported that mean expression levels of BMP15 mRNA are lower than GDF9 mRNA in denuded mouse oocytes (Crawford & McNatty, 2012) in contrast to the similar ratio of BMP15 and GDF9 produced by the oocyte in sheep (Crawford & McNatty, 2012). Indeed, ewes with spontaneous homozygous mutations in the *Bmp15* or *Gdf9* gene have high similarity in their infertile phenotype (McNatty et al., 2005). Ewes with spontaneous heterozygous mutations in *Bmp15* or *Gdf9* also share the increased fecundity with an augmented ovulation rate and increased litters of twins and triplets, due to an amplified sensitivity to LH and development of secondary follicles followed by an increased number of antral follicles (Galloway et al., 2000; Hanrahan et al., 2004; Juengel, Hudson, Whiting, & McNatty, 2004), showing that the dosage of BMP15 or GDF9 in the oocyte can modulate fertility (Galloway et al., 2000; Moore et al., 2004).

6. OOCYTE ACTIONS ON GCs

The evidence of a critical function of the oocyte in promoting proliferation of GCs in preantral and antral follicles dates back to the early 1990s with the studies of Vanderhyden, Cohen, and Morley (1993) and Vanderhyden, Telfer, and Eppig (1992). The oocyte was proved to induce GC proliferation through the production of a soluble factor that was acting in a paracrine fashion on GCs, most likely via follicular fluid. Experiments with mouse CCs from COCs or GCs from preantral follicles cultured with oocyte-conditioned medium suggested that no direct contact between the oocyte and somatic cells was required to proliferate (Vanderhyden et al., 1992). However, the ability of the oocyte to promote GC proliferation and differentiation is highly influenced by the developmental stage of the oocyte, besides the developmental state of GC itself (Gilchrist, Ritter, & Armstrong, 2001). In particular, the meiotic status of the oocyte determines the mitogenic influence on GC proliferation. During meiotic competence acquisition the oocyte highly promotes GC proliferation and starts losing this capability at the metaphase I of the meiosis. Coculturing experiments with meiotically arrested GV oocytes and mural GCs gave the highest GC proliferation rate that was gradually decreased when GCs were cultured with maturing oocytes, ovulated oocytes, or zygotes (Gilchrist et al., 2001). In fact, meiotically growing incompetent oocytes, even if they were proved to express BMP15 and GDF9 (Dube et al., 1998; McGrath et al., 1995), are less effective in regulating follicular cell function than fully grown GV-stage oocytes (Gilchrist et al., 2001). When oocytes from secondary follicles were paired with somatic cells from newborn mouse ovaries and implanted beneath the renal capsules of ovariectomized female mice, a double rate of follicle growth along with GC differentiation was observed after only 9 days from implantation. In this study, the oocytes in the newly formed and developed follicles were able to resume meiosis, be fertilized, and complete the preimplantation stages of development in vitro (Eppig, Wigglesworth, & Pendola, 2002).

OSFs modulate the production of factors that control ovarian functions in promoting follicular growth, including inhibin (Lanuza, Groome, Baranao, & Campo, 1999), luteinizing hormone receptor (Eppig, Pendola, & Wigglesworth, 1998; Eppig, Wigglesworth, Pendola, & Hirao, 1997), and kit ligand (Joyce, Pendola, Wigglesworth, & Eppig, 1999), and contribute to the regulation of gene expression (Paradis et al., 2010; Regassa et al., 2011), metabolism (Sugiura et al., 2007), and apoptosis (Hussein, 2005). As for apoptosis, the level of CC apoptosis was increased when the oocytes were removed from the COCs but reversed when the oocytes were added back to the CC culture (Hussein, 2005), suggesting that OSFs play a role to prevent CC apoptosis. Oocyte-secreted BMP15 and BMP6 as well as theca cell-secreted BMP7 (Erickson & Shimasaki, 2003) were shown to protect spontaneous CC apoptosis, but GDF9 has no significant effect (Hussein, 2005). This is likely to be due to the different signaling pathways, namely SMAD1/5/8 for BMP15, BMP6, and BMP7, and SMAD2/3 for GDF9.

7. OSF EFFECTS ON THE OOCYTE DEVELOPMENTAL COMPETENCE

GDF9 and BMP15 have critical roles in the ovary for the transition after primary stage of follicle development, but they also control later events in follicle development. The absence of GDF9 or BMP15 in mice, as previously described, brings to a phenotype of infertility or reduced fertility, respectively, due to misregulation of GCs and low developmental competence of the oocyte (Su et al., 2004). However, the treatment of COCs with recombinant GDF9, BMP15, or denuded oocytes during IVM led to higher rates of blastocyst formation and fetal yield after IVM and IVF of COCs in cattle (Dey et al., 2012; Hussein et al., 2011, 2006; Su, Hu, et al., 2014; Su, Wang, et al., 2014), in mice (Glister, Groome, & Knight, 2003; Sudiman et al., 2014; Yeo et al., 2008), in goats (Romaguera et al., 2010), and in pigs (Gomez et al., 2012). BMP15 in particular seems to interact with amphiregulin (one of EGF-like growth factors) to enhance oocyte developmental competence, via a prolonged gap-junctional coupling between the oocyte and CCs that would allow the passage of metabolites to the oocyte (Sugimura et al., 2014). The implication of these results resides in the improvement of IVM techniques, in which compromised oocytes from IVM cycles might benefit from treatment with OSFs to restore a good developmental competence. In fact, IVM oocytes might not be able to secrete oocyte factors that are able to sustain oocyte meiotic maturation, as demonstrated by the evidence that murine IVM COCs lack a processed form of BMP15 compared with in vivo matured COCs (Sugimura et al., 2014). The synergistic effects of BMP15 and GDF9 in the improvement of oocyte meiotic competence might be related to a better nuclear and cytoplasmic maturation, the prevention of zona pellucida hardening, a correct fertilization pattern, and the higher expression of glutathione peroxidase 1 in the oocyte (Dey et al., 2012).

8. GDF9 AND BMP15 IN DIZYGOTIC TWINNING, POI, AND PCOS

As stated earlier, GDF9 and BMP15 can influence female reproduction. Published data that have been accumulated so far indicate that nonsynonymous mutations in the *GDF9* and *BMP15* genes are correlated with dizygotic (DZ) twinning (Montgomery et al., 2004; Palmer et al., 2006; Zhao et al., 2008) (Table 1) as well as gynecological diseases such as primary ovarian insufficiency (POI) (Bouilly et al., 2016; Di Pasquale, Beck-Peccoz, & Persani, 2004; Di Pasquale et al., 2006; Dixit et al., 2005, 2006; Ferrarini et al., 2013; Kovanci et al., 2007; Laissue et al., 2006; Ledig, Ropke, Haeusler, Hinney, & Wieacker, 2008; Mayer, Fouquet, Pugeat, & Misrahi, 2017; Rossetti et al., 2009; Tiotiu et al., 2010; Wang et al., 2013, 2010; Zhao et al., 2007) (Table 2) and polycystic ovary syndrome (PCOS) (Liu et al., 2011; Mehdizadeh, Sheikhha, Kalantar, Aali, & Ghanei, 2016; Wang et al., 2010) (Table 3).

8.1 DZ Twinning

The spontaneous conception of DZ twinning rate is influenced by a combination of male and female reproductive factors as it reflects the frequency of double ovulation, the successful fertilization, and the survival of the

Table 1 Prevalence of Carriers With BMP15 and GDF9 Nonsynonymous Variants in Cohorts of Mothers of DZ Twins and Respective Controls

Variants			No. of Cases/Total			
DNA	**Protein**	**Domains**	**DZ Twinning**	**Normal**	**Countries**	**References**
BMP15 in DZ twinning						
c.520C>T	P174S	Proregion	2/1693	0/1512	Australia	Zhao et al. (2008)
c.581T>C	F194S	Proregion	5/1693	11/1512	Australia	Zhao et al. (2008)
c.931G>A	A311T	Mature region	3/1693	0/1512	Australia	Zhao et al. (2008)
c.1175G>C	R392T	Mature region	1/1693	0/1512	Australia	Zhao et al. (2008)
Total			11/1693	11/1512	$P=0.8328$ by Fisher's exact test	
GDF9 in DZ twinning						
c.207_210delATCT	L69LfsTer88	Proregion	1/449	NA	Australia	Montgomery et al. (2004)
c.307C>T	P103S	Proregion	30/1693	13/1512	Australia	Palmer et al. (2006)*
c.362C>T	T121L	Proregion	4/1693	1/1512	Australia	Palmer et al. (2006)
c.392_393insT	Frame shift	Proregion	NA	NA	Australia	Palmer et al. (2006)
c.1121C>T	P374L	Mature region	4/1693	0/1512	Australia	Palmer et al. (2006)*
c.1268_1269delAA	Frame shift	Mature region	4/1693	0/1512	Australia	Palmer et al. (2006)*
c.1360C>T	R454C	Mature region	22/1693	12/1512	Australia	Palmer et al. (2006)
Total			64/1693	26/1512	$P=0.0006$ by Fisher's exact test	

Asterisks denote a significance ($P<0.05$) that was calculated for the likelihood ratio test (MENDEL binomial link measured genotype model) (Palmer et al., 2006). *NA* denotes unavailable data, and thus the respective frequency numbers were excluded from the calculation. The total number of women analyzed in the same study was only counted once.

Table 2 Prevalence of Carriers With BMP15 and GDF9 Nonsynonymous Variants in Cohorts With POI and Respective Controls

Variants			No. of Cases/Total			
DNA	**Protein**	**Domains**	**POI**	**Normal**	**Countries**	**References**
BMP15 in POI						
c.13A>C	S5R	Signal peptide	1/300	0/216	Italy	Rossetti et al. (2009)
c.13A>C	S5R	Signal peptide	4/100	NA	France	Bouilly et al. (2016)
c.151_152delGA	E51IfsTer27	Proregion	NA	NA	France	Mayer et al. (2017)
c.181C>T	R61W	Proregion	2/133	0/197	India	Dixit et al. (2006)
c.182G>A	R61E	Proregion	1/133	0/197	India	Dixit et al. (2006)
c.189_198delAGGGCATTCAinsTG	E64AfsTer12	Proregion	NA	NA	France	Mayer et al. (2017)
c.202C>T	R68W	Proregion	2/300	0/216	Italy	Rossetti et al. (2009)
c.202C>T	R68W	Proregion	1/160	0/181	Italy	Di Pasquale et al. (2006)
c.226C>T	R76C	Proregion	3/133	0/197	India	Dixit et al. (2006)
c.227G>A	R76H	Proregion	1/133	0/197	India	Dixit et al. (2006)
c.242A>G	H81R	Proregion	1/50	0/214	Belgium	Tiotiu et al. (2010)
c.308A>G	N103S	Proregion	13/133	10/197	India	Dixit et al. (2006)
c.308A>G	N103S	Proregion	11/160	16/181	Italy	Di Pasquale et al. (2006)
c.308A>G	N103S	Proregion	7/50	25/214	Belgium	Tiotiu et al. (2010)
c.308A>G	N103S	Proregion	NA	NA	France	Bouilly et al. (2016)

c.413G>A	R138H	Proregion	1/300	0/216	Italy	Rossetti et al. (2009)
c.443T>C	L148P	Proregion	1/203	0/54	France	Laissue et al. (2006)
c.443T>C	L148P	Proregion	2/300	0/216	Italy	Rossetti et al. (2009)
c.443T>C	L148P	Proregion	1/50	1/214	Belgium	Tiotiu et al. (2010)
c.443T>C	L148P	Proregion	2/100	NA	France	Bouilly et al. (2016)
c.538G>A	A180T	Proregion	1/133	0/197	India	Dixit et al. (2006)
c.538G>A	A180T	Proregion	2/203	0/54	France	Laissue et al. (2006)
c.538G>A	A180T	Proregion	5/160	0/181	Italy	Di Pasquale et al. (2006)
c.538G>A	A180T	Proregion	2/20	1/93	Germany	Ledig et al. (2008)
c.538G>A	A180T	Proregion	6/300	0/216	Italy	Rossetti et al. (2009)
c.538G>A	A180T	Proregion	2/50	2/214	Belgium	Tiotiu et al. (2010)
c.538G>A	A180T	Proregion	1/50	0/150	Italy	Ferrarini et al. (2013)
c.538G>A	A180T	Proregion	NA	NA	France	Bouilly et al. (2016)
c.538G>T+539C>T	A180F/S+V	Proregion	1/133	0/197	India	Dixit et al. (2006)
c.581T>C	F194S	Proregion	1/50	1/214	Belgium	Tiotiu et al. (2010)
c.588T>A	N196K	Proregion	1/133	0/197	India	Dixit et al. (2006)
c.595G>A	G199R	Proregion	1/50	0/214	Belgium	Tiotiu et al. (2010)
c.598C>T	H200Y	Proregion	5/100	1/100	China	Wang et al. (2010)
c.617G>A	R206H	Proregion	1/133	0/197	India	Dixit et al. (2006)

Continued

Table 2 Prevalence of Carriers With BMP15 and GDF9 Nonsynonymous Variants in Cohorts With POI and Respective Controls—cont'd

Variants			No. of Cases/Total			
DNA	**Protein**	**Domains**	**POI**	**Normal**	**Countries**	**References**
c.631C > T	E211X	Proregion	1/133	0/197	India	Dixit et al. (2006)
c.661T > C	W221R	Proregion	1/133	0/197	India	Dixit et al. (2006)
c.704A > G	Y235C	Proregion	NA	NA	Italy	Di Pasquale et al. (2004)
c.727A > G	I243G	Proregion	1/133	0/197	India	Dixit et al. (2006)
c.788insTCT	Ins263L	Proregion	4/133	4/197	India	Dixit et al. (2006)
c.788insTCT	Ins263L	Proregion	10/203	9/54	France	Laissue et al. (2006)
c.788insTCT	Ins263L	Proregion	2/160	5/95	Italy	Di Pasquale et al. (2006)
c.788insTCT	Ins263L	Proregion	17/300	5/216	Italy	Rossetti et al. (2009)
c.788insTCT	Ins263L	Proregion	3/50	5/214	Belgium	Tiotiu et al. (2010)
c.788insTCT	Ins263L	Proregion	5/100	15/100	China	Wang et al. (2010)
c.985C > T	R329C	Mature region	1/100	0/100	China	Wang et al. (2010)
None	—	—	0/92	0/76	China	Zhang, Shi, Wang, and Chen (2007)
None	—	—	0/38	0/51	New Zealand	Chand, Ponnampalam, Harris, Winship, and Shelling (2006)
None	—	—	0/15	0/3	Japan	Takebayashi et al. (2000)
Total			122/1161	100/1335	$P = 0.0170$ by Fisher's exact test	

GDF9 in POI						
c.169G > T	D57Y	Proregion	8/100	11/100	China	Wang et al. (2010)
c.169G > T	D57Y	Proregion	6/139	2/152	China	Wang et al. (2013)
c.199A > C	K67E	Proregion	4/127	0/220	India	Dixit et al. (2005)
c.307C > T	P103S	Proregion	1/60	0/60	USA	Kovanci et al. (2007)
c.307C > T	P103S	Proregion	1/100	NA	France	Bouilly et al. (2016)
c.436C > T	R146C	Proregion	1/100	1/96	China	Zhao et al. (2007)
c.436C > T	R146C	Proregion	3/139	0/152	China	Wang et al. (2013)
c.442A > G	T148A	Proregion	NA	NA	France	Bouilly et al. (2016)
c.557C > A	S186Y	Proregion	1/203	0/54	France	Laissue et al. (2006)
c.646G > A	V216M	Proregion	2/127	0/220	India	Dixit et al. (2005)
c.712A > G	T238A	Proregion	1/100	0/96	China	Zhao et al. (2007)
c.1283G > C	S428T	Mature region	2/100	2/96	China	Zhao et al. (2007)
c.1360C > T	R454C	Mature region	1/100	NA	France	Bouilly et al. (2016)
None	—	—	0/50	0/150	Italy	Ferrarini et al. (2013)
None	—	—	0/15	0/3	Japan	Takebayashi et al. (2000)
Total			29/794	16/835	$P = 0.0484$ by Fisher's exact test	

NA denotes unavailable data, and thus the respective frequency numbers were excluded from the calculation. The total number of women analyzed in the same study was only counted once.

Table 3 Prevalence of Carriers With BMP15 and GDF9 Nonsynonymous Variants in Cohorts With PCOS and Respective Controls

Variants			No. of Cases/ Total			
DNA	**Protein**	**Domains**	**PCOS**	**Normal**	**Countries**	**References**
BMP15 in PCOS						
c.34C>G	L12V	Signal peptide	1/216	0/200	China	Liu et al. (2011)
c.109G>C	A37P	Proregion	1/216	0/200	China	Liu et al. (2011)
c.169C>G	P57A	Proregion	1/216	0/200	China	Liu et al. (2011)
c.288G>C	R96S	Proregion	1/216	0/200	China	Liu et al. (2011)
c.308A>G	N103S	Proregion	1/216	0/200	China	Liu et al. (2011)
c.308A>G	N103S	Proregion	2/70	NA	Iran	Mehdizadeh et al. (2016)
c.598C>T	H200Y	Proregion	1/216	2/200	China	Liu et al. (2011)
c.788insTCT	Ins263L	Proregion	21/216	30/200	China	Liu et al. (2011)
None	—	—	0/38	0/3	Japan	Takebayashi et al. (2000)
Total			27/254	32/203	$P=0.1665$ by Fisher's exact test	
GDF9 in PCOS						
c.15C>G	N5L	Signal peptide	1/216	0/200	China	Wang et al. (2010)
c.118T>G	L40V	Proregion	4/216	0/200	China	Wang, Zhou et al. (2010)
c.133A>G	M45V	Proregion	1/216	0/200	China	Wang, Zhou et al. (2010)
c.169G>T	D57Y	Proregion	12/216	24/200	China	Wang, Zhou et al. (2010)

Table 3 Prevalence of Carriers With BMP15 and GDF9 Nonsynonymous Variants in Cohorts With PCOS and Respective Controls—cont'd

Variants			No. of Cases/ Total			
DNA	**Protein**	**Domains**	**PCOS**	**Normal**	**Countries**	**References**
c.436C>T	R146C	Proregion	2/216	0/200	China	Wang, Zhou et al. (2010)
c.1025A>T	Y342F	Mature region	1/216	0/200	China	Wang, Zhou et al. (2010)
c.1275C>A	S425R	Mature region	1/216	0/200	China	Wang, Zhou et al. (2010)
c.1283G>C	S428T	Mature region	2/216	0/200	China	Wang, Zhou et al. (2010)
None	—	—	0/38	0/3	Japan	Takebayashi et al. (2000)
Total			24/254	24/203	$P=0.5422$ by Fisher's exact test	

NA denotes unavailable data, and thus the respective frequency numbers were excluded from the calculation. The total number of women analyzed in the same study was only counted once.

zygote (Tong & Short, 1998). Of GDF9 and BMP15 as potentially associated genes with DZ twinning is the genetic influence in the *GDF9* gene more frequent than that in the *BMP15* gene (Hoekstra et al., 2008). The first evidence of association between DZ twinning and GDF9 mutation was the finding of a loss-of-function mutation (c.207_210delATCT, p. L69LfsTer88) in two sisters with spontaneous DZ twins (Montgomery et al., 2004) (Table 1). This mutation alters the reading frame for the protein introducing a stop codon at codon position 88 of exon 1. Other mutations in the *GDF9* gene with insertion/deletions (c.392–393insT and c.1268–1269delAA) or missense codons (P103S and T121L in the proregion; P374L and R454C in the mature region) were found (Palmer et al., 2006). Variants P103S and R454C were also reported in POI women (see below), supporting the hypothesis that mutations causative of DZ twinning may be candidates for the development of POI as well and vice versa (Inagaki & Shimasaki, 2010). In biochemical and functional studies for in vitro GC proliferation, it was demonstrated that mutations P103S and P374L abrogated mature GDF9 expression, suggesting a 50% decrease in

GDF9 levels in heterozygous women carrying these mutations (Inagaki & Shimasaki, 2010; Simpson et al., 2014). Pooling the prevalence data of GDF9 variants in Table 1 provides a significant difference ($P=0.0006$) between DZ twinning and controls by Fisher's exact test (https://www.graphpad.com/quickcalcs/contingency1.cfm). In this regard, Palmer et al. previously predicted the frequency differences of individual variants in the *GDF9* gene found in mothers of DZ twins and normal controls by contingency χ^2 testing with χ^2 and asymptotic P values estimated, allowing for the family nature of the data as implemented in MENDEL binomial link measured genotype model (Palmer et al., 2006). They found that the frequencies of the variants P103S, P374L, and c.1268_1269delAA (frame shift) were significantly higher ($P<0.05$, denoted by asterisks in Table 1) in mothers of DZ twins than the control (Palmer et al., 2006), supporting the results by Fisher's exact test.

In contrast, Zhao et al. identified several mutations and deletions/insertions in the *BMP15* gene in mothers of DZ twins, three of which (P174S, A311T, and R392T) were not detected in 1512 controls (Zhao et al., 2008). However, no significant evidence has been provided in these variations in DZ twinning (Zhao et al., 2008). Fisher's exact test ($P=0.8328$) also supports their conclusion (Table 1).

It was shown that mothers of DZ twins reach menopause significantly earlier than those of monozygotic twins, with a higher number of diagnoses for POI (Gosden et al., 2007; Martin, Healey, Pangan, Heath, & Turner, 1997), highlighting the correlation between the faster depletion of follicle reserve with multiple ovulation and the early ovarian insufficiency (Moore et al., 2004). Given that reduced levels of BMP15 and/or GDF9 in ewes are associated with increased ovulation rate and litter size, women with GDF9 (and possibly BMP15) mutations may have an increased number of dominant follicles, resulting in an increased likelihood of bearing DZ twins.

8.2 Primary Ovarian Insufficiency

POI is a dysfunction of the ovary caused by heterogeneous factors (viral infections, autoimmune disorders, genetic disorders, iatrogenic causes), among which the genetic component plays a substantial role, as the synergistic action of different mutations may underlie the development of this phenotype (Rossetti, Ferrari, Bonomi, & Persani, 2017; Tucker, Grover, Bachelot, Touraine, & Sinclair, 2016). Relatively a few genes are considered involved in the development of syndromic and nonsyndromic POI (Cox & Liu, 2014; Fortuno & Labarta, 2014). Approximately 1% of women are affected by POI presenting prolonged amenorrhea (4–6 months) before the age of 40, which is a major cause of infertility, and they show elevated plasma gonadotropin levels. *BMP15* and *GDF9* gene variants were found to have a high incidence on the POI phenotype, namely 0%–12% prevalence of BMP15 variants (Dixit et al., 2006; Ma et al., 2015; Persani, Rossetti, Di Pasquale, Cacciatore, & Fabre, 2014) and 1%–4% prevalence of GDF9 variants (Dixit et al., 2006; Laissue et al., 2006; Ma et al., 2015; Qin, Jiao, Simpson, & Chen, 2015) in certain ethnicities.

The first identified human BMP-15 mutation associated with hypergonadotropic ovarian failure due to ovarian dysgenesis was a heterozygous Y235C missense mutation (Di Pasquale et al., 2004). Thereafter, a number of nonsynonymous mutations in the *BMP15* and *GDF9* genes were identified in women with POI as follows: (i) *BMP15*, S5R, E51IfsTer27, R61W, R61E, E64AfsTer12, R68W, R76C, R76H, H81R, N103S, R138H, L148P, A180T, A180F/S+V, F194S, N196K, G199R, H200Y, R206H, E211X, W221R, Y235C, I243G, Ins263L, and R329C; and (ii) *GDF9*, D57Y, K67E, P103S, R146C, T148A, S186Y, V216M, T238A, S428T, and R454C (Table 2).

The biochemical and biological properties of these variants have been extensively studied. For example, BMP15 variants of R68W, R138H, L148P (Rossetti et al., 2009), R76C, and R206H (Inagaki & Shimasaki, 2010) located in the proregion lead to a reduced production of mature BMP15 in an in vitro assay, showing that reduced biological effects may be due to an impaired processing or a decreased precursor stability (Inagaki & Shimasaki, 2010; Rossetti et al., 2009). Ins263L might not have a functional abnormality since it is present also in control women and might represent, together with N103S, a polymorphism with no functional significance (Bouilly et al., 2016; Di Pasquale et al., 2006; Dixit et al., 2006; Laissue et al., 2006; Rossetti et al., 2009; Tiotiu et al., 2010; Wang et al., 2010).

A study from Dixit and coworkers in 2006 found a homozygous mutation, E211X, in the proregion of BMP15 associated with POI (Dixit et al., 2006). This variant is likely to represent a knockout of the *BMP15* gene, with the production of a prematurely truncated protein that leads to the lack of the mature protein. The patient carrying the mutation was infertile and presented ovarian hypoplasia, together with other developmental defects. A study from 2010 on Chinese POI patients showed the identification of a missense mutation in BMP15 in the mature region (R329C) that might impair the correct folding and final three-dimensional structure of BMP15 subunit as well as the formation of homo- or heterodimers with GDF9 (Wang et al., 2010). A more recent study was published on a family with two sisters who exhibited to be compound heterozygous with E51IfsTer27 inherited from their father and E64AfsTer12 from their mother (Mayer et al., 2017). Due to frame-shift mutations, both of these mutant BMP15 proteins are prematurely truncated in the proregion, precluding mature BMP15 production. Thus, a deletion in the two alleles of the *BMP15* gene in two sisters revealed a human "knockout-like" effect, which resulted in primary amenorrhea in one sister and primo-secondary amenorrhea in the other. Moreover, hormonal analysis showed high FSH and LH, very low estradiol, and low but detectable AMH. Five years later, FSH and LH levels were elevated, estradiol level lowered, and AMH became undetectable, indicating a complete loss of growing follicles. Interestingly, their brother, hemizygous for BMP15, showed asthenozoospermia but refused further investigations.

As described earlier, an increasing number of BMP15 variants have been identified in women with POI in different ethnicities. However, the statistical significance of the mutations in POI has not been provided in each study due to low numbers of patients and controls examined. In this regard, Persani et al. performed an excellent work in which they put together all BMP15 variants identified by January 2014 (the date in which the manuscript was submitted) by different laboratories in different ethnicities and found that the prevalence of nonsynonymous mutations in the POI cohorts is 10-fold higher than in the control populations ($P<0.0001$ by Fisher's exact test), thus supporting the pathogenic role of these rare variants (Persani et al., 2014). We repeated this calculation with all nonsynonymous BMP15 variants identified by March 2017. As a result, the P value of the difference between the POI and normal cohorts in the prevalence of nonsynonymous mutations of BMP15 is 0.0170 (Table 2). However, if the relatively common variants, N103S and Ins263L, are excluded from the calculation, the P value would be <0.0001, which is consistent with the previous report by Persani et al. (2014). Collectively, the mutation rate in the *BMP15* gene in women with POI is significantly higher than the normal control.

Variants K67E, P103S, and S428T rose to misfolded GDF9 proproteins when they were expressed in vitro, resulting in significantly reduced or nearly abrogated mature proteins (Inagaki & Shimasaki, 2010; Simpson et al., 2014). It is of interest that variants S186Y, V216M, and T238A, located in the proregion of GDF9, have reduced association with the mature domain as compared with the wild-type control, thus the availability of the mature protein dimers is increased (Simpson et al., 2014). The variants P103S and R454C are common in mothers of DZ twins as described earlier (Palmer et al., 2006). Another GDF9 variant, R146C, was found in 3 out of 139 Chinese women with diminished ovarian reserve but absent in the control population ($n=152$) (Wang et al., 2013). This mutation reduced the levels of mature GDF9 protein in transfected cells presumably through disruption of its three-dimensional structure. Moreover, variant R146C reduced the ability of the wild-type GDF9 to stimulate GC proliferation and to activate SMAD2 signaling, supporting the hypothesis that the R146C variant is deleterious and thus could have pathogenic effects (Wang et al., 2013). Putting all GDF9 variants detected in women with POI together, the GDF9 mutation rate in POI appeared to be significantly higher than control women ($P=0.0484$) (Table 2).

Women with nonsyndromic POI showed heterozygous mutations not only in *BMP15* or *GDF9* but also in other key genes of ovarian function, such as *NOBOX, FOXL2, SOHLH1, FIGLA, GALT, STAG3, HFM1, SYCE1, MCM8, MCM9, SCM1β, REC8, LHX8*, and many more, involved in cell functions such as regulation of transcription, meiosis, and DNA repair (Bouali et al., 2016; Bouilly et al., 2016; Qin et al., 2015). A recent study on whole-exome sequencing of Chinese POI patients showed a novel homozygous truncating variant in the *NOBOX* gene (chr7:144098161delC) that impaired severely the transcriptional activation of the *GDF9* gene in functional analyses, suggesting that a loss-of-function effect on NOBOX transcriptional activity could be associated with POI via reduced GDF9 (Li et al., 2017).

8.3 Polycystic Ovary Syndrome

PCOS affects 5%–10% of women in their reproductive age and is considered the most common endocrinological disorder characterized by anovulation, ovarian cysts, elevated plasma level of androgens, hirsutism, insulin resistance, obesity, and menstrual irregularities that lead to infertility (Dunaif, 1997; Fauser et al., 2012; Franks, 1995; Legro et al., 1998). Genetic factors have been shown to influence the impaired follicular development that contribute to PCOS pathogenesis, and the dysregulation of oocyte BMP15 and GDF9 expression in women with PCOS has been reported; however, a consistent conclusion with regard to the dysregulated levels of BMP15 and GDF9 in PCOS remains to be lead (de Resende et al., 2012; Teixeira Filho et al., 2002; Wei et al., 2014; Wei, Liang, Fang, & Zhang, 2011; Zhao et al., 2010).

Nonsynonymous mutations in the *BMP15* gene were also identified in PCOS patients (L12V, A37P, P57A, R96S, N103S, H200Y, and Ins263L) (Liu et al., 2011; Mehdizadeh et al., 2016) (Table 3). Several of these mutations (L12V, A37P, P57A, R96S, and N103S) in the *BMP15* gene were found in PCOS, but not normal, women (Liu et al., 2011). However, the prevalence of all these BMP15 variants in women with PCOS is not significantly different from that in normal controls (Table 3). Nonsynonymous mutations were also found in the *GDF9* gene in PCOS patients (N5L, L40V, M45V, D57Y, R146C, Y342F, S425R, and S428T) (Wang et al., 2010) with no significant difference in mutation rates between PCOS and normal women (Table 3). Despite the presence of a variety of studies indicating a correlation between genetic variations in key genes of ovarian function and PCOS development, other evidence exists in support of the absence of such associations between BMP15 and GDF9 polymorphisms and PCOS susceptibility (Sproul, Jones, Mathur, Azziz, & Goodarzi, 2010; Takebayashi et al., 2000).

9. PERSPECTIVES

Most of the mutations in the *GDF9* and *BMP15* genes identified in mothers of DZ twins and women with POI and PCOS are located in the proregion of the proprotein. As the proregion is necessary for dimerization of the mature protein (Hogan, 1996), mutations in the proregion may cause the impaired processing of the proproteins by forming misfolded proprotein dimers and thereby negatively impact the production of functional mature protein dimers. This impairment could occur in various combinations of GDF9 or BMP15 subunits (i.e., homodimers of the mutant GDF9, heterodimers of the mutant GDF9 and the wild-type GDF9, heterodimers of the mutant GDF9 and the wild-type BMP15, homodimers of the mutant BMP15, heterodimers of the mutant BMP15 and the wild-type BMP15, and heterodimers of the mutant BMP15 and the wild-type GDF9).

Misfolded proteins are degraded in the endoplasmic reticulum (ER) by the process termed ER-associated degradation (McCracken & Brodsky, 1996), which is necessary for protein quality control in the eukaryotic secretory pathway to ensure that only properly folded proteins transit through cellular organelles (Ellgaard & Helenius, 2001). These pathological events caused by adoption of nonnative protein conformations lead to a large range of conditions recognized as conformational diseases (Convertino, Das, & Dokholyan, 2016; Kopito & Ron, 2000). In the control process of misfolded proteins, chaperones control the folding status of proteins by preventing and reversing protein aggregation together with ATP-dependent proteases (Weibezahn, Bukau, & Mogk, 2004). As a therapeutic strategy for conformational diseases, cell-permeable pharmacological chaperones are expected to selectively provide a molecular scaffold for misfolded GDF9 and BMP15 proproteins to recover their native folding, leading to the secretion of functional mature proteins.

ACKNOWLEDGMENTS

We apologize to researchers whose work could not be cited in this review due to space constrains. This work was supported by the NIH Grant 2P50HD012303-34A1 awarded in part to S.S. and by the UC San Diego Department of Reproductive Medicine Fund 60121B awarded to S.S.

REFERENCES

Al-Musawi, S. L., Walton, K. L., Heath, D., Simpson, C. M., & Harrison, C. A. (2013). Species differences in the expression and activity of bone morphogenetic protein 15. *Endocrinology*, *154*, 888–899.

Anderson, E., & Albertini, D. F. (1976). Gap junctions between the oocyte and companion follicle cells in the mammalian ovary. *Journal of Cell Biology*, *71*, 680–686.

Behrouzi, A., Colazo, M. G., & Ambrose, D. J. (2016). Alterations in bone morphogenetic protein 15, growth differentiation factor 9, and gene expression in granulosa cells in preovulatory follicles of dairy cows given porcine LH. *Theriogenology*, *85*, 1249–1257.

Bouali, N., Francou, B., Bouligand, J., Lakhal, B., Malek, I., Kammoun, M., et al. (2016). NOBOX is a strong autosomal candidate gene in Tunisian patients with primary ovarian insufficiency. *Clinical Genetics*, *89*, 608–613.

Bouilly, J., Beau, I., Barraud, S., Bernard, V., Azibi, K., Fagart, J., et al. (2016). Identification of multiple gene mutations accounts for a new genetic architecture of primary ovarian insufficiency. *The Journal of Clinical Endocrinology and Metabolism*, *101*, 4541–4550.

Buccione, R., Schroeder, A. C., & Eppig, J. J. (1990). Interactions between somatic cells and germ cells throughout mammalian oogenesis. *Biology of Reproduction*, *43*, 543–547.

Carabatsos, M. J., Elvin, J., Matzuk, M. M., & Albertini, D. F. (1998). Characterization of oocyte and follicle development in growth differentiation factor-9-deficient mice. *Developmental Biology*, *204*, 373–384.

Chand, A. L., Ponnampalam, A. P., Harris, S. E., Winship, I. M., & Shelling, A. N. (2006). Mutational analysis of BMP15 and GDF9 as candidate genes for premature ovarian failure. *Fertility and Sterility*, *86*, 1009–1012.

Convertino, M., Das, J., & Dokholyan, N. V. (2016). Pharmacological chaperones: Design and development of new therapeutic strategies for the treatment of conformational diseases. *ACS Chemical Biology*, *11*, 1471–1489.

Cox, L., & Liu, J. H. (2014). Primary ovarian insufficiency: An update. *International Journal of Women's Health*, *6*, 235–243.

Crawford, J. L., & McNatty, K. P. (2012). The ratio of growth differentiation factor 9: Bone morphogenetic protein 15 mRNA expression is tightly co-regulated and differs between species over a wide range of ovulation rates. *Molecular and Cellular Endocrinology, 348*, 339–343.

de Resende, L. O., Vireque, A. A., Santana, L. F., Moreno, D. A., de Sa Rosa, E. S. A. C., Ferriani, R. A., et al. (2012). Single-cell expression analysis of BMP15 and GDF9 in mature oocytes and BMPR2 in cumulus cells of women with polycystic ovary syndrome undergoing controlled ovarian hyperstimulation. *Journal of Assisted Reproduction and Genetics, 29*, 1057–1065.

Dey, S. R., Deb, G. K., Ha, A. N., Lee, J. I., Bang, J. I., Lee, K. L., et al. (2012). Coculturing denuded oocytes during the in vitro maturation of bovine cumulus oocyte complexes exerts a synergistic effect on embryo development. *Theriogenology, 77*, 1064–1077.

Di Pasquale, E., Beck-Peccoz, P., & Persani, L. (2004). Hypergonadotropic ovarian failure associated with an inherited mutation of human bone morphogenetic protein-15 (BMP15) gene. *American Journal of Human Genetics, 75*, 106–111.

Di Pasquale, E., Rossetti, R., Marozzi, A., Bodega, B., Borgato, S., Cavallo, L., et al. (2006). Identification of new variants of human BMP15 gene in a large cohort of women with premature ovarian failure. *The Journal of Clinical Endocrinology and Metabolism, 91*, 1976–1979.

Dixit, H., Rao, L. K., Padmalatha, V., Kanakavalli, M., Deenadayal, M., Gupta, N., et al. (2005). Mutational screening of the coding region of growth differentiation factor 9 gene in Indian women with ovarian failure. *Menopause, 12*, 749–754.

Dixit, H., Rao, L. K., Padmalatha, V. V., Kanakavalli, M., Deenadayal, M., Gupta, N., et al. (2006). Missense mutations in the BMP15 gene are associated with ovarian failure. *Human Genetics, 119*, 408–415.

Dong, J., Albertini, D. F., Nishimori, K., Kumar, T. R., Lu, N., & Matzuk, M. (1996). Growth differentiation factor-9 is required during early ovarian folliculogenesis. *Nature, 383*, 531–535.

Dragovic, R. A., Ritter, L. J., Schulz, S. J., Amato, F., Thompson, J. G., Armstrong, D. T., et al. (2007). Oocyte-secreted factor activation of SMAD 2/3 signaling enables initiation of mouse cumulus cell expansion. *Biology of Reproduction, 76*, 848–857.

Dube, J. L., Wang, P., Elvin, J., Lyons, K. M., Celeste, A. J., & Matzuk, M. M. (1998). The bone morphogenetic protein 15 gene is X-linked and expressed in oocytes. *Molecular Endocrinology, 12*, 1809–1817.

Duffy, D. M. (2003). Growth differentiation factor-9 is expressed by the primate follicle throughout the periovulatory interval. *Biology of Reproduction, 69*, 725–732.

Dunaif, A. (1997). Insulin resistance and the polycystic ovary syndrome: Mechanism and implications for pathogenesis. *Endocrine Reviews, 18*, 774–800.

Ellgaard, L., & Helenius, A. (2001). ER quality control: Towards an understanding at the molecular level. *Current Opinion in Cell Biology, 13*, 431–437.

Eppig, J. J. (2001). Oocyte control of ovarian follicular development and function in mammals. *Reproduction, 122*, 829–838.

Eppig, J. J., Pendola, F. L., & Wigglesworth, K. (1998). Mouse oocytes suppress cAMP-induced expression of LH receptor mRNA by granulosa cells in vitro. *Molecular Reproduction and Development, 49*, 327–332.

Eppig, J. J., Wigglesworth, K., & Pendola, F. (2002). The mammalian oocyte orchestrates the rate of ovarian follicular development. *Proceedings of the National Academy of Sciences of the United States of America, 99*, 2890–2894.

Eppig, J. J., Wigglesworth, K., Pendola, F., & Hirao, Y. (1997). Murine oocytes suppress expression of luteinizing hormone receptor messenger ribonucleic acid by granulosa cells. *Biology of Reproduction, 56*, 976–984.

Erickson, G. F., & Shimasaki, S. (2003). The spatiotemporal expression pattern of the bone morphogenetic protein family in rat ovary cell types during the estrous cycle. *Reproductive Biology and Endocrinology*, *1*, 9.
Fauser, B. C., Tarlatzis, B. C., Rebar, R. W., Legro, R. S., Balen, A. H., Lobo, R., et al. (2012). Consensus on women's health aspects of polycystic ovary syndrome (PCOS): The Amsterdam ESHRE/ASRM-sponsored 3rd PCOS Consensus Workshop Group. *Fertility and Sterility*, *97*, 28–38, e25.
Ferrarini, E., Russo, L., Fruzzetti, F., Agretti, P., De Marco, G., Dimida, A., et al. (2013). Clinical characteristics and genetic analysis in women with premature ovarian insufficiency. *Maturitas*, *74*, 61–67.
Fortuno, C., & Labarta, E. (2014). Genetics of primary ovarian insufficiency: A review. *Journal of Assisted Reproduction and Genetics*, *31*, 1573–1585.
Franks, S. (1995). Polycystic ovary syndrome. *New England Journal of Medicine*, *333*, 853–861.
Galloway, S. M., McNatty, K. P., Cambridge, L. M., Laitinen, M. P. E., Juengel, J. L., Jokiranta, T. S., et al. (2000). Mutations in an oocyte-derived growth factor gene (BMP15) cause increased ovulation rate and infertility in a dosage-sensitive manner. *Nature Genetics*, *25*, 279–283.
Gilchrist, R. B., Lane, M., & Thompson, J. G. (2008). Oocyte-secreted factors: Regulators of cumulus cell function and oocyte quality. *Human Reproduction Update*, *14*, 159–177.
Gilchrist, R. B., Ritter, L. J., & Armstrong, D. T. (2001). Mouse oocyte mitogenic activity is developmentally coordinated throughout folliculogenesis and meiotic maturation. *Developmental Biology*, *240*, 289–298.
Gilchrist, R. B., Ritter, L. J., & Armstrong, D. T. (2004). Oocyte-somatic cell interactions during follicle development in mammals. *Animal Reproduction Science*, *82–83*, 431–446.
Gilula, N. B., Epstein, M. L., & Beers, W. H. (1978). Cell-to-cell communication and ovulation. A study of the cumulus-oocyte complex. *The Journal of Cell Biology*, *78*, 58–75.
Glister, C., Groome, N. P., & Knight, P. G. (2003). Oocyte-mediated suppression of follicle-stimulating hormone- and insulin-like growth factor-induced secretion of steroids and inhibin-related proteins by bovine granulosa cells in vitro: Possible role of transforming growth factor alpha. *Biology of Reproduction*, *68*, 758–765.
Gomez, M. N., Kang, J. T., Koo, O. J., Kim, S. J., Kwon, D. K., Park, S. J., et al. (2012). Effect of oocyte-secreted factors on porcine in vitro maturation, cumulus expansion and developmental competence of parthenotes. *Zygote*, *20*, 135–145.
Goodenough, D. A., Simon, A. M., & Paul, D. L. (1999). Gap junctional intercellular communication in the mouse ovarian follicle. *Novartis Foundation Symposium*, *219*, 226–235. discussion 235–240.
Gosden, R. G., Treloar, S. A., Martin, N. G., Cherkas, L. F., Spector, T. D., Faddy, M. J., et al. (2007). Prevalence of premature ovarian failure in monozygotic and dizygotic twins. *Human Reproduction*, *22*, 610–615.
Hanrahan, J. P., Gregan, S. M., Mulsant, P., Mullen, M., Davis, G. H., Powell, R., et al. (2004). Mutations in the genes for oocyte-derived growth factors GDF9 and BMP15 are associated with both increased ovulation rate and sterility in Cambridge and Belclare sheep (*Ovis aries*). *Biology of Reproduction*, *70*, 900–909.
Hayashi, M., McGee, E. A., Min, G., Klein, C., Rose, U. M., van Duin, M., et al. (1999). Recombinant growth differentiation factor-9 (GDF-9) enhances growth and differentiation of cultured early ovarian follicles. *Endocrinology*, *140*, 1236–1244.
Hoekstra, C., Zhao, Z. Z., Lambalk, C. B., Willemsen, G., Martin, N. G., Boomsma, D. I., et al. (2008). Dizygotic twinning. *Human Reproduction Update*, *14*, 37–47.
Hogan, B. L. (1996). Bone morphogenetic proteins: Multifunctional regulators of vertebrate development. *Genes & Development*, *10*, 1580–1594.
Hussein, M. R. (2005). Apoptosis in the ovary: Molecular mechanisms. *Human Reproduction Update*, *11*, 162–177.

Hussein, T. S., Sutton-McDowall, M. L., Gilchrist, R. B., & Thompson, J. G. (2011). Temporal effects of exogenous oocyte-secreted factors on bovine oocyte developmental competence during IVM. *Reproduction, Fertility, and Development*, *23*, 576–584.

Hussein, T. S., Thompson, J. G., & Gilchrist, R. B. (2006). Oocyte-secreted factors enhance oocyte developmental competence. *Developmental Biology*, *296*, 514–521.

Inagaki, K., & Shimasaki, S. (2010). Impaired production of BMP-15 and GDF-9 mature proteins derived from proproteins with mutations in the proregion. *Molecular and Cellular Endocrinology*, *328*, 1–7.

Israel, D. I., Nove, J., Kerns, K. M., Moutsatsos, I. K., & Kaufman, R. J. (1992). Expression and characterization of bone morphogenetic protein-2 in Chinese hamster ovary cells. *Growth Factors*, *7*, 139–150.

Jones, W. K., Richmond, E. A., White, K., Sasak, H., Kusmik, W., Smart, J., et al. (1994). Osteogenic protein-1 (OP-1) expression and processing in Chinese hamster ovary cells: Isolation of a soluble complex containing the mature and pro-domains of OP-1. *Growth Factors*, *11*, 215–225.

Joyce, I. M., Pendola, F. L., Wigglesworth, K., & Eppig, J. J. (1999). Oocyte regulation of kit ligand expression in mouse ovarian follicles. *Developmental Biology*, *214*, 342–353.

Juengel, J. L., Hudson, N. L., Whiting, L., & McNatty, K. P. (2004). Effects of immunization against bone morphogenetic protein 15 and growth differentiation factor 9 on ovulation rate, fertilization, and pregnancy in ewes. *Biology of Reproduction*, *70*, 557–561.

Kaivo-Oja, N., Bondestam, J., Kamarainen, M., Koskimies, J., Vitt, U., Cranfield, M., et al. (2003). Growth differentiation factor-9 induces Smad2 activation and inhibin B production in cultured human granulosa-luteal cells. *Journal of Clinical Endocrinology and Metabolism*, *88*, 755–762.

Kopito, R. R., & Ron, D. (2000). Conformational disease. *Nature Cell Biology*, *2*, E207–E209.

Kovanci, E., Rohozinski, J., Simpson, J. L., Heard, M. J., Bishop, C. E., & Carson, S. A. (2007). Growth differentiating factor-9 mutations may be associated with premature ovarian failure. *Fertility and Sterility*, *87*, 143–146.

Laissue, P., Christin-Maitre, S., Touraine, P., Kuttenn, F., Ritvos, O., Aittomaki, K., et al. (2006). Mutations and sequence variants in GDF9 and BMP15 in patients with premature ovarian failure. *European Journal of Endocrinology*, *154*, 739–744.

Laitinen, M., Vuojolainen, K., Jaatinen, R., Ketola, I., Aaltonen, J., Lehtonen, E., et al. (1998). A novel growth differentiation factor-9 (GDF-9) related factor is co-expressed with GDF-9 in mouse oocytes during folliculogenesis. *Mechanisms of Development*, *78*, 135–140.

Lanuza, G. M., Groome, N. P., Baranao, J. L., & Campo, S. (1999). Dimeric inhibin A and B production are differentially regulated by hormones and local factors in rat granulosa cells. *Endocrinology*, *140*, 2549–2554.

Ledig, S., Ropke, A., Haeusler, G., Hinney, B., & Wieacker, P. (2008). BMP15 mutations in XX gonadal dysgenesis and premature ovarian failure. *American Journal of Obstetrics and Gynecology*, *198*(84), e81–85.

Legro, R. S., Spielman, R., Urbanek, M., Driscoll, D., Strauss, J. F., 3rd, & Dunaif, A. (1998). Phenotype and genotype in polycystic ovary syndrome. *Recent Progress in Hormone Research*, *53*, 217–256.

Li, Q., Agno, J. E., Edson, M. A., Nagaraja, A. K., Nagashima, T., & Matzuk, M. M. (2011). Transforming growth factor beta receptor type 1 is essential for female reproductive tract integrity and function. *PLoS Genetics*, *7*, e1002320.

Li, Q., Rajanahally, S., Edson, M. A., & Matzuk, M. M. (2009). Stable expression and characterization of N-terminal tagged recombinant human bone morphogenetic protein 15. *Molecular Human Reproduction*, *15*, 779–788.

Li, L., Wang, B., Zhang, W., Chen, B., Luo, M., Wang, J., et al. (2017). A homozygous NOBOX truncating variant causes defective transcriptional activation and leads to primary ovarian insufficiency. *Human Reproduction*, *32*, 248–255.

Liao, W. X., Moore, R. K., Otsuka, F., & Shimasaki, S. (2003). Effect of intracellular interactions on the processing and secretion of bone morphogenetic protein-15 (BMP-15) and growth and differentiation factor-9: Implication of the aberrant ovarian phenotype of BMP-15 mutant sheep. *Journal of Biological Chemistry*, *278*, 3713–3719.

Liao, W. X., Moore, R. K., & Shimasaki, S. (2004). Functional and molecular characterization of naturally occurring mutations in the oocyte-secreted factors BMP-15 and GDF-9. *Journal of Biological Chemistry*, *279*, 17391–17396.

Lin, J. Y., Pitman-Crawford, J. L., Bibby, A. H., Hudson, N. L., McIntosh, C. J., Juengel, J. L., et al. (2012). Effects of species differences on oocyte regulation of granulosa cell function. *Reproduction*, *144*, 557–567.

Ling, N., Ying, S.-Y., Ueno, N., Shimasaki, S., Esch, F., Hotta, M., et al. (1986). Pituitary FSH is released by a heterodimer of the ß-subunits from the two forms of inhibin. *Nature*, *321*, 779–782.

Liu, J., Wang, B., Wei, Z., Zhou, P., Zu, Y., Zhou, S., et al. (2011). Mutational analysis of human bone morphogenetic protein 15 in Chinese women with polycystic ovary syndrome. *Metabolism*, *60*, 1511–1514.

Ma, L., Chen, Y., Mei, S., Liu, C., Ma, X., Li, Y., et al. (2015). Single nucleotide polymorphisms in premature ovarian failure-associated genes in a Chinese Hui population. *Molecular Medicine Reports*, *12*, 2529–2538.

Martin, N. G., Healey, S. C., Pangan, T. S., Heath, A. C., & Turner, G. (1997). Do mothers of dizygotic twins have earlier menopause? A role for fragile X? *American Journal of Medical Genetics*, *69*, 114–116.

Matzuk, M. M., Burns, K. H., Viveiros, M. M., & Eppig, J. J. (2002). Intercellular communication in the mammalian ovary: Oocytes carry the conversation. *Science*, *296*, 2178–2180.

Mayer, A., Fouquet, B., Pugeat, M., & Misrahi, M. (2017). BMP15 "knockout-like" effect in familial premature ovarian insufficiency with persistent ovarian reserve. *Clinical Genetics*, *92*, 208–212.

Mazerbourg, S., Klein, C., Roh, J., Kaivo-Oja, N., Mottershead, D. G., Korchynskyi, O., et al. (2004). Growth differentiation factor-9 signaling is mediated by the type I receptor, activin receptor-like kinase 5. *Molecular Endocrinology*, *18*, 653–665.

McCracken, A. A., & Brodsky, J. L. (1996). Assembly of ER-associated protein degradation in vitro: Dependence on cytosol, calnexin, and ATP. *Journal of Cell Biology*, *132*, 291–298.

McGrath, S. A., Esquela, A. F., & Lee, S.-J. (1995). Oocyte-specific expression of growth/differentiation factor-9. *Molecular Endocrinology*, *9*, 131–136.

McIntosh, C. J., Lun, S., Lawrence, S., Western, A. H., McNatty, K. P., & Juengel, J. L. (2008). The proregion of mouse BMP15 regulates the cooperative interactions of BMP15 and GDF9. *Biology of Reproduction*, *79*, 889–896.

McMahon, H. E., Sharma, S., & Shimasaki, S. (2008). Phosphorylation of bone morphogenetic protein-15 and growth and differentiation factor-9 plays a critical role in determining agonistic or antagonistic functions. *Endocrinology*, *149*, 812–817.

McNatty, K. P., Juengel, J. L., Reader, K. L., Lun, S., Myllymaa, S., Lawrence, S. B., et al. (2005a). Bone morphogenetic protein 15 and growth differentiation factor 9 co-operate to regulate granulosa cell function. *Reproduction*, *129*, 473–480.

McNatty, K. P., Juengel, J. L., Reader, K. L., Lun, S., Myllymaa, S., Lawrence, S. B., et al. (2005b). Bone morphogenetic protein 15 and growth differentiation factor 9 co-operate to regulate granulosa cell function in ruminants. *Reproduction*, *129*, 481–487.

McNatty, K. P., Lawrence, S., Groome, N. P., Meerasahib, M. F., Hudson, N. L., Whiting, L., et al. (2006). Meat and livestock association plenary lecture 2005. Oocyte signalling molecules and their effects on reproduction in ruminants. *Reproduction, Fertility, and Development, 18*, 403–412.

McNatty, K. P., Moore, L. G., Hudson, N. L., Quirke, L. D., Lawrence, S. B., Reader, K., et al. (2004). The oocyte and its role in regulating ovulation rate: A new paradigm in reproductive biology. *Reproduction, 128*, 379–386.

McNatty, K. P., Smith, P., Moore, L. G., Reader, K., Lun, S., Hanrahan, J. P., et al. (2005). Oocyte-expressed genes affecting ovulation rate. *Molecular and Cellular Endocrinology, 234*, 57–66.

McPherron, A. C., & Lee, S.-J. (1993). GDF-3 and GDF-9: Two new members of the transforming growth factor-ß superfamily containing a novel pattern of cysteines. *Journal of Biological Chemistry, 268*, 3444–3449.

Mehdizadeh, A., Sheikhha, M. H., Kalantar, S. M., Aali, B. S., & Ghanei, A. (2016). Mutation analysis of exon1 of bone morphogenetic protein-15 gene in Iranian patients with polycystic ovarian syndrome. *International Journal of Reproductive Biomedicine (Yazd, Iran), 14*, 527–532.

Meno, C., Ito, Y., Saijoh, Y., Matsuda, Y., Tashiro, K., Kuhara, S., et al. (1997). Two closely-related left-right asymmetrically expressed genes, lefty-1 and lefty-2: Their distinct expression domains, chromosomal linkage and direct neutralizing activity in xenopus embryos. *Genes to Cells, 2*, 513–524.

Mester, B., Ritter, L. J., Pitman, J. L., Bibby, A. H., Gilchrist, R. B., McNatty, K. P., et al. (2015). Oocyte expression, secretion and somatic cell interaction of mouse bone morphogenetic protein 15 during the peri-ovulatory period. *Reproduction, Fertility, and Development, 27*, 801–811.

Montgomery, G. W., Zhao, Z. Z., Marsh, A. J., Mayne, R., Treloar, S. A., James, M., et al. (2004). A deletion mutation in GDF9 in sisters with spontaneous DZ twins. *Twin Research, 7*, 548–555.

Moore, R. K., Erickson, G. F., & Shimasaki, S. (2004). Are BMP-15 and GDF-9 primary determinants of ovulation quota in mammals? *Trends in Endocrinology and Metabolism, 15*, 356–361.

Moore, R. K., Otsuka, F., & Shimasaki, S. (2003). Molecular basis of bone morphogenetic protein-15 signaling in granulosa cells. *Journal of Biological Chemistry, 278*, 304–310.

Mottershead, D. G., Pulkki, M. M., Muggalla, P., Pasternack, A., Tolonen, M., Myllymaa, S., et al. (2008). Characterization of recombinant human growth differentiation factor-9 signaling in ovarian granulosa cells. *Molecular and Cellular Endocrinology, 283*, 58–67.

Mottershead, D. G., Ritter, L. J., & Gilchrist, R. B. (2012). Signalling pathways mediating specific synergistic interactions between GDF9 and BMP15. *Molecular Human Reproduction, 18*, 121–128.

Mottershead, D. G., Sugimura, S., Al-Musawi, S. L., Li, J. J., Richani, D., White, M. A., et al. (2015). Cumulin, an oocyte-secreted heterodimer of the transforming growth factor-beta family, is a potent activator of granulosa cells and improves oocyte quality. *The Journal of Biological Chemistry, 290*, 24007–24020.

Otsuka, F., Moore, R. K., & Shimasaki, S. (2001). Biological function and cellular mechanism of bone morphogenetic protein-6 in the ovary. *Journal of Biological Chemistry, 276*, 32889–32895.

Palmer, J. S., Zhao, Z. Z., Hoekstra, C., Hayward, N. K., Webb, P. M., Whiteman, D. C., et al. (2006). Novel variants in growth differentiation factor 9 in mothers of dizygotic twins. *The Journal of Clinical Endocrinology and Metabolism, 91*, 4713–4716.

Pangas, S. A., & Matzuk, M. M. (2004). Genetic models for transforming growth factor beta superfamily signaling in ovarian follicle development. *Molecular and Cellular Endocrinology*, *225*, 83–91.
Pangas, S. A., & Matzuk, M. M. (2005). The art and artifact of GDF9 activity: Cumulus expansion and the cumulus expansion-enabling factor. *Biology of Reproduction*, *73*, 582–585.
Paradis, F., Moore, H. S., Pasternak, J. A., Novak, S., Dyck, M. K., Dixon, W. T., et al. (2010). Pig preovulatory oocytes modulate cumulus cell protein and gene expression in vitro. *Molecular and Cellular Endocrinology*, *320*, 87–96.
Park, J. Y., Su, Y. Q., Ariga, M., Law, E., Jin, S. L., & Conti, M. (2004). EGF-like growth factors as mediators of LH action in the ovulatory follicle. *Science*, *303*, 682–684.
Peng, J., Li, Q., Wigglesworth, K., Rangarajan, A., Kattamuri, C., Peterson, R. T., et al. (2013). Growth differentiation factor 9: Bone morphogenetic protein 15 heterodimers are potent regulators of ovarian functions. *Proceedings of the National Academy of Sciences of the United States of America*, *110*, E776–785.
Persani, L., Rossetti, R., Di Pasquale, E., Cacciatore, C., & Fabre, S. (2014). The fundamental role of bone morphogenetic protein 15 in ovarian function and its involvement in female fertility disorders. *Human Reproduction Update*, *20*, 869–883.
Pulkki, M. M., Mottershead, D. G., Pasternack, A. H., Muggalla, P., Ludlow, H., van Dinther, M., et al. (2012). A covalently dimerized recombinant human bone morphogenetic protein-15 variant identifies bone morphogenetic protein receptor type 1B as a key cell surface receptor on ovarian granulosa cells. *Endocrinology*, *153*, 1509–1518.
Qin, Y., Jiao, X., Simpson, J. L., & Chen, Z. J. (2015). Genetics of primary ovarian insufficiency: New developments and opportunities. *Human Reproduction Update*, *21*, 787–808.
Reader, K. L., Heath, D. A., Lun, S., McIntosh, C. J., Western, A. H., Littlejohn, R. P., et al. (2011). Signalling pathways involved in the cooperative effects of ovine and murine GDF9+BMP15-stimulated thymidine uptake by rat granulosa cells. *Reproduction*, *142*, 123–131.
Regassa, A., Rings, F., Hoelker, M., Cinar, U., Tholen, E., Looft, C., et al. (2011). Transcriptome dynamics and molecular cross-talk between bovine oocyte and its companion cumulus cells. *BMC Genomics*, *12*, 57.
Requena, A., Cruz, M., Agudo, D., Pacheco, A., & Garcia-Velasco, J. A. (2016). Type of gonadotropin during controlled ovarian stimulation affects the endocrine profile in follicular fluid and apoptosis rate in cumulus cells. *European Journal of Obstetrics, Gynecology, and Reproductive Biology*, *197*, 142–146.
Ritter, L. J., Sugimura, S., & Gilchrist, R. B. (2015). Oocyte induction of EGF responsiveness in somatic cells is associated with the acquisition of porcine oocyte developmental competence. *Endocrinology*, *156*, 2299–2312.
Roh, J.-S., Bondestam, J., Mazerbourg, S., Kaivo-Oja, N., Groome, N., Ritvos, O., et al. (2003). Growth differentiation factor-9 stimulates inhibin production and activates Smad2 in cultured rat granulosa cells. *Endocrinology*, *144*, 172–178.
Romaguera, R., Morato, R., Jimenez-Macedo, A. R., Catala, M., Roura, M., Paramio, M. T., et al. (2010). Oocyte secreted factors improve embryo developmental competence of COCs from small follicles in prepubertal goats. *Theriogenology*, *74*, 1050–1059.
Rossetti, R., Di Pasquale, E., Marozzi, A., Bione, S., Toniolo, D., Grammatico, P., et al. (2009). BMP15 mutations associated with primary ovarian insufficiency cause a defective production of bioactive protein. *Human Mutation*, *30*, 804–810.
Rossetti, R., Ferrari, I., Bonomi, M., & Persani, L. (2017). Genetics of primary ovarian insufficiency. *Clinical Genetics*, *91*, 183–198.

Saito, S., Yano, K., Sharma, S., McMahon, H. E., & Shimasaki, S. (2008). Characterization of the post-translational modification of recombinant human BMP-15 mature protein. *Protein Science*, *17*, 362–370.

Shimasaki, S., Moore, R. K., Otsuka, F., & Erickson, G. F. (2004). The bone morphogenetic protein system in mammalian reproduction. *Endocrine Reviews*, *25*, 72–101.

Simpson, C. M., Robertson, D. M., Al-Musawi, S. L., Heath, D. A., McNatty, K. P., Ritter, L. J., et al. (2014). Aberrant GDF9 expression and activation are associated with common human ovarian disorders. *The Journal of Clinical Endocrinology and Metabolism*, *99*, E615–624.

Simpson, C. M., Stanton, P. G., Walton, K. L., Chan, K. L., Ritter, L. J., Gilchrist, R. B., et al. (2012). Activation of latent human GDF9 by a single residue change (Gly 391 Arg) in the mature domain. *Endocrinology*, *153*, 1301–1310.

Sproul, K., Jones, M. R., Mathur, R., Azziz, R., & Goodarzi, M. O. (2010). Association study of four key folliculogenesis genes in polycystic ovary syndrome. *British Journal of Obstetrics and Gynaecology*, *117*, 756–760.

Su, J., Hu, G., Wang, Y., Liang, D., Gao, M., Sun, H., et al. (2014). Recombinant human growth differentiation factor-9 improves oocyte reprogramming competence and subsequent development of bovine cloned embryos. *Cellular Reprogramming*, *16*, 281–289.

Su, Y. Q., Sugiura, K., Li, Q., Wigglesworth, K., Matzuk, M. M., & Eppig, J. J. (2010). Mouse oocytes enable LH-induced maturation of the cumulus-oocyte complex via promoting EGF receptor-dependent signaling. *Molecular Endocrinology*, *24*, 1230–1239.

Su, J., Wang, Y., Zhang, L., Wang, B., Liu, J., Luo, Y., et al. (2014). Oocyte-secreted factors in oocyte maturation media enhance subsequent development of bovine cloned embryos. *Molecular Reproduction and Development*, *81*, 341–349.

Su, Y. Q., Wu, X., O'Brien, M. J., Pendola, F. L., Denegre, J. N., Matzuk, M. M., et al. (2004). Synergistic roles of BMP15 and GDF9 in the development and function of the oocyte-cumulus cell complex in mice: Genetic evidence for an oocyte-granulosa cell regulatory loop. *Developmental Biology*, *276*, 64–73.

Sudiman, J., Ritter, L. J., Feil, D. K., Wang, X., Chan, K., Mottershead, D. G., et al. (2014). Effects of differing oocyte-secreted factors during mouse in vitro maturation on subsequent embryo and fetal development. *Journal of Assisted Reproduction and Genetics*, *31*, 295–306.

Sugimura, S., Ritter, L. J., Rose, R. D., Thompson, J. G., Smitz, J., Mottershead, D. G., et al. (2015). Promotion of EGF receptor signaling improves the quality of low developmental competence oocytes. *Developmental Biology*, *403*, 139–149.

Sugimura, S., Ritter, L. J., Sutton-McDowall, M. L., Mottershead, D. G., Thompson, J. G., & Gilchrist, R. B. (2014). Amphiregulin co-operates with bone morphogenetic protein 15 to increase bovine oocyte developmental competence: Effects on gap junction-mediated metabolite supply. *Molecular Human Reproduction*, *20*, 499–513.

Sugiura, K., Su, Y. Q., Diaz, F. J., Pangas, S. A., Sharma, S., Wigglesworth, K., et al. (2007). Oocyte-derived BMP15 and FGFs cooperate to promote glycolysis in cumulus cells. *Development*, *134*, 2593–2603.

Tagliabracci, V. S., Engel, J. L., Wen, J., Wiley, S. E., Worby, C. A., Kinch, L. N., et al. (2012). Secreted kinase phosphorylates extracellular proteins that regulate biomineralization. *Science*, *336*, 1150–1153.

Takebayashi, K., Takakura, K., Wang, H. Q., Kimura, F., Kasahara, K., & Noda, Y. (2000). Mutation analysis of the growth differentiation factor-9 and -9b genes in patients with premature ovarian failure and polycystic ovary syndrome. *Fertility and Sterility*, *74*, 976–979.

Teixeira Filho, F. L., Baracat, E. C., Lee, T. H., Suh, C. S., Matsui, M., Chang, R. J., et al. (2002). Aberrant expression of growth differentiation factor-9 in oocytes of women with

polycystic ovary syndrome. *The Journal of Clinical Endocrinology and Metabolism*, *87*, 1337–1344.
Thies, R. S., Chen, T. J., Davies, M. V., Tomkinson, K. N., Pearson, A. A., Shakey, Q. A., et al. (2000). GDF-8 propeptide binds to GDF-8 and antagonizes biological activity by inhibiting GDF-8 receptor binding. *Growth Factors*, *18*, 251–259.
Tiotiu, D., Alvaro Mercadal, B., Imbert, R., Verbist, J., Demeestere, I., De Leener, A., et al. (2010). Variants of the BMP15 gene in a cohort of patients with premature ovarian failure. *Human Reproduction*, *25*, 1581–1587.
Tong, S., & Short, R. V. (1998). Dizygotic twinning as a measure of human fertility. *Human Reproduction*, *13*, 95–98.
Tucker, E. J., Grover, S. R., Bachelot, A., Touraine, P., & Sinclair, A. H. (2016). Premature ovarian insufficiency: New perspectives on genetic cause and phenotypic spectrum. *Endocrine Reviews*, *37*, 609–635.
Vanderhyden, B. C., Cohen, J. N., & Morley, P. (1993). Mouse oocytes regulate granulosa cell steroidogenesis. *Endocrinology*, *133*, 423–426.
Vanderhyden, B. C., Telfer, E. E., & Eppig, J. J. (1992). Mouse oocytes promote proliferation of granulosa cells from preantral and antral follicles in vitro. *Biology of Reproduction*, *46*, 1196–1204.
Vitt, U. A., Mazerbourg, S., Klein, C., & Hsueh, A. J. (2002). Bone morphogenetic protein receptor type II is a receptor for growth differentiation factor-9. *Biology of Reproduction*, *67*, 473–480.
Wang, T. T., Ke, Z. H., Song, Y., Chen, L. T., Chen, X. J., Feng, C., et al. (2013). Identification of a mutation in GDF9 as a novel cause of diminished ovarian reserve in young women. *Human Reproduction*, *28*, 2473–2481.
Wang, E. A., Rosen, V., D'Alessandro, J. S., Bauduy, M., Cordes, P., Harada, T., et al. (1990). Recombinant human bone morphogenetic protein induces bone formation. *Proceedings of the National Academy of Sciences of the United States of America*, *87*, 2220–2224.
Wang, B., Wen, Q., Ni, F., Zhou, S., Wang, J., Cao, Y., et al. (2010). Analyses of growth differentiation factor 9 (GDF9) and bone morphogenetic protein 15 (BMP15) mutation in Chinese women with premature ovarian failure. *Clinical Endocrinology*, *72*, 135–136.
Wang, B., Zhou, S., Wang, J., Liu, J., Ni, F., Yan, J., et al. (2010). Identification of novel missense mutations of GDF9 in Chinese women with polycystic ovary syndrome. *Reproductive Biomedicine Online*, *21*, 344–348.
Wei, L. N., Huang, R., Li, L. L., Fang, C., Li, Y., & Liang, X. Y. (2014). Reduced and delayed expression of GDF9 and BMP15 in ovarian tissues from women with polycystic ovary syndrome. *Journal of Assisted Reproduction and Genetics*, *31*, 1483–1490.
Wei, L. N., Liang, X. Y., Fang, C., & Zhang, M. F. (2011). Abnormal expression of growth differentiation factor 9 and bone morphogenetic protein 15 in stimulated oocytes during maturation from women with polycystic ovary syndrome. *Fertility and Sterility*, *96*, 464–468.
Weibezahn, J., Bukau, B., & Mogk, A. (2004). Unscrambling an egg: Protein disaggregation by AAA+ proteins. *Microbial Cell Factories*, *3*, 1.
Wu, Y. T., Tang, L., Cai, J., Lu, X. E., Xu, J., Zhu, X. M., et al. (2007). High bone morphogenetic protein-15 level in follicular fluid is associated with high quality oocyte and subsequent embryonic development. *Human Reproduction*, *22*, 1526–1531.
Yan, C., Wang, P., DeMayo, J., DeMayo, F. J., Elvin, J. A., Carino, C., et al. (2001). Synergistic roles of bone morphogenetic protein 15 and growth differentiation factor 9 in ovarian function. *Molecular Endocrinology*, *15*, 854–866.
Yeo, C. X., Gilchrist, R. B., Thompson, J. G., & Lane, M. (2008). Exogenous growth differentiation factor 9 in oocyte maturation media enhances subsequent embryo development and fetal viability in mice. *Human Reproduction*, *23*, 67–73.

Yoshino, O., McMahon, H. E., Sharma, S., & Shimasaki, S. (2006). A unique preovulatory expression pattern plays a key role in the physiological functions of bmp-15 in the mouse. *Proceedings of the National Academy of Sciences of the United States of America*, *103*, 10678–10683.

Zhang, P., Shi, Y. H., Wang, L. C., & Chen, Z. J. (2007). Sequence variants in exons of the BMP-15 gene in Chinese patients with premature ovarian failure. *Acta Obstetricia et Gynecologica Scandinavica*, *86*, 585–589.

Zhao, Z. Z., Painter, J. N., Palmer, J. S., Webb, P. M., Hayward, N. K., Whiteman, D. C., et al. (2008). Variation in bone morphogenetic protein 15 is not associated with spontaneous human dizygotic twinning. *Human Reproduction*, *23*, 2372–2379.

Zhao, S. Y., Qiao, J., Chen, Y. J., Liu, P., Li, J., & Yan, J. (2010). Expression of growth differentiation factor-9 and bone morphogenetic protein-15 in oocytes and cumulus granulosa cells of patients with polycystic ovary syndrome. *Fertility and Sterility*, *94*, 261–267.

Zhao, H., Qin, Y., Kovanci, E., Simpson, J. L., Chen, Z.-J., & Rajkovic, A. (2007). Analyses of GDF9 mutation in 100 Chinese women with premature ovarian failure. *Fertility and Sterility*, *88*, 1474–1476.

CHAPTER THIRTEEN

The Menstrual Cycle Influences Emotion but Has Limited Effect on Cognitive Function

Inger Sundström-Poromaa[1]
Department of Women's and Children's Health, Uppsala University, Uppsala, Sweden
[1]Corresponding author: e-mail address: inger.sundstrom@kbh.uu.se

Contents

Abstract

From a psychological perspective, the menstrual cycle has been a research topic for more than 50 years. The most recent menstrual cycle research has been driven by an increased interest in sex differences in neuroscience, and the urge to understand sex disparities in prevalence, clinical presentation, and treatment response in psychiatric or neurologic disorders. Indeed, the menstrual cycle is an excellent model of ovarian steroid influence on emotion, behavior, and cognition.

This review summarizes the emotion-related and cognitive findings of methodologically sound menstrual cycle studies. In particular, the review is devoted to the sex hormone-induced emotional disturbances in women with premenstrual dysphoric disorder, a subgroup of women responding with enhanced sensitivity to the normal fluctuations in endogenous hormone levels during the menstrual cycle. In addition, emotion processing and cognitive findings across the menstrual cycle in healthy women are also discussed.

The overall conclusion is that that menstrual cycle differences in sexually dimorphic cognitive tasks are small and difficult to replicate. Emotion-related changes are more consistently found and are better associated with progesterone and the luteal phase, than with estradiol.

Vitamins and Hormones, Volume 107
ISSN 0083-6729
https://doi.org/10.1016/bs.vh.2018.01.016

1. INTRODUCTION

From a psychological perspective, the menstrual cycle has been a research topic over the past 60–70 years. In fact, the entry of women into the work force can be traced by the various research questions that have been posed over time, with the overarching theme to elucidate if women are equally skilled at work, independent of which menstrual cycle phase they are in. Hence, in the 1950s to the 1970s, most attention was given to tasks that would be necessary to master as a secretary or a clerk, including attention, finger tapping, simple arithmetic tests, and some memory tasks (Sommer, 1973). Already in 1973, the first review on menstrual cycle influence on cognitive and perceptual-motor performance was published, at that point including 33 scientific papers. Not surprisingly, the review concluded that no menstrual cycle-related changes in any of these work-related tasks were evident, but the review also pointed out numerous methodological concerns with menstrual cycle research (Sommer, 1973). The concerns that were raised included lack of hormonal measures and incorrect dating of menstrual cycle stage. These methodological problems have gradually improved over time, and nowadays most menstrual cycle-related studies include saliva or serum levels of estradiol and progesterone, for correct dating of cycle phase. In the early 2000s, studies on female flight simulator performance were published, presumably reflecting the fact that women at this point also had entered male-dominated areas such the aviation business (Mumenthaler, O'Hara, Taylor, Friedman, & Yesavage, 2001a; Mumenthaler, O'Hara, Taylor, Friedman, & Yesavage, 2001b). Even later, reports on menstrual problems in deployed military women have emerged (Deuster, Powell-Dunford, Crago, & Cuda, 2011; Manski, Grindlay, Burns, Holt, & Grossman, 2014).

The past 25 years of menstrual cycle research has been driven by an increased interest in sex differences in neuroscience, and the urge to understand sex differences in prevalence, clinical presentation, and treatment response in psychiatric or neurologic disorders. From this point of view, menstrual cycle studies have sometimes been performed to explain the greater variability noted in women, especially in areas where sex differences have been difficult to establish. In addition, the menstrual cycle has also attracted interest in itself as an excellent and ecological model of ovarian steroid influence on emotion, behavior, and cognition. Indeed, there is plenty of evidence to suggest that the ovarian steroids, estradiol and progesterone, should influence these aspects of brain function.

2. SEX HORMONE EFFECTS IN THE BRAIN

Estradiol and progesterone are both highly lipophilic and easily pass through the blood–brain barrier. The estradiol receptors (ERα and ERβ) and the progesterone receptors (PRA and PRB) are highly expressed in brain areas associated with reproduction, but also found in areas of importance for cognitive function and emotional processing (for review, see Brinton et al., 2008; Gruber, Tschugguel, Schneeberger, & Huber, 2002). The estradiol receptors are expressed in the human hypothalamus, amygdala, hippocampus, claustrum, and the cerebral cortex (Osterlund, Gustafsson, Keller, & Hurd, 2000; Osterlund, Keller, & Hurd, 2000), with the densest expression in the cerebral cortex being found in the temporal parts (Osterlund, Gustafsson, et al., 2000; Osterlund, Keller, et al., 2000). While the progesterone receptors have not yet been mapped in the human brain, animal data suggest that progesterone receptors are also distributed throughout the amygdala, hippocampus, hypothalamus, thalamus, and the frontal cortex (Guerra-Araiza, Cerbon, Morimoto, & Camacho-Arroyo, 2000; Guerra-Araiza, Villamar-Cruz, Gonzalez-Arenas, Chavira, & Camacho-Arroyo, 2003). Additional membrane-bound receptors have emerged as potential mediators of rapid nongenomic effects of estradiol and progesterone. As to estradiol, the G protein-coupled estrogen receptors have been described in brain areas such as the hippocampus, hypothalamus, cortex, and substantia nigra in the rodent brain (Hazell et al., 2009). Two different families of membrane-bound progesterone receptors have been described, among which receptors belonging to the progesterone receptor membrane component 1 (PGRMC1) family are relatively abundant in rodent forebrain areas (Petersen et al., 2013). Yet another route by which progesterone may influence emotional and behavioral circuits in the brain is via its' metabolization to GABAergic neurosteroids. The most well-known progesterone-derived neurosteroid is allopregnanolone, which, by binding to the GABA-A receptor, has sedative, anxiolytic, anticonvulsant, neuroprotective properties, and memory-impairing effects (Brunton, 2015; Kask, Backstrom, Nilsson, & Sundstrom-Poromaa, 2008; Melcangi, Panzica, & Garcia-Segura, 2011).

Adding further complexity to the picture, both estradiol and progesterone act as modulators for other, classical neurotransmitters, such as the cholinergic, serotonergic, and dopaminergic systems (Comasco, Frokjaer, & Sundstrom-Poromaa, 2014). These neurotransmitters have been preliminarily investigated in relation to estrogen treatment in postmenopausal women and seem to be biological mediators of positive effects of estrogens (Henderson & Greicius, 2010). Besides the joint influence by estradiol and cholinergic neurotransmission on cognitive performance (Craig et al., 2009; Smith, Minoshima, Kuhl, & Zubieta, 2001), estradiol and progesterone have widespread interactions with the serotonin neurotransmitter system, of relevance for emotional processing and mood (Bethea, Reddy, Tokuyama, Henderson, & Lima, 2009). In humans, perhaps the most consistent finding is the estradiol-induced upregulation of serotonin 2A receptors in frontal brain areas (Frokjaer et al., 2010; Kugaya et al., 2003; Moses et al., 2000; Moses-Kolko et al., 2003). Further, in female cynomolgus monkeys an effect of menstrual cycle phase on dopamine D2 receptor availability in the caudate nucleus and putamen has been demonstrated (Czoty et al., 2009).

Estradiol is mostly associated with positive effects, both in terms of mood and cognition (Comasco et al., 2014; Sherwin, 2012). Most of the human data on estradiol effects have been derived from studies on estrogen replacement in postmenopausal women or from experimental studies using gonadotropin-releasing hormone (GnRH) agonist suppression of estradiol (Comasco et al., 2014; Sherwin, 2012). From these studies, it is generally established that estrogen therapy in postmenopausal women is associated with improved well-being (Kaunitz & Manson, 2015), although this effect in part may be due to improved sleep and vasomotor symptoms. The usefulness of estrogen for treatment of depression is debated (Craig, 2016; Studd, 2016), and at present the antidepressant effect seems to be restricted only to premenopausal women (Soares, 2014). Imaging studies in postmenopausal women suggest that estrogen treatment seems to enhance brain reactivity in frontocingulate regions during cognitive functioning, although in many cases no difference in cognitive performance was present (Comasco et al., 2014).

Progesterone has been implicated in negative mood effects, especially during the menstrual cycle. Synthetic progestogens also seem to play a role for the adverse mood effects, typically irritability and depression, noted by many hormonal contraceptive users (Gingnell et al., 2013; Lundin et al., 2017). The role of the progesterone-derived neurosteroids is, however, somewhat more complicated. There is evidence to suggest that allopregnanolone may have protective and mood stabilizing effects at certain periods in the female reproductive life cycle, such as pregnancy and the postpartum period (Hellgren, Akerud, Skalkidou, Backstrom, & Sundstrom-Poromaa, 2014; Kanes et al., 2017; Osborne et al., 2017).

However, at other times in women's lives, neurosteroids may have anxiogenic and depressogenic effects, which may depend on dose, duration of exposure, and innate sensitivity (Andreen, Sundstrom-Poromaa, Bixo, Nyberg, & Backstrom, 2006; Backstrom, Bixo, & Stromberg, 2015; Gulinello, Gong, & Smith, 2002).

3. EMOTIONAL ASPECTS OF THE MENSTRUAL CYCLE: PREMENSTRUAL DYSPHORIC DISORDER

A recent review concluded that the scientific findings of menstrual cycle-induced emotional effects are far more consistent than the cognitive effects (Sundstrom Poromaa & Gingnell, 2014). This may not come as a surprise, as many women complain of emotional problems in relation to the menstrual cycle, most commonly in the premenstrual (luteal) phase. Among women in childbearing ages, approximately 2%–10% are afflicted by severe premenstrual symptoms, and 2%–5% fulfill criteria for premenstrual dysphoric disorder (PMDD) (O'Brien et al., 2011). Subthreshold PMDD may be even more prevalent, as population-based epidemiological studies suggest that 18% of women experience at least one mood symptom in the luteal phase of the menstrual cycle (Wittchen, Becker, Lieb, & Krause, 2002). PMDD, now categorized as a mood disorder in the diagnostic and statistical manual of mental disorders (DSM-5), is defined by the onset of functionally impairing or distressing mood and physical symptoms in the late luteal phase of ovulatory menstrual cycles. Depression, irritability, mood lability, and anxiety are the most commonly reported symptoms by the women (Epperson et al., 2012; Yonkers, O'Brien, & Eriksson, 2008), and symptom severity during the luteal phase is clearly in the same range as in women with moderate depression, with suicidal thoughts and ideations not uncommonly described (Pilver, Libby, & Hoff, 2013). Further, it should be noted that PMDD disease burden generally exceeds that of major depression, simply because of the larger number of symptomatic days throughout fertile life (Halbreich, Borenstein, Pearlstein, & Kahn, 2003).

PMDD is defined by the relation to the late luteal phase of the menstrual cycle. In fact, PMDD is a prototypical female-specific mood disorder in which symptom onset and offset is associated with endogenous progesterone levels (Segebladh, Borgstrom, Nyberg, Bixo, & Sundstrom-Poromaa, 2009; Sundstrom, Nyberg, Bixo, Hammarback, & Backstrom, 1999), and the disorder is an important model for the understanding of how sex hormones influence mood and anxiety in women. As progesterone is only present in the luteal phase, PMDD is commonly regarded as a disorder caused by the variation in (or mere presence of) progesterone levels. Research in support of this includes findings of symptom relief during GnRH agonist-induced anovulatory cycles (Wyatt, Dimmock, Ismail, Jones, & O'Brien, 2004), the reinstatement of symptoms when add-back progesterone is administered together with GnRH agonists (Segebladh et al., 2009), and findings of progestagen-induced mood symptoms in postmenopausal women (Andreen, Bixo, Nyberg, Sundstrom-Poromaa, & Backstrom, 2003; Andreen et al., 2005, 2006) and combined oral contraceptive users (Lundin et al., 2017). As said, progesterone is highly lipophilic and easily passes through the blood–brain barrier. Not only so, animal studies and postmortem studies in reproductive and postmenopausal women indicate that progesterone is accumulated in the brain, with the highest concentrations found in the amygdala (Bixo, Andersson, Winblad, Purdy, & Backstrom, 1997). Given the high number of progesterone receptors in areas of importance for emotion processing, such as the amygdala, it is thus not surprising that progesterone treatment increases amygdala reactivity in healthy women (van Wingen et al., 2008). In women with PMDD amygdala reactivity is highly correlated with endogenous progesterone levels, already at much lower concentrations than those needed to elicit an amygdala response in healthy women (Gingnell, Morell, Bannbers, Wikstrom, & Sundstrom Poromaa, 2012). Although the exact mechanism by which progesterone precipitates the symptoms of PMDD is unknown, interactions with the serotonin (Jovanovic et al., 2006; Marjoribanks, Brown, O'Brien, & Wyatt, 2013) and the GABAergic systems (Epperson et al., 2002; Sundstrom Poromaa, Smith, & Gulinello, 2003) are plausible. Yet, another line of research suggest that GABA-active progesterone metabolites, such as allopregnanolone, may be responsible for precipitating the premenstrual symptoms, and that allopregnanolone antagonists may alleviate symptom severity (Bixo et al., 2017).

While it is commonly accepted that progesterone is the symptom-provoking hormone in PMDD, hormone interventions have suggested that also estradiol plays a role. Segebladh and colleagues treated PMDD women with GnRH agonists for hormone suppression and evaluated the return of symptoms when women were exposed to three different hormonal treatments. The combination of a high estrogen dose (and progesterone) was more symptom provoking than a low estrogen dose together with progesterone, suggesting that estradiol dose or concentration also plays a role for the symptom surfacing (Segebladh et al., 2009). Possibly, as one of the most important aspects of estrogen action, in the brain and elsewhere, is to upregulate progesterone receptors, increased availability to estrogen may thus result in a larger number of progesterone receptors for progesterone to act upon. In addition, other studies also using GnRH-agonist intervention have demonstrated that women with PMDD may experience depression and anxiety, solely due to estrogen-only exposure (Schmidt, Nieman, Danaceau, Adams, & Rubinow, 1998). Finally, women with premenstrual disorders experience their most intense symptoms in the late luteal phase, when progesterone levels are declining, not at the progesterone peak (Nevatte et al., 2013). Thus, the final symptom provocation may be the result of progesterone withdrawal, a concept known to induce anxiety-like behavior in rodents (Gulinello et al., 2002).

Not surprisingly, emotion processing is impaired in the luteal phase of women with PMDD. They show impaired facial emotion recognition performance and a negative bias in the luteal phase, meaning that neutral faces commonly are being misjudged as sad (Rubinow, Smith, Schenkel, Schmidt, & Dancer, 2007), and display more behavioral impulsivity and greater difficulties in emotion regulation and in socioemotional functioning (Petersen et al., 2016). These findings are corroborated by functional neuroimaging studies, where the neural correlates of PMDD include impaired prefrontal cortex top-down control of the emotions that are generated in the limbic system (Comasco & Sundstrom-Poromaa, 2015). These findings are consistent with the functional neuroanatomy that typifies a range of anxiety disorders (Etkin & Wager, 2007), but what is unique for these women is that these features are mostly, but not always, confined to the symptomatic luteal phase of the menstrual cycle. However, it should be noted that PMDD is presumably not only due to biased emotion processing in the luteal phase, as psychophysiology studies have suggested heightened arousal in PMDD women at this time point (Epperson et al., 2007; Kask, Gulinello, Backstrom, Geyer, & Sundstrom-Poromaa, 2008). Altogether, these studies provide evidence of a distorted affective perception and impaired anxiety regulation in PMDD. Cognitive disability is also integrated with negative emotion processing and psychiatric symptoms such as anxiety and depression (Okon-Singer, Hendler, Pessoa, & Shackman, 2015). Greater cognitive impairment is seen in patients with anxiety as well as depression in several domains such as processing speed, attention, reasoning, and verbal knowledge (Lam, Kennedy, McIntyre, & Khullar, 2014; Okon-Singer et al., 2015). As to PMDD, most studies have failed to demonstrate cognitive deficits (Protopopescu et al., 2008; Reed, Levin, & Evans, 2008; Sundstrom Poromaa & Gingnell, 2014; Sveinsdottir, Lundman, & Norberg, 1999), although preliminary data suggest alterations in cognitive–emotional processes such as attentional bias (Eggert, Kleinstauber, Hiller, & Witthoft, 2017; Slyepchenko et al., 2017). Thus, it cannot be excluded that hormone-sensitive cognitive deficits are trigger factors of PMDD. To date there is only one neuroimaging study investigating cognitive processing in PMDD, showing a greater late luteal phase recruitment of the left insula in women with PMDD (Bannbers et al., 2012).

Parenthetically, selective serotonin reuptake inhibitors (SSRIs) are effective first-line therapy for PMDD, supporting the notion that progesterone-induced alterations in serotonergic signaling is an important pathophysiological mechanism for the disorder. The usefulness of the SSRIs in treating PMDD has been demonstrated in numerous individual trials and is supported a meta-analysis of 29 randomized, placebo-controlled clinical trials (Marjoribanks, Brown, O'Brien, & Wyatt, 2013). In addition, the short onset of action of SSRIs in women with PMDD, manifesting within the first days of treatment, enables the limited use of SSRIs in the symptomatic luteal phase (Yonkers et al., 2015). However, treatments targeting the symptom-triggering factor, i.e., progesterone, seem equally efficient, but their clinical usefulness is limited due to short- and long-term side effects. On such treatment is the GnRH agonists, which have long been used to treat PMDD (Segebladh et al., 2009; Wyatt et al., 2004). GnRH agonist treatment leads to complete suppression of endogenous estradiol and progesterone levels, and this suppression is associated with improvement in both psychological and physical symptoms (Wyatt et al., 2004). However, side effects are problematic and often lead to discontinuation of treatment, with vasomotor symptoms in the short run, being the most common and bothersome one (Segebladh et al., 2009).

4. EMOTIONAL ASPECTS OF THE MENSTRUAL CYCLE: NATURALLY CYCLING HEALTHY WOMEN

While PMDD represents the far end of the spectrum of menstrual cycle-induced emotional disturbances, it is also clear that progesterone-induced emotion-processing disturbances are found in healthy women. Again, this may not be surprising given the relatively high proportion of women with subthreshold PMDD in the population (Wittchen et al., 2002), and the potentially even greater number of women with subclinical premenstrual symptoms (Sveindottir & Backstrom, 2000). With this in mind, it is unclear as to what extent premenstrual disorders may have influenced the overall results in studies of healthy, naturally cycling, women. While most studies have used psychiatric interviews to exclude ongoing depressive and anxiety disorders, only one study utilized daily symptom scoring for diagnosis of PMDD and consequently, excluded PMDD patients in the healthy female group (Rubinow et al., 2007). In addition, the genetic makeup of healthy women may also play a role for sex hormone influence on emotion processing and contribute to variability (Hamstra, de Kloet, Quataert, Jansen, & Van der Does, 2017).

Emotion processing involves the detection and evaluation of salient stimuli, and includes components of attention, emotional arousal, and regulation of arousal. The ability to identify one's emotional response, and how those emotions are experienced and expressed is referred to as emotion regulation. Given the above caveats, findings are relatively consistent in targeting progesterone as the most important hormone for emotion processing and regulation in healthy women. One of the most common tasks to probe emotion processing is emotion recognition, typically by using emotional faces as stimuli, and a number of studies have addressed emotion recognition across the menstrual cycle phases. Importantly, women have greater early automatic visual processing than men, and this sex difference is most pronounced in mid-luteal women (Lusk, Carr, Ranson, Bryant, & Felmingham, 2015). In spite of this, better emotion recognition accuracy is noted in the follicular phase compared to the mid-luteal phase independent of emotion stimuli (Derntl, Hack, Kryspin-Exner, & Habel, 2013; Derntl, Kryspin-Exner, Fernbach, Moser, & Habel, 2008; Derntl, Windischberger, et al., 2008), or specifically for sad faces (Guapo et al., 2009). Thus, emotion recognition accuracy appear poorer in the progesterone-dominated luteal phase, and this seems most relevant for negative emotional stimuli (Derntl, Kryspin-Exner, et al., 2008). Indeed, these menstrual cycle studies are further corroborated by a report on decreased facial recognition accuracy upon acute progesterone administration (van Wingen et al., 2007). Women also demonstrate a greater tendency to perceive fearful expressions as more intense if progesterone levels are high (Conway et al., 2007), and respond faster to negative or aversive stimuli in the mid-luteal phase (Derntl et al., 2013; Gasbarri, Pompili, d'Onofrio, Cifariello, et al., 2008; Masataka & Shibasaki, 2012). Further, reaction times to many of the facial expressions is positively correlated with progesterone levels (Kamboj, Krol, & Curran, 2015). However, a menstrual cycle influence on negative bias in emotion recognition, i.e., the tendency to perceive neutral stimuli as more negative has not been demonstrated in healthy women, in contrast to women with PMDD (Rubinow et al., 2007).

The sex hormone variations across the menstrual cycle seem also to influence emotional memory, although findings are not entirely consistent. Emotional memory depends on hypothalamus–pituitary–adrenal (HPA) axis hormones and sympathetic activity, and a sex-related difference has been suggested (Andreano & Cahill, 2009). Again, progesterone, and the luteal phase has attracted most attention, presumably because of the close relationship between progesterone and the HPA axis (Fajer, Holzbauer, & Newport, 1971; Frye, 2007), and previous findings on increased cortisol reactivity to stress in the luteal phase (Kirschbaum, Kudielka, Gaab, Schommer, & Hellhammer, 1999), however see (Maki et al., 2015). The only longitudinal study performed thus far report on decreased recognition for negative items in the luteal phase (Bayer, Schultz, Gamer, & Sommer, 2014), whereas cross-sectional studies have demonstrated no difference across cycle phases (Felmingham, Fong, & Bryant, 2012; Maki et al., 2015), enhanced memory for emotional items (Ertman, Andreano, & Cahill, 2011), or worse memory of positive items in the luteal phase (Pompili, Arnone, D'Amico, Federico, & Gasbarri, 2016). However, enhanced memory for peripheral details of an emotional story has been linked to the luteal phase (Nielsen, Ahmed, & Cahill, 2013) and emotional memory correlated positively with progesterone levels sampled at the time of encoding (Ertman et al., 2011).

Emotional memory can be modulated by manipulations of the HPA axis, for instance by cortisol treatment or stress tests. Cortisol treatment leads to impaired emotional verbal memory, but no difference in the cortisol-induced memory impairment has been noted between menstrual cycle phases (Kuhlmann & Wolf, 2005), possibly because the HPA axis manipulation overrides any potential effects of the menstrual cycle. Similarly, while one study reported that women with high progesterone levels had greater memory recall for negative images if subjected to posttraining physical stress (Felmingham et al., 2012), other studies have found no difference across cycle phases in memory performance following stress (Andreano, Arjomandi, & Cahill, 2008; Maki et al., 2015). Poststress cortisol response has been correlated with memory retrieval in the late follicular phase (Maki et al., 2015) as well in the luteal phase (Andreano et al., 2008).

Intense emotional events may be followed by spontaneous intrusive recollections (SIRs), from a clinical point of view most importantly noted in individuals suffering from posttraumatic stress disorder. Indeed, the menstrual cycle stage during which a trauma occurs may play a role for intrusive recollections in trauma patients (Bryant et al., 2011) and in healthy women (Ferree, Kamat, & Cahill, 2011; Soni, Curran, & Kamboj, 2013). These flashbacks or SIRs appear to be more common if the trauma or experimental exposure to aversive stimuli occurred in the luteal phase (Bryant et al., 2011; Ferree et al., 2011; Soni et al., 2013), and again, progesterone levels were positively correlated with the frequency of SIR (Ferree et al., 2011).

Fear conditioning is often used as a model for the formation of emotional memories and may be manipulated in standardized manners (LeDoux, 2003). For instance, by using a 2-day fear conditioning paradigm consisting of fear conditioning, and extinction learning on the first day, and extinction recall and fear renewal on the second day, researchers have presented relatively consistent results as to the estradiol involvement in consolidation or maintenance of fear extinction (Graham & Milad, 2013; Merz et al., 2012; Milad et al., 2010; Zeidan et al., 2011). Fear acquisition and fear extinction per se, however, appear not to be influenced by hormonal state or menstrual cycle phase. Thus, less recovery of fear upon fear renewal is noted in women during the late follicular phase, when estradiol levels are high, than at other stages of the menstrual cycle (Graham & Milad, 2013; Milad et al., 2010; Zeidan et al., 2011). Similarly, single dose estradiol administration during the early follicular phase leads to enhanced consolidation of fear extinction (Graham & Milad, 2013), while no effect of progesterone was noted (Milad et al., 2010; Zeidan et al., 2011). From these findings it may be suggested that maintenance of fear extinction, which is an important goal of cognitive behavioral therapy, appears superior if applied in the follicular rather than in the luteal phase.

Neuroimaging techniques, such as functional magnetic resonance imaging, are useful tools to gather further insight into emotion processing. The effects of the menstrual cycle and other hormone interventions during emotional tasks were recently summarized, both in healthy women (Toffoletto, Lanzenberger, Gingnell, Sundstrom-Poromaa, & Comasco, 2014) and in women with PMDD (Comasco & Sundstrom-Poromaa, 2015). The overall conclusion from these two reviews is, unfortunately, that findings thus far are too disparate to allow for proper meta-analyses. However, some tentative conclusions may be drawn. A pattern of increased amygdala reactivity to negative emotional stimuli in the luteal phase appears in several studies (Andreano & Cahill, 2010; Bayer et al., 2014; Gingnell et al., 2012). A sensitivity in the amygdala to increases in progesterone is also supported by the increased reactivity in the amygdala, fusiform gyrus, inferior frontal gyri, cerebellar vermis, and supplementary motor area observed after acute progesterone administration (van Wingen et al., 2008). Being one of the core structures in the fear network (Shin & Liberzon, 2010) the sensitivity in the amygdala to changes in ovarian steroid hormones across the menstrual cycle fits nicely with the increased susceptibility to anxiety and depressive symptoms in the luteal phase. In addition, both naturally fluctuating and exogenously administered hormones have been found to affect the functional connections of the brain, with findings indicating higher connectivity with higher concentrations of estradiol and progesterone (Peper, van den Heuvel, Mandl, Hulshoff Pol, & van Honk, 2011). In line with this, a one-subject longitudinal study demonstrated sex hormone-related effects on intrinsic functional connectivity, with progesterone being associated with global network connectivity of the dorsolateral prefrontal and sensorimotor cortex, and their connectivity with the hippocampus (Arelin et al., 2015). While these findings remain to be replicated, they suggest a progesterone-dependent plasticity mechanism in the resting state connectivity, which may be of relevance in emotional processing and women with PMDD.

Clearly, all of these findings point toward altered emotion processing across the menstrual cycle, which is in line with the clinically relevant emotional disturbances that are reported by many women. Further studies in this area could help explain the underlying mechanisms of emotional problems in the luteal phase and aid in better nosology, early detection, and, potentially, new treatment options. However, longitudinal studies are awaited, and preferably authors should investigate to which extent also PMDD (or subthreshold PMDD) influences their results.

5. COGNITIVE PERFORMANCE ACROSS THE MENSTRUAL CYCLE

Great interest has been devoted to exploring cognitive effects throughout the menstrual cycle. In this area, estrogen is the most studied hormone, in keeping with the commonly reported association between estrogen treatment and larger hippocampal volume, along with enhanced hippocampus function, in postmenopausal women (see review Comasco et al., 2014). Further, the hippocampus has been a candidate substrate for the hormonal effects because of its relation to neurogenesis and synaptic plasticity, and the abundance of estrogen receptors (Daniel, 2013; Gibbs, 2010;

Voytko, Tinkler, Browne, & Tobin, 2009; Wnuk, Korol, & Erickson, 2012). Indeed, gray matter morphology seems altered by the menstrual cycle in healthy women. Volumetric differences in within-subject study designs indicate gray matter volume to be increased in the hippocampus in the estrogen-dominated late follicular phase relative to the early follicular and mid-luteal phases (Lisofsky et al., 2015), in the right fusiform/parahippocampal gyrus during the early follicular compared to the mid-luteal phase (Pletzer et al., 2010), or in the right anterior hippocampus during the late follicular compared to late luteal phase (Protopopescu et al., 2008). In spite of the imaging findings suggesting enhanced hippocampal volume during the estrogen-dominated phase of the menstrual cycle, the cognitive findings across the menstrual cycle are thus far sparse.

The predominant hypotheses in the field of menstrual cycle-related cognitive changes have been that: (1) sexually dimorphic cognitive abilities/skills that favor men are improved during phases with low estrogen and progesterone levels such as the early follicular phase, (2) sexually dimorphic cognitive abilities/skills that favor women are improved during phases with increased estrogen and/or progesterone such as the late follicular phase and mid-luteal phase. This hypothesis was postulated in the early 1990s and has been tested in a number of studies. However, a recent review of the field, including a meta-analytic approach, found no support for any of these hypotheses (Sundstrom Poromaa & Gingnell, 2014).

5.1 Visuospatial Ability

5.1.1 Mental Rotation

According to the predominant hypothesis, sexually dimorphic cognitive abilities that favor men are improved during phases with low estrogen and progesterone levels. A typical skill where men outperform women is visuospatial ability (Andreano & Cahill, 2009), and the most commonly used test by which visuospatial ability has been tested is mental rotation (Andreano & Cahill, 2009). Overall, 18 studies on menstrual cycle influence on mental rotation have been published, of which 10 have incorporated hormonal measures. The majority of studies have been negative, i.e., have not been able to find any difference in mental rotation accuracy across the menstrual cycle (Dietrich et al., 2001; Epting & Overman, 1998; Gordon & Lee, 1993; Griksiene & Ruksenas, 2011; Halari et al., 2005; Kozaki & Yasukouchi, 2009; Mordecai, Rubin, & Maki, 2008; Schoning et al., 2007; Zhu, Kelly, Curry, Lal, & Joseph, 2015), whereas a minority have been able to substantiate the hypothesis of superior performance in the early follicular phase (Courvoisier et al., 2013; Hausmann, Slabbekoorn, Van Goozen, Cohen-Kettenis, & Gunturkun, 2000; Maki, Rich, & Rosenbaum, 2002). While it can be argued that some of these studies had significant methodological concerns, these methodological problems are equally spread between the studies with negative and positive results. For instance, two of the positive studies included sample sizes that were extremely small (Hausmann et al., 2000), one positive study used an unbalanced longitudinal design which may have opened up for training effects (Courvoisier et al., 2013), one negative study reported on a composite score for visuospatial tasks which may have precluded the detection of more direct effects on mental rotation (Gordon & Lee, 1993), and one negative study used a task that may have been too easy (Epting & Overman, 1998). Finally, three of the studies were neuroimaging studies, and it cannot be excluded that behavioral measures in the scanner may differ from that obtained in pure behavioral studies (Dietrich et al., 2001; Schoning et al., 2007; Zhu et al., 2015). However, even if the studies with methodological concerns are disregarded, three out of the four remaining studies were unable to detect any menstrual cycle differences in mental rotation performance (Griksiene & Ruksenas, 2011; Kozaki & Yasukouchi, 2009; Maki et al., 2002; Mordecai et al., 2008). Six of the mental rotation studies provided sufficient information for inclusion in a meta-analytic approach (Epting & Overman, 1998; Hausmann et al., 2000; Kozaki & Yasukouchi, 2009; Maki et al., 2002; Mordecai et al., 2008; Schoning et al., 2007), which yielded an standardized mean difference in error rate of 1.61 (95% CI −0.35 to 3.57, ns), *at present* suggesting no favor of an early follicular phase improvement in mental rotation performance. In addition to these studies, Hampson and colleagues evaluated mental rotation in women with low (approximately corresponding to early follicular phase levels) and high (approximately corresponding to late follicular levels) estradiol levels, regardless if subjects had been assessed in the follicular or luteal phases of the menstrual cycle (Hampson, Levy-Cooperman, & Korman, 2014). While this may be a biologically sound approach, it also serves as an example that perhaps no predictive menstrual cycle-related effects in visuospatial abilities exist, as low estradiol levels can be found during the early follicular phase, postovulation, the late luteal phase, and during anovulatory cycles.

The neural correlates of mental rotation has been evaluated in relation to the menstrual cycle by two studies (Dietrich et al., 2001; Schoning et al., 2007). Both studies report changes in brain reactivity across the menstrual cycle and an increased reactivity in Brodmann area (BA) 39, or the angular gyrus, during the presence of high levels of estradiol. The angular gyrus is involved not only in verbal processing but also in spatial judgment (Seghier, 2013), and according to the authors, the increased reactivity may reflect an increased need to recruit this area to solve the task at hand during the luteal phase (Dietrich et al., 2001; Schoning et al., 2007). Further, the right hemisphere is considered the dominant hemisphere in visuospatial processing, and estradiol induced improved mental rotation performance seems associated with increased recruitment of the left hemisphere, in turn suggesting reduced functional asymmetry during the task (Zhu et al., 2015).

5.1.2 Other Visuospatial Tasks

Besides mental rotation, spatial tests can also broadly be categorized into tasks that evaluate spatial perception and spatial visualization, and among the latter, navigation tests and object location test are included (Linn & Petersen, 1985). Notably, while men outperform women on most tasks reflecting visuospatial ability, a female advantage has been noted for tasks that can be verbalized (Andreano & Cahill, 2009). Again, the hypothesis being tested in most studies is that visuospatial task performance should be superior in the early follicular phase in comparison with high-hormone menstrual cycle phases.

However, except for two studies from the same group reporting improved performance on visuospatial performance during the early follicular phase, or at times of low estradiol levels (Hampson, 1990a; Hampson et al., 2014), the majority of studies have not been able to establish any menstrual cycle influence on tests of visuospatial memory or ability, or have reported findings in contrast with the hypothesis (Epting & Overman, 1998; Gordon & Lee, 1993; Halari et al., 2005; Hausmann et al., 2000; Mordecai et al., 2008; Phillips & Sherwin, 1992; Solis-Ortiz & Corsi-Cabrera, 2008; Weis, Hausmann, Stoffers, & Sturm, 2011). Again, a number of methodological concerns have been identified in the studies; several studies suffer from low power (Hausmann et al., 2000; Solis-Ortiz & Corsi-Cabrera, 2008), and two studies report on a visuospatial composite score (Gordon & Lee, 1993; Hampson, 1990a).

More recently, two studies on spatial navigation have been published (Hussain, Hanafi, Konishi, Brake, & Bohbot, 2016; Scheuringer & Pletzer, 2017). Spatial navigation in complex environments engages multiple memory systems, and cognitive strategies used to find a target location include the egocentric perspective (or landmark-based strategy), and the hippocampus-dependent allocentric (or spatial) strategy

(Best, White, & Minai, 2001). Typically, men take a more allocentric perspective, whereas women take a more egocentric perspective and rely more on a landmark-based strategy in navigation tasks (Scheuringer & Pletzer, 2017). While no menstrual cycle differences have been reported in navigation performance, i.e., number of trials needed to reach the target or the number of errors made in the task, two studies have suggested that landmark-based strategies are more commonly used, or more successful, in the luteal phase than in the follicular phase (Hussain et al., 2016; Scheuringer & Pletzer, 2017).

It may also be argued that certain math tests involve components of visuospatial ability. Two studies have evaluated such tests across the menstrual cycle (Becker, Creutzfeldt, Schwibbe, & Wuttke, 1982; Pletzer, Kronbichler, Ladurner, Nuerk, & Kerschbaum, 2011). Although both studies reported on superior performance in the follicular phase compared to the luteal phase, the study by Becker and colleagues had an unbalanced longitudinal design, and Pletzer and colleagues did not distinguish between the early and late follicular phase, hence the role of low estradiol was not entirely captured (Becker et al., 1982; Pletzer et al., 2011). Performance on simple mathematical calculations such a subtraction and multiplication appears not to differ across cycle phases (Hampson, 1990a).

5.2 Verbal Tasks

A female advantage for verbal fluency and verbal memory is well documented (Gulinello et al., 2002). Accordingly, the predominant hypothesis for menstrual cycle studies on verbal fluency and memory has thus been that women should perform better at these tasks during time periods of high estradiol levels, i.e., during the late follicular phase or mid-luteal phase, than at phases with low hormone levels such as the early follicular phase.

However, the hypothesis has not been substantiated in the majority of studies. Whereas two studies report on improved verbal fluency and verbal memory in the late follicular (Hussain et al., 2016) or mid-luteal phases (Maki et al., 2002), most studies find no menstrual cycle phase differences (Gordon & Lee, 1993; Griksiene & Ruksenas, 2011; Halari et al., 2005; Hampson, 1990b; Hampson et al., 2014; Hatta & Nagaya, 2009; Jacobs & D'Esposito, 2011; Konrad et al., 2008; Mordecai et al., 2008; Phillips & Sherwin, 1992; Scheuringer & Pletzer, 2017), or contrary findings of decreased performance in the late follicular phase

(Solis-Ortiz & Corsi-Cabrera, 2008). Again, a number of studies have methodological flaws including low power (Konrad et al., 2008; Rosenberg & Park, 2002; Solis-Ortiz & Corsi-Cabrera, 2008), designs that open up for learning effects (Rosenberg & Park, 2002; Solis-Ortiz & Corsi-Cabrera, 2008), or used unwisely chosen time frames for their assessments leaving little room for hormonal changes (Halari et al., 2005). The most frequently used tasks include verbal fluency and verbal recall, but tests reflecting semantic retrieval and implicit verbal memory have also been employed. Two imaging studies have evaluated verbal tasks across the menstrual cycle. In line with the behavioral results, Rumberg and colleagues found no difference in brain activation during verb generation between women examined in the early follicular and late follicular/early luteal phase (Rumberg et al., 2010). Dietrich, using a similar task, on the other hand reported on increased activation of a number of language-related areas (BA 45/46, 6, and 40) during the late follicular phase in comparison with the early follicular phase (Dietrich et al., 2001).

Two studies have evaluated verbal working memory, which also taps into prefrontal dopaminergic function. The Rosenberg study, albeit small in sample size, is one of few studies which demonstrated improved performance during phases of the menstrual cycle that are characterized by high estrogen levels (Rosenberg & Park, 2002). In addition, while the overall cycle effect was negative, Jacobs and D'Esposito, in fact, demonstrated that verbal working memory task performance was modulated by an interaction between *COMT* Val158Met and estradiol levels (Jacobs & D'Esposito, 2011). In the presence of high late follicular phase estradiol levels, an improved cognitive performance was found in Val/Val carriers (which putatively is associated with lower frontal dopamine levels), whereas a deteriorated performance was found in Met/Met carriers in comparison with the performance during the early follicular phase (Bixo et al., 2017). Thus, verbal working memory seems to be one verbal task where estradiol load is important; however, the genetic makeup of women may also influence the outcome.

In summary, the majority of menstrual cycle studies on cognitive function has failed to demonstrate robust phase-dependent findings. Nevertheless, some concluding remarks are important for interpretation of these findings. First, although the menstrual cycle studies on verbal memory were mostly negative, this does not rule out, or contradict, the possibility of an estradiol influence on cognitive function. A number of imaging studies have suggested increased hippocampal volume during phases of high estradiol levels (Lisofsky et al., 2015; Pletzer et al., 2010; Protopopescu et al., 2008), and it is an important mission to determine if the gray matter volume changes translate to menstrual cycle-induced changes in cognitive function. Clearly, it is expected that tasks that target estrogen-dependent areas and functions, of sufficient difficulty, are needed to unmask menstrual cycle changes. The latter is especially relevant, as menstrual cycle studies, by definition, include reproductive-aged young women with high cognitive capacity leading to roof effects. Further, the hormonal changes across the menstrual cycle may simply be too swift for detection of an altered cognitive performance. Estrogen deficiency, as stressed in the hypothesis on sexually dimorphic tests, is only found during a few days of the early follicular phase, and not all women will have estradiol levels in the postmenopausal range during this phase of the cycle. However, if estrogen deficiency is maintained for longer periods in young women, such as in studies using GnRH agonist intervention, estradiol suppression has been reported to be associated with impaired verbal memory performance (Craig et al., 2007). Finally, verbal memory is a relatively common cognitive outcome in randomized clinical trials on estrogen treatment in postmenopausal women, and possibly the only cognitive task where there is some evidence that estrogen treatment may have a beneficial effect. According to a recent review, there is some evidence that estrogen treatment protects verbal memory in surgically postmenopausal women, whereas it has no effect when initiated more than a decade after the menopause (Sherwin, 2012).

6. CONCLUSION

The menstrual cycle remains an intriguing, natural experiment of relevance to many researchers in medical and psychological disciplines. While the earliest reports on menstrual cycle findings were devoted to explore the suitability of women in work life and areas dominated by males (Sommer, 1973), the past years research appear driven by the increasing interest in sex influences on neurobiology. However, despite its' immediate appeal and accessibility, menstrual cycle studies require skillful and meticulous handling (Becker et al., 2005), and some of the positive findings have in many cases turned out to be notoriously difficult to replicate.

Further studies on the menstrual cycle modulation of emotional processing are needed, as findings from such studies are relevant for women's mental health, and the improvement thereof. Premenstrual syndrome and

PMDD are relatively common in women of fertile ages and represent a great burden for afflicted women, often associated with impaired social or work-related functioning. Besides the emotion-processing disturbances noted in women with PMDD, this review has emphasized that the luteal phase is associated with impaired emotion recognition accuracy (Conway et al., 2007; Derntl et al., 2013; Derntl, Kryspin-Exner, et al., 2008; Gasbarri, Pompili, D'Onofrio, Abreu, & Tavares, 2008; Guapo et al., 2009) and enhanced emotional memory. Emotional events that occur during the luteal phase more often result in intrusive recollections (Ferree et al., 2011; Soni et al., 2013), and traumatic flashback memories are more common when the trauma takes place in the luteal phase (Bryant et al., 2011). In addition, a number of studies have pinpointed progesterone as the driving factor for these findings; progesterone levels correlated with emotional memory (Ertman et al., 2011) and positively predicted intrusive memories (Ferree et al., 2011). A number of imaging studies have also reported on increased luteal phase reactivity in core structures of the fear network such as the amygdala (Andreano & Cahill, 2010; Bayer, Bandurski, & Sommer, 2013; Gingnell et al., 2014, 2012) and the anterior cingulate cortex (Amin, Epperson, Constable, & Canli, 2006; Protopopescu et al., 2005), and progesterone appears a key factor for the increased amygdala reactivity (Gingnell et al., 2014, 2012; van Wingen et al., 2008). Taken together, these findings suggest that progesterone, or at least the combined effect of luteal phase estradiol and progesterone, has the ability to influence various aspects of emotional processing, which may affect the clinical presentation of emotional disturbances in the luteal phase.

As to cognitive function, the best evidence suggest that differences across the menstrual cycle in sexually dimorphic tasks, such as mental rotation, visuospatial ability, verbal memory, and verbal fluency, are small and difficult to replicate. This finding is partly in line with previous reviews which have either suggested no influence of the menstrual cycle (Sommer, 1973) or some influence although at the same time emphasizing that such changes would not be clinically relevant (Sherwin, 2012). However, given the recent reports on menstrual cycle-induced changes in hippocampal volume (Lisofsky et al., 2015; Pletzer et al., 2010; Protopopescu et al., 2008), further research in this area is mandated. However, future studies should potentially not pursue the hypothesis of sexually dimorphic tasks, but instead focus on more demanding tasks, by which menstrual cycle performance changes can be revealed also in the young women.

REFERENCES

Amin, Z., Epperson, C. N., Constable, R. T., & Canli, T. (2006). Effects of estrogen variation on neural correlates of emotional response inhibition. *NeuroImage*, *32*(1), 457–464.

Andreano, J. M., Arjomandi, H., & Cahill, L. (2008). Menstrual cycle modulation of the relationship between cortisol and long-term memory. *Psychoneuroendocrinology*, *33*(6), 874–882.

Andreano, J. M., & Cahill, L. (2009). Sex influences on the neurobiology of learning and memory. *Learning & Memory*, *16*(4), 248–266.

Andreano, J. M., & Cahill, L. (2010). Menstrual cycle modulation of medial temporal activity evoked by negative emotion. *NeuroImage*, *53*(4), 1286–1293.

Andreen, L., Bixo, M., Nyberg, S., Sundstrom-Poromaa, I., & Backstrom, T. (2003). Progesterone effects during sequential hormone replacement therapy. *European Journal of Endocrinology*, *148*(5), 571–577.

Andreen, L., Sundstrom-Poromaa, I., Bixo, M., Andersson, A., Nyberg, S., & Backstrom, T. (2005). Relationship between allopregnanolone and negative mood in postmenopausal women taking sequential hormone replacement therapy with vaginal progesterone. *Psychoneuroendocrinology*, *30*(2), 212–224.

Andreen, L., Sundstrom-Poromaa, I., Bixo, M., Nyberg, S., & Backstrom, T. (2006). Allopregnanolone concentration and mood—A bimodal association in postmenopausal women treated with oral progesterone. *Psychopharmacology*, *187*(2), 209–221.

Arelin, K., Mueller, K., Barth, C., Rekkas, P. V., Kratzsch, J., Burmann, I., et al. (2015). Progesterone mediates brain functional connectivity changes during the menstrual cycle—A pilot resting state MRI study. *Frontiers in Neuroscience*.

Backstrom, T., Bixo, M., & Stromberg, J. (2015). GABAA receptor-modulating steroids in relation to women's behavioral health. *Current Psychiatry Reports*, *17*(11), 92.

Bannbers, E., Gingnell, M., Engman, J., Morell, A., Comasco, E., Kask, K., et al. (2012). The effect of premenstrual dysphoric disorder and menstrual cycle phase on brain activity during response inhibition. *Journal of Affective Disorders*, *142*(1–3), 347–350.

Bayer, J., Bandurski, P., & Sommer, T. (2013). Differential modulation of activity related to the anticipation of monetary gains and losses across the menstrual cycle. *The European Journal of Neuroscience*, *38*(10), 3519–3526.

Bayer, J., Schultz, H., Gamer, M., & Sommer, T. (2014). Menstrual-cycle dependent fluctuations in ovarian hormones affect emotional memory. *Neurobiology of Learning and Memory*, *110*, 55–63.

Becker, J. B., Arnold, A. P., Berkley, K. J., Blaustein, J. D., Eckel, L. A., Hampson, E., et al. (2005). Strategies and methods for research on sex differences in brain and behavior. *Endocrinology*, *146*(4), 1650–1673.

Becker, D., Creutzfeldt, O. D., Schwibbe, M., & Wuttke, W. (1982). Changes in physiological, EEG and psychological parameters in women during the spontaneous menstrual cycle and following oral contraceptives. *Psychoneuroendocrinology*, 7(1), 75–90.

Best, P. J., White, A. M., & Minai, A. (2001). Spatial processing in the brain: The activity of hippocampal place cells. *Annual Review of Neuroscience*, *24*, 459–486.

Bethea, C. L., Reddy, A. P., Tokuyama, Y., Henderson, J. A., & Lima, F. B. (2009). Protective actions of ovarian hormones in the serotonin system of macaques. *Frontiers in Neuroendocrinology*, *30*(2), 212–238.

Bixo, M., Andersson, A., Winblad, B., Purdy, R. H., & Backstrom, T. (1997). Progesterone, 5alpha-pregnane-3,20-dione and 3alpha-hydroxy-5alpha-pregnane-20-one in specific regions of the human female brain in different endocrine states. *Brain Research*, *764*(1–2), 173–178.

Bixo, M., Ekberg, K., Poromaa, I. S., Hirschberg, A. L., Jonasson, A. F., Andreen, L., et al. (2017). Treatment of premenstrual dysphoric disorder with the GABAA receptor modulating steroid antagonist Sepranolone (UC1010)-A randomized controlled trial. *Psychoneuroendocrinology*, *80*, 46–55.

Brinton, R. D., Thompson, R. F., Foy, M. R., Baudry, M., Wang, J., Finch, C. E., et al. (2008). Progesterone receptors: Form and function in brain. *Frontiers in Neuroendocrinology, 29*(2), 313–339.
Brunton, P. J. (2015). Programming the brain and behaviour by early-life stress: A focus on neuroactive steroids. *Journal of Neuroendocrinology, 27*(6), 468–480.
Bryant, R. A., Felmingham, K. L., Silove, D., Creamer, M., O'Donnell, M., & McFarlane, A. C. (2011). The association between menstrual cycle and traumatic memories. *Journal of Affective Disorders, 131*(1–3), 398–401.
Comasco, E., Frokjaer, V. G., & Sundstrom-Poromaa, I. (2014). Functional and molecular neuroimaging of menopause and hormone replacement therapy. *Frontiers in Neuroscience, 8*, 388.
Comasco, E., & Sundstrom-Poromaa, I. (2015). Neuroimaging the menstrual cycle and premenstrual dysphoric disorder. *Current Psychiatry Reports, 17*(10), 77.
Conway, C. A., Jones, B. C., DeBruine, L. M., Welling, L. L., Law Smith, M. J., Perrett, D. I., et al. (2007). Salience of emotional displays of danger and contagion in faces is enhanced when progesterone levels are raised. *Hormones and Behavior, 51*(2), 202–206.
Courvoisier, D. S., Renaud, O., Geiser, C., Paschke, K., Gaudy, K., & Jordan, K. (2013). Sex hormones and mental rotation: An intensive longitudinal investigation. *Hormones and Behavior, 63*(2), 345–351.
Craig, M. C. (2016). HRT should be considered as first line therapy for perimenopausal depression: AGAINST: More clinical trials are needed. *BJOG: An International Journal of Obstetrics and Gynaecology, 123*(6), 1011.
Craig, M. C., Fletcher, P. C., Daly, E. M., Rymer, J., Brammer, M., Giampietro, V., et al. (2009). The interactive effect of the cholinergic system and acute ovarian suppression on the brain: An fMRI study. *Hormones and Behavior, 55*(1), 41–49.
Craig, M. C., Fletcher, P. C., Daly, E. M., Rymer, J., Cutter, W. J., Brammer, M., et al. (2007). Gonadotropin hormone releasing hormone agonists alter prefrontal function during verbal encoding in young women. *Psychoneuroendocrinology, 32*(8–10), 1116–1127.
Czoty, P. W., Riddick, N. V., Gage, H. D., Sandridge, M., Nader, S. H., Garg, S., et al. (2009). Effect of menstrual cycle phase on dopamine D2 receptor availability in female cynomolgus monkeys. *Neuropsychopharmacology: Official Publication of the American College of Neuropsychopharmacology, 34*(3), 548–554.
Daniel, J. M. (2013). Estrogens, estrogen receptors, and female cognitive aging: The impact of timing. *Hormones and Behavior, 63*(2), 231–237.
Derntl, B., Hack, R. L., Kryspin-Exner, I., & Habel, U. (2013). Association of menstrual cycle phase with the core components of empathy. *Hormones and Behavior, 63*(1), 97–104.
Derntl, B., Kryspin-Exner, I., Fernbach, E., Moser, E., & Habel, U. (2008). Emotion recognition accuracy in healthy young females is associated with cycle phase. *Hormones and Behavior, 53*(1), 90–95.
Derntl, B., Windischberger, C., Robinson, S., Lamplmayr, E., Kryspin-Exner, I., Gur, R. C., et al. (2008). Facial emotion recognition and amygdala activation are associated with menstrual cycle phase. *Psychoneuroendocrinology, 33*(8), 1031–1040.
Deuster, P. A., Powell-Dunford, N., Crago, M. S., & Cuda, A. S. (2011). Menstrual and oral contraceptive use patterns among deployed military women by race and ethnicity. *Women & Health, 51*(1), 41–54.
Dietrich, T., Krings, T., Neulen, J., Willmes, K., Erberich, S., Thron, A., et al. (2001). Effects of blood estrogen level on cortical activation patterns during cognitive activation as measured by functional MRI. *NeuroImage, 13*(3), 425–432.
Eggert, L., Kleinstauber, M., Hiller, W., & Witthoft, M. (2017). Emotional interference and attentional processing in premenstrual syndrome. *Journal of Behavior Therapy and Experimental Psychiatry, 54*, 77–87.

Epperson, C. N., Haga, K., Mason, G. F., Sellers, E., Gueorguieva, R., Zhang, W., et al. (2002). Cortical gamma-aminobutyric acid levels across the menstrual cycle in healthy women and those with premenstrual dysphoric disorder: A proton magnetic resonance spectroscopy study. *Archives of General Psychiatry*, *59*(9), 851–858.

Epperson, C. N., Pittman, B., Czarkowski, K. A., Stiklus, S., Krystal, J. H., & Grillon, C. (2007). Luteal-phase accentuation of acoustic startle response in women with premenstrual dysphoric disorder. *Neuropsychopharmacology*, *32*(10), 2190–2198.

Epperson, C. N., Steiner, M., Hartlage, S. A., Eriksson, E., Schmidt, P. J., Jones, I., et al. (2012). Premenstrual dysphoric disorder: Evidence for a new category for DSM-5. *The American Journal of Psychiatry*, *169*(5), 465–475.

Epting, L. K., & Overman, W. H. (1998). Sex-sensitive tasks in men and women: A search for performance fluctuations across the menstrual cycle. *Behavioral Neuroscience*, *112*(6), 1304–1317.

Ertman, N., Andreano, J. M., & Cahill, L. (2011). Progesterone at encoding predicts subsequent emotional memory. *Learning & Memory*, *18*(12), 759–763.

Etkin, A., & Wager, T. D. (2007). Functional neuroimaging of anxiety: A meta-analysis of emotional processing in PTSD, social anxiety disorder, and specific phobia. *The American Journal of Psychiatry*, *164*(10), 1476–1488.

Fajer, A. B., Holzbauer, M., & Newport, H. M. (1971). The contribution of the adrenal gland to the total amount of progesterone produced in the female rat. *The Journal of Physiology*, *214*(1), 115–126.

Felmingham, K. L., Fong, W. C., & Bryant, R. A. (2012). The impact of progesterone on memory consolidation of threatening images in women. *Psychoneuroendocrinology*, *37*(11), 1896–1900.

Ferree, N. K., Kamat, R., & Cahill, L. (2011). Influences of menstrual cycle position and sex hormone levels on spontaneous intrusive recollections following emotional stimuli. *Consciousness and Cognition*, *20*(4), 1154–1162.

Frokjaer, V. G., Erritzoe, D., Juul, A., Nielsen, F. A., Holst, K., Svarer, C., et al. (2010). Endogenous plasma estradiol in healthy men is positively correlated with cerebral cortical serotonin 2A receptor binding. *Psychoneuroendocrinology*, *35*(9), 1311–1320.

Frye, C. A. (2007). Progestins influence motivation, reward, conditioning, stress, and/or response to drugs of abuse. *Pharmacology, Biochemistry, and Behavior*, *86*(2), 209–219.

Gasbarri, A., Pompili, A., D'Onofrio, A., Abreu, C. T., & Tavares, M. C. (2008). Working memory for emotional facial expressions: Role of estrogen in humans and non-human primates. *Reviews in the Neurosciences*, *19*(2–3), 129–148.

Gasbarri, A., Pompili, A., d'Onofrio, A., Cifariello, A., Tavares, M. C., & Tomaz, C. (2008). Working memory for emotional facial expressions: Role of the estrogen in young women. *Psychoneuroendocrinology*, *33*(7), 964–972.

Gibbs, R. B. (2010). Estrogen therapy and cognition: A review of the cholinergic hypothesis. *Endocrine Reviews*, *31*(2), 224–253.

Gingnell, M., Ahlstedt, V., Bannbers, E., Wikstrom, J., Sundstrom-Poromaa, I., & Fredrikson, M. (2014). Social stimulation and corticolimbic reactivity in premenstrual dysphoric disorder: A preliminary study. *Biology of Mood & Anxiety Disorders*, *4*(1), 3.

Gingnell, M., Engman, J., Frick, A., Moby, L., Wikstrom, J., Fredrikson, M., et al. (2013). Oral contraceptive use changes brain activity and mood in women with previous negative affect on the pill-a double-blinded, placebo-controlled randomized trial of a levonorgestrel-containing combined oral contraceptive. *Psychoneuroendocrinology*, *38*(7), 1133–1144.

Gingnell, M., Morell, A., Bannbers, E., Wikstrom, J., & Sundstrom Poromaa, I. (2012). Menstrual cycle effects on amygdala reactivity to emotional stimulation in premenstrual dysphoric disorder. *Hormones and Behavior*, *62*(4), 400–406.

Gordon, H. W., & Lee, P. A. (1993). No difference in cognitive performance between phases of the menstrual cycle. *Psychoneuroendocrinology*, *18*(7), 521–531.
Graham, B. M., & Milad, M. R. (2013). Blockade of estrogen by hormonal contraceptives impairs fear extinction in female rats and women. *Biological Psychiatry*, *73*(4), 371–378.
Griksiene, R., & Ruksenas, O. (2011). Effects of hormonal contraceptives on mental rotation and verbal fluency. *Psychoneuroendocrinology*, *36*(8), 1239–1248.
Gruber, C. J., Tschugguel, W., Schneeberger, C., & Huber, J. C. (2002). Production and actions of estrogens. *The New England Journal of Medicine*, *346*(5), 340–352.
Guapo, V. G., Graeff, F. G., Zani, A. C., Labate, C. M., dos Reis, R. M., & Del-Ben, C. M. (2009). Effects of sex hormonal levels and phases of the menstrual cycle in the processing of emotional faces. *Psychoneuroendocrinology*, *34*(7), 1087–1094.
Guerra-Araiza, C., Cerbon, M. A., Morimoto, S., & Camacho-Arroyo, I. (2000). Progesterone receptor isoforms expression pattern in the rat brain during the estrous cycle. *Life Sciences*, *66*(18), 1743–1752.
Guerra-Araiza, C., Villamar-Cruz, O., Gonzalez-Arenas, A., Chavira, R., & Camacho-Arroyo, I. (2003). Changes in progesterone receptor isoforms content in the rat brain during the oestrous cycle and after oestradiol and progesterone treatments. *Journal of Neuroendocrinology*, *15*(10), 984–990.
Gulinello, M., Gong, Q. H., & Smith, S. S. (2002). Progesterone withdrawal increases the alpha4 subunit of the GABA(A) receptor in male rats in association with anxiety and altered pharmacology—A comparison with female rats. *Neuropharmacology*, *43*(4), 701–714.
Halari, R., Hines, M., Kumari, V., Mehrotra, R., Wheeler, M., Ng, V., et al. (2005). Sex differences and individual differences in cognitive performance and their relationship to endogenous gonadal hormones and gonadotropins. *Behavioral Neuroscience*, *119*(1), 104–117.
Halbreich, U., Borenstein, J., Pearlstein, T., & Kahn, L. S. (2003). The prevalence, impairment, impact, and burden of premenstrual dysphoric disorder (PMS/PMDD). *Psychoneuroendocrinology*, *28*(Suppl. 3), 1–23.
Hampson, E. (1990a). Estrogen-related variations in human spatial and articulatory-motor skills. *Psychoneuroendocrinology*, *15*(2), 97–111.
Hampson, E. (1990b). Variations in sex-related cognitive abilities across the menstrual cycle. *Brain and Cognition*, *14*(1), 26–43.
Hampson, E., Levy-Cooperman, N., & Korman, J. M. (2014). Estradiol and mental rotation: Relation to dimensionality, difficulty, or angular disparity? *Hormones and Behavior*, *65*(3), 238–248.
Hamstra, D. A., de Kloet, E. R., Quataert, I., Jansen, M., & Van der Does, W. (2017). Mineralocorticoid receptor haplotype, estradiol, progesterone and emotional information processing. *Psychoneuroendocrinology*, *76*, 162–173.
Hatta, T., & Nagaya, K. (2009). Menstrual cycle phase effects on memory and Stroop task performance. *Archives of Sexual Behavior*, *38*(5), 821–827.
Hausmann, M., Slabbekoorn, D., Van Goozen, S. H., Cohen-Kettenis, P. T., & Gunturkun, O. (2000). Sex hormones affect spatial abilities during the menstrual cycle. *Behavioral Neuroscience*, *114*(6), 1245–1250.
Hazell, G. G., Yao, S. T., Roper, J. A., Prossnitz, E. R., O'Carroll, A. M., & Lolait, S. J. (2009). Localisation of GPR30, a novel G protein-coupled oestrogen receptor, suggests multiple functions in rodent brain and peripheral tissues. *The Journal of Endocrinology*, *202*(2), 223–236.
Hellgren, C., Akerud, H., Skalkidou, A., Backstrom, T., & Sundstrom-Poromaa, I. (2014). Low serum allopregnanolone is associated with symptoms of depression in late pregnancy. *Neuropsychobiology*, *69*(3), 147–153.

Henderson, V. W., & Greicius, M. D. (2010). Functional magnetic resonance imaging and estrogen effects on the brain: Cautious interpretation of a BOLD finding. *Menopause, 17*(4), 669–671.

Hussain, D., Hanafi, S., Konishi, K., Brake, W. G., & Bohbot, V. D. (2016). Modulation of spatial and response strategies by phase of the menstrual cycle in women tested in a virtual navigation task. *Psychoneuroendocrinology, 70*, 108–117.

Jacobs, E., & D'Esposito, M. (2011). Estrogen shapes dopamine-dependent cognitive processes: Implications for women's health. *The Journal of Neuroscience: The Official Journal of the Society for Neuroscience, 31*(14), 5286–5293.

Jovanovic, H., Cerin, A., Karlsson, P., Lundberg, J., Halldin, C., & Nordstrom, A. L. (2006). A PET study of 5-HT1A receptors at different phases of the menstrual cycle in women with premenstrual dysphoria. *Psychiatry Research, 148*(2–3), 185–193.

Kamboj, S. K., Krol, K. M., & Curran, H. V. (2015). A specific association between facial disgust recognition and estradiol levels in naturally cycling women. *PLoS One, 10*(4), e0122311.

Kanes, S. J., Colquhoun, H., Doherty, J., Raines, S., Hoffmann, E., Rubinow, D. R., et al. (2017). Open-label, proof-of-concept study of brexanolone in the treatment of severe postpartum depression. *Human Psychopharmacology, 32*(2).

Kask, K., Backstrom, T., Nilsson, L. G., & Sundstrom-Poromaa, I. (2008). Allopregnanolone impairs episodic memory in healthy women. *Psychopharmacology, 199*(2), 161–168.

Kask, K., Gulinello, M., Backstrom, T., Geyer, M. A., & Sundstrom-Poromaa, I. (2008). Patients with premenstrual dysphoric disorder have increased startle response across both cycle phases and lower levels of prepulse inhibition during the late luteal phase of the menstrual cycle. *Neuropsychopharmacology, 33*(9), 2283–2290.

Kaunitz, A. M., & Manson, J. E. (2015). Management of menopausal symptoms. *Obstetrics and Gynecology, 126*(4), 859–876.

Kirschbaum, C., Kudielka, B. M., Gaab, J., Schommer, N. C., & Hellhammer, D. H. (1999). Impact of gender, menstrual cycle phase, and oral contraceptives on the activity of the hypothalamus-pituitary-adrenal axis. *Psychosomatic Medicine, 61*(2), 154–162.

Konrad, C., Engelien, A., Schoning, S., Zwitserlood, P., Jansen, A., Pletziger, E., et al. (2008). The functional anatomy of semantic retrieval is influenced by gender, menstrual cycle, and sex hormones. *Journal of Neural Transmission, 115*(9), 1327–1337.

Kozaki, T., & Yasukouchi, A. (2009). Sex differences on components of mental rotation at different menstrual phases. *The International Journal of Neuroscience, 119*(1), 59–67.

Kugaya, A., Epperson, C. N., Zoghbi, S., van Dyck, C. H., Hou, Y., Fujita, M., et al. (2003). Increase in prefrontal cortex serotonin 2A receptors following estrogen treatment in postmenopausal women. *The American Journal of Psychiatry, 160*(8), 1522–1524.

Kuhlmann, S., & Wolf, O. T. (2005). Cortisol and memory retrieval in women: Influence of menstrual cycle and oral contraceptives. *Psychopharmacology, 183*(1), 65–71.

Lam, R. W., Kennedy, S. H., Mclntyre, R. S., & Khullar, A. (2014). Cognitive dysfunction in major depressive disorder: Effects on psychosocial functioning and implications for treatment. *Canadian Journal of Psychiatry, 59*(12), 649–654.

LeDoux, J. (2003). The emotional brain, fear, and the amygdala. *Cellular and Molecular Neurobiology, 23*(4–5), 727–738.

Linn, M. C., & Petersen, A. C. (1985). Emergence and characterization of sex differences in spatial ability: A meta-analysis. *Child Development, 56*(6), 1479–1498.

Lisofsky, N., Martensson, J., Eckert, A., Lindenberger, U., Gallinat, J., & Kuhn, S. (2015). Hippocampal volume and functional connectivity changes during the female menstrual cycle. *NeuroImage, 118*, 154–162.

Lundin, C., Danielsson, K. G., Bixo, M., Moby, L., Bengtsdotter, H., Jawad, I., et al. (2017). Combined oral contraceptive use is associated with both improvement and worsening of mood in the different phases of the treatment cycle-A double-blind, placebo-controlled randomized trial. *Psychoneuroendocrinology*, *76*, 135–143.

Lusk, B. R., Carr, A. R., Ranson, V. A., Bryant, R. A., & Felmingham, K. L. (2015). Early visual processing is enhanced in the midluteal phase of the menstrual cycle. *Psychoneuroendocrinology*, *62*, 343–351.

Maki, P. M., Mordecai, K. L., Rubin, L. H., Sundermann, E., Savarese, A., Eatough, E., et al. (2015). Menstrual cycle effects on cortisol responsivity and emotional retrieval following a psychosocial stressor. *Hormones and Behavior*, *74*, 201–208.

Maki, P. M., Rich, J. B., & Rosenbaum, R. S. (2002). Implicit memory varies across the menstrual cycle: Estrogen effects in young women. *Neuropsychologia*, *40*(5), 518–529.

Manski, R., Grindlay, K., Burns, B., Holt, K., & Grossman, D. (2014). Reproductive health access among deployed U.S. servicewomen: A qualitative study. *Military Medicine*, *179*(6), 645–652.

Marjoribanks, J., Brown, J., O'Brien, P. M., & Wyatt, K. (2013). Selective serotonin reuptake inhibitors for premenstrual syndrome. *Cochrane Database of Systematic Reviews*, *6*, CD001396.

Masataka, N., & Shibasaki, M. (2012). Premenstrual enhancement of snake detection in visual search in healthy women. *Scientific Reports*, *2*, 307.

Melcangi, R. C., Panzica, G., & Garcia-Segura, L. M. (2011). Neuroactive steroids: Focus on human brain. *Neuroscience*, *191*, 1–5.

Merz, C. J., Tabbert, K., Schweckendiek, J., Klucken, T., Vaitl, D., Stark, R., et al. (2012). Oral contraceptive usage alters the effects of cortisol on implicit fear learning. *Hormones and Behavior*, *62*(4), 531–538.

Milad, M. R., Zeidan, M. A., Contero, A., Pitman, R. K., Klibanski, A., Rauch, S. L., et al. (2010). The influence of gonadal hormones on conditioned fear extinction in healthy humans. *Neuroscience*, *168*(3), 652–658.

Mordecai, K. L., Rubin, L. H., & Maki, P. M. (2008). Effects of menstrual cycle phase and oral contraceptive use on verbal memory. *Hormones and Behavior*, *54*(2), 286–293.

Moses, E. L., Drevets, W. C., Smith, G., Mathis, C. A., Kalro, B. N., Butters, M. A., et al. (2000). Effects of estradiol and progesterone administration on human serotonin 2A receptor binding: A PET study. *Biological Psychiatry*, *48*(8), 854–860.

Moses-Kolko, E. L., Berga, S. L., Greer, P. J., Smith, G., Cidis Meltzer, C., & Drevets, W. C. (2003). Widespread increases of cortical serotonin type 2A receptor availability after hormone therapy in euthymic postmenopausal women. *Fertility and Sterility*, *80*(3), 554–559.

Mumenthaler, M. S., O'Hara, R., Taylor, J. L., Friedman, L., & Yesavage, J. A. (2001a). Influence of the menstrual cycle on flight simulator performance after alcohol ingestion. *Journal of Studies on Alcohol*, *62*(4), 422–433.

Mumenthaler, M. S., O'Hara, R., Taylor, J. L., Friedman, L., & Yesavage, J. A. (2001b). Relationship between variations in estradiol and progesterone levels across the menstrual cycle and human performance. *Psychopharmacology*, *155*(2), 198–203.

Nevatte, T., O'Brien, P. M., Backstrom, T., Brown, C., Dennerstein, L., Endicott, J., et al. (2013). ISPMD consensus on the management of premenstrual disorders. *Archives of Women's Mental Health*, *16*(4), 279–291.

Nielsen, S. E., Ahmed, I., & Cahill, L. (2013). Sex and menstrual cycle phase at encoding influence emotional memory for gist and detail. *Neurobiology of Learning and Memory*, *106*, 56–65.

O'Brien, P. M., Backstrom, T., Brown, C., Dennerstein, L., Endicott, J., Epperson, C. N., et al. (2011). Towards a consensus on diagnostic criteria, measurement and trial design of the premenstrual disorders: The ISPMD Montreal consensus. *Archives of Women's Mental Health*, *14*(1), 13–21.

Okon-Singer, H., Hendler, T., Pessoa, L., & Shackman, A. J. (2015). The neurobiology of emotion-cognition interactions: Fundamental questions and strategies for future research. *Frontiers in Human Neuroscience, 9*, 58.
Osborne, L. M., Gispen, F., Sanyal, A., Yenokyan, G., Meilman, S., & Payne, J. L. (2017). Lower allopregnanolone during pregnancy predicts postpartum depression: An exploratory study. *Psychoneuroendocrinology, 79*, 116–121.
Osterlund, M. K., Gustafsson, J. A., Keller, E., & Hurd, Y. L. (2000). Estrogen receptor beta (ERbeta) messenger ribonucleic acid (mRNA) expression within the human forebrain: Distinct distribution pattern to ERalpha mRNA. *The Journal of Clinical Endocrinology and Metabolism, 85*(10), 3840–3846.
Osterlund, M. K., Keller, E., & Hurd, Y. L. (2000). The human forebrain has discrete estrogen receptor alpha messenger RNA expression: High levels in the amygdaloid complex. *Neuroscience, 95*(2), 333–342.
Peper, J. S., van den Heuvel, M. P., Mandl, R. C., Hulshoff Pol, H. E., & van Honk, J. (2011). Sex steroids and connectivity in the human brain: A review of neuroimaging studies. *Psychoneuroendocrinology, 36*(8), 1101–1113.
Petersen, S. L., Intlekofer, K. A., Moura-Conlon, P. J., Brewer, D. N., Del Pino Sans, J., & Lopez, J. A. (2013). Novel progesterone receptors: Neural localization and possible functions. *Frontiers in Neuroscience, 7*, 164.
Petersen, N., London, E. D., Liang, L., Ghahremani, D. G., Gerards, R., Goldman, L., et al. (2016). Emotion regulation in women with premenstrual dysphoric disorder. *Archives of Women's Mental Health, 19*(5), 891–898.
Phillips, S. M., & Sherwin, B. B. (1992). Variations in memory function and sex steroid hormones across the menstrual cycle. *Psychoneuroendocrinology, 17*(5), 497–506.
Pilver, C. E., Libby, D. J., & Hoff, R. A. (2013). Premenstrual dysphoric disorder as a correlate of suicidal ideation, plans, and attempts among a nationally representative sample. *Social Psychiatry and Psychiatric Epidemiology, 48*(3), 437–446.
Pletzer, B., Kronbichler, M., Aichhorn, M., Bergmann, J., Ladurner, G., & Kerschbaum, H. H. (2010). Menstrual cycle and hormonal contraceptive use modulate human brain structure. *Brain Research, 1348*, 55–62.
Pletzer, B., Kronbichler, M., Ladurner, G., Nuerk, H. C., & Kerschbaum, H. (2011). Menstrual cycle variations in the BOLD-response to a number bisection task: Implications for research on sex differences. *Brain Research, 1420*, 37–47.
Pompili, A., Arnone, B., D'Amico, M., Federico, P., & Gasbarri, A. (2016). Evidence of estrogen modulation on memory processes for emotional content in healthy young women. *Psychoneuroendocrinology, 65*, 94–101.
Protopopescu, X., Butler, T., Pan, H., Root, J., Altemus, M., Polanecsky, M., et al. (2008). Hippocampal structural changes across the menstrual cycle. *Hippocampus, 18*(10), 985–988.
Protopopescu, X., Pan, H., Altemus, M., Tuescher, O., Polanecsky, M., McEwen, B., et al. (2005). Orbitofrontal cortex activity related to emotional processing changes across the menstrual cycle. *Proceedings of the National Academy of Sciences of the United States of America, 102*(44), 16060–16065.
Reed, S. C., Levin, F. R., & Evans, S. M. (2008). Changes in mood, cognitive performance and appetite in the late luteal and follicular phases of the menstrual cycle in women with and without PMDD (premenstrual dysphoric disorder). *Hormones and Behavior, 54*(1), 185–193.
Rosenberg, L., & Park, S. (2002). Verbal and spatial functions across the menstrual cycle in healthy young women. *Psychoneuroendocrinology, 27*(7), 835–841.
Rubinow, D. R., Smith, M. J., Schenkel, L. A., Schmidt, P. J., & Dancer, K. (2007). Facial emotion discrimination across the menstrual cycle in women with premenstrual dysphoric disorder (PMDD) and controls. *Journal of Affective Disorders, 104*(1–3), 37–44.

Rumberg, B., Baars, A., Fiebach, J., Ladd, M. E., Forsting, M., Senf, W., et al. (2010). Cycle and gender-specific cerebral activation during a verb generation task using fMRI: Comparison of women in different cycle phases, under oral contraception, and men. *Neuroscience Research, 66*(4), 366–371.

Scheuringer, A., & Pletzer, B. (2017). Sex differences and menstrual cycle dependent changes in cognitive strategies during spatial navigation and verbal fluency. *Frontiers in Psychology, 8*, 381.

Schmidt, P. J., Nieman, L. K., Danaceau, M. A., Adams, L. F., & Rubinow, D. R. (1998). Differential behavioral effects of gonadal steroids in women with and in those without premenstrual syndrome. *The New England Journal of Medicine, 338*(4), 209–216.

Schoning, S., Engelien, A., Kugel, H., Schafer, S., Schiffbauer, H., Zwitserlood, P., et al. (2007). Functional anatomy of visuo-spatial working memory during mental rotation is influenced by sex, menstrual cycle, and sex steroid hormones. *Neuropsychologia, 45*(14), 3203–3214.

Segebladh, B., Borgstrom, A., Nyberg, S., Bixo, M., & Sundstrom-Poromaa, I. (2009). Evaluation of different add-back estradiol and progesterone treatments to gonadotropin-releasing hormone agonist treatment in patients with premenstrual dysphoric disorder. *American Journal of Obstetrics & Gynecology, 201*(2), 139 e1–8.

Seghier, M. L. (2013). The angular gyrus: Multiple functions and multiple subdivisions. *The Neuroscientist, 19*(1), 43–61.

Sherwin, B. B. (2012). Estrogen and cognitive functioning in women: Lessons we have learned. *Behavioral Neuroscience, 126*(1), 123–127.

Shin, L. M., & Liberzon, I. (2010). The neurocircuitry of fear, stress, and anxiety disorders. *Neuropsychopharmacology, 35*(1), 169–191.

Slyepchenko, A., Lokuge, S., Nicholls, B., Steiner, M., Hall, G. B., Soares, C. N., et al. (2017). Subtle persistent working memory and selective attention deficits in women with premenstrual syndrome. *Psychiatry Research, 249*, 354–362.

Smith, Y. R., Minoshima, S., Kuhl, D. E., & Zubieta, J. K. (2001). Effects of long-term hormone therapy on cholinergic synaptic concentrations in healthy postmenopausal women. *The Journal of Clinical Endocrinology and Metabolism, 86*(2), 679–684.

Soares, C. N. (2014). Mood disorders in midlife women: Understanding the critical window and its clinical implications. *Menopause, 21*(2), 198–206.

Solis-Ortiz, S., & Corsi-Cabrera, M. (2008). Sustained attention is favored by progesterone during early luteal phase and visuo-spatial memory by estrogens during ovulatory phase in young women. *Psychoneuroendocrinology, 33*(7), 989–998.

Sommer, B. (1973). The effect of menstruation on cognitive and perceptual-motor behavior: A review. *Psychosomatic Medicine, 35*(6), 515–534.

Soni, M., Curran, V. H., & Kamboj, S. K. (2013). Identification of a narrow post-ovulatory window of vulnerability to distressing involuntary memories in healthy women. *Neurobiology of Learning and Memory, 104*, 32–38.

Studd, J. (2016). HRT should be considered as first line therapy for perimenopausal depression: FOR: Estrogens are the first line treatment for perimenopausal women. *BJOG: An International Journal of Obstetrics and Gynaecology, 123*(6), 1011.

Sundstrom, I., Nyberg, S., Bixo, M., Hammarback, S., & Backstrom, T. (1999). Treatment of premenstrual syndrome with gonadotropin-releasing hormone agonist in a low dose regimen. *Acta Obstetricia et Gynecologica Scandinavica, 78*(10), 891–899.

Sundstrom Poromaa, I., & Gingnell, M. (2014). Menstrual cycle influence on cognitive function and emotion processing-from a reproductive perspective. *Frontiers in Neuroscience, 8*, 380.

Sundstrom Poromaa, I., Smith, S., & Gulinello, M. (2003). GABA receptors, progesterone and premenstrual dysphoric disorder. *Archives of Women's Mental Health, 6*(1), 23–41.

Sveindottir, H., & Backstrom, T. (2000). Prevalence of menstrual cycle symptom cyclicity and premenstrual dysphoric disorder in a random sample of women using and not using oral contraceptives. *Acta Obstetricia et Gynecologica Scandinavica, 79*(5), 405–413.

Sveinsdottir, H., Lundman, B., & Norberg, A. (1999). Women's perceptions of phenomena they label premenstrual tension: Normal experiences reflecting ordinary behaviour. *Journal of Advanced Nursing, 30*(4), 916–925.

Toffoletto, S., Lanzenberger, R., Gingnell, M., Sundstrom-Poromaa, I., & Comasco, E. (2014). Emotional and cognitive functional imaging of estrogen and progesterone effects in the female human brain: A systematic review. *Psychoneuroendocrinology, 50*, 28–52.

van Wingen, G., van Broekhoven, F., Verkes, R. J., Petersson, K. M., Backstrom, T., Buitelaar, J., et al. (2007). How progesterone impairs memory for biologically salient stimuli in healthy young women. *The Journal of Neuroscience: The Official Journal of the Society for Neuroscience, 27*(42), 11416–11423.

van Wingen, G. A., van Broekhoven, F., Verkes, R. J., Petersson, K. M., Backstrom, T., Buitelaar, J. K., et al. (2008). Progesterone selectively increases amygdala reactivity in women. *Molecular Psychiatry, 13*(3), 325–333.

Voytko, M. L., Tinkler, G. P., Browne, C., & Tobin, J. R. (2009). Neuroprotective effects of estrogen therapy for cognitive and neurobiological profiles of monkey models of menopause. *American Journal of Primatology, 71*(9), 794–801.

Weis, S., Hausmann, M., Stoffers, B., & Sturm, W. (2011). Dynamic changes in functional cerebral connectivity of spatial cognition during the menstrual cycle. *Human Brain Mapping, 32*(10), 1544–1556.

Wittchen, H. U., Becker, E., Lieb, R., & Krause, P. (2002). Prevalence, incidence and stability of premenstrual dysphoric disorder in the community. *Psychological Medicine, 32*(1), 119–132.

Wnuk, A., Korol, D. L., & Erickson, K. I. (2012). Estrogens, hormone therapy, and hippocampal volume in postmenopausal women. *Maturitas, 73*(3), 186–190.

Wyatt, K. M., Dimmock, P. W., Ismail, K. M., Jones, P. W., & O'Brien, P. M. (2004). The effectiveness of GnRHa with and without 'add-back' therapy in treating premenstrual syndrome: A meta analysis. *BJOG: An International Journal of Obstetrics and Gynaecology, 111*(6), 585–593.

Yonkers, K. A., Kornstein, S. G., Gueorguieva, R., Merry, B., Van Steenburgh, K., & Altemus, M. (2015). Symptom-onset dosing of sertraline for the treatment of premenstrual dysphoric disorder: A randomized clinical trial. *JAMA Psychiatry, 72*(10), 1037–1044.

Yonkers, K. A., O'Brien, P. M., & Eriksson, E. (2008). Premenstrual syndrome. *Lancet (London, England), 371*(9619), 1200–1210.

Zeidan, M. A., Igoe, S. A., Linnman, C., Vitalo, A., Levine, J. B., Klibanski, A., et al. (2011). Estradiol modulates medial prefrontal cortex and amygdala activity during fear extinction in women and female rats. *Biological Psychiatry, 70*(10), 920–927.

Zhu, X., Kelly, T. H., Curry, T. E., Jr., Lal, C., & Joseph, J. E. (2015). Altered functional brain asymmetry for mental rotation: Effect of estradiol changes across the menstrual cycle. *Neuroreport, 26*(14), 814–819.

CHAPTER FOURTEEN

Developmental Programming of Ovarian Functions and Dysfunctions

Muraly Puttabyatappa, Vasantha Padmanabhan[1]
University of Michigan, Ann Arbor, MI, United States
[1]Corresponding author: e-mail address: vasantha@umich.edu

Contents

Abstract

The pathophysiological mechanisms underlying the origin of several ovarian pathologies remain unclear. In addition to the genetic basis, developmental insults are gaining

Vitamins and Hormones, Volume 107
ISSN 0083-6729
https://doi.org/10.1016/bs.vh.2018.01.017

attention as a basis for the origin of these pathologies. Such early insults include maternal over or under nutrition, stress, and exposure to environmental chemicals. This chapter reviews the development and physiological function of the ovary, the known ovarian pathologies, the developmental check points of ovarian differentiation impacted by developmental insults, the role played by steroidal and metabolic factors as mediaries, the epigenetic mechanisms via which these mediaries induce their effects, and the knowledge gaps for targeting future studies to ultimately aid in the development of improved treatments.

1. INTRODUCTION

Ovary is the female gonad that is present as a pair in most animals and is responsible for generation of female gametes and production of hormones that regulate reproductive functions. Although the existence of the ovary was probably recognized long before the time of Aristotle, the first detailed descriptions came from Leonardo Da Vinci, Andreas Vesalius, Gabriele Falloppio, and others during the 16th century with Vesalius pointing to the existence of follicles within the ovary (Correia, 1998). Later in the 17th century, William Harvey stated "ex ovo de vivum" pointing to everything coming from the egg or the ovum (Albertini, 2015; Correia, 1998). The concept put forth by Niels Stensen and Regnier de Graaf that the entire follicle was the egg that underwent fertilization and then released was later refuted by von Baer in the 19th century who described the ovum within the follicle as the fertilizing unit (Correia, 1998; Pangas & Rajkovic, 2015). Since then there has been an explosion of knowledge relative to ovarian structure, function, and its role in sexuality and regeneration of life. Considering the key role played by the ovary in housing and supporting the oocyte—the germline of the next generation, and several ovarian pathologies contributing to reduced fecundity have been identified, it is imperative to recognize modifiable factors that contribute to the development of such pathologies.

1.1 Ovarian Structure

In most mammals, ovaries are present in the upper pelvic cavity on either side of the body held at place by a mesentery connected to a broad ligament. Structurally ovary is made of inner medulla, which contains vascular elements (blood and lymphatic vessels) and an outer cortex containing the ovarian follicles, the functional unit of the ovary. The cortex is enclosed in a capsule called tunica albuginea. The structure of the ovarian follicle varies depending on the stage of follicular differentiation (discussed in subsequent sections). A fully developed follicular unit is made up of the female gamete, ovum, and the surrounding somatic cells comprised of an avascular layer containing granulosa cells, the site of estrogen production and theca cell layer, the androgen producing unit, where the nerve innervations and blood supply terminate.

1.2 Establishment of the Ovarian Reserve/Pool

Gonads develop in the urogenital ridge adjacent to the primitive embryonic kidney and exist initially in a bipotential or sexually undifferentiated state (Capel, 2000). Presence or absence of testis-determining factor (TDF; also known as sex-determining region Y [SRY]) gene in the somatic cells (Capel, 2000; Goodfellow & Lovell-Badge, 1993) dictates whether the bipotential gonad differentiates into a testes or ovary, respectively. A ovary-determining gene (Z gene) has been postulated but is yet to be identified (Blecher & Erickson, 2007). Although the current dogma is that ovarian development occurs by default or in a passive manner, gene expression studies point to active regulation of ovarian differentiation involving differential expression of various genes including Wnt4 (Yao, 2005).

The bipotential primordial germ cells (PGCs), which form the ova in the females, originate from the endoderm of the yolk sac and migrate into the genital ridge and undergo further proliferation (Hummitzsch et al., 2015; Lin & Capel, 2015). The coelomic epithelium also proliferates and infiltrates the mesenchyma of the genital ridge to form the primitive sex cords that surround the PGCs. The mesenchymal cells that are part of the mesonephros also migrate into the gonad at this time through the medulla (Smith, Wilhelm, & Rodgers, 2014). Following sex determination, the bipotential gonad in the male immediately organizes into a testis, while the ovaries in the female show little structural differentiation. By the 7th week of gestation in human female fetus, primitive sex cords degenerate and are replaced by cortical sex cords (Knobil & Neill, 1998). Around this time, the glomeruli and ductal cells that are part of the migrating mesonephros undergo regression and the remnant remains to form the ovarian rete (Smith et al., 2014). Final stages of ovarian development involve involution of medulla and the development of cortex where sex cords split into clusters within which the germ cells develop into oocyte and the cord cells form the granulosa cells. The mesonephric rete cells that are initially located in the ovarian medulla increase in size through proliferation and migrate into the cortex. It has been suggested that theca cells originate from these mesenchymal mesonephric cells later on as follicular differentiation progresses (Smith et al., 2014).

The PGCs first mitotically divide to form oogonia, which undergo first meiosis to form the primary oocyte, which occurs at around 9th week of fetal life in the human (Albertini, 2015). Primary oocytes stay arrested at the diplotene stage until ovulation. The process of oocyte maturation is closely associated with follicular development. Upon formation of primary oocytes, they are surrounded by the cord-derived, epithelial origin, squamous cell layer called pregranulosa cells to form the primordial follicles. The "naked oocytes" that are not surrounded by the pregranulosa cell layer undergo atresia. In the humans, complete ovarian differentiation occurs in utero, but in altricial species such as rodents differentiation gets completed during postnatal life (McGee & Hsueh, 2000). In the rodents, the PGCs migrate into gonadal ridge during late embryonic period and develop into oogonia such that at birth only sex cords and oogonia are present. Primordial follicles in the rodents develop between days 1 and 2 after birth in contrast to the humans where primordial follicles appear at 16 weeks of fetal life.

Unlike in the testes where spermatogonium undergoes mitotic renewal throughout reproductive life, the mitotic activity of the oogonia stops before or around birth depending on the species, resulting in the number of oocytes formed in each species being set by the time of birth or early postnatal life. This premise has been challenged recently from studies in mice with the suggestion that oocyte renewal occurs even after formation of initial pool (Johnson, Canning, Kaneko, Pru, & Tilly, 2004) and the stem cells that participate in this renewal are of bone marrow origin (Johnson et al., 2005). Although other investigators have failed to confirm these findings (Zhang et al., 2012), recent studies suggest presence of germline stem cells in the ovary (Wang et al., 2014). The relevance of these findings as it relates to formation of functional ovarian follicles remains unclear (Hummitzsch et al., 2015). As such, the prevailing dogma is that, by birth ovaries have established a finite set of primordial follicles that constitutes the follicular pool or ovarian reserve (Fig. 1) from which to draw from throughout the lifespan of the female. Developmentally at about 20 weeks of gestation an estimated 6–7 million germ cells are present in the human ovaries reducing down to about 1–2 million follicles at birth and ~250,000 follicles at puberty. By 35 years of age only 25,000 follicles remain reducing further down to 2500 follicles by 50 years of age (Gougeon, Ecochard, & Thalabard, 1994; Richardson, Senikas, & Nelson, 1987). By the time women are

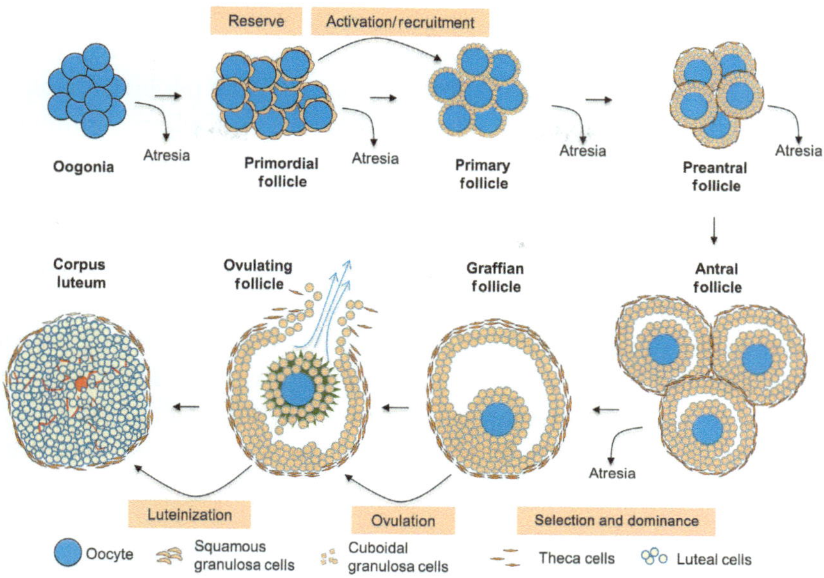

Fig. 1 Schematic showing the different stages of follicular differentiation and the developmental check points, potential targets for developmental insults.

65 years of age virtually no follicles are present in both ovaries (Costoff & Mahesh, 1975). This reduction in follicular numbers has been shown to involve programmed cell death or apoptosis (Escobar, Vázquez-Nin, & Echeverría, 2011; Tilly, 1996).

1.3 Follicular Activation or Recruitment

The vast majority of the primordial follicles from the follicular pool remain in a quiescent state with few follicles from this pool emerging to develop further during each reproductive cycle. The fate of those emerging from the pool during each cycle is continued differentiation or atresia (Fig. 1). Follicular differentiation from primordial pool begins with follicular activation (also referred to as initial recruitment) and occurs in a gonadotropin-independent manner. Follicular activation is regulated by locally produced growth factors, hormones, and cytokines (reviewed in McLaughlin & McIver, 2009) such as kit ligand, fibroblast growth factor, transforming growth factor (TGF) α, leukemia inhibitory factor, bone morphogenetic protein (BMP) 4, antimullerian hormone (AMH), TGFβ, and androgens (Durlinger et al., 1999; Fortune, 2003; Skinner, 2005). While the exact mechanism of activation is not completely understood, gene knockout studies provide some insights. For instance, premature follicular activation following gene deletion of forkhead box O3a (FOXO3A) and protein kinase B (alpha serine/threonine-protein kinase, AKT) inhibitor, phosphatase and tensin homolog (PTEN) (Castrillon, Miao, Kollipara, Horner, & DePinho, 2003; Reddy et al., 2008) suggest that the primordial follicular pool is under inhibitory control and activation occurs when the inhibitory factors decrease or stimulatory factors increase. As such, the balance between activators (kit ligand, fibroblast growth factor, TGFα, leukemia inhibitory factor, and BMP4) and inhibitors (AMH and TGFβ) appears to drive follicular activation (Gougeon, 2004). The activation process involves intracellular signaling pathways such as phosphoinositide 3-kinase (PI3K), AKT, and mammalian target of rapamycin (reviewed in Sobinoff, Sutherland, & McLaughlin, 2013) and leads to the formation of primary follicles that is characterized by cuboidal-shaped granulosa cell layer.

1.4 Follicular Selection and Dominance

Further differentiation of primary follicle into preantral follicle involves acquisition of theca layer around the granulosa cell layer and granulosa cell division through mitosis to form multiple layers of granulosa cells (the theca layer remains as a single layer at this stage) (Fig. 1). The primary follicles are evident in the human fetal ovary at 20 weeks and antral follicles by 32 weeks of gestation. In contrast, primary follicles appear by days 3–5 and antral follicles by days 15–17 after birth in the rodents (McGee & Hsueh, 2000).

The follicular growth from primary to preantral stage is considered to be gonadotropin independent. Identification of follicle-stimulating hormone (FSH) receptors during early stages of follicular development (Touraine et al., 1999) and the finding that loss of FSH affects preantral follicle development (McGee & Hsueh, 2000) go against this premise and suggest that gonadotropins do influence preantral follicular development. The other players that influence follicular growth during this transition are TGFβ, growth differentiation factor (GDF) 9, BMP family members, estrogens, androgens, insulin, and insulin-like growth factor (IGF) 1 (Zeleznik, 2004). As the preantral follicle acquires multiple granulosa cell layers, they become less sensitive to oocyte-secreted growth factors and transition from the gonadotropin-independent to gonadotropin-dependent state. Furthermore, there is a reduction in the amount of AMH produced by the growing follicles, resulting in increased sensitivity of granulosa cells to FSH (Dewailly et al., 2016) consistent with them becoming more gonadotropin responsive.

In the humans, constant pools of preantral follicles are always present such that it has been referred to as the second reserve of follicles (Monniaux et al., 2014).

Preantral follicles differentiate into antral follicles (Fig. 1) with the appearance of antrum, a cavity filled with follicular fluid containing granulosa cell secretions and diffusions from blood vessels in the theca layer. At this stage, theca layer also differentiates into inner layer of theca interna and the outer layer of theca externa. Within the follicle, oocyte gets surrounded by granulosa cells called corona radiata and is suspended in the antral fluid held by a stalk of granulosa cells called the cumulus oophorus (Fig. 1). Following the onset of menses, a rise in FSH secretion leads to cyclic recruitment (McGee & Hsueh, 2000) of a small cohort of antral follicles that gets selected to grow further.

Antral follicular development was believed to occur either continuously or through a single recruitment process where the privileged follicle(s) continue to develop further (Baerwald, Adams, & Pierson, 2012). However, advancement in noninvasive ultrasound monitoring technology has shown that follicular development in the human and domestic animals occurs in waves, which involves synchronous growth of a group of antral follicles occurring at regular intervals during the menstrual/estrous cycle (Baerwald et al., 2012). The ovulatory follicle originates from the final wave with all other follicles undergoing demise through atretic process, although exceptions have been reported (Baerwald, Adams, & Pierson, 2003).

Depending on ovulatory quota of the species, one or more follicles emerge to become dominant to ovulate during the ovulatory wave. The rest of the follicles from the cohort termed nondominant or subordinate undergo atresia (Zeleznik, 2001). The dominant follicles due to their increased sensitivity to withstand the low FSH milieu, a consequence of negative feedback from increasing levels of estradiol produced by the growing follicles, continue their growth. The members of the IGF family and their binding proteins have a role in determining follicular selection and/or dominance (Fortune, Rivera, & Yang, 2004).

Follicular growth is also associated with increased secretion of steroidal hormones, androgens, and estrogens by theca and granulosa cells, respectively, as suggested by the two-cell two-gonadotropin hypothesis (Liu & Hsueh, 1986). Increasing estradiol levels promote expression of luteinizing hormone (LH) receptors in granulosa cells thus allowing them to switch their dependence to LH from FSH thereby preparing the follicle, now known as preovulatory or Graafian follicle, to respond to the ovulatory LH surge.

Ovulation involves expulsion of the oocyte (Fig. 1) that undergoes final stages of maturation and resumption of meiosis and germinal vesicle breakdown. The follicular somatic cells left behind after ovulation transform into the endocrine structure called corpus luteum through a process called luteinization (Fig. 1). This involves neovascularization of the granulosa cell layer which along with theca cells invade the antrum where they undergo differentiation into luteal cells. Luteinization is also associated with switch in the major hormone production from estradiol to progesterone. Both the ovulatory process and luteinization are intricately regulated by members of various gene families such as proteases that breakdown follicular wall to allow oocyte expulsion, epidermal growth factors that promote cumulus granulosa cell expansion and proteases that breakdown follicular wall to allow oocyte expulsion (Conti, Hsieh, Zamah, & Oh, 2012; Richards, 2007; Russell & Robker, 2007; Stouffer, Xu, & Duffy, 2007).

1.5 Corpus Luteum

Corpus luteum is a transient endocrine structure that produces progesterone required to support uterine development, implantation, and pregnancy. The corpus luteum ceases function and structurally involutes through the process known as luteolysis either at the end of a nonfecund reproductive ovarian cycle or when placenta develops to take over its function in the human or at the end of pregnancy in other species. The process of luteolysis may depend on the species as in some species they undergo functional and structural luteolysis and involute into connective tissue laden corpus albicans, while in other species such as rodents, they only undergo functional luteolysis, leading to many corpora lutea from multiple cycles being present in the ovary (Smith & Meidan, 2014; Stouffer, Bishop, Bogan, Xu, & Hennebold, 2013).

2. OVARIAN PATHOLOGIES

Ovarian dysfunctions underlie many infertility disorders. There are several ovarian pathologies some of genetic origin and others stemming from genetic susceptibility and environmental interactions. Ovarian pathologies include primary ovary insufficiency (POI), Turner's syndrome, fragile X syndrome, polycystic ovary syndrome (PCOS), ovarian cysts, ovarian torsion, hyperthecosis, luteomas, and ovarian cancers. Although a case can be made for some of these to have extraovarian origin, this chapter will discuss conditions that have major changes within the ovary and discuss briefly the influence of extraovarian factors in the pathogenesis of the ovarian condition. These conditions emerge due to defects in one or more check points of follicular developmental trajectory such as ovarian reserve, activation, selection, dominance, ovulation, or corpus luteum formation (Table 1).

2.1 Premature Ovarian Insufficiency

The POI condition is characterized by decline in ovarian function due to premature depletion of follicles that result in an earlier than average menopause affecting approximately 1%–2% of women under 40 years of age (Monniaux et al., 2014). This could be the result of chromosomal abnormalities such as mutation in the fragile X mental retardation 1 (*FMR1*) gene (Qin, Jiao, Simpson, & Chen, 2015), autoimmune disorders (Silva et al., 2014), or outcome of chemo/radiotherapy (Thomakos, Trachana, Koutroumpa, Rodolakis, & Gavalas, 2015). Additionally, environmental xenobiotic exposures can cause premature menopause potentially due to POI (Grindler et al., 2015; Hoyer & Keating, 2014). The POI in autoimmune disorders may also be associated with endocrine disorders such as hypoparathyroidism and hypoadrenalism (Silva et al., 2014; Wemeau, Proust-Lemoine, Ryndak, & Vanhove, 2013). The POI may be the result of increased atresia of early follicular stages or premature activation of the follicular pool leading to its depletion.

2.2 Turner's Syndrome

Turner syndrome, also known as gonadal dysgenesis, is a genetic disorder arising from chromosomal abnormalities containing a missing X chromosome (45,X karyotype) (Shah, Sivapalan, Gibbons, Tempest, & Griffin, 2003). It occurs with a frequency of 0.02% of live births and is characterized by gonadal dysgenesis with presence of gonadal streaks, lack of pubertal development, amenorrhea, and hypergonadotropic hypogonadism along with premature ovarian failure (Venkatesh, Suresh, & Tsutsumi, 2014). Within the ovary, accelerated loss of ovarian follicles by apoptosis begins from 18 weeks of gestation and the rate of depletion varies such that 45,X karyotype phenotypes range from complete loss of follicles to mosaic karyotypes showing follicular development (Novak, Kokai, Popovic, Ludoski, & Jurukovski, 1995).

2.3 Fragile X Syndrome

Fragile X syndrome (FXS), also known at Martin–Bell syndrome, characterized by developmental delays, learning disabilities, and behavioral problems

Table 1 Common Ovarian Pathologies and the Stage of Follicular Development Affected

	Stages in Ovarian Follicular Development							
Ovarian Pathologies	**Reserve Establishment**	**Reserve Depletion**	**Activation/ Recruitment**	**Selection**	**Dominance**	**Ovulation**	**Corpus Luteum**	**Phenotype**
Turner's syndrome	↓	?						Loss of oocytes
Fragile X syndrome	?	↑	B*	B*	B*	B*	B*	Reduced primordial follicle pool; reduced follicular development
Primary ovarian insufficiency	?	↓	↑					Reduced primordial follicle pool; premature loss of follicle pool
Polycystic ovary syndrome			↑	B*	B*	B*	B*	Multifollicular appearance; absence of preovulatory follicles; absence of corpus luteum
Follicular cysts					B	B		Large diameter cyst
Luteal cysts						B*	B	Fluid-filled corpus luteum
Luteomas							↑*	Brownish-yellow nodular mass
Hyperreactio luteinalis							↑*	Bilateral cystic ovarian enlargement
Germ cell tumors	?							Ovarian cancer
Stromal cell tumors				?	?	?		Ovarian cancer

↑, increase; ↓, decrease; ?, unknown; *, predicted based on the ovarian phenotype observed; *B*, blocked.

also manifest an ovarian phenotype of POI. In these subjects, POI is caused by either a decrease in initial pool of follicles or through an accelerated rate of atresia (Wittenberger et al., 2007). Other factors that contribute to decreased ovarian function in FXS cannot be ruled out as evidenced from the decreased levels of circulating inhibin B, inhibin A, and progesterone seen in these patients many years preceding POI (Welt, Smith, & Taylor, 2004).

2.4 Polycystic Ovary Syndrome

PCOS is the most common infertility disorder affecting about 5.5%–19.9% of women of reproductive age (Lizneva et al., 2016). It is characterized by functional hyperandrogenism (clinical and biochemical), menstrual irregularities, chronic anovulation, polycystic ovaries, and reduced fertility (Conway et al., 2014; Dumesic et al., 2015). Majority of patients with PCOS are at increased risk of developing type 2 diabetes mellitus (T2DM) and cardiovascular diseases (Diamanti-Kandarakis & Dunaif, 2012; Moran, Norman, & Teede, 2015). The etiology of PCOS is unknown. Multiple factors including genetics and early developmental exposures have been implicated (Puttabyatappa, Cardoso, & Padmanabhan, 2016). While high familial aggregation (Legro, Driscoll, Strauss, Fox, & Dunaif, 1998) and the observations among Dutch twins that heritability for PCOS is 0.79 (Vink, Sadrzadeh, Lambalk, & Boomsma, 2006) provide evidence in support of genetic susceptibility, studies with congenital adrenal hyperplasia patients (Barnes et al., 1994) and animal models (Abbott et al., 2008; Maliqueo, Benrick, & Stener-Victorin, 2014; Padmanabhan & Veiga-Lopez, 2013a) implicate prenatal androgen excess in the development of PCOS.

The ovarian phenotype of PCOS subjects is the multifollicular appearance, stemming from accumulation of antral follicles, which could originate from enhanced recruitment and/or decreased depletion. While longitudinal studies characterizing developmental ontogeny of ovarian defects in PCOS subjects is lacking, cross-sectional studies showing a decrease in number of primordial follicles and a corresponding increase in number of growing follicles are supportive of enhanced recruitment (Webber et al., 2003). The prevailing endocrine milieu, namely, hyperandrogenic and hyperinsulinemic status of PCOS subjects, either by increasing FSH action and/or decreasing the negative regulators of follicular activation might contribute to the enhanced follicular recruitment (Dumesic & Richards, 2013). Another contributing factor for development of multifollicular ovarian morphology is follicular arrest that may involve failure to select follicle pool to differentiate further (Franks, Stark, & Hardy, 2008), low FSH to stimulate follicular growth (Rebar et al., 1976), and/or reduced rate of atresia (Webber et al., 2003). While follicular arrest has been postulated (Franks, Mason, & Willis, 2000), serial ultrasonographic studies are yet to be undertaken in these subjects to prove or disprove this premise.

2.5 Ovarian Cysts

Ovarian cysts are a common problem in women of reproductive age (Kozak, Owings, & Hall, 2005) and domestic animals such as dairy cattle (Kesler & Garverick, 1982). In United States alone a quarter million women per year have been diagnosed with ovarian cysts (Kozak et al., 2005). Commonly observed ovarian cysts include functional cysts, luteinized unruptured follicles (LUF), and enlarged follicles. Functional ovarian cysts are observed during the menstrual/estrous cycle and include follicular and corpus luteum cysts. Follicular cysts are ovarian follicles that fail to ovulate, persist for more than two cycles, and develop 30 mm or more in diameter. They usually regress within days or weeks of their development. Hemorrhagic cysts are follicular cysts that are filled with blood through extravasation. Although defects in hypothalamic and pituitary secretions primarily contribute to the follicular cyst development, intraovarian contributors cannot be ruled out (Ortega, Marelli, et al., 2015). Corpus luteum cysts form when the follicular remnant after ovulation reseals a fluid-filled cavity without infiltration of the luteal cells. LUFs are ovarian cysts where the follicle that reached preovulatory stage, fails to ovulate, entraps the oocyte with rest of the follicle undergoing luteinization. Enlarged follicles are similar to follicular cysts but fail to persist for more than two cycles.

2.6 Luteomas and Hyperreactio Luteinalis

These are the two most common ovarian pathologies that are apparent during pregnancy and frequently cause hyperandrogenic state and induce maternal virilization (Joshi & Dunaif, 1995; Phelan & Conway, 2011). Pregnancy luteomas usually appear as a solid ovarian mass that are benign and in up to 50% of cases present in both ovaries. These appear as brownish-yellow nodular mass made of lutein cells with internal hemorrhagic deposits. This condition is common among Afro-Caribbean ethnic subgroup and other predisposing factors are maternal age, multiparity, and PCOS (Joshi & Dunaif, 1995). Growth in these luteomas is promoted by human chorionic gonadotropin (hCG) and they usually regress after childbirth very rarely requiring surgical interventions (Wang, Su, Liu, Chang, & Chen, 2005).

Hyperreactio luteinalis (HL) is a relatively rare condition characterized by bilateral cystic ovarian enlargement and excessive androgen production during pregnancy (Joshi & Dunaif, 1995). The causes for this condition are not known but similar to luteoma arise due to increased responsiveness to hCG with the cyst made up mostly of theca lutein cells (Bradshaw, Santos-Ramos, Rawlins, MacDonald, & Parker, 1986). Conditions with elevated levels of hCG, such as multiple pregnancy, trophoblastic diseases, or chronic kidney disease, can induce the development of HL (Phelan & Conway, 2011).

2.7 Ovarian Cancers

According to the National Cancer Institute (NCI), ovarian cancer is the fifth leading cause of cancer death in women (Al-Alem & Curry, 2015). Ovarian cancers can originate in different cell types and include epithelial, sex cord stromal, and germ cell tumors (Bast, Hennessy, & Mills, 2009). Among these the epithelial cell cancers that originate from ovarian surface or reproductive tract epithelium accounts for about 90% of the ovarian cancers (Al-Alem & Curry, 2015). Risk factors include genetics and increased number of ovulatory cycles from nulliparity and early menarche with late menopause (Rossing, Daling, Weiss, Moore, & Self, 1994). The ovarian surface epithelial cell tumor can develop due to "incessant ovulation" leading to inflammatory reaction, oxidative stress, and release of growth factors that damage the ovarian surface epithelium and promote their growth culminating in the development of tumor (Fathalla, 1971).

Gonadal stromal tumors comprise a heterogeneous group of benign and malignant cancers that include granulosa cell tumor (GCT) and Sertoli–Leydig cell tumor (SLCT) (Karnezis, Cho, Gilks, Pearce, & Huntsman, 2017). There are juvenile and adult types of GCT depending on the affected age group and involve somatic missense mutation in the forkhead box L2 gene. Majority of these tumors are estrogenic although androgenic or progestogenic varieties have been reported (Foulkes, Gore, & McCluggage, 2016). The SLCTs observed in young- to middle-aged women are rare tumors involving dicer 1, ribonuclease III (DICER1) mutations that promote Sertoli and Leydig cell differentiation from granulosa cells and often present androgenic phenotypes (Karnezis et al., 2017).

Ovarian germ cell tumors (OGCTs) are derived from PGCs and comprised of a diverse group of histologic subtypes that can be either benign or malignant. Malignant OGCTs, which include dysgerminomas and non-dysgerminomas are rare, accounting for only 1% of all ovarian tumors with an average age of onset of age <25 years (Cools, Wolffenbuttel, Drop, Oosterhuis, & Looijenga, 2011).

2.8 Ovarian Torsion

Adnexal torsion is one of a few gynecologic surgical emergencies and refers to rotation of the ovary at its pedicle and consequent occlusion of the ovarian artery and/or vein. Misdiagnosis or delay in treatment can have permanent sequelae including loss of an ovary with effect on future fertility, peritonitis, and even death (Sasaki & Miller, 2014). Adnexal torsion commonly occurs in the pregnant women while other predisposing causes are benign ovarian cysts (follicular, luteal, or theca lutein cysts), cystadenomas, mature dermoid cysts, malignant tumor, pelvic inflammatory disease, or endometriosis (Ricci et al., 2016). Torsion of a normal ovary is uncommon but can occur in adolescents (Spinelli, Piscioneri, & Strambi, 2015).

3. BASIS OF OVARIAN PATHOLOGIES

Originally, most of the noninfectious disease states were thought to be due to aberrant changes in the organism's genetic information (DNA sequence) resulting from mutations and deletions or gene fusion, amplification, or tandem duplication (Tang & Ho, 2007) thus providing the genetic basis for the origin of the disease. In recent years, developmental exposure to adverse insults during critical periods of organ differentiation is also getting attention as one of the leading causes. In the context of ovarian pathologies, as discussed earlier, the critical period of ovarian differentiation occurs during prenatal life in precocial species (human, monkeys, sheep, cow) and extends to postnatal life in altricial species (rats and mouse) (Padmanabhan & Veiga-Lopez, 2013b). Insults occurring during these critical windows result in adaptive changes that could have either positive or negative consequences, a concept known as the developmental origins of health and disease (DoHAD) (Barker, 2004). The concept was first proposed based on epidemiological data from the 1944 to 1945 Dutch famine cohort showing maternal starvation during gestation correlates with an increased risk for cardiovascular and metabolic diseases in the offspring (Ravelli, Stein, & Susser, 1976). Since then it has become quite apparent that varying developmental insults such as maternal undernutrition, overnutrition, stress, lifestyle, disease states, and exposures to environmental chemicals contribute to the development of adult reproductive and metabolic pathologies (Padmanabhan, Cardoso, & Puttabyatappa, 2016). The genetic and environmental bases for the development of various ovarian pathologies are addressed later.

3.1 Genetic

Although the genetically determined ovarian disorders such as Turner syndrome and FXS are relatively infrequent, other ovarian conditions such as ovarian cancers, POI, and PCOS that are more frequent appear to also have a genetic contribution (Franks & McCarthy, 2004). The genetic basis for ovarian defects for the most part stem from chromosomal aberrations (Turners and FXS) although polymorphism and phenotypes with multifactorial inheritance (POI, PCOS, and ovarian cancers) have been reported (Brugh, Maduro, & Lamb, 2003; Franks & McCarthy, 2004).

Turner syndrome characterized by POI stems from a missing X chromosome (45,X karyotype) (Shah et al., 2003). In most cases the single X chromosome is of maternal origin (Hassold, Benham, & Leppert, 1988). 47,XXX is another condition with chromosomal abnormality that has an extra X chromosome of maternal origin and is associated with early onset of menopause (Hassold et al., 1996). Additionally cases with four or more X chromosomes have been reported where the severity of the symptoms increases with the number of X chromosomes (Shah et al., 2003). These chromosomal abnormalities involving extra X chromosomes are also characterized by POI suggestive of a strong genetic basis. Gene mutations in the FMR1 gene are shown to be responsible for the POI in FXS patients (Qin et al., 2015). Gene mutations and gene polymorphisms are implicated in considering the genetic basis for the origin of PCOS and ovarian cancers and techniques such as genome-wide association studies have mapped various susceptibility genes for PCOS (Dunaif, 2016; Puttabyatappa et al., 2016; Zhao, Lv, Li, & Chen, 2016) and ovarian cancers (Longacre et al., 2016).

3.2 Developmental Insults and Programming Agents

The impact of early developmental insults as they relate to various ovarian developmental check points that stem from maternal undernutrition, overnutrition, stress, and exposure to environmental chemicals are discussed later and summarized in Table 2.

Table 2 Developmental Insults and the Stage of Follicular Development Affected

	Stages in Ovarian Follicular Development							
Maternal Insult	**Reserve Establishment**	**Reserve Depletion**	**Activation/ Recruitment**	**Selection**	**Dominance**	**Ovulation**	**Corpus Luteum**	**Remarks**
Undernutrition	↓ (Bernal, Vickers, Hampton, Poynton, & Sloboda, 2010; Chan, Tsoulis, & Sloboda, 2015)		↑* (Mossa et al., 2013)		B* (Khorram, Keen-Rinehart, Chuang, Ross, & Desai, 2015)	B* (Chan et al., 2015; Khorram et al., 2015)		Increased oocyte apoptosis; decrease in primordial and increase in other follicle classes; appearance of cysts
Overnutrition	↓* (Tsoulis et al., 2016)		↑* (Cheong, Sadek, Bruce, Macklon, & Cagampang, 2014)		B* (Cheong et al., 2014)	B* (Cheong et al., 2014)		Increase in oocyte apoptosis; decrease in primordial and increase in other follicle classes; follicular developmental defects

Stress	↓* (Ashworth, George, Hogg, Lai, & Brunton, 2016; Barra, Cruz, Mayerhofer, Paredes, & Lara, 2014)		↑* (Ashworth et al., 2016; Barra et al., 2014; Hulas-Stasiak, Dobrowolski, & Tomaszewska, 2016; Ristic et al., 2008)				Increased oocyte apoptosis; decrease in primordial and increase in other follicle classes
EDC	↓	↑	↑	B	B	B	See below for specific EDC examples
Bisphenol A	↓* (Adewale, Jefferson, Newbold, & Patisaul, 2009; Karavan & Pepling, 2012)	↓*	↓* (Karavan & Pepling, 2012)	B* (Markey, Coombs, Sonnenschein, & Soto, 2003; Newbold, Jefferson, & Padilla-Banks, 2007)	B* (Markey et al., 2003; Newbold et al., 2007)	B* (Markey et al., 2003; Newbold et al., 2007)	Reduced germ nest breakdown; increased primordial follicles at early age with increased antral and cystic follicles at older age

Continued

Table 2 Developmental Insults and the Stage of Follicular Development Affected—cont'd

Maternal Insult	Stages in Ovarian Follicular Development							
	Reserve Establishment	Reserve Depletion	Activation/ Recruitment	Selection	Dominance	Ovulation	Corpus Luteum	Remarks
Methoxychlor		↑*	↑* (Armenti, Zama, Passantino, & Uzumcu, 2008)	B* (Armenti et al., 2008)	B* (Armenti et al., 2008)	B* (Armenti et al., 2008)		Increased number of preantral and antral follicles; reduced corpus luteum; numerous cystic follicles
Diethylstilbesterol	↓ (Kim et al., 2009)	↓*	↓* (Karavan & Pepling, 2012; Yamamoto et al., 2003)	B* (Newbold, Bullock, & Mc Lachlan, 1983)	B* (Newbold et al., 1983)			Reduced germ cell nest breakdown; increased primordial follicles at early age with increased antral and cystic follicles at older age
Phthalates	↓ (Zhang et al., 2014)	↑*	↑ (Moyer & Hixon, 2012)	B* (Moyer & Hixon, 2012)	B* (Moyer & Hixon, 2012)			Reduced germ cell nest breakdown; increased preantral and antral follicles

↑, increase; ↓, decrease; *, indicates the affected follicular developmental stage is predicted based on the phenotype observed; *B*, blocked.

3.2.1 Maternal Undernutrition

The concept that maternal undernutrition through either caloric or protein reduction leads to in utero growth restriction (IUGR) and adult onset metabolic and cardiovascular diseases (Barker & Osmond, 1988), also extends to ovarian pathologies. In general, clinical and experimental studies provide evidence that nutritional deficiency during early-life is associated with a decline in ovarian follicular reserve and changes in ovulation rates (Chan et al., 2015). For instance, one study in rats found maternal undernutrition that lasted throughout the pregnancy and lactation resulted in reduced ovarian reserve (Bernal et al., 2010). Similarly, undernourishment during the first trimester in cows, a precocial species as human, also resulted in reduced ovarian reserve in the offspring, based on their reduction in their antral follicle count (Mossa et al., 2013). Another study in rodents found maternal undernutrition confined to just the second half of the pregnancy resulted in large ovarian cysts, reduced number of corpora lutea, and elevated circulating testosterone (T) (Khorram et al., 2015). The differences in phenotypic outcomes in the two rodent studies (Bernal et al., 2010; Khorram et al., 2015) may relate to differences in the timing of undernutrition. To what extent such outcomes are a function of the IUGR induced by maternal undernutrition is unclear.

3.2.2 Maternal Overnutrition

Overnutrition, the opposite end of the nutritional spectrum from undernutrition, and consequent obesity in women is also associated with increased reproductive risks manifested as irregular menstrual cycles and anovulation and PCOS (Naderpoor et al., 2015). Obesity is also a risk factor for abnormal ovarian development in the offspring. For instance, neonatal offspring of rats fed high-fat diet became obese, had fewer oocytes at embryonic day 20, and showed reduced AMH signaling at postnatal day 4 that favor increased follicular activation (Tsoulis et al., 2016). As adults these offspring manifested increased follicular atresia, reduced FSH responsiveness, and high AMH levels in antral follicles indicative of a compromised follicular environment. Maternal high-fat diet also produced deleterious effects in mice evidenced as obesity, reduced primordial, antral, and Graafian follicular numbers along with increased expression levels of FOXO3A and GDF9 (Cheong et al., 2014). Similar studies with precocial species that focus on ovarian pathologies are lacking.

Considering maternal obesity also leads to offspring obesity (Zambrano & Nathanielsz, 2013), it is unclear to what extent the ovarian disruptions seen in the offspring are a function of the obese phenotype of the offspring themselves, which is likely to replicate the maternal environment of the obese mothers. Reduced adipokine and increased inflammatory markers are hallmarks of obese mothers (Jungheim, Travieso, Carson, & Moley, 2012), aspects that are likely to be detrimental to ovarian function (Boots & Jungheim, 2015; Campos et al., 2015). If so, the impact of maternal obesity on oocyte metabolism, meiosis, and maturity will also be seen in the obese F1 generation, setting up a self-perpetuating vicious cycle. The effect of obesity on the ovary may be reflected as lipid accumulation in oocytes (Robker et al., 2009), a detriment to oocyte competence (Wu, Norman, & Robker, 2011). For instance, increases in follicular fluid free fatty acid concentrations in women undergoing IVF have been found to be associated with poor oocyte quality (Jungheim et al., 2011) and saturated free fatty acids have been shown to induce apoptosis of human granulosa cells (Mu et al., 2001). Multigenerational studies focusing on the impact of maternal obesity on offspring ovarian reserve and function along with offspring metabolic phenotype are required to determine if ovarian dysfunctions are a reflection of repetitive obese phenotype reflecting the prevailing metabolic environment or transgenerationally programmed.

3.2.3 Maternal Stress

Maternal stress can arise from major life events bereavement, catastrophic events, depression, or anxiety (Dunkel Schetter, 2011) involving classic endocrine response of increased glucocorticoid secretion that could underlie abnormal ovarian programming. Programming by maternal undernutrition discussed earlier may also be secondary to the stress response they manifest (Whirledge & Cidlowski, 2010). Stress can also elevate sympathetic neurotransmitters such as norepinephrine (Nugent et al., 2012) and influence ovarian function (Lara et al., 2002).

Animal models of maternal stress have provided evidence supportive of maladaptive programming of offspring ovarian differentiation and function. For instance, gestational exposure of rats to cold stress was found to reduce primordial follicle formation in the offspring either through reduced germinal nest breakdown or increase in oocyte apoptosis indicative of reduced ovarian reserve as well as decrease in the number of growing follicles suggestive of defects in the follicular activation/recruitment process (Barra et al., 2014). Likewise, prenatal social stress in pigs (Ashworth et al., 2011) and rats (Ashworth et al., 2016) resulted in fewer primordial and more primary follicles, respectively, stressing the impact of maternal stress in programming offspring ovarian reserve and follicular activation. To what extent the impact of other stressors such as heat and bright light involve ovarian developmental defects in delaying vaginal opening and increasing estrus cycle length in the offspring (Politch & Herrenkohl, 1984) remains to be ascertained.

Because glucocorticoid secretion is the main endocrine response during stress, animal models administered glucocorticoids during gestation help address the effects of stress hormones on the ovary. For instance, prenatal dexamethasone, a synthetic glucocorticoid, treatment resulted in a decrease in primordial follicular number in spiny mouse (Hulas-Stasiak et al., 2016) and rat (Ristic et al., 2008). Because maternal stress also leads to maternal and fetal androgen excess as evidenced from increases in anogenital distance (Barrett & Swan, 2015; Herrenkohl & Scott, 1984) and elevations in circulating levels of T (Barrett & Swan, 2015; Takahashi, Turner, & Kalin, 1998), another well-established programming agent (Hakim, Padmanabhan, & Vyas, 2016), to what extent androgens are involved in mediating the ovarian response in stress models remains to be addressed.

3.2.4 Environmental Chemical Exposures

Targeted studies in animals have provided enough evidence that developmental exposure to endocrine disrupting chemicals (EDCs), chemicals with endocrine-like functions (Diamanti-Kandarakis et al., 2009; Gore et al., 2015), affect offspring ovarian development and function. Maternal exposure to common EDCs such as methoxychlor (MXC), diethylstilbestrol (DES), bisphenol A (BPA), phthalates, and genistein have an impact on the ovary in various animal models culminating in reduced ovarian reserve, POI, and/or development of cystic follicles (Crain et al., 2008; Hannon & Flaws, 2015; Zama & Uzumcu, 2010). For instance, perinatal (gestational day 19 to postnatal day 7) exposure to MXC in the rat and gestational (days 9–16 or 9–20) exposure to BPA in mice increased the number of preantral and antral follicles, induced cystic follicles, and reduced the number of corpus luteum (Armenti et al., 2008; Markey et al., 2003; Newbold, Jefferson, & Padilla-Banks, 2009), indicative of defects at the level of follicular activation/recruitment, selection, dominance, and ovulation. Other studies found neonatal exposure to DES in mice (Kim et al., 2009) and BPA in rats (Adewale et al., 2009) lead to the development of multioocyte follicles indicative of defects in the establishment of the ovarian reserve. Among the phthalates, neonatal diethylhexyl phthalate (DEHP) treatment affected germ cell nest breakdown (Zhang et al., 2014), while gestational (days 17–19) mono-2-ethylhexylphthalate (MEHP) exposure increased number of preantral and antral follicles in the offspring (Moyer & Hixon, 2012), indicating that the former negatively affected the establishment of ovarian reserve and the latter activation/recruitment process suggestive of early depletion of ovarian reserve and potential for premature reproductive aging. These and other observations that are reviewed elsewhere (Hannon & Flaws, 2015; Peretz et al., 2014; Zama & Uzumcu, 2010) point to the profound impact of perinatal EDC exposure in programming ovarian defects (Nilsson & Skinner, 2015).

Although association studies point to detrimental effects of EDC exposure on ovarian function in humans (Bhattacharya & Keating, 2012; Craig, Wang, & Flaws, 2011; Hoyer & Keating, 2014), the impact of developmental exposures to EDCs on ovarian function in the offspring has not been studied. Findings that (1) EDCs are present in the human maternal circulation (Inoue et al., 2004; Mazdai, Dodder, Abernathy, Hites, & Bigsby, 2003; Peretz et al., 2014; Sathyanarayana, Calafat, Liu, & Swan, 2008), (2) EDCs such as BPA are associated with excess androgen—a well-documented developmental programming agent (Zhou, Liu, Liao, Han, & Liu, 2008), (3) observations of reduced pregnancy rates in daughters of mothers with high serum dichlorodiphenyltrichloroethane (DDT) compounds (Cohn et al., 2003), and (4) a steady decline in fertility in the developed world with increase in the manufacture and use of chemicals (Woodruff, 2011) indicate that developmental exposures to EDC are a threat to female reproductive function. To what extent this is manifested at the ovarian level, is a research gap to focus future studies upon.

4. MECHANISM OF PROGRAMMED DISRUPTIONS

Although the types of ovarian defects induced by the different developmental insults (undernutrition, overnutrition, stress, EDCs) reflect the timing of exposure as it relates to critical periods of ovarian differentiation, common mediators, and mechanisms appear to be involved (Padmanabhan et al., 2016). The common programming mediators include steroidal and metabolic factors.

4.1 Steroidal and Metabolic Factors in Programming Ovarian Defects

Hormones, both steroidal and metabolic, play an important role in organizing the structure/function of various organ systems including the ovary.

Among the steroidal hormones estradiol, T and glucocorticoids of gonadal and/or adrenal origin are the major programmers of reproductive dysfunctions. The fact that the undifferentiated gonad develops into a testes or ovary depending on the presence or absence of T during the critical period of gonadal differentiation—former driving the development of testes and the latter formation of ovaries (Capel, 2000; Jost, 1983)—highlights the importance of steroid hormones as programming agents of ovarian differentiation and dysfunctions in offspring.

The role played by abnormal maternal exposure to steroids in the development of ovarian pathologies in the offspring is evident from animal studies involving administration of agonist, receptor antagonist, steroid synthesis inhibitors, or EDCs with steroidogenic potential during the developmental window (Table 3). For instance, creation of a hypoestrogenic milieu in fetal baboon by maternal treatment with letrozole—an aromatase inhibitor—from mid- to late-gestation leads to a reduction in primordial follicular numbers that was prevented by estradiol cotreatment (Zachos et al., 2002), thus stressing the role played by estradiol in setting the ovarian reserve. In altricial species such as mice, where this window occurs during postnatal life, administration of native (Chen et al., 2007), or synthetic (Karavan & Pepling, 2012) estradiol during neonatal life also reduces ovarian reserve. Treatment with DES, a potent synthetic estrogen, from days 7 to 21 of gestation increases the percentages of primary and secondary ovarian follicles in the pups (Yamamoto et al., 2003) and induces inflammation, medullary tubule-like structures in the interstitial compartment, and intra- and paraovarian cysts in the 10-month offspring (Newbold et al., 1983).

While studies testing the effects of maternal exposure to excess estradiol or deficiency at the ovarian level are few, a large number of studies involving both altricial and precocial models have investigated the effects of maternal exposure to T in the context of developmental origin of PCOS (Abbott et al., 2008; Maliqueo et al., 2014; Padmanabhan & Veiga-Lopez, 2013a). Among the animal models studied, Suffolk sheep model of gestational T exposure is the only model that has undertaken serial ultrasonographic studies and extensively investigated the ovary at the organismal, tissue, and molecular level to relate phenotypic changes with underlying molecular changes (Cardoso, Puttabyatappa, & Padmanabhan, 2015; Padmanabhan & Veiga-Lopez, 2013b). T excess from days 30 to 90 of gestation in sheep doubled the number of primordial follicles in day 90 T-treated fetuses, supportive of advancement in follicular differentiation (Comim, Hardy, Robinson, & Franks, 2015), produced a multifollicular ovarian morphology

Table 3 Steroidal Influence on the Stage of Follicular Development

Animal Model			Stages in Ovarian Follicular Development							
Species	**Treatment**	**References**	**Reserve Establishment**	**Reserve Depletion**	**Activation/ Recruitment**	**Selection**	**Dominance**	**Ovulation**	**Corpus Luteum**	**Remarks**
Rhesus macaques	Testosterone GD 40–55/65/75 (term 164)	Abbott, Barnett, Bruns, and Dumesic (2005)				B*	B*	B*		Anovulation; polyfollicular ovarian morphology
Rhesus macaques	Testosterone GD 100/115–124/139 (term 164)	Abbott et al. (2005)				B*	B*	B*		Anovulation; polyfollicular ovarian morphology
Baboon	Letrozole GD 100–164 (term 184)	Zachos, Billiar, Albrecht, and Pepe (2002)	↓	↓						Ovarian outer cortex has fewer follicles and more interfollicular nests in fetuses. Adult ovarian phenotype not assessed
Sufflock sheep	Testosterone GD 30–90 (term 147)	Padmanabhan and Veiga-Lopez (2013b)		↑	↑	B	B	B	B	Multifollicular phenotype; reduced primordial and increase in other follicular classes; anovulation

Sufflock sheep	Testosterone GD 60–90 (term 147)	Steckler, Roberts, Doop, Lee, and Padmanabhan (2007) and Veiga-Lopez, Moeller, Abbott, and Padmanabhan (2016)					↓*	Altered intrafollicular steroids; reduced fecundity
Sufflock sheep	DHT GD 30–90 (term 147)	Padmanabhan and Veiga-Lopez (2013b)	↑	↑				Decrease in percentage of primordial follicles with corresponding increase in other follicular classes in fetal ovary
Scottish Greyface × Texel sheep	Testosterone GD 62–102 (term 147)	Hogg, McNeilly, and Duncan (2011) and Hogg et al. (2012)						No changes in ovarian morphology or follicular counts at fetal and 11 months of age

Continued

Table 3 Steroidal Influence on the Stage of Follicular Development—cont'd

Animal Model			Stages in Ovarian Follicular Development							
Species	Treatment	References	Reserve Establishment	Reserve Depletion	Activation/ Recruitment	Selection	Dominance	Ovulation	Corpus Luteum	Remarks
Rat (Sprague Dawley)	Testosterone GD 16–19 (term 21)	Wu et al. (2010)			↑*	B*	B*			Increase in preantral and antral follicles; fewer preovulatory follicles and corpus luteum
Rat (Sprague Dawley)	DHT GD 16–19 (term 21)	Wu et al. (2010)			↑*	B*	B*			Increase in preantral and antral follicles; fewer preovulatory follicles and corpus luteum
Mouse	DHT GD 16–18 (term 20)	Roland, Nunemaker, Keller, and Moenter (2010)						B*		Longer estrous cycle
Rat# (Wistar)	DHT 21–111 days PN	Manneras et al. (2007)				B*	B*	B*		Multiple cystic follicles
Rat# (Wistar)	Letrozole 21–111 days PN	Manneras et al. (2007)				B*	B*	B*		Multiple cystic follicles
Rat# (Wistar)	Letrozole 34–55 days PN	Kafali, Iriadam, Ozardali, and Demir (2004)				B*	B*	B*		Multiple cystic follicles

Rat# (Wistar)	Letrozole 56–77 days PN	Baravalle, Salvetti, Mira, Pezzone, and Ortega (2006)				B*	B*	B*	Multiple cystic follicles
Mouse#	Letrozole 28–63 days PN	Kauffman et al. (2015)				B*	B*	B*	Multiple cystic follicles
Rat# (Wistar)	Estradiol valerate 14 days PN	Rosa, Guimaraes, Padmanabhan, and Lara (2003)				B*	B*	B*	Multiple cystic follicles
Mouse	Estradiol 1–4d PN	Karavan and Pepling (2012) and Chen, Jefferson, Newbold, Padilla-Banks, and Pepling (2007)	↓	↓	↓				Impaired germ cell nest breakdown; increased primordial follicle numbers
Mouse (CD1)	Ethinyl estradiol 1–4 d PN	Karavan and Pepling (2012)	↓	↓	↓				Impaired germ cell nest breakdown and increased primordial follicle numbers

DHT, dihydrotestosterone; *GD*, gestational day; *PN*, postnatal. ↑, increase; ↓, decrease; *, indicates the affected follicular developmental stage is predicted based on the phenotype observed; #, indicates animal models with treatments outside the ovarian developmental window; *B*, blocked.

due to accumulation of antral follicles, increased recruitment, induced follicular persistence, and induced ovulatory and corpus luteum dysfunctions (Cardoso et al., 2015; Padmanabhan & Veiga-Lopez, 2013b), defects also reported in women with PCOS (Dumesic & Richards, 2013; Padmanabhan & Veiga-Lopez, 2013b).

Serial ultrasonographic studies in this sheep model found multifollicular appearance to be the result of increased follicular recruitment and follicular persistence (Manikkam, Steckler, Welch, Inskeep, & Padmanabhan, 2006; Veiga-Lopez, Wurst, Steckler, Ye, & Padmanabhan, 2014). Of relevance, a multifollicular phenotype is also evident in prenatal T-treated monkeys (Abbott et al., 2008), although serial ultrasonographic studies are lacking to assess contribution from recruitment and persistence. As opposed to the arrested follicular phenotype of women with PCOS (Franks et al., 2008), prenatal T-treated monkeys (Abbott et al., 2008), and sheep (Padmanabhan & Veiga-Lopez, 2013b), the follicles in prenatal T-treated rats appear to be cystic in nature (Maliqueo et al., 2014).

Examination of key mediators of ovarian function at the tissue level in the 30–90 days T-treated sheep model of PCOS has yielded mechanistic insights (Fig. 2) linking molecular/protein changes with functional changes. Importantly, use of ovarian sections from the same set of animals for assessment of various molecular/protein markers has helped obtain an integrative picture of mechanisms underlying the ovarian pathology. These investigations indicate that increased androgen receptor expression (which promotes

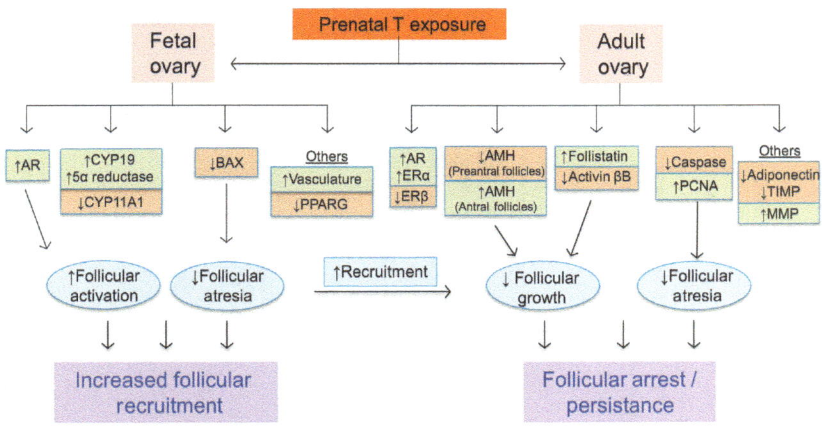

Fig. 2 Fetal and adult ovarian changes that account for enhanced follicular recruitment and follicular persistence evidenced in prenatal T-treated sheep. Increases in gene/protein markers are denoted in *green* and decreases in *orange*.

follicular activation) (Ortega, Salvetti, & Padmanabhan, 2009), in the absence of changes in B-cell lymphoma 2 (BCL2, which instills resistance to atresia) and reduced BCL2-associated X protein (Salvetti, Ortega, Veiga-Lopez, & Padmanabhan, 2012) in early growing follicles are the likely contributors to the enhanced follicular recruitment. On the contrary, increased expression levels of follistatin, reduced activin βB (West, Foster, Evans, Robinson, & Padmanabhan, 2001), and increased AMH protein expression (Veiga-Lopez, Ye, & Padmanabhan, 2012) in preantral and antral follicles likely create a net negative FSH milieu, thus contributing to the follicular arrest and follicular persistence. In addition, reduced adiponectin—an insulin sensitizer (Ortega, Rey, Velazquez, & Padmanabhan, 2010), alterations in (1) the apoptotic machinery manifested as decreased BCL2 and caspase 3 in granulosa/theca cells of antral follicles (Salvetti et al., 2012), (2) intrafollicular androgen/estrogen ratio the result of reduced granulosa aromatase (CYP19A1) and thecal 17α hydroxylase (CYP17A1) (Padmanabhan, Salvetti, Matiller, & Ortega, 2014), (3) steroid receptor balance expressed as an increase in estrogen receptor α (ESR1) and decrease in ESR 2 coupled with increased androgen receptor expression in granulosa cells of antral follicle (Ortega et al., 2009), and (4) ratio of metalloproteases to its inhibitor (Puttabyatappa et al., 2017) in antral follicles, as well as disruptions in ovarian stromal vasculature (Ortega, Veiga-Lopez, et al., 2015) are all consistent with follicular arrest and/or block in the progression of preovulatory follicle development.

While ovarian stimulation studies found that exogenous gonadotropins are capable of rescuing ovulatory function in prenatal T-treated (days 30–90 of fetal life) sheep (Steckler, Lee, Ye, Inskeep, & Padmanabhan, 2008), similar studies using an even less severe phenotype (T treatment from days 60 to 90 of fetal life) found that ovarian defects continue to persist in the developing preovulatory follicles and involve impairment in the final maturation step and intrafollicular steroid milieu (Veiga-Lopez et al., 2016). Disruptions in intrafollicular steroidal milieu accompanied by a reduction in oocyte competence have also been evidenced in prenatal T-treated monkeys (Dumesic et al., 2003), a finding that could potentially explain the reduced fecundity observed in prenatal T-treated sheep (Steckler et al., 2007).

Other studies with Scottish Greyface × Texel sheep found T treatment from gestational days 60 to 90 or 62 to 102 did not produce multifollicular ovarian phenotype (Hogg et al., 2012), indicating that the critical window for programming this phenotype spans from days 30 to 60 of gestation, when gonadal differentiation occurs. However, in vitro studies of follicles from these prenatal T-treated sheep provided evidence in support of increased thecal androgen production (Hogg et al., 2012). In addition, these investigators found that a single injection of T directly into the fetus on day 60 altered the expression pattern of key genes involved in steroid biosynthesis and action (Hogg et al., 2011). The functional significance of the findings from the direct fetal T administration and late gestational Scottish Greyface × Texel T exposure models is unclear, as adult reproductive phenotypes remain yet to be characterized.

As T can be aromatized to estradiol and gestational T treatment increases maternal and fetal T and estradiol concentrations (Abi Salloum, Veiga-Lopez, Abbott, Burant, & Padmanabhan, 2015), the ovarian changes detailed earlier could be programmed by androgenic and estrogenic pathways. Comparative studies with prenatal dihydrotestosterone (DHT, a non-aromatizable form of T)-treated sheep suggest that the programming of follicular recruitment is androgen dependent, while the follicular persistence and multifollicular appearance are estrogen dependent (Padmanabhan & Veiga-Lopez, 2011). However, because DHT can be metabolized to estrogen agonist 3β-diol (Handa, Pak, Kudwa, Lund, & Hinds, 2008) and exert its action via estrogen receptor beta, one needs to exert caution in interpreting these findings relative to the type of steroid serving as programming agent.

As opposed to the information available on prenatal programming of ovarian structure/function by gonadal steroids, the role played by glucocorticoids in the developmental programming of offspring ovarian defects is yet to be elucidated. However, because dexamethasone has been shown to increase oocyte apoptosis in fetal ovaries in vitro (Poulain et al., 2012), the potential for stress-induced increases in glucocorticoids having an effect in establishing the ovarian reserve cannot be overlooked.

Relating the findings from prenatal T and estradiol-treated models to undernutrition, overnutrition, stress, and EDC exposure models discussed in Section 3, there are commonalities in phenotypic outcomes. For instance, the impact of prenatal undernourishment, overnourishment, or stress on follicular activation/recruitment, etc. parallel that seen after prenatal exposure to excess T, and thus may relate to the elevated androgen levels seen also in these models (Barrett & Swan, 2015; Mossa et al., 2013; Whyte et al., 2007).

Similarly, the impact of developmental exposure to EDCs on ovarian function may relate to their potential to serve as steroidomimetics and/or alter steroid biosynthesis (Diamanti-Kandarakis et al., 2009; Gore, 2008). EDCs with steroidomimetic capability include (1) phytoestrogens, dioxins, MXC, DDT, and endosulfan that affect estrogen signaling pathways; (2)

vinclozolin and DDT metabolites that affect androgen receptor actions; and (3) BPA, phthalates, and polychlorinated biphenyl (PCB) that affect both pathways (Cooper & Kavlock, 1997).

In stark contrast to the data available relative to steroidal programming of ovarian function, information relative to metabolic programming of offspring ovarian perturbations is limiting. Clearly maternal undernutrition (Fowden & Forhead, 2004) and overnutrition (Holemans, Caluwaerts, Poston, & Van Assche, 2004) lead to metabolic alterations manifested as hyperinsulinemia, hyperleptinemia, hyperglycemia, dyslipidemia, etc. It needs to be recognized that some of the changes observed in prenatal T-treated sheep model discussed earlier may have also been subject to programming influences via maternal hyperinsulinemia induced by gestational T treatment (Abi Salloum et al., 2015). Because hyperinsulinemic and hyperandrogenic states often coexist during pregnancy in other disorders such as PCOS, obesity, and preeclampsia, it is difficult to pinpoint the specific metabolic/steroidal contributions programming the ovarian pathology. Because insulin is required for establishment of ovarian reserve (Isik et al., 2012) metabolic contributions to programming of ovarian dysfunctions are likely to exist. The programing effects of other factors associated with metabolic disorders such as adipokines, chemokines, amino acids, fatty acids, and glucose (Heerwagen, Miller, Barbour, & Friedman, 2010; Padmanabhan et al., 2016) that have been shown to have direct effects at the ovarian level, have not been investigated in the context of developmental origin of ovarian pathologies. Furthermore, elevated reactive oxygen species generated through oxidative stress in metabolic disorders (Dong, Zheng, Ford, Nathanielsz, & Ren, 2013) due to their potential to damage cell components, such as proteins, lipids, and DNA (Valko et al., 2007), can also have an impact on the developing ovary. Interventional studies targeting the metabolic factor in question, such as gestational insulin sensitizer treatment and follow-up of ovarian function in offspring are very much needed to address the role played by metabolic factors in programming ovarian dysfunctions in the offspring.

4.2 Epigenetic Mechanisms

The mechanisms by which steroidal and metabolic mediators program the ovarian developmental trajectory likely involve epigenetic modulations. Epigenetic changes involve heritable changes in the gene function without altering the DNA sequence. These occur through methylation of DNA, histone modifications, and noncoding RNA expression. Methylation of DNA occurs at cytosine residues in CpG dinucleotides (Ehrlich & Lacey, 2013). Histone modification involves acetylation, methylation, phosphorylation, palmitoylation, and ubiquitination of amino acid residues (Turner, 2011). Noncoding RNAs include micro, small, and long RNA that do not code for any proteins but regulate gene expression of target genes through posttranscriptional processing (Amaral & Mattick, 2008). These epigenetic modifications can reduce (or silence) expression or degrade mRNA posttranscription (Heard & Martienssen, 2014) of keys genes involved in programming ovarian function.

Relatively little is known in terms of epigenetic programming of ovarian disorders with most studies pointing to associations than causality. For instance, the programming of ovarian dysfunction in the prenatal T-treated sheep model discussed earlier involves epigenetic alterations as evidenced from increased expression of microRNAs in fetal day 65 ovaries, several of which are predicted to target *ESR1*, *CYP19*, and Sry-related high-mobility-group box (*SOX*) family (Luense, Veiga-Lopez, Padmanabhan, & Christenson, 2011), genes of importance to gonadal development. Further evidence that epigenetic alterations are involved in programming ovarian disorders comes from studies with EDCs (Huo et al., 2015; Zama & Uzumcu, 2010). For instance, gestational BPA treatment was found to increase expression of microRNAs that regulate SOX family genes, kit ligand, and insulin-related genes that regulate ovarian follicular differentiation and insulin sensitivity (Veiga-Lopez, Luense, Christenson, & Padmanabhan, 2013). Because silencing kit ligand, a growth factor that maintains primordial follicular pool, would negatively affect the ovarian reserve (Gougeon, 2004), the ovarian defects observed in gestational (9–16 days) BPA exposed mice (Newbold et al., 2009) may be secondary to the impact on the microRNA that regulates kit ligand expression. Similarly, exposures to MXC, BPA, and DEHP have been found to alter DNA methyl transferase (DNMT), an enzyme involved in methylating gene promoters (Cruz, Foster, Paredes, Yi, & Uzumcu, 2014). In other studies with ovaries from perinatal MXC-exposed rats, targeted genome-wide methylation analysis revealed hypermethylation of multiple genes involved in estrogen receptor, PTEN, and insulin-mediated cell signaling along with increase in DNMT expression (Zama & Uzumcu, 2013). The accelerated follicle loss observed in these rats (Zama & Uzumcu, 2013) may therefore result from silencing of genes involved in PTEN signaling such as Foxo3, likely due to removal of the inhibitory action on AKT (Sobinoff et al., 2013). Because BPA exposure has been shown to reduce histone methylation in the mouse follicles in vitro (Trapphoff, Heiligentag, El Hajj, Haaf, & Eichenlaub-Ritter, 2013), the involvement of histone modifications in the developmental programming of ovarian dysfunctions by BPA and other EDCs cannot be ruled out.

Importantly, epigenetic changes induced by developmental insults can be transferred to subsequent generations (Miska & Ferguson-Smith, 2016). A role for epigenetics in the etiology of ovarian disease and transgenerational transmission is evident at the level of epigenome and transcriptome in the F3 granulosa cells following ancestral exposure to fungicide, pesticide mixture, plastic mixture, dioxin, and a hydrocarbon mixture spanning the window of embryonic gonadal sex determination (Skinner, 2014). These changes appear to underlie the disease phenotypes manifested as increase in ovarian cysts and reduction in primordial follicle pool in F1 and F3 offspring (Nilsson & Skinner, 2015). In addition, crossing the ancestrally MXC exposed F3 females with wild-type male rats produced similar ovarian defects (reduction in primordial pools and polycystic ovaries) in F4 females, indicating that the epigenetic transfer occurs through female germline (Manikkam, Haque, Guerrero-Bosagna, Nilsson, & Skinner, 2014). These emerging data indicate genome-wide epigenetic alterations, which indeed occurs in response to developmental insults and point to the need for developing intervention strategies such as use of dietary supplements (methyl donors like folic acid) to prevent epigenetic alterations and development of ovarian pathologies.

5. CONCLUSIONS

As discussed earlier, in addition to genetic basis, developmental insults also contribute to the establishment of many ovarian pathologies. These insults, albeit of varying origin, target several developmental check points such as establishment of ovarian reserve, activation/recruitment, selection, dominance, ovulation, and formation of corpus luteum with the final phenotypic outcome likely dictated by the developmental time point of insult. All these developmental insults work via common mediaries such as steroidal and metabolic factors to induce gene-specific epigenetic changes that alter the expression pattern of genes and consequently the ovarian developmental trajectory culminating in adult pathologies. With many such insults, it is difficult to distinguish the contribution from steroidal and metabolic mediaries in programming the pathologies as they often coexist. Interventions targeting the specific steroidal or metabolic factors are necessary to identify the specific mediator or determine if both metabolic and steroidal factors synergize in the programming of the pathology.

Although the focus of this chapter has been on the impact of developmental insults, recent evidence suggests that a subsequent insult (a second hit) has the potential to amplify or unmask defects induced by the first insult (Puttabyatappa et al., 2016). For instance, postnatal obesity in sheep amplifies the reproductive defects (Steckler, Herkimer, Dumesic, & Padmanabhan, 2009) and postnatal estradiol increases the incidences of subluteal cycles (Veiga-Lopez et al., 2014) in prenatal T-treated sheep. Such studies are lacking at the ovarian level. Considering the potential for interaction between developmental insults and subsequent changes in lifestyle, diet, and environmental exposures, this is an important avenue for future investigation. In addition, because susceptibility to a given insult is likely to differ from individual to individual, focus on gene × environment interactions should be taken into consideration in probing these pathologies.

It also needs to be recognized that for the most part impacts of the various developmental insults at the level of the ovary have come mostly from altricial species. While this has been crucial for addressing underlying mechanisms such as transgenerational transfer of pathology to subsequent generations and testing intervention strategies, similar studies are required in precocial species such as sheep, cow, and nonhuman primates for effective human translation. This is important as ovarian differentiation in these species gets completed in utero similar to humans, as opposed to them getting completed postnatally in altricial models.

Since targeted mechanistic studies are not possible in human at the ovarian level, there is also a need for development of biomarkers for recognizing early onset of developmentally programmed ovarian pathologies. Currently, AMH and antral follicular count are widely used to assess recruitment defects and declining ovarian reserve in patients undergoing assisted reproductive technologies procedures (Tal & Seifer, 2017). Noninvasive ultrasonography tools and circulating biomarkers are very much needed for early detection of ovarian dysfunctions in "at risk" individuals.

Effective intervention strategies are required to prevent ovarian pathologies originating or amplified by developmental insults. Because use of pharmacological agents that block common endocrine programmers such as T and insulin has the potential to serve as programming agents themselves, intervention strategies need to focus on lifestyle changes. When environmental endocrine disrupting chemicals are identified as the culprit, focus needs to be not only on lifestyle changes but also at the regulatory level to reduce exposures.

ACKNOWLEDGMENT

This work was supported by National Institutes of Health Grant P01 HD44232.

REFERENCES

Abbott, D. H., Barnett, D. K., Bruns, C. M., & Dumesic, D. A. (2005). Androgen excess fetal programming of female reproduction: A developmental aetiology for polycystic ovary syndrome? *Human Reproduction Update, 11*, 357–374.

Abbott, D. H., Barnett, D. K., Levine, J. E., Padmanabhan, V., Dumesic, D. A., Jacoris, S., et al. (2008). Endocrine antecedents of polycystic ovary syndrome in fetal and infant prenatally androgenized female rhesus monkeys. *Biology of Reproduction, 79*, 154–163.

Abi Salloum, B., Veiga-Lopez, A., Abbott, D. H., Burant, C. F., & Padmanabhan, V. (2015). Developmental programming: Exposure to testosterone excess disrupts steroidal and metabolic environment in pregnant sheep. *Endocrinology, 156*, 2323–2337.

Adewale, H. B., Jefferson, W. N., Newbold, R. R., & Patisaul, H. B. (2009). Neonatal bisphenol-a exposure alters rat reproductive development and ovarian morphology without impairing activation of gonadotropin-releasing hormone neurons. *Biology of Reproduction, 81*, 690–699.

Al-Alem, L., & Curry, T. E., Jr. (2015). Ovarian cancer: Involvement of the matrix metalloproteinases. *Reproduction, 150*, R55–R64.

Albertini, D. F. (2015). *Chapter 2—The mammalian oocyte, Knobil and Neill's physiology of reproduction* (4th ed., pp. 59–97). San Diego: Academic Press.

Amaral, P. P., & Mattick, J. S. (2008). Noncoding RNA in development. *Mammalian Genome, 19*, 454–492.

Armenti, A. E., Zama, A. M., Passantino, L., & Uzumcu, M. (2008). Developmental methoxychlor exposure affects multiple reproductive parameters and ovarian folliculogenesis and gene expression in adult rats. *Toxicology and Applied Pharmacology, 233*, 286–296.

Ashworth, C. J., George, S. O., Hogg, C. O., Lai, Y. T., & Brunton, P. J. (2016). Sex-specific prenatal stress effects on the rat reproductive axis and adrenal gland structure. *Reproduction, 151*, 709–717.

Ashworth, C. J., Hogg, C. O., Hoeks, C. W., Donald, R. D., Duncan, W. C., Lawrence, A. B., et al. (2011). Pre-natal social stress and post-natal pain affect the developing pig reproductive axis. *Reproduction, 142*, 907–914.

Baerwald, A. R., Adams, G. P., & Pierson, R. A. (2003). Characterization of ovarian follicular wave dynamics in women. *Biology of Reproduction, 69*, 1023–1031.

Baerwald, A. R., Adams, G. P., & Pierson, R. A. (2012). Ovarian antral folliculogenesis during the human menstrual cycle: A review. *Human Reproduction Update, 18*, 73–91.

Baravalle, C., Salvetti, N. R., Mira, G. A., Pezzone, N., & Ortega, H. H. (2006). Microscopic characterization of follicular structures in letrozole-induced polycystic ovarian syndrome in the rat. *Archives of Medical Research, 37*, 830–839.

Barker, D. J. (2004). The developmental origins of adult disease. *Journal of the American College of Nutrition, 23*, 588S–595S.

Barker, D., & Osmond, C. (1988). Low birth weight and hypertension. *BMJ [British Medical Journal], 297*, 134.

Barnes, R. B., Rosenfield, R. L., Ehrmann, D. A., Cara, J. F., Cuttler, L., Levitsky, L. L., et al. (1994). Ovarian hyperandrogynism as a result of congenital adrenal virilizing disorders: Evidence for perinatal masculinization of neuroendocrine function in women. *The Journal of Clinical Endocrinology and Metabolism, 79*, 1328–1333.

Barra, R., Cruz, G., Mayerhofer, A., Paredes, A., & Lara, H. E. (2014). Maternal sympathetic stress impairs follicular development and puberty of the offspring. *Reproduction, 148*, 137–145.

Barrett, E. S., & Swan, S. H. (2015). Stress and androgen activity during fetal development. *Endocrinology, 156*, 3435–3441.
Bast, R. C., Jr., Hennessy, B., & Mills, G. B. (2009). The biology of ovarian cancer: New opportunities for translation. *Nature Reviews. Cancer, 9*, 415–428.
Bernal, A. B., Vickers, M. H., Hampton, M. B., Poynton, R. A., & Sloboda, D. M. (2010). Maternal undernutrition significantly impacts ovarian follicle number and increases ovarian oxidative stress in adult rat offspring. *PLoS One, 5*, e15558.
Bhattacharya, P., & Keating, A. F. (2012). Impact of environmental exposures on ovarian function and role of xenobiotic metabolism during ovotoxicity. *Toxicology and Applied Pharmacology, 261*, 227–235.
Blecher, S. R., & Erickson, R. P. (2007). Genetics of sexual development: A new paradigm. *American Journal of Medical Genetics. Part A, 143A*, 3054–3068.
Boots, C. E., & Jungheim, E. S. (2015). Inflammation and human ovarian follicular dynamics. *Seminars in Reproductive Medicine, 33*, 270–275.
Bradshaw, K. D., Santos-Ramos, R., Rawlins, S. C., MacDonald, P. C., & Parker, C. R., Jr. (1986). Endocrine studies in a pregnancy complicated by ovarian theca lutein cysts and hyperreactio luteinalis. *Obstetrics and Gynecology, 67*, 66S–69S.
Brugh, V. M., 3rd, Maduro, M. R., & Lamb, D. J. (2003). Genetic disorders and infertility. *The Urologic Clinics of North America, 30*, 143–152.
Campos, D. B., Albornoz, M., Papa, P. C., Palin, M. F., Bordignon, V., & Murphy, B. D. (2015). Relationship between adiponectin and fertility in the female pig. *Reproduction, Fertility, and Development, 27*, 458–470.
Capel, B. (2000). The battle of the sexes. *Mechanisms of Development, 92*, 89–103.
Cardoso, R. C., Puttabyatappa, M., & Padmanabhan, V. (2015). Steroidogenic versus metabolic programming of reproductive neuroendocrine, ovarian and metabolic dysfunctions. *Neuroendocrinology, 102*, 226–237.
Castrillon, D. H., Miao, L., Kollipara, R., Horner, J. W., & DePinho, R. A. (2003). Suppression of ovarian follicle activation in mice by the transcription factor Foxo3a. *Science, 301*, 215–218.
Chan, K. A., Tsoulis, M. W., & Sloboda, D. M. (2015). Early-life nutritional effects on the female reproductive system. *The Journal of Endocrinology, 224*, R45–R62.
Chen, Y., Jefferson, W. N., Newbold, R. R., Padilla-Banks, E., & Pepling, M. E. (2007). Estradiol, progesterone, and genistein inhibit oocyte nest breakdown and primordial follicle assembly in the neonatal mouse ovary in vitro and in vivo. *Endocrinology, 148*, 3580–3590.
Cheong, Y., Sadek, K. H., Bruce, K. D., Macklon, N., & Cagampang, F. R. (2014). Diet-induced maternal obesity alters ovarian morphology and gene expression in the adult mouse offspring. *Fertility and Sterility, 102*, 899–907.
Cohn, B. A., Cirillo, P. M., Wolff, M. S., Schwingl, P. J., Cohen, R. D., Sholtz, R. I., et al. (2003). DDT and DDE exposure in mothers and time to pregnancy in daughters. *Lancet, 361*, 2205–2206.
Comim, F. V., Hardy, K., Robinson, J., & Franks, S. (2015). Disorders of follicle development and steroidogenesis in ovaries of androgenised foetal sheep. *The Journal of Endocrinology, 225*, 39–46.
Conti, M., Hsieh, M., Zamah, A. M., & Oh, J. S. (2012). Novel signaling mechanisms in the ovary during oocyte maturation and ovulation. *Molecular and Cellular Endocrinology, 356*, 65–73.
Conway, G., Dewailly, D., Diamanti-Kandarakis, E., Escobar-Morreale, H. F., Franks, S., Gambineri, A., et al. (2014). The polycystic ovary syndrome: A position statement from the European Society of Endocrinology. *European Journal of Endocrinology, 171*, P1–P29.

Cools, M., Wolffenbuttel, K. P., Drop, S. L., Oosterhuis, J. W., & Looijenga, L. H. (2011). Gonadal development and tumor formation at the crossroads of male and female sex determination. *Sexual Development, 5*, 167–180.

Cooper, R. L., & Kavlock, R. J. (1997). Endocrine disruptors and reproductive development: A weight-of-evidence overview. *The Journal of Endocrinology, 152*, 159–166.

Correia, C. P. (1998). *The ovary of Eve egg and sperm and preformation* (p. xxiii). Chicago, Ill.: University of Chicago Press. 396 p.

Costoff, A., & Mahesh, V. B. (1975). Primordial follicles with normal oocytes in the ovaries of postmenopausal women. *Journal of the American Geriatrics Society, 23*, 193–196.

Craig, Z. R., Wang, W., & Flaws, J. A. (2011). Endocrine-disrupting chemicals in ovarian function: Effects on steroidogenesis, metabolism and nuclear receptor signaling. *Reproduction, 142*, 633–646.

Crain, D. A., Janssen, S. J., Edwards, T. M., Heindel, J., Ho, S. M., Hunt, P., et al. (2008). Female reproductive disorders: The roles of endocrine-disrupting compounds and developmental timing. *Fertility and Sterility, 90*, 911–940.

Cruz, G., Foster, W., Paredes, A., Yi, K. D., & Uzumcu, M. (2014). Long-term effects of early-life exposure to environmental oestrogens on ovarian function: Role of epigenetics. *Journal of Neuroendocrinology, 26*, 613–624.

Dewailly, D., Robin, G., Peigne, M., Decanter, C., Pigny, P., & Catteau-Jonard, S. (2016). Interactions between androgens, FSH, anti-Mullerian hormone and estradiol during folliculogenesis in the human normal and polycystic ovary. *Human Reproduction Update, 22*, 709–724.

Diamanti-Kandarakis, E., Bourguignon, J. P., Giudice, L. C., Hauser, R., Prins, G. S., Soto, A. M., et al. (2009). Endocrine-disrupting chemicals: An endocrine society scientific statement. *Endocrine Reviews, 30*, 293–342.

Diamanti-Kandarakis, E., & Dunaif, A. (2012). Insulin resistance and the polycystic ovary syndrome revisited: An update on mechanisms and implications. *Endocrine Reviews, 33*, 981–1030.

Dong, M., Zheng, Q., Ford, S. P., Nathanielsz, P. W., & Ren, J. (2013). Maternal obesity, lipotoxicity and cardiovascular diseases in offspring. *Journal of Molecular and Cellular Cardiology, 55*, 111–116.

Dumesic, D. A., Oberfield, S. E., Stener-Victorin, E., Marshall, J. C., Laven, J. S., & Legro, R. S. (2015). Scientific statement on the diagnostic criteria, epidemiology, pathophysiology, and molecular genetics of polycystic ovary syndrome. *Endocrine Reviews, 36*, 487–525.

Dumesic, D. A., & Richards, J. S. (2013). Ontogeny of the ovary in polycystic ovary syndrome. *Fertility and Sterility, 100*, 23–38.

Dumesic, D. A., Schramm, R. D., Bird, I. M., Peterson, E., Paprocki, A. M., Zhou, R., et al. (2003). Reduced intrafollicular androstenedione and estradiol levels in early-treated prenatally androgenized female rhesus monkeys receiving follicle-stimulating hormone therapy for in vitro fertilization. *Biology of Reproduction, 69*, 1213–1219.

Dunaif, A. (2016). Perspectives in polycystic ovary syndrome: From hair to eternity. *The Journal of Clinical Endocrinology and Metabolism, 101*, 759–768.

Dunkel Schetter, C. (2011). Psychological science on pregnancy: Stress processes, biopsychosocial models, and emerging research issues. *Annual Review of Psychology, 62*, 531–558.

Durlinger, A. L., Kramer, P., Karels, B., de Jong, F. H., Uilenbroek, J. T., Grootegoed, J. A., et al. (1999). Control of primordial follicle recruitment by anti-Mullerian hormone in the mouse ovary. *Endocrinology, 140*, 5789–5796.

Ehrlich, M., & Lacey, M. (2013). DNA methylation and differentiation: Silencing, upregulation and modulation of gene expression. *Epigenomics, 5*, 553–568.

Escobar, M., Vázquez-Nin, G., & Echeverría, O. (2011). *Follicular cells, cell death in mammalian ovary* (pp. 185–200). Springer Netherlands.

Fathalla, M. F. (1971). Incessant ovulation—A factor in ovarian neoplasia? *Lancet, 2*, 163.

Fortune, J. (2003). The early stages of follicular development: Activation of primordial follicles and growth of preantral follicles. *Animal Reproduction Science, 78*, 135–163.

Fortune, J. E., Rivera, G. M., & Yang, M. Y. (2004). Follicular development: The role of the follicular microenvironment in selection of the dominant follicle. *Animal Reproduction Science, 82-83*, 109–126.

Foulkes, W. D., Gore, M., & McCluggage, W. G. (2016). Rare non-epithelial ovarian neoplasms: Pathology, genetics and treatment. *Gynecologic Oncology, 142*, 190–198.

Fowden, A. L., & Forhead, A. J. (2004). Endocrine mechanisms of intrauterine programming. *Reproduction, 127*, 515–526.

Franks, S., Mason, H., & Willis, D. (2000). Follicular dynamics in the polycystic ovary syndrome. *Molecular and Cellular Endocrinology, 163*, 49–52.

Franks, S., & McCarthy, M. (2004). Genetics of ovarian disorders: Polycystic ovary syndrome. *Reviews in Endocrine & Metabolic Disorders, 5*, 69–76.

Franks, S., Stark, J., & Hardy, K. (2008). Follicle dynamics and anovulation in polycystic ovary syndrome. *Human Reproduction Update, 14*, 367–378.

Goodfellow, P. N., & Lovell-Badge, R. (1993). SRY and sex determination in mammals. *Annual Review of Genetics, 27*, 71–92.

Gore, A. C. (2008). Developmental programming and endocrine disruptor effects on reproductive neuroendocrine systems. *Frontiers in Neuroendocrinology, 29*, 358–374.

Gore, A. C., Chappell, V. A., Fenton, S. E., Flaws, J. A., Nadal, A., Prins, G. S., et al. (2015). EDC-2: The endocrine society's second scientific statement on endocrine-disrupting chemicals. *Endocrine Reviews, 36*, 593–602.

Gougeon, A. (2004). Dynamics of human follicular growth: Morphologic, dynamic, and functional aspects. *The Ovary, 2*, 25–43.

Gougeon, A., Ecochard, R., & Thalabard, J. C. (1994). Age-related changes of the population of human ovarian follicles: Increase in the disappearance rate of non-growing and early-growing follicles in aging women. *Biology of Reproduction, 50*, 653–663.

Grindler, N. M., Allsworth, J. E., Macones, G. A., Kannan, K., Roehl, K. A., & Cooper, A. R. (2015). Persistent organic pollutants and early menopause in U.S. women. *PLoS One, 10*, e0116057.

Hakim, C., Padmanabhan, V., & Vyas, A. K. (2016). Gestational hyperandrogenism in developmental programming. *Endocrinology, 158*, 199–212.

Handa, R. J., Pak, T. R., Kudwa, A. E., Lund, T. D., & Hinds, L. (2008). An alternate pathway for androgen regulation of brain function: Activation of estrogen receptor beta by the metabolite of dihydrotestosterone, 5alpha-androstane-3beta,17beta-diol. *Hormones and Behavior, 53*, 741–752.

Hannon, P. R., & Flaws, J. A. (2015). The effects of phthalates on the ovary. *Frontiers in Endocrinology (Lausanne), 6*, 8.

Hassold, T., Abruzzo, M., Adkins, K., Griffin, D., Merrill, M., Millie, E., et al. (1996). Human aneuploidy: Incidence, origin, and etiology. *Environmental and Molecular Mutagenesis, 28*, 167–175.

Hassold, T., Benham, F., & Leppert, M. (1988). Cytogenetic and molecular analysis of sex-chromosome monosomy. *American Journal of Human Genetics, 42*, 534–541.

Heard, E., & Martienssen, R. A. (2014). Transgenerational epigenetic inheritance: Myths and mechanisms. *Cell, 157*, 95–109.

Heerwagen, M. J., Miller, M. R., Barbour, L. A., & Friedman, J. E. (2010). Maternal obesity and fetal metabolic programming: A fertile epigenetic soil. *American Journal of Physiology. Regulatory, Integrative and Comparative Physiology, 299*, R711–R722.

Herrenkohl, L. R., & Scott, S. (1984). Prenatal stress and postnatal androgen: Effects on reproduction in female rats. *Experientia*, *40*, 101–103.
Hogg, K., McNeilly, A. S., & Duncan, W. C. (2011). Prenatal androgen exposure leads to alterations in gene and protein expression in the ovine fetal ovary. *Endocrinology*, *152*, 2048–2059.
Hogg, K., Young, J. M., Oliver, E. M., Souza, C. J., McNeilly, A. S., & Duncan, W. C. (2012). Enhanced thecal androgen production is prenatally programmed in an ovine model of polycystic ovary syndrome. *Endocrinology*, *153*, 450–461.
Holemans, K., Caluwaerts, S., Poston, L., & Van Assche, F. A. (2004). Diet-induced obesity in the rat: A model for gestational diabetes mellitus. *American Journal of Obstetrics and Gynecology*, *190*, 858–865.
Hoyer, P. B., & Keating, A. F. (2014). Xenobiotic effects in the ovary: Temporary versus permanent infertility. *Expert Opinion on Drug Metabolism & Toxicology*, *10*, 511–523.
Hulas-Stasiak, M., Dobrowolski, P., & Tomaszewska, E. (2016). Prenatally administered dexamethasone impairs folliculogenesis in spiny mouse offspring. *Reproduction, Fertility, and Development*, *28*, 1038–1048.
Hummitzsch, K., Anderson, R. A., Wilhelm, D., Wu, J., Telfer, E. E., Russell, D. L., et al. (2015). Stem cells, progenitor cells, and lineage decisions in the ovary. *Endocrine Reviews*, *36*, 65–91.
Huo, X., Chen, D., He, Y., Zhu, W., Zhou, W., & Zhang, J. (2015). Bisphenol-A and female infertility: A possible role of gene-environment interactions. *International Journal of Environmental Research and Public Health*, *12*, 11101–11116.
Inoue, K., Okada, F., Ito, R., Kato, S., Sasaki, S., Nakajima, S., et al. (2004). Perfluorooctane sulfonate (PFOS) and related perfluorinated compounds in human maternal and cord blood samples: Assessment of PFOS exposure in a susceptible population during pregnancy. *Environmental Health Perspectives*, *112*, 1204–1207.
Isik, S., Ozcan, H. N., Ozuguz, U., Tutuncu, Y. A., Berker, D., Alimli, A. G., et al. (2012). Evaluation of ovarian reserve based on hormonal parameters, ovarian volume, and antral follicle count in women with type 2 diabetes mellitus. *The Journal of Clinical Endocrinology and Metabolism*, *97*, 261–269.
Johnson, J., Bagley, J., Skaznik-Wikiel, M., Lee, H. J., Adams, G. B., Niikura, Y., et al. (2005). Oocyte generation in adult mammalian ovaries by putative germ cells in bone marrow and peripheral blood. *Cell*, *122*, 303–315.
Johnson, J., Canning, J., Kaneko, T., Pru, J. K., & Tilly, J. L. (2004). Germline stem cells and follicular renewal in the postnatal mammalian ovary. *Nature*, *428*, 145–150.
Joshi, R., & Dunaif, A. (1995). Ovarian disorders of pregnancy. *Endocrinology and Metabolism Clinics of North America*, *24*, 153–169.
Jost, A. (1983). Genetic and hormonal factors in sex differentiation of the brain. *Psychoneuroendocrinology*, *8*, 183–193.
Jungheim, E. S., Macones, G. A., Odem, R. R., Patterson, B. W., Lanzendorf, S. E., Ratts, V. S., et al. (2011). Associations between free fatty acids, cumulus oocyte complex morphology and ovarian function during in vitro fertilization. *Fertility and Sterility*, *95*, 1970–1974.
Jungheim, E. S., Travieso, J. L., Carson, K. R., & Moley, K. H. (2012). Obesity and reproductive function. *Obstetrics and Gynecology Clinics of North America*, *39*, 479–493.
Kafali, H., Iriadam, M., Ozardali, I., & Demir, N. (2004). Letrozole-induced polycystic ovaries in the rat: A new model for cystic ovarian disease. *Archives of Medical Research*, *35*, 103–108.
Karavan, J. R., & Pepling, M. E. (2012). Effects of estrogenic compounds on neonatal oocyte development. *Reproductive Toxicology*, *34*, 51–56.

Karnezis, A. N., Cho, K. R., Gilks, C. B., Pearce, C. L., & Huntsman, D. G. (2017). The disparate origins of ovarian cancers: Pathogenesis and prevention strategies. *Nature Reviews. Cancer*, *17*, 65–74.

Kauffman, A. S., Thackray, V. G., Ryan, G. E., Tolson, K. P., Glidewell-Kenney, C. A., Semaan, S. J., et al. (2015). A novel letrozole model recapitulates both the reproductive and metabolic phenotypes of polycystic ovary syndrome in female mice. *Biology of Reproduction*, *93*, 69.

Kesler, D. J., & Garverick, H. A. (1982). Ovarian cysts in dairy cattle: A review. *Journal of Animal Science*, *55*, 1147–1159.

Khorram, O., Keen-Rinehart, E., Chuang, T. D., Ross, M. G., & Desai, M. (2015). Maternal undernutrition induces premature reproductive senescence in adult female rat offspring. *Fertility and Sterility*, *103*, 291–298, e292.

Kim, H., Nakajima, T., Hayashi, S., Chambon, P., Watanabe, H., Iguchi, T., et al. (2009). Effects of diethylstilbestrol on programmed oocyte death and induction of polyovular follicles in neonatal mouse ovaries. *Biology of Reproduction*, *81*, 1002–1009.

Knobil, E., & Neill, J. D. (1998). *Encyclopedia of reproduction*. San Diego: Academic Press.

Kozak, L. J., Owings, M. F., & Hall, M. J. (2005). National Hospital Discharge Survey: 2002 annual summary with detailed diagnosis and procedure data. *Vital and Health Statistics*, *13*, 1–199.

Lara, H. E., Dorfman, M., Venegas, M., Luza, S. M., Luna, S. L., Mayerhofer, A., et al. (2002). Changes in sympathetic nerve activity of the mammalian ovary during a normal estrous cycle and in polycystic ovary syndrome: Studies on norepinephrine release. *Microscopy Research and Technique*, *59*, 495–502.

Legro, R. S., Driscoll, D., Strauss, J. F., 3rd, Fox, J., & Dunaif, A. (1998). Evidence for a genetic basis for hyperandrogenemia in polycystic ovary syndrome. *Proceedings of the National Academy of Sciences of the United States of America*, *95*, 14956–14960.

Lin, Y. T., & Capel, B. (2015). Cell fate commitment during mammalian sex determination. *Current Opinion in Genetics & Development*, *32*, 144–152.

Liu, Y. X., & Hsueh, A. J. (1986). Synergism between granulosa and theca-interstitial cells in estrogen biosynthesis by gonadotropin-treated rat ovaries: Studies on the two-cell, two-gonadotropin hypothesis using steroid antisera. *Biology of Reproduction*, *35*, 27–36.

Lizneva, D., Suturina, L., Walker, W., Brakta, S., Gavrilova-Jordan, L., & Azziz, R. (2016). Criteria, prevalence, and phenotypes of polycystic ovary syndrome. *Fertility and Sterility*, *106*, 6–15.

Longacre, M., Snyder, N. A., Housman, G., Leary, M., Lapinska, K., Heerboth, S., et al. (2016). A comparative analysis of genetic and epigenetic events of breast and ovarian cancer related to tumorigenesis. *International Journal of Molecular Sciences*, *17*, 759.

Luense, L. J., Veiga-Lopez, A., Padmanabhan, V., & Christenson, L. K. (2011). Developmental programming: Gestational testosterone treatment alters fetal ovarian gene expression. *Endocrinology*, *152*, 4974–4983.

Maliqueo, M., Benrick, A., & Stener-Victorin, E. (2014). Rodent models of polycystic ovary syndrome: Phenotypic presentation, pathophysiology, and the effects of different interventions. *Seminars in Reproductive Medicine*, *32*, 183–193.

Manikkam, M., Haque, M. M., Guerrero-Bosagna, C., Nilsson, E. E., & Skinner, M. K. (2014). Pesticide methoxychlor promotes the epigenetic transgenerational inheritance of adult-onset disease through the female germline. *PLoS One*, *9*, e102091.

Manikkam, M., Steckler, T. L., Welch, K. B., Inskeep, E. K., & Padmanabhan, V. (2006). Fetal programming: Prenatal testosterone treatment leads to follicular persistence/luteal defects; partial restoration of ovarian function by cyclic progesterone treatment. *Endocrinology*, *147*, 1997–2007.

Manneras, L., Cajander, S., Holmang, A., Seleskovic, Z., Lystig, T., Lonn, M., et al. (2007). A new rat model exhibiting both ovarian and metabolic characteristics of polycystic ovary syndrome. *Endocrinology, 148*, 3781–3791.

Markey, C. M., Coombs, M. A., Sonnenschein, C., & Soto, A. M. (2003). Mammalian development in a changing environment: Exposure to endocrine disruptors reveals the developmental plasticity of steroid-hormone target organs. *Evolution & Development, 5*, 67–75.

Mazdai, A., Dodder, N. G., Abernathy, M. P., Hites, R. A., & Bigsby, R. M. (2003). Polybrominated diphenyl ethers in maternal and fetal blood samples. *Environmental Health Perspectives, 111*, 1249–1252.

McGee, E. A., & Hsueh, A. J. (2000). Initial and cyclic recruitment of ovarian follicles. *Endocrine Reviews, 21*, 200–214.

McLaughlin, E. A., & McIver, S. C. (2009). Awakening the oocyte: Controlling primordial follicle development. *Reproduction, 137*, 1–11.

Miska, E. A., & Ferguson-Smith, A. C. (2016). Transgenerational inheritance: Models and mechanisms of non-DNA sequence-based inheritance. *Science, 354*, 59–63.

Monniaux, D., Clement, F., Dalbies-Tran, R., Estienne, A., Fabre, S., Mansanet, C., et al. (2014). The ovarian reserve of primordial follicles and the dynamic reserve of antral growing follicles: What is the link? *Biology of Reproduction, 90*, 85.

Moran, L. J., Norman, R. J., & Teede, H. J. (2015). Metabolic risk in PCOS: Phenotype and adiposity impact. *Trends in Endocrinology and Metabolism, 26*, 136–143.

Mossa, F., Carter, F., Walsh, S. W., Kenny, D. A., Smith, G. W., Ireland, J. L., et al. (2013). Maternal undernutrition in cows impairs ovarian and cardiovascular systems in their offspring. *Biology of Reproduction, 88*, 92.

Moyer, B., & Hixon, M. L. (2012). Reproductive effects in F1 adult females exposed in utero to moderate to high doses of mono-2-ethylhexylphthalate (MEHP). *Reproductive Toxicology, 34*, 43–50.

Mu, Y. M., Yanase, T., Nishi, Y., Tanaka, A., Saito, M., Jin, C. H., et al. (2001). Saturated FFAs, palmitic acid and stearic acid, induce apoptosis in human granulosa cells. *Endocrinology, 142*, 3590–3597.

Naderpoor, N., Shorakae, S., Joham, A., Boyle, J., De Courten, B., & Teede, H. J. (2015). Obesity and polycystic ovary syndrome. *Minerva Endocrinologica, 40*, 37–51.

Newbold, R. R., Bullock, B. C., & Mc Lachlan, J. A. (1983). Exposure to diethylstilbestrol during pregnancy permanently alters the ovary and oviduct. *Biology of Reproduction, 28*, 735–744.

Newbold, R. R., Jefferson, W. N., & Padilla-Banks, E. (2007). Long-term adverse effects of neonatal exposure to bisphenol A on the murine female reproductive tract. *Reproductive Toxicology, 24*, 253–258.

Newbold, R. R., Jefferson, W. N., & Padilla-Banks, E. (2009). Prenatal exposure to bisphenol a at environmentally relevant doses adversely affects the murine female reproductive tract later in life. *Environmental Health Perspectives, 117*, 879–885.

Nilsson, E. E., & Skinner, M. K. (2015). Environmentally Induced Epigenetic Transgenerational Inheritance of Reproductive Disease. *Biology of Reproduction, 93*, 145.

Novak, A. Z., Kokai, G. K., Popovic, V. P., Ludoski, M. D., & Jurukovski, V. A. (1995). Interphase cytogenetics on paraffin-embedded sections of ovary for detection of genomic constitution in a patient with Turner's syndrome and chromosomal mosaicism. *Human Genetics, 95*, 293–298.

Nugent, B. M., Tobet, S. A., Lara, H. E., Lucion, A. B., Wilson, M. E., Recabarren, S. E., et al. (2012). Hormonal programming across the lifespan. *Hormone and Metabolic Research, 44*, 577–586.

Ortega, H. H., Marelli, B. E., Rey, F., Amweg, A. N., Diaz, P. U., Stangaferro, M. L., et al. (2015). Molecular aspects of bovine cystic ovarian disease pathogenesis. *Reproduction, 149*, R251–R264.

Ortega, H. H., Rey, F., Velazquez, M. M., & Padmanabhan, V. (2010). Developmental programming: Effect of prenatal steroid excess on intraovarian components of insulin signaling pathway and related proteins in sheep. *Biology of Reproduction, 82*, 1065–1075.

Ortega, H. H., Salvetti, N. R., & Padmanabhan, V. (2009). Developmental programming: Prenatal androgen excess disrupts ovarian steroid receptor balance. *Reproduction, 137*, 865–877.

Ortega, H. H., Veiga-Lopez, A., Sreedharan, S., del Lujan Velazquez, M. M., Salvetti, N. R., & Padmanabhan, V. (2015). Developmental programming: Does prenatal steroid excess disrupt the ovarian VEGF system in sheep? *Biology of Reproduction, 93*, 58.

Padmanabhan, V., Cardoso, R. C., & Puttabyatappa, M. (2016). Developmental programming, a pathway to disease. *Endocrinology, 157*, 1328–1340.

Padmanabhan, V., Salvetti, N. R., Matiller, V., & Ortega, H. H. (2014). Developmental programming: Prenatal steroid excess disrupts key members of intraovarian steroidogenic pathway in sheep. *Endocrinology, 155*, 3649–3660.

Padmanabhan, V., & Veiga-Lopez, A. (2011). Developmental origin of reproductive and metabolic dysfunctions: Androgenic versus estrogenic reprogramming. *Seminars in Reproductive Medicine, 29*, 173–186.

Padmanabhan, V., & Veiga-Lopez, A. (2013a). Animal models of the polycystic ovary syndrome phenotype. *Steroids, 78*, 734–740.

Padmanabhan, V., & Veiga-Lopez, A. (2013b). Sheep models of polycystic ovary syndrome phenotype. *Molecular and Cellular Endocrinology, 373*, 8–20.

Pangas, S. A., & Rajkovic, A. (2015). *Chapter 21—Follicular development: Mouse, sheep, and human models, Knobil and Neill's physiology of reproduction* (4th ed., pp. 947–995). San Diego: Academic Press.

Peretz, J., Vrooman, L., Ricke, W. A., Hunt, P. A., Ehrlich, S., Hauser, R., et al. (2014). Bisphenol a and reproductive health: Update of experimental and human evidence, 2007-2013. *Environmental Health Perspectives, 122*, 775–786.

Phelan, N., & Conway, G. S. (2011). Management of ovarian disease in pregnancy. *Best Practice & Research. Clinical Endocrinology & Metabolism, 25*, 985–992.

Politch, J. A., & Herrenkohl, L. R. (1984). Effects of prenatal stress on reproduction in male and female mice. *Physiology & Behavior, 32*, 95–99.

Poulain, M., Frydman, N., Duquenne, C., N'Tumba-Byn, T., Benachi, A., Habert, R., et al. (2012). Dexamethasone induces germ cell apoptosis in the human fetal ovary. *The Journal of Clinical Endocrinology and Metabolism, 97*, E1890–E1897.

Puttabyatappa, M., Irwin, A., Martin, J. D., Mesquita, M., Veiga-Lopez, A., & Padmanabhan, V. (2017). Developmental programming: Gestational exposure to excess testosterone alters expression of ovarian matrix metalloproteases and their target proteins. *Reproductive Sciences*. 1933719117697127, https://doi.org/10.1177/1933719117697127.

Puttabyatappa, M., Cardoso, R. C., & Padmanabhan, V. (2016). Effect of maternal PCOS and PCOS-like phenotype on the offspring's health. *Molecular and Cellular Endocrinology, 435*, 29–39.

Qin, Y., Jiao, X., Simpson, J. L., & Chen, Z. J. (2015). Genetics of primary ovarian insufficiency: New developments and opportunities. *Human Reproduction Update, 21*, 787–808.

Ravelli, G. P., Stein, Z. A., & Susser, M. W. (1976). Obesity in young men after famine exposure in utero and early infancy. *The New England Journal of Medicine, 295*, 349–353.

Rebar, R., Judd, H. L., Yen, S. S., Rakoff, J., Vandenberg, G., & Naftolin, F. (1976). Characterization of the inappropriate gonadotropin secretion in polycystic ovary syndrome. *The Journal of Clinical Investigation, 57*, 1320–1329.

Reddy, P., Liu, L., Adhikari, D., Jagarlamudi, K., Rajareddy, S., Shen, Y., et al. (2008). Oocyte-specific deletion of PTEN causes premature activation of the primordial follicle pool. *Science, 319*, 611–613.

Ricci, Z. J., Oh, S. K., Stein, M. W., Kaul, B., Flusberg, M., Chernyak, V., et al. (2016). Solid organ abdominal ischemia, part I: Clinical features, etiology, imaging findings, and management. *Clinical Imaging, 40*, 720–731.

Richards, J. S. (2007). Genetics of ovulation. *Seminars in Reproductive Medicine, 25*, 235–242.

Richardson, S. J., Senikas, V., & Nelson, J. F. (1987). Follicular depletion during the menopausal transition: Evidence for accelerated loss and ultimate exhaustion. *The Journal of Clinical Endocrinology and Metabolism, 65*, 1231–1237.

Ristic, N., Nestorovic, N., Manojlovic-Stojanoski, M., Filipovic, B., Sosic-Jurjevic, B., Milosevic, V., et al. (2008). Maternal dexamethasone treatment reduces ovarian follicle number in neonatal rat offspring. *Journal of Microscopy, 232*, 549–557.

Robker, R. L., Akison, L. K., Bennett, B. D., Thrupp, P. N., Chura, L. R., Russell, D. L., et al. (2009). Obese women exhibit differences in ovarian metabolites, hormones, and gene expression compared with moderate-weight women. *The Journal of Clinical Endocrinology and Metabolism, 94*, 1533–1540.

Roland, A. V., Nunemaker, C. S., Keller, S. R., & Moenter, S. M. (2010). Prenatal androgen exposure programs metabolic dysfunction in female mice. *The Journal of Endocrinology, 207*, 213–223.

Rosa, E. S. A., Guimaraes, M. A., Padmanabhan, V., & Lara, H. E. (2003). Prepubertal administration of estradiol valerate disrupts cyclicity and leads to cystic ovarian morphology during adult life in the rat: Role of sympathetic innervation. *Endocrinology, 144*, 4289–4297.

Rossing, M. A., Daling, J. R., Weiss, N. S., Moore, D. E., & Self, S. G. (1994). Ovarian tumors in a cohort of infertile women. *The New England Journal of Medicine, 331*, 771–776.

Russell, D. L., & Robker, R. L. (2007). Molecular mechanisms of ovulation: Co-ordination through the cumulus complex. *Human Reproduction Update, 13*, 289–312.

Salvetti, N. R., Ortega, H. H., Veiga-Lopez, A., & Padmanabhan, V. (2012). Developmental programming: Impact of prenatal testosterone excess on ovarian cell proliferation and apoptotic factors in sheep. *Biology of Reproduction, 87*, 22, 1–10.

Sasaki, K. J., & Miller, C. E. (2014). Adnexal torsion: Review of the literature. *Journal of Minimally Invasive Gynecology, 21*, 196–202.

Sathyanarayana, S., Calafat, A. M., Liu, F., & Swan, S. H. (2008). Maternal and infant urinary phthalate metabolite concentrations: Are they related? *Environmental Research, 108*, 413–418.

Shah, K., Sivapalan, G., Gibbons, N., Tempest, H., & Griffin, D. K. (2003). The genetic basis of infertility. *Reproduction, 126*, 13–25.

Silva, C. A., Yamakami, L. Y., Aikawa, N. E., Araujo, D. B., Carvalho, J. F., & Bonfa, E. (2014). Autoimmune primary ovarian insufficiency. *Autoimmunity Reviews, 13*, 427–430.

Skinner, M. K. (2005). Regulation of primordial follicle assembly and development. *Human Reproduction Update, 11*, 461–471.

Skinner, M. K. (2014). Endocrine disruptor induction of epigenetic transgenerational inheritance of disease. *Molecular and Cellular Endocrinology, 398*, 4–12.

Smith, G. W., & Meidan, R. (2014). Ever-changing cell interactions during the life span of the corpus luteum: Relevance to luteal regression. *Reproductive Biology, 14*, 75–82.

Smith, P., Wilhelm, D., & Rodgers, R. J. (2014). Development of mammalian ovary. *The Journal of Endocrinology, 221*, R145–R161.

Sobinoff, A. P., Sutherland, J. M., & McLaughlin, E. A. (2013). Intracellular signalling during female gametogenesis. *Molecular Human Reproduction, 19*, 265–278.

Spinelli, C., Piscioneri, J., & Strambi, S. (2015). Adnexal torsion in adolescents: Update and review of the literature. *Current Opinion in Obstetrics & Gynecology*, *27*, 320–325.
Steckler, T. L., Herkimer, C., Dumesic, D. A., & Padmanabhan, V. (2009). Developmental programming: Excess weight gain amplifies the effects of prenatal testosterone excess on reproductive cyclicity—implication for polycystic ovary syndrome. *Endocrinology*, *150*, 1456–1465.
Steckler, T. L., Lee, J. S., Ye, W., Inskeep, E. K., & Padmanabhan, V. (2008). Developmental programming: Exogenous gonadotropin treatment rescues ovulatory function but does not completely normalize ovarian function in sheep treated prenatally with testosterone. *Biology of Reproduction*, *79*, 686–695.
Steckler, T., Roberts, E., Doop, D., Lee, T., & Padmanabhan, V. (2007). Developmental programming in sheep: Administration of testosterone during 60–90 days of pregnancy reduces breeding success and pregnancy outcome. *Theriogenology*, *67*, 459–467.
Stouffer, R. L., Bishop, C. V., Bogan, R. L., Xu, F., & Hennebold, J. D. (2013). Endocrine and local control of the primate corpus luteum. *Reproductive Biology*, *13*, 259–271.
Stouffer, R. L., Xu, F., & Duffy, D. M. (2007). Molecular control of ovulation and luteinization in the primate follicle. *Frontiers in Bioscience*, *12*, 297–307.
Takahashi, L. K., Turner, J. G., & Kalin, N. H. (1998). Prolonged stress-induced elevation in plasma corticosterone during pregnancy in the rat: Implications for prenatal stress studies. *Psychoneuroendocrinology*, *23*, 571–581.
Tal, R., & Seifer, D. B. (2017). Ovarian reserve testing: A user's guide. *American Journal of Obstetrics and Gynecology*, *217*, 129–140.
Tang, W. Y., & Ho, S. M. (2007). Epigenetic reprogramming and imprinting in origins of disease. *Reviews in Endocrine & Metabolic Disorders*, *8*, 173–182.
Thomakos, N., Trachana, S. P., Koutroumpa, I., Rodolakis, A., & Gavalas, N. G. (2015). Molecular aspects and clinical methods for preserving ovarian reserves in women receiving cancer treatment. *Clinical and Experimental Obstetrics & Gynecology*, *42*, 416–425.
Tilly, J. L. (1996). The molecular basis of ovarian cell death during germ cell attrition, follicular atresia, and luteolysis. *Frontiers in Bioscience*, *1*, d1–d11.
Touraine, P., Beau, I., Gougeon, A., Meduri, G., Desroches, A., Pichard, C., et al. (1999). New natural inactivating mutations of the follicle-stimulating hormone receptor: Correlations between receptor function and phenotype. *Molecular Endocrinology*, *13*, 1844–1854.
Trapphoff, T., Heiligentag, M., El Hajj, N., Haaf, T., & Eichenlaub-Ritter, U. (2013). Chronic exposure to a low concentration of bisphenol A during follicle culture affects the epigenetic status of germinal vesicles and metaphase II oocytes. *Fertility and Sterility*, *100*, 1758–1767, e1751.
Tsoulis, M. W., Chang, P. E., Moore, C. J., Chan, K. A., Gohir, W., Petrik, J. J., et al. (2016). Maternal high-fat diet-induced loss of fetal oocytes is associated with compromised follicle growth in adult rat offspring. *Biology of Reproduction*, *94*, 94.
Turner, B. M. (2011). Environmental sensing by chromatin: An epigenetic contribution to evolutionary change. *FEBS Letters*, *585*, 2032–2040.
Valko, M., Leibfritz, D., Moncol, J., Cronin, M. T., Mazur, M., & Telser, J. (2007). Free radicals and antioxidants in normal physiological functions and human disease. *The International Journal of Biochemistry & Cell Biology*, *39*, 44–84.
Veiga-Lopez, A., Luense, L. J., Christenson, L. K., & Padmanabhan, V. (2013). Developmental programming: Gestational bisphenol-A treatment alters trajectory of fetal ovarian gene expression. *Endocrinology*, *154*, 1873–1884.
Veiga-Lopez, A., Moeller, J., Abbott, D. H., & Padmanabhan, V. (2016). Developmental programming: Rescuing disruptions in preovulatory follicle growth and steroidogenesis from prenatal testosterone disruption. *Journal of Ovarian Research*, *9*, 39.

Veiga-Lopez, A., Wurst, A. K., Steckler, T. L., Ye, W., & Padmanabhan, V. (2014). Developmental programming: Postnatal estradiol amplifies ovarian follicular defects induced by fetal exposure to excess testosterone and dihydrotestosterone in sheep. *Reproductive Sciences*, *21*, 444–455.

Veiga-Lopez, A., Ye, W., & Padmanabhan, V. (2012). Developmental programming: Prenatal testosterone excess disrupts anti-Müllerian hormone expression in preantral and antral follicles. *Fertility and Sterility*, *97*, 748–756.

Venkatesh, T., Suresh, P. S., & Tsutsumi, R. (2014). New insights into the genetic basis of infertility. *The Application of Clinical Genetics*, 7, 235–243.

Vink, J. M., Sadrzadeh, S., Lambalk, C. B., & Boomsma, D. I. (2006). Heritability of polycystic ovary syndrome in a Dutch twin-family study. *The Journal of Clinical Endocrinology and Metabolism*, *91*, 2100–2104.

Wang, H., Jiang, M., Bi, H., Chen, X., He, L., Li, X., et al. (2014). Conversion of female germline stem cells from neonatal and prepubertal mice into pluripotent stem cells. *Journal of Molecular Cell Biology*, *6*, 164–171.

Wang, Y. C., Su, H. Y., Liu, J. Y., Chang, F. W., & Chen, C. H. (2005). Maternal and female fetal virilization caused by pregnancy luteomas. *Fertility and Sterility*, *84*, 509.

Webber, L. J., Stubbs, S., Stark, J., Trew, G. H., Margara, R., Hardy, K., et al. (2003). Formation and early development of follicles in the polycystic ovary. *Lancet*, *362*, 1017–1021.

Welt, C. K., Smith, P. C., & Taylor, A. E. (2004). Evidence of early ovarian aging in fragile X premutation carriers. *The Journal of Clinical Endocrinology & Metabolism*, *89*, 4569–4574.

Wemeau, J. L., Proust-Lemoine, E., Ryndak, A., & Vanhove, L. (2013). Thyroid autoimmunity and polyglandular endocrine syndromes. *Hormones (Athens, Greece)*, *12*, 39–45.

West, C., Foster, D. L., Evans, N. P., Robinson, J., & Padmanabhan, V. (2001). Intrafollicular activin availability is altered in prenatally-androgenized lambs. *Molecular and Cellular Endocrinology*, *185*, 51–59.

Whirledge, S., & Cidlowski, J. A. (2010). Glucocorticoids, stress, and fertility. *Minerva Endocrinologica*, *35*, 109–125.

Whyte, J., Alexenko, A., Davis, A., Ellersieck, M., Fountain, E., & Rosenfeld, C. (2007). Maternal diet composition alters serum steroid and free fatty acid concentrations and vaginal pH in mice. *Journal of Endocrinology*, *192*, 75–81.

Wittenberger, M. D., Hagerman, R. J., Sherman, S. L., McConkie-Rosell, A., Welt, C. K., Rebar, R. W., et al. (2007). The FMR1 premutation and reproduction. *Fertility and Sterility*, *87*, 456–465.

Woodruff, T. J. (2011). Bridging epidemiology and model organisms to increase understanding of endocrine disrupting chemicals and human health effects. *The Journal of Steroid Biochemistry and Molecular Biology*, *127*, 108–117.

Wu, X. Y., Li, Z. L., Wu, C. Y., Liu, Y. M., Lin, H., Wang, S. H., et al. (2010). Endocrine traits of polycystic ovary syndrome in prenatally androgenized female Sprague-Dawley rats. *Endocrine Journal*, *57*, 201–209.

Wu, L. L., Norman, R. J., & Robker, R. L. (2011). The impact of obesity on oocytes: Evidence for lipotoxicity mechanisms. *Reproduction, Fertility, and Development*, *24*, 29–34.

Yamamoto, M., Shirai, M., Sugita, K., Nagai, N., Miura, Y., Mogi, R., et al. (2003). Effects of maternal exposure to diethylstilbestrol on the development of the reproductive system and thyroid function in male and female rat offspring. *The Journal of Toxicological Sciences*, *28*, 385–394.

Yao, H. H. (2005). The pathway to femaleness: Current knowledge on embryonic development of the ovary. *Molecular and Cellular Endocrinology*, *230*, 87–93.

Zachos, N. C., Billiar, R. B., Albrecht, E. D., & Pepe, G. J. (2002). Developmental regulation of baboon fetal ovarian maturation by estrogen. *Biology of Reproduction*, *67*, 1148–1156.

Zama, A. M., & Uzumcu, M. (2010). Epigenetic effects of endocrine-disrupting chemicals on female reproduction: An ovarian perspective. *Frontiers in Neuroendocrinology, 31*, 420–439.

Zama, A. M., & Uzumcu, M. (2013). Targeted genome-wide methylation and gene expression analyses reveal signaling pathways involved in ovarian dysfunction after developmental EDC exposure in rats. *Biology of Reproduction, 88*, 52.

Zambrano, E., & Nathanielsz, P. W. (2013). Mechanisms by which maternal obesity programs offspring for obesity: Evidence from animal studies. *Nutrition Reviews, 71*(Suppl. 1), S42–S54.

Zeleznik, A. J. (2001). Follicle selection in primates: "Many are called but few are chosen". *Biology of Reproduction, 65*, 655–659.

Zeleznik, A. J. (2004). The physiology of follicle selection. *Reproductive Biology and Endocrinology, 2*, 31.

Zhang, T., Li, L., Qin, X. S., Zhou, Y., Zhang, X. F., Wang, L. Q., et al. (2014). Di-(2-ethylhexyl) phthalate and bisphenol A exposure impairs mouse primordial follicle assembly in vitro. *Environmental and Molecular Mutagenesis, 55*, 343–353.

Zhang, H., Zheng, W., Shen, Y., Adhikari, D., Ueno, H., & Liu, K. (2012). Experimental evidence showing that no mitotically active female germline progenitors exist in postnatal mouse ovaries. *Proceedings of the National Academy of Sciences of the United States of America, 109*, 12580–12585.

Zhao, H., Lv, Y., Li, L., & Chen, Z. J. (2016). Genetic studies on polycystic ovary syndrome. *Best Practice & Research. Clinical Obstetrics & Gynaecology, 37*, 56–65.

Zhou, W., Liu, J., Liao, L., Han, S., & Liu, J. (2008). Effect of bisphenol A on steroid hormone production in rat ovarian theca-interstitial and granulosa cells. *Molecular and Cellular Endocrinology, 283*, 12–18.

FURTHER READING

Padmanabhan, V., Cardoso, R. C., & Puttabyatappa, M. (2016). Developmental programming, a pathway to disease. *Endocrinology, 157*, 1328–1340.

Richards, J. S., & Pangas, S. A. (2010). The ovary: basic biology and clinical implications. *Journal of clinical investigation, 120*, 963–972.

CHAPTER FIFTEEN

The Ovarian Hyperstimulation Syndrome

Zeev Blumenfeld[1]
Reproductive Endocrinology, Faculty of Medicine, Technion—Israel Institute of Technology, Haifa, Israel
[1]Corresponding author: e-mail addresses: bzeev@technion.ac.il; z_blumenfeld@rambam.health.gov.il

Contents

Abstract

The ovarian hyperstimulation syndrome (OHSS) is an iatrogenic syndrome, which may, infrequently, become severe and evenly fatal. Until the last decade, the worldwide popularity of assisted reproductive technology (ART) and ovulation induction/controlled ovarian hyperstimulation employed in ART for either in vivo fertilization or in vitro fertilization, in the last two decades, has been accompanied by an increase in the cases of OHSS. In the last decade, the substitution of the human chorionic gonadotropin trigger with the gonadotropin-releasing hormone agonist trigger has decreased the incidence of this syndrome but did not eliminate it completely. We summarize here the classification, pathophysiology, and treatment of the OHSS.

Vitamins and Hormones, Volume 107
ISSN 0083-6729
https://doi.org/10.1016/bs.vh.2018.01.018

1. INTRODUCTION

The ovarian hyperstimulation syndrome (OHSS) is a syndrome comprising marked ovarian enlargement, high concentration of sex steroids, and extravascular exudate accumulation (Golan, Ron-El, et al., 1989; Navot, Bergh, et al., 1992, 1996; Rabau, Serr, et al., 1967; Schenker, 1993; Schenker & Weinstein, 1978). This exudate is generated by an increased vascular permeability and massive shift of fluid and protein from the intravascular fluid and protein from the intravascular compartment to the peritoneal cavity and other compartments (Navot et al., 1992). The resulting depletion in intravascular volume may lead to hemoconcentration, decreased perfusion of many vital organs, and propensity for thromboembolic phenomena. These changes usually accompany exaggerated ovarian steroidogenesis, secretion of vascular endothelial growth factor (VEGF), ovarian enlargement, and potentially multiple system failure, including renal, hepatic, and respiratory dysfunctions.

The worldwide popularity of assisted reproductive technology (ART) and ovulation induction/controlled ovarian hyperstimulation (COH) employed in ART for either in vivo fertilization or in vitro fertilization (IVF), in the last two decades, has been accompanied by an increase in the cases of OHSS (Blankenstein, Shalev, et al., 1987; Navot, Bergh, et al., 1996).

The incidence of severe OHSS ranges between 0.1% and 2%, and that of moderate OHSS has been reported to range between 3% and 7% of the women undergoing ovulation induction and COH by human menopausal gonadotropins (hMGs) and hCG (Navot et al., 1996). In the last two decades, the substitution of hCG by GnRH agonists (GnRHas), as triggering of final follicular maturation, before ovulation and follicular aspiration, has significantly decreased the incidence and severity of OHSS in clinical practice (Dosouto, Haahr, & Humaidan, 2017; Humaidan et al., 2016; Thomsen & Humaidan, 2015). According to the latest European Society of Human Reproduction and Embryology report, the incidence of OHSS ranges from 0.18% to 1.40% in European countries (Thomsen & Humaidan, 2015). However, OHSS is still subject to substantial underreporting. New triggering protocols using GnRHa have shown to be similar to the gold standard hCG trigger with regard to the reproductive outcome, but with a significant decrease in OHSS (Thomsen & Humaidan, 2015). More recently, promising results were published for the use of kisspeptin to induce final oocyte maturation (Kasum et al., 2017). Although until now no study has been reported in OHSS prone patients, one may speculate that the risk of OHSS might be even further reduced after kisspeptin trigger (Thomsen & Humaidan, 2015). It seems that GnRHa trigger is currently the best existing tool for the prevention of OHSS while maintaining acceptable reproductive outcomes (Thomsen & Humaidan, 2015).

2. CLASSIFICATION OF OHSS

Historically, OHSS was classified into three categories and six grades corresponding to mild, moderate, or severe forms (Golan et al., 1989; Humaidan et al., 2016; Navot et al., 1992, 1996; Zivi, Simon, & Laufer, 2010). The mild form is characterized by high serum 17-estradiol (E_2) and progesterone (P_4), namely "chemical OHSS." The ovaries are only mildly enlarged, less than 5 cm by diameter. In the moderate form, bloating, abdominal distension, and discomfort accompany nausea, vomiting, and diarrhea. The ovaries are more enlarged, between 5 and 12 cm in diameter. In the severe form, the ovarian diameters are frequently larger than 12 cm, usually surrounded by massive ascites, less frequently with hydrothorax and/or pericardial effusion, and not infrequently, renal dysfunction and hemoconcentration may occur.

More recent modifications to the original classification suggested including mild ascites detected by transvaginal sonography (TVS) and added a critical, life-threatening form of the syndrome to the severe form. In the revised classification, generalized edema (anasarca) and liver dysfunction (Balasch, Carmona, et al., 1990; Humaidan et al., 2016; Ryley, Forman, et al., 1990; Sueldo, Price, et al., 1988; Younis, Zeevi, et al., 1988; Zivi et al., 2010) are considered additional signs of severe OHSS. Adult respiratory distress syndrome (Zosmer, Katz, et al., 1987), tense ascites, severe hemoconcentration with an hematocrit above 55%, and leukocytosis above 35,000 cell/mm^3 are signs of the critical, life-endangering form of OHSS (Humaidan et al., 2016; Ryley et al., 1990; Sueldo et al., 1988; Younis et al., 1988; Zivi et al., 2010).

Since follicular aspiration for egg retrieval, in IVF/ART, may affect ovarian size, this parameter has been encountered as nondefinitive for the classification of the severity of OHSS accompanying ART/IVF. Indeed, in cases of OHSS after IVF, massive ascites may frequently exist in the presence of only moderate ovarian enlargement (<10 cm) (Humaidan et al., 2016; Navot et al., 1996; Zivi et al., 2010).

3. PATHOPHYSIOLOGY

Although the exact pathophysiological mechanisms have not been unequivocally elucidated, the underlying mechanism responsible for the clinical manifestations of OHSS appears to be an increase in the capillary permeability of various blood vessels (Kwik & Maxwell, 2016; Navot et al., 1996; Nelson, 2017; Polishuk & Schenker, 1969; Rizk, Aboulghar, Smitz, & Ron-El, 1997). This increased vascular permeability brings about a rapid and intensive shift of fluid from the intravascular space to the "third compartment," associated with acute hemodynamic abnormalities and a wide spectrum of clinical presentations (Kwik & Maxwell, 2016; Nelson, 2017; Rizk et al., 1997; Tollan, Holst, et al., 1990). Several systems and mediators have been suggested to be involved in the pathophysiological process of the OHSS (Delbaere, Bergmann, et al., 1997; Kwik & Maxwell, 2016; Miller et al., 2016; Navot et al., 1996; Nelson, 2017; Orvieto & Ben-Rafael, 1998; Rizk et al., 1997):

a. Ovarian renin–angiotensin system (OVRAS).
b. Prostaglandins.
c. Histamines.
d. Cytokines.
e. VEGF/vascular permeability factor (VPF).
f. Pigment epithelium-derived factor (PEDF) (Fig. 1).

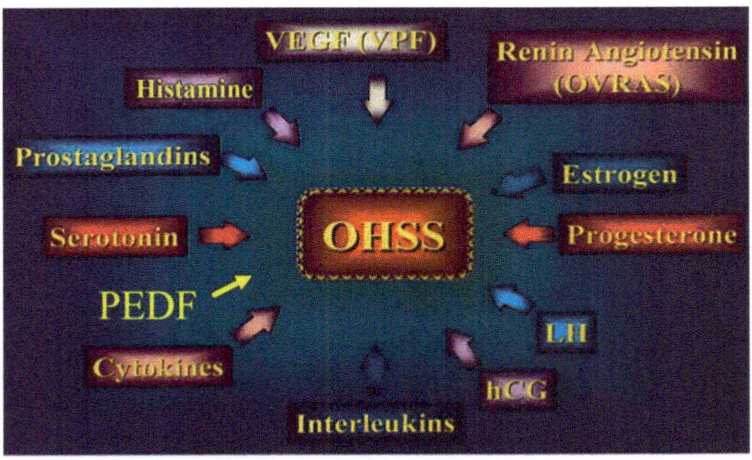

Fig. 1 The multiple mediators associated with OHSS.

4. THE OVARIAN RENIN–ANGIOTENSIN SYSTEM

Renin and its high molecular weight precursor, prorenin, circulate in plasma, the latter at levels up to 10-folds higher than active renin (Glorioso, Atlas, et al., 1986; Navot et al., 1996; Palumbo, Ávila, & Naftolin, 2016; Sealey, Atlas, et al., 1980, 1985). In addition to its traditional and abundant origin in the juxtaglomerular cells in the wall of the afferent arterioles in the kidney, prorenin is also generated by the ovary, placenta, and other tissues. Angiotensinogen is cleaved by renin, an acid protease, to form angiotensin I (AI). This decapeptide, AI, is cleaved by angiotensin-converting enzyme to angiotensin II (AII). Sealey et al. (1980, 1985) described the synchronized periodicity of prorenin with LH and P_4 in the menstrual cycle. In parallel to the midcycle LH surge, there is a significant increase in prorenin, and a second increase parallel to the midluteal P_4, the latter accompanied also by a rise in active renin (Palumbo et al., 2016; Sealey et al., 1980, 1985). Plasma prorenin also increases after hCG administration in hMG/hCG ovulation induction and COH, and follicular fluid (FF) prorenin levels in such patients are about 10-folds higher than plasma prorenin concentrations after hMG/hCG stimulation (Bergh & Navot, 1992; Fernandez, Tarlatzis, et al., 1985; Fernandez, Twickler, et al., 1985; Palumbo et al., 2016). The midcycle prorenin rise is related to the number of preovulatory follicles detected by TVS (Blumenfeld, Yoffe, & Bronshtein, 1991), and the degree of increase in early gestational prorenin is correlated to the number of corpora lutea paralleling the increase in endogenous hCG secretion. The ovarian source of systemic prorenin is substantiated by the lack of increase in prorenin in pregnancies in patients with premature ovarian failure, achieved by egg donation (Navot et al., 1996; Palumbo et al., 2016).

Navot and coworkers (Navot et al., 1996; Navot, Margalioth, et al., 1987) found a direct correlation between plasma renin concentrations and the degree of OHSS, hypothesizing that through AII secretion, the OVRAS plays a cardinal role in the pathophysiology of OHSS. Ong, Eisen, et al. (1991) confirmed this hypothesis by finding elevated plasma renin activity and plasma aldosterone concentrations in women who developed OHSS after COH, persisting in spite of significant infusion of plasma expanders.

Although AII is primarily a pressor agent, its ability to increase the vascular permeability and to induce angiogenesis suggests an important pathophysiological role in the accumulation of postovulatory peritoneal fluid in spontaneous cycles and in OHSS (Palumbo et al., 2016). In addition to increasing the capillary permeability, AII also increases the capillary filtration pressure by constricting the postcapillary venules. A possible contractile response of the larger arterioles to AII may cause the separation of endothelial cells, leading to a further increase in vascular permeability (Palumbo et al., 2016). The described mechanisms may act in concert to cause a marked decrease in blood volume and an increase in the extravascular fluid and flow of lymphatic fluid (Palumbo et al., 2016).

Thus, AII as an intraovarian peptide (Culler, Tarlatzis, et al., 1986; Palumbo et al., 2016), able to promote angiogenesis (Fernandez, Tarlatzis, et al., 1985; Fernandez, Twickler, et al., 1985), increase vascular permeability (Robertson & Khairallah, 1972), and stimulate the release of prostaglandins (Yoshimura, Karube, et al., 1993), is one of the factors most probably involved in the pathophysiological mechanisms of OHSS (Palumbo et al., 2016). However, the ovarian vs renal origin of the elevated plasma renin activity during OHSS is controversial (Delbaere et al., 1997). On one hand, the elevated plasma renin activity was not normalized by plasma volume expansion during OHSS, further suggesting a nonrenal origin (Ong et al., 1991); on the other hand, it correlated well with the plasma levels of other volume-dependent vasoactive substances such as norepinephrine and antidiuretic hormone, supporting a renal origin (Balasch, Arroyo, et al., 1994).

Delbaere, Bergmann, et al. (1994) have shown a marked elevation of AII immunoreactivity in the ascites of three women who developed severe OHSS, strongly suggesting an ovarian origin for the ascites. The demonstration of a very high renin concentration, the majority of which was prorenin, in the peritoneal fluid of severe OHSS, Rosenberg, Mckenzie, et al. (1994) lended further support to the theory of a marked stimulation of the OVRAS in OHSS (Palumbo et al., 2016). Delbaere et al. (1997) have confirmed their preliminary observation (Delbaere et al., 1994), supporting an ovarian origin of the AII in the ascites of OHSS. The majority of ascitic AII immunoreactivity was generated by AII, most of which (>70%) attributed to a bioactive peptide (Delbaere et al., 1997; Palumbo et al., 2016). This study (Delbaere et al., 1997) has also suggested that ascitic AII does not contribute significantly to plasma AII, and that the latter mainly reflects activation of the renal renin–angiotensin system. Since the highest levels of AII immunoreactivity were found in the peritoneal fluid of pregnant women with OHSS, one may assume and speculate that pregnancy acts synergistically with gonadotropins to maximally stimulate the OVRAS, which in turn could participate either directly or indirectly in the pathogenesis of the OHSS and can reflect its severity (Palumbo et al., 2016). Although increased levels of AII are consistently present in the ascites of OHSS patients, they are not sufficient for the induction of ascites. However, AII may act as a mediator of ascites formation, synergistically with other factors such as VEGF (McClure, Healy, et al., 1994) or cytokines (Revel, Barak, et al., 1996) that have also been involved in the pathophysiological mechanism of OHSS (Brinsden, Wada, et al., 1995; Morris & Paulson, 1994; Palumbo et al., 2016).

4.1 Prostaglandins

Prostaglandins have been traditionally put forward as key mediators in the pathophysiology of the OHSS (Kumar, Sait, Sharma, & Kumar, 2011; Navot et al., 1996; Schenker & Polishuk, 1976a, 1976b, 1976c, 1976d). Prostaglandins may induce the OHSS either directly or indirectly, through their enhanced production of AII. The severely increased pharmacological levels of estrogens may stimulate prostaglandin production (Kumar et al., 2011). Although earlier studies (Erlik, Naot, et al., 1979; Katz, Lancet, et al., 1984; Schenker & Polishuk, 1976a, 1976b, 1976c, 1976d) have enthusiastically employed indomethacin, a prostaglandin inhibitor, as a therapeutic measure in cases of severe OHSS, later human trials of indomethacin for the treatment of OHSS have been disappointing (Borenstein, Elhalah, et al., 1989; Kumar et al., 2011). Moreover, despite its ability to decrease ovarian prostaglandin synthesis in rabbits (Pride, Yuen, et al., 1986), indomethacin failed to prevent ascites formation and ovarian enlargement even in pharmacological doses.

4.2 Histamines

Histamines have been also suggested as possible mediators of the OHSS (Delvigne & Rozenberg, 2002; Knox, Dowd, et al., 1975; Kumar et al., 2011). Pride, Ho Yuen, et al. (1984) using a rabbit model evaluated treatment with H_1 receptor blockers in OHSS. Although initial results were encouraging in inhibiting ascites and ovarian cysts formation (Gergly, Paldi, et al., 1976), later studies were disappointing. Erlik et al. (1979) did not find any difference in histamine levels in rabbits with OHSS or their controls. Similarly, Zaides, Friedman, et al. (1983), using two different H_1 receptor antagonists, could not find any clues suggesting a role to histamine or serotonin in the pathophysiological mechanism of the OHSS. Therefore, currently histamines are not believed to play a cardinal role in the pathogenesis of OHSS (Delvigne & Rozenberg, 2002; Kumar et al., 2011).

4.3 Cytokines

A promising and exiting area in gonadal physiology suggesting new possible mediators of OHSS appears to be the interaction between the immune system and the ovary (Kumar et al., 2011; Navot et al., 1996; Orvieto, Achiron, Margalit, & Ben-Rafael, 1998, 2014; Shi et al., 2015). Thus, the ovarian function in general and the ovulatory process more specifically may be regulated and mediated by various cytokines. The cytokines are polypeptides synthesized by activated immunoregulatory and mesenchymal cells. The interleukins have been more particularly studied (Adashi, 1989, 1996; Kumar et al., 2011; Shi et al., 2015) as possible modulators of various intraovarian regulatory mechanisms. Although macrophages were found in the ovarian stroma, other components of the immune system undergo significant variations of cell content throughout the different phases of the normal menstrual cycle (Hume, Halpin, et al., 1984; Kumar et al., 2011; Shi et al., 2015).

In the late follicular phase, mast cells massively infiltrate the ovarian stroma, undergoing degranulation in response to the midcycle gonadotropin surge (Nakamura, Smith, et al., 1987). This degranulation is followed by migration of polymorphonuclear (PMN) cells, "T" lymphocytes, and macrophages. Indeed, tissue macrophages and monocytes have been identified in 5%–15% of the cell population aspirated with the FF during egg retrieval for IVF/ART (Loukides, Loy, et al., 1990). This apparent acute inflammatory reaction is probably secondary to the increased follicular neovascularization occurring during ovulation, associated with the secretion and/or release of follicular chemotactic factors. Furthermore, the oocyte itself may also produce cytokines (Zolti, Ben-Rafael, et al., 1991).

Interleukin-1 (IL-1) is an activated macrophage-derived cytokine previously known as lymphocyte-activating factor. Various tissues such as epithelial, vascular, lymphoid, epidermal, and granulosa cells produce IL-1 in addition to its classical site of production, the macrophage. IL-1 exists in two structurally related forms, IL-1α and IL-1β, the latter being biologically more potent. Both these IL-1 suppress the in vitro luteinization of porcine and murine granulosa cells (Gottschall, Katsuura, et al., 1989; Mori, Masatsune, et al., 1989; Shi et al., 2015), yet stimulating their proliferation. Moreover, IL-1β can act synergistically with hCG to stimulate in vitro thecal progesterone production. On the other hand, IL-1β is a potent stimulator of prostaglandin production by bovine luteal cells (Adashi, 1996; Shi et al., 2015).

Hurwitz, Loukides, et al. (1992) have shown that the intraovarian IL-1 system is highly compartmentalized and can modulate the intraovarian regulation of cell differentiation and folliculogenesis. In addition to IL-1α and IL-1β, a third actor in the IL-1 gene family is the high-affinity IL-1 receptor antagonist inhibiting both IL-1 and is characteristic of the tight control over the production and activity exerted on IL-1 (Dinarello, 1991; Revel et al., 1996; Shi et al., 2015). During inflammation, the three members of the IL-1 gene family are expressed and their products compete for IL-1 receptor occupancy and promote PMN cell accumulation. When experimentally administered, IL-1 induces many local and systemic effects, including prominent vascular leakage (Dinarello, 1991). In addition to the demonstration of IL-1, its receptor and receptor antagonist in the ovary, the expression of IL-1β is upregulated by gonadotropins, suggesting a physiological role in ovulation and a possible pathophysiologic mediator mechanism for hMG-/hCG-associated or induced increased ovarian vascular permeability (Revel et al., 1996).

Another cytokine, IL-6, was initially described as a type of interferon purified with hepatocyte and hybrid growth factor (Adashi, 1996; Shi et al., 2015). The mRNA of IL-6 is produced in vivo in two self-limiting angiogenetic processes: the neovascularization associated with folliculogenesis and the formation of the capillary network in the uterine decidua after embryo nidation (Motro, Itin, et al., 1990). Geva, Lessing, et al. (1997) attempted to determine the possible predictive role of IL-2, IL-6, and tumor necrosis factor-α (TNF-α) in the development of OHSS. They have found significantly higher levels of IL-6 in the FF of those IVF patients who later developed OHSS as compared to non-OHSS controls, either high or low responders. It seems therefore that the FF concentration of IL-6 on the day of egg retrieval may positively predict the development of OHSS in IVF/ART patients (Shi et al., 2015). No correlation was found (Geva et al., 1997) between the FF concentrations of IL-2, IL-6, and TNF-α and serum E_2 levels or the number of retrieved oocytes. Therefore, cryopreservation of all embryos without fresh embryo transfer (ET) is recommended in IVF cycles where high levels of IL-6 in the FF are predictive of OHSS. Alternatively limiting the number of transferred embryos to one, in all other IVF cycles, may minimize the risk of OHSS, in addition to minimizing multiple gestations.

Orvieto and Ben-Rafael (1996, 1998) have suggested a role for IL-2 as a possible culprit of OHSS. Orvieto et al. (2014) have suggested that the normal negative effect of hCG on peripheral blood mononuclear cells' IL-2 production is lost in patients who suffered OHSS and the resulting dysregulation of IL-2 expression and suppressor of cytokine signaling (SOCS) activation may be a possible mediator of OHSS. These investigators (Orvieto et al., 2014) have therefore speculated that patients who are prone of developing OHSS may have a genetic paradoxical immune response to hCG, with IL-2 instead of SOCS-1 dominance, leading to systemic inflammatory response and consequently developing OHSS.

Revel et al. (1996) have measured the peritoneal fluid concentrations of cytokines and nitrites in women with severe OHSS. The OHSS patients had significantly higher peritoneal fluid levels of IL-6 (3523 vs 30 pg/mL), TNF-α (14 vs 4.2 pg/mL), and IL-8 (1695 vs 900 pg/mL) vs controls. In the serum, only IL-6 levels were significantly higher (375 vs 11 pg/mL) (Revel et al., 1996). On the contrary, nitrite levels were significantly lower in the ascitic fluid of OHSS patients (0.5 vs 34 nmol/mL). IL-1 levels were higher and IL-1 receptor antagonist levels were lower in OHSS patients, suggesting potentially increased biological potency of IL-1 (Revel et al., 1996). Each of the cytokines TNF-α, IL-1, IL-2, IL-6, and IL-8 has been shown to induce changes in vascular permeability, speculating that their combined effects may be additive or even synergistic (Andus, Gross, et al., 1992; Burke-Gaffney & Keenan, 1993; Elliot & Finn, 1993; Funke, Prummer, et al., 1994; Shi et al., 2015). High levels of IL-1, TNF-α, and IL-6 were found in the ascitic fluid of patients with ovarian carcinoma (Kutteh & Kutteh, 1992). Whereas IL-6, IL-8, and TNF-α levels were significantly increased in the peritoneal fluid of OHSS patients as compared to controls, the IL-1 levels were nonsignificantly increased (Revel et al., 1996). IL-6, a ubiquitous cytokine produced by many cells including endothelial cells, macrophages, lymphocytes, epithelial cells, and fibroblasts, is induced by shock-triggering cytokines such as IL-1 and TNF-α. Among its numerous effects, IL-6 is involved in angiogenesis and hyperpermeability. Several studies (Friedlander, de Mola, et al., 1995; Geva et al., 1997; Orvieto & Ben-Rafael, 1998) have confirmed the finding of increased IL-6 levels in OHSS ascitic fluid and serum.

TNF-α, previously known as cachectin, is a 17-kDa polypeptide produced mainly by stimulated macrophages (Revel et al., 1996). It is an extremely potent cytokine, acting as an endogenous mediator of inflammatory, immune, and host defense mechanisms. More than any other cytokine, TNF-α has been linked to pathophysiological processes characterizing various acute human diseases and is a recognized mediator of vascular permeability (Chistyakova, Remizova, Gazieva, & Chermyaninova, 2014).

Whereas the localized, restricted, effects of TNF-α may be beneficial, the systemic and general release of TNF-α may be devastating and rapidly lethal, such as in the capillary leak syndrome (Chistyakova et al., 2014). In the ovary, TNF-α stimulates cell growth, participates in folliculogenesis, and may even promote epithelial ovarian cancer (Terranova, Sancho-Tello, et al., 1993). The ovary can synthesize TNF-α, the compound was detected in the oocyte, granulosa cells, and ovarian mononuclear cells, and high-affinity receptors were found in granulosa cells. The ovarian production and release of TNF-α may be stimulated by gonadotropins (Chistyakova et al., 2014). Thus, not only the existence, synthesis, and reception mechanisms for TNF-α and IL-1 were elucidated in the ovary but also its physiological and/or pathophysiological mechanisms of stimulation, suggesting another possible link between hMG/hCG stimulation and OHSS-associated ovarian vascular hyperpermeability (Barak, Mordel, et al., 1992; Chistyakova et al., 2014).

Interleukin 8 is a neutrophil chemotactic factor produced by lipopolysaccharide-stimulated human blood monoleukocytes (Chen et al., 2010; Revel et al., 1996). In vitro, IL-8 stimulates chemotaxis, augments lysosomal enzyme release, upregulates Mac-1, and complements receptor expression. In vivo, IL-8 induces PMN chemotaxis and increases vascular permeability (Chen et al., 2010; Larsen, Anderson, et al., 1989). Since massive and significant changes occur in the distribution of white blood cells around and following ovulation and because some of the elements responsible for hyperpermeability may be induced by IL-8, a potential role for this cytokine in ovarian physiology and in the OHSS mechanism has been put forward (Chen et al., 2010; Revel et al., 1996).

Endothelium-derived relaxing factor has been identified as NO. This important messenger molecule can mediate many and diverse functions including vasodilatation, neurotransmission, and cytotoxic activities (Ekerhovd et al., 2001). Different cells such as endothelial cells, macrophages, and neurons produce NO from original. In addition to its ubiquitous distribution in various tissues and organs, NO has been shown to be involved in physiological mechanisms in the ovary (Ben-Shlomo, Kokia, Jackson, Adashi, & Payne, 1994; Ekerhovd et al., 2001), and to modulate vascular permeability (Kubes & Granger, 1992). The continuous release of NO from the endothelium is probably an intrinsic modulator of blood perfusion in various tissues, possibly affecting the interactions between inflammatory cells and vascular endothelium (Ekerhovd et al., 2001; Gardiner, Compton, et al., 1990). Inhibition of NO production by vascular endothelium leads to a reversible increase in microvascular efflux (Ekerhovd et al., 2001;

Kubes & Granger, 1992). Thus, NO preserves endothelial function and inhibition of NO synthase promotes leukocytes and platelet adhesion, bringing about endothelial dysfunction and increased permeability (Ekerhovd et al., 2001; Kubes & Granger, 1992). Endothelial leukocyte adhesion molecules are rapidly generated by endothelial cells upon stimulation with TNF-α and IL-1, significantly mediating PMN adhesion (Revel et al., 1996). Since the continuous local release of NO may serve as a possible scavenger mechanism of small amounts of superoxide released by the endothelial cells, inhibition of NO synthase may lead to the accumulation of superoxide radicals, which may increase microvascular permeability, an alternative possible mechanism in OHSS (Ekerhovd et al., 2001; Kubes & Granger, 1992; Revel et al., 1996).

Indeed, the regulation of NO synthase and IL-1 system may be closely related. TNF-α has been also shown to downregulate endothelial NO synthase mRNA by shortening its half-life (Yoshizumi & Perrella, 1993). On the contrary, another study (Burke-Gaffney & Keenan, 1993) has shown that inhibition of vascular permeability caused by TNF-α is not mediated by NO. Revel et al. (1996) were the first to demonstrate a significant decrease in nitrite level in the ascitic fluid of patients with severe OHSS, in keeping with a previous report linking low nitrites to increased vascular permeability (Filep, Foldes-Filep, et al., 1993).

Several explanatory mechanisms have been put forward to the observation of very low nitrites in OHSS:

(i) decreased NO synthase production in response to increased TNF-α concentrations, which have been shown to destabilize NO synthase mRNA (Nathan & Xie, 1993),

(ii) similar to preeclampsia (Seligman, Buyon, et al., 1994), an endogenous inhibition to NO synthesis may exist, and

(iii) possibly, decreased availability of L-arginine or other cofactors of NO synthase may result in decreased NO secretion (Revel et al., 1996).

The increased levels of inflammatory cytokines (IL-1, IL-4, IL-6, IL-8, and TNF-α) may mediate the severely increased capillary permeability in OHSS, possibly through the NO and/or the VEGF system (Ekerhovd et al., 2001; Orvieto et al., 1998; Revel et al., 1996).

4.4 Vascular Endothelial Growth Factor/Vascular Permeability Factor (VEGF/VPF)

Increased capillary permeability induced by FF and peritoneal fluid from women undergoing COH for IVF is due in part to VEGF, suggesting an ovarian origin for the OHSS (Goldsman, Pedram, et al., 1995; Kransnow, Berga, Guzick, Zeleznik, & Yeo, 1996; Levin et al., 1998). The VEGF/VPF involvement in the pathogenesis of OHSS offers an attractive theory, but clinical applications of these findings were not available until recently (Abramov, Barak, et al., 1997; Dourron & Williams, 1996; Kransnow et al., 1996; Levin et al., 1998). The cytokine VEGF (VPF) is a heparin-binding glycoprotein possessing potent angiogenic, endothelial cell-specific mitogenic, and vascular permeability-enhancing activities (Abramov et al., 1997; Ferrera, Houck, et al., 1991; Ferrara & Davis-Smyth, 1997; Kransnow et al., 1996; Levin et al., 1998). VEGF plays important roles in inflammation and in normal and pathological angiogenesis, a process that is associated with wound healing, embryogenesis, growth, and metastasis of neoplasms (Abramov et al., 1997; Levin et al., 1998). A few preliminary reports have suggested a possible role for VEGF/VPF in the pathogenesis of capillary leakage in the OHSS (Abramov et al., 1997; Kransnow et al., 1996; Levin et al., 1998; McClure et al., 1994; Neulen et al., 1995; Robertson, Selleck, et al., 1995). VEGF/VPF acts directly on endothelial cells to induce proliferation and angiogenesis, expressed in vitro by luteinized human granulosa cells (Ferrara & Davis-Smyth, 1997, Ferrara, Houck, et al., 1992), and is demonstrated in biologically significant concentrations in FF (Kransnow et al., 1996; Levin et al., 1998). High concentrations of VEGF/VPF increase vascular permeability and lead to extravasation of plasma, followed by pronounced hemoconcentration (Kransnow et al., 1996; Levin et al., 1998; Senger, Galli, et al., 1983). Similarly, high concentrations of VEGF/VPF were found in the ascitic fluid from patients with severe OHSS (Abramov et al., 1997; Levin et al., 1998). In addition to the demonstration that VEGF/VPF may be an important capillary permeability factor in the ascites formation in OHSS, 70% of the capillary permeability activity in OHSS was neutralized by recombinant human VEGF/VPF antiserum (Abramov et al., 1997; Levin et al., 1998; McClure et al., 1994). A humanized version of a high-affinity anti-VEGF/VPF monoclonal antibody, which retains the same affinity and efficacy as the original murine antibody, has been generated and offers clinical therapeutic endeavors for treatment of severe cases of OHSS (Ferrara & Davis-Smith, 1997; Levin et al., 1998; Meldrum, 2012). Abramov et al. (1997) have detected high levels of VEGF/VPF in the plasma and ascitic fluid of seven patients with severe OHSS upon hospitalization, decreasing significantly in the plasma along clinical resolution of the syndrome. The significant correlation between VEGF/VPF levels and certain biological characteristics of capillary leakage and hemoconcentration, such as elevated hematocrit and white blood cells count, is in keeping with the suggested theory of VEGF/VPF in the pathophysiological mechanism of increased ovarian vascular permeability in OHSS. The peritoneal fluid levels of VEGF/VPF were higher than those in the plasma, suggesting a local secretion from the ovaries and/or mesothelial surfaces or an active transport of this factor through ovarian and/or mesothelial vessels (Abramov et al., 1997).

4.5 Pigment Epithelium-Derived Factor

In the last few years, PEDF has emerged as a new factor and possible mediator of OHSS pathophysiology (Bar-Joseph, Ben-Ami, Ron-El, Shalgi, & Chuderland, 2016; Chuderland et al., 2013; Miller et al., 2016; Miller, Chuderland, Ron-El, Shalgi, & Ben-Ami, 2015).

PEDF is a secreted 50-kDa glycoprotein that belongs to the non-inhibitory members of the serine protease inhibitors, being a natural angiogenesis inhibitor with neurotrophic and immune-modulating properties (Craword, Fitchev, Veliceasa, & Volpert, 2013). PEDF inhibits the stimulatory activity of potent proangiogenic factors and can inhibit VEGF effect on endothelial cell activity (Becerra & Notario, 2013). Shalgi's group (Bar-Joseph et al., 2016; Chuderland et al., 2013; Miller et al., 2016, 2015) has found that PEDF is expressed in human ovaries, produced and secreted by granulosa cells in a hormonally dependent manner. PEDF has an anti-angiogenic effect, as shown by its in vitro ability to inhibit proliferation, migration, and tube formation of human umbilical vein endothelial cells (Ninio-Many, Grossman, Shomron, Chuderland, & Shalgi, 2013). hCG can decrease PEDF expression, both in vivo and in vitro, and the expression of PEDF is opposite to VEGF expression (Ninio-Many et al., 2013). It seems, therefore, that the balance between PEDF and VEGF regulates the physiological ovarian function, and the "fine tuning" between these two antagonistic regulators may be modulated, at least in part, by hCG or LH (Chuderland et al., 2013). The group of Ruth Shalgi has provided evidence, in several experiments, that correlates the downregulation of OHSS symptoms caused by PEDF with both angiogenic and inflammatory pathways (Bar-Joseph et al., 2016; Chuderland et al., 2013; Miller et al., 2016, 2015). These investigators (Miller et al., 2016) have demonstrated that expression of IL-6 and IL-8 increased in lysophosphatidic acid (LPA)-stimulated human primary granulosa cells ($P < 0.01$) as a model of OHSS. Coadministration of recombinant PEDF decreased the expression of

LPA-induced IL-6 and IL-8 mRNA and protein by four- and two- to five-fold, respectively (Miller et al., 2016). Similarly, IL-8-stimulated human granulosa cells augmented the expression of VEGF mRNA, and costimulation with hCG induced a significantly higher increase in the expression of VEGF mRNA ($P < 0.001$), which was significantly inhibited by administration of PEDF (Miller et al., 2016). In vivo, injection of PEDF to OHSS-induced mice decreased the ovarian expression of IL-6 ($P < 0.01$) and ameliorated the severity of OHSS (Miller et al., 2016). Thus, PEDF seems to play a dual role in OHSS pathophysiology; it can antagonize both the inflammatory and the angiogenic mechanisms generating OHSS (Bar-Joseph et al., 2016; Chuderland et al., 2013; Miller et al., 2016, 2015).

4.6 The Origin of Ascites in OHSS

Not only the pathophysiological mechanisms of the OHSS are not completely clear and unequivocally elucidated; even the main source and origin of the ascitic fluid in this syndrome are still equivocal. According to Schenker (1997), the sudden body fluid shift is generated by increased capillary permeability, especially of the ovarian vessels, as has been previously demonstrated in an experimental study using intravenous Evans blue and pentamine sky blue dyes (Schenker & Polishuk, 1976a, 1976b, 1976c, 1976d). The volume of the peritoneal fluid is correlated to cyclic ovarian activity; it is consistently low in normal men, postmenopausal women, and women on oral contraceptives (Donnez, Langerock, et al., 1982; Koninckx, Heyns, et al., 1980; Koninckx, Renaer, et al., 1980). Since the production of the peritoneal fluid is independent of tubal patency and uterine or tubal presence, its presumed origin is either ovarian or peritoneal. The marked degree of peritoneal permeability seen in OHSS is reflected to a much lesser degree in the variations of peritoneal fluid volume observed in normal cycling women (Navot et al., 1996). Furthermore, the concentrations of sex steroids in the peritoneal fluid vary throughout the menstrual cycle. After ovulation, E_2 and P_4 concentrations in the peritoneal fluid are significantly higher than plasma levels (Koninckx, Heyns, et al., 1980; Koninckx, Renaer, et al., 1980). Before vascularization, the secretion of the early corpus luteum is preferentially directed toward the peritoneal cavity, presumably from the thecal capillaries and the ovulatory ostium. A suggested mechanism for these cyclic peritoneal fluid changes may be related to the marked increase in the capillary permeability induced by intraovarian angiogenesis (Jakob, Jentrsch, et al., 1977). During the first trimester of pregnancy, the concentrations of E_2 and P_4 were significantly higher in the peritoneal fluid than in serum (Delbaere et al., 1997). The peritoneal fluid is an ovarian exudation product (Koninckx, Heyns, et al., 1980; Koninckx, Renaer, et al., 1980). Steroid levels in the peritoneal fluid have not been reported during early pregnancy except for the incidental finding of elevated levels of E_2 and P_4 in the peritoneal fluid of a pregnant woman at 5 weeks' gestation (Maatjuis, Look, et al., 1978). The concentration gradient for E_2 and P_4 between the peritoneal fluid and the serum suggests a direct exchange between the corpus luteum and the peritoneal cavity that may continue during early gestation. Indeed, in early pregnancy, the levels of immunoreactive AII are higher in the peritoneal fluid than in plasma, or in the peritoneal fluid during a normal, nonpregnant ovulatory cycle (Delbaere et al., 1997; Delbaere, Bergmann, et al., 1996). Kim-Bjrklund, Landgren, et al. (1991) have measured the peritoneal fluid volume and levels of steroid hormones and gonadotropins in peritoneal fluid of normal and norethisterone-treated women. Whereas E_2 and P_4 levels were higher in the peritoneal fluid than in plasma throughout the luteal phase, the opposite was detected for gonadotropin concentrations. The latter observation contradicts those of a previous study (Bouckacrt, Evers, et al., 1986), whereby the levels of FSH and LH in the peritoneal fluid, during the luteal phase, were higher than or equal to those in serum. In an attempt to assess the direct ovarian contribution to ascites formation in experimental OHSS in the rabbit, Yarali and Fleige-Zahradka (1993) have microsurgically isolated the ovaries from the peritoneal cavity. This model was a modification of a similar experiment done almost half a century ago (Polishuk & Schenker, 1969), where rabbit's ovaries were extra-peritonealized by subcutaneous displacement. Both studies found no decrease in the ascitic volume in rabbits induced to develop the OHSS by ovarian displacement or enclosure, concluding that the main contribution to the ascites in OHSS is increased transudation across extra ovarian serosal surfaces. If their conclusion is correct, i.e., the main source of ascites is peritoneal and not ovarian, the concentrations of the ovarian secreted hormones would be higher, or similar, in the serum compared to the respective concentrations in the ascitic fluid. Therefore, we measured the concentrations of E_2, P_4, testosterone (T), hCG, prolactin (PRL), insulin-like growth factor-1 (IGF-1), and IGF-binding proteins (IGF-BP_3) in the ascitic fluid aspirated by abdominal paracentesis from 14 patients with severe OHSS for clinical indications such as tense ascites, dyspnea, and/or oliguria (Blumenfeld, Shabadash, & Shen-Or, 1998). Our objective was to deduct from the "concentration cascade" of the various sex hormones on the main secretion origin of the ascitic fluid in these patients with severe OHSS (Blumenfeld et al., 1998). Our working hypothesis was that in case the main contribution is from ovarian vessels, the levels of the gonadal secreted hormones, E_2, P_4, and T, would be higher in the ascitic fluid, whereas the pituitary or chorionic generated hormones would be similar or higher in the serum. The mean ± SEM of the various hormones in the serum and ascitic fluid is summarized in Table 1.

Whereas the mean ± SE concentrations of E_2, P_4, and T were significantly higher in the ascitic fluid than the respective concentrations in serum (Fig. 2), the concentrations of hCG, PRL, IGF-1, and IGF-BP_3 (Fig. 3) were not different between the two compartments. The FSH and LH concentrations were similarly below 1 IU/L in both compartments, not different from each other, respectively.

The significantly higher concentrations of the ovarian-derived sex steroids (E_2, P_4, T) in the ascitic fluid vs serum, unlike the pituitary (FSH, LH, PRL), trophoblast (hCG), or decidual (PRL)-derived, or liver (IGF-1, IGF-BP_3)-generated nongonadal-derived hormones/proteins, point toward an "ovarian leakage" as the main source of peritoneal fluid accumulation in human OHSS. These findings contradict those of previous studies (Yarali & Fleige-Zahradka, 1993) in the rabbit model, suggesting that the main source of ascites in OHSS is from extra gonadal blood vessels.

4.7 Treatment of OHSS

The best treatment of OHSS is its avoidance. Although OHSS can seldom occur in spontaneous pregnancy, the vast majority of OHSS occurs after

Table 1 The Mean ± SEM Concentration of the Various Hormones in Serum and Ascitic Fluid in 14 Women With Severe OHSS

Hormone	Serum	Ascitic Fluid	*P*
E_2 (pmol/L)	14,918 ± 1978	28,552 ± 3412	<0.01
P_4 (nmol/L)	598 ± 98	933 ± 154	<0.05
Testosterone (nmol/L)	7.99 ± 1.7	17.2 ± 1.9	<0.014
hCG (IU/L)	1549 ± 822	599 ± 347	NS
PRL (mU/L)	1070 ± 185	1004 ± 126	NS
IGF-1 (nmol/L)	9 ± 1	9.5 ± 1.5	NS
IGF-BP_3 (g/L)	1.4 ± 0.1	1.45 ± 0.1	NS

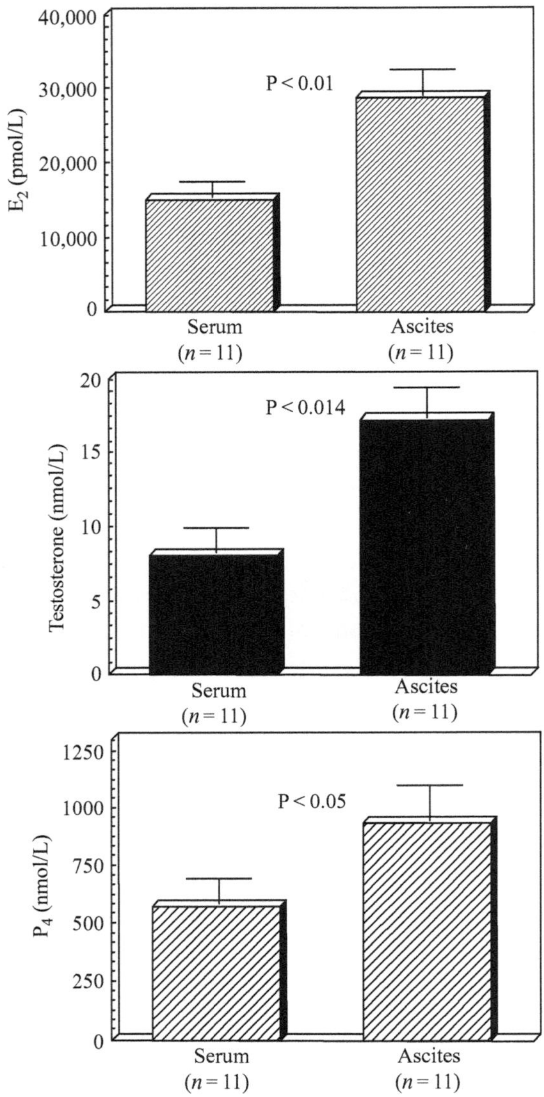

Fig. 2 The significantly higher (mean ± SE) concentrations of the ovarian-derived sex hormones: estradiol (E_2), progesterone (P_4), and testosterone (T) in the ascitic fluid from women with severe OHSS as compared to serum levels.

ovulation induction or controlled ovarian stimulation (COS) for in vivo or in vitro fertilization. For the latter, the last decade protocols using GnRH antagonists for prevention of premature LH rise, and GnRHa's as triggering final follicular maturation have significantly decreased the incidence of OHSS, although not completely eliminating it (Blumenfeld, 2015a, 2015b;

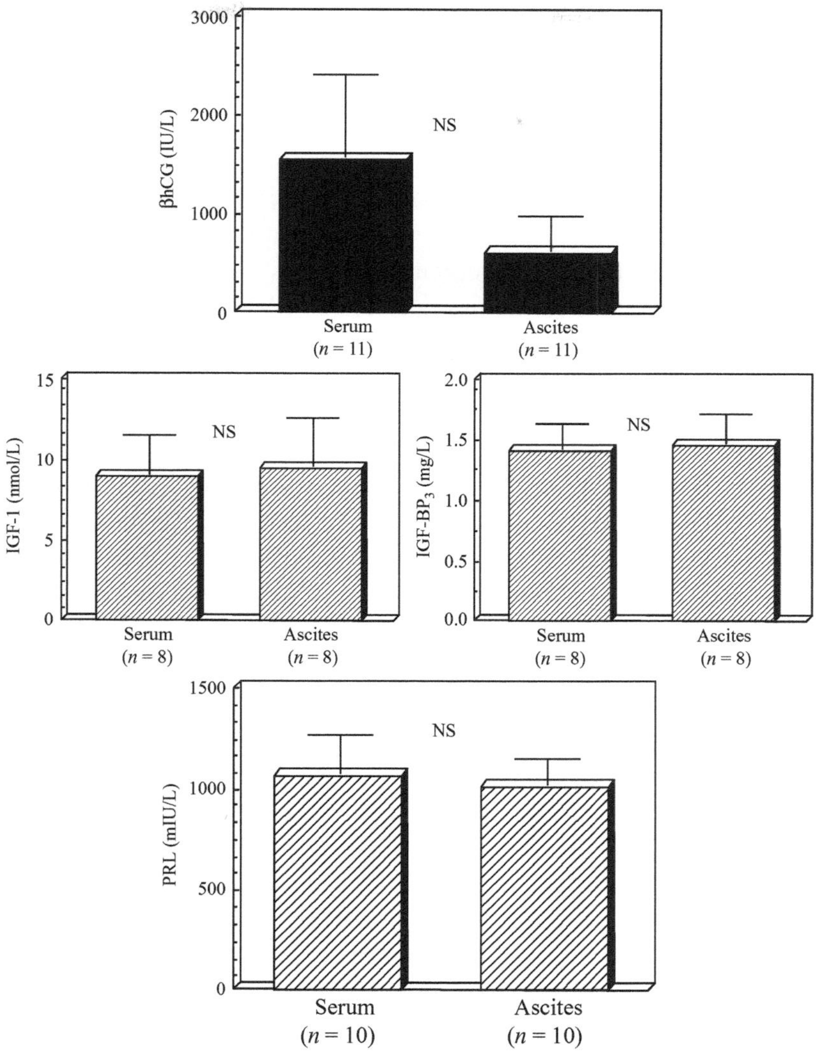

Fig. 3 The concentrations of the nonovarian-derived hormones/proteins: hCG, IGF-I, IGF-BP3, and PRL are not significantly different between serum and ascitic fluid in OHSS patients.

Humaidan et al., 2016; Meldrum, 2012). Whenever there is a risk of OHSS, it is advisable to consider the "freeze all" option and postpone ET to a future cycle, where no ovarian stimulation is experienced (Blumenfeld, 2015a, 2015b; Humaidan et al., 2016; Meldrum, 2012). In COS for in vivo conception, it is recommended to use very low doses of gonadotropins to generate monofollicular maturation. However, even with a careful attitude, it cannot be guaranteed that always only one follicle will ovulate, and even then, a young, lean, and PCOS patient may eventually develop OHSS (Blumenfeld, 2015a, 2015b; Humaidan et al., 2016; Meldrum, 2012). Whereas the injected gonadotropins also stimulate the small to medium size follicles that contribute most to the OHSS, it cannot be completely avoidable (Blumenfeld, 2015a, 2015b; Humaidan et al., 2016; Meldrum, 2012). Women with a low clearance rate of FSH may be at particular risk of OHSS, therefore tapering of the FSH dose after a dominant follicle is selected may be an important preventive routine for minimizing OHSS (Meldrum, 2012; Mizunuma et al., 1991). Decreasing the trigger dose of hCG may also be a preventive measure, since hCG is the prime stimulator of VEGF (Meldrum, 2012; Shapiro et al., 2005). Reduced hCG clearance may also increase the risk of OHSS (Meldrum, 2012; Shapiro et al., 2005). Whereas hCG has a long half-life time, one may consider using native GnRH (Blumenfeld, Lang, Amit, Kahana, & Yoffe, 1994) or GnRHa, instead of hCG for final follicular maturation and ovulation (Blumenfeld, 2015a, 2015b; Humaidan et al., 2016; Meldrum, 2012). The theoretical advantage of native GnRH over GnRHa is that the former induces LH surge without the associated luteolysis induced by the agonist (Blumenfeld et al., 1994). However, in patients who are at high risk of OHSS, the GnRHa-induced luteolysis may be beneficial for decreasing the risk, despite the disadvantage of necessary extensive luteal support with E+P (Blumenfeld, 2015a, 2015b; Humaidan et al., 2016; Meldrum, 2012). Using a dual trigger with GnRHa followed by 1000–1500 IU of hCG may rescue luteal function but OHSS cannot be completely avoided (Blumenfeld, 2015a, 2015b; Eskew & Omurtag, 2017; Fatemi et al., 2014; Humaidan, 2012; Humaidan et al., 2016; Meldrum, 2012).

The use of metformin, 1500–2000 mg/day for 6 weeks before and in parallel to COS, is another means for minimizing OHSS and increasing pregnancy rate in PCOS patients (Costello, Chapman, & Conway, 2006; Meldrum, 2012).

A large RCT has demonstrated that low dose aspirin can significantly reduce severe OHSS (Varnagy et al., 2010), possibly through its effect on platelet activation (Meldrum, 2012).

In cases where the estradiol rises severely, ceasing gonadotropin administration, the so-called coasting, may also decrease the risk of OHSS (Meldrum, 2012).

Dopamine agonists, such as cabergoline, a VEGF antagonist, may reduce the incidence of moderate or severe OHSS in women at high risk of OHSS (moderate quality evidence), without any apparent effect on pregnancy outcome (Tang, Mourad, Zhai, & Hart, 2016). Administration of cabergoline

(Dostinex®), 0.5 mg/day, orally for 7 days starting on hCG-triggering day in high-risk patients can reduce early OHSS and its severity in GnRHa IVF cycles, but cannot prevent the incidence of OHSS (Jellad et al., 2017). However, dopamine agonists might increase the risk of adverse events, such as gastrointestinal symptoms (Tang et al., 2016).

However, despite all the modalities to prevent OHSS, occasionally it may occur. In severe cases, necessitating hospitalization, a very efficient therapeutic measure is paracentesis of the ascitic fluid, abdominally or transvaginally, once, or repeated a few times, if necessary (Aboulghar et al., 1993; Aboulghar, Mansour, Serour, & Amin, 1990; Meldrum, 2012; Smith, Hacker, & Alper, 2009; Thaler, Yoffe, Kaftory, & Brandes, 1981). This efficient treatment, initiated in 1981, eliminates the vasoactive factors, such as VEGF, in the ascitic fluid, diminishes the mechanical pressure on the diaphragm, and significantly increases diuresis (Aboulghar et al., 1990, 1993; Meldrum, 2012; Smith et al., 2009; Thaler et al., 1981). Paracentesis may be effective, as an outpatient procedure, transvaginally, minimizing hospitalization (Meldrum, 2012; Smith et al., 2009).

Whereas hemoconcentration and high estrogen levels may increase the risk of thromboembolic phenomena, adding low molecular weight heparin, 40–80 mg/day, tailored to patient's BMI, and risk of thrombosis, is a preventive, suggested treatment. Additionally, intensive critical care, including intravenous crystalloids and plasma expanders, such as albumin, to maintain adequate renal perfusion and minimize the critical risk of renal "shutdown" (Carter, Petrie, Sadighi, & Skene, 2015; Kumar et al., 2011; Meldrum, 2012).

Finally, although many effective agents have been introduced in minimizing and treating OHSS, it has not been completely eradicated. One of the great modern medicine endeavors in the 21st century is to really establish an OHSS-free fertility clinic.

ACKNOWLEDGMENT

The assistance of Ruth Blumenfeld in preparation of the figures is thankfully acknowledged.

REFERENCES

Aboulghar, M. A., Mansour, R. T., Serour, G. I., & Amin, Y. (1990). Ultrasonically guided vaginal aspiration of ascites in the treatment of severe ovarian hyperstimulation syndrome. *Fertility and Sterility*, *53*, 933–935.

Aboulghar, M. A., Mansour, R. T., Serour, G. I., Sattar, M. A., Amin, Y. M., & Elattar, I. (1993). Management of severe ovarian hyperstimulation syndrome by ascitic fluid aspiration and intensive intravenous fluid therapy. *Obstetrics and Gynecology*, *81*, 108–111.

Abramov, Y., Barak, V., et al. (1997). Vascular endothelial growth factor plasma levels correlate to the clinical picture in severe ovarian hyperstimulation syndrome. *Fertility and Sterility*, *67*, 261–265.
Adashi, E. Y. (1989). Cytokine-mediated regulation of ovarian function: Encounters of a third kind. *Endocrinology*, *124*, 7043–7045.
Adashi, E. Y. (1996). The ovarian follicular apparatus. In E. Y. Adashi, J. A. Rock, & Z. Rosenwaks (Eds.), *Reproductive endocrinology, surgery, and technology* (pp. 17–40). Philadelphia: Lippincott-Raven Publishers.
Andus, T., Gross, V., et al. (1992). Evidence for the production of high amounts of IL-6 in the peritoneal cavity of patients with ascites. *Journal of Hepatology*, *15*, 378–381.
Balasch, J., Arroyo, V., et al. (1994). Neurohormonal and hemodynamic changes in severe cases of the ovarian hyperstimulation syndrome. *Annals of Internal Medicine*, *121*, 27–33.
Balasch, J., Carmona, F., et al. (1990). Acute prenatal failure and liver dysfunction in a patient with severe ovarian hyperstimulation syndrome. *Human Reproduction*, *5*, 348–351.
Barak, V., Mordel, N., et al. (1992). The correlation of IL-1 and TNF to oestradiol, progesterone and testosterone levels in periovulatory follicular fluid of in-vitro fertilization patients. *Human Reproduction*, *7*, 462–464.
Bar-Joseph, H., Ben-Ami, I., Ron-El, R., Shalgi, R., & Chuderland, D. (2016). Pigment epithelium-derived factor regulation by human chorionic gonadotropin in granulosa cells. *Reproduction*, *151*, 179–185.
Becerra, S. P., & Notario, V. (2013). The effects of PEDF on cancer biology: Mechanisms of action and therapeutic potential. *Nature Reviews. Cancer*, *13*, 258–271.
Ben-Shlomo, I., Kokia, E., Jackson, M. J., Adashi, E. Y., & Payne, D. W. (1994). Interleukin-1 beta stimulates nitrite production in the rat ovary: Evidence for heterologous cell-cell interaction and for insulin-mediated regulation of the inducible isoform of nitric oxide synthase. *Biology of Reproduction*, *51*, 310–318.
Bergh, P. A., & Navot, D. (1992). Ovarian hyperstimulation syndrome—A review of pathophysiology. *Journal of Assisted Reproduction and Genetics*, *9*, 429–438.
Blankenstein, J., Shalev, J., et al. (1987). Ovarian hyperstimulation syndrome: Prediction by number and size of preovulatory ovarian follicles. *Fertility and Sterility*, *47*, 597–602.
Blumenfeld, Z. (2015a). Why more is less and less is more when it comes to ovarian stimulation. *Journal of Assisted Reproduction and Genetics*, *32*, 1713–1719.
Blumenfeld, Z. (2015b). Does freeze all policy eliminate OHSS??? "It ain't necessarily so…". *Journal of Assisted Reproduction and Genetics*, *32*, 1571.
Blumenfeld, Z., Lang, N., Amit, A., Kahana, L., & Yoffe, N. (1994). Native gonadotropin-releasing hormone for triggering follicular maturation in polycystic ovary syndrome patients undergoing human menopausal gonadotropin ovulation induction. *Fertility and Sterility*, *62*, 456–460.
Blumenfeld, Z., Shabadash, V., & Shen-Or, Z. (1998). In *The origin of ascites in the ovarian hyperstimulation syndrome is mainly ovarian in human Presented at the 44th annual meeting of the Society for Gynecologic Investigation, March 19–22, 1997, San-Diego, CA* (pp. 102–110). J.S.G.I.—suppl. 1997; 4: 123A (abstr. # 175). Assist. Reprod. Reviews 8.
Blumenfeld, Z., Yoffe, N., & Bronshtein, M. (1991). Transvaginal sonography in infertility and reproductive endocrinology. *Obstetrical & Gynecological Survey*, *46*, 36–49.
Borenstein, R., Elhalah, U., et al. (1989). Severe ovarian hyperstimulation syndrome: A reevaluated therapeutic approach. *Fertility and Sterility*, *51*, 791–795.
Bouckacrt, P. X. G. M., Evers, J. L. H., et al. (1986). Patterns of changes in glycoproteins, polypeptides and steroids in the peritoneal fluid of women during the preovulatory phase of the menstrual cycle. *The Journal of Clinical Endocrinology and Metabolism*, *62*, 293–299.
Brinsden, P. R., Wada, I., et al. (1995). Diagnosis, prevention and management of ovarian hyperstimulation syndrome. *British Journal of Obstetrics and Gynaecology*, *102*, 767–772.

Burke-Gaffney, A., & Keenan, A. K. (1993). Modulation of human endothelial cell permeability by combinations of cytokines Interleukin-1 alpha/beta, tumor necrosis factor-alpha and interferon. *Immunopharmacology, 25,* 1–9.

Carter, R., Petrie, K., Sadighi, A., & Skene, H. (2015). Ovarian hyperstimulation syndrome on the acute medical unit: A problem-based review. *Acute Medicine, 14,* 21–27.

Chen, S. U., Chou, C. H., Lin, C. W., Lee, H., Wu, J. C., Lu, H. F., et al. (2010). Signal mechanisms of vascular endothelial growth factor and interleukin-8 in ovarian hyperstimulation syndrome: Dopamine targets their common pathways. *Human Reproduction, 25,* 757–767.

Chistyakova, G. N., Remizova, I. I., Gazieva, I. A., & Chermyaninova, O. V. (2014). Immunological and hemostasiological disorders in women with ovarian hyperstimulation syndrome. *Gynecological Endocrinology, 30*(Suppl. 1), 39–42.

Chuderland, D., Ben-Ami, I., Kaplan-Kraicer, R., Grossman, H., Ron-El, R., & Shalgi, R. (2013). The role of pigment epithelium-derived factor in the pathophysiology and treatment of ovarian hyperstimulation syndrome in mice. *The Journal of Clinical Endocrinology and Metabolism, 98,* E258–66.

Costello, M. F., Chapman, M., & Conway, U. (2006). A systematic review and meta-analysis of randomized controlled trials on metformin co-administration during gonadotrophin ovulation induction or IVF in women with polycystic ovary syndrome. *Human Reproduction, 21,* 1387–1399.

Craword, S. E., Fitchev, P., Veliceasa, D., & Volpert, O. V. (2013). The many facets of PEDF in drug discovery and disease: A diamond in the rough or split personality disorder? *Expert Opinion on Drug Discovery, 8,* 769–792.

Culler, M. D., Tarlatzis, B. C., et al. (1986). Angiotensin II-like immunoreactivity in human ovarian follicular fluid. *The Journal of Clinical Endocrinology and Metabolism, 62,* 613–615.

Delbaere, A., Bergmann, P. J. M., et al. (1994). Angiotensin II immunoreactivity is elevated in ascites during severe ovarian hyperstimulation syndrome: Implications for pathophysiology and clinical management. *Fertility and Sterility, 62,* 731–737.

Delbaere, A., Bergmann, P. J. M., et al. (1996). Preovulatory elevation of angiotensin II in the peritoneal fluid during the human menstrual cycle. *The Journal of Clinical Endocrinology and Metabolism, 81,* 2810–2815.

Delbaere, A., Bergmann, P. J. M., et al. (1997). Increased angiotensin II in ascites during severe ovarian hyperstimulation syndrome: Role of early pregnancy and ovarian gonadotropin stimulation. *Fertility and Sterility, 67,* 1038–1045.

Delvigne, A., & Rozenberg, S. (2002). Systematic review of data concerning etiopathology of ovarian hyperstimulation syndrome. *International Journal of Fertility and Women's Medicine, 47,* 211–226 [Review].

Dinarello, C. A. (1991). IL-1 and IL-1 antagonist. *Blood, 77,* 1627–1652.

Donnez, J., Langerock, S., et al. (1982). Peritoneal fluid volume and 17-estradiol and progesterone concentrations in ovulatory, anovulatory, and postmenopausal women. *Obstetrics and Gynecology, 59,* 687–692.

Dosouto, C., Haahr, T., & Humaidan, P. (2017). Gonadotropin-releasing hormone agonist (GnRHa) trigger—State of the art. *Reproductive Biology, 17,* 1–8 [Review].

Dourron, N. E., & Williams, D. B. (1996). Prevention and treatment of ovarian hyperstimulation syndrome. *Seminars in Reproductive Endocrinology, 14,* 355–365.

Ekerhovd, E., Enskog, A., Caidahl, K., Klintland, N., Nilsson, L., Brännström, M., et al. (2001). Plasma concentrations of nitrate during the menstrual cycle, ovarian stimulation and ovarian hyperstimulation syndrome. *Human Reproduction, 16,* 1334–1339.

Elliot, M. J., & Finn, A. H. (1993). Interaction between neutrophils and endothelium. *The Annals of Thoracic Surgery, 56,* 1503–1508.

Erlik, Y., Naot, Y., et al. (1979). Histamine levels in ovarian hyperstimulation syndrome. *Obstetrics and Gynecology, 53,* 580–582.

Eskew, A. M., & Omurtag, K. R. (2017). A review on ovarian hyperstimulation syndrome (OHSS) management strategies. *Minerva Endocrinologica*. https://doi.org/10.23736/S0391-1977.17.02638-4 [Epub ahead of print].
Fatemi, H. M., Popovic-Todorovic, B., Humaidan, P., Kol, S., Banker, M., Devroey, P., et al. (2014). Severe ovarian hyperstimulation syndrome after gonadotropin-releasing hormone (GnRH) agonist trigger and "freeze-all" approach in GnRH antagonist protocol. *Fertility and Sterility*, *101*, 1008–1011.
Fernandez, L. A., Tarlatzis, B. C., et al. (1985). Renin-like activity in ovarian follicular fluid. *Fertility and Sterility*, *44*, 219–223.
Fernandez, L. A., Twickler, J., et al. (1985). Neovascularization produced by angiotensin II. *The Journal of Laboratory and Clinical Medicine*, *105*, 141–145.
Ferrara, N., & Davis-Smyth, T. (1997). The biology of vascular endothelial growth factor. *Endocrine Reviews*, *18*(1), 4–25, Review.
Ferrara, N., Houck, K., et al. (1992). Molecular and biological properties of the vascular endothelial growth factor family of proteins. *Endocrine Reviews*, *13*, 18–32.
Ferrera, N., Houck, K., et al. (1991). The vascular endothelial growth factor family of polypeptides. *The Journal of Cell Biology*, *47*, 211–218.
Filep, J. G., Foldes-Filep, E., et al. (1993). Nitric oxide modulates vascular permeability in the rat coronary circulation. *British Journal of Pharmacology*, *8*, 323–326.
Friedlander, M. A., de Mola, J. R. L., et al. (1995). Elevated levels of interleukin-6 in ascites and serum from women with ovarian hyperstimulation syndrome. *Fertility and Sterility*, *60*, 826–833.
Funke, I., Prummer, O., et al. (1994). Capillary leak syndrome associated with elevated IL-2 serum levels after allogeneic bone marrow transplantation. *Annals of Hematology*, *68*, 49–52.
Gardiner, S. M., Compton, A. M., et al. (1990). Control of regional blood flow by EDRF. *Hypertension*, *15*, 485–492.
Gergly, R. Z., Paldi, E., et al. (1976). Treatment of ovarian hyperstimulation syndrome by antihistamine. *Obstetrics and Gynecology*, *47*, 83–85.
Geva, E., Lessing, J. B., et al. (1997). Elevated levels of interleukin-6 in the follicular fluid at the time of oocyte retrieval for in vitro fertilization may predict the development of early-form ovarian hyperstimulation syndrome. *Fertility and Sterility*, *68*, 133–137.
Glorioso, N., Atlas, S. A., et al. (1986). Prorenin in high concentrations in human follicular fluid. *Science*, *233*, 1422–1424.
Golan, A., Ron-El, R., et al. (1989). Ovarian hyperstimulation syndrome: An update review. *Obstetrical & Gynecological Survey*, *44*(6), 430–440.
Goldsman, M. P., Pedram, A., et al. (1995). Increased capillary permeability induced by human follicular fluid: A hypothesis for an ovarian origin of the hyperstimulation syndrome. *Fertility and Sterility*, *63*, 268–272.
Gottschall, P. E., Katsuura, G., et al. (1989). Interleukin-1 is more potent than interleukin-1 in suppressing follicle-stimulation hormone-induced differentiation of ovarian granulosa cells. *Biochemical and Biophysical Research Communications*, *163*, 764–777.
Humaidan, P. (2012). Agonist trigger: What is the best approach? Agonist trigger and low dose hCG. *Fertility and Sterility*, *97*, 529–530.
Humaidan, P., Nelson, S. M., Devroey, P., Coddington, C. C., Schwartz, L. B., Gordon, K., et al. (2016). Ovarian hyperstimulation syndrome: Review and new classification criteria for reporting in clinical trials. *Human Reproduction*, *31*, 1997–2004.
Hume, D. A., Halpin, D., et al. (1984). The mononuclear phagocyte system of the mouse defined by immunohistochemical localization of antigen F4/80: Macrophages of endocrine organs. *Proceedings of the National Academy of Sciences of the United States of America*, *81*, 4174–4177.

Hurwitz, A., Loukides, J., et al. (1992). Human intraovarian IL-1 system: Highly compartmentalized and hormonally dependent regulation of the genes encoding IL-1, its receptor and its receptor antagonist. *The Journal of Clinical Investigation*, *89*, 1746–1754.

Jakob, W., Jentrsch, K. D., et al. (1977). Demonstration of angiogenesis activity in the corpus luteum of cattle. *Experimental Pathology*, *13*, 231–236.

Jellad, S., Haj Hassine, A., Basly, M., Mrabet, A., Chibani, M., & Rachdi, R. (2017). Vascular endothelial growth factor antagonist reduces the early onset and the severity of ovarian hyperstimulation syndrome. *Journal de Gynécologie, Obstétrique et Biologie de la Reproduction*, *46*, 87–91.

Kasum, M., Franulić, D., Čehić, E., Orešković, S., Lila, A., & Ejubović, E. (2017). Kisspeptin as a promising oocyte maturation trigger for in vitro fertilisation in humans. *Gynecological Endocrinology*, *33*, 583–587. https://doi.org/10.1080/09513590.2017.1309019 [Epub ahead of print].

Katz, Z., Lancet, M., et al. (1984). Absence of teratogenicity of indomethacin in ovarian hyperstimulation syndrome. *International Journal of Fertility*, *29*, 186–188.

Kim-Bjrklund, T., Landgren, B. M., et al. (1991). Peritoneal fluid volume and levels of steroid hormones and gonadotropins in peritoneal fluid of normal and norethisterone-treated women. *Human Reproduction*, *6*, 1233–1237.

Knox, G. E., Dowd, A., et al. (1975). Antihistamine blockade of the ovarian hyperstimulation syndrome. II. Possible role of antigen-antibody complexes in the pathogenesis of the syndrome. *Fertility and Sterility*, *26*, 418–421.

Koninckx, P. R., Heyns, W., et al. (1980). Biochemical characterization of peritoneal fluid in women during the menstrual cycle. *The Journal of Clinical Endocrinology and Metabolism*, *51*, 1239–1244.

Koninckx, P. R., Renaer, M., et al. (1980). Origin of peritoneal fluid in women: An ovarian exudation product. *British Journal of Obstetrics and Gynaecology*, *87*, 177–183.

Kransnow, J. S., Berga, S. L., Guzick, D. S., Zeleznik, A. J., & Yeo, K. T. (1996). Vascular permeability factor and vascular endothelial growth factor in ovarian hyperstimulation syndrome: A preliminary report. *Fertility and Sterility*, *65*, 552–555.

Kubes, P., & Granger, D. N. (1992). Nitric oxide modulates microvascular permeability. *The American Journal of Physiology*, *262*, 611–615.

Kumar, P., Sait, S. F., Sharma, A., & Kumar, M. (2011). Ovarian hyperstimulation syndrome. *Journal of Human Reproductive Sciences*, *4*, 70–75.

Kutteh, W. H., & Kutteh, C. C. (1992). Quantitation of tumor necrosis factor-alpha, IL-1beta and IL-6 in the effusion of ovarian epithelial neoplasms. *American Journal of Obstetrics and Gynecology*, *167*, 1864–1869.

Kwik, M., & Maxwell, E. (2016). Pathophysiology, treatment and prevention of ovarian hyperstimulation syndrome. *Current Opinion in Obstetrics & Gynecology*, *28*, 236–241.

Larsen, G. G., Anderson, A. O., et al. (1989). The neutrophil activating protein (NAP-1) is also chemotactic for T lymphocytes. *Science*, *243*, 1464–1466.

Levin, E. R., Rosen, G. F., Cassidenti, D. L., Yee, B., Meldrum, D., Wisot, A., et al. (1998). Role of vascular endothelial cell growth factor in ovarian hyperstimulation syndrome. *The Journal of Clinical Investigation*, *102*, 1978–1985.

Loukides, J. A., Loy, R. A., et al. (1990). Human follicular fluids contain tissue macrophages. *The Journal of Clinical Endocrinology and Metabolism*, *71*, 1363–1367.

Maatjuis, J. B., Look, P. F. A., et al. (1978). Changes in volume, total protein and ovarian steroid concentrations of peritoneal fluid throughout the human menstrual cycle. *The Journal of Endocrinology*, *76*, 123–133.

McClure, N., Healy, D. L., et al. (1994). Vascular endothelial growth factor as a capillary permeability agent in ovarian hyperstimulation syndrome. *Lancet*, *344*, 235–236.

Meldrum, D. R. (2012). Preventing severe OHSS has many different facets. *Fertility and Sterility*, *97*, 536–538.

Miller, I., Chuderland, D., Grossman, H., Ron-El, R., Ben-Ami, I., & Shalgi, R. (2016). The dual role of PEDF in the pathogenesis of OHSS: Negating both angiogenic and inflammatory pathways. *The Journal of Clinical Endocrinology and Metabolism, 101*, 4699–4709.

Miller, I., Chuderland, D., Ron-El, R., Shalgi, R., & Ben-Ami, I. (2015). GnRH agonist triggering modulates PEDF to VEGF ratio inversely to hCG in granulosa cells. *The Journal of Clinical Endocrinology and Metabolism, 100*, E1428–36.

Mizunuma, H., Takagi, T., Yamada, K., Andoh, K., Ibuki, Y., & Igarashi, M. (1991). The role of endogenous gonadotropin release in the etiology of ovarian enlargement during purified urinary follicle-stimulating hormone therapy. *Fertility and Sterility, 55*, 66–72.

Mori, T., Masatsune, F., et al. (1989). Modulation of porcine granulosa cell functions by interleukin-1. *Steroids, 54*, 543–552.

Morris, R. S., & Paulson, R. J. (1994). Ovarian derived prorenin-angiotensin cascade in human reproduction. *Fertility and Sterility, 62*, 1105–1114.

Motro, B., Itin, A., et al. (1990). Pattern of interleukin 6 gene expression in vivo suggests a role for this cytokine in angiogenesis. *Cell Biology, 87*, 3092–3096.

Nakamura, Y., Smith, M., et al. (1987). Increased number of mast cells in the dominant follicle of the cow: Relationships among luteal, stromal, and hilar regions. *Biology of Reproduction, 37*, 546–549.

Nathan, C., & Xie, Q. W. (1993). Regulation of biosynthesis of nitric oxide. *The Journal of Biological Chemistry, 269*, 13725–13728.

Navot, D., Bergh, P. A., et al. (1992). Ovarian hyperstimulation syndrome in novel reproductive technologies: Prevention and treatment. *Fertility and Sterility, 58*, 249–260.

Navot, D., Bergh, P. A., et al. (1996). The ovarian hyperstimulation syndrome. In E. Y. Adashi, J. A. Rock, & Z. Rosenwaks (Eds.), *Reproductive endocrinology, surgery, and technology* (pp. 2225–2232). Philadelphia: Lippincott-Raven Publishers.

Navot, D., Margalioth, E. J., et al. (1987). Direct correlation between plasma renin activity and severity of the ovarian hyperstimulation syndrome. *Fertility and Sterility, 48*, 57–61.

Nelson, S. M. (2017). Prevention and management of ovarian hyperstimulation syndrome. *Thrombosis Research, 151*(Suppl. 1), S61–S64.

Neulen, J., Yan, Z., Raczek, S., Weindel, K., Keck, C., Weich, H. A., et al. (1995). Human chorionic gonadotropin-dependent expression of vascular endothelial growth factor/vascular permeability factor in human granulosa cells: Importance in ovarian hyperstimulation syndrome. *The Journal of Clinical Endocrinology and Metabolism, 80*, 1967–1971.

Ninio-Many, L., Grossman, H., Shomron, N., Chuderland, D., & Shalgi, R. (2013). microRNA-125a-3p reduces cell proliferation and migration by targeting Fyn. *Journal of Cell Science, 126*, 2867–2876.

Ong, A. C. M., Eisen, V., et al. (1991). The pathogenesis of the ovarian hyperstimulation syndrome (OHS): A possible role for ovarian renin. *Clinical Endocrinology, 34*, 43–49.

Orvieto, R., Achiron, A., Margalit, R., & Ben-Rafael, Z. (1998). The role of intravenous immunoglobulin in the prevention of severe ovarian hyperstimulation syndrome. *Journal of Assisted Reproduction and Genetics, 15*(1), 46–49.

Orvieto, R., & Ben-Rafael, Z. (1996). The immune system in severe ovarian hyperstimulation syndrome. *Israel Journal of Medical Sciences, 32*, 1180–1182.

Orvieto, R., & Ben-Rafael, Z. (1998). Ovarian hyperstimulation syndrome: A new insight into an old enigma. *Journal of the Society for Gynecologic Investigation, 5*, 110–113.

Orvieto, R., Dratviman-Storobinsky, O., Lantsberg, D., Haas, J., Mashiach, R., & Cohen, Y. (2014). Interleukin-2 and SOCS-1 proteins involvement in the pathophysiology of severe ovarian hyperstimulation syndrome—A preliminary proof of concept. *Journal of Ovarian Research, 7*, 106.

Palumbo, A., Ávila, J., & Naftolin, F. (2016). The ovarian renin-angiotensin system (OVRAS): A major factor in ovarian function and disease. *Reproductive Sciences, 23*, 1644–1655.
Polishuk, W. Z., & Schenker, J. G. (1969). Ovarian overstimulation syndrome. *Fertility and Sterility, 20*, 443–450.
Pride, S. M., Ho Yuen, B., et al. (1984). Clinical, endocrinological and intraovarian prostaglandin-F response to H-1 receptor blockade in the ovarian hyperstimulation syndrome. *American Journal of Obstetrics and Gynecology, 148*, 670–674.
Pride, S. M., Yuen, B. H., et al. (1986). Relationship of gonadotropin-releasing hormone, danazol and prostaglandin blockade to ovarian enlargement ascites formation of the ovarian hyperstimulation syndrome in the rabbit. *American Journal of Obstetrics and Gynecology, 154*, 1155–1160.
Rabau, E., Serr, D. M., et al. (1967). Human menopausal gonadotropins for anovulation and sterility. *American Journal of Obstetrics and Gynecology, 98*, 92–98.
Revel, A., Barak, V., et al. (1996). Characterization of intraperitoneal cytokines and nitrites in women with severe ovarian hyperstimulation syndrome. *Fertility and Sterility, 66*, 66–71.
Rizk, B., Aboulghar, M., Smitz, J., & Ron-El, R. (1997). The role of vascular endothelial growth factor and interleukins in the pathogenesis of severe ovarian hyperstimulation syndrome. *Human Reproduction Update, 3*, 255–266 [Review].
Robertson, A. L., & Khairallah, P. A. (1972). Effects of angiotensin II and some analogues on vascular permeability in the rabbit. *Circulation Research, 31*, 923–931.
Robertson, D., Selleck, K., et al. (1995). Urinary vascular endothelial growth factor concentrations in women undergoing gonadotropin treatment. *Human Reproduction, 10*, 2478–2482.
Rosenberg, M. E., Mckenzie, J. K., et al. (1994). Increased ascitic fluid prorenin in the ovarian hyperstimulation syndrome. *American Journal of Kidney Diseases, 23*, 427–429.
Ryley, N. G., Forman, R., et al. (1990). Liver abnormality in ovarian hyperstimulation syndrome. *Human Reproduction, 5*, 938–943.
Schenker, J. G. (1993). Prevention and treatment of ovarian hyperstimulation. *Human Reproduction, 8*, 653–659.
Schenker, J. G. (1997). Clinical aspects and pathophysiology of ovarian hyperstimulation syndrome. *Israel Journal of Obstetrics and Gynecology, 8*, 114–121.
Schenker, J. G., & Polishuk, W. Z. (1976a). The role of prostaglandins in ovarian hyperstimulation syndrome. *European Journal of Obstetrics, Gynecology, and Reproductive Biology, 6*, 47–50.
Schenker, J. G., & Polishuk, W. Z. (1976b). In M. Tichner & J. Pilch (Eds.), *Vol. 4. An experimental model of ovarian hyperstimulation syndrome. Proceedings of the international congress animal reproduction* (p. 635). Krakow: Drukarnia Naukowa.
Schenker, J. G., & Polishuk, W. Z. (1976c). Role of prostaglandin in ovarian hyperstimulation syndrome. *Obstetrical & Gynecological Survey, 31*, 742–744.
Schenker, J. G., & Polishuk, W. Z. (1976d). Ovarian hyperstimulation syndrome: Clinical and experimental data. *International Congress Series, 396*, 204–209.
Schenker, J. G., & Weinstein, D. (1978). Ovarian hyperstimulation syndrome: A current survey. *Fertility and Sterility, 30*, 255–268.
Sealey, J. E., Atlas, S. A., et al. (1980). Protein and large molecular weight forms of renin. *Endocrine Reviews, 1*, 365–391.
Sealey, J. E., Atlas, S. A., et al. (1985). Cyclical secretion of prorenin during the menstrual cycle: Synchronization with luteinizing hormone and progesterone. *Proceedings of the National Academy of Sciences of the United States of America, 82*, 8705–8709.
Seligman, S. P., Buyon, J. P., et al. (1994). The role of nitric oxide in the pathogenesis of preeclampsia. *American Journal of Obstetrics and Gynecology, 171*, 944–948.

Senger, D. R., Galli, S. J., et al. (1983). Tumor cells secrete a vascular permeability factor that promotes accumulation of ascites fluid. *Science*, *219*, 983–985.
Shapiro, B. S., Daneshmand, S. T., Garner, F. C., Aguirre, M., Ross, R., & Morris, S. (2005). Effects of ovulatory serum concentration of human chorionic gonadotropin on the incidence of ovarian hyperstimulation syndrome and success rates for in vitro fertilization. *Fertility and Sterility*, *84*, 93–98.
Shi, S. L., Peng, Z. F., Yao, G. D., Jin, H. X., Song, W. Y., Yang, H. Y., et al. (2015). Expression of CD11c+HLA-DR+dendritic cells and related cytokines in the follicular fluid might be related to pathogenesis of ovarian hyperstimulation syndrome. *International Journal of Clinical and Experimental Pathology*, *8*, 15133–15137.
Smith, L. P., Hacker, M. R., & Alper, M. M. (2009). Patients with severe ovarian hyperstimulation syndrome can be managed safely with aggressive outpatient transvaginal paracentesis. *Fertility and Sterility*, *92*, 1953–1959.
Sueldo, C. E., Price, H. M., et al. (1988). Liver dysfunction in ovarian hyperstimulation syndrome. A case report. *The Journal of Reproductive Medicine*, *33*, 387–390.
Tang, H., Mourad, S., Zhai, S. D., & Hart, R. J. (2016). Dopamine agonists for preventing ovarian hyperstimulation syndrome. *Cochrane Database of Systematic Reviews*, *11*. CD008605.
Terranova, P. F., Sancho-Tello, M., et al. (1993). TNF- and ovarian function. In E. Y. Adashi & C. K. Leung (Eds.), *The ovary* (pp. 395–411). New York: Raven Press Ltd.
Thaler, I., Yoffe, N., Kaftory, J. K., & Brandes, J. M. (1981). Treatment of ovarian hyperstimulation syndrome: The physiologic basis for a modified approach. *Fertility and Sterility*, *36*, 110–113.
Thomsen, L., & Humaidan, P. (2015). Ovarian hyperstimulation syndrome in the 21st century: The role of gonadotropin-releasing hormone agonist trigger and kisspeptin. *Current Opinion in Obstetrics & Gynecology*, *27*, 210–214.
Tollan, A., Holst, N., et al. (1990). Transcapillary fluid dynamic during ovarian stimulation for in vitro fertilization. *American Journal of Obstetrics and Gynecology*, *162*, 554–558.
Varnagy, A., Bodis, J., Manfai, Z., Wilhelm, F., Busznyak, C., & Koppan, M. (2010). Low-dose aspirin therapy to prevent ovarian hyperstimulation syndrome. *Fertility and Sterility*, *93*, 2281–2284.
Yarali, H., & Fleige-Zahradka, B. G. (1993). The ascites in the ovarian hyperstimulation syndrome does not originate from the ovary. *Fertility and Sterility*, *59*, 657–661.
Yoshimura, Y., Karube, M., et al. (1993). Locally produced angiotensin II induces ovulation by stimulating prostaglandin production in in vitro perfused rabbit ovaries. *Endocrinology*, *133*, 1609–1616.
Yoshizumi, M., & Perrella, M. A. (1993). TNF downregulates an endothelial nitric oxide synthase mRNA. *Circulation Research*, *73*, 205–209.
Younis, J. S., Zeevi, D., et al. (1988). Transient liver function tests abnormalities in ovarian hyperstimulation syndrome. *Fertility and Sterility*, *50*, 176–178.
Zaides, I., Friedman, M., et al. (1983). Serotonin and the ovarian hyperstimulation syndrome. *European Journal of Obstetrics, Gynecology, and Reproductive Biology*, *15*, 55–61.
Zivi, E., Simon, A., & Laufer, N. (2010). Ovarian hyperstimulation syndrome: Definition, incidence, and classification. *Seminars in Reproductive Medicine*, *28*, 441–447.
Zolti, M., Ben-Rafael, Z., et al. (1991). Cytokine involvement in oocyte and early embryos. *Fertility and Sterility*, *56*, 265–272.
Zosmer, A., Katz, Z., et al. (1987). Adult respiratory distress syndrome complicating ovarian hyperstimulation syndrome. *Fertility and Sterility*, *47*, 524–526.

FURTHER READING

Borges, E., Jr., Braga, D. P., Setti, A. S., Vingris, L. S., Figueira, R. C., & Iaconelli, A., Jr. (2016). Strategies for the management of OHSS: Results from freezing-all cycles. *JBRA Assisted Reproduction*, *20*, 8–12.

Medical Research International, Society for Assisted Reproductive Technology (SART), American Fertility Society. (1992). In vivo fertilization–embryo transfer (IVF-ET) in the United States: 1990 results from the IVF-ET Registry. *Fertility and Sterility*, *57*, 15–24.

Oriowo, M. A. (2004). Ex vivo reactivity of the ovarian vascular bed to noradrenaline and carbachol during ovarian hyperstimulation syndrome. *Medical Principles and Practice*, *13*(3), 115–121.

CHAPTER SIXTEEN

From Follicular Development and Ovulation to Ovarian Cancers: An Unexpected Journey

JoAnne S. Richards[1]
Baylor College of Medicine, Houston, TX, United States
Dan L. Duncan Cancer Center, Baylor College of Medicine, Houston, TX, United States
Center for Reproductive Medicine, Baylor College of Medicine, Houston, TX, United States
[1]Corresponding author: e-mail address: joanner@bcm.edu

Contents

Abstract

Follicular development and ovulation are complex development processes that are regulated by multiple, interacting pathways and cell types. The oocyte, cumulus cells, granulosa cells, and theca cells communicate to impact follicular development and ovulation. Many hormones and cytokines control intracellular regulatory networks and transcription factors, some of which are cell type specific. Molecular biology approaches and mutant mouse models have contributed immensely to our knowledge of what genes and signaling cascades impact each stage of follicular development and ovulation, and how the alteration of gene expression profiles and the activation of specific signaling pathways can impact ovarian cancer development in ovarian surface epithelial cells as well as granulosa cells. This chapter explores how pathways controlling normal follicle development and ovulation can be diverted to abnormal development.

Vitamins and Hormones, Volume 107
ISSN 0083-6729
https://doi.org/10.1016/bs.vh.2018.01.019

1. INTRODUCTION

Research in my laboratory has focused on the hormonal and cellular events that control ovulation, specifically the factors and pathways that drive cumulus cell–oocyte complex (COC) expansion and follicle rupture. As a consequence of generating mutant mouse models to analyze specific signaling pathways related to ovulation, we obtained mice that unexpectedly developed epithelial ovarian cancer (EOC). Furthermore, in analyzing signaling factors that impact follicular development, we generated mutant mice that developed granulosa cell tumors (GCTs). This chapter will provide an overview of how we got from studying follicular development and ovulation to our current studies on ovarian cancer.

2. EARLY STUDIES ON OVULATION

Our venture into understanding the molecular basis of ovulation began with (1) our efforts in 1989–93 to characterize and clone prostaglandin synthase (PTGS or cyclooxygenase), the rate-limiting enzyme in prostaglandin biosynthesis that was presumed, at that time, to be encoded by a single gene, and (2) our collaborations with Dr. Larry Espey. To clone *Ptgs*, we were generating antibodies in rabbits against purified seminal vesicle PTGS to use for screening our ovarian-derived lambda phage expression library. All efforts appeared futile until we carefully analyzed protein samples from preovulatory follicles, ovulating follicles, and corpora lutea using several different antibodies generated in different rabbits (Wong & Richards, 1991, 1992). Remarkably, one antibody recognized a protein (band) induced by hCG in preovulatory follicles that was not present before hCG or in corpora. Using the same protein samples, another antibody recognized a protein (band) of slightly different molecular size that was present in all samples (Fig. 1). These results provided the first immunological evidence that there might be two distinct (but related?) enzymes capable of synthesizing prostaglandins. Purification of the inducible PTGS protein and N-terminal sequencing further documented that this protein was distinct from the deduced sequence of a PTGS gene that had been cloned (Sirois, Levy, Simmons, & Richards, 1993; Sirois, Simmons, & Richards, 1992). Concurrent with these observations, a cyclooxygenase-like gene was cloned from chicken embryo fibroblasts in response to the Rous sarcoma virus (Xie, Chipman, Robertson, Erikson, & Simmons, 1991; Xie, Merrill, Bradshaw, & Simmons, 1993). This gene encoded a protein similar to that

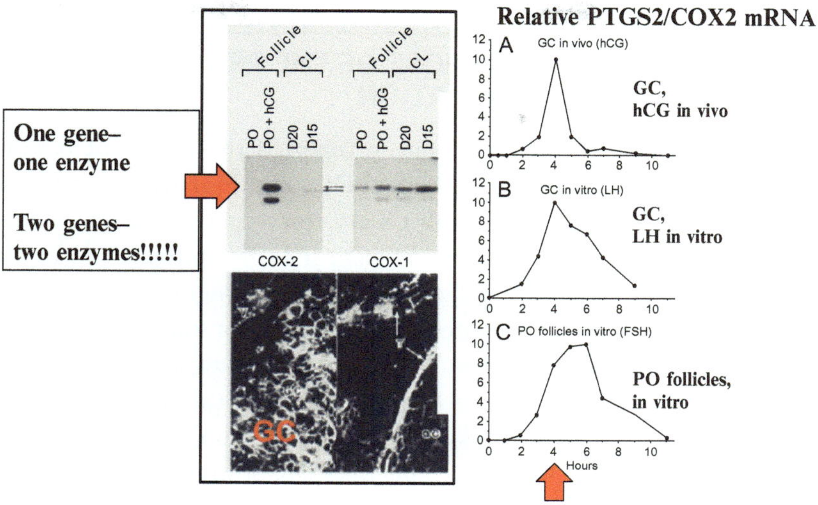

Fig. 1 Rapid but transient induction of prostaglandin synthase 2 (cyclooxygenase 2, COX2) in granulosa cells of preovulatory follicles by LH/hCG. *Adapted from Wong, W. Y. W., & Richards, J. S. (1991). Evidence for two antigenically distinct molecular weight variants of prostaglandin H synthase in the rat ovary.* Molecular Endocrinology, 5, *1269–1279; Sirois, J., Simmons, D. L., & Richards, J. S. (1992). Hormonal regulation of messenger ribonucleic acid encoding a novel isoform of prostaglandin endoperoxide H synthase in rat preovulatory follicles.* The Journal of Biological Chemistry, 267, *11586–11592.*

induced by LH in granulosa cells. Suddenly there was both immunological and genetic evidence for two distinct genes. The inducible gene is now known to encode PTGS2 (COX2) and the original, noninducible, gene to encode PTGS1 (COX1). Further studies showed that the *Ptgs2* gene was induced by multiple pathways and required an E-box region in its promoter (Morris & Richards, 1993, 1996). Ultimately, the knockout of *Ptgs2* and *Ep2* in mice verified its essential role in ovulation (Richards, Liu, & Shimada, 2015).

Using differential display expression analyses, Dr. Espey made seminal discoveries that identified numerous genes that were induced rapidly but transiently by hCG in ovulating rat follicles (Espey, Ujoka, et al., 2000; Espey, Yoshioka, Russell, Robker, et al., 2000; Espey, Yoshioka, Russell, Ujioka, et al., 2000; Espey, Yoshioka, Ujioka, Fujii, & Richards, 2001). These genes included: a disintegrin and metalloproteinase with thrombospondin-like repeats 1 (*Adamts1*), tumor necrosis-stimulated gene 6 (*Tsg6*; also known as tumor necrosis factor alpha-induced protein 6, *Tnfaip6*), epiregulin (*Ereg*), and early growth regulator 1 (*Egr1*) that eventually helped to unravel key factors in ovulation (Espey & Richards, 2002).

3. UNRAVELING THE ROLE OF THE PROGESTERONE RECEPTOR IN OVULATION

In 1991, Kelly Mayo's laboratory reported the rapid but transient induction of the progesterone receptor (PGR) by LH in ovulating follicles in mice (Park & Mayo, 1991). This observation combined with evidence that the *Pgr* knockout (PRKO) mice were infertile and failed to ovulate (Lydon et al., 1995) (Fig. 2) highlighted PGR as another signaling pathway controlling ovulation and supported previous inhibitor studies using the steroidogenic antagonist, epostane (Tanaka, Espey, Kawano, & Okamura, 1991). In addition, we were able to show that both the A and B forms of PGR were induced by LH or forskolin and PMA in granulosa cells in culture (Natraj & Richards, 1993). By analyzing the PGR promoter we were able to document that this gene was not induced by estradiol in granulosa cells as occurs in many other tissues but rather was dependent on SP1 binding sites within the promoter (Sriraman, Sharma, & Richards, 2003). Furthermore, forskolin and progesterone but not progesterone alone activated a glucocorticoid (progesterone) promoter–reporter construct indicating the coordinate

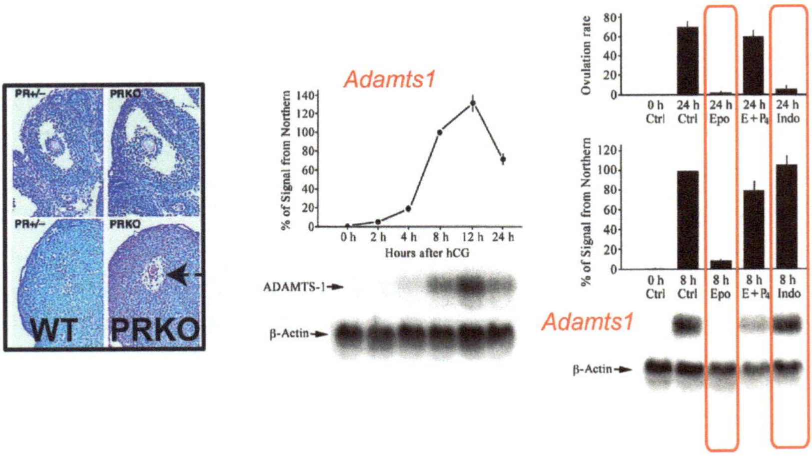

Fig. 2 *Pgr* null (PRKO) mice are infertile and *Adamts1* is a PGR-regulated gene. *Adapted from Lydon et al., 1996; Robker, R. L., Russell, D. L., Espey, L.L., Lydon, J. P., O'Malley, B.W., & Richards, J. S. (2000). Progesterone-regulated genes in the ovulation process: ADAMTS-1 and cathepsin L proteases.* Proceedings of the National Academy of Sciences of the United States of America, 97, *4689–4694; PMID:10781075; Espey, L.L., Yoshioka, S., Russell, D.L., Robker, R. L., Fujii, S., & Richards, J. S. (2000). Ovarian expression of a disintegrin metalloproteinase with thrombospondin motifs during ovulation in the gonadotropin-primed immature rat.* Biology of Reproduction, 62, *1090–1095.*

induction and activation of PGR. That the PGR antagonist RU486 did not block the induction of PGR indicated that progesterone did not regulate the induction of its own receptor (Natraj & Richards, 1993). By analyzing the PGR knockout mice and gene promoter activation, we documented that *Adamts1* and cathepsin L, synaptin protein 25 (*Snap 25*), *Adam8*, endothelin 2 (*Edn2*), and cGMP-dependent protein kinase (*Pkg*) are progesterone and PGR target genes in granulosa cells of ovulating follicles (Doyle, Russell, Sriraman, & Richards, 2004; Robker et al., 2000; Shimada et al., 2007; Sriraman et al., 2008; Sriraman, Lohmann, Mulders, & Richards, 2006; Sriraman, Rudd, Lohmann, Mulders, & Richards, 2006) (Fig. 2). We have also shown that PRA and PRB regulate the induction of distinct genes in cultured granulosa cells (Sriraman, Sinha, & Richards, 2010). Because PGR null mice fail to ovulate but corpora lutea form, the role of PGR and progesterone appears to be primarily related to events that impact follicle rupture.

4. UNRAVELING THE ROLES OF INFLAMMATION AND PROSTAGLANDINS IN OVULATION

The mystery of ovulation continued to unravel with evidence that expression (*Tsg6, Tnfaip6*) was induced by LH (Yoshioka et al., 2000) and was disrupted in COCs of *Ptgs2* and *Ep2* null mice (Ochsner, Russell, & Richards, 2003). Impaired expression of *Tnfaip6* in vivo or disrupting its functions in culture blocked COC expansion (Ochsner, Day, Breyer, Gomer, & Richards, 2003; Ochsner, Russell, et al., 2003). Microarray analyses of COCs collected from mouse ovaries and oviducts at selected time intervals before and after hCG-induced ovulation revealed many genes that were rapidly and selectively induced in COCs, including hyaluronic acid synthase 2 (*Has2*), pentraxin 3 (*Ptx3*), *amphiregulin* (*Areg*), betacellulin (*Btc*), Cbp/P300 interacting transactivator with Glu/Asp-rich carboxy-terminal domain 4 (*Cited4*), *interleukin 6* (*Il6*), myelin basic protein (*Mbp*), and tenascin C (*Tnc*) (Hernandez-Gonzalez et al., 2006). These factors are critical COC expansion, a process by which the hyaluronan-rich matrix is synthesized and stabilized (Liu & Richards, 2008; Richards et al., 2015; Richards & Pangas, 2010). These results indicated that factors controlling *Ptgs2* expression and COC expansion were distinct from those controlling the induction of *Pgr* and follicle rupture during ovulation (Espey & Richards, 2002).

But exactly how LH/hCG or forskolin/PMA were inducing these genes and pathways did not come to light until the induction of *Ereg* was fully appreciated and evidence implicating the EGF receptor pathway emerged from EGF receptor hypomorphic mice (Conti, Hsieh, Park, & Su, 2005; Hsieh & Conti, 2005; Park et al., 2004). These results were rapidly connected with data showing LH/hCG induced the EGF-like factors EREG, AREG, and BTC, that in turn, activated the EGF receptors leading to phosphorylation of the extracellular regulated kinases, ERK1/2 (MAPK3/1) (Shimada, Gonzalez-Robayna, Hernandez-Gonzalez, & Richards, 2006; Shimada, Hernandez-Gonzalez, Gonzalez-Robayna, & Richards, 2005) (Figs. 3 and 4). Moreover, mice in which the genes encoding *Erk1/2* were depleted selectively in granulosa cells failed to ovulate, COC expansion did not occur and the FSH program of gene expression was not suppressed (Fan, Liu, Shimada, et al., 2009). Their observations solidified the key role of the EGF-like factor, EGF receptor, ERK1/2 pathway in mediating LH induction of ovulation (Fig. 5). More recent studies have shown that CITED4 and C/EBP α/β are downstream targets of ERK1/2 and PKA that modify histone acetylation and ovulation, respectively (Fan, Liu, Johnson, & Richards, 2011; Zhang et al., 2014) (Fig. 6). Additionally, COCs cultured in the presence of AREG or EREG expanded as well or better than COCs cultured in the presence of FSH or forskolin (Fig. 5) and that expansion was dependent in part on calpain-mediated cell movement (Kawashima et al., 2012; Shimada et al., 2005). IL6 and MAPK14 also impact COC expansion (Liu, Fan, Wang, & Richards, 2010; Liu & Richards, 2008). Thus, the EGF-like factors could completely bypass the need for LH and cAMP

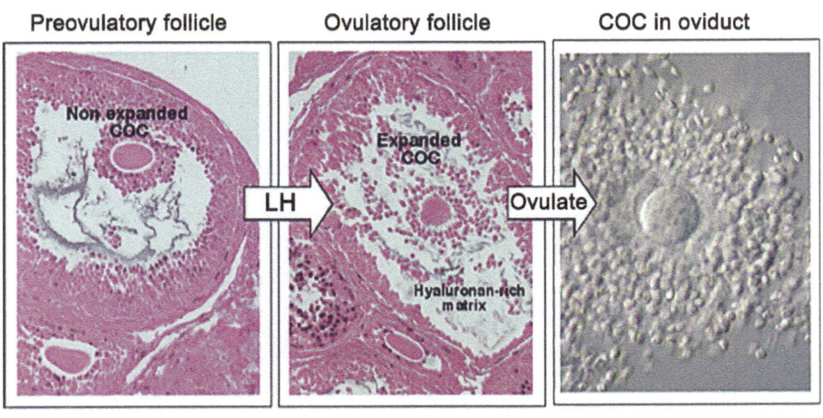

Fig. 3 Cumulus cell–oocyte complex (COC) expansion is induced by LH.

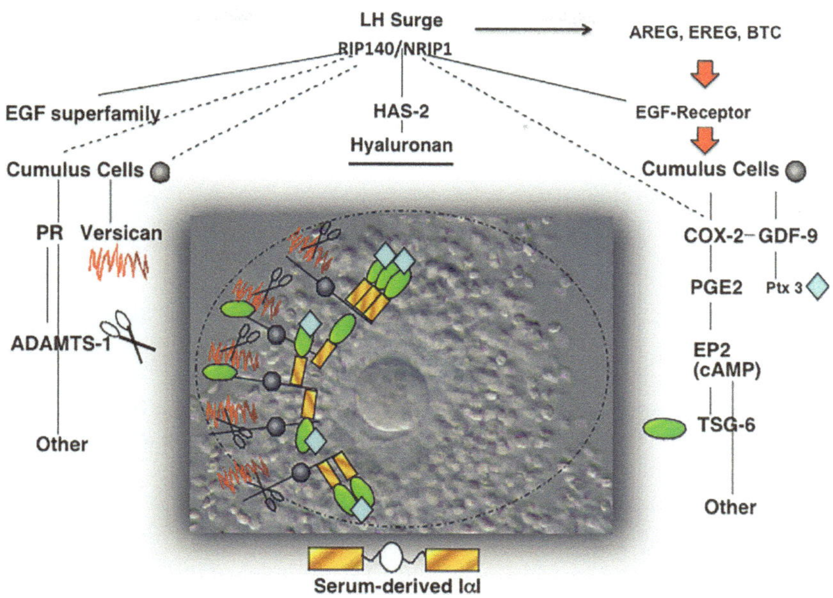

Fig. 4 The LH surge induces the expression of many signaling pathways and factors that lead to the production and stabilization of the matrix that controls COC expansion.

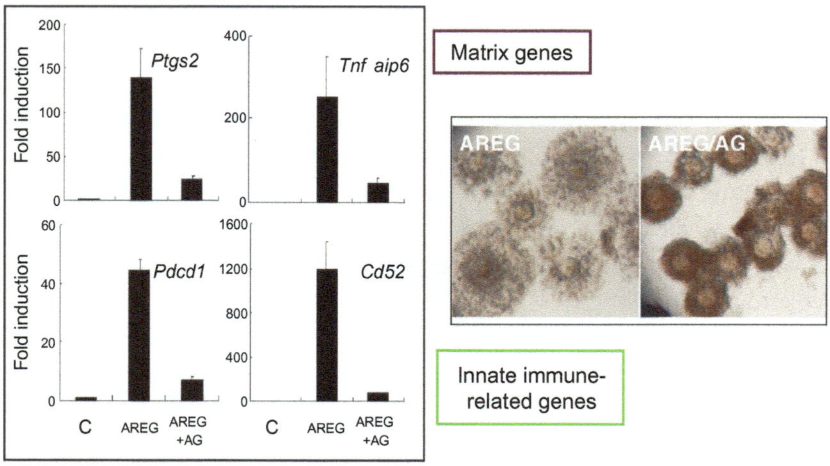

Fig. 5 AREG induces genes in COCs in an EGF receptor-dependent manner. *Adapted from Shimada, M., Gonzalez-Robayna, I., Hernandez-Gonzalez, I., & Richards, J.S. (2006). Paracrine and autocrine regulation of EGF-like factors in cumulus oocyte complexes and granulosa cells: Key role for prostaglandin synthase 2 (Ptgs2) and progesterone receptor (Pgr),* Molecular Endocrinology, 20, *348–364, 16543407.*

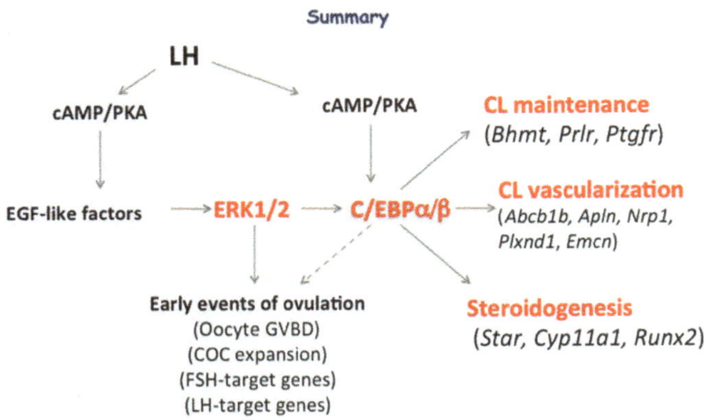

Fig. 6 Signaling pathways downstream of the LH surge that activate the transcription factors, C/EBPα and C/EBPβ, that control luteinization (corpus luteum (CL) maintenance and vascularization) and ovulation.

production. Thus, the binding of LH to the LH receptors on mural granulosa cells initiates the expression of the EGF-like factors that can then activate the EGF receptors in cumulus cells as well as in granulosa cells, allowing for the complete coordination of the ovulating follicle by the surge of LH.

5. PCOS AND OVULATION

Ovulation is impaired in women with polycystic ovarian syndrome (Dumesic & Richards, 2013). The underlying causes of PCOS are multifaceted but are manifested by elevated circulating androgens, a systemic proinflammatory condition and ovulation failure. Obesity appears to aggravate the PCOS condition, in part, by enhancing inflammation not only at the systemic level but also in ovulating follicles of women undergoing hormonal treatments for in vitro fertilization (Adams et al., 2016). These observations reinforce the impact of the immune system and inflammation in ovulation.

6. THE PORTAL TO EOC, p53, AND STEROID HORMONES

The central role of the EGF receptor pathway in mediating LH-induced ovulation indicated that a RAS/ERK1/2 pathway exerted a major effect on preovulatory granulosa cells. The power of this pathway in ovulating follicles led us to ask what role RAS might play if activated at earlier stages of follicular development. To address this, mutant mice were generated to express an active form of active RAS, $KRAS^{G12D}$, prematurely in granulosa cells of growing follicles using *Amhr2-cre* (Fan, Shimada, et al., 2008). The phenotype of $Kras^{G12D}$;*Amhr2-cre* expressing mice was striking and unexpected. Specifically, follicular development was derailed at an early stage. The abnormal follicles were depleted of oocytes, ceased to grow, did not differentiate, and did not die. Equally striking was the elevated presence of PTEN protein in granulosa cells of the abnormal follicles as revealed by immunohistochemistry and Western blots (Fan, Shimada, et al., 2008).

To determine the effects of PTEN in the mouse ovary and if we could reverse the effects of elevated PTEN in the abnormal follicles of the $KRAS^{G12D}$ mutant mice, we either depleted the *Pten* gene alone (Fan, Liu, & Richards, 2008) or in the $Kras^{G12D}$;*Amhr2-Cre* mutant background (Fan, Liu, Paquet, et al., 2009). Depletion of *Pten* alone enhanced ovulation and extended the life span of corpora lutea (Fan, Liu, et al., 2008). Depletion of *Pten* in the *Kras* mutant mice (*Pten*;$Kras^{G12D}$;*Amhr2-Cre*) did not prevent the development of abnormal follicles. Rather follicular development was arrested as in the $Kras^{G12D}$;*Amhr2-Cre* mice. However, the *Pten*/*Kras* double mutant mice developed large tumors derived from the ovarian surface epithelium (Fan, Liu, Paquet, et al., 2009) (Fig. 7). The tumors appeared early, occurred with 100% penetrance, and invaded the ovarian stroma. Thus, we suddenly had a mouse model of ovarian cancer in which follicular development and follicular steroid hormone biosynthesis was disrupted giving us an

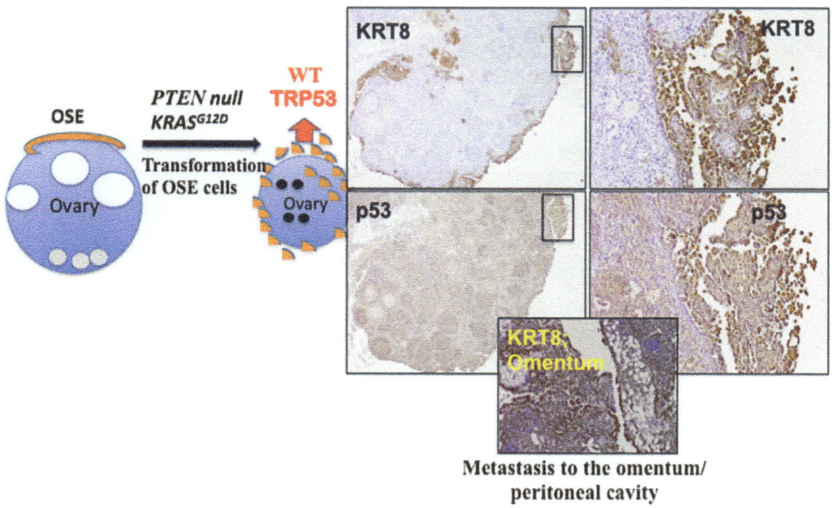

Fig. 7 The *Pten;*$Kras^{G12D}$*-Amhr2-Cre* mice develop serous adenocarcinomas with 100% penetrance at an early age but NO granulosa cell tumors are formed.

ideal model in which to determine the roles of steroid hormones on ovarian cancer progression. These mice became our entryway into the complex world of ovarian cancer characterized by epithelial cell transformation, genomic instability, oncogene expression, metastasis, and immune evasion. Ovarian cancer is also characterized by >96% mutations in the tumor protein, TP53 (Cole et al., 2016). Some of these mutations lead to a loss of p53 activity; others lead to a gain of oncogenic function. The gain-of-function (GOF) mutants contribute to genomic instability, metastasis, and drug resistance (Muller & Vousden, 2014; Vousden & Privies, 2009). Using our mutant mouse model we have investigated the functions of wild-type (WT) p53, loss of p53, and GOF p53 mutant alleles in the presence or absence of steroids.

To determine the role of p53 in the tumor-bearing double mutant mice, we made triple mutant mice in which the mouse *Trp53* gene was disrupted in the *Pten;Kras*^{G12D}*;Amhr2-Cre* mutant mice (*Trp53;Pten;Kras*^{G12D}*;Amhr2-Cre*) (Mullany, Liu, King, Wong, & Richards, 2012; Mullany et al., 2014). Abnormal follicles continued to be observed in the 53 null triple mutant mice, but to our surprise, the depletion of p53 did not cause more aggressive tumor growth as might be expected based on HGSCs in women. Rather only small lesions were observed, metastasis appeared minimal and large tumors never developed (Fig. 8). However, when the triple mutant mice lacking p53 were exposed to estradiol (implants), massive tumors developed and metastasized to the peritoneal cavity (Fig. 9). Progesterone (implants) blocked the effects of estradiol (Mullany et al., 2014). By contrast, when the double mutant mice that expressed (WT) p53 were exposed to estradiol

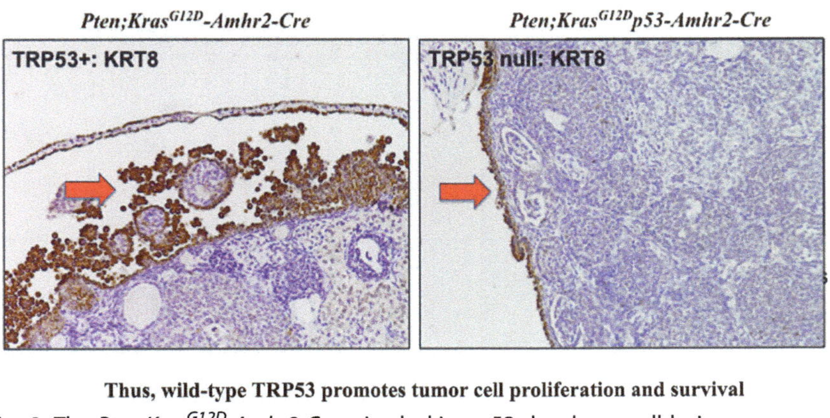

Fig. 8 The *Pten;Kras*^{G12D}*-Amhr2-Cre* mice lacking p53 develop small lesions.

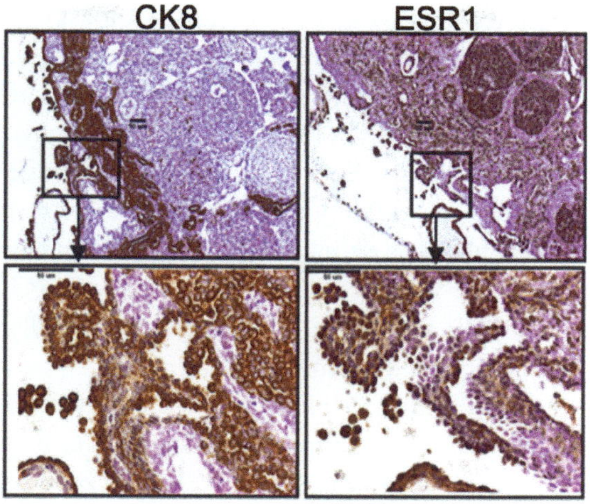

Fig. 9 Estrogen induces proliferation and invasion of *Pten/Kras/p53* null tumor cells into the ovarian stroma. *Adapted from Mullany, L.K., Liu, Z., Wong, K.K., Deneke, V., Ren, Y. A., Herron, A., et al., (2014). Tumor repressor protein 53 and steroid hormones provide a new paradigm for ovarian cancer metastases.* Molecular Endocrinology, 28*(1), 127–137 PMID: 24264574; PMCID: PMC23874458.*

(implants), the effects were minimal. Microarray analyses of cells derived from the tumors revealed marked differences in gene profiles of the triple mutant mice without or with estradiol. These results indicate that epithelial cell-derived ovarian cancers can be sensitive to steroids hormones and that responsiveness to steroids appears to depend, in part, on the functional status of p53.

To analyze the impact of specific GOF p53 mutants and steroid hormones on tumor growth, mice expressing the $p53^{R172H}$ mutant allele were bred into the *Pten;Kras*^{G12D};*Amhr2-Cre* double mutant mice (Ren et al., 2016). Tumors in mice heterozygous for the p53-R172H mutant and WTp53 alleles exhibited similar morphology and gene expression profiles as observed in the double mutant mice expressing WT p53 alleles. However, the p53-R172H/WT expressing tumors metastasized to the omentum and estradiol promoted proliferation. Unexpectedly, the p53-R172H/WT heterozygous mutant mice also developed mucinous tumor-like structures at 12 weeks of age with 80% penetrance. These mice represent the first model of mucinous tumors. Of note, some mucinous HGSCs in women express heterozygous WT and mutant p53 alleles, indicating that the presence of heterozygous alleles exerts functions distinct from WT or homozygous mutant alleles.

Indeed, the triple mutant mice expressing homozygous p53-R172H/R172H mutant alleles showed early signs of lesions and metastasis to the omentum within 4 weeks (Ren et al., 2016). Unfortunately, these mice die prematurely of other causes. However, cell lines derived from lesions of homozygous R172H tumors have allowed us to determine that these cells do not activate a p53 promoter–reporter construct, fail to respond to nutlin-3A that activates WT p53, and exhibit a gene profile distinct from tumor cell lines derived from mice with WT p53 or heterozygous R172H/WT p53 alleles. When these cells lines are injected intraperitoneally into syngenic mice, they grow rapidly on the omentum.

To relate our mouse studies to HGSCs in women, we are currently analyzing human ovarian cancer cell lines expressing WT, null, or specific GOF (R175H, R248Q, or R273H) p53 mutant alleles. These cell lines exhibit distinct morphologies, gene expression profiles, and responses to cytotoxic drugs in culture and xenografts in athymic mice in vivo.

7. LINKS TO GCT FORMATION

Components of the WNT/FZD pathway are differentially expressed in the murine ovary (Boyer et al., 2010; Hsieh, Johnson, Greenberg, & Richards, 2002). WNT4 has been shown to impact specification of the female gonad (Vainio, Heikkila, Kispert, Chin, & McMahon, 1999), is expressed in granulosa cells and corpora lutea (Hsieh et al., 2002), and exerts its activity via a canonical pathway leading to the activation of beta-catenin (CTNNB1) (Boyer et al., 2010). WNT5a antagonizes the effects of WNT4 in the ovary (Abedini et al., 2016; Boyer et al., 2010). Expression of dominant stable CTNNB1 during follicle development using *Cyp19a1-Cre* (*Ctnnb1((Ex3)fl/fl);Cyp19-Cre*) enhanced granulosa cell responses to FSH, including proliferation and expression of *Cyp19a1* but reduced LH-mediated ovulation and luteinization, in part, by reducing LH receptor levels in granulosa cells and LH-mediated induction of *Areg* and *Ereg* (Fan et al., 2010). Expression of a dominant stable form of CTNNB1 in mice using another recombinase *Amhr2-Cre* (*Ctnnb1((Ex3)fl/fl);Amhr2-cre*) impaired follicular development and eventually led to GCT formation with 57% penetrance at 7.5 months of age (Boerboom et al., 2005). However, the role of CTNNB1 is complex because this pathway has also been reported to increase LH receptor expression in granulosa cells in culture (Law, Weck, Kyriss, Nilson, & Hunzicker-Dunn, 2013). Thus, WNT signaling impacts granulosa cell proliferation, differentiation, and tumor formation but much remains to be understood.

As mentioned earlier, the expression of $Kras^{G12D}$ alone in granulosa cells potently blocks follicular development at an early stage leading to the formation of abnormal structures that lack oocytes and in which granulosa cells do not divide or differentiate (Fan, Shimada, et al., 2008). In stark contrast, the expression of $Kras^{G12D}$ in the *Ctnnb1((Ex3)fl/fl);Amhr2-cre* mutant mice markedly enhances granulosa cells tumor growth (Boerboom et al., 2005; Boerboom, White, Dalle, Courty, & Richards, 2006; Richards et al., 2012). Likewise, disruption of *Pten* in granulosa cells expressing activated CTNNB1 leads to a similar, almost identical GCT phenotype and gene expression profile to that of observed in the $Kras^{G12D}$;*Ctnnb1((Ex3)fl/fl); Amhr2-cre* mice (Richards et al., 2012). These observations indicate that activated CTNNB1 drives tumor formation and the gene expression profiles, whereas either the expression of mutant $Kras^{G12D}$ or depletion of *Pten* facilitates the actions of CTNNB1 (Richards et al., 2012). It is likely that $KRAS^{G12D}$ is activating the PI3K pathway in the GCTs, whereas $KRAS^{G12D}$ appears to be inhibiting this pathways in granulosa cells of the $Kras^{G12D}$;*Amhr2-Cre* mice where PTEN levels are strikingly high (Fan, Shimada, et al., 2008). What switches the effects of $KRAS^{G12D}$ is not yet clear.

GCTs have also been observed in mutant mice lacking inhibin (Matzuk, Finegold, Su, Hsueh, & Bradley, 1992), SMADs1/5(Middlebrook, Eldin, Li, Shivasankaran, & Pangas, 2009), or FOXO1/3 (Liu, Castrillon, Zhou, & Richards, 2013; Liu et al., 2015). Inhibin is a member of the TGFβ super family. Its major function is to antagonize the actions of activin on activin-specific receptors. Thus, in mice lacking inhibin, the expression and activity of activin are unopposed leading to persistent high levels of phospho-SMAD2 and phosphor-SMAD3 that act to enhance proliferation. SMAD1 and SMAD5 are downstream targets of bone morphogenic proteins (BMPs) that are also members of the TGFβ superfamily. BMPs bind and activate their own specific receptors that are distinct from those activated by activin. Activation of the BMP pathway leads to the phosphorylation of SMAD1/5 that preferentially mediate differentiation and suppress the effects of SMAD2/3. Thus, depletion of SMAD1/5 also facilitates activin-driven events including the persistent phosphorylation and activation of the SMAD2/3 pathway and proliferation and GCT formation.

FOXO1 and FOXO3 are members of the Forkhead boxO family of transcription factors that play many diverse roles in many different tissues

(Accili & Arden, 2004; Calnan & Brunet, 2008). In granulosa cells, FOXO1 impacts genes regulated by FSH (Herndon, Law, Donaubauer, Kyriss, & Hunzicker-Dunn, 2016) that are involved in steriodogenesis (Liu et al., 2009), cell survival, and apoptosis by interacting with the activin and BMP pathways (Liu et al., 2013). FOXO3 levels increase in corpora lutea (Fan, Liu, et al., 2008). Depletion of the genes encoding *Foxo1* and *Foxo3* in granulosa cells leads to GCT formation with low penetrance (Liu et al., 2013, 2015). However, when *Pten* is also depleted in the *Foxo1/3;Cyp19-cre* mutant mice, GCTs occur frequently and appear early, perhaps due to the loss of apoptosis-mediating events and/or differentiation to luteal cells (Fan, Liu, et al., 2008; Liu et al., 2015) (Fig. 10). The tumors in these mice also exhibit persistent high levels of phospho-SMAD3. In addition, the tumors express low levels of *Fshr*, *Amh*, *Nrob1*, and *Wt1* mRNA but retain normal levels of WNT4, FOXL2, and GATA4 mRNA and protein that are markers

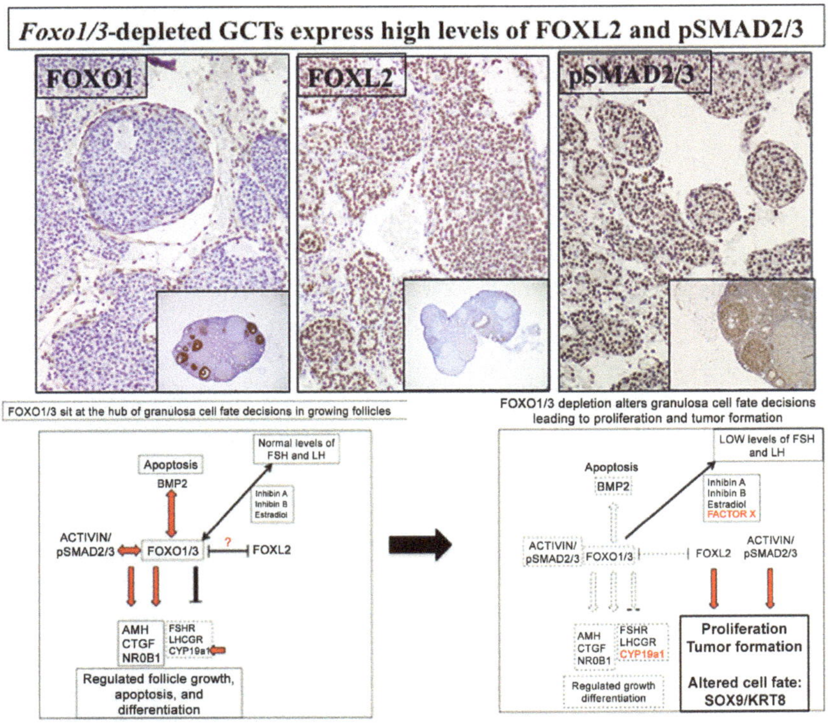

Fig. 10 Depletion of FOXO1/3 in granulosa cells leads to granulosa cell tumor formation and changes in granulosa cell fate decisions that are driven, in part, by FOXL2, activin mediated phosphorylation of SMAD2/3 and the maintenance of WNT4 and GATA4 expression.

of granulosa cell specification and proliferation but not differentiation (Liu et al., 2015). We conclude that FOXO1 and FOXO3 control granulosa cell fate decisions and proliferation by interacting with multiple signaling pathways.

8. SUMMARY

Follicular development and ovulation are complex development processes that are regulated by multiple, interacting pathways and cell types. Molecular biology approaches and mutant mouse models have contributed immensely to our knowledge of what genes and signaling cascades impact each stage of follicular development and ovulation, and how the alteration of gene expression profiles and the activation of specific signaling pathways can impact ovarian cancer development in epithelial as well as granulosa cells. Much remains to be learned. Hopefully, a greater understanding of ovarian biology will improve approaches to fertility regulation and the management of ovarian cancer.

ACKNOWLEDGMENTS

Supported in part by NIH-HD-076980 and NCI-181808.

REFERENCES

Abedini, A., Zamberlam, G., Lapointe, E., Tourigny, C., Boyer, A., Paquet, M., et al. (2016). WNT5a is required for normal ovarian follicle development and antagonizes gonadotropin responsiveness in granulosa cells by suppressing canonical WNT signaling. *The FASEB Journal*, *30*(4), 1534–1547. 26667040.

Accili, D., & Arden, K. C. (2004). FoxOs at the crossroads of cellular metabolism, differentiation and transformation. *Cell*, *117*, 421–426. 15137936.

Adams, J., Liu, Z., Ren, Y. A., Wun, W.-S., Zhou, W., Kenigsberg, S., et al. (2016). Enhanced inflammatory transcriptome in granulosa cells of women with polycystic ovarian syndrome. *The Journal of Clinical Endocrinology and Metabolism*, *101*, 3459–3468. 27228368.

Boerboom, D., Paquet, M., Hsieh, M., Liu, J., Jamin, S. P., Behringer, R. R., et al. (2005). Misregulated Wnt/b-catenin signaling leads to ovarian granulosa cell tumor development. *Cancer Research*, *65*, 9206–9215. 16230381.

Boerboom, D., White, L. D., Dalle, S., Courty, J., & Richards, J. S. (2006). Dominant-stable β-catenin expression causes cell fate alterations and Wnt signaling antagonist expression in a murine granulosa cell model. *Cancer Research*, *66*, 1964–1973. 16488995.

Boyer, A., Lapointe, E., Zheng, X., Cowan, R. G., Quirk, S. M., DeMayo, F. J., et al. (2010). WNT4 is required for normal ovarian follicle development and female fertility. *FASEB Journal: Official Publication of the Federation of American Societies for Experimental Biology*, *24*, 3010–3025. 20371632. PMCID:PMC22909279.

Calnan, D. R., & Brunet, A. (2008). The FoxO code. *Oncogene*, *27*, 2276–2288.

Cole, A. J., Dwight, T., Gill, A. J., Dickson, K.-A., Zhu, Y., Clarkson, A., et al. (2016). Assessing mutant p53 in primary high-grade serous ovarian cancer using immunohistochemistry and massively parallel sequencing [Article]. *6*, 26191. PMID: 27189670 https://doi.org/10.1038/srep26191. https://http://www.nature.com/articles/srep26191 - supplementary-information.

Conti, M., Hsieh, M., Park, J.-Y., & Su, Y.-Q. (2005). Role of the EGF network in ovarian follicles. *Molecular Endocrinology*, *20*, 715–723.

Doyle, K. M. H., Russell, D. L., Sriraman, V., & Richards, J. S. (2004). Coordinate transactivation of the ADAMTS-1 gene by luteinizing hormone and progesterone receptor. *Molecular Endocrinology*, *18*, 2463–2478.

Dumesic, D. A., & Richards, J. S. (2013). Ontogeny of the ovary in polycystic ovary syndrome. *Fertility and Sterility*, *100*(1), 23–38. PMID: 23472949 (https://www.ncbi.nlm.nih.gov/pubmed/23472949,)https://doi.org/10.1016/j.fertnstert.2013.02.011.

Espey, L. L., & Richards, J. S. (2002). Temporal and spatial patterns of ovarian gene transcription following an ovulatory dose of gonadotropin in the rat. *Biology of Reproduction*, *67*, 1662–1670.

Espey, L. L., Ujoka, T., Russell, D. L., Skelsey, M., Vladu, B., Robker, R. L., et al. (2000). Induction of early growth response protein-1 (Egr-1) gene expression in the rat ovary in response to an ovulatory dose of hCG. *Endocrinology*, *141*, 2385–2391.

Espey, L. L., Yoshioka, S., Russell, D. L., Robker, R. L., Fujii, S., & Richards, J. S. (2000). Ovarian expression of a disintegrin metalloproteinase with thrombospondin motifs during ovulation in the gonadotropin-primed immature rat. *Biology of Reproduction*, *62*, 1090–1095.

Espey, L. L., Yoshioka, S., Russell, D. L., Ujioka, T., Vladu, B., Skelsey, M., et al. (2000). Characterization of ovarian carbonyl reductase gene expression during ovulation in the gonadotropin-primed immature rat. *Biology of Reproduction*, *62*, 390–397.

Espey, L. L., Yoshioka, S., Ujioka, T., Fujii, S., & Richards, J. S. (2001). 3a-Hydroxysteroid dehydrogenase mRNA transcription in the immature rat ovary in response to an ovulatory dose of gonadotropin. *Biology of Reproduction*, *65*, 72–78.

Fan, H. Y., Liu, Z., Johnson, P. F., & Richards, J. S. (2011). CCAAT/enhancer-binding proteins (C/EBP)-{alpha} and -{beta} are essential for ovulation, luteinization and expression of key target genes. *Molecular Endocrinology (Baltimore, Md)*, *25*, 253–268. 21177758. PMCID:PMC23386543.

Fan, H. Y., Liu, Z., Paquet, M., Wang, J., Lydon, J. P., DeMayo, F. J., et al. (2009). Cell type specific targeted mutation of Kras and Pten document proliferation arrest in granulosa cells versus oncogenic insult in ovarian surface epithelial cells. *Cancer Research*, *69*, 6463–6472. 19679546. PMCID: PMC12741085.

Fan, H. Y., Liu, Z., & Richards, J. S. (2008). Targeted disruption of *Pten* in ovarian granulosa cells enhances ovulation and extends the life span of luteal cells. *Molecular Endocrinology*, *22*, 2128–2140. 18606860. PMCID: PMC12631369.

Fan, H. Y., Liu, Z., Shimada, M., Sterneck, E., Johnson, P. F., Hedrick, S. M., et al. (2009). MAPK3/1 (ERK1/2) in ovarian granulosa cells are essential for female fertility. *Science (New York, N.Y.)*, *324*, 938–941. 19443782. PMC2847890.

Fan, H. Y., O'Connor, A., Shitanaka, M., Shimada, M., Liu, Z., & Richards, J. S. (2010). Beta-catenin (CTNNB1) promotes preovulatory follicular development but represses LH-mediated ovulation and luteinization. *Molecular Endocrinology (Baltimore, Md.)*, *24*, 1794–1804. 20610534.

Fan, H. Y., Shimada, M., Liu, Z., Cahill, N., Noma, N., Wu, Y., et al. (2008). Selective expression of $Kras^{G12D}$ in granulosa cells of the mouse ovary causes defects in follicular development and ovulation. *Development (Cambridge, England)*, *135*, 2127–2137. 18506027. PMCID:PMC13541831.

Hernandez-Gonzalez, I., Gonzalez-Robayna, I., Shimada, M., Wayne, C. M., Ochsner, S. A., White, L., et al. (2006). Gene expression profiles of cumulus cell oocyte complexes during ovulation reveal cumulus cells express neuronal and immune-related genes: Does this expand their role in the ovulation process? *Molecular Endocrinology, 20*(6), 1300–1321. 16455817.

Herndon, M. K., Law, N. C., Donaubauer, E. M., Kyriss, B., & Hunzicker-Dunn, M. (2016). Forkhead box O member FOXO1 regulates the majority of follicle-stimulating hormone responsive genes in ovarian granulosa cells. *Molecular and Cellular Endocrinology, 434*, 116–126. PMID:27328024. https://doi.org/10.1016/j.mce.2016.06.020.

Hsieh, M., & Conti, M. (2005). G-protein-coupled receptor signaling and the EGF network in endocrine systems. *Trends in Endocrinology and Metabolism, 16*, 3320–3326.

Hsieh, M., Johnson, M., Greenberg, N. M., & Richards, J. S. (2002). Regulated expression of Wnt and Frizzled signals in the rodent ovary. *Endocrinology, 143*, 898–908.

Kawashima, I., Liu, Z., Mullany, L. K., Mihara, T., Richards, J. S., & Shimada, M. (2012). EGF-ike factors induce expansion of the cumulus cell-oocyte complexes by activating calpain-mediated cell movement. *Endocrinology, 153*, 3949–3959. 22673225. PMCID: PMC23404342.

Law, N. C., Weck, J., Kyriss, B., Nilson, J. H., & Hunzicker-Dunn, M. (2013). Lhcgr expression in granulosa cells: Roles for PKA-phosphorylated β-catenin, TCF3, and FOXO1. *Molecular Endocrinology (Baltimore, Md.), 27*(8), 1295–1310. https://doi.org/10.1210/me.2013-1025. 23754802.

Liu, Z., Castrillon, D. H., Zhou, W., & Richards, J. S. (2013). FOXO1/3 depletion in granulosa cells alters follicle growth, death and regulation of pituitary FSH. *Molecular Endocrinology (Baltimore, Md.), 27*(2), 238–252. 23322722.

Liu, Z., Fan, H. Y., Wang, Y., & Richards, J. S. (2010). Targeted disruption of Mapk14 (p38MAPKa) in granulosa cells and cumulus cells causes cell-specific changes in gene expression profiles that rescue cell-oocyte complex expansion and maintain fertility. *Molecular Endocrinology, 24*, 1794–1804.

Liu, Z., Ren, Y. A., Pangas, S. A., Adams, J., Zhou, W., Castrillon, D. H., et al. (2015). FOXO1/3 and PTEN depletion in granulosa cells promotes ovarian granulosa cell tumor development. *Molecular Endocrinology, 29*(7), 1006–1024. https://doi.org/10.1210/me.2015-1103. 26061565.

Liu, Z., & Richards, J. S. (2008). IL6: An autocrine regulator of the mouse cumulus cell-oocyte complex expansion process. *Endocrinology, 150*, 3360–3368. 19299453. PMCID: PMC12703543.

Liu, Z., Rudd, M. D., Hernandez-Gonzalez, I., Gonzalez-Robayna, I., Fan, H. Y., Zeleznik, A. J., et al. (2009). FSH and FOXO1 regulate genes in the sterol/steroid and lipid biosynthesis pathways in granulosa cells. *Molecular Endocrinology (Baltimore, Md.), 23*, 649–661. 19196834. PMCID:PMC12675958.

Lydon, J. P., DeMayo, F., Funk, C. R., Mani, S. K., Hughes, A. R., Montgomery, C. A., et al. (1995). Mice lacking progesterone receptor exhibit reproductive abnormalities. *Genes & Development, 9*, 2266–2278.

Matzuk, M. M., Finegold, M., Su, J.-G. J., Hsueh, A. J. W., & Bradley, A. (1992). a-Inhibin is a tumour-suppressor gene with gonadal specificity in mice. *Nature, 360*, 313–319. PMID:1448148.

Middlebrook, B. S., Eldin, K., Li, X., Shivasankaran, S., & Pangas, S. A. (2009). Smad1-Smad5 ovarian conditional knockout mice develop a disease profile similar to the juvenile form of human granulosa cells tumors. *Endocrinology, 150*, 5208–5217. 19819941.

Morris, J. K., & Richards, J. S. (1993). Hormonal induction of luteinization and prostaglandin endoperoxide synthase-2 involves multiple cellular signaling pathways. *Endocrinology, 133*, 770–779.

Morris, J. K., & Richards, J. S. (1996). An E-box region within the prostaglandin endoperoxide synthase 2 (PGS-2) promoter is required for transcription in rat ovarian granulosa cells. *The Journal of Biological Chemistry, 271*, 16633–16643.
Mullany, L. K., Liu, Z., King, E. R., Wong, K.-K., & Richards, J. S. (2012). Wild-type tumor repressor protein 53 (TRP53) promotes ovarian cancer cell survival. *Endocrinology, 153*(4), 1638–1648. 22396451. PMCID:PMC23320246.
Mullany, L. K., Liu, Z., Wong, K. K., Deneke, V., Ren, Y. A., Herron, A., et al. (2014). Tumor repressor protein 53 and steroid hormones provide a new paradigm for ovarian cancer metastases. *Molecular Endocrinology (Baltimore, Md.), 28*(1), 127–137. https://doi.org/10.1210/me.2013-1308. 24264574. PMCID: PMC23874458.
Muller, P. A. J., & Vousden, K. H. (2014). Mutant p53 in cancer: New functions and therapeutic opportunities. *Cancer Cell, 25*(3), 304–317. PMID:24651012 https://doi.org/10.1016/j.ccr.2014.01.021.
Natraj, U., & Richards, J. S. (1993). Hormonal regulation, localization and functional activity of the progesterone receptor in granulosa cells of rat preovulatory follicles. *Endocrinology, 133*, 761–769.
Ochsner, S. A., Day, A. J., Breyer, R. M., Gomer, R. H., & Richards, J. S. (2003). Disrupted function of tumor necrosis stimulated gene 6 blocks cumulus cell-oocyte complex function. *Endocrinology, 144*, 4376–4387.
Ochsner, S. A., Russell, D. L., & Richards, J. S. (2003). Decreased expression of tumor necrosis factor-a-stimulated gene 6 in cumulus cells of the cyclooxygenase-2 and EP2 null mice. *Endocrinology, 144*, 1008–1019.
Park, O.-K., & Mayo, K. (1991). Transient expression of progesterone receptor messenger RNA in ovarian granulosa cells after the preovulatory luteinizing hormone surge. *Molecular Endocrinology, 5*, 967–978. PMID:1840636.
Park, J.-Y., Su, Y.-Q., Ariga, M., Law, E., Jin, S.-L. C., & Conti, M. (2004). EGF-like growth factors as mediators of LH action in the ovulatory follicle. *Science, 290*, 395–398. 14726596.
Ren, Y. A., Mullany, L. K., Liu, Z., Herron, A. J., Wong, K.-K., & Richards, J. S. (2016). Mutant p53 promotes epithelial ovarian cancer by regulating tumor differentiation, metastasis, and responsiveness to steroid hormones. *Cancer Research, 76*, 2202–2208. 26964623.
Richards, J. S., Fan, H. Y., Liu, Z., Tsoi, M., Lague, M. N., Boyer, A., et al. (2012). Either Kras activation or Pten loss similarly enhances the dominant-stable CTNNB1-induced genetic program to promote granulosa cell tumor development in the ovary and testis. *Oncogene, 31*, 1504–1520. https://doi.org/10.1038/onc.2011.1341 21860425.
Richards, J. S., Liu, Z., & Shimada, M. (2015). Ovulation. In T. M. Plant & A. J. Zeleznik (Eds.), *Knobil and Neill's physiology of reproduction: Vol. 1* (4th ed., pp. 997–1021). Waltham, MA: Academic Press.
Richards, J. S., & Pangas, S. A. (2010). The ovary: Basic biology and clinical implications. *The Journal of Clinical Investigation, 120*, 963–972. 20364094. PMCID:PMC22846061.
Robker, R. L., Russell, D. L., Espey, L. L., Lydon, J. P., O'Malley, B. W., & Richards, J. S. (2000). Progesterone-regulated genes in the ovulation process: ADAMTS-1 and cathepsin L proteases. *Proceedings of the National Academy of Sciences of the United States of America, 97*, 4689–4694. 10781075.
Shimada, M., Gonzalez-Robayna, I., Hernandez-Gonzalez, I., & Richards, J. S. (2006). Paracrine and autocrine regulation of EGF-like factors in cumulus oocyte complexes and granulosa cells: Key role for prostaglandin synthase 2 (*Ptgs2*) and progesterone receptor (*Pgr*). *Molecular Endocrinology, 20*, 348–364. 16543407.
Shimada, M., Hernandez-Gonzalez, I., Gonzalez-Robayna, I., & Richards, J. S. (2005). Cumulus-oocyte complexes (COCs) express the EGF-like factor Amphiregulin that impacts not only induction of COX2 in these cells but also other genes associated with COC function. *Biology of Reproduction*. Abstract #57.

Shimada, M., Yanai, Y., Okazaki, T., Yamashita, Y., Sriraman, V., Wilson, M. C., et al. (2007). Synaptosomal associated protein 25 gene expression is hormonally regulated during ovulation and is involved in cytokine/chemokine exocytosis from granulosa cells. *Molecular Endocrinology, 21*, 2487–2502.

Sirois, J., Levy, L. O., Simmons, D. L., & Richards, J. S. (1993). Characterization and hormonal regulation of the promoter of the rat prostaglandin endoperoxide synthase 2 gene in granulosa cells. *The Journal of Biological Chemistry, 286*, 12199–12206.

Sirois, J., Simmons, D. L., & Richards, J. S. (1992). Hormonal regulation of messenger ribonucleic acid encoding a novel isoform of prostaglandin endoperoxide H synthase in rat preovulatory follicles. *The Journal of Biological Chemistry, 267*, 11586–11592.

Sriraman, V., Eichenlaub-Ritter, U., Bartsch, J. W., Rittger, A., Mulders, S. M., & Richards, J. S. (2008). Regulated expression of ADAM8 (a disintegrin and metalloprotease domain 8) in the mouse ovary: Evidence for a regulatory role of luteinizing hormone, progesterone receptor, and epidermal growth factor-like growth factors. *Biology of Reproduction, 78*(6), 1038–1048.

Sriraman, V., Lohmann, S., Mulders, S., & Richards, J. S. (2006). Induction of cGKII by LH and PR in the mouse ovary. *Molecular Endocrinology, 20*, 348–361 (in press).

Sriraman, V., Rudd, M. D., Lohmann, S. M., Mulders, S. M., & Richards, J. S. (2006). Cyclic guanosine 5′-monophosphate-dependent protein kinase II is induced by luteinizing hormone and progesterone receptor-dependent mechanisms in granulosa cells and cumulus oocyte complexes of ovulating follicles. *Molecular Endocrinology, 20*(2), 348–361.

Sriraman, V., Sharma, S. C., & Richards, J. S. (2003). Transactivation of the progesterone receptor gene in granulosa cells: Evidence that Sp1/Sp3 binding sites in the proximal promoter play a key role in luteinizing hormone inducibility. *Molecular Endocrinology, 17*, 436–449.

Sriraman, V., Sinha, M., & Richards, J. S. (2010). Progesterone receptor-induced gene expression in primary mouse granulosa cell cultures. *Biology of Reproduction, 82*(2), 402–412.

Tanaka, N., Espey, L. L., Kawano, T., & Okamura, H. (1991). Comparison of inhibitory actions of indomethacin and epostane on ovulation in rats. *The American Journal of Physiology, 260*, E170–E174.

Vainio, S., Heikkila, M., Kispert, A., Chin, N., & McMahon, A. P. (1999). Female development in mammals is regulated by Wnt-4 signalling. *Nature, 397*(6718), 405–409. https://doi.org/10.1038/17068.

Vousden, K. H., & Privies, C. (2009). Blinded by the light: The growing complexity of p53. *Cell, 137*, 413–431. 19410540.

Wong, W. Y. W., & Richards, J. S. (1991). Evidence for two antigenically distinct molecular weight variants of prostaglandin H synthase in the rat ovary. *Molecular Endocrinology, 5*, 1269–1279.

Wong, W. Y., & Richards, J. S. (1992). Induction of prostaglandin H synthase in rat preovulatory follicles by gonadotropin-releasing hormone. *Endocrinology, 130*, 3512–3521.

Xie, W. L., Chipman, J. G., Robertson, D. L., Erikson, R. L., & Simmons, D. L. (1991). Expression of a mitogen-responsive gene encoding prostaglandin synthase is regulated by mRNA splicing. *Proceedings of the National Academy of Sciences of the United States of America, 88*(7), 2692–2696. PMID:1849272.

Xie, W. L., Merrill, J. R., Bradshaw, W. S., & Simmons, D. L. (1993). Structural determination and promoter analysis of the chicken mitogen-inducible prostaglandin G/H synthase gene and genetic mapping of the murine homolog. *Archives of Biochemistry and Biophysics, 300*(1), 247–252. PMID:8424659 https://doi.org/10.1006/abbi.1993.1034.

Yoshioka, S., Ochsner, S., Russell, D. L., Ujioka, T., Fujii, S., Richards, J. S., et al. (2000). Expression of tumor necrosis factor-stimulated gene-6 in the rat ovary in response to an ovulatory dose of gonadotropin. *Endocrinology*, *141*, 4114–4119.

Zhang, Y.-L., Xia, Y., Yu, C., Richards, J. S., Liu, J., & Fan, H.-Y. (2014). CBP-CITED4 is required for luteinizing hormone-triggered target gene expression during ovulation. *MHR: Basic Science of Reproductive Medicine*, *20*(9), 850–860. https://doi.org/10.1093/molehr/gau040. 24878634.

CHAPTER SEVENTEEN

Human Thymidylate Synthase Inhibitors Halting Ovarian Cancer Growth

Stefania Ferrari*, Leda Severi*, Cecilia Pozzi†, Antonio Quotadamo*, Glauco Ponterini*, Lorena Losi*, Gaetano Marverti*,[1], Maria Paola Costi*,[1]

*University of Modena and Reggio Emilia, Modena, Italy
†University of Siena, Siena, Italy
[1]Corresponding author: e-mail addresses: gaetano.marverti@unimore.it; mariapaola.costi@unimore.it

Contents

Abstract

Human thymidylate synthase (hTS) has an important role in DNA biosynthesis, thus it is essential for cell survival. TS is involved in the folate pathways, specifically in the de novo pyrimidine biosynthesis. Structure and functions are intimately correlated, account for cellular activity and, in a broader view, with in vivo mechanisms. hTS is a target for anticancer agents, some of which are clinical drugs. The understanding of the detailed mechanism of TS inhibition by currently used drugs and of the interaction with the mechanism of action of other anticancer agents can suggest new perspective of TS inhibition able to improve the anticancer effect and to overcome drug resistance. TS-targeting drugs in therapy today are inhibitors that bind at the active site and that mostly resemble the substrates. Nonsubstrate analogs offer an opportunity for allosteric binding and novel mode of inhibition in the cancer cells. This chapter illustrates the relationship among the large number of hTS actions at molecular and clinical levels,

Vitamins and Hormones, Volume 107
ISSN 0083-6729
https://doi.org/10.1016/bs.vh.2017.12.002

its role as a target for ovarian cancer therapy, in particular in cases of overexpression of hTS and other folate proteins such as those induced by platinum drug treatments, and address the potential combination of TS inhibitors with other suitable anticancer agents.

ABBREVIATIONS

5-FU	5-fluorouracil
α-FR	α-folate receptor
BRCA1	breast cancer type 1 susceptibility protein
BRCA2	breast cancer type 2 susceptibility protein
CDK1	cyclin-dependent kinase 1
CDK12	cyclin-dependent kinase 12
CSMD3	CUB and sushi domain-containing protein 3
DHF	dihydrofolate
DHFR	dihydrofolate reductase
dTMP	2′-deoxythymidine-5′-monophosphate
dUMP	2′-deoxyuridine-5′-monophosphate
ENTPD1	ectonucleoside triphosphate diphosphohydrolase 1
ENTPD3	ectonucleoside triphosphate diphosphohydrolase 3
ENTPD8	ectonucleoside triphosphate diphosphohydrolase 8
FAT3	protocadherin fat 3
FPGS	folylpoly-γ-glutamate synthetase
GABRA6	gamma-aminobutyric acid receptor subunit alpha-6
GARFTase	glycinamide ribonucleotide formyltransferase
HNC	head and neck cancer
hTS	human thymidylate synthase
MRP1	multidrug resistance-associated protein 1
MRP2	multidrug resistance-associated protein 2
mTHF	5,10-methylene-5,6,7,8,-tetrahydrofolate
MTHFD1	C-1-tetrahydrofolate synthase, cytoplasmic
MTHFR	methylenetetrahydrofolate reductase
MTX	methotrexate
NF1	neurofibromin
NHL	non-Hodgkin lymphoma
NSCLC	non-small-cell lung carcinoma
OC	ovarian cancer
OS	overall survival
PCNA	proliferating cell nuclear antigen
PDX	pralatrexate
PFS	progression-free survival
PLD	pegylated liposomal doxorubicin
PMX	pemetrexed
PTCL	peripheral T-cell lymphoma
RB1	retinoblastoma-associated protein
RFC	reduced folate carriers
RTX	raltitrexed
SHMT1	serine hydroxymethyltransferase, cytosolic

TCGA The Cancer Genomic Atlas
TOP2A DNA topoisomerase 2-alpha
TS or TYMS thymidylate synthase

Thymidylate synthase (TS) (E.C. 2.1.1.45) is a highly conserved enzyme, essential for cell survival due to its important role in DNA biosynthesis (Carreras & Santi, 1995; Garg et al., 2010). The methylation reaction that it catalyzes provides the only de novo source of thymidylate (dTMP, 2′-deoxythymidine-5′-monophosphate). The TS role and interaction with the other metabolic proteins of the cellular environment can be described at growing level of complexity as reported in Fig. 1. Staring from the catalytic reaction, hTS belong to the carbon metabolic network, more specifically, of the thymidylate synthesis cycle involved in the de novo pyrimidine biosynthesis (Fig. 1, KEGGS). It is possible to identify the network that characterizes the hTS action taking advantage of the STRING database https://string-db.org/ (Szklarczyk et al., 2017) that allows access to the protein–protein interaction network centered on hTS. STRING is a database of known and predicted protein–protein interactions that includes direct (physical) and indirect (functional) associations. These predicted interactions

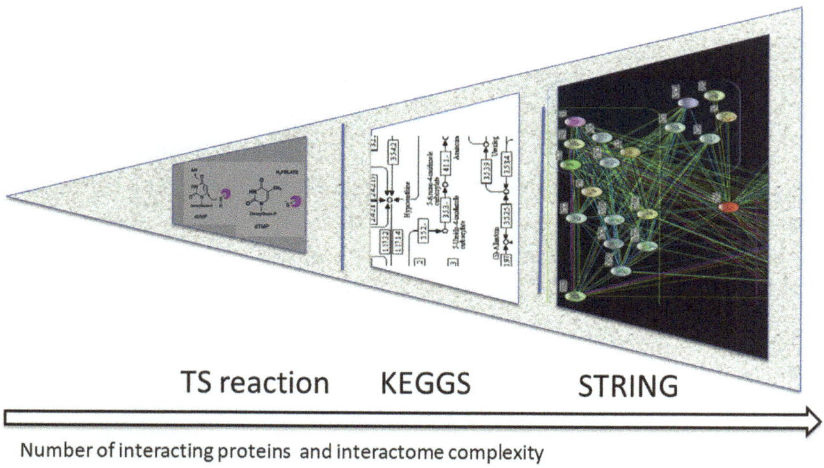

Fig. 1 Representation of the growing complexity of the metabolic role and interactome of hTS with the other components of the cellular environments. Two software names as examples are given to indicate that study of the interactome complexity can be described with the help of specific softwares on the basis of metadata and prediction (KEGGS and STRING).

stem from computational prediction, from knowledge transfer between organisms, and from interactions aggregated from other (primary) databases. Thus, it is possible to point out other proteins that have been experimentally demonstrated to be connected with hTS and are modulated as a result of hTS inhibition. This approach focuses on the metabolic interactions that are modulated by the methylating protein and allows a larger view of their biological effects (Fig. 1).

hTS is a target for anticancer agents, some of which are clinical drugs. The TS-targeting drugs in therapy are inhibitors that bind at the active site and that mostly resemble either the substrate 2′-deoxyuridine-5′-monophosphate (dUMP), being classified as dUMP-binding site inhibitors, or the cofactor 5,10-methylene-5,6,7,8,-tetrahydrofolate (mTHF), named folate-binding site inhibitors. The main drawback of these therapies is related to drug toxicity and resistance development, a phenomenon due to multiple effects but often related to overexpression of TS and some folate-related proteins. Thus, other strategies to target hTS are presently being pursued (Cardinale et al., 2011; Garg et al., 2010). In order to avoid to interfere with TS binding with the TSmRNA, thus triggering hTS upregulation, investigational drugs are now designed by focusing on allosteric-binding sites, such as the dimeric interface. hTS is itself an oncogenic protein and its high levels can favor cancer development. Therefore, counteracting the drug resistance effect may avoid further complications in cancer treatment (Cardinale et al., 2011). At the discovery levels, the new compounds that target hTS may exploit metabolic pathways and solicit unexpected cellular responses that may help to understand the roles of the protein in cells.

This chapter illustrates the relationship between the structure and the function of hTS, its role as a target for ovarian cancer therapy, in particular in cases of overexpression of hTS and folate protein such as those induced by platinum drugs treatments. Overall this chapter underlines the importance of the probing role of inhibitors in increasing our understanding of hTS mechanisms in ovarian cancer cells shedding light on the effects of biomolecular interconnections. TS-targeting drugs in ovarian cancer therapy are also discussed in the attempt of bridging the gaps between the knowledge in the preclinical and clinical mechanisms, the anticancer efficacy, and the complex interaction with the human body.

1. THYMIDYLATE SYNTHASE STRUCTURAL INSIGHTS

Due to its important catalytic role, TS is a highly conserved enzyme. Different TS states and conformations were so far suggested, based on experimental evidence: monomer/dimer, active/inactive, open/closed (Deschamps, Réty, Bareille, & Leulliot, 2017; Genovese et al., 2010). However, a definitive characterization of their roles and of the equilibria that regulate them has not yet been achieved. X-ray crystallography and mutagenesis studies have been often addressed to this purpose. The thymidylate synthase crystallographic 3D structure is reported in Figs. 2 and 3, where the domains and catalytic residues of main importance are reported. Each monomer of the human enzyme is composed of 313 residues. Its topology is an α/β-fold with 7 α-helices and 10 β-strands, arranged in three layers: a six-stranded mixed β-sheet that forms the monomer–monomer interface, a long α-helix across the sheet flanked by two shorter helices, and a mixed layer containing the remaining four helices and two antiparallel two-stranded β-sheets. The dimer contains two active sites, one per monomer. Each site is formed by residues of only one monomer except for two arginine residues from the second monomer that participate in the binding of the phosphate group in the first monomer (Fig. 2A). The residues of the active site are: Arg50, Thr51, Lys77, Phe80, Glu87, Ile108, Trp109, Asn112, Tyr135, Leu192, Pro193, Cys195, His196, Gln214, Arg215, Ser216, Gly217, Asp218, Leu221, Gly222, Val223, Phe225, Asn226, His256, Tyr258, Ile307, Lys308, Met309, Met311, Ala312, and Arg175' and Arg176' (the apostrophe indicates that the residues belong to the opposite monomer) (Almog, Waddling, Maley, Maley, & Van Roey, 2001; Ferrari, Losasso, & Costi, 2008).

The N-terminus region has been suggested to play a role for protein stability and in active–inactive equilibria. Two other regions, 99–129 and 144–158, whose conformations differ in the active and the inactive states, participate to the closure of the active site upon ternary complex formation (Brunn, Dibrov, Kao, Ghassemian, & Hermann, 2014; Chen, Jansson, Sim, Larsson, & Nordlund, 2017; Gibson, Celeste, Lovelace, & Lebioda, 2011; Lovelace, Gibson, & Lebioda, 2007; Phan, Koli, et al., 2001; Sayre et al., 2001).

From 2001, 50 X-ray crystallographic structures that include the apoenzyme, binary and ternary complexes, and the inactive and active forms of human thymidylate synthase (hTS) were released on the Protein Data Bank (https://www.rcsb.org). Twenty-seven of them are structures of the apoenzyme (Almog et al., 2001; Brunn et al., 2014; Cardinale et al., 2011; Chen et al., 2017; Gibson, Lovelace, & Lebioda, 2008; Huang et al., 2010; Lovelace et al., 2009; Lovelace, Minor, & Lebioda, 2005; Phan, Steadman, et al., 2001; Salo-Ahen et al., 2015; and the PDB IDs 4O1U and 4O1X). Almost all of them show the enzyme in the inactive

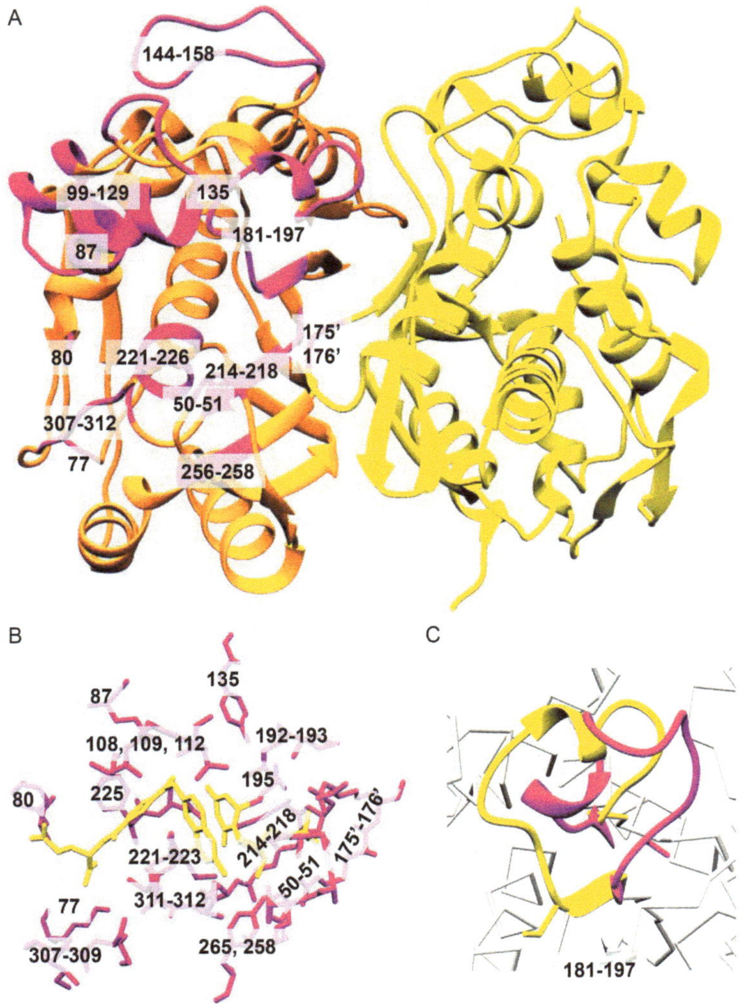

Fig. 2 Thymidylate synthase crystallographic 3D structure. (A) Dimer of TS (PDB ID: 1HVY) with the two monomers colored in *orange* and *yellow*. Within the first monomer, the regions 99–129, 144–158,181–199 and other residues, lining the active site are colored in *purple*; two residues, 175′ and 176′, from the opposite monomer are part of the active site of this monomer. (B) Detailed view of the active site of TS (PDB ID: 1HVY). In *yellow* are the substrate dUMP and the cofactor analog raltitrexed; residues delimiting the active site are in *purple*. (C) Detailed view of the active-site loop, 181–197, in the active (*yellow*) and inactive (*purple*) conformation of TS.

conformation. The exceptions are few mutated proteins that were selected to stabilize the active conformation (Almog et al., 2001; Chen et al., 2017; Gibson et al., 2008; Huang et al., 2010) and two recent structures obtained in low-salt conditions (Brunn et al., 2014; Deschamps et al., 2017).

In the active conformation, the catalytic Cys195 lies inside the active site and the enzyme can interact with its physiological substrates, dUMP and mTHF, or its substrate-mimicking inhibitors (Fig. 3A). The inactive conformation is characterized by the active-site loop, residues 181–197 (Fig. 2C), twisted by approximately 180 degrees with respect to its orientation in the active form. Thus, the sulfhydryl group of the catalytic nucleophile, Cys195, is more than 10Å away from its position in the active conformation, shifted toward the subunit interface. Movement of other two regions 99–129 and 144–158 (Fig. 2A) also accompanies this active-site loop conformation change; in particular the region 99–129 that is almost disordered in the inactive conformation becomes ordered in the active state. The native protein exists in equilibrium between the two conformational states. As shown later, TS inhibitors presently used in chemotherapy are substrates analogs and bind to the active conformation of hTS thus stabilizing it (Fig. 2) (Berger, Berger, & Lebioda, 2004).

Fourteen X-ray crystallographic structures of hTS are binary complexes (Cardinale et al., 2011; Carosati et al., 2012; Chen et al., 2017; Gibson et al., 2011; Huang et al., 2010; Lovelace et al., 2007; Tochowicz et al., 2015; and the PDB IDs: 4G6W, 4GD7, 4G20, 5WRN): five with the substrate or a substrate analogs in which TS is in the active conformation, and the remaining ones complexed with other ligands in which, with the exception of PDB ID: 3OB7, the enzyme is in the inactive conformation. The only exception is represented by a complex of R163K human TS and glutathione (PDB ID: 3OB7) obtained by soaking of the mutated enzyme stabilized in the active conformation, with the ligand bound in the substrate-binding site (Gibson et al., 2011). The structures of these binary complexes show the existence of different, allosteric, binding sites with respect to the substrate- and cofactor-binding sites. Nine X-ray crystallographic structures are those of the ternary complex of hTS with dUMP and five active-site binding antifolate TS inhibitors such as raltitrexed, methotrexate, nolatrexed, pemetrexed, and CBK11533 (Almog et al., 2001; Chen et al., 2017; Phan, Koli, et al., 2001; Sayre et al., 2001; Almqvist et al., 2016). One structure refers to a ternary complex with a ligand, 1-(5-methoxy-1H-indol-3-yl)-*N*,*N*-dimethyl-methanamine, that binds at the dimer-interface of the active structure (Fig. 3A) (Chen et al., 2017). However, this ligand does not inhibit the catalytic function of the protein.

While binding of dUMP in the binary complex does not induce important protein rearrangements, dUMP binding to form ternary complexes of the enzyme with active-site ligands induce a closure of the active site. This goes together with movements of the C-terminus region and various protein

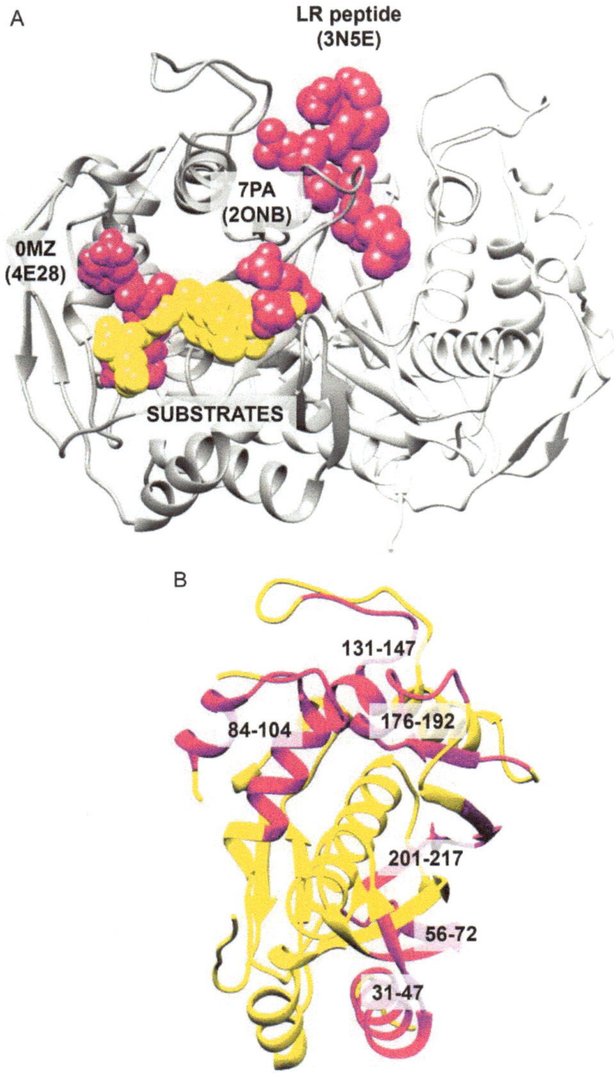

Fig. 3 TS regions interacting with substrates, monomer, and TSmRNA. (A) The active site and the other binding sites as shown from crystallographic studies are presented. Dimer of TS (PDB ID: 3N5E) with the two monomer colored in *gray* and *white*. The active site is indicated by the substrate dUMP and the cofactor analog raltitrexed, represented in *sphere* and colored in *yellow* (taken from PDB ID: 5X5Q). The other three known binding sites are indicated by the LR peptide (PDB ID: 3N5E), compound 0MZ (taken from PDB ID: 4E28), compound 7PA (taken from PDB ID: 2ONB), represented in *sphere* and colored in *purple*. (B) Monomer of TS (PDB ID: 1HW3) colored in *yellow*. In *purple* are indicated the regions suggested to be involved in the mRNA binding: 31–47, 56–72, 84–104, 131–147, 176–192, 201–217.

segments lining the substrate-binding sites. Moreover, in the human enzyme the two specific regions, 99–129 and 144–158, contribute to the closure of the active site following binding of the substrates (Almog et al., 2001; Chen et al., 2017; Sayre et al., 2001).

2. TS FUNCTION

2.1 TS Catalytic Function

The catalytic activity is performed by TS in its homodimeric active state, through the catalytic Cys195 residue and assisted by the cofactor 5,10-methylene-tetrahydrofolate (mTHF) (Carreras & Santi, 1995). The active-site cysteine residue gives Michael addition to the 6-position of dUMP to give an enolate, which attacks mTHF, thus forming a ternary complex (Fig. 4). The enzyme-catalyzed removal of the C5 proton leads to the beta-elimination of THF. Oxidation of the latter gives dihydrofolate (DHF) and the enzyme-bound thymidylate enolate. Finally, the reverse of the first step releases the active-site cysteine residue and produces dTMP (Carreras & Santi, 1995). A great amount of mutagenesis, enzymatic kinetic, and crystallographic studies have been addressed to clarify the TS catalytic mechanism. Finer-Moore, Santi, and Stroud (2003) reviewed all these studies and proposed that, in addition to Cys195, other few conserved

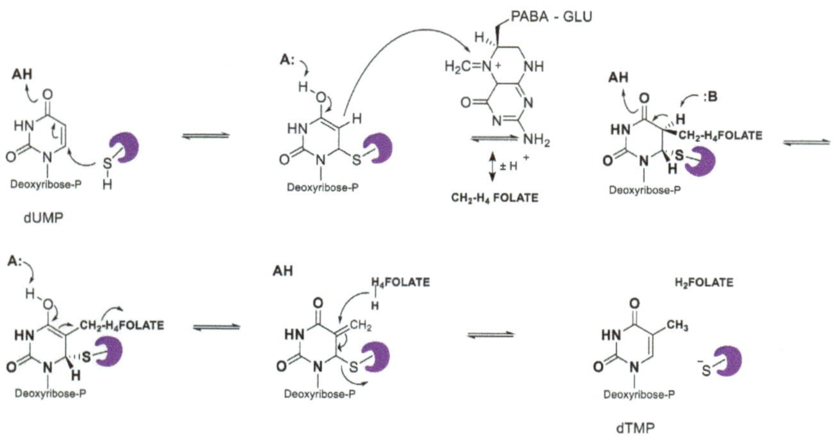

Fig. 4 Thymidylate synthase catalysis. The –SH group from Cys195 in human thymidylate synthase reacts with 2′-deoxyuridine-5′-monophosphate (dUMP) and 5,10-methylentetrahydrofolate (CH_2-H_4FOLATE) to produce 2′-deoxythymidine-5′-monophosphate (dTMP) and dihydrofolate (H_2FOLATE). PABA-GLU state for *para*-aminobenzoic acid and glutamic acid moieties of the folate cofactor.

residues were essential in the catalysis. Arg215, adjacent to the active-site cysteine, was postulated to activate the cysteine through electrostatic interactions. Although its role has not been proven, the variant Arg215Lys, among all the tested ones, is the only one with a value of k_{cat} close to that of the wild-type enzyme. This is likely due to the fact that lysine has electrostatic properties similar to those of arginine. His196 and Asn226 are proposed to modulate the orientation and pK_a of ordered water molecules acting as a general acid during nucleophilic attack of the catalytic cysteine at the C6 atom of dUMP. Asp218 has been proposed to be the acid that catalyzes protonation of N5 of the cofactor, while Tyr135 favors removal of a proton from the C5 atom of dUMP thus catalyzing the breakdown of the covalent ternary complex. Glu87 catalyzes hydrogen transfers to and from the nucleotide pyrimidine ring during different chemical steps. Asn226 has also a critical role in substrate selectivity. Most of the other highly conserved residues in the enzyme, Arg50, Trp109, Arg175, Gln214, Ser216, His256, Tyr258, contribute to catalysis mainly driving conformational changes in the enzyme and helping the different actors of the reactions to assume the optimal orientations to react (Ferrari et al., 2008; Finer-Moore et al., 2003). Inhibitors able to interact with them or to affect their contribution to the reaction, e.g., by binding in an allosteric site, may force and stabilize the enzyme in a conformational state, such as the inactive one, that cannot afford catalysis. TS is describe as a half-of-the-sites activity enzyme under physiological conditions, i.e., the two active sites of a protein dimer can work for a catalytic event one at a time (Anderson, O'Neil, DeLano, & Stroud, 1999; Weichsel, Montfort, Cieśla, & Maley, 1995). Some residues are believed to be involved in monomer–monomer communications responsible for a negative cooperativity between the substrate bindings at the two active pockets in a dimer. These are Gnl214, Arg215, and Ser216 that interact with dUMP and influence Tyr213 (Fig. 2). The latter residue interacts with its counterpart in the opposite monomer. However, crystallographic studies can rarely reproduce such structural details because of the experimental conditions required to obtain the crystallographic structures.

2.2 TSmRNA Interaction

TS is an mRNA-binding protein too. By binding an mRNA, it may act to coordinately regulate the cellular expression of several genes and, in doing so, may function as a critical regulator of several key aspects of cell cycle control, including apoptosis and chemosensitivity (Liu et al., 2002). TS was observed to bind its own mRNA, a behavior proposed to act as a feedback inhibition mechanism that regulates TS levels (Chu et al., 1991). Also, the native protein has been reported to bind several other mRNAs such as p53 and c-myc (Phan, Steadman, et al., 2001).

The autoregulation of the TS translational efficiency has been shown to occur through the high-affinity interaction (of the order of 1–3 nM) of the TS protein with either or both of the two *cis*-acting elements on its cognate mRNA. The first of these two regions includes the translational start sequence in a stable stem–loop structure and occurs at nucleotides (nt) 81–110 and the second lies between nt 450 and 520 in the protein-coding region (Lin, 2003; Liu et al., 2002; Voeller, Zajac-Kaye, Fisher, & Allegra, 2002). In vitro and in vivo studies have shown that each site could function independently of the other. However, both elements are required for the complete translational autoregulatory effects of TS.

There is growing evidence that the RNA-binding activity of TS is affected by its state of ligand occupancy. When TS is free of ligands, the RNA-binding activity is maximal, resulting in translational repression of TSmRNA (Lin, 2003). However, when TS is bound to either of its physiologic substrates or to substrate-analog inhibitors the in vitro binding of hTS to TS mRNA is reduced, and so is the repression by TS of TSmRNA translation (Lin, 2003; Phan, Steadman, et al., 2001). The net effect of ligand binding is abrogation of translational repression, which results in synthesis of new TS protein (Lin, 2003). As a critical chemotherapeutic target, intracellular levels of TS are an important determinant of sensitivity to TS inhibitors. This phenomenon ultimately results in emergence of drug resistance by the tumor. Since exposure to the TS inhibitors currently used in clinical practice leads to increased levels of intracellular TS activity, e.g., from about two- to fourfold within 24 h after exposure to fluoropyrimidines, the intracellular level of TS in the malignant tissue of patients may be used to predict response to therapies. The levels of TSmRNA do not change significantly, because the increase is predominantly at the protein level. Reduction of TS expression levels might be beneficial during cancer therapy with TS-inhibiting drugs which ablate binding of the protein to its mRNA. In particular, compounds that stabilize the TS-binding RNA motif, and thereby inhibit ribosomal initiation, might be used in combination with existing TS-targeting drugs to overcome tumor resistance development during chemotherapy (Brunn, Garcia Sega, Kao, & Hermann, 2012; Garg et al., 2010).

The native protein exists in two conformational states, an active and an inactive one. All TS inhibitors presently used in chemotherapy are substrate analogs and have shown to stabilize the active conformation of hTS and to abrogate the binding of the protein to its mRNA (Berger et al., 2004). In order to exert its function of protein expression regulation, TS must be ligands-free and, presumably, in the inactive state. Although the residues involved in the mRNA binding have not been identified with certainty, the residues in the surface region 84–104 are believed to be involved (Brunn et al., 2014). Also, peptides corresponding to the TS amino acid sequences 31–47, 56–72, 131–147, 176–192, and 201–217 from the monomer–monomer interface region were found to bind mRNA as well (Voeller et al., 2002). In particular, crystal structure analysis reveals that in the active conformation of the ternary complex of the enzyme with the substrates, the Gly84-Lys104 helix–loop–helix motif is obstructed by the Lys107-Glu128 peptide which adopts an ordered, predominantly helical structure. This may prevent protein binding to RNA (Brunn et al., 2014) (Fig. 3B).

2.3 Thymidylate Synthase Interactome Inside STRING

The catalytic function of TS is part of the thymidylate synthase cycle, within the pyrimidine metabolism and one carbon pool by folate pathways. This last is also connected to the folate biosynthesis pathway that in human cell is implicated in the regeneration of the active form of reduced folates (http://www.kegg.jp/). Folate allows a rapid division of tumor cells, since it is an essential compound for purine and pyrimidine biosyntheses. It is supplied from diet and delivered inside human cells by folate transporters. Several kinds of tumors have been shown to express high levels of folate receptor α (FRα) on the cell surface (Assaraf, Leamon, & Reddy, 2014).

With the aim of understanding the connections among TS and the proteins involved in other metabolic process, we employed the STRING software, version 10.5 with 0.9 highest confidence, https://string-db.org/ (Szklarczyk et al., 2017). The software allows expanding the knowledge on the interactions of a protein of interest by exploiting its database of experimentally determined and predicted protein–protein interactions.

The functional partners predicted by STRING are related to DNA biosynthesis metabolic processes such as TK1 and TK2 (Fig. 5). Moreover, the included proteins are involved in biological processes other than the catalytic function of hTS through the STRING tool. An example is provided by proteins ENTPD3, ENTPD8, and ENTPD1 that are implicated in nucleoside phosphate metabolic processes. Also, TS is related to MTHFR,

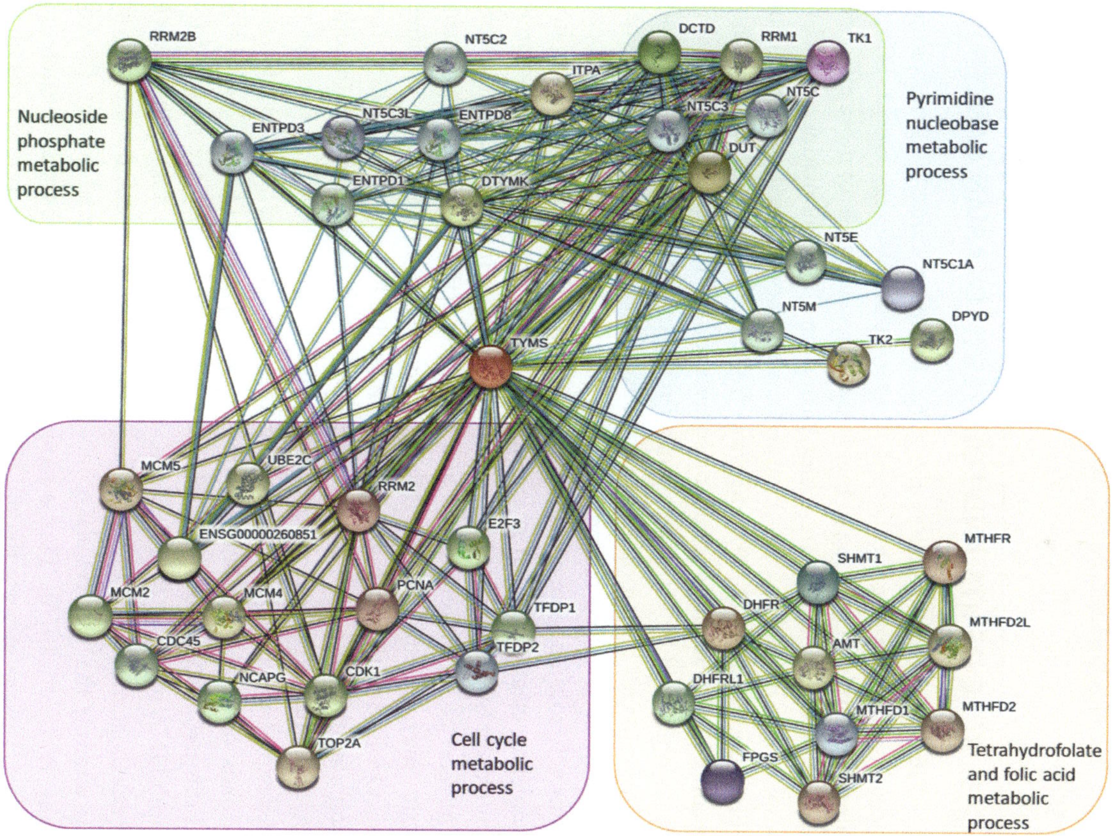

Fig. 5 Network of thymidylate synthase (TYMS) and 50 protein connectors obtained through String software (0.9 as highest confidence value). Thymidylate synthase is involved in several biological processes: nucleoside phosphate (proteins in *green square*), pyrimidine nucleobase (*blue square*), cell cycle (*violet square*), tetrahydrofolate and folic acid metabolic process (*orange square*).

DHFR, MTHFD1, SHMT1, i.e., the most important proteins in folic acid biological processes, and to several cell-cycle-related proteins (PCNA, CDK1, TOP2A). More attention needs to be devoted to the understanding of the complex proteins modulation that happens when TS is active. This is important to understand the difference when a disease status is developed in the cells.

3. ROLE OF hTS AND FOLATES IN OVARIAN CANCER

Ovarian cancer (OC) is the most common cause of gynecologic cancer mortality among women of all ages in the Western world. Efforts to improve survival continue to focus on the development of more effective systemic therapies, which are mainly based on the combination of cyclophosphamide or paclitaxel and a platinum compound, such as cisplatin or carboplatin (Muggia, 2009) and more recently on targeted therapy and immunotherapy as single agents or combinations. The use of drug combinations is justified by the occurrence of resistant cell populations in tumors and the resultant limitation of the efficacy of platinum drugs. The mechanisms behind acquired resistance to cisplatin and its derivatives have been extensively studied in many different tumor cell lines but they are still not completely clear, although it is evident that the process is multifactorial. While most of the mechanisms of resistance may remain unchanged, overexpression of DNA replication and increased activity of folate cycle and repair enzymes are almost always encountered in resistant cell lines (Lai, Ozols, Smyth, Young, & Hamilton, 1988; Masuda et al., 1988), this being a significant component of enhanced DNA repair capacity. In addition, resistant cells often respond to cisplatin treatment with an increase in gene expression for DNA repair enzymes and enzymes necessary for the synthesis of thymidine, such as thymidylate synthase (TS, EC 2.1.1.45) and dihydrofolate reductase (DHFR, 1.5.1.3), since they are required for a rapid DNA turnover (Scanlon & Kashani-Sabet, 1988). Cisplatin interacts selectively at the tumor membrane and inhibits methionine uptake (Fig. 6) thus altering the intracellular concentrations of reduced folates (Krebs, Hems, & Tyler, 1976). This is likely the mechanism by which sublethal doses of cisplatin increase the mTHF levels and make the tumor cells more sensitive to fluoropyrimidines (Scanlon, Newman, Lu, & Priest, 1986).

On the other hand, the characterization of ovarian cancer tissues that is reported in The Cancer Genomic Atlas (TCGA, https://cancergenome.nih.gov/) contains the list of mutated genes in cancer with respect to normal

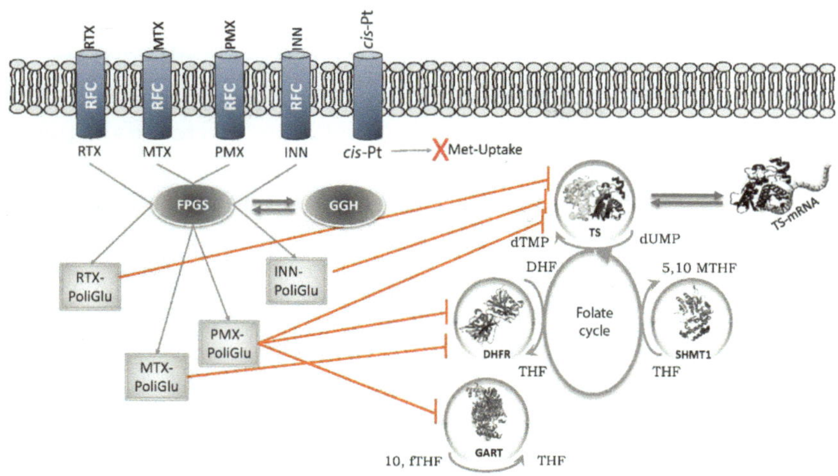

Fig. 6 Main targets involved in antifolates targeting thymidylate synthase and the folate cycle. The RFC transporters are reported. *cis*-Pt drug transporter and membrane-related effect of blocking methionine uptake. (Folate receptors for folic acid transport and lower involvement in antifolates transport are not shown.)

tissues and suggests a nonleading role of the TS gene mutation in OC development. Even if TS is not considered a first-line therapy target in OC treatment, allosteric TS-targeting drugs can be used in combination to prevent drug resistance development or can be used to reduce high TS levels in platinum drug-resistant OC. This concept has been applied in different cases at the cellular level (Marverti et al., 2012, 2010, 2013; Rytelewski et al., 2013).

3.1 Insides From TCGA

TCGA researchers measured comprehensively genomic and epigenomic abnormalities on clinically annotated high-grade serous ovarian cancer (HGS-OvCa) samples (Cancer Genome Atlas Research Network, 2011). They focused on the identification of molecular abnormalities able to influence pathophysiology, to affect outcome, or to constitute therapeutic targets. They reported the analysis of 489 clinically annotated stage II–IV HGS-OvCa samples and the corresponding normal DNA. Samples were surgically resected before systemic treatment, the patients received a platinum agent and, almost all, a taxane. According to the data, the median progression-free survival (PFS) and overall survival of the cohort was similar to those published in other trials (Bookman et al., 2009; Muggia et al., 2000). The researchers were interested in finding mutations that might be important in HGS-OvCa pathophysiology. So, they searched for non-synonymous or splice site mutations present at significantly increased frequencies relative to background and predicted the mutations' impacts on protein function. The genes that were found altered were TP53, mutated in nearly 96% of the analyzed samples, BRCA1 and BRCA2, mutated in 11%–12% of the cases, and other six genes recurrently mutated in the selected patients, RB1, NF1, FAT3, CSMD3, GABRA6, and CDK12. Searching of the TCGA repository (https://portal.gdc.cancer.gov/) for information on the mutations of the TS gene (TYMS) afforded three cases of single nucleotide substitution. Their consequences are two mutated proteins, Met219Arg and Asp69Gly, with a medium predicted impact on protein function and a nonfunctional protein truncated at the catalytic residue Cys195. From the structure–activity relationship analysis, the Met219Arg mutation occurs in the core of the monomer but outside the active site, while the Asp69Gly mutation occurs at the surface of the monomer and, therefore, the protein and its activity can be only slightly affected by the mutation and it is predicted to be viable. The number of clinical cases is limited, but the broad characterization triggers the exploration of the role of some important proteins that could be developed also as anticancer targets. Some of them are connected to TS such as those involved in cell replication, DNA repair. The most of the interactome remains to be explored.

3.2 Thymidylate Synthase Inhibition

The most important class of anti-TS drugs used in the anticancer therapy are nucleotide analogs such as 5-fluorouracil (5-FU), the prodrug of the corresponding nucleotide, and dUMP analogues such as capecitabine and tegafur, that are activated to 5-fluoro-deoxyuridine-monophosphate and reduced folate cofactor analogues, known as antifolates, such as methotrexate (MTX), pemetrexed (PMX), raltitrexed (RTX), and pralatrexate (PDX). These antifolates perturb the folate metabolic pathway by causing a marked decrease in the production of dTMP, a thymidylate deficiency and consequent imbalances in the cellular nucleotide pools. The final effect is slowing of cancer cells and tumor growth due to impairment of DNA replication and repair (Assaraf, 2007; Chu, Callender, Farrell, & Schmitz, 2003; Gonen & Assaraf, 2012). Furthermore, their cytotoxic activity is due to perturbation of additional metabolic pathways, such as the de novo purine and pyrimidine syntheses as well as DNA and RNA biosynthesis.

5-FU has been used in combination with cisplatin in the human OC line A2780 in culture, and it was shown that clinically achievable exposures to these drugs produced synergistic toxicity toward these cells (Table 1). As reported earlier, the TS protein exerts a tight control over its own synthesis during the cell cycle by binding to its own mRNA in a negative autoregulatory manner (Chu & Allegra, 1996). A number of studies with cultured cell lines, tumor models, and clinical specimens have shown that classical TS inhibitors that mimic the substrate and cofactor interactions disrupt the TSmRNA–protein complex and restore the mRNA translational efficiency (Chu & Allegra, 1993, 1996; Chu et al., 1993; Keyomarsi, Samet, Molnar, & Pardee, 1993), thus accounting for the occurrence of resistance to TS-targeted compounds. Recent clinical and preclinical reports demonstrate that levels of expression of TS gene and a few other genes represent important predictors of response and survival for patients treated with drugs including cisplatin and 5-FU (Theisen et al., 2008; Wang et al., 2008).

RTX, MTX, and PMX undergo polyglutamation by folylpoly-γ-glutamate synthetase (FPGS), and the polyglutamated forms of these agents are stronger inhibitors of folate-dependent enzymes. Short- or long-term exposure to the antifolates induces the decrease of the activity of the polyglutamilating protein FPGS thus reducing the antifolates activity (Jansen & Pieters, 1998). In addition, MRP1 and MRP2 overexpression is associated with resistance to short-term exposure of polyglutamatable antifolate drugs in human OC cells (Hooijberg et al., 1999).

To overcome the above-mentioned problems due to classical hTS inhibitors, novel strategies aiming to inhibit TS overexpression and restore chemosensitivity to TS inhibitors have been developed and effectively tested (Cardinale et al., 2011; Garg et al., 2010), including strategies exploiting the endocytotic properties of folate receptors (Leamon, 2008). All data of hTS inhibition and EC_{50} in ovarian cancer cells are reported in Table 1.

3.2.1 Anti-TS Drugs in Therapy and Their Use in Ovarian Cancer Treatment

Over 1600 clinical trials have been established all over the world for the evaluation of the therapeutic use of fluoropyrimidines, PMX, RTX, MTX, and PDX for the possible treatment of several different solid tumors in Phases II, III, and IV. TS-targeted agents are used in clinical practice to treat numerous solid and hematological malignancies, either alone or as anchor drugs in combination treatment regimens (Wilson, Danenberg, Johnston, Lenz, & Ladner, 2014). Fluoropyrimidines, PMX, and PDX are in clinical studies either alone or in combinations to treat ovarian cancers.

Fig. 6 shows the main targets of the antifolates actions.

Table 1 IC_{50} and EC_{50} hTS Inhibitors

Name	Structure	IC_{50} hTS	EC_{50} Ovarian Cell Line
5FU	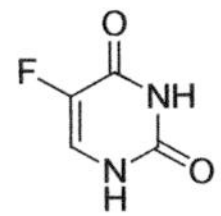	1.86 μM (El-Naggar, Abou-El-Regal, El-Metwally, Sherbiny, & Eissa, 2017)	A2780 cells 5.1 ± 0.4 μM (Carosati et al., 2012) OVCAR-8 cells 24(h) = 31 μM (Huehls et al., 2011)
FdUMP	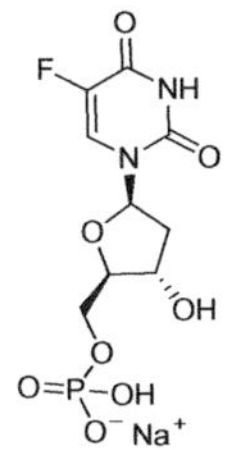	K_i FdUMP = 3.1 + 0.3 nM (Honeywell, Ruiz van Haperen, Veerman, Smid, & Peters, 2015)	OVCAR-8 cells 24(h) = 12 μM (Huehls et al., 2011)
Capecitabine	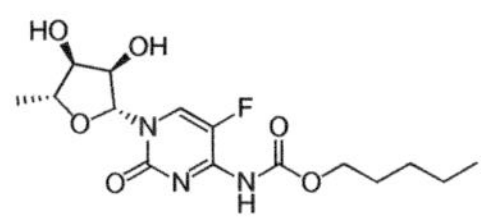	Prodrug of 5FU by thymidine phosphorylase	
Gemcitabine	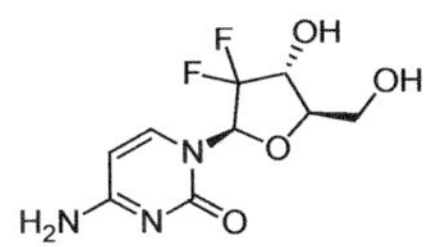	K_i dFdUMP (metabolite gemcitabine = 132 + 40 μM (Honeywell et al., 2015)	dFdC (μM) EC50 A2780 Incubation time (h) 4(h) = 0.007 ± 0.001 24(h) = 0.002 ± 0.0006 72(h) = 0.002 ± 0.001

Pemetrexed	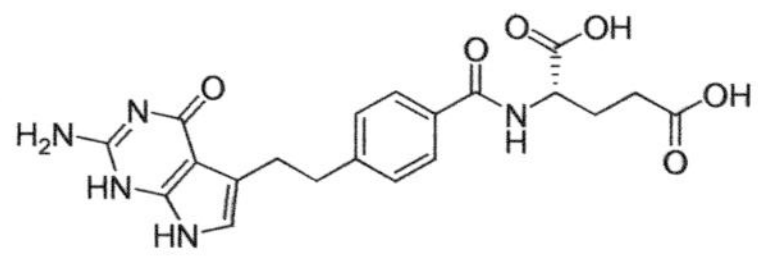	$K_i = 109\,nM$ (Curtin & Hughes, 2001; Gangjee, Jain Jaclyn, Guo, Queener, & Kisliuk, 2010)	EC50 (nmol/L) IGROV1 57.7 ± 3 SKOV3 77.5 ± 5.5 TOV112D 48.9 ± 5.4 (Hou et al., 2017)
Raltitrexed	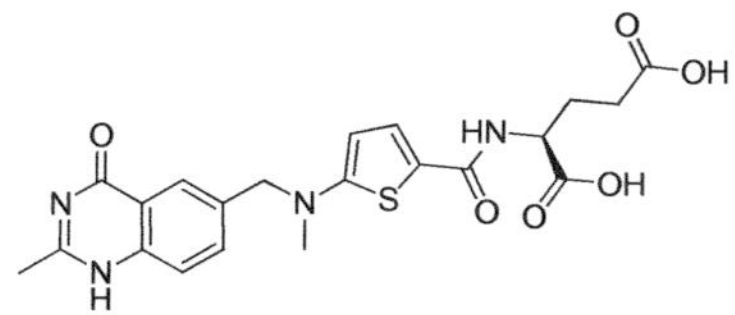	0.38 μM (Gangjee et al., 2010)	A2780 cells Incubation time (h) RTX (μM) 4(h) = 1.2 ± 0.4 24(h) = 0.041 ± 0.011 72(h) = 0.0055 ± 0.0005 Growth inhibition EC50 (μM) SKOV3 ovarian 0.0023 HX62 ovarian 0.0049 PXN94 ovarian 0.032 CH1 ovarian 0.017 A2780 ovarian 0.022 41M ovarian 0.039 (Honeywell et al., 2015; Jackman et al., 2002)
Plevitrexed (ZD9331)	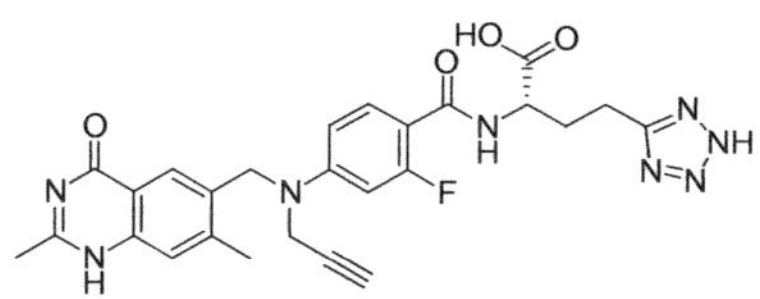	$K_i = 0.4\,nM$ (Marsham et al., 1999)	Growth inhibition EC50 (μM) SKOV3 ovarian 0.098 HX62 ovarian 0.024 PXN94 ovarian 0.081 CH1 ovarian 0.068 A2780 ovarian 0.068 41M ovarian 15 (Jackman et al., 2002)

Continued

Table 1 IC_{50} and EC_{50} hTS Inhibitors—cont'd

Name	Structure	IC_{50} hTS	EC_{50} Ovarian Cell Line
GW1843	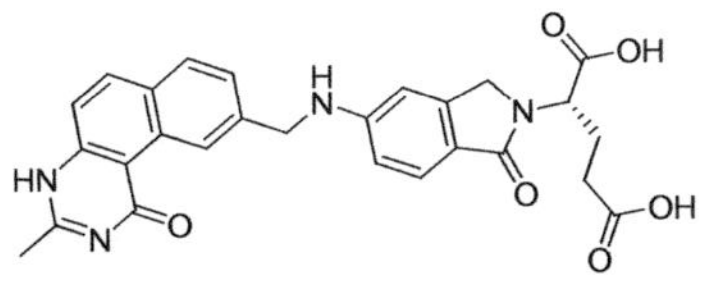	0.032 µM (Zhang et al., 2011)	2008 cells 1.1 ± 0.2 nM 2008/MRP1 cells 4.0 ± 0.3 nM 2008/MRP2 cells 1.4 ± 0.2 nM (Hooijberg et al., 1999)
BGC945	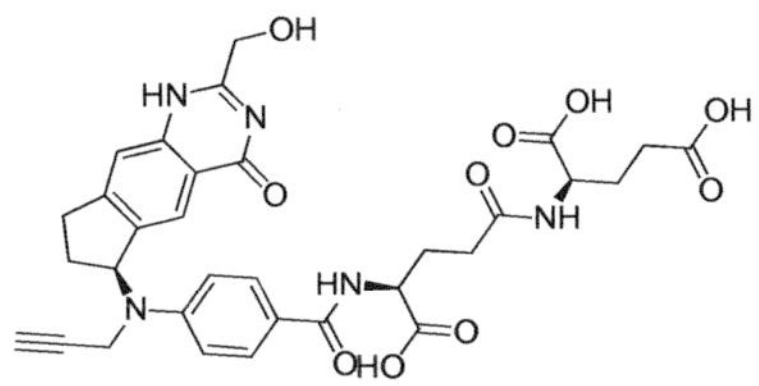	$K_i = 1.2$ (nmol/L) (Gibbs et al., 2005)	IGROV-1 cells 0.090 ± 0.01 µM (−FA) 3.7 ± 1.1 µM (+FA) (Jackman et al., 2002)
Compound **5**	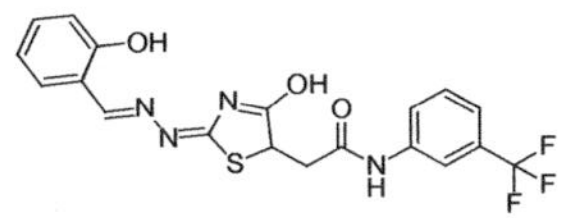	$K_i = 1.3$ µM (Carosati et al., 2012)	A2780 cells 19.4 µM (Carosati et al., 2012)
Nolatrexed (AG337)	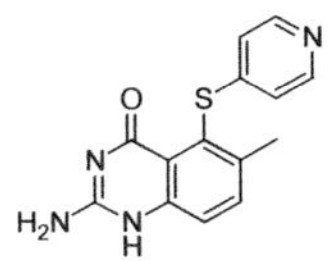	$K_i = 11$ nM (Webber et al., 1996)	2008 cells 0.65 µM (Raymond, Djelloul, Buquet-Fagot, Mester, & Gespach, 1996)
1. $R = CH_3$ 2. $R = CH_2CH_3$	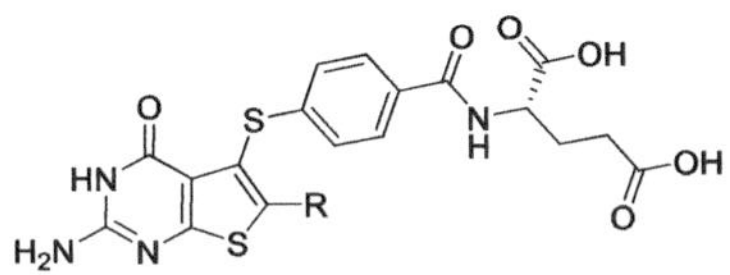	1. 0.04 µM 2. 0.054 µM (Gangjee et al., 2009)	1. OVCAR-8 $> 1*10^{-4}$ M 2. OVCAR-8 $5.79*10^{-7}$ (Gangjee et al., 2009)

3.2.1.1 5-FU, Capecitabine, and Tegafur

5-FU is largely used alone or in combination regimens for the treatment of colorectal, gastric, pancreas, head and neck, and metastatic breast tumors, among other malignancies.

Capecitabine, the orally administered prodrug of 5-FU, is often used as an alternative to 5-FU injection. The appearance of capecitabine in clinical use has rendered treatment easier to administer, with a different but manageable toxicity profile. It is a prodrug synthesized with the aim to provide prolonged drug exposure and tissue selectivity. Indeed, its chemical structure makes capecitabine a substrate for tissue-specific enzymatic activity found predominantly in liver tissue and malignancies (Budman, 2000; Budman et al., 1998). The drug is hepatically metabolized and ultimately converted into 5-FU at the tissue level and shows a comparable efficacy to 5-FU (Popa & Shah, 2013). As such, capecitabine is progressively replacing infusional 5-FU in the treatment of many cancers. However, 5-FU and capecitabine have particular relevance for the standard of care of metastatic colorectal cancer (mCRC). Combination regimens consisting of a protracted infusion of 5-FU modulated by leucovorin and oxaliplatin or by leucovorin and irinotecan form the backbone to which targeted agents are added in common clinical practice (Tournigand et al., 2004). Capecitabine can be used as an alternative to FU, particularly in combination with oxaliplatin (CAPOX or XELOX) (Cassidy et al., 2008).

5-FU possesses several undesired toxic effects (Jensen & Sørensen, 2006; Saif, Shah, & Shah, 2009; Speyer et al., 1980). The toxicity primarily associated with capecitabine is to the mucosa and skin, with less stomatitis than bolus 5-FU, and comparable rates of diarrhea (Hoff, Cassidy, & Schmoll, 2001; Van Cutsem et al., 2001).

The first Phase II trial of single-agent capecitabine included patients with epithelial ovarian cancer (EOC) who had received prior platinum-based chemotherapy and relapsed within 12 months of their last chemotherapy (Pisano et al., 2009; Vasey, McMahon, Paul, Reed, & Kaye, 2003). The median PFS was 3.7 months with an overall survival (OS) of 8.0 months. After 6 months of treatment, 28% remained without evidence of disease progression and 62% were still alive (Vasey et al., 2003). A second Phase II trial enrolled patients who had received a median of four prior chemotherapy regimens for relapsed ovarian carcinoma, including both platinum and taxane regimens (Rischin et al., 2004). Using the intent-to-treat population, the estimated median PFS and median OS were 2.3 and 7.1 months, respectively, without grade 4 toxicity experienced (Rischin et al., 2004).

Few studies have evaluated combination therapy with capecitabine in ovarian cancer. A small Phase I/II trial of epirubicin–carboplatin–capecitabine in 11 ovarian cancer patients who had relapsed after platinum therapy evaluated the maximum tolerated dose of the combination, while tumor response was a secondary end point (Rothermundt et al., 2006). Tegafur and S-1 (an oral combination of tegafur and two biochemical modulators) are bioactivated to 5-FU via 5′-hydroxylation mediated by cytochrome P-450 (probably P-450 2A6), and subsequently transformed to 5-FU (Ikeda et al., 2000) and used in different cancer types (Ishikawa, 2008; Wisher, 2012).

3.2.1.2 Pralatrexate

PDX (Folotyn) was first approved by FDA in 2009 for the treatment of relapsed or refractory peripheral T-cell lymphoma (PTCL), which comprises a subgroup of rare and aggressive non-Hodgkin lymphomas that develop from T cells in different stages of maturity outside the thymus. Having higher affinity for the reduced folate carrier (RFC) and FPGS when compared to MTX, PDX is more efficiently transported than MTX and, once inside the cell, it is efficiently polyglutamated by FPGS (Sirotnak, DeGraw, Colwell, & Piper, 1998; Visentin, Unal, Zhao, & Goldman, 2013). PDX has demonstrated variable degrees of efficacy in PTCL, with response rates differing among the multiple subtypes of the disease. While Phase III studies are still to be completed, early clinical trials indicate that PDX is a promising new therapeutic agent for PTCL (Rodd, Ververis, & Karagiannis, 2012). At present, 27 studies from all over the world have been registered in the US NIH database of publicly and privately supported clinical studies (source: https://www.clinicaltrials.gov/). To further improve the efficacy of PDX, the combination with other drugs has also been exploited (Fury et al., 2006).

The still ongoing Phase II (ClinicalTrials.gov Identifier: NCT01188876) is evaluating the highest dose of PDX that can be given safely in combination with carboplatin in platinum-sensitive ovarian, fallopian, or primary peritoneal cancer.

3.2.1.3 Pemetrexed

Pemetrexed (PMX, Alimta, LY231514, Eli Lilly, Indianapolis, IN) is an antifolate, antineoplastic agent that exerts its action by inhibiting TS, DHFR, and glycinamide ribonucleotide formyltransferase (GARFTase), all folate-dependent enzymes involved in the de novo biosynthesis of thymidine and purine nucleotides (Liani et al., 2003). PMX enters the cancer cells via the RFC that are expressed in many cancers including ovarian cancer (Westerhof et al., 1995; Wu, Gunning, & Ratnam, 1999).

PMX is approved for the treatment of NSCLC and metastatic NSCLC, mesothelioma and pleural mesothelioma. It is currently involved, alone or in combination with other drugs, in Phase II clinical trials for the treatment of gynecological tumors (cervical cancer, endometrial cancer, metastatic breast cancer, and ovarian cancer), colorectal cancer, gastric cancer, and other solid tumors. Phase III studies, instead, are active on NSCLC and mesothelioma. PMX has also been evaluated as a novel chemotherapeutic for head and neck cancer (HNC), in combination with other agents and/or radiotherapy and results are promising in the first-line treatment setting, having a similar range of activity to those of other active agents (Argiris, Pennella, Koustenis, Hossain, & Obasaju, 2013).

Phase I and II studies indicate that the addition of PMX to platinum, gemcitabine, or PLD (pegylated liposomal doxorubicin) is associated with a high incidence of grade 3 and 4 toxicity (e.g., neutropenia, thrombocytopenia, febrile neutropenia) (Hensley et al., 2008; Matulonis et al., 2008; Richards et al., 2011; Sehouli et al., 2012).

In patients with platinum-resistant EOC, an activity of PMX administered as a single agent has been reported, although based on small nonrandomized trials (Miller et al., 2009; Vergote et al., 2009). In these studies, the response is at least comparable with those of agents currently in use (PLD 12.3%, gemcitabine 13%–14%, topotecan 6.5%–12.4%) (Gordon et al., 2001; Gordon et al., 2013; Lund, Hansen, Theilade, Hansen, & Neijt, 1994; Shapiro et al., 1996).

PMX is the only TS-targeting drug in EOC clinical trials and has demonstrated increased PFS and overall survival (OS) (Sehouli et al., 2012). Guidelines from the National Comprehensive Cancer Network (NCCN) support the use of PMX as a single agent for patients with recurrent ovarian cancer (Morgan et al., 2012). This recommendation was based upon clinical trial evidence that PMX showed some activity in a Phase II trial in patients with platinum-resistant ovarian cancer (Miller et al., 2009). Other clinical trials with PMX in combination therapy have been recently completed. In this regard, the combination of PMX with gemcitabine has been tested on patients with recurrent platinum-sensitive or platinum-resistant ovarian cancer (ClinicalTrials.gov Identifier: NCT00055432). A Phase II study determined the effects (good and bad) of the combination of PMX and carboplatin on patients with recurrent ovarian, peritoneal, or fallopian tube cancers (ClinicalTrials.gov Identifier: NCT00230542). The now completed Phase II Trial of PMX plus Bevacizumab, a monoclonal antibody that acts as an angiogenesis inhibitor against the vascular endothelial growth factor, ClinicalTrials.gov Identifier: NCT00868192) has been undertaken to determine the effects of this combination on recurrent ovarian and primary peritoneal carcinoma by looking at progression and survival at 6 months.

3.2.2 Investigational Drugs to Treat OC

3.2.2.1 Classical Active-Site Inhibitors

BGC945 is a cyclopenta[*g*]quinazoline-based, thymidylate synthase inhibitor specifically transported into α-folate receptor (α-FR)-overexpressing tumors (Table 1). It has been shown to inhibit the growth of human ovarian IGROV-1 cancer cells as well as other cell lines. BGC945 selectively inhibits thymidylate synthase in α-FR-overexpressing tumors and should cause very low toxicity to humans at therapeutic doses (Gibbs et al., 2005).

AGF94 and AGF154 are a novel 6-substituted 2-amino-4-oxo-pyrrolo [2,3-*d*]pyrimidine scaffold and a thieno side chain (i.e., AGF94) with a high level of selectivity for folate receptors (FR) and PCFT over RFC. Both AGF94 and AGF154 inhibited proliferation of IGROV1 and SKOV3 EOC cells in vitro (Hou et al., 2017) (Table 1). Cytotoxicity was directly attributable to inhibition of de novo purine nucleotide biosynthesis by GARFTase, the first folate-dependent step. These results were further tested in vivo, whereby AGF94 and AGF154 exhibited similar antitumor efficacies toward early-stage SKOV3 EOC xenografts with modest toxicity, thus reflecting their tumor-selective uptake by FR and PCFT over RFC (Wang et al., 2015).

Plevitrexed (ZD9331, BGC9331) is a rationally designed, orally bioavailable, nonpolyglutamatable quinazoline antifolate that is a selective inhibitor of TS with potent antineoplastic activity and the ability to overcome antifolate resistance due to decreased FPGS activity. It also features a lower toxicity than polyglutamatable TS inhibitors, presumably due to a lesser cellular retention. Plevitrexed is primarily transported into cells via RFC and selectively binds to the folate-binding site of TS with high affinity.

Plevitrexed was assessed and in preclinical studies against a panel of human ovarian cancer cell lines (Table 1) and in several Phase II clinical studies for the treatment of various solid cancers including ovarian cancer (Rader et al., 2003).

Another novel investigational drug is GW1843 (BW1843, 1843U89, OSI-7904, GSL7904L), a very potent antifolate inhibitor of TS (Gonen & Assaraf, 2012). GW1843 is efficiently transported into human cells via

RFC and, unlike RTX and MTX, retains high cytotoxic activity also in FPGS-deficient human ovarian cancer cell lines (Jackman et al., 1995). Through a virtual screening approach against hTS, Costi et al. have identified a thiazole derivative (5) that binds the inactive conformer of hTS, in contrast to other known hTS inhibitors (Table 1). Nolatrexed (AG337) derivatives have been studied by introducing a novel heterocyclic 2-amino-4-oxo-5-substituted thieno[2,3-*d*]pyrimidines with methyl or ethyl group in 6-position by Gangjee et al. (Table 1).

In addition, the homologation of the 6-methyl to the 6-ethyl was instrumental in increasing the potency as well as the spectrum of tumor inhibition in culture. Thus, the in vitro tumor cell inhibitory activity of **2** was much superior to that of **1**. Possible explanation for this increased in vivo activity could be the increased lipophilicity of **2** compared to **1**, which could facilitate passive diffusion and/or active transport of **2** into tumor cells (Table 1).

3.2.2.2 Interface Inhibitors

One of the major concerns in the application of classical active-site inhibitors of hTS for the treatment of ovarian cancer (OC) is the rapid onset of drug resistance. To overcome this problem, a few years ago a library of octapeptides targeting the dimer interface of hTS was developed (Cardinale et al., 2011). These shorter sequences where derived from a 20-mer peptide able to inhibit hTS ($IC_{50} = 30 \pm 6\ \mu M$) whose sequence corresponds to a β-hairpin structure (residues 198–217) at the dimer interface of the enzyme. Smaller octapeptides, better suited to be modified into therapeutic agents, were derived by fragmentation of the 20-mer. Among them, the LR peptide (primary sequence: LSCQLYQR) was the most active inhibitor ($K_i = 26 \pm 2\ \mu M$). The X-ray crystal structure of the LR/enzyme complex revealed that the peptide binds at a previously undescribed allosteric site at the interface of the diinactive protein dimer (PDB ID: 3N5E). LR occupies an inverted-cone-shaped cleft in the interface area and its three amino-terminal residues are buried by the side chains of Met190', Ala191', and Leu192' (Fig. 7). The peptide is anchored to the enzyme through a reversible covalent disulfide bridge between its Cys^3 and the hTS Cys180. The peptide Gln^4, Leu^5, and Arg^8 residues are accommodated in lipophilic pockets, whereas Tyr^6 and Gln^7 are oriented toward the solvent-exposed protein surface (Fig. 7). Its novel mechanism of inhibition relies on the stabilization of the enzyme inactive conformation and was confirmed by a combination of kinetic analysis and biophysical and computational techniques (Cardinale et al., 2011). Although less effective than the existing drugs that target the active site of hTS such as 5-FU, the LR peptide

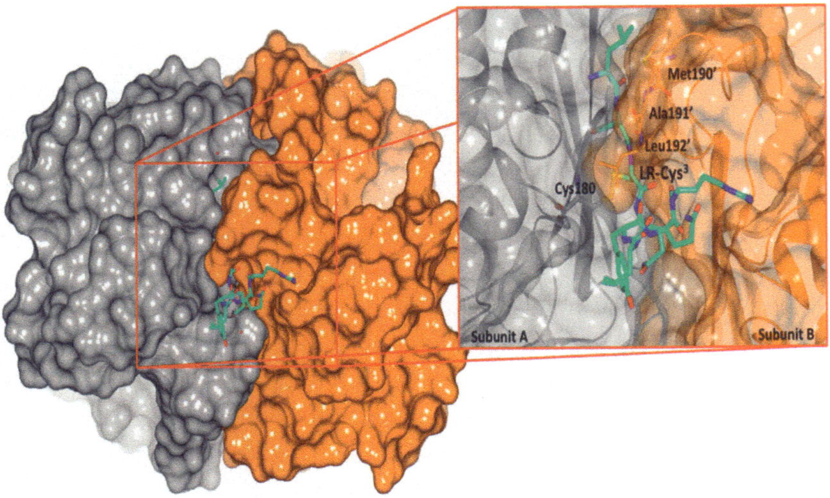

Fig. 7 X-ray crystal structure of the diinactive hTS dimer (represented as solid surface, subunit A and B are colored *gray* and *orange*, respectively) in complex with the peptide LR (in *sticks*, *cyan* carbon atoms) (PDB ID: 3N5E). A detailed view of the LR-binding pocket, an allosteric site stabilizing the inactive enzyme conformation, is shown in the *inset*. The LR peptide is anchored to the enzyme through a disulfide bond connecting the LR-Cys^3 to the hTS Cys180 (subunit A). The three amino-terminal residues of the peptide are buried by the side chains of Met190′, Ala191′, and Leu192′ belonging to the B subunit of the enzyme.

inhibited the growth of both cisplatin-sensitive and -resistant OC cell lines (with IC_{50} values after 48 h of 4.2 ± 0.5 μM and 4.7 ± 0.8 μM in A2780 and A2780/CP cell lines, respectively). At variance with 5-FU that induces TS overexpression presumably because of the loss of regulation of TSmRNA translation due to formation of TS–FdUMP–mTHF ternary complexes, treatment with the peptides did not lead to increased cellular levels of TS. Indeed, the extent of TS expression remained constant in drug-resistant cell lines, whereas a decreased amount of protein (∼−25%, by Western immunoblot analysis) was observed in sensitive cells. Furthermore, the concomitant decrease in DHFR intracellular levels was also reported (more pronounced in sensitive cells), a fact reasonably related to lowering of the dihydrofolate cellular pool due to TS inhibition. Thereafter, the mechanism of inhibition of these peptides and their ability to bind at this newly discovered allosteric site were also demonstrated by the investigation of various hTS interface variants in which residues proven to play a crucial role in stabilizing the enzyme quaternary structure were mutated into alanine (Salo-Ahen et al., 2015; Tochowicz et al., 2015). The inhibition potency of the peptides was drastically reduced when the mutated residues, such as Leu198 and Phe59, were in the proximity of the LR-binding pocket, whereas the mutation of residues located on the opposite side of the enzyme interface, such as Lys47 and Tyr202, did not affect their activity (Tochowicz et al., 2015). The X-ray crystal structure of the hTS variant K47A complexed with the CG peptide (primary sequence: CQLYQRSG) showed the inhibitor bound in the hydrophobic groove forming the allosteric site in which also the LR peptide was accommodated. Notably, the amino-terminal cysteine of the CG peptide was covalently anchored to the thiol moiety of Cys180 on the dimer interface, an interaction proved to be crucial also for LR binding. Like most peptides, those developed against hTS could not cross the cell membrane without using specific delivery systems (Cardinale et al., 2011). To overcome this problem, solid lipid nanoparticles were developed to encapsulate and deliver the hydrophilic LR peptide. The formulation of specific nanoparticles prepared in the absence of squalene, demonstrated to be responsible for the ineffectiveness of the carrier, resulted effective on cisplatin-resistant OC cell lines (19% cell growth viability inhibition at 0.2 μM LR) (Sacchetti et al., 2015). Later on, pH-sensitive liposomes were developed for peptide delivery and compared with conventional ones in cell growth inhibition experiments on cisplatin-resistant cancer cells (Sacchetti et al., 2017). At variance with conventional liposomes that require longer incubation times to convey the peptide inside the cells, acid-pH-triggered liposomal delivery reduced significantly the time required for cellular release and allowed LR activity to be observed after only 24 h of incubation. Despite this effect in accelerating the "switch on" of the drug activity, no enhancement of the peptide efficacy was induced. On the other hand, the evaluation of the LR stability showed only a modest cellular half-life (around 1.4 h) in a cisplatin-resistant epithelial OC cell line (Cannazza et al., 2014). The cysteine thiol moiety of the peptide, that plays a key role in LR binding to hTS, is a reactive metabolic function reasonably constituting a source of structural liability for the peptide. A new series of LR derivatives was generated through a SAR approach to improve its affinity for the target enzyme and its antiproliferative activity on OC cells (Pelà et al., 2014). The main structural features accounting for peptide binding and allosteric inhibition were elucidated through a wide range of modifications including N-terminal acetylation, C-terminal amidation, N- and C-terminal truncation, D- and alanine scanning. The modification that produced the largest potency enhancement was the chirality inversion of the glutamine residue in position 4 (peptide [D-Gln^4]LR). The cytotoxicity studies, performed on four cisplatin-sensitive and -resistant OC cell lines, demonstrated the improved effectiveness of

[D-Gln^4]LR compared to both LR and 5-FU (25%–30% and less than 50% cell survival of sensitive and resistant lines, respectively). The evaluation of both TS and DHFR cellular expression after treatment with [D-Gln^4]LR showed that the amount of the former enzyme did not change significantly whereas the DHFR levels decreased. This modulation pattern is the same observed also for the forefather LR but differs from that displayed by 5-FU whose administration leads to the onset of TS overexpression and drug resistance. The variation of the cellular modulation patterns of TS and DHFR represents an additional proof of the different mode of action of the peptides with respect to classical hTS inhibitors.

Later on, the cellular behavior of the peptides LR and [D-Gln^4]LR was investigated using a combination of proteomic/mass spectrometry differential analysis of the cytoplasmic proteins in treated and untreated OC cells and the resulting protein data sets were compared with those obtained from the same OC cell lines exposed to a classical active site inhibitor of hTS, PMX (Genovese et al., 2014). A common set of six proteins similarly modulated after the treatment with both peptides was identified in three different OC cell lines. It included proteins belonging to the folate cycle and the purine metabolism (TS, DHFR, ATIC, and GART) and to the proteasome-HSP90-TRAP1 functional cluster (HSP90AA1 and TRAP1). On the contrary, the antifolate hTS inhibitor PMX, that also interacts to a lower extent with DHRF and other folate-related enzymes, produced modulations of both TS and DHFR as well as of the stress-related proteins TRAP1 and HSP90AA1 opposite to those of the peptides. Even if the detailed machinery by which the peptides work at the cellular level is yet to be fully clarified, these results clearly confirmed that LR and [D-Gln^4]LR act through a new mode of action that diverges from the one characterizing classical hTS inhibitors. Notably, the peptides were proven to be effective even on PMX-resistant OC cell lines.

Very recently, a fragment screening campaign performed on hTS by differential scanning fluorimetry led to the identification of the small compound F13 (1-(5-methoxy-1H-indol-3-yl)-*N*,*N*-dimethyl-methanamine). The fragment was shown to bind at the hTS dimer interface but, in contrast to the peptides that bound to the diinactive dimer, it interacted with the diactive enzyme form (PDB ID: 5X69) (Chen et al., 2017). Assays of the inhibition activity on the isolated enzyme showed a low potency toward hTS (inhibition data not reported), while the cellular inhibition profile has not yet been reported.

As of today, the peptides LR and [D-Gln^4]LR, which specifically target the interface of hTS and inhibit OC cell growth through a new mechanism of action, represent an innovative and valuable weapon to fight anticancer drug resistance and reduce the overall drug toxicity, thus expanding the anticancer pipeline.

3.2.2.3 Compounds Conjugates and Other Compounds Targeting the Folate Receptor

Most of the drugs that act on folate metabolism enzymes are predominantly transported into the cell via the reduced folate carrier (RFC). The RFC is expressed in many normal tissues, and this limits the use of antifolate drugs because of unwanted side effects. In OC cells, RFC plays a dominant role in folate uptake at physiological concentrations and α-FR overexpression does not appear to be determinant for the folate cellular demand (Corona, Giannini, Fabris, Toffoli, & Boiocchi, 1998). Folic acid has an essential role for the one carbon transfer process in DNA synthesis. Additionally, folate is crucial for nucleotide synthesis and methylation reactions. Cancer cells are strongly dependent on folate. Folic acid analogues block cellular proliferation by blocking one-carbon transfer and methylation reactions. Additionally, they form active polyglutamated derivates that inhibit enzymes in the folate metabolism. Folate uptake is due to α- and β-folate receptors (FR), but only α-FR (FOLR1) significant expression is mostly found in epithelial cells of uterus, placenta, choroid plexus, retina, and kidney. Most serous ovarian cancers, but especially high grade, high stage cancers, overexpress α-FR and show higher expression when the cancer cells are resistant to platinum drugs thus suggesting a functional correlation with resistance to platinum therapy (Toffoli et al., 1997, p. 199; Toffoli et al., 1998). The observation of high α-FR expression suggested a role as a therapeutic target for primary and recurrent ovarian cancers. Folate metabolism impairment by coadministration of the thymidylate synthase inhibitor PMX and farletuzumab, a monoclonal folate receptor-targeting antibody, showed improved PFS rates. Therefore, simultaneous administration of chemotherapeutic agents with folate receptor inhibitors has prolonged survival in Phase II trials (Walters, Arend, Armstrong, Naumann, & Alvarez, 2013) especially in platinum-sensitive recurrent OC patients (Sehouli et al., 2012). Consequently, α-FR offers a novel target for both selective imaging and therapeutic strategies including the development of cytotoxic agents such as BGC945, and α-FR-targeted thymidylate synthase inhibitor (Gibbs et al., 2005). Hence, α-FR more and more proves to be a useful biological target for disease management and personalized medicine as well as for imaging and diagnostic purposes. In this respect, it was recently demonstrated that α-FR can be efficiently harnessed with folic acid conjugated to fluorescein isothiocyanate (folate-FITC) for intraoperative tumor-specific fluorescence imaging in patients harboring ovarian cancer (van Dam et al., 2011).

A randomized Phase II trial termed PRECEDENT used PLD with or without Vintafolide (EC145) against recurrent platinum-resistant ovarian cancer. Vintafolide is a folic acid-linked desacetylvinblastine hydrazide, i.e., a Vinca alkaloid conjugate, which disrupts microtubule polymerization (Herzog et al., 2016; Naumann & Coleman, 2011; Naumann et al., 2013). It binds to the FRα with high affinity (0.47 relative to that of folic acid, K_d 0.1–1 nmol/L for α-FR) and inhibits tumor cell growth at nanomolar concentrations.

The drug combination regimen of Vintafolide and platinum compounds showed a significant PFS advantage over standard therapy in terms of PFS (2.7–5.0 months), particularly with ovarian cancer patients harboring tumors and metastases that were FR-positive. Consequently, for personalized medicine, a new Phase III clinical study, called PROCEED, was launched to evaluate the effectiveness and safety of Vintafolide in combination with PLD for the treatment of women with OC whose tumors developed resistance to standard platinum-based chemotherapy (Reddy et al., 2014). However, the PROCEED trial was terminated before completion (Matsumoto, Onda, & Yaegashi, 2015). Several FR-targeted agents are currently in development and clinically tested.

Farletuzumab (MORAb-003) is a humanized α-FR-targeted monoclonal antibody derived from the murine antibody, LK26 (Teng, Xie, Teng, & Lee, 2012). In preclinical studies, farletuzumab elicited robust antibody-dependent cellular cytotoxicity and complement-dependent cytotoxicity, thus inhibiting the growth of human ovarian tumor xenografts (Teng et al., 2012). Farletuzumab combined with carboplatin and taxane may enhance the response rate and duration of response in patients with platinum-sensitive ovarian cancer with first relapse after remission duration of 6–18 months (Konner et al., 2010). Based on these encouraging findings, a Phase III study was undertaken in patients with platinum-sensitive recurrent ovarian cancer (Walters et al., 2013). The FAR131 trial did not prove efficacy for patients with platinum-sensitive ovarian cancer (PSOC, defined as a PFI of $\geq$6 months), in terms of the primary endpoint of PFS. Aside from Farletuzumab, other antibodies have been developed to target FR, and tested clinically. Similar to studies exploiting Vintafolide, a Phase III, open-label, randomized study (ClinicalTrials.gov Identifier: NCT02631876) was designed to compare the safety and efficacy of Mirvetuximab soravtansine, also known as IMGN853, an α-FR-targeting

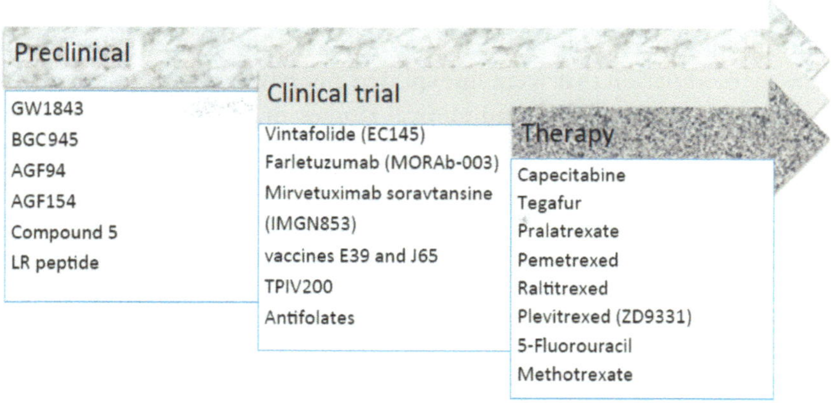

Fig. 8 Development pipeline for ovarian cancer. The drugs listed are divided on the basis of general anticancer development stage. Clinical trial development is specifically for ovarian cancer. They have hTS as one of the target.

antibody-drug conjugate, to that of selected single-agent chemotherapies in women with platinum-resistant α-FR-positive advanced EOC, and other pelvic cancers. The antibody serves to specifically target the highly potent microtubule inhibitor maytansinoid DM4 to the α-FR-positive cancer cells. In addition, vaccines against FR have been produced and evaluated, such as the folate-binding protein vaccines E39 and J65 involved in the Phase Ib trial (ClinicalTrials.gov Identifier: NCT02019524) for patients with breast or ovarian cancer diagnosis who have been treated and are without evidence of disease. A Phase II clinical trial (NCT02764333) is testing on patients with Pt-resistant ovarian cancer the safety and effectiveness of two investigational drugs, TPIV200/huFR-1 (also called TPIV200), a vaccine consisting of proteins from α-FR mixed with GM-CSF, and durvalumab (MEDI4736).

The pipeline of the drugs described is reported in Fig. 8 where the different inhibitors and drugs are distributed in the preclinical phase, in the clinical studies, and in therapy.

4. CONCLUSIONS AND PERSPECTIVES

Thymidylate synthase is one of the best-known enzymes and one of the most relevant targets for anticancer therapy. Despite the enormous amount of work performed, hTS roles have to be fully clarified, in particular the different conformations and monomers/dimer forms and the interactome that emerges from metadata and prediction based on interaction analysis. Allosteric inhibitors binding at the protein dimeric interface trigger cellular pathway modulation that were unexplored and pave the way to novel drug discovery studies. Cellular and clinical studies on ovarian cancer can be promising if the TS-targeting inhibitor is administered in combinations at the therapy initiation to prevent high TS level developed after first- or second-line therapy and combat specifically platinum drug resistance. A much stronger interplay between preclinical and clinical studies is needed to better understand the observed biological effects triggered by the hTS protein.

ACKNOWLEDGMENT

This work was financially supported by the Italian Association for Cancer Research AIRC IG10474 and AIRC2015 IG 16977.

REFERENCES

Almog, R., Waddling, C. A., Maley, F., Maley, G. F., & Van Roey, P. (2001). Crystal structure of a deletion mutant of human thymidylate synthase Δ (7-29) and its ternary complex with Tomudex and dUMP. *Protein Science*, *10*(5), 988–996. https://doi.org/10.1110/ps.47601.

Almqvist, H., Axelsson, H., Jafari, R., Dan, C., Mateus, A., Haraldsson, M., et al. (2016). CETSA screening identifies known and novel thymidylate synthase inhibitors and slow intracellular activation of 5-fluorouracil. *Nature Communications*, 7, 11040. https://doi.org/10.1038/ncomms11040.

Anderson, A. C., O'Neil, R. H., DeLano, W. L., & Stroud, R. M. (1999). The structural mechanism for half-the-sites reactivity in an enzyme, thymidylate synthase, involves a relay of changes between subunits. *Biochemistry*, *38*(42), 13829–13836.

Argiris, A., Pennella, E., Koustenis, A., Hossain, A. M., & Obasaju, C. K. (2013). Pemetrexed in head and neck cancer: A systematic review. *Oral Oncology*, *49*(6), 492–501. https://doi.org/10.1016/j.oraloncology.2013.01.007.

Assaraf, Y. G. (2007). Molecular basis of antifolate resistance. *Cancer Metastasis Reviews*, *26*(1), 153–181. https://doi.org/10.1007/s10555-007-9049-z.

Assaraf, Y. G., Leamon, C. P., & Reddy, J. A. (2014). The folate receptor as a rational therapeutic target for personalized cancer treatment. *Drug Resistance Updates*, *17*(4–6), 89–95. https://doi.org/10.1016/j.drup.2014.10.002.

Berger, S. H., Berger, F. G., & Lebioda, L. (2004). Effects of ligand binding and conformational switching on intracellular stability of human thymidylate synthase. *Biochimica et Biophysica Acta*, *1696*(1), 15–22.

Bookman, M. A., Brady, M. F., McGuire, W. P., Harper, P. G., Alberts, D. S., Friedlander, M., et al. (2009). Evaluation of new platinum-based treatment regimens in advanced-stage ovarian cancer: A phase III trial of the gynecologic cancer intergroup. *Journal of Clinical Oncology: Official Journal of the American Society of Clinical Oncology*, *27*(9), 1419–1425. https://doi.org/10.1200/JCO.2008.19.1684.

Brunn, N. D., Dibrov, S. M., Kao, M. B., Ghassemian, M., & Hermann, T. (2014). Analysis of mRNA recognition by human thymidylate synthase. *Bioscience Reports*, *34*(6), 905–913. https://doi.org/10.1042/BSR20140137.

Brunn, N. D., Garcia Sega, E., Kao, M. B., & Hermann, T. (2012). Targeting a regulatory element in human thymidylate synthase mRNA. *Chembiochem*, *13*(18), 2738–2744. https://doi.org/10.1002/cbic.201200603.

Budman, D. R. (2000). Capecitabine. *Investigational New Drugs*, *18*(4), 355–363.

Budman, D. R., Meropol, N. J., Reigner, B., Creaven, P. J., Lichtman, S. M., Berghorn, E., et al. (1998). Preliminary studies of a novel oral fluoropyrimidine carbamate: Capecitabine. *Journal of Clinical Oncology: Official Journal of the American Society of Clinical Oncology*, *16*(5), 1795–1802. https://doi.org/10.1200/JCO.1998.16.5.1795.

Cancer Genome Atlas Research Network. (2011). Integrated genomic analyses of ovarian carcinoma. *Nature*, *474*(7353), 609–615. https://doi.org/10.1038/nature10166.

Cannazza, G., Cazzato, A. S., Marraccini, C., Pavesi, G., Pirondi, S., Guerrini, R., et al. (2014). Internalization and stability of a thymidylate synthase peptide inhibitor in ovarian cancer cells. *Journal of Medicinal Chemistry*, *57*(24), 10551–10556. https://doi.org/10.1021/jm501397h.

Cardinale, D., Guaitoli, G., Tondi, D., Luciani, R., Henrich, S., Salo-Ahen, O. M. H., et al. (2011). Protein-protein interface-binding peptides inhibit the cancer therapy target human thymidylate synthase. *Proceedings of the National Academy of Sciences*, *108*(34), E542–E549. https://doi.org/10.1073/pnas.1104829108.

Carosati, E., Tochowicz, A., Marverti, G., Guaitoli, G., Benedetti, P., Ferrari, S., et al. (2012). Inhibitor of ovarian cancer cells growth by virtual screening: A new Thiazole derivative targeting human thymidylate synthase. *Journal of Medicinal Chemistry*, *55*(22), 10272–10276. https://doi.org/10.1021/jm300850v.

Carreras, C. W., & Santi, D. V. (1995). The catalytic mechanism and structure of thymidylate synthase. *Annual Review of Biochemistry*, *64*, 721–762. https://doi.org/10.1146/annurev.bi.64.070195.003445.

Cassidy, J., Clarke, S., Díaz-Rubio, E., Scheithauer, W., Figer, A., Wong, R., et al. (2008). Randomized phase III study of capecitabine plus oxaliplatin compared with fluorouracil/folinic acid plus oxaliplatin as first-line therapy for metastatic colorectal cancer. *Journal of Clinical Oncology: Official Journal of the American Society of Clinical Oncology*, *26*(12), 2006–2012. https://doi.org/10.1200/JCO.2007.14.9898.

Chen, D., Jansson, A., Sim, D., Larsson, A., & Nordlund, P. (2017). Structural analyses of human thymidylate synthase reveal a site that may control conformational switching between active and inactive states. *The Journal of Biological Chemistry*, *292*(32), 13449–13458. https://doi.org/10.1074/jbc.M117.787267.

Chu, E., & Allegra, C. J. (1993). Regulation of thymidylate synthase in human colon cancer cells treated with 5-fluorouracil and interferon-gamma. *Advances in Experimental Medicine and Biology*, *339*, 143–151 [discussion 165–167].

Chu, E., & Allegra, C. J. (1996). The role of thymidylate synthase as an RNA binding protein. *BioEssays*, *18*(3), 191–198. https://doi.org/10.1002/bies.950180306.

Chu, E., Callender, M. A., Farrell, M. P., & Schmitz, J. C. (2003). Thymidylate synthase inhibitors as anticancer agents: From bench to bedside. *Cancer Chemotherapy and Pharmacology*, *52*(Suppl. 1), S80–89. https://doi.org/10.1007/s00280-003-0625-9.

Chu, E., Koeller, D. M., Casey, J. L., Drake, J. C., Chabner, B. A., Elwood, P. C., et al. (1991). Autoregulation of human thymidylate synthase messenger RNA translation by thymidylate synthase. *Proceedings of the National Academy of Sciences of the United States of America*, *88*(20), 8977–8981.

Chu, E., Voeller, D., Koeller, D. M., Drake, J. C., Takimoto, C. H., Maley, G. F., et al. (1993). Identification of an RNA binding site for human thymidylate synthase. *Proceedings of the National Academy of Sciences of the United States of America*, *90*(2), 517–521.

Corona, G., Giannini, F., Fabris, M., Toffoli, G., & Boiocchi, M. (1998). Role of folate receptor and reduced folate carrier in the transport of 5-methyltetrahydrofolic acid in human ovarian carcinoma cells. *International Journal of Cancer*, *75*(1), 125–133.

Curtin, N. J., & Hughes, A. N. (2001). Pemetrexed disodium, a novel antifolate with multiple targets. *Lancet Oncology*, *2*(5), 298–306.

Deschamps, P., Réty, S., Bareille, J., & Leulliot, N. (2017). Crystal structure of the active form of native human thymidylate synthase in the absence of bound substrates. *Acta Crystallographica. Section F, Structural Biology Communications*, *73*(6), 336–341. https://doi.org/10.1107/S2053230X17007233.

El-Naggar, A. M., Abou-El-Regal, M. M., El-Metwally, S. A., Sherbiny, F. F., & Eissa, I. H. (2017). Synthesis, characterization and molecular docking studies of thiouracil derivatives as potent thymidylate synthase inhibitors and potential anticancer agents. *Molecular Diversity*, *21*(4), 967–983. https://doi.org/10.1007/s11030-017-9776-1.

Ferrari, S., Losasso, V., & Costi, M. P. (2008). Sequence-based identification of specific drug target regions in the thymidylate synthase enzyme family. *ChemMedChem*, *3*(3), 392–401. https://doi.org/10.1002/cmdc.200700215.

Finer-Moore, J. S., Santi, D. V., & Stroud, R. M. (2003). Lessons and conclusions from dissecting the mechanism of a bisubstrate enzyme: Thymidylate synthase mutagenesis, function, and structure. *Biochemistry*, *42*(2), 248–256.

Fury, M. G., Krug, L. M., Azzoli, C. G., Sharma, S., Kemeny, N., Wu, N., et al. (2006). A phase I clinical pharmacologic study of pralatrexate in combination with probenecid in adults with advanced solid tumors. *Cancer Chemotherapy and Pharmacology*, *57*(5), 671–677. https://doi.org/10.1007/s00280-005-0080-x.

Gangjee, A., Jain Jaclyn, H. D., Guo, P. X., Queener, S. F., & Kisliuk, R. L. (2010). 2,4-Diamino-5-methyl-6-substituted arylthio-furo[2,3-d]pyrimidines as novel classical and nonclassical antifolates as potential dual thymidylate synthase and dihydrofolate reductase inhibitors. *Bioorganic & Medicinal Chemistry*, *18*, 953–961.

Gangjee, A., Li, W., Kisliuk, R. L., Cody, V., Pace, J., Piraino, J., et al. (2009). Design, synthesis, and X-ray crystal structure of classical and nonclassical 2-amino-4-oxo-5-substituted-6-ethylthieno[2,3-d]pyrimidines as dual thymidylate synthase and dihydrofolate reductase inhibitors and as potential antitumor agents. *Journal of Medicinal Chemistry*, *52*(15), 4892–4902. https://doi.org/10.1021/jm900490a.

Garg, D., Henrich, S., Salo-Ahen, O. M. H., Myllykallio, H., Costi, M. P., & Wade, R. C. (2010). Novel approaches for targeting thymidylate synthase to overcome the resistance and toxicity of anticancer drugs. *Journal of Medicinal Chemistry*, *53*, 6539–6549.

Genovese, F., Ferrari, S., Guaitoli, G., Caselli, M., Costi, M. P., & Ponterini, G. (2010). Dimer-monomer equilibrium of human thymidylate synthase monitored by fluorescence resonance energy transfer. *Protein Science*, *19*, 1023–1030. https://doi.org/10.1002/pro.379.

Genovese, F., Gualandi, A., Taddia, L., Marverti, G., Pirondi, S., Marraccini, C., et al. (2014). Mass spectrometric/bioinformatic identification of a protein subset that characterizes the cellular activity of anticancer peptides. *Journal of Proteome Research*, *13*(11), 5250–5261. https://doi.org/10.1021/pr500510v.

Gibbs, D. D., Theti, D. S., Wood, N., Green, M., Raynaud, F., Valenti, M., et al. (2005). BGC 945, a novel tumor-selective thymidylate synthase inhibitor targeted to alpha-folate receptor-overexpressing tumors. *Cancer Research*, *65*(24), 11721–11728. https://doi.org/10.1158/0008-5472.CAN-05-2034.

Gibson, L. M., Celeste, L. R., Lovelace, L. L., & Lebioda, L. (2011). Structures of human thymidylate synthase R163K with dUMP, FdUMP and glutathione show asymmetric ligand binding. *Acta Crystallographica Section D Biological Crystallography*, *67*(1), 60–66. https://doi.org/10.1107/S0907444910044732.

Gibson, L. M., Lovelace, L. L., & Lebioda, L. (2008). The R163K mutant of human thymidylate synthase is stabilized in an active conformation: Structural asymmetry and reactivity of cysteine 195. *Biochemistry*, *47*(16), 4636–4643. https://doi.org/10.1021/bi7019386.

Gonen, N., & Assaraf, Y. G. (2012). Antifolates in cancer therapy: Structure, activity and mechanisms of drug resistance. *Drug Resistance Updates: Reviews and Commentaries in Antimicrobial and Anticancer Chemotherapy*, *15*(4), 183–210. https://doi.org/10.1016/j.drup.2012.07.002.

Gordon, A. N., Fleagle, J. T., Guthrie, D., Parkin, D. E., Gore, M. E., & Lacave, A. J. (2001). Recurrent epithelial ovarian carcinoma: A randomized phase III study of pegylated liposomal doxorubicin versus topotecan. *Journal of Clinical Oncology: Official Journal of the American Society of Clinical Oncology*, *19*(14), 3312–3322. https://doi.org/10.1200/JCO.2001.19.14.3312.

Gordon, M. S., Rosen, L. S., Mendelson, D., Ramanathan, R. K., Goldman, J., Liu, L., et al. (2013). A phase 1 study of TRC102, an inhibitor of base excision repair, and pemetrexed in patients with advanced solid tumors. *Investigational New Drugs*, *31*(3), 714–723. https://doi.org/10.1007/s10637-012-9876-9.

Hensley, M. L., Larkin, J., Fury, M., Gerst, S., Tai, D. F., Sabbatini, P., et al. (2008). A phase I trial of pemetrexed plus gemcitabine given biweekly with B-vitamin support in solid tumor malignancies or advanced epithelial ovarian cancer. *Clinical Cancer Research: An Official Journal of the American Association for Cancer Research*, *14*(19), 6310–6316. https://doi.org/10.1158/1078-0432.CCR-08-0338.

Herzog, T. J., Kutarska, E., Bidzińsk, M., Symanowski, J., Nguyen, B., Rangwala, R. A., et al. (2016). Adverse event profile by folate receptor status for vintafolide and pegylated liposomal doxorubicin in combination, versus pegylated liposomal doxorubicin alone, in platinum-resistant ovarian cancer: Exploratory analysis of the Phase II PRECEDENT trial. *International Journal of Gynecological Cancer: Official Journal of the International Gynecological Cancer Society*, *26*(9), 1580–1585. https://doi.org/10.1097/IGC.0000000000000806.

Hoff, P. M., Cassidy, J., & Schmoll, H. J. (2001). The evolution of fluoropyrimidine therapy: From intravenous to oral. *The Oncologist*, *6*(Suppl. 4), 3–11.

Honeywell, R. J., Ruiz van Haperen, V. W., Veerman, G., Smid, K., & Peters, G. J. (2015). Inhibition of thymidylate synthase by 2′,2′-difluoro-2′-deoxycytidine (Gemcitabine) and its metabolite 2′,2′-difluoro-2′-deoxyuridine. *International Journal of Biochemistry & Cell Biology*, *60*, 73–81. https://doi.org/10.1016/j.biocel.2014.12.010.

Hooijberg, J. H., Broxterman, H. J., Kool, M., Assaraf, Y. G., Peters, G. J., Noordhuis, P., et al. (1999). Antifolate resistance mediated by the multidrug resistance proteins MRP1 and MRP2. *Cancer Research*, *59*(11), 2532–2535.

Hou, Z., Gattoc, L., O'Connor, C., Yang, S., Wallace-Povirk, A., George, C., et al. (2017). Dual targeting of epithelial ovarian cancer via folate receptor α and the proton-coupled folate transporter with 6-substituted Pyrrolo[2,3-d]pyrimidine Antifolates. *Molecular Cancer Therapeutics*, *16*(5), 819–830. https://doi.org/10.1158/1535-7163.MCT-16-0444.

Huang, X., Gibson, L. M., Bell, B. J., Lovelace, L. L., Peña, M. M. O., Berger, F. G., et al. (2010). Replacement of Val3 in human thymidylate synthase affects its kinetic properties and intracellular stability. *Biochemistry*, *49*(11), 2475–2482. https://doi.org/10.1021/bi901457e.

Huehls, A. M., Wagner, J. M., Huntoon, C. J., Geng, L., Erlichman, C., Patel, A. G., et al. (2011). Poly(ADP-ribose) polymerase inhibition synergizes with 5-fluorodeoxyuridine but not 5-fluorouracil in ovarian cancer cells. *Cancer Research*, *71*(14), 4944–4954. https://doi.org/10.1158/0008-5472.

Ikeda, K., Yoshisue, K., Matsushima, E., Nagayama, S., Kobayashi, K., Tyson, C. A., et al. (2000). Bioactivation of tegafur to 5-fluorouracil is catalyzed by cytochrome P-450 2A6 in human liver microsomes in vitro. *Clinical Cancer Research: An Official Journal of the American Association for Cancer Research*, *6*(11), 4409–4415.

Ishikawa, T. (2008). Chemotherapy with enteric-coated tegafur/uracil for advanced hepatocellular carcinoma. *World Journal of Gastroenterology, 14*(18), 2797–2801.

Jackman, A. L., Kelland, L. R., Kimbell, R., Brown, M., Gibson, W., Aherne, G. W., et al. (1995). Mechanisms of acquired resistance to the quinazoline thymidylate synthase inhibitor ZD1694 (Tomudex) in one mouse and three human cell lines. *British Journal of Cancer, 71*(5), 914–924.

Jackman, A. L., Melin, C. J., Kimbell, R., Brunton, L., Aherne, G. W., Theti, D. S., et al. (2002). A rationale for the clinical development of the thymidylate synthase inhibitor ZD9331 in ovarian and other solid tumours. *Biochimica Biophysica Acta, 1587*(2–3), 215–223.

Jansen, G., & Pieters, R. (1998). The role of impaired transport in (pre)clinical resistance to methotrexate: Insights on new antifolates. *Drug Resistance Updates: Reviews and Commentaries in Antimicrobial and Anticancer Chemotherapy, 1*(3), 211–218.

Jensen, S. A., & Sørensen, J. B. (2006). Risk factors and prevention of cardiotoxicity induced by 5-fluorouracil or capecitabine. *Cancer Chemotherapy and Pharmacology, 58*(4), 487–493. https://doi.org/10.1007/s00280-005-0178-1.

Keyomarsi, K., Samet, J., Molnar, G., & Pardee, A. B. (1993). The thymidylate synthase inhibitor, ICI D1694, overcomes translational detainment of the enzyme. *The Journal of Biological Chemistry, 268*(20), 15142–15149.

Konner, J. A., Bell-McGuinn, K. M., Sabbatini, P., Hensley, M. L., Tew, W. P., Pandit-Taskar, N., et al. (2010). Farletuzumab, a humanized monoclonal antibody against folate receptor alpha, in epithelial ovarian cancer: A phase I study. *Clinical Cancer Research: An Official Journal of the American Association for Cancer Research, 16*(21), 5288–5295. https://doi.org/10.1158/1078-0432.CCR-10-0700.

Krebs, H. A., Hems, R., & Tyler, B. (1976). The regulation of folate and methionine metabolism. *Biochemical Journal, 158*(2), 341–353.

Lai, G.-M., Ozols, R. F., Smyth, J. F., Young, R. C., & Hamilton, T. C. (1988). Enhanced DNA repair and resistance to cisplatin in human ovarian cancer. *Biochemical Pharmacology, 37*(24), 4597–4600. https://doi.org/10.1016/0006-2952(88)90325-5.

Leamon, C. P. (2008). Folate-targeted drug strategies for the treatment of cancer. *Current Opinion in Investigational Drugs (London, England: 2000), 9*(12), 1277–1286.

Liani, E., Rothem, L., Bunni, M. A., Smith, C. A., Jansen, G., & Assaraf, Y. G. (2003). Loss of folylpoly-gamma-glutamate synthetase activity is a dominant mechanism of resistance to polyglutamylation-dependent novel antifolates in multiple human leukemia sublines. *International Journal of Cancer, 103*(5), 587–599. https://doi.org/10.1002/ijc.10829.

Lin, X. (2003). Role of cysteine amino acid residues on the RNA binding activity of human thymidylate synthase. *Nucleic Acids Research, 31*(16), 4882–4887. https://doi.org/10.1093/nar/gkg678.

Liu, J., Schmitz, J. C., Lin, X., Tai, N., Yan, W., Farrell, M., et al. (2002). Thymidylate synthase as a translational regulator of cellular gene expression. *Biochimica et Biophysica Acta, 1587*(2–3), 174–182.

Lovelace, L. L., Gibson, L. M., & Lebioda (2007). Cooperative inhibition of human thymidylate synthase by mixtures of active site binding and allosteric inhibitors. *Biochemistry, 46*(10), 2823–2830. https://doi.org/10.1021/bi061309j.

Lovelace, L. L., Johnson, S. R., Gibson, L. M., Bell, B. J., Berger, S. H., & Lebioda, L. (2009). Variants of human thymidylate synthase with loop 181-197 stabilized in the inactive conformation. *Protein Science, 18*(8), 1628–1636. https://doi.org/10.1002/pro.171.

Lovelace, L. L., Minor, W., & Lebioda, L. (2005). Structure of human thymidylate synthase under low-salt conditions. *Acta Crystallographica Section D, Biological Crystallography, 61*(Pt. 5), 622–627. https://doi.org/10.1107/S0907444905005895.

Lund, B., Hansen, O. P., Theilade, K., Hansen, M., & Neijt, J. P. (1994). Phase II study of gemcitabine (2′,2′-difluorodeoxycytidine) in previously treated ovarian cancer patients. *Journal of the National Cancer Institute*, *86*(20), 1530–1533.

Marsham, P. R., Wardleworth, J. M., Boyle, F. T., Hennequin, L. F., Kimbell, R., Brown, M., et al. (1999). Design and synthesis of potent non-polyglutamatable quinazoline antifolate thymidylate synthase inhibitors. *Journal of Medicinal Chemistry*, *42*, 3809–3820.

Marverti, G., Guaitoli, G., Ligabue, A., Frassineti, C., Monti, M. G., Lombardi, P., et al. (2012). Distamycin A and derivatives as synergic drugs in cisplatin-sensitive and -resistant ovarian cancer cells. *Amino Acids*, *42*(2–3), 641–653. https://doi.org/10.1007/s00726-011-1039-3.

Marverti, G., Ligabue, A., Guerrieri, D., Paglietti, G., Piras, S., Costi, M. P., et al. (2010). Spermidine/spermine N1-acetyltranferase modulation by novel folate cycle inhibitors in cisplatin-sensitive and -resistant human ovarian cancer cell lines. *Gynecologic Oncology*, *117*(2), 202–210. https://doi.org/10.1016/j.ygyno.2009.11.030.

Marverti, G., Ligabue, A., Lombardi, P., Ferrari, S., Monti, M. G., Frassineti, C., et al. (2013). Modulation of the expression of folate cycle enzymes and polyamine metabolism by berberine in cisplatin-sensitive and -resistant human ovarian cancer cells. *International Journal of Oncology*, *43*(4), 1269–1280. https://doi.org/10.3892/ijo.2013.2045.

Masuda, H., Ozols, R. F., Lai, G. M., Fojo, A., Rothenberg, M., & Hamilton, T. C. (1988). Increased DNA repair as a mechanism of acquired resistance to cis-diamminedichloroplatinum (II) in human ovarian cancer cell lines. *Cancer Research*, *48*(20), 5713–5716.

Matsumoto, K., Onda, T., & Yaegashi, N. (2015). Pharmacotherapy for recurrent ovarian cancer: Current status and future perspectives. *Japanese Journal of Clinical Oncology*, *45*(5), 408–410. https://doi.org/10.1093/jjco/hyv014.

Matulonis, U. A., Horowitz, N. S., Campos, S. M., Lee, H., Lee, J., Krasner, C. N., et al. (2008). Phase II study of carboplatin and pemetrexed for the treatment of platinum-sensitive recurrent ovarian cancer. *Journal of Clinical Oncology: Official Journal of the American Society of Clinical Oncology*, *26*(35), 5761–5766. https://doi.org/10.1200/JCO.2008.17.0282.

Miller, D. S., Blessing, J. A., Krasner, C. N., Mannel, R. S., Hanjani, P., Pearl, M. L., et al. (2009). Phase II evaluation of pemetrexed in the treatment of recurrent or persistent platinum-resistant ovarian or primary peritoneal carcinoma: A study of the gynecologic oncology group. *Journal of Clinical Oncology: Official Journal of the American Society of Clinical Oncology*, *27*(16), 2686–2691. https://doi.org/10.1200/JCO.2008.19.2963.

Morgan, R. J., Alvarez, R. D., Armstrong, D. K., Burger, R. A., Castells, M., Chen, L., et al. (2012). Ovarian cancer, version 3.2012. *Journal of the National Comprehensive Cancer Network: JNCCN*, *10*(11), 1339–1349.

Muggia, F. (2009). Platinum compounds 30 years after the introduction of cisplatin: Implications for the treatment of ovarian cancer. *Gynecologic Oncology*, *112*(1), 275–281. https://doi.org/10.1016/j.ygyno.2008.09.034.

Muggia, F. M., Braly, P. S., Brady, M. F., Sutton, G., Niemann, T. H., Lentz, S. L., et al. (2000). Phase III randomized study of cisplatin versus paclitaxel versus cisplatin and paclitaxel in patients with suboptimal stage III or IV ovarian cancer: A gynecologic oncology group study. *Journal of Clinical Oncology*, *18*(1), 106–115. https://doi.org/10.1200/JCO.2000.18.1.106.

Naumann, R. W., & Coleman, R. L. (2011). Management strategies for recurrent platinum-resistant ovarian cancer. *Drugs*, *71*(11), 1397–1412. https://doi.org/10.2165/11591720-000000000-00000.

Naumann, R. W., Coleman, R. L., Burger, R. A., Sausville, E. A., Kutarska, E., Ghamande, S. A., et al. (2013). PRECEDENT: A randomized phase II trial comparing vintafolide (EC145) and pegylated liposomal doxorubicin (PLD) in combination versus PLD alone in patients with platinum-resistant ovarian cancer. *Journal of Clinical Oncology: Official Journal of the American Society of Clinical Oncology*, *31*(35), 4400–4406. https://doi.org/10.1200/JCO.2013.49.7685.

Pelà, M., Saxena, P., Luciani, R., Santucci, M., Ferrari, S., Marverti, G., et al. (2014). Optimization of peptides that target human thymidylate synthase to inhibit ovarian cancer cell growth. *Journal of Medicinal Chemistry*, *57*(4), 1355–1367. https://doi.org/10.1021/jm401574p.

Phan, J., Steadman, D. J., Koli, S., Ding, W. C., Minor, W., Dunlap, R. B., et al. (2001). Structure of human thymidylate synthase suggests advantages of chemotherapy with noncompetitive inhibitors. *Journal of Biological Chemistry*, *276*(17), 14170–14177. https://doi.org/10.1074/jbc.M009493200.

Phan, J., Koli, S., Minor, W., Dunlap, R. B., Berger, S. H., & Lebioda, L. (2001). Human thymidylate synthase is in the closed conformation when complexed with dUMP and raltitrexed, an antifolate drug. *Biochemistry*, *40*(7), 1897–1902.

Pisano, C., Morabito, A., Sorio, R., Breda, E., Lauria, R., Gebbia, V., et al. (2009). A phase II study of capecitabine in the treatment of ovarian cancer resistant or refractory to platinum therapy: A multicentre Italian trial in ovarian cancer (MITO-6) trial. *Cancer Chemotherapy and Pharmacology*, *64*(5), 1021–1027. https://doi.org/10.1007/s00280-009-0958-0.

Popa, E. C., & Shah, M. A. (2013). Capecitabine in the treatment of esophageal and gastric cancers. *Expert Opinion on Investigational Drugs*, *22*(12), 1645–1657. https://doi.org/10.1517/13543784.2013.842974.

Rader, J. S., Clarke-Pearson, D., Moore, M., Carson, L., Holloway, R., Kao, M.-S., et al. (2003). A phase II study to determine the efficacy and tolerability of intravenous ZD9331 in heavily pretreated patients with ovarian cancer. *Gynecologic Oncology*, *91*(2), 318–325.

Raymond, E., Djelloul, S., Buquet-Fagot, C., Mester, J., & Gespach, C. (1996). Synergy between the non-classical thymidylate synthase inhibitor AG337 (Thymitaq) and cisplatin in human colon and ovarian cancer cells. *Anticancer Drugs*, 7(7), 752–757.

Reddy, J. A., Dorton, R., Bloomfield, A., Nelson, M., Vetzel, M., Guan, J., et al. (2014). Rational combination therapy of vintafolide (EC145) with commonly used chemotherapeutic drugs. *Clinical Cancer Research: An Official Journal of the American Association for Cancer Research*, *20*(8), 2104–2114. https://doi.org/10.1158/1078-0432.CCR-13-2423.

Richards, D. A., Loesch, D., Vukelja, S. J., Wu, H., Hyman, W. J., Nieves, J., et al. (2011). Phase I study of pemetrexed and pegylated liposomal doxorubicin in patients with refractory breast, ovarian, primary peritoneal, or fallopian tube cancer. *Investigational New Drugs*, *29*(5), 963–970. https://doi.org/10.1007/s10637-010-9414-6.

Rischin, D., Phillips, K.-A., Friedlander, M., Harnett, P., Quinn, M., Richardson, G., et al. (2004). A phase II trial of capecitabine in heavily pre-treated platinum-resistant ovarian cancer. *Gynecologic Oncology*, *93*(2), 417–421. https://doi.org/10.1016/j.ygyno.2004.01.037.

Rodd, A. L., Ververis, K., & Karagiannis, T. C. (2012). Safety and efficacy of pralatrexate in the management of relapsed or refractory peripheral T-cell lymphoma. *Clinical Medical Insights Oncology*, *6*, 305–314. https://doi.org/10.4137/CMO.S8536.

Rothermundt, C., Hubner, R., Ahmad, T., Gibbens, I., Keyzor, C., Habeshaw, T., et al. (2006). Combination chemotherapy with carboplatin, capecitabine and epirubicin (ECarboX) as second- or third-line treatment in patients with relapsed ovarian cancer: A phase I/II trial. *British Journal of Cancer*, *94*(1), 74–78. https://doi.org/10.1038/sj.bjc.6602879.

Rytelewski, M., Ferguson, P. J., Maleki Vareki, S., Figueredo, R., Vincent, M., & Koropatnick, J. (2013). Inhibition of BRCA2 and thymidylate synthase creates multidrug sensitive tumor cells via the induction of combined "complementary lethality" *Molecular Therapy Nucleic Acids*, *2* e78 https://doi.org/10.1038/mtna.2013.7.

Sacchetti, F., D'Arca, D., Genovese, F., Pacifico, S., Maretti, E., Hanuskova, M., et al. (2017). Conveying a newly designed hydrophilic anti-human thymidylate synthase peptide to cisplatin resistant cancer cells: Are pH-sensitive liposomes more effective than conventional ones? *Drug Development and Industrial Pharmacy*, *43*(3), 465–473. https://doi.org/10.1080/03639045.2016.1262870.

Sacchetti, F., Marraccini, C., D'Arca, D., Pelà, M., Pinetti, D., Maretti, E., et al. (2015). Enhanced anti-hyperproliferative activity of human thymidylate synthase inhibitor peptide by solid lipid nanoparticle delivery. *Colloids and Surfaces. B, Biointerfaces*, *136*, 346–354. https://doi.org/10.1016/j.colsurfb.2015.09.040.

Saif, M. W., Shah, M. M., & Shah, A. R. (2009). Fluoropyrimidine-associated cardiotoxicity: Revisited. *Expert Opinion on Drug Safety*, *8*(2), 191–202. https://doi.org/10.1517/14740330902733961.

Salo-Ahen, O. M. H., Tochowicz, A., Pozzi, C., Cardinale, D., Ferrari, S., Boum, Y., et al. (2015). Hotspots in an obligate homodimeric anticancer target. Structural and functional effects of interfacial mutations in human thymidylate synthase. *Journal of Medicinal Chemistry*, *58*(8), 3572–3581. https://doi.org/10.1021/acs.jmedchem.5b00137.

Sayre, P. H., Finer-Moore, J. S., Fritz, T. A., Biermann, D., Gates, S. B., MacKellar, W. C., et al. (2001). Multi-targeted antifolates aimed at avoiding drug resistance form covalent closed inhibitory complexes with human and Escherichia coli thymidylate synthases. *Journal of Molecular Biology*, *313*(4), 813–829. https://doi.org/10.1006/jmbi.2001.5074.

Scanlon, K. J., & Kashani-Sabet, M. (1988). Elevated expression of thymidylate synthase cycle genes in cisplatin-resistant human ovarian carcinoma A2780 cells. *Proceedings of the National Academy of Sciences of the United States of America*, *85*(3), 650–653.

Scanlon, K. J., Newman, E. M., Lu, Y., & Priest, D. G. (1986). Biochemical basis for cisplatin and 5-fluorouracil synergism in human ovarian carcinoma cells. *Proceedings of the National Academy of Sciences of the United States of America*, *83*(23), 8923–8925.

Sehouli, J., Alvarez, A. M., Manouchehrpour, S., Ghatage, P., Szczylik, C., Zimmermann, A., et al. (2012). A phase II trial of pemetrexed in combination with carboplatin in patients with recurrent ovarian or primary peritoneal cancer. *Gynecologic Oncology*, *124*(2), 205–209. https://doi.org/10.1016/j.ygyno.2011.09.007.

Shapiro, J. D., Millward, M. J., Rischin, D., Michael, M., Walcher, V., Francis, P. A., et al. (1996). Activity of gemcitabine in patients with advanced ovarian cancer: Responses seen following platinum and paclitaxel. *Gynecologic Oncology*, *63*(1), 89–93. https://doi.org/10.1006/gyno.1996.0284.

Sirotnak, F. M., DeGraw, J. I., Colwell, W. T., & Piper, J. R. (1998). A new analogue of 10-deazaaminopterin with markedly enhanced curative effects against human tumor xenografts in mice. *Cancer Chemotherapy and Pharmacology*, *42*(4), 313–318. https://doi.org/10.1007/s002800050823.

Speyer, J. L., Collins, J. M., Dedrick, R. L., Brennan, M. F., Buckpitt, A. R., Londer, H., et al. (1980). Phase I and pharmacological studies of 5-fluorouracil administered intraperitoneally. *Cancer Research*, *40*(3), 567–572.

Szklarczyk, D., Morris, J. H., Cook, H., Kuhn, M., Wyder, S., Simonovic, M., et al. (2017). The STRING database in 2017: Quality-controlled protein–protein association networks, made broadly accessible. *Nucleic Acids Research*, *45*(D1), D362–D368. https://doi.org/10.1093/nar/gkw937.

Teng, L., Xie, J., Teng, L., & Lee, R. J. (2012). Clinical translation of folate receptor-targeted therapeutics. *Expert Opinion on Drug Delivery*, *9*(8), 901–908. https://doi.org/10.1517/17425247.2012.694863.

Theisen, J., Danenberg, K., Ott, K., Becker, K., Danenberg, P., Stein, H. J., et al. (2008). Predictors of response and survival for neoadjuvant treated patients with esophageal adenocarcinoma. *Diseases of the Esophagus, 21*(7), 601–606. https://doi.org/10.1111/j.1442-2050.2008.00820.x.

Tochowicz, A., Santucci, M., Saxena, P., Guaitoli, G., Trande, M., Finer-Moore, J., et al. (2015). Alanine mutants of the interface residues of human thymidylate synthase decode key features of the binding mode of allosteric anticancer peptides. *Journal of Medicinal Chemistry, 58*(2), 1012–1018. https://doi.org/10.1021/jm5011176.

Toffoli, G., Cernigoi, C., Russo, A., Gallo, A., Bagnoli, M., & Boiocchi, M. (1997). Overexpression of folate binding protein in ovarian cancers. *International Journal of Cancer, 74*(2), 193–198.

Toffoli, G., Russo, A., Gallo, A., Cernigoi, C., Miotti, S., Sorio, R., et al. (1998). Expression of folate binding protein as a prognostic factor for response to platinum-containing chemotherapy and survival in human ovarian cancer. *International Journal of Cancer, 79*(2), 121–126.

Tournigand, C., André, T., Achille, E., Lledo, G., Flesh, M., Mery-Mignard, D., et al. (2004). FOLFIRI followed by FOLFOX6 or the reverse sequence in advanced colorectal cancer: A randomized GERCOR study. *Journal of Clinical Oncology: Official Journal of the American Society of Clinical Oncology, 22*(2), 229–237. https://doi.org/10.1200/JCO.2004.05.113.

Van Cutsem, E., Twelves, C., Cassidy, J., Allman, D., Bajetta, E., Boyer, M., et al. (2001). Oral capecitabine compared with intravenous fluorouracil plus leucovorin in patients with metastatic colorectal cancer: Results of a large phase III study. *Journal of Clinical Oncology: Official Journal of the American Society of Clinical Oncology, 19*(21), 4097–4106. https://doi.org/10.1200/JCO.2001.19.21.4097.

van Dam, G. M., Themelis, G., Crane, L. M. A., Harlaar, N. J., Pleijhuis, R. G., Kelder, W., et al. (2011). Intraoperative tumor-specific fluorescence imaging in ovarian cancer by folate receptor-α targeting: First in-human results. *Nature Medicine, 17*(10), 1315–1319. https://doi.org/10.1038/nm.2472.

Vasey, P. A., McMahon, L., Paul, J., Reed, N., & Kaye, S. B. (2003). A phase II trial of capecitabine (Xeloda) in recurrent ovarian cancer. *British Journal of Cancer, 89*(10), 1843–1848. https://doi.org/10.1038/sj.bjc.6601381.

Vergote, I., Calvert, H., Kania, M., Kaiser, C., Zimmermann, A. H., & Sehouli, J. (2009). A randomised, double-blind, phase II study of two doses of pemetrexed in the treatment of platinum-resistant, epithelial ovarian or primary peritoneal cancer. *European Journal of Cancer (Oxford, England: 1990), 45*(8), 1415–1423. https://doi.org/10.1016/j.ejca.2008.12.013.

Visentin, M., Unal, E. S., Zhao, R., & Goldman, I. D. (2013). The membrane transport and polyglutamation of pralatrexate: A new-generation dihydrofolate reductase inhibitor. *Cancer Chemotherapy and Pharmacology, 72*(3), 597–606. https://doi.org/10.1007/s00280-013-2231-9.

Voeller, D. M., Zajac-Kaye, M., Fisher, R. J., & Allegra, C. J. (2002). The identification of thymidylate synthase peptide domains located in the interface region that bind thymidylate synthase mRNA. *Biochemical and Biophysical Research Communications, 297*(1), 24–31.

Walters, C. L., Arend, R. C., Armstrong, D. K., Naumann, R. W., & Alvarez, R. D. (2013). Folate and folate receptor alpha antagonists mechanism of action in ovarian cancer. *Gynecologic Oncology, 131*(2), 493–498. https://doi.org/10.1016/j.ygyno.2013.07.080.

Wang, L., Liu, B., Wang, T., Ding, Y., Qian, X., & Zhao, Y. (2008). Detection of cell-free ERCC1 and thymidylate synthase (TS) mRNA in malignant effusions and its association with anticancer drug sensitivity. *Anticancer Research, 28*(2A), 1085–1091.

Wang, L., Wallace, A., Raghavan, S., Deis, S. M., Wilson, M. R., Yang, S., et al. (2015). 6-Substituted pyrrolo[2,3-d]pyrimidine thienoyl regioisomers as targeted antifolates for folate receptor α and the proton-coupled folate transporter in human tumors. *Journal of Medicinal Chemistry*, *58*(17), 6938–6959. https://doi.org/10.1021/acs.jmedchem.5b00801.

Webber, S., Bartlett, C. A., Boritzki, T. J., Hillard, J. A., Howland, E. F., Johnston, A. L., et al. (1996). AG337, a novel lipophilic thymidylate synthase inhibitor: In vitro and in vivo preclinical studies. *Cancer Chemotherapy and Pharmacology*, *37*(6), 509–517.

Weichsel, A., Montfort, W. R., Cieśla, J., & Maley, F. (1995). Promotion of purine nucleotide binding to thymidylate synthase by a potent folate analogue inhibitor, 1843U89. *Proceedings of the National Academy of Sciences of the United States of America*, *92*(8), 3493–3497.

Westerhof, G. R., Schornagel, J. H., Kathmann, I., Jackman, A. L., Rosowsky, A., Forsch, R. A., et al. (1995). Carrier- and receptor-mediated transport of folate antagonists targeting folate-dependent enzymes: Correlates of molecular-structure and biological activity. *Molecular Pharmacology*, *48*(3), 459–471.

Wilson, P. M., Danenberg, P. V., Johnston, P. G., Lenz, H.-J., & Ladner, R. D. (2014). Standing the test of time: Targeting thymidylate biosynthesis in cancer therapy. *Nature Reviews. Clinical Oncology*, *11*(5), 282–298. https://doi.org/10.1038/nrclinonc.2014.51.

Wisher, D. (2012). Martindale: The complete drug reference. 37th ed. *Journal of the Medical Library Association: JMLA*, *100*(1), 75–76. https://doi.org/10.3163/1536-5050.100.1.018.

Wu, M., Gunning, W., & Ratnam, M. (1999). Expression of folate receptor type alpha in relation to cell type, malignancy, and differentiation in ovary, uterus, and cervix. *Cancer Epidemiology, Biomarkers & Prevention: A Publication of the American Association for Cancer Research, cosponsored by the American Society of Preventive Oncology*, *8*(9), 775–782.

Zhang, X., Zhou, X., Kisliuk, R. L., Piraino, J., Cody, V., & Gangjee, A. (2011). Design, synthesis, biological evaluation and X-ray crystal structure of novel classical 6,5,6-tricyclic benzo[4,5]thieno[2,3-d]pyrimidines as dual thymidylate synthase and dihydrofolate reductase inhibitors. *Bioorganic & Medicinal Chemistry*, *19*, 3585–3594.

CHAPTER EIGHTEEN

Characterization of the Ovarian Tumor Peptidome

Tao Liu, Karin D. Rodland, Richard D. Smith[1]
Pacific Northwest National Laboratory, Richland, WA, United States
[1]Corresponding author: e-mail address: rds@pnnl.gov

Contents

Abstract

Aberrant degradation of proteins is associated with many pathological states, including cancers. Mass spectrometric analysis of the tumor peptidome has the potential to provide biological insights on proteolytic processing in cancer. However, attempts to use the tumors peptidome information in cancer research have been fairly limited to date, largely due to the lack of effective approaches for robust peptidomics identification and quantification, and the prevalence of confounding factors and biases associated with sample handling and processing. To address this need, we have recently developed an effective and robust analytical platform as well as a novel informatics approach for comprehensive analyses of tissue peptidomes. The ability of this new peptidomics pipeline for high-throughput, comprehensive, and quantitative peptidomics analysis, as well as the suitability of clinical ovarian tumor samples

Vitamins and Hormones, Volume 107
ISSN 0083-6729
https://doi.org/10.1016/bs.vh.2018.01.020

with postexcision delay limited to less than 60 min before freezing for peptidomics analysis, has been demonstrated. These initial analyses set a stage for further determination of molecular details and functional significance of the peptidomic activities in ovarian cancer.

1. INTRODUCTION

By avoiding the digestion of proteins into peptides, as applied in conventional bottom-up proteomics, peptidomics can provide otherwise unobtainable insights on posttranslational modifications (PTMs) and endogenous proteolytic products. For example, our recent study of blood plasma peptidome revealed strikingly different peptidomics profiles between samples from breast cancer (BrCa) patients and healthy controls, despite the very similar bottom-up proteomic profiles, showing the potential of peptidomics to reveal endogenous proteolytic processing information to which conventional bottom-up proteomics is effectively blind (Shen, Tolic, et al., 2010). Recent studies in cell lines have also identified proteolytic products of intracellular and intercellular proteins that appear to be stable bioactive molecules with explicit roles in cellular signaling pathways, rather than simple transient protein degradation products (Cunha et al., 2008; Gelman, Sironi, Castro, Ferro, & Fricker, 2011). Therefore, peptidomics provides important insights into the activity of various endogenous proteases, as well as potential information on other biologically active peptides (Lopez-Otin & Overall, 2002; Villanueva et al., 2006). Broad- and large-scale peptidomics studies using advanced liquid chromatography-tandem mass spectrometry (LC-MS/MS) have been increasingly conducted to characterize endogenous peptides in different biological samples, including cell lines (Fricker, Gelman, Castro, Gozzo, & Ferro, 2012; Gelman et al., 2011), body fluids (Fiedler et al., 2007; Villanueva et al., 2006), as well as tissues (Xu et al., 2015), for biomarker screening or clinically related studies.

Interpreting the significance of the peptidomic components identified from clinical tissue samples in a biological context is also of biomedical interest, and particularly the ability to differentiate between specific regulated events that convey biological information and more general protein degradation. Identifying the natural proteases and their associated pathways in clinical tissue samples and exploring the biological significance of the protease/substrate interaction has the potential to provide novel insights into the mechanisms of molecular tumor pathology (Shahinian et al., 2013).

To achieve this goal, it is desirable that the peptidomics pipeline provides comprehensive peptidome coverage and robust quantification, which is dependent on each step of the pipeline, including peptide extraction and/or enrichment/fractionation, separation, LC-MS data acquisition, as well as the subsequent informatics analysis (Tinoco & Saghatelian, 2011). Other issues related to peptidomics analysis of clinical samples, e.g., the ability to efficiently extract the full repertoire of peptides from complicated clinical tissue samples while maintaining the "original" status of the samples (e.g., avoiding potential confounding factors), also need to be considered. For example, postmortem stability is potentially problematic for peptidomic studies (Zhu & Desiderio, 1993), as are issues, associated with the handling of clinical samples, including degradation associated with postexcision delay, making it difficult to characterize the "true" peptidome of clinical tumor samples.

We have recently developed and applied a highly effective, label-free quantitative peptidomics pipeline to characterize the peptidomes in human ovarian cancer (OvCa) tumor samples. The ability of the platform to achieve highly reproducible measurements and comprehensive peptidomic coverage was demonstrated (Wu et al., 2015; Xu et al., 2015). Compared to the conventional LC-MS/MS-based pipeline, the use of a novel LC-MS-based informatics approach, informed quantification (IQ), provided not only significantly improved peptidome coverage but also dramatically reduced "missing values" across different sample analyses. In addition, analyses of tumor samples undergoing controlled postexcision delay of 5–60 min showed little or no effect of warm ischemia time on peptidomes from the OvCa tumors, an important observation supporting the utility of peptidomic approaches. In line with other observations, the proteasome was observed as the potential major contributor to human OvCa tumor peptidomes (Xu et al., 2015). These results demonstrate that robust quantitative peptidomics analysis can be effectively performed on clinical tumor samples using our peptidomics platform. We anticipate continued application of this platform to further determine molecular details and functional significance of the peptidomic activities in cancers.

2. MATERIALS AND METHODS

2.1 Ovarian Tumor Samples

Ovarian tumor tissues (FIGO stage IIIC or IV) were collected from three patients with high-grade serous ovarian carcinoma under IRB-approved protocols. Prior to performing primary tumor resection and before any compromise to the vascular supply, a portion of ovarian tumor attached to the omentum was rapidly resected. The tumor specimen was immediately dissected into four contiguous and adjacent specimens strips each no larger than $10 \times 3 \times 3$ mm, and placed into cryovials and frozen in liquid nitrogen at four different time points (0, 5, 30, and 60 min, at room temperature). All specimens were then stored in −80°C freezers until cryopulverization using a Covaris CP02 Cryoprep device (Covaris, Woburn, MA). All procedures were carried out on dry ice to maintain tissue in a powdered, frozen state.

2.2 Sample Preparation

For peptide extraction, extraction buffer containing 0.25% acetic acid and protease inhibitor cocktail (Sigma-Aldrich, St. Louis, MO) was added into approximately 20 mg (wet weight) pulverized OvCa tumor samples at a ratio of 1:10 (w:v). Mixed samples were homogenized on ice for 1 min, sonicated with three short bursts of 10 s followed by intervals of 5 min for cooling on ice bed, and centrifuged at 4°C, 14,000 × *g* for 30 min. The supernatant was filtered in an Amicon Ultracel 30-kDa MWCO filter tube (Millipore, Billerica, MA) by centrifugation (4°C, 8000 × *g*) to remove peptides larger than 30 kDa. The flow-through was concentrated in Speed-Vac (Thermo Fisher Scientific, Waltham, MA), and its peptidome content was measured using the BCA protein assay kit (Pierce, Rockford, IL).

2.3 LC-MS/MS Analysis

The peptidomics samples were analyzed using nanoACQUITY UPLC® system (Waters Corporation, Milford, MA) coupled on-line to a LTQ Orbitrap Velos mass spectrometer (Thermo Fisher Scientific, Waltham, MA). A 110 cm × 75 μm i.d. (flow rate 200 nL/min) fused-silica capillary column packed with 3-μm Jupiter C18 bonded particles (Phenomenex, Torrance, CA) was used. Mobile phases consisted of 0.1% formic acid in water (A) and 0.1% formic acid acetonitrile (B) were operated with effective gradient profile of (min:%B): 0:1, 6:8, 60:12, 225:35, 291:45, 300:95. The LTQ Orbitrap Velos mass spectrometer was operated in the data-dependent mode acquiring high-resolution collision-induced dissociation (CID) scans ($R = 15{,}000$, 5×10^4 target ions) after each full MS scan ($R = 60{,}000$, 1×10^6 target ions) for the top six most abundant ions (400–2000 m/z). An isolation window of 2 Th and a normalized collision energy of 35 were used for CID. The dynamic exclusion time was 60 s.

2.4 Data Analysis

Peptide identification and quantification were carried out using the IQ approach as described previously (Wu et al., 2015; Xu et al., 2015). Briefly, a comprehensive peptide database was first created by combining confident identifications (<1% false identification rate; precursor mass error less than 10 ppm) derived from independent search of MS/MS spectra from all analyses using MS-GF+ (Kim & Pevzner, 2014), MS-Align+ (Liu et al., 2012), and the Unique Sequence Tags (Shen, Liu, et al., 2010) algorithms, respectively, against the human NCBI database (Build 37.3). IQ uses the *m*/*z* values of the theoretical isotopic profile derived based on the peptide sequences that were included in the peptide database to guide the extraction of the observed isotopic profile from the summed mass spectra in each individual analysis. Least-squares fitting of the theoretical isotopic profile on the observed profile is then performed (Jaitly et al., 2009). After alignment of mass and the LC elution time, the processed data were filtered by fit score (<0.1), normalized elution time tolerance (<2.5%), and mass accuracy (<10 ppm), followed by manual validation to eliminate false positives (Slysz et al., 2014). The abundance from different charge states is added up for the specific peptide for quantification. For comparison, the data sets were also analyzed using the PNNL-developed accurate mass and time (AMT) tag approach (Smith et al., 2002; Zimmer, Monroe, Qian, & Smith, 2006) as previously described (Wu et al., 2015). The identified peptides from the IQ and AMT analyses were finally grouped to nonredundant protein lists using IDpicker3 (Holman, Ma, & Tabb, 2012).

The intensities of peptides were first Log10 transformed and then median normalized across all samples. Pearson correlation, unsupervised hierarchical clustering, and ANOVA and moderated F-test (for evaluating changes in the peptidome over time using a kinetics-based regression model) were performed in *R* using corresponding R packages (Xu et al., 2015); the resulting *P*-values were adjusted for multiple testing using the Benjamini–Hochberg method.

The ProteinCoverageSummarizer tool developed in-house and corresponding protein databases were used to map the unique peptide sequences into proteins. The frequency of various amino acids within the cleavage site (P6-1, P1-6′) from all identified peptides was analyzed (Schechter & Berger, 1968). For each peptide, N- and C-terminal cleavage sites were analyzed as PX_Up and PX_Down, respectively, the frequencies of which were plotted together for individual P sites. Frequencies of amino acids at all

12 P sites were analyzed to identify potential proteases based on the protease cleavage rules which were adopted from the comprehensive publication by Keil (1992). Gene List Analysis was conducted using Panther (http://www.pantherdb.org/) to obtain biological information (e.g., function, pathway) of the precursor proteins.

3. RESULTS AND DISCUSSION

3.1 Robust Characterization of the Tumor Peptidome

A total of 5756 distinct peptides ranging from 500 to ~6000 Da (Fig. 1A) were identified and quantified (label-free) from analyses of 12 ovarian cancer tumor tissue samples (from three patients, each with four time points). The number of identified peptides was highly consistent across samples (4952 ± 285; CV = 5.8%) (Fig. 1B), demonstrating the good, consistent peptidome coverage of the analytical platform. These peptides were mapped into 974 distinct precursor proteins (peptidome peptides, their precursor proteins, and original peptide intensities from each ovarian tumor sample are available in supplemental table S2 of Xu et al., 2015); >60% of these proteins have more than one peptidome peptides identified (Fig. 1C). To our knowledge, our study represents the broadest characterization and first quantitative analysis of human cancer tumor peptidomes. This entire quantitative peptidomics platform encompassing sample preparation, LC-MS analysis, and data analysis provides a foundation for future peptidomics applications in biological and translational studies in cancer.

3.2 IQ Analysis Significantly Improves Data Quality

The OvCa peptidomics studies benefitted significantly from data analysis afforded by the new IQ approach. As an example, we compared the IQ analysis results of the OvCa data (LC-MS data) to those obtained through conventional analyses such as MS-GF+ (MS/MS data) and the AMT tag strategy (LC-MS data) (Wu et al., 2015). As expected, MS-GF+ suffers from the undersampling issue of typical data-dependent acquisition of MS/MS data and hence provided much lower peptidome coverage and less consistent results for each individual analysis (Fig. 2A). IQ and AMT tag analyses both use a database consisting of peptides identified from database search of all MS/MS spectra from the entire peptidomics datasets, yet IQ provided higher peptidome coverage, and more importantly, less missing data across the entire datasets (Fig. 2B). We attribute this to IQ using all isotopic peaks

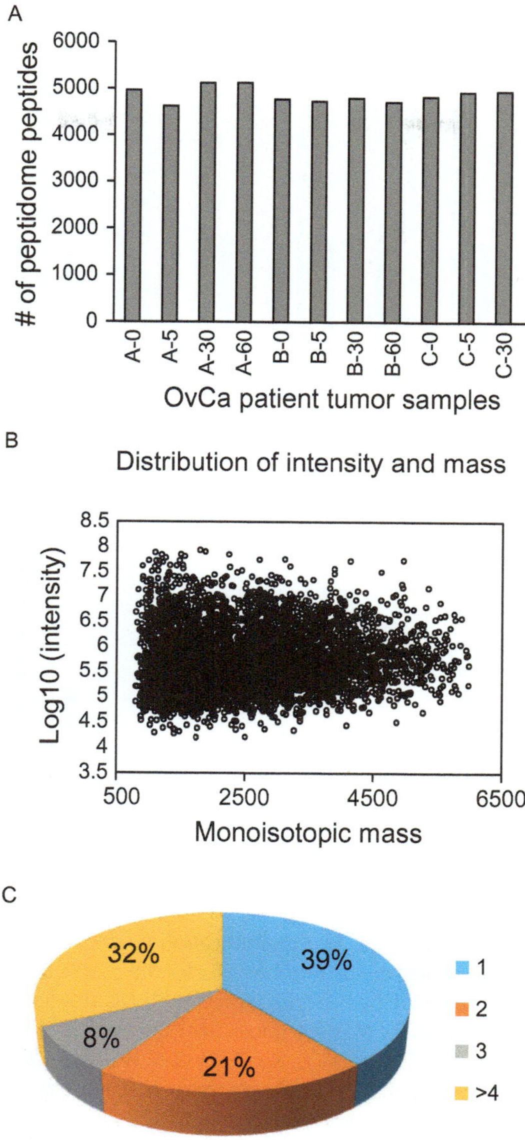

Fig. 1 Summary of peptide identifications from OvCa tumor samples. Bar graphs show the numbers of identified peptides from each OvCa tumor (A). The intensities of identified peptides (Log10 values) from OvCa tumor samples were plotted against their monoisotopic masses (Da), showing a similar peptide intensity range but different mass ranges (B). The distributions of the precursor proteins in terms of number of unique peptidome peptide identifications are shown in pie graph (C). *Modified with permission from Xu, Z., Wu, C., Xie, F., Slysz, G. W., Tolic, N., Monroe, M. E., et al. (2015). Comprehensive quantitative analysis of ovarian and breast cancer tumor peptidomes.* Journal of Proteome Research, 14, *422–433, available at http://pubs.acs.org/doi/full/10.1021/pr500840w. These figures are licensed under American Chemical Society (ACS) AuthorChoice, and further permission requests should be directed to the ACS.*

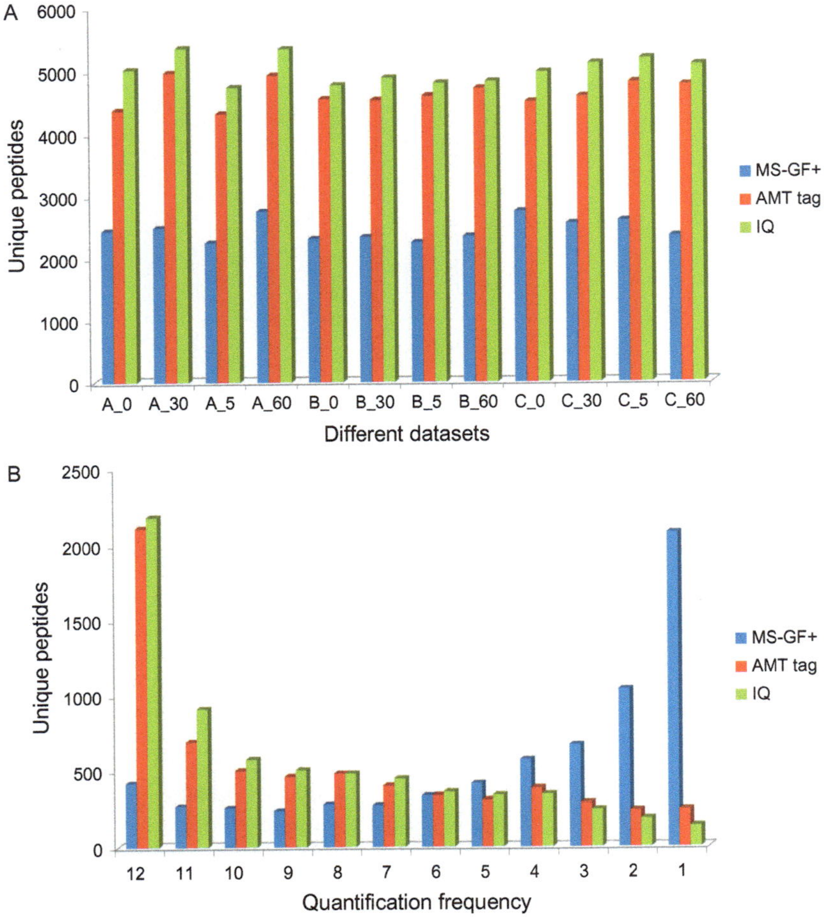

Fig. 2 Comparison of identification and quantification results from MS-GF+, AMT tag, and IQ analyses. (A) Comparison of number of identified unique peptides in each sample for MS-GF+ (*blue*), AMT (*red*), and IQ (*green*). (B) Distribution of quantification frequency among all samples for MS-GF+ (*blue*), AMT tag (*red*), and IQ (*green*). *Adapted with permission from Wu, C., Monroe, M. E., Xu, Z., Slysz, G. W., Payne, S. H., Rodland, K. D., et al. (2015). An optimized informatics pipeline for mass spectrometry-based peptidomics.* Journal of the American Society for Mass Spectrometry, 26, *2002–2008. These figures are licensed under Springer, and further permission requests should be directed to Springer.*

as opposed to the AMT tag method (Smith et al., 2002; Zimmer et al., 2006) which uses individually deisotoped spectra, resulting in improved sensitivity, better distinguished overlapping features, more accurate quantification, and better reproducibility.

3.3 Negligible Effects From Postexcision Delay

Pearson correlation analysis, unsupervised hierarchical clustering analysis, principle component analysis (PCA), and a kinetics-based regression analysis were performed to assess the ischemic effect on the OvCa tumor peptidome. Despite considerable overlap (95.4%) of the precursor proteins in all three patients (supplemental table S2 of Xu et al., 2015), the pairwise peptide intensities of all three patients were poorly correlated (Fig. 3A). In contrast, peptides of the four time point samples from the same patient showed much

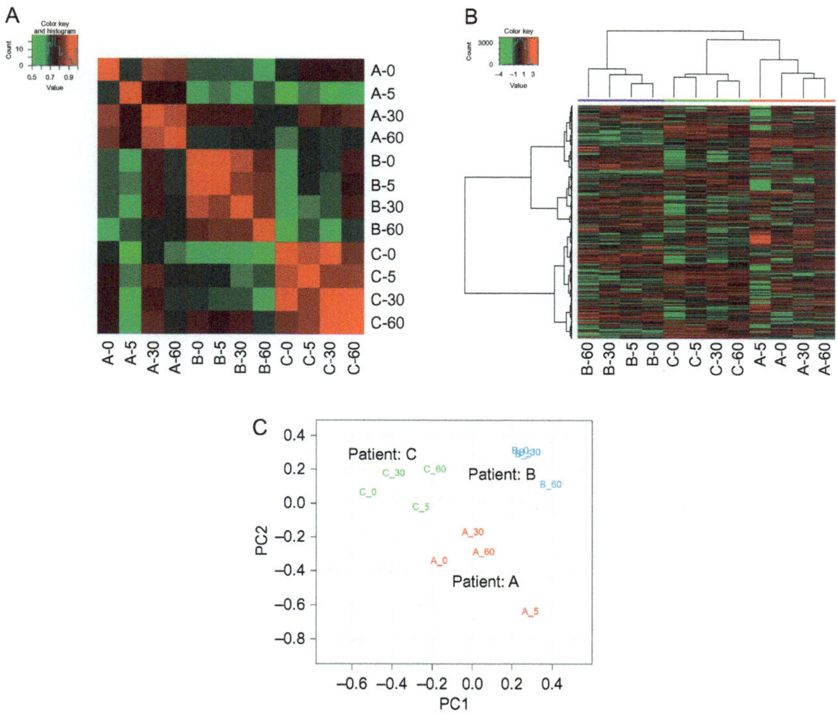

Fig. 3 Analysis of the OvCa peptidome data demonstrated minimal ischemic effect within 60 min of postexcision delay. (A) Pearson correlation, (B) unsupervised hierarchical clustering, and (C) PCA analyses of all OvCa tumor samples (four time point samples from each of the three patients). It is evident that potential ischemic effect on tumor peptidome as a result of up to 60 min postexcision delay is much smaller than patient heterogeneity. *Modified with permission from Xu, Z., Wu, C., Xie, F., Slysz, G. W., Tolic, N., Monroe, M. E., et al. (2015). Comprehensive quantitative analysis of ovarian and breast cancer tumor peptidomes.* Journal of Proteome Research, 14, *422–433, available at http://pubs.acs.org/doi/full/10.1021/pr500840w. These figures are licensed under American Chemical Society (ACS) AuthorChoice, and further permission requests should be directed to the ACS.*

better correlation, supporting good reproducibility of the analytical platform and indicating that any potential ischemic effect on the peptidome is much smaller than the patient-to-patient difference. The clustering analysis (Fig. 3B) and PCA analysis (Fig. 3C) showed similar trends: the peptide intensity patterns changed significantly across the three patients, and the four time point samples for each patient had more similar peptidome profiles and were clustered together. The kinetics-based regression analysis results indicated that none of the peptides was significantly changing across the three patients as a result of 60-min ischemic stress (data not shown). Overall, results from all these analyses indicated that the ischemic effect contributing to variations in the OvCa tumor peptidome within the first 60 min after excision, if any, is much less significant than that of patient heterogeneity. This is consistent with our previous observations from bottom-up proteomics analysis of the same set of OvCa tumor time course samples (Mertins et al., 2014), i.e., the overall protein abundance does not change significantly with up to 60 min postexcision delay.

Peptidomics profiles have been previously reported as capable of discriminating cancer types such as breast, prostate, and bladder cancer, and healthy controls (Villanueva et al., 2006). We have also shown even two patient-derived xenografts derived from the same subtype of breast cancer (basal) can be clearly separated by their peptidomics profiles (Xu et al., 2015). In the case of OvCa, tumor peptidome samples from different patients were also clearly separated. These results suggest that tumor peptidomics profiles are sensitive to human pathophysical status and human heterogeneity, and not affected by the ischemia caused by postexcision delay (up to 60 min as tested), and hence could provide highly valuable information on endogenous proteolytic events for cancer studies.

3.4 No Strong Correlation Between the OvCa Peptidome Peptides and the Abundance/Stability of Their Precursor Proteins

The OvCa peptidome precursor proteins were compared to the list of the most abundant proteins from our previous bottom-up study of the same OvCa tumors (Mertins et al., 2014). Of the 974 nonredundant protein groups identified in OvCa peptidome, 238 were among the 500 most abundant proteins in the OvCa tumors, and 34 of these 238 proteins were among the 50 most abundant proteins. Further comparing the distributions of the spectral counts of the peptidome precursor proteins and the entire list of protein identifications in the bottom-up studies (9498 proteins) showed that more than 25% of the OvCa peptidome precursor proteins are relatively low abundance proteins in the proteome. Moreover, the relative abundances for the same precursor protein in the proteome and peptidome do not correlate well ($R^2 < 0.2$; results not shown). There are many potentially interesting low abundance precursor proteins that were identified with multiple peptidome peptide identifications. By comparing the OvCa peptidome precursor proteins with the 600 proteins in A549 cell and 6000 proteins in Hela cell with the highest turnover rates (Boisvert et al., 2012; Doherty, Hammond, Clague, Gaskell, & Beynon, 2009), we found 67 matched proteins in the A549 cell study and 684 matched proteins in the Hela cell study. Only 5 precursor proteins were among the top 50 unstable proteins in A549 cell, and only 53 were among the top 500 category in Hela cell. Overall, our data suggest that there is no strong correlation between the peptides identified in OvCa peptidome and the abundance/stability of the proteins from which these peptides are derived.

The biological processes in which the precursor proteins are involved were analyzed using gene ontology. Approximately 25% of the 974 precursor proteins in ovarian cancer, representing the most prevalent proteolytic substrates, were participating in metabolic processes. The second largest group of peptidome precursor proteins (~15%) was relevant to the cellular processes including cellular component morphogenesis, neural development, and immune responses. The rest of the precursor proteins were participating in processes such as transport, cell communication, reproduction, cellular component organization, and apoptosis, counting for two-thirds of total biological processes. However, none of these represent more than 10% of the total biological processes (Xu et al., 2015). These results suggest that protein degradation is a global event in OvCa tumors.

3.5 Chymotrypsin Activity From the Proteasome Subunit Is Likely the Major Contributor to the OvCa Peptidome

For peptides identified from OvCa samples, the most frequently observed amino acids at P1 position were Leu, Lys, and Phe (Fig. 4A). When normalized to the relative abundance of each amino acid within all human proteome, these amino acids were still the most common in the P1 position (Fig. 4B). Predominant amino acids at P1′ position of the cleavage sites were Leu, Val, Phe, Ala, and Ser (Fig. 4C and D). This suggests the involvement of chymotrypsin that preferentially cleaves proteins at Phe, Tyr, and Trp with high specificity and at Leu, Met, and His with lower specificity (Keil, 1992), for producing the OvCa peptidome peptides. Similar analyses

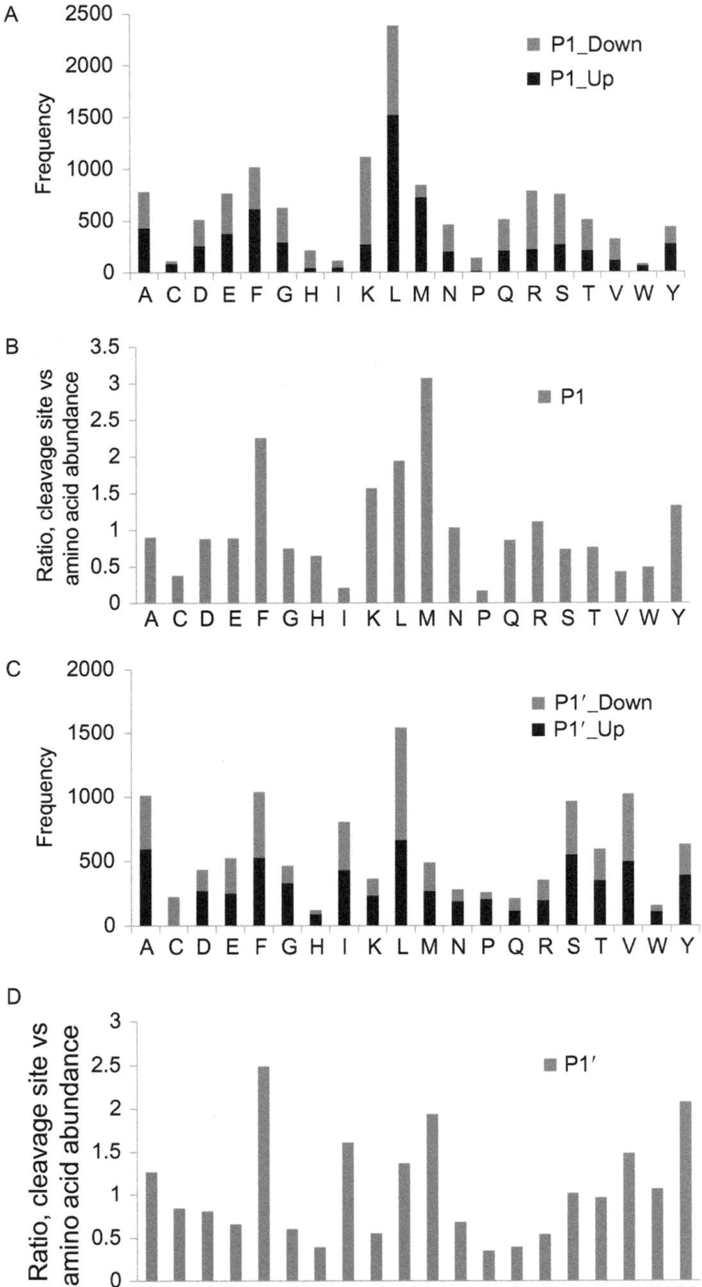

Fig. 4 Cleavage specificity analysis at P1 and P1′ sites for OvCa peptidome peptides. (A) Number of peptides undergoing cleavages is shown on the *y*-axis for all P1 amino acids. P1_Down, downstream P1 site on the peptides; P1_Up, upstream P1 site.

of the residues performed for other sites (P2-6, P2-6′) suggest the involvement of other proteases such as enterokinase and pepsin (data not shown). The percentages of terminal and internal peptides in OvCa peptidome were also examined, and the results showed that 82% of the 5756 peptides were internal peptides while 17% were terminal ones. In addition, 54% of the peptides were within the length range of 3–23 amino acids.

One of the major intracellular protein degradation pathways is mediated by the proteasome, which generally cleaves proteins after hydrophobic residues by its chymotrypsin-like β5 subunit, and after basic residues by its trypsin-like β2 subunit (Voges, Zwickl, & Baumeister, 1999; Zwickl, Voges, & Baumeister, 1999). The length of the proteasomal cleavage products ranges from 3 to ~23 amino acid residues (Glickman & Ciechanover, 2002; Wenzel, Eckerskorn, Lottspeich, & Baumeister, 1994). In addition, proteasomal degradation is expected to produce a larger number of internal fragments, whereas the proteolytic products of other cytoplasmic proteases are terminal ones (Nixon et al., 1994). OvCa peptidome consisted of peptides rich in Leu, Lys, and Phe at P1 position, more than half of the peptides fell within the range of proteasome products, and the majority of which were internal peptides. These analyses suggest that the chymotrypsin activity, which is possibly from the proteasome subunit, is the major, but not the only, contributor to the OvCa peptidome.

We further examined the cleavage specificities of amino acid residues at the N-terminus and C-terminus of identified OvCa peptidome peptides. Ala and Ser were the most frequently observed amino acids at the starting position of the peptides, which were derived from N-terminus of precursor proteins with the first Met cleaved off; for the other N-terminal, all C-terminal, and overall peptidome peptides, no amino acid was observed with prominent frequency (data not shown). Based on the *MEROPS*

(B) P1 amino acid frequency in (A) was adjusted to the relative abundance of each amino acid in all human proteins. Ratio of 1.0 indicates that corresponding amino acid present in the cleavage site has the same frequency in overall human protein amino acid composition, while >1.0 suggests an amino acid found more frequently in the cleavage site than elsewhere in the peptides. (C) As in (A) except for the P1′ position. (D) As in (B), except for the P1′ position. *Modified with permission from Xu, Z., Wu, C., Xie, F., Slysz, G. W., Tolic, N., Monroe, M. E., et al. (2015). Comprehensive quantitative analysis of ovarian and breast cancer tumor peptidomes.* Journal of Proteome Research, 14, *422–433, available at http://pubs.acs.org/doi/full/10.1021/pr500840w. These figures are licensed under American Chemical Society (ACS) AuthorChoice, and further permission requests should be directed to the ACS.*

database search (http://merops.sanger.ac.uk/index.shtml), an uncharacterized methionyl dipeptidase (Van Damme et al., 2009) (Met-Xaa dipeptidase) which most frequently cleaves Ala and Ser after Met matched the observed cleavage pattern, suggesting the involvement of N-terminal Methionine Excision (NME) in the generation of OvCa peptidome. In addition, ~51% of the Met at P1 position of all OvCa peptidome peptides was contributed by the NME process. Therefore, in addition to the contribution of proteasome subunit's low chymotrypsin specificity for Met, NME process gives rise to the distinctive prevalence of Met at P1 position of all OvCa peptidome peptides (Fig. 4B).

NME is an important proteolytic pathway regulating the diversity of N-terminal amino acids of proteins, conserved from prokaryotes to eukaryotes. Although Met is the first amino acid of newly synthesized proteins, it is often removed from the mature proteins for nonbulky N-terminal residue to avoid immediate degradation by ubiquitin–proteasome system (Sherman, Stewart, & Tsunasawa, 1985). Besides being part of the protein maturing mechanism, recent data suggest that NME also serves to regulate protein's half-life based on N-end rule at posttranslational level throughout development, tumorigenesis and in response to abiotic stress. For example, retained Met, Ser, Ala, Thr, Val, or Gly at N-terminus generally stabilize proteins for half-lives longer than 20 h, while proteins with N-terminal Phe, Leu, Asp, Lys, or Arg have half-lives of 3 min or less in mammals (Varshavsky, 1996). Therefore, NME is tightly regulated with other mechanisms including ubiquitin–proteasome pathway to determine the fate of proteins under various circumstances (Giglione, Boularot, & Meinnel, 2004). Our data revealed the prominent contribution of NME for the generation of OvCa tumor peptidome peptides. Based on the N-end rule, those proteins starting with nonbulky residues such as Ser, Ala, Thr, and Val are generally stabilized for longer half-lives instead of undergoing proteolytic process, suggesting that NME contributes possibly at translational level rather than serving as a switch for protein degradation at posttranslational level in OvCa tumor cells.

4. CONCLUSIONS

The study of low-molecular-weight bioreactive peptides, protein degradation products, and small intact proteins (i.e., the peptidome) is often criticized as being less sensitive, less reproducible, and as vulnerable to a lack of controls in sample collection and preparation (Fricker, Lim, Pan, & Che,

2006; Leichtle, Dufour, & Fiedler, 2013). Herein, we developed and evaluated a robust, efficient, and quantitative peptidomics platform and demonstrated its utility for in-depth characterization of the human ovarian tumor peptidome. Our study also indicates that warm ischemia effects are negligible up to 60 min of postexcision delay in OvCa tumor peptidome, providing a basis for proper tumor specimen processing, handling, and selection for tumor peptidomics studies. Furthermore, we have shown that protein degradation is a global event in OvCa tumors, and that chymotrypsin and trypsin activities of proteasome subunits are the major contributors to the OvCa tumor peptidome. The observation that the cancer peptidome differs in a controlled, but unspecified, manner between the known subtypes of BrCa and between BrCa and OvCa (Xu et al., 2015) suggests that there are cancer-specific changes in the regulation of normal proteolytic processes. The observation that varying times of warm ischemia, from 0 to 60 min, introduced less variability in the peptidome profile than that observed between subjects in actual clinical samples of OvCa suggests that these cancer-specific proteolytic process are tightly regulated, and not merely a nonspecific increase in protein degradation. These observations suggest that a more extensive and tightly controlled analysis of tumor peptidomes, including the use of normal control tissues and clinical metadata, has the potential to provide new and potentially useful insights regarding cancer biology.

ACKNOWLEDGMENTS

Portions of this work were supported by the Grant U24CA160019, from the National Cancer Institute Clinical Proteomic Tumor Analysis Consortium (CPTAC), National Institutes of Health Grant P41GM103493, and Department of Defense Interagency Agreement MIPR2DO89M2058. The experimental work described herein was performed in the Environmental Molecular Sciences Laboratory, a national scientific user facility sponsored by the DOE and located at Pacific Northwest National Laboratory, which is operated by Battelle Memorial Institute for the DOE under Contract DE-AC05-76RL0 1830.

REFERENCES

Boisvert, F. M., Ahmad, Y., Gierlinski, M., Charriere, F., Lamont, D., Scott, M., et al. (2012). A quantitative spatial proteomics analysis of proteome turnover in human cells. *Molecular & Cellular Proteomics, 11,* M111 011429.

Cunha, F. M., Berti, D. A., Ferreira, Z. S., Klitzke, C. F., Markus, R. P., & Ferro, E. S. (2008). Intracellular peptides as natural regulators of cell signaling. *The Journal of Biological Chemistry, 283,* 24448–24459.

Doherty, M. K., Hammond, D. E., Clague, M. J., Gaskell, S. J., & Beynon, R. J. (2009). Turnover of the human proteome: Determination of protein intracellular stability by dynamic SILAC. *Journal of Proteome Research*, *8*, 104–112.

Fiedler, G. M., Baumann, S., Leichtle, A., Oltmann, A., Kase, J., Thiery, J., et al. (2007). Standardized peptidome profiling of human urine by magnetic bead separation and matrix-assisted laser desorption/ionization time-of-flight mass spectrometry. *Clinical Chemistry*, *53*, 421–428.

Fricker, L. D., Gelman, J. S., Castro, L. M., Gozzo, F. C., & Ferro, E. S. (2012). Peptidomic analysis of HEK293T cells: Effect of the proteasome inhibitor epoxomicin on intracellular peptides. *Journal of Proteome Research*, *11*, 1981–1990.

Fricker, L. D., Lim, J., Pan, H., & Che, F. Y. (2006). Peptidomics: Identification and quantification of endogenous peptides in neuroendocrine tissues. *Mass Spectrometry Reviews*, *25*, 327–344.

Gelman, J. S., Sironi, J., Castro, L. M., Ferro, E. S., & Fricker, L. D. (2011). Peptidomic analysis of human cell lines. *Journal of Proteome Research*, *10*, 1583–1592.

Giglione, C., Boularot, A., & Meinnel, T. (2004). Protein N-terminal methionine excision. *Cellular and Molecular Life Sciences*, *61*, 1455–1474.

Glickman, M. H., & Ciechanover, A. (2002). The ubiquitin-proteasome proteolytic pathway: Destruction for the sake of construction. *Physiological Reviews*, *82*, 373–428.

Holman, J. D., Ma, Z. Q., & Tabb, D. L. (2012). Identifying proteomic LC-MS/MS data sets with Bumbershoot and IDPicker. *Current Protocols in Bioinformatics*, 1–15, Chapter 13, Unit 13.17.

Jaitly, N., Mayampurath, A., Littlefield, K., Adkins, J. N., Anderson, G. A., & Smith, R. D. (2009). Decon2LS: An open-source software package for automated processing and visualization of high resolution mass spectrometry data. *BMC Bioinformatics*, *10*, 87.

Keil, B. (1992). *Specificity of proteolysis*. Berlin Heidelberg: Springer-Verlag.

Kim, S., & Pevzner, P. A. (2014). MS-GF+ makes progress towards a universal database search tool for proteomics. *Nature Communications*, *5*, 5277.

Leichtle, A. B., Dufour, J. F., & Fiedler, G. M. (2013). Potentials and pitfalls of clinical peptidomics and metabolomics. *Swiss Medical Weekly*, *143*, w13801.

Liu, X., Sirotkin, Y., Shen, Y., Anderson, G., Tsai, Y. S., Ting, Y. S., et al. (2012). Protein identification using top-down. *Molecular & Cellular Proteomics*, *11*, M111 008524.

Lopez-Otin, C., & Overall, C. M. (2002). Protease degradomics: A new challenge for proteomics. *Nature Reviews. Molecular Cell Biology*, *3*, 509–519.

Mertins, P., Yang, F., Liu, T., Mani, D. R., Petyuk, V. A., Gillette, M. A., et al. (2014). Ischemia in tumors induces early and sustained phosphorylation changes in stress kinase pathways but does not affect global protein levels. *Molecular & Cellular Proteomics*, *13*, 1690–1704.

Nixon, R. A., Saito, K. I., Grynspan, F., Griffin, W. R., Katayama, S., Honda, T., et al. (1994). Calcium-activated neutral proteinase (calpain) system in aging and Alzheimer's disease. *Annals of the New York Academy of Sciences*, *747*, 77–91.

Schechter, I., & Berger, A. (1968). On the active site of proteases. 3. Mapping the active site of papain; specific peptide inhibitors of papain. *Biochemical and Biophysical Research Communications*, *32*, 898–902.

Shahinian, H., Loessner, D., Biniossek, M. L., Kizhakkedathu, J. N., Clements, J. A., Magdolen, V., et al. (2013). Secretome and degradome profiling shows that Kallikrein-related peptidases 4, 5, 6, and 7 induce TGFbeta-1 signaling in ovarian cancer cells. *Molecular Oncology*, *8*, 68–82.

Shen, Y., Liu, T., Tolic, N., Petritis, B. O., Zhao, R., Moore, R. J., et al. (2010). Strategy for degradomic-peptidomic analysis of human blood plasma. *Journal of Proteome Research*, *9*, 2339–2346.

Shen, Y., Tolic, N., Liu, T., Zhao, R., Petritis, B. O., Gritsenko, M. A., et al. (2010). Blood peptidome-degradome profile of breast cancer. *PLoS One, 5,* e13133.

Sherman, F., Stewart, J. W., & Tsunasawa, S. (1985). Methionine or not methionine at the beginning of a protein. *BioEssays, 3,* 27–31.

Slysz, G. W., Steinke, L., Ward, D. M., Klatt, C. G., Clauss, T. R., Purvine, S. O., et al. (2014). Automated data extraction from in situ protein-stable isotope probing studies. *Journal of Proteome Research, 13,* 1200–1210.

Smith, R. D., Anderson, G. A., Lipton, M. S., Pasa-Tolic, L., Shen, Y., Conrads, T. P., et al. (2002). An accurate mass tag strategy for quantitative and high-throughput proteome measurements. *Proteomics, 2,* 513–523.

Tinoco, A. D., & Saghatelian, A. (2011). Investigating endogenous peptides and peptidases using peptidomics. *Biochemistry, 50,* 7447–7461.

Van Damme, P., Maurer-Stroh, S., Plasman, K., Van Durme, J., Colaert, N., Timmerman, E., et al. (2009). Analysis of protein processing by N-terminal proteomics reveals novel species-specific substrate determinants of granzyme B orthologs. *Molecular & Cellular Proteomics, 8,* 258–272.

Varshavsky, A. (1996). The N-end rule: Functions, mysteries, uses. *Proceedings of the National Academy of Sciences of the United States of America, 93,* 12142–12149.

Villanueva, J., Shaffer, D. R., Philip, J., Chaparro, C. A., Erdjument-Bromage, H., Olshen, A. B., et al. (2006). Differential exoprotease activities confer tumor-specific serum peptidome patterns. *The Journal of Clinical Investigation, 116,* 271–284.

Voges, D., Zwickl, P., & Baumeister, W. (1999). The 26S proteasome: A molecular machine designed for controlled proteolysis. *Annual Review of Biochemistry, 68,* 1015–1068.

Wenzel, T., Eckerskorn, C., Lottspeich, F., & Baumeister, W. (1994). Existence of a molecular ruler in proteasomes suggested by analysis of degradation products. *FEBS Letters, 349,* 205–209.

Wu, C., Monroe, M. E., Xu, Z., Slysz, G. W., Payne, S. H., Rodland, K. D., et al. (2015). An optimized informatics pipeline for mass spectrometry-based peptidomics. *Journal of the American Society for Mass Spectrometry, 26,* 2002–2008.

Xu, Z., Wu, C., Xie, F., Slysz, G. W., Tolic, N., Monroe, M. E., et al. (2015). Comprehensive quantitative analysis of ovarian and breast cancer tumor peptidomes. *Journal of Proteome Research, 14,* 422–433.

Zhu, X., & Desiderio, D. M. (1993). Methionine enkephalin-like immunoreactivity, substance P-like immunoreactivity and beta-endorphin-like immunoreactivity postmortem stability in rat pituitary. *Journal of Chromatography, 616,* 175–187.

Zimmer, J. S., Monroe, M. E., Qian, W. J., & Smith, R. D. (2006). Advances in proteomics data analysis and display using an accurate mass and time tag approach. *Mass Spectrometry Reviews, 25,* 450–482.

Zwickl, P., Voges, D., & Baumeister, W. (1999). The proteasome: A macromolecular assembly designed for controlled proteolysis. *Philosophical Transactions of the Royal Society of London. Series B, Biological Sciences, 354,* 1501–1511.

MIX
Papier | Fördert
gute Waldnutzung
FSC® C083411

Kolibri 360 GmbH
Ferdinand-Jühlke-Straße 7
99095 Erfurt, Deutschland
produktsicherheit@kolibri360.de